AMERICAN
PROMETHEUS

A contributing editor at *The Nation*, Kai Bird is the author of several biographies. He lives in Kathmandu, Nepal, with his wife and son.

Martin J. Sherwin is University Professor of History at George Mason University. He is author of *A World Destroyed: Hiroshima and Its Legacies*. He and his wife live in Washington, D.C.

AMERICAN PROMETHEUS

The Triumph and Tragedy of

J. ROBERT OPPENHEIMER

KAI BIRD *and*

MARTIN J. SHERWIN

Atlantic Books
London

For Susan Goldmark and Susan Sherwin
and in memory of
Angus Cameron
and
Jean Mayer

First published in the United States of America in 2005 by Alfred A. Knopf, a division of Random House, Inc., New York.

First published in hardback in Great Britain in 2008 by Atlantic Books, an imprint of Grove Atlantic Ltd.

This paperback edition published in Great Britain in 2009 by Atlantic Books.

A portion of this work previously appeared in a slightly different form in *the Nation*.

11

A CIP catalogue record for this book is available from the British Library.

ISBN 978 1 84354 705 1

Printed and bound in Great Britain by CPI Group (UK) Ltd, Croydon, CR0 4YY

Atlantic Books
An imprint of Grove Atlantic Ltd
Ormond House
26–27 Boswell Street
London WC1N 3JZ

Modern Prometheans have raided Mount Olympus again and have brought back for man the very thunderbolts of Zeus.

—*Scientific Monthly*,
September 1945

Prometheus stole fire and gave it to men. But when Zeus learned of it, he ordered Hephaestus to nail his body to Mount Caucasus. On it Prometheus was nailed and kept bound for many years. Every day an eagle swooped on him and devoured the lobes of his liver, which grew by night.

—Apollodorus, *The Library*, book 1:7,
second century b.c.

CONTENTS

Contents

PREFACE

ROBERT OPPENHEIMER'S life—his career, his reputation, even his sense of self-worth—suddenly spun out of control four days before Christmas in 1953. "I can't believe what is happening to me," he exclaimed, staring through the window of the car speeding him to his lawyer's Georgetown home in Washington, D.C. There, within a few hours, he had to confront a fateful decision. Should he resign from his government advisory positions? Or should he fight the charges contained in the letter that Lewis Strauss, chairman of the Atomic Energy Commission (AEC), had handed to him out of the blue earlier that afternoon? The letter informed him that a new review of his background and policy recommendations had resulted in his being declared a security risk, and went on to delineate thirty-four charges ranging from the ridiculous—"it was reported that in 1940 you were listed as a sponsor of the Friends of the Chinese People"—to the political—"in the autumn of 1949, and subsequently, you strongly opposed the development of the hydrogen bomb."

Curiously, ever since the atomic bombings of Hiroshima and Nagasaki, Oppenheimer had been harboring a vague premonition that something dark and ominous lay in wait for him. A few years earlier, in the late 1940s, at a time when he had achieved a veritably iconic status in American society as the most respected and admired scientist and public policy adviser of his generation—even being featured on the covers of *Time* and *Life* magazines—he had read Henry James' short story "The Beast in the Jungle." Oppenheimer was utterly transfixed by this tale of obsession and tormented egotism in which the protagonist is haunted by a premonition that he was "being kept for something rare and strange, possibly prodigious and terrible, that was sooner or later to happen." Whatever it was, he knew that it would "overwhelm" him.

As the tide of anticommunism rose in postwar America, Oppenheimer became increasingly aware that "a beast in the jungle" was stalking him. His appearances before Red-hunting congressional investigative committees, the FBI taps on his home and office phones, the scurrilous stories about his political past and policy recommendations planted in the press made him feel like a hunted man. His left-wing activities during the 1930s in Berkeley, combined with his postwar resistance to the Air Force's plans for massive strategic bombing with nuclear weapons—plans he called genocidal—had angered many powerful Washington insiders, including FBI Director J. Edgar Hoover and Lewis Strauss.

That evening, at the Georgetown home of Herbert and Anne Marks, he contemplated his options. Herbert was not only his lawyer but one of his closest friends. And Herbert's wife, Anne Wilson Marks, had once been his secretary at Los Alamos. That night Anne observed that he seemed to be in an "almost despairing state of mind." Yet, after much discussion, Oppenheimer concluded, perhaps as much in resignation as conviction, that no matter how stacked the deck, he could not let the charges go unchallenged. So, with Herb's guidance, he drafted a letter addressed to "Dear Lewis." In it Oppenheimer noted that Strauss had encouraged him to resign. "You put to me as a possibly desirable alternative that I request termination of my contract as a consultant to the [Atomic Energy] Commission, and thereby avoid an explicit consideration of the charges. . . ." Oppenheimer said he had earnestly considered this option. But "[u]nder the circumstances," he continued, "this course of action would mean that I accept and concur in the view that I am not fit to serve this government, that I have now served for some twelve years. This I cannot do. If I were thus unworthy I could hardly have served our country as I have tried, or been the Director of our Institute [for Advanced Study] in Princeton, or have spoken, as on more than one occasion I have found myself speaking, in the name of our science and our country."

By the end of the evening, Robert was exhausted and despondent. After several drinks, he retired upstairs to the guest bedroom. A few minutes later, Anne, Herbert and Robert's wife, Kitty, who had accompanied him to Washington, heard a "terrible crash." Racing upstairs, they found the bedroom empty and the bathroom door closed. "I couldn't get it open," Anne said, "and I couldn't get a response from Robert."

He had collapsed on the bathroom floor, and his unconscious body was blocking the door. They gradually forced it open, pushing Robert's limp form to one side. When he revived, "he sure was mumbly," Anne recalled. He said he had taken one of Kitty's prescription sleeping pills. "Don't let him go to sleep," a doctor warned over the phone. So for almost an hour, until the doctor arrived, they walked Robert back and forth, coaxing him to swallow sips of coffee.

Robert's "beast" had pounced; the ordeal that would end his career of public service, and, ironically, both enhance his reputation and secure his legacy, had begun.

• • • •

THE ROAD ROBERT TRAVELED from New York City to Los Alamos, New Mexico—from obscurity to prominence—led him to participation in the great struggles and triumphs, in science, social justice, war, and Cold War, of the twentieth century. His journey was guided by his extraordinary intelligence, his parents, his teachers at the Ethical Culture School, and his youthful experiences. Professionally, his development began in the 1920s in Germany where he learned quantum physics, a new science that he loved and proselytized. In the 1930s, at the University of California, Berkeley, while building the most prominent center for its study in the United States, he was moved by the consequences of the Great Depression at home and the rise of fascism abroad to work actively with friends—many of them fellow travelers and communists—in the struggle to achieve economic and racial justice. Those years were some of the finest of his life. That they were so easily used to silence his voice a decade later is a reminder of how delicately balanced are the democratic principles we profess, and how carefully they must be guarded.

The agony and humiliation that Oppenheimer endured in 1954 were not unique during the McCarthy era. But as a defendant, he was incomparable. He was America's Prometheus, "the father of the atomic bomb," who had led the effort to wrest from nature the awesome fire of the sun for his country in time of war. Afterwards, he had spoken wisely about its dangers and hopefully about its potential benefits and then, near despair, critically about the proposals for nuclear warfare being adopted by the military and promoted by academic strategists: "What are we to make of a civilization which has always regarded ethics as an essential part of human life [but] which has not been able to talk about the prospect of killing almost everybody except in prudential and game-theoretical terms?"

In the late 1940s, as U.S.-Soviet relations deteriorated, Oppenheimer's persistent desire to raise such tough questions about nuclear weapons greatly troubled Washington's national security establishment. The return of the Republicans to the White House in 1953 elevated advocates of massive nuclear retaliation, such as Lewis Strauss, to positions of power in Washington. Strauss and his allies were determined to silence the one man who they feared could credibly challenge their policies.

In assaulting his politics and his professional judgments—his life and his values really—Oppenheimer's critics in 1954 exposed many aspects of his character: his ambitions and insecurities, his brilliance and naïveté, his determination and fearfulness, his stoicism and his bewilderment. Much was revealed in the more than one thousand densely printed pages of the transcript of the AEC's Personnel Security Hearing Board, *In the Matter of J. Robert Oppenheimer;* and yet the hearing transcript reveals how little his antagonists had been able to pierce through the emotional armor this complex man had constructed around himself since his early years. *American Prometheus* explores the enigmatic personality behind that armor as it follows Robert from his childhood on New York's Upper West Side at the turn of the twentieth century to his death in 1967. It is a deeply personal biography researched and written in the belief that a person's public behavior and his policy decisions (and in Oppenheimer's case perhaps even his science) are guided by the private experiences of a lifetime.

A QUARTER CENTURY in the making, *American Prometheus* is based on many thousands of records gathered from archives and personal collections in this country and abroad. It draws on Oppenheimer's own massive collection of papers in the Library of Congress, and on thousands of pages of FBI records accumulated over more than a quarter century of surveillance. Few men in public life have been subjected to such scrutiny. Readers will "hear" his words, captured by FBI recording devices and transcribed. And yet, because even the written record tells only part of the truth of a man's life, we have also interviewed nearly a hundred of Oppenheimer's closest friends, relatives and colleagues. Many of the individuals interviewed in the 1970s and 1980s are no longer alive. But the stories they told leave behind a nuanced portrait of a remarkable man who led us into the nuclear age and struggled, unsuccessfully—as we have continued to struggle—to find a way to eliminate the danger of nuclear war.

Oppenheimer's story also reminds us that our identity as a people remains intimately connected with the culture of things nuclear. "We have had the bomb on our minds since 1945," E. L. Doctorow has observed. "It was first our weaponry and then our diplomacy, and now it's our economy. How can we suppose that something so monstrously powerful would not, after forty years, compose our identity? The great golem we have made against our enemies is our culture, our bomb culture—its logic, its faith, its vision." Oppenheimer tried valiantly to divert us from that bomb culture by containing the nuclear threat he had helped to set loose. His most impressive effort was a plan for the international control of atomic energy, which became known as the Acheson-Lilienthal Report (but was in fact conceived and largely written by Oppenheimer). It remains a singular model for rationality in the nuclear age.

Cold War politics at home and abroad, however, doomed the plan, and America, along with a growing list of other nations, embraced the bomb for the next half century. With the end of the Cold War, the danger of nuclear annihilation seemed to pass, but in another ironic twist, the threat of nuclear war and nuclear terrorism is probably more imminent in the twenty-first century than ever before.

In the post-9/11 era, it is worth recalling that at the dawn of the nuclear age, the father of the atomic bomb warned us that it was a weapon of indiscriminate terror that instantly had made America more vulnerable to wanton attack. When he was asked in a closed Senate hearing in 1946 "whether three or four men couldn't smuggle units of an [atomic] bomb into New York and blow up the whole city," he responded pointedly, "Of course it could be done, and people could destroy New York." To the follow-up question of a startled senator, "What instrument would you use to detect an atomic bomb hidden somewhere in a city?" Oppenheimer quipped, "A screwdriver [to open each and every crate or suitcase]." The only defense against nuclear terrorism was the elimination of nuclear weapons.

Oppenheimer's warnings were ignored—and ultimately, he was silenced. Like that rebellious Greek god Prometheus—who stole fire from Zeus and bestowed it upon humankind, Oppenheimer gave us atomic fire. But then, when he tried to control it, when he sought to make us aware of its terrible dangers, the powers-that-be, like Zeus, rose up in anger to punish him. As Ward Evans, the dissenting member of the Atomic Energy Commission's hearing board, wrote, denying Oppenheimer his security clearance was "a black mark on the escutcheon of our country."

AMERICAN
PROMETHEUS

PROLOGUE

PRINCETON, NEW JERSEY, February 25, 1967: Despite the menacing weather and bitter cold that chilled the Northeast, six hundred friends and colleagues—Nobel laureates, politicians, generals, scientists, poets, novelists, composers and acquaintances from all walks of life—gathered to recall the life and mourn the death of J. Robert Oppenheimer. Some knew him as their gentle teacher and affectionately called him "Oppie." Others knew him as a great physicist, a man who in 1945 had become the "father" of the atomic bomb, a national hero and an emblem of the scientist as public servant. And everyone remembered with deep bitterness how, just nine years later, the new Republican administration of President Dwight D. Eisenhower had declared him a security risk—making Robert Oppenheimer the most prominent victim of America's anticommunist crusade. And so they came with heavy hearts to remember a brilliant man whose remarkable life had been touched by triumph as well as tragedy.

The Nobelists included such world-renowned physicists as Isidor I. Rabi, Eugene Wigner, Julian Schwinger, Tsung Dao Lee and Edwin McMillan. Albert Einstein's daughter, Margot, was there to honor the man who had been her father's boss at the Institute for Advanced Study. Robert Serber—a student of Oppenheimer's at Berkeley in the 1930s and a close friend and veteran of Los Alamos—was there, as was the great Cornell physicist Hans Bethe, the Nobelist who had revealed the inner workings of the sun. Irva Denham Green, a neighbor from the tranquil Caribbean island of St. John, where the Oppenheimers had built a beach cottage as a refuge after his public humiliation in 1954, sat elbow to elbow with powerful luminaries of America's foreign policy establishment: lawyer and perennial presidential adviser John J. McCloy; the Manhattan Project's military chief, General Leslie R. Groves; Secretary of the Navy Paul Nitze; Pulitzer Prize–winning historian Arthur Schlesinger, Jr.; and Senator Clifford Case of New Jersey. To represent the White House, President Lyndon B. Johnson sent his scientific adviser, Donald F. Hornig, a Los Alamos veteran who had been with Oppenheimer at "Trinity," the test on July 16, 1945, of the first atomic bomb. Sprinkled among the scientists and Washington's power elite were men of literature and culture: the poet Stephen Spender, the novelist John O'Hara, the composer Nicholas Nabokov and George Balanchine, the director of the New York City Ballet.

Oppenheimer's widow, Katherine "Kitty" Puening Oppenheimer, sat in the front row at Princeton University's Alexander Hall for what many would remember as a subdued, bittersweet memorial service. Sitting with her were their daughter, Toni, age twenty-two, and their son, Peter, age twenty-five. Robert's younger brother, Frank Oppenheimer, whose own career as a physicist had been destroyed during the McCarthyite maelstrom, sat next to Peter.

Strains of Igor Stravinsky's *Requiem Canticles,* a work Robert Oppenheimer had heard for the first time, and admired, in this very hall the previous autumn, filled the auditorium. And then Hans Bethe—who had known Oppenheimer for three decades—gave the first of three eulogies. "He did more than any other man," Bethe said, "to make American theoretical physics great. . . . He was a leader. . . . But he was not domineering, he never dictated what should be done. He brought out the best in us, like a good host with his guests. . . ." At Los Alamos, where he directed thousands in a putative race against the Germans to build the atomic bomb, Oppenheimer had transformed a pristine mesa into a laboratory and forged a diverse group of scientists into an efficient team. Bethe and other veterans of Los Alamos knew that without Oppenheimer the primordial "gadget" they had built in New Mexico would never have been finished in time for its use in the war.

Henry DeWolf Smyth, a physicist and Princeton neighbor, gave the second eulogy. In 1954, Smyth had been the only one of five commissioners of the Atomic Energy Commission (AEC) who had voted to restore Oppenheimer's security clearance. As a witness to the star-chamber "security hearing" Oppenheimer had endured, Smyth fully comprehended the travesty that had been committed: "Such a wrong can never be righted; such a blot on our history never erased. . . . We regret that his great work for his country was repaid so shabbily. . . ."

Finally, it was the turn of George Kennan, veteran diplomat and ambassador, the father of America's postwar containment policy against the Soviet Union, and a longtime friend and colleague of Oppenheimer's at the Institute for Advanced Studies. No man had stimulated Kennan's thinking about the myriad dangers of the nuclear age more than Oppenheimer. No man had been a better friend, defending his work and providing him a refuge at the Institute when Kennan's dissenting views on America's militarized Cold War policies made him a pariah in Washington.

"On no one," Kennan said, "did there ever rest with greater cruelty the dilemmas evoked by the recent conquest by human beings of a power over nature out of all proportion to their moral strength. No one ever saw more clearly the dangers arising for humanity from this mounting disparity. This anxiety never shook his faith in the value of the search for truth in all its forms, scientific and humane. But there was no one who more passionately desired to be useful in averting the catastrophes to which the development of the weapons of mass destruction threatened to lead. It was the interests of mankind that he had in mind here; but it was as an American, and through the medium of this national community to which he belonged, that he saw his greatest possibilities for pursuing these aspirations.

"In the dark days of the early fifties, when troubles crowded in upon him from many sides and when he found himself harassed by his position at the center of controversy, I drew his attention to the fact that he would be welcome in a hundred academic centers abroad and asked him whether he had not thought of taking residence outside this country. His answer, given to me with tears in his eyes: 'Damn it, I happen to love this country.' "*

ROBERT OPPENHEIMER WAS AN ENIGMA, a theoretical physicist who displayed the charismatic qualities of a great leader, an aesthete who cultivated ambiguities. In the decades after his death, his life became shrouded in controversy, myth and mystery. For scientists, like Dr. Hideki Yukawa, Japan's first Nobelist, Oppenheimer was "a symbol of the tragedy of the modern nuclear scientist." To liberals, he became the most prominent martyr of the McCarthyite witch-hunt, a symbol of the right wing's unprincipled animus. To his political enemies, he was a closet communist and a proven liar.

He was, in fact, an immensely human figure, as talented as he was complex, at once brilliant and naïve, a passionate advocate for social justice and a tireless government adviser whose commitment to harnessing a runaway nuclear arms race earned him powerful bureaucratic enemies. As his friend Rabi said, in addition to being "very wise, he was very foolish."

The physicist Freeman Dyson saw deep and poignant contradictions in Robert Oppenheimer. He had dedicated his life to science and rational thought. And yet, as Dyson observed, Oppenheimer's decision to participate in the creation of a genocidal weapon was "a Faustian bargain if there ever was one. . . . And of course we are still living with it. . . ." And like Faust, Robert Oppenheimer tried to renegotiate the bargain—and was cut

---

*Kennan was deeply moved by Oppenheimer's emphatic reaction. In 2003, at Kennan's hundredth-birthday party, he retold this story—and this time there were tears in *his* eyes.

down for doing so. He had led the effort to unleash the power of the atom, but when he sought to warn his countrymen of its dangers, to constrain America's reliance on nuclear weapons, the government questioned his loyalty and put him on trial. His friends compared this public humiliation to the 1633 trial of another scientist, Galileo Galilei, by a medieval-minded church; others saw the ugly spectre of anti-Semitism in the event and recalled the ordeal of Captain Alfred Dreyfus in France in the 1890s.

But neither comparison helps us to understand Robert Oppenheimer the man, his accomplishments as a scientist and the unique role he played as an architect of the nuclear era. This is the story of his life.

PART ONE

CHAPTER ONE

"He Received Every New Idea as Perfectly Beautiful"

I was an unctuous, repulsively good little boy.

ROBERT OPPENHEIMER

IN THE FIRST DECADE of the twentieth century, science initiated a second American revolution. A nation on horseback was soon transformed by the internal combustion engine, manned flight and a multitude of other inventions. These technological innovations quickly changed the lives of ordinary men and women. But simultaneously an esoteric band of scientists was creating an even more fundamental revolution. Theoretical physicists across the globe were beginning to alter the way we understand space and time. Radioactivity was discovered in 1896, by the French physicist Henri Becquerel. Max Planck, Marie Curie and Pierre Curie and others provided further insights into the nature of the atom. And then, in 1905, Albert Einstein published his special theory of relativity. Suddenly, the universe appeared to have changed.

Around the globe, scientists were soon to be celebrated as a new kind of hero, promising to usher in a renaissance of rationality, prosperity and social meritocracy. In America, reform movements were challenging the old order. Theodore Roosevelt was using the bully pulpit of the White House to argue that good government in alliance with science and applied technology could forge an enlightened new Progressive Era.

Into this world of promise was born J. Robert Oppenheimer, on April 22, 1904. He came from a family of first- and second-generation German immigrants striving to be American. Ethnically and culturally Jewish, the Oppenheimers of New York belonged to no synagogue. Without rejecting their Jewishness they chose to shape their identity within a uniquely American offshoot of Judaism—the Ethical Culture Society—that celebrated rationalism and a progressive brand of secular humanism. This was at the same time an innovative approach to the quandaries any immigrant to America faced—and yet for Robert Oppenheimer it reinforced a lifelong ambivalence about his Jewish identity.

As its name suggests, Ethical Culture was not a religion but a way of life that promoted social justice over self-aggrandizement. It was no accident that the young boy who would become known as the father of the atomic era was reared in a culture that valued independent inquiry, empirical exploration and the free-thinking mind—in short, the values of science. And yet, it was the irony of Robert Oppenheimer's odyssey that a life devoted to social justice, rationality and science would become a metaphor for mass death beneath a mushroom cloud.

ROBERT'S FATHER, Julius Oppenheimer, was born on May 12, 1871, in the German town of Hanau, just east of Frankfurt. Julius' father, Benjamin Pinhas Oppenheimer, was an untutored peasant and grain trader who had been raised in a hovel in "an almost medieval German village," Robert later reported. Julius had two brothers and three sisters. In 1870, two of Benjamin's cousins by marriage emigrated to New York. Within a few years these two young men—named Sigmund and Solomon Rothfeld—joined another relative, J. H. Stern, to start a small company to import men's suit linings. The company did extremely well serving the city's flourishing new trade in ready-made clothing. In the late 1880s, the Rothfelds sent word to Benjamin Oppenheimer that there was room in the business for his sons.

Julius arrived in New York in the spring of 1888, several years after his older brother Emil. A tall, thin-limbed, awkward young man, he was put to work in the company warehouse, sorting bolts of cloth. Although he brought no monetary assets to the firm and spoke not a word of English, he was determined to remake himself. He had an eye for color and in time acquired a reputation as one of the most knowledgeable "fabrics" men in the city. Emil and Julius rode out the recession of 1893, and by the turn of the century Julius was a full partner in the firm of Rothfeld, Stern & Company. He dressed to fit the part, always adorned in a white high-collared shirt, a conservative tie and a dark business suit. His manners were as immaculate as his dress. From all accounts, Julius was an extremely likeable young man. "You have a way with you that just invites confidence to the highest degree," wrote his future wife in 1903, "and for the best and finest reasons." By the time he turned thirty, he spoke remarkably good English, and, though completely self-taught, he had read widely in American and European history. A lover of art, he spent his free hours on weekends roaming New York's numerous art galleries.

It may have been on one such occasion that he was introduced to a young painter, Ella Friedman, "an exquisitely beautiful" brunette with finely chiseled features, "expressive gray-blue eyes and long black lashes," a slender figure—and a congenitally unformed right hand. To hide this deformity, Ella always wore long sleeves and a pair of chamois gloves. The glove covering her right hand contained a primitive prosthetic device with a spring attached to an artificial thumb. Julius fell in love with her. The Friedmans, of Bavarian Jewish extraction, had settled in Baltimore in the 1840s. Ella was born in 1869. A family friend once described her as "a gentle, exquisite, slim, tallish, blue-eyed woman, terribly sensitive, extremely polite; she was always thinking what would make people comfortable or happy." In her twenties, she spent a year in Paris studying the early Impressionist painters. Upon her return she taught art at Barnard College. By the time she met Julius, she was an accomplished enough painter to have her own students and a private rooftop studio in a New York apartment building.

All this was unusual enough for a woman at the turn of the century, but Ella was a powerful personality in many respects. Her formal, elegant demeanor struck some people upon first acquaintance as haughty coolness. Her drive and discipline in the studio and at home seemed excessive in a woman so blessed with material comforts. Julius worshipped her, and she returned his love. Just days before their marriage, Ella wrote to her fiancé: "I do so want you to be able to enjoy life in its best and fullest sense, and you will help me take care of you? To take care of someone whom one really loves has an indescribable sweetness of which a whole lifetime cannot rob me. Good-night, dearest."

On March 23, 1903, Julius and Ella were married and moved into a sharp-gabled stone house at 250 West 94th Street. A year later, in the midst of the coldest spring on record, Ella, thirty-four years old, gave birth to a son after a difficult pregnancy. Julius had already settled on naming his firstborn Robert; but at the last moment, according to family lore, he decided to add a first initial, "J," in front of "Robert." Actually, the boy's birth certificate reads "Julius Robert Oppenheimer," evidence that Julius had decided to name the boy after himself. This would be unremarkable—except that naming a baby after any *living* relative is contrary to European Jewish tradition. In any case, the boy would always be called Robert and, curiously, he in turn always insisted that his first initial stood for nothing at all. Apparently, Jewish traditions played no role in the Oppenheimer household.

Sometime after Robert's arrival, Julius moved his family to a spacious eleventh-floor apartment at 155 Riverside Drive, overlooking the Hudson River at West 88th Street. The apartment, occupying an entire floor, was exquisitely decorated with fine European furniture. Over the years, the Oppenheimers also acquired a remarkable collection of French Postimpressionist and Fauvist paintings chosen by Ella. By the time Robert was a young man, the collection included a 1901 "blue period" painting by Pablo Picasso entitled *Mother and Child,* a Rembrandt etching, and paintings by Edouard Vuillard, André Derain and Pierre-Auguste Renoir. Three Vincent Van Gogh paintings—*Enclosed Field with Rising Sun* (Saint-Remy, 1889), *First Steps (After Millet)* (Saint-Remy, 1889) and *Portrait of Adeline Ravoux* (Auvers-sur-Oise, 1890)—dominated a living room wallpapered in gilted gold. Sometime later they acquired a drawing by Paul Cézanne and a painting by Maurice de Vlaminck. A head by the French sculptor Charles Despiau rounded out this exquisite collection.*

Ella ran the household to exacting standards. "Excellence and purpose" was a constant refrain in young Robert's ears. Three live-in maids kept the apartment spotless. Robert had a Catholic Irish nursemaid named Nellie Connolly, and later, a French governess who taught him a little French. German, on the other hand, was not spoken at home. "My mother didn't talk it well," Robert recalled, "[and] my father didn't believe in talking it." Robert would learn German in school.

On weekends, the family would go for drives in the countryside in their Packard, driven by a gray-uniformed chauffeur. When Robert was eleven or twelve, Julius bought a substantial summer home at Bay Shore, Long Island, where Robert learned to sail. At the pier below the house, Julius moored a forty-foot sailing yacht, christened the *Lorelei,* a luxurious craft outfitted with all the amenities. "It was lovely on that bay," Robert's brother, Frank, would later recall fondly. "It was seven acres . . . a big vegetable garden and lots and lots of flowers." As a family friend later observed, "Robert was doted on by his parents. . . . He had everything he wanted; you might say he was brought up in luxury." But despite this, none of his childhood friends thought him spoiled. "He was extremely generous with money and material things," recalled Harold Cherniss. "He was not a spoiled child in any sense."

By 1914, when World War I broke out in Europe, Julius Oppenheimer was a very prosperous businessman. His net worth certainly totaled more than several hundred thousand dollars—which made him the equivalent of a multimillionaire in current dollars. By all accounts, the Oppenheimer marriage was a loving partnership. But Robert's friends were always struck by their contrasting personalities. "He [Julius] was jolly German-Jewish,"

---

*The Oppenheimers spent a small fortune on these works of art. In 1926, for instance, Julius paid $12,900 for Van Gogh's *First Steps (After Millet).*

recalled Francis Fergusson, one of Robert's closest friends. "Extremely likeable. I was surprised that Robert's mother had married him because he seemed such a hearty and laughing kind of person. But she was very fond of him and handled him beautifully. They were very fond of each other. It was an excellent marriage."

Julius was a conversationalist and extrovert. He loved art and music and thought Beethoven's *Eroica* symphony "one of the great masterpieces." A family friend, the philosopher George Boas, later recalled that Julius "had all the sensitiveness of both his sons." Boas thought him "one of the kindest men I ever knew." But sometimes, to the embarrassment of his sons, Julius would burst out singing at the dinner table. He enjoyed a good argument. Ella, by contrast, sat quietly and never joined in the banter. "She [Ella] was a very delicate person," another friend of Robert's, the distinguished writer Paul Horgan, observed, ". . . highly attenuated emotionally, and she always presided with a great delicacy and grace at the table and other events, but [she was] a mournful person."

Four years after Robert's birth, Ella bore another son, Lewis Frank Oppenheimer, but the infant soon died, a victim of stenosis of the pylorus, a congenital obstruction of the opening from the stomach to the small intestine. In her grief, Ella thereafter always seemed physically more fragile. Because young Robert himself was frequently ill as a child, Ella became overly protective. Fearing germs, she kept Robert apart from other children. He was never allowed to buy food from street vendors, and instead of taking him to get a haircut in a barber shop Ella had a barber come to the apartment.

Introspective by nature and never athletic, Robert spent his early childhood in the comfortable loneliness of his mother's nest on Riverside Drive. The relationship between mother and son was always intense. Ella encouraged Robert to paint—he did landscapes—but he gave it up when he went to college. Robert worshipped his mother. But Ella could be quietly demanding. "This was a woman," recalled a family friend, "who would never allow anything unpleasant to be mentioned at the table."

Robert quickly sensed that his mother disapproved of the people in her husband's world of trade and commerce. Most of Julius's business colleagues, of course, were first-generation Jews, and Ella made it clear to her son that she felt ill-at-ease with their "obtrusive manners." More than most boys, Robert grew up feeling torn between his mother's strict standards and his father's gregarious behavior. At times, he felt ashamed of his father's spontaneity—and at the same time he would feel guilty that he felt ashamed. "Julius's articulate and sometimes noisy pride in Robert annoyed him greatly," recalled a childhood friend. As an adult, Robert gave his friend and former teacher Herbert Smith a handsome engraving of the scene in Shakespeare's *Coriolanus* where the hero is unclasping his mother's hand and throwing her to the ground. Smith was sure that Robert was sending him a message, acknowledging how difficult it had been for him to separate from his own mother.

When he was only five or six, Ella insisted that he take piano lessons. Robert dutifully practiced every day, hating it all the while. About a year later, he fell sick and his mother characteristically suspected the worst, perhaps a case of infantile paralysis. Nursing him back to health, she kept asking him how he felt until one day he looked up from his sickbed and grumbled, "Just as I do when I have to take piano lessons." Ella relented, and the lessons ended.

In 1909, when Robert was only five, Julius took him on the first of four transatlantic crossings to visit his grandfather Benjamin in Germany. They made the trip again two years later; by then Grandfather Benjamin was seventy-five years old, but he left an indelible impression on his grandson. "It was clear," Robert recalled, "that one of the great joys in life for him was reading, but he had probably hardly been to school." One day, while watching Robert play with some wooden blocks, Benjamin decided to give him an encyclopedia of architecture. He also gave him a "perfectly conventional" rock collection consisting of a box with perhaps two dozen rock samples labeled in German. "From then on," Robert later recounted, "I became, in a completely childish way, an ardent mineral collector." Back home in New York, he persuaded his father to take him on rock-hunting expeditions along the Palisades. Soon the apartment on Riverside Drive was crammed with Robert's rocks, each neatly labeled with its scientific name. Julius encouraged his son in this solitary hobby, plying him with books on the subject. Long afterward, Robert recounted that he had no interest in the geological origins of his rocks, but was fascinated by the structure of crystals and polarized light.

From the ages of seven through twelve, Robert had three solitary but all-consuming passions: minerals, writing and reading poetry, and building with blocks. Later he would recall that he occupied his time with these activities "not because they were something I had companionship in or because they had any relation to school—but just for the hell of it." By the age of twelve, he was using the family typewriter to correspond with a number of well-known local geologists about the rock formations he had studied in Central Park. Not aware of his youth, one of these correspondents nominated Robert for membership in the New York Mineralogical Club, and soon thereafter a letter arrived inviting him to deliver a lecture before the club. Dreading the thought of having to talk to an audience of adults, Robert begged his father to explain that they had invited a twelve-year-old. Greatly amused, Julius encouraged his son to accept this honor. On the designated evening, Robert showed up at the club with his parents, who proudly introduced their son as "J. Robert Oppenheimer." The startled audience of geologists and amateur rock collectors burst out laughing when he stepped up to the podium; a wooden box had to be found for him to stand on so that the audience could see more than the shock of his wiry black hair sticking up above the lectern. Shy and awkward, Robert nevertheless read his prepared remarks and was given a hearty round of applause.

Julius had no qualms about encouraging his son in these adult pursuits. He and Ella knew they had a "genius" on their hands. "They adored him, worried about him and protected him," recalled Robert's cousin Babette Oppenheimer. "He was given every opportunity to develop along the lines of his own inclinations and at his own rate of speed." One day, Julius gave Robert a professional-quality microscope which quickly became the boy's favorite toy. "I think that my father was one of the most tolerant and human of men," Robert would remark in later years. "His idea of what to do for people was to let them find out what they wanted." For Robert, there was no doubt about what he wanted; from an early age, he lived within the world of books and science. "He was a dreamer," said Babette Oppenheimer, "and not interested in the rough-and-tumble life of his age group . . . he was often teased and ridiculed for not being like other fellows." As he grew older, even his mother on occasion worried about her son's "limited interest" in play and children his own age. "I know she kept trying to get me to be more like other boys, but with indifferent success."

In 1912, when Robert was eight years old, Ella gave birth to another son, Frank Friedman Oppenheimer, and thereafter much of her attention shifted to the new baby. At some point, Ella's mother moved into the Riverside apartment and lived with the family until she died when Frank was a young teenager. The eight years separating the boys left few opportunities for sibling rivalry. Robert later thought he had been not only an elder brother but also perhaps "father to him because of that age difference." Frank's early childhood was as nurturing, if not more so, than Robert's. "If we had some enthusiasm," Frank recalled, "my parents would cater to it." In high school, when Frank showed an interest in reading Chaucer, Julius promptly went out and bought him a 1721 edition of the poet's works. When Frank expressed a desire to play the flute, his parents hired one of America's greatest flutists, George Barère, to give him private lessons. Both boys were excessively pampered—but as the firstborn, only Robert acquired a certain conceit. "I repaid my parents' confidence in me by developing an unpleasant ego," Robert later confessed, "which I am sure must have affronted both children and adults who were unfortunate enough to come into contact with me."

IN SEPTEMBER 1911, soon after returning from his second visit to Grandfather Benjamin in Germany, Robert was enrolled in a unique private school. Years earlier, Julius had become an active member of the Ethical Culture Society. He and Ella had been married by Dr. Felix Adler, the Society's leader and founder, and, beginning in 1907, Julius served as a trustee of the Society. There was no question but that his sons would receive their primary and secondary education at the Society's school on Central Park West. The school's motto was "Deed, not Creed." Founded in 1876, the Ethical Culture Society inculcated in its members a commitment to social action and humanitarianism: "Man must assume responsibility for the direction of his life and destiny." Although an outgrowth of American Reform Judaism, Ethical Culture was itself a "non-religion," perfectly suited to upper-middle-class German Jews, most of whom, like the Oppenheimers, were intent on assimilating into American society. Felix Adler and his coterie of talented teachers promoted this process and would have a powerful influence in the molding of Robert Oppenheimer's psyche, both emotionally and intellectually.

The son of Rabbi Samuel Adler, Felix Adler had, with his family, emigrated to New York from Germany in 1857, when he was only six years old. His father, a leader of the Reform Judaism movement in Germany, had come to head Temple Emanu-El, the largest Reform congregation in America. Felix might easily have succeeded his father, but as a young man he returned to Germany for his university studies and there he was exposed to radical new notions about the universality of God and man's responsibilities to society. He read Charles Darwin, Karl Marx and a host of German philosophers, including Felix Wellhausen, who rejected the traditional belief in the Torah as divinely inspired. Adler returned to his father's Temple Emanu-El in 1873 and preached a sermon on what he called the "Judaism of the Future." To survive in the modern age, the younger Adler argued, Judaism must renounce its "narrow spirit of exclusion." Instead of defining themselves by their biblical identity as the "Chosen People," Jews should distinguish themselves by their social concern and their deeds on behalf of the laboring classes.

Within three years, Adler led some four hundred congregants of Temple Emanu-El out of the established Jewish community. With the financial support of Joseph Seligman and other wealthy Jewish businessmen of German origin, he founded a new movement that he called "Ethical Culture." Meetings were held on Sunday mornings, at which Adler lectured; organ music was played but there were no prayers and no other religious ceremonies. Beginning in 1910, when Robert was six years old, the Society assembled in a handsome meeting house at 2 West 64th Street. Julius Oppenheimer attended the dedication ceremonies for the new building in 1910. The auditorium featured hand-carved oak paneling, beautiful stained-glass windows and a Wicks pipe organ in the balcony. Distinguished speakers like W. E. B. DuBois and Booker T. Washington, among many other prominent public personalities, were welcomed in this ornate auditorium.

"Ethical Culture" was a reformist Judaic sect. But the seeds of this particular movement had clearly been planted by elite efforts to reform and integrate upper-class Jews into German society in the nineteenth century. Adler's radical notions of Jewish identity struck a popular chord among wealthy Jewish businessmen in New York precisely because these men were grappling with a rising tide of anti-Semitism in nineteenth-century American life. Organized, institutional discrimination against Jews was a relatively recent phenomenon; since the American Revolution, when deists like Thomas Jefferson had insisted on a radical separation of organized religion from the state, American Jews had experienced a sense of tolerance. But after the stock market crash of 1873, the atmosphere in New York began to change. Then, in the summer of 1877, the Jewish community was scandalized when Joseph Seligman, the wealthiest and most prominent Jew of German origin in New York, was rudely turned away, as a Jew, from the Grand Union Hotel in Saratoga, New York. Over the next few years, the doors of other elite institutions, not only hotels but social clubs and preparatory private schools, suddenly slammed shut against Jewish membership.

Thus, by the end of the 1870s, Felix Adler's Ethical Culture Society provided New York Jewish society with a timely vehicle for dealing with this mounting bigotry. Philosophically, Ethical Culture was as deist and republican as the Founding Fathers' revolutionary principles. If the revolution of 1776 had brought with it an emancipation of American Jews, well, an apt response to nativist Christian bigotry was to become more American— more republican—than the Americans. These Jews would take the next step to assimilation, but they would do it, so to speak, as deist Jews. In Adler's view, the notion of Jews as a nation was an anachronism. Soon he began creating the institutional structures that would make it practical for his adherents to lead their lives as "emancipated Jews."

Adler insisted that the answer to anti-Semitism was the global spread of intellectual culture. Interestingly, Adler criticized Zionism as a withdrawal into Jewish particularism: "Zionism itself is a present-day instance of the segregating tendency." For Adler, the future for Jews lay in America, not

Palestine: "I fix my gaze steadfastly on the glimmering of a fresh morning that shines over the Alleghenies and the Rockies, not on the evening glow, however tenderly beautiful, that broods and lingers over the Jerusalem hills."

To transform his Weltanschauung into reality, Adler founded in 1880 a tuition-free school for the sons and daughters of laborers called the Workingman's School. In addition to the usual subjects of arithmetic, reading and history, Adler insisted that his students should be exposed to art, drama, dance and some kind of training in a technical skill likely to be of use in a society undergoing rapid industrialization. Every child, he believed, had some particular talent. Those who had no talent for mathematics might possess extraordinary "artistic gifts to make things with their hands." For Adler, this insight was the "ethical seed—and the thing to do is to cultivate these various talents." The goal was a "better world," and thus the school's mission was to "train reformers." As the school evolved, it became a showcase of the progressive educational reform movement, and Adler himself fell under the influence of the educator and philosopher John Dewey and his school of American pragmatists.

While not a socialist, Adler was spiritually moved by Marx's description in *Das Kapital* of the plight of the industrial working class. "I must square myself," he wrote, "with the issues that socialism raises." The laboring classes, he came to believe, deserved "just remuneration, constant employment, and social dignity." The labor movement, he later wrote, "is an ethical movement, and I am with it, heart and soul." Labor leaders reciprocated these sentiments; Samuel Gompers, head of the new American Federation of Labor, was a member of the New York Society for Ethical Culture.

Ironically, by 1890 the school had so many students that Adler felt compelled to subsidize the Ethical Culture Society's budget by admitting some tuition-paying students. At a time when many elite private schools were closing their doors to Jews, scores of prosperous Jewish businessmen were clamoring to have their children admitted to the Workingman's School. By 1895, Adler had added a high school and renamed the school the Ethical Culture School. (Decades later, it was renamed the Fieldston School.) By the time Robert Oppenheimer enrolled in 1911, only about ten percent of the student body came from a working-class background. But the school nevertheless retained its liberal, socially responsible outlook. These sons and daughters of the relatively prosperous patrons of the Ethical Culture Society were infused with the notion that they were being groomed to reform the world, that they were the vanguard of a highly modern ethical gospel. Robert was a star student.

Needless to say, Robert's adult political sensibilities can easily be traced to the progressive education he received at Felix Adler's remarkable school. Throughout the formative years of his childhood and education, he was surrounded by men and women who thought of themselves as catalysts for a better world. In the years between the turn of the century and the end of World War I, Ethical Culture members served as agents of change on such politically charged issues as race relations, labor rights, civil liberties and environmentalism. In 1909, for instance, such prominent Ethical Culture members as Dr. Henry Moskowitz, John Lovejoy Elliott, Anna Garlin Spencer and William Salter helped to found the National Association for the Advancement of Colored People (NAACP). Dr. Moskowitz similarly played an important role in the garment workers' strikes that occurred between 1910 and 1915. Other Ethical Culturists helped to found the National Civil Liberties Bureau, a forerunner of the American Civil Liberties Union. Though they shunned notions of class struggle, members of the Society were pragmatic radicals committed to playing an active role in bringing about social change. They believed that a better world required hard work, persistence and political organization. In 1921, the year Robert graduated from the Ethical Culture high school, Adler exhorted his students to develop their "ethical imagination," to see "things not as they are, but as they might be."*

Robert was fully aware of Adler's influence not only on himself but on his father. And he was not above teasing Julius about it. At seventeen, he wrote a poem on the occasion of his father's fiftieth birthday that included the line "and after he came to America, he swallowed Dr. Adler like morality compressed."

Like many Americans of German background, Dr. Adler was deeply saddened and conflicted when America was drawn into World War I. Unlike another prominent member of the Ethical Culture Society, Oswald Garrison Villard, editor of *The Nation* magazine, Adler was not a pacifist. When a German submarine sank the British passenger ship *Lusitania,* he supported the arming of American merchant ships. While he opposed American entry into the conflict, when the Wilson Administration declared war in April 1917, Adler urged his congregation to give its "undivided allegiance" to America. At the same time, he declared that he could not label Germany the only guilty party. As a critic of the German monarchy, at the war's end he

*Decades later, Robert's classmate Daisy Newman recalled: "When his idealism got him into difficulties, I felt this was the logical outcome of our superb training in ethics. A faithful pupil of Felix Adler and John Lovejoy Elliott would have been obliged to act in accordance with his conscience, however unwise his choice might be." (Newman ltr. to Alice K. Smith, 2/17/77, Smith correspondence, Sherwin collection.)

welcomed the downfall of imperial rule and the collapse of the Austro-Hungarian Empire. But as a fierce anticolonialist, he openly deplored the hypocrisy of a victors' peace that appeared only to strengthen the British and French empires. Naturally, his critics accused him of pro-German sentiments. As a trustee of the Society and as a man who deeply admired Dr. Adler, Julius Oppenheimer likewise felt conflicted about the European war and his identity as a German-American. But there is no evidence of how young Robert felt about the conflict. His teacher at school in ethical studies, however, was John Lovejoy Elliott, who remained a fierce critic of American entry into the war.

Born in 1868 to an Illinois family of abolitionists and freethinkers, Elliott became a beloved figure in the progressive humanist movement of New York City. A tall, affectionate man, Elliott was the pragmatist who put into practice Adler's Ethical Culture principles. He built one of the country's most successful settlement houses, the Hudson Guild, in New York's poverty-stricken Chelsea district. A lifelong trustee of the ACLU, Elliott was politically and personally fearless. When two Austrian leaders of the Ethical Culture Society in Vienna were arrested by Hitler's Gestapo in 1938, Elliott—at the age of seventy—went to Berlin and spent several months negotiating with the Gestapo for their release. After paying a bribe, Elliott succeeded in spiriting the two men out of Nazi Germany. When he died in 1942, the ACLU's Roger Baldwin eulogized him as "a witty saint . . . a man who so loved people that no task to aid them was too small."

It was to this "witty saint" that the Oppenheimer brothers were exposed throughout the years of their weekly dialogues in ethics class. Years later, when they were young men, Elliott wrote their father: "I did not know how close I could get to your boys. Along with you, I am glad and grateful for them." Elliott taught ethics in a Socratic-style seminar where students discussed specific social and political issues. Education in Life Problems was a required course for all of the high school students. Often he would pose a personal moral dilemma for his students, such as asking them if they had a choice between a job teaching or a job that paid more working in Wrigley's chewing gum factory—which would they choose? During Robert's years at the school, some of the topics vigorously debated included the "Negro problem," the ethics of war and peace, economic inequality and understanding "sex relations." In his senior year, Robert was exposed to an extended discussion on the role of "the State." The curriculum included a "short catechism of political ethics," including "the ethics of loyalty and treason." It was an extraordinary education in social relations and world affairs, an education that planted deep roots in Robert's psyche—and one that would produce a bountiful harvest in the decades to come.

· · ·

"I WAS AN UNCTUOUS, repulsively good little boy," Robert remembered. "My life as a child did not prepare me for the fact that the world is full of cruel and bitter things." His sheltered home life had offered him "no normal, healthy way to be a bastard." But it had created an inner toughness, even a physical stoicism, that Robert himself may not have recognized.

Anxious to get him out of doors and among boys his own age, Julius decided to send Robert, at the age of fourteen, to a summer camp. For most of the other boys there, Camp Koenig was a mountain paradise of fun and camaraderie. For Robert, it was an ordeal. Everything about him made him a target for the cruelties young adolescents delight in inflicting on those who are shy, sensitive or different. The other boys soon began calling him "Cutie" and teased him mercilessly. But Robert refused to fight back. Shunning athletics, he walked the trails, collecting rocks. He made one friend, who recalled that Robert was obsessed that summer with the writings of George Eliot. The novelist's major work, *Middlemarch,* appealed to him greatly, perhaps because it explored so thoroughly a topic he found so mysterious: the life of the inner mind in relation to the making and breaking of human relationships.

Then, however, Robert made the mistake of writing his parents that he was glad he had come to camp because the other boys were teaching him the facts of life. This prompted a quick visit by the Oppenheimers, and subsequently the camp director announced a crackdown on the telling of salacious stories. Inevitably, Robert was fingered for tattling, and so one night he was carried off to the camp icehouse, stripped and knocked about. As a final humiliation, the boys doused his buttocks and genitals with green paint. Robert was then left naked and locked inside the icehouse for the night. His one friend later said of this incident that Robert had been "tortured." Robert suffered this gross degradation in stoic silence; he neither left the camp nor complained. "I don't know how Robert stuck out those remaining weeks," said his friend. "Not many boys would have—or could have—but Robert did. It must have been hell for him." As his friends often discovered, Robert's seemingly brittle and delicate shell actually disguised a stoic personality built of stubborn pride and determination, a characteristic that would reappear throughout his life.

Back in school, Robert's highbrow personality was nurtured by the Ethical Culture School's attentive teachers, all of whom had been carefully selected by Dr. Adler as models of the progressive education movement. When Robert's math teacher, Matilda Auerbach, noticed that he was bored and restless, she sent him to the library to do independent work, and later he was allowed to explain to his fellow students what he had learned. His Greek and Latin instructor, Alberta Newton, recalled that he was a delight to teach: "He received every new idea as perfectly beautiful." He read Plato and Homer in Greek, and Caesar, Virgil and Horace in Latin.

Robert always excelled. As early as third grade, he was doing laboratory experiments and by the time he was ten years old, in fifth grade, he was studying physics and chemistry. So clearly eager was Robert to study the sciences that the curator at the American Museum of Natural History agreed to tutor him. As he had skipped several grades, everyone regarded him as precocious—and sometimes too precious. When he was nine, he was once overheard telling an older girl cousin, "Ask me a question in Latin and I will answer you in Greek."

Robert's peers thought him distant at times. "We were thrown together a lot," said a childhood acquaintance, "and yet we were never close. He was usually preoccupied with whatever he was doing or thinking." Another classmate recalled him sitting laconically in class, "exactly as though he wasn't getting enough to eat or drink." Some of his peers thought him "rather gauche . . . he didn't really know how to get along with other children." Robert himself was painfully aware of the costs of knowing so much more than his classmates. "It's no fun," he once told a friend, "to turn the pages of a book and say, 'Yes, yes, of course, I know that.' " Jeanette Mirsky knew Robert well enough in their senior year to think of him as a "special friend." She never thought of him as shy in the usual sense, only distant. He bore a certain "hubris," she thought, of the kind that carries with it the seeds of its own destruction. Everything about Robert's personality—from his abrupt, jerky way of walking to such little things as the making of a salad dressing—displayed, she thought, "a great need to declare his preeminence."

Throughout his high school years, Robert's "homeroom" teacher was Herbert Winslow Smith, who had joined the English department in 1917 after receiving his master's degree from Harvard. A man of remarkable intellect, Smith was well on his way to obtaining a doctorate when he was recruited to teach. He was so taken by his initial experience at Ethical Culture that he never went back to Cambridge. Smith would spend his entire career at Ethical Culture, eventually becoming the school's principal. Barrel-chested and athletic, he was a warm and gentle teacher who somehow always managed to find out what each student was most curious about and then relate it to the topic at hand. After his lectures, students invariably could be found lingering around his desk, trying to squeeze a little more conversation out of their teacher. Though Robert's first passion was clearly science, Smith stoked his literary interests; he thought Robert already had a "magnificent prose style." Once, after Robert wrote an entertaining essay on oxygen, Smith suggested, "I think your vocation is to be a science writer." Smith would become Robert's friend and counselor. He was "very, very kind to his students," Francis Fergusson recalled. "He took on Robert and me and various other people . . . saw them through their troubles and advised them what to do next."

Robert had his breakthrough year as a junior, when he took a course in physics with Augustus Klock. "He was marvelous," Robert said. "I got so excited after the first year, I arranged to spend the summer working with him setting up equipment for the following year, when I would then take chemistry. We must have spent five days a week together; once in a while we would even go off on a mineral hunting junket as a reward." He began to experiment with electrolytes and conduction. "I loved chemistry so deeply. . . . Compared to physics, it starts right in the heart of things and very soon you have that connection between what you see and a really very sweeping set of ideas which could exist in physics but is very much less likely to be accessible." Robert would always feel indebted to Klock for having set him on the road to science. "He loved the bumpy contingent nature of the way in which you actually find out something, and he loved the excitement that he could stir up in young people."

Even fifty years later, Jane Didisheim's memories of Robert were particularly vivid. "He blushed extraordinarily easily," she recalled. He seemed "very frail, very pink-cheeked, very shy, and very brilliant of course. Very quickly everybody admitted that he was different from all the others and superior. As far as studies were concerned he was good in everything. . . ."

The sheltered atmosphere of the Ethical Culture School was ideal for an unusually awkward adolescent polymath. It allowed Robert to shine when and where he wished—and protected him from those social challenges with which he was not yet prepared to cope. And yet, this same cocoon of security offered by the school may help to explain his prolonged adolescence. He was permitted to remain a child, and allowed to grow gradually out of his immaturity rather than being wrenched abruptly from it. At sixteen or seventeen he had only one real friend, Francis Fergusson, a scholarship boy from New Mexico who became his classmate during their senior year. By the time Fergusson met him in the fall of 1919, Robert was just coasting. "He was just sort of playing around and trying to find something to keep himself occupied," recalled Fergusson. In addition to courses in history, English literature, math and physics, Robert enrolled in Greek, Latin, French and German. "He still took straight A's." He would graduate as the valedictorian of his class.

Besides hiking and rock-collecting, Robert's chief physical activity was sailing. By all accounts, he was an audacious, expert sailor who pushed his boat to the edge. As a young boy he had honed his skills on several smaller boats, but when he turned sixteen, Julius bought him a twenty-eight-foot sloop. He christened it the *Trimethy,* a name derived from the chemical compound trimethylene dioxide. He loved sailing in summer storms, racing the boat against the tides through the inlet at Fire Island and straight out into the Atlantic. With his younger brother, Frank, hunkered down in the cockpit, Robert would stand with the tiller between his legs, screaming gleefully into the wind as he tacked the boat back into Long Island's Great South Bay. His parents could not reconcile such impetuous behavior with the Robert they knew as a shy introvert. Invariably, Ella found herself standing at the window of their Bay Shore home, searching for a trace of the *Trimethy* on the skyline. More than once, Julius felt compelled to chase the *Trimethy* back to port in a motor launch, reprimanding Robert for the risks he was taking with his own and others' lives. "Roberty, Roberty . . . ," he would say, shaking his head. Robert, however, was unabashed; indeed, he never failed to display absolute confidence in his mastery over wind and sea. He knew the full measure of his skill and saw no reason to cheat himself of what was clearly an emotionally liberating experience. Still, if not foolhardy, his behavior in stormy seas struck some friends as an example of Robert's deeply ingrained arrogance, or perhaps a not very surprising extension of his inner resiliency. He had an irresistible urge to flirt with danger.

Fergusson would never forget the first time he sailed with Robert. The two friends had both just turned seventeen. "It was a blowy day in spring— very chilly—and the wind made little waves all over the bay," Fergusson said, "and there was rain in the air. It was a little bit scary to me, because I didn't know whether he could do it or not. But he did; he was already a pretty skilled sailor. His mother was watching from the upstairs window and probably having palpitations of all kinds. But he had induced her to let him go. She worried, but she put up with it. We got thoroughly soaked, of course, with the wind and the waves. But I was very impressed."

ROBERT GRADUATED FROM THE Ethical Culture School in the spring of 1921, and that summer Julius and Ella took their sons for the summer to Germany. Robert struck out on his own for a few weeks on a prospecting field trip among some of the old mines near Joachimsthal, northeast of Berlin. (Ironically, just two decades later, the Germans would be mining uranium from this site for their atomic bomb project.) After camping out in rugged conditions, he returned with a suitcase full of rock specimens and what turned out to be a near-fatal case of trench dysentery. Shipped home on a stretcher, he was ill and bedridden long enough to force the postponement of his enrollment at Harvard that autumn. Instead, his parents compelled him to remain at home, recuperating from the dysentery and a subsequent case of colitis. The latter would plague him for the rest of his days, aggravated by a stubborn appetite for spicy foods. He was not a good patient. It was a long winter, cooped up in the New York apartment, and at times he acted boorishly, locking himself in his room and brushing aside his mother's ministrations.

By the spring of 1922, Julius thought the boy well enough to get him out of the house. To this end, he urged Herbert Smith to take Robert with him that summer on a trip to the Southwest. The Ethical Culture School teacher had made a similar trip with another student the previous summer, and Julius thought a Western adventure would help to harden his son. Smith agreed; he was taken aback, however, when Robert approached him in private shortly before their departure with a strange proposition: Would Smith agree to let him travel under the name "Smith" as his younger brother? Smith rejected the suggestion out of hand, and couldn't help but think that some part of Robert was uncomfortable with being identifiably Jewish. Robert's classmate Francis Fergusson later speculated, similarly, that his friend may have felt self-conscious about "his Jewishness and his wealth, and his eastern connections, and [that] his going to New Mexico was partly to escape from that." Another classmate, Jeanette Mirsky, also thought Robert felt some unease about his Jewishness. "We all did," said Mirsky. Yet just a few years later, at Harvard, Robert seemed much more relaxed about his Jewish heritage, telling a friend of Scotch-Irish ancestry, "Well, neither one of us came over on the *Mayflower.*"

STARTING OUT IN THE SOUTH, Robert and Smith gradually made their way across to the mesas of New Mexico. In Albuquerque, they stayed with Fergusson and his family. Robert enjoyed their company and the visit cemented a lifelong friendship. Fergusson introduced Robert to another Albuquerque boy their age, Paul Horgan, an equally precocious boy who later had a successful career as a writer. Horgan happened also to be bound for Harvard, as was Fergusson. Robert liked Horgan and found himself mesmerized by the dark-haired, blue-eyed beauty of Horgan's sister Rosemary. Frank Oppenheimer said that his brother later confided in him that he had been strongly attracted to Rosemary.

When they got to Cambridge and continued to hang out together, Horgan quipped that they were "this great troika" of "polymaths." But New

Mexico had brought out new attitudes and interests in Robert. In Albuquerque, Horgan's first impressions of Robert were particularly vivid: ". . . he combined incredibly good wit and gaiety and high spirits. . . . he had this lovely social quality that permitted him to enter into the moment very strongly, wherever it was and whenever it was."

From Albuquerque, Smith took Robert—and his two friends Paul and Francis—twenty-five miles northeast of Santa Fe to a dude ranch called Los Pinos, run by a twenty-eight-year-old Katherine Chaves Page. This charming and yet imperious young woman would become a lifelong friend. But first there was an infatuation—Robert was intensely attracted to Katherine, who was then newly married. The previous year she had been desperately ill and, seemingly on her deathbed, she had married an Anglo, Winthrop Page, a man her father's age. And then she hadn't died. Page, a businessman in Chicago, rarely spent any time in the Pecos.

The Chaveses were an aristocratic hidalgo family with deep roots in the Spanish Southwest. Katherine's father, Don Amado Chaves, had built the handsome ranch house near the village of Cowles with a majestic view of the Pecos River looking north to the snowcapped Sangre de Cristo mountain range. Katherine was the "reigning princess" of this realm, and, to his delight, Robert found himself to be her "favorite" courtier. She became, according to Fergusson, "his very good friend. . . . He would bring her flowers all the time and he would flatter her to death whenever he saw her."

That summer, Katherine taught Robert horseback-riding and soon had him exploring the surrounding pristine wilderness on rides that sometimes lasted five or six days. Smith was astonished by the boy's stamina and gritty resilience on horseback. Despite his lingering ill-health and fragile appearance, Robert clearly relished the physical challenges of horseback-riding as much as he had enjoyed skirting the edge of danger in his sailboat. One day they were riding back from Colorado and Robert insisted he wanted to take a snow-laden trail over the highest pass in the mountains. Smith was certain that trail could easily expose them to death by freezing—but Robert was dead set on going anyway. Smith proposed they toss a coin to decide the issue. "Thank God I won," Smith recalled. "I don't know how I'd have got out of it if I hadn't." He thought such foolhardiness on Robert's part bordered on the suicidal. In all his dealings with Robert, Smith sensed that this was a boy who wouldn't allow the prospect of death to "keep him from doing something he much wanted to do."

Smith had known Robert since he was fourteen, and the boy had always been physically delicate and somehow emotionally vulnerable. But now, seeing him in the rugged mountains, camping out in spartan conditions, Smith began to wonder whether Robert's persistent colitis might be psychosomatic. It occurred to him that these episodes invariably came on when Robert heard someone making "disparaging" remarks about Jews. Smith thought he had developed the habit of "kicking an intolerable fact under the rug." It was a psychological mechanism, Smith thought, that "when it was carried to its most dangerous, got him into trouble."

Smith was also well caught up on the latest Freudian theories of child development, and he concluded from Robert's relaxed campfire conversations that the boy had pronounced oedipal issues. "I never heard a murmur of criticism on Robert's part of [his] mother," Smith recalled. "He was certainly critical enough of [his] father."

As an adult, Robert clearly loved his father, deferred to him and indeed, until his father's death, went to extraordinary lengths to accommodate him, introduce him to his friends and generally make room for him in his life. But Smith sensed that as a particularly shy and sensitive child, Robert was profoundly mortified by his father's sometimes maladroit affability. Robert told Smith one night around the campfire about the icehouse incident at Camp Koenig—which of course had been prompted by his father's overreaction to his letter home about the sex talk at camp. As an adolescent, he had become increasingly self-conscious about his father's garment business, which he no doubt saw as a traditional Jewish trade. Smith later recalled that once on that 1922 Western trip, he had turned to Robert as they were packing up and asked him to fold a jacket for his suitcase. "He looked at me sharply," Smith recalled, "and said, 'Oh yes. The tailor's son would know how to do that, wouldn't he?' "

Such outbursts aside, Smith thought Robert grew emotionally in stature and confidence during their time together on the Los Pinos ranch. He knew Katherine Page could take a great deal of credit for this. Her friendship was extremely important to Robert. The fact that Katherine and her aristocratic hidalgo friends could accept this insecure New York Jewish boy in their midst was somehow a watershed event in Robert's inner life. To be sure, he knew he was accepted inside the forgiving womb of the Ethical Culture community in New York. But here was approbation from people he liked outside his own world. "For the first time in his life," Smith thought, ". . . [Robert] found himself loved, admired, sought after." It was a feeling Robert cherished, and in the years ahead he would learn to cultivate the social skills required to call up such admiration on demand.

One day he, Katherine and a few others from Los Pinos took packhorses out and, starting from the village of Frijoles west of the Rio Grande, they rode south and ascended the Pajarito (Little Bird) Plateau, which rises to a height of over 10,000 feet. They rode through the Valle Grande, a canyon inside the Jemez Caldera, a bowl-shaped volcanic crater twelve miles wide.

Turning northeast, they then rode four miles and came upon another canyon which took its Spanish name from the cottonwood trees that bordered a stream trickling through the valley: Los Alamos. At the time, the only human habitation for many miles consisted of a spartan boys' school, the Los Alamos Ranch School.

Los Alamos, the physicist Emilio Segré would later write when he saw it, was "beautiful and savage country." Patches of grazing meadows broke up dense pine and juniper forests. The ranch school stood atop a two-mile-long mesa bounded on the north and south by steep canyons. When Robert first visited the school in 1922, there were only some twenty-five boys enrolled, most of them the sons of newly affluent Detroit automobile manu-facturers. They wore shorts throughout the year and slept on unheated sleeping porches. Each boy was responsible for tending a horse, and pack trips into the nearby Jemez mountains were frequent. Robert admired the setting—so obviously a contrast to his Ethical Culture environment—and in years to come would repeatedly find his way back to this desolate mesa.

Robert came away from that summer love-struck with the stark desert/mountain beauty of New Mexico. When, some months later, he heard that Smith was planning another trip to "Hopi country," Robert wrote him: "Of course I am insanely jealous. I see you riding down from the mountains to the desert at that hour when thunderstorms and sunsets capari-son the sky; I see you in the Pecos . . . spending the moonlight on Grass Mountain."

CHAPTER TWO

"His Separate Prison"

The notion that I was traveling down a clear track would be wrong.

ROBERT OPPENHEIMER

IN SEPTEMBER 1922, Robert Oppenheimer enrolled at Harvard. Although the university awarded him a fellowship, he didn't accept it "because I could get along well without the money." In lieu of the scholarship, Harvard gave him a volume of Galileo's early writings. He was assigned a single room in Standish Hall, a freshman dormitory facing the Charles River. At nineteen, Robert was an oddly handsome young man. Every feature of his body was of an extreme. His fine pale skin was drawn taut across high cheekbones. His eyes were the brightest pale blue, but his eyebrows were glossy black. He wore his coarse, kinky black hair long on top, but short at the sides—so he seemed even taller than his lanky five feet, ten-inch frame. He weighed so little—never more than 130 pounds—that he gave an impression of flimsiness. His straight Roman nose, thin lips and large, almost pointed ears accentuated an image of exaggerated delicacy. He spoke in fully grammatical sentences with the kind of ornate European politeness his mother had taught him. But as he talked, his long, thin hands made his gestures seem somehow contorted. His appearance was mesmerizing, and slightly bizarre.

His behavior in Cambridge over the next three years did nothing to soften the impression his appearance gave of a studious, socially inept and immature young man. As surely as New Mexico had opened up Robert's personality, Cambridge drove him back to his former introversion. At Harvard his intellect thrived, but his social development floundered; or so it seemed to those who knew him. Harvard was an intellectual bazaar filled with delights for the mind. But it offered Robert none of the careful guidance and devoted nurturing of his Ethical Culture experience. He was on his own, and so he retreated into the security his powerful intellect assured. He seemed incapable of not flaunting his eccentricities. His diet often consisted of little more than chocolate, beer and artichokes. Lunch was often just a "black and tan"—a piece of toast slathered with peanut butter and topped with chocolate syrup. Most of his classmates thought him diffident. Fortunately, both Francis Fergusson and Paul Horgan were also at Harvard that year, so he had at least two soul mates. But he made very few new friends. One was Jeffries Wyman, a Boston Brahmin who was beginning graduate studies in biology. "He [Robert] found social adjustment very difficult," Wyman recalled, "and I think he was often very unhappy. I suppose he was lonely and felt he didn't fit in well. . . . We were good friends, and he had some other friends, but there was something that he lacked . . . because our contacts were largely, I should say wholly, on an intellectual basis."

Introverted and intellectual, Robert was already reading such dark-spirited writers as Chekhov and Katherine Mansfield. His favorite Shakespearean character was Hamlet. Horgan recalled years later that "Robert had bouts of melancholy, deep, deep depressions as a youngster. He would seem to be incommunicado emotionally for a day or two at a time. That happened while I was staying with him once or twice, and I was very distressed, had no idea what was causing it."

Sometimes Robert's flair for the intellectual went beyond the merely ostentatious. Wyman recalled a hot spring day when Oppenheimer walked into his room and said, "What intolerable heat. I have been spending all afternoon lying on my bed reading Jeans' *Dynamical Theory of Gases*. What else can one do in weather like this?" (Forty years later, Oppenheimer still had in his possession a weathered and salt-encrusted copy of James Hopwood Jeans' book *Electricity and Magnetism*.)

In the spring of Robert's freshman year, he formed a friendship with Frederick Bernheim, a pre-med student who had graduated from the Ethical Culture School a year after him. They shared an interest in science, and with Fergusson about to leave for England on a Rhodes Scholarship, Robert soon anointed Bernheim as his new best friend. Unlike most college-age men—who tend to have many acquaintances and few deep friendships—Robert's friendships were few and intense.

In September 1923, at the beginning of their sophomore year, he and Bernheim decided to share adjacent rooms in an old house at 60 Mount Auburn Street, close to the offices of the *Harvard Crimson*. Robert decorated his room with an oriental rug, oil paintings and etchings he brought from home, and insisted on making tea from a charcoal-fired Russian samovar. Bernheim was more amused than annoyed by his friend's eccentricities: "He wasn't a comfortable person to be around, in a way, because he always gave the impression that he was thinking very deeply about things. When we roomed together he would spend evenings locked in his room, trying to do something with Planck's constant or something like that. I had visions of him suddenly bursting forth as a great physicist, and here I was just trying to get through Harvard."

Bernheim thought Robert was something of a hypochondriac. "He went to bed with an electric pad every night, and one day it started to smoke." Robert woke up and ran to the bathroom with the burning pad. He then went back to sleep, unaware that the pad was still burning. Bernheim recalled having to put the thing out before it burned the house down. Living with Robert was always "a little bit of a strain," Bernheim noted, "because you had to more or less adjust to his standards or moods—he was really the dominant one." Difficult or not, Bernheim roomed with Robert for their two remaining years at Harvard and credited him with inspiring his later career in medical research.

Only one other Harvard student dropped by their Mt. Auburn Street quarters with any regularity. William Clouser Boyd had met Robert in chemistry class one day and took an instant liking to him. "We had lots of interests in common aside from science," he recalled. They both tried to write poetry, sometimes in French, and short stories imitative of Chekhov. Robert always called him "Clowser," purposely misspelling his middle name. "Clowser" often joined Robert and Fred Bernheim on occasional weekend expeditions to Cape Ann, an hour's drive northeast of Boston. Robert didn't yet know how to drive, so the boys would go in Bernheim's Willys Overland and spend the night at an inn in Folly Cove outside of Gloucester where the food was particularly good. Boyd would finish Harvard in three years, and, like Robert, he worked hard to do it. But while Robert obviously spent many long hours in his room studying, Boyd remembers that "he was pretty careful not to let you catch him at it." He thought Robert could run circles around him intellectually. "He had a very quick mind. For instance, when someone would propose a problem, he would give two or three wrong answers, followed by the right one, before I could think of a single answer."

The one thing Boyd and Oppenheimer did not have in common was music. "I was very fond of music," Boyd recalled, "but once a year he would go to an opera, with me and Bernheim usually, and he'd leave after the first act. He just couldn't take any more." Herbert Smith had also noticed this peculiarity, and once said to Robert, "You're the only physicist I've ever known who wasn't also musical."

INITIALLY, ROBERT WAS NOT SURE which academic path he should choose. He took a variety of unrelated courses, including philosophy, French literature, English, introductory calculus, history and three chemistry courses (qualitative analysis, gas analysis and organic chemistry). He briefly considered architecture, but because he had loved Greek in high school he also toyed with the thought of becoming a classicist or even a poet or painter. "The notion that I was traveling down a clear track," he recalled, "would be wrong." But within months he settled on his first passion, chemistry, as a major. Determined to graduate in three years, he took the maximum number of allowed courses, six. But each semester he also managed to audit two or three others. With virtually no social life, he studied long hours—though he made an effort to hide the fact because somehow it was important to him that his brilliance appear effortless. He read all three thousand pages of Gibbon's classic history, *The Decline and Fall of the Roman Empire.* He also read widely in French literature and began writing poetry, a few examples of which appeared in *Hound and Horn,* a student journal. "When I am inspired," he wrote Herbert Smith, "I jot down verses. As you so neatly remarked, they aren't either meant or fit for anyone's perusal, and to force their masturbatic excesses on others is a crime. But I shall stuff them in a drawer for a while and, if you want to see them, send them off." That year, T. S. Eliot's *The Waste Land* was published, and when Robert read it, he instantly identified with the poet's sparse existentialism. His own poetry dwelt with themes of sadness and loneliness. Early in his tenure at Harvard, he wrote these lines:

> *The dawn invests our substance with desire*
> *And the slow light betrays us, and our wistfulness:*
> *When the celestial saffron*
> *Is faded and grown colourless,*
> *And the sun*
> *Gone sterile, and the growing fire*
> *Stirs us to waken,*
> *We find ourselves again*
> *Each in his separate prison*
> *Ready, hopeless*
> *For negotiation*
> *With other men.*

Harvard's political culture in the early 1920s was decidedly conservative. Soon after Robert's arrival, the university imposed a quota to restrict the number of Jewish students. (By 1922, the Jewish student population had risen to twenty-one percent.) In 1924, the *Harvard Crimson* reported on its front page that the university's former president Charles W. Eliot had publicly declared it "unfortunate" that growing numbers of the "Jewish race" were intermarrying with Christians. Few such marriages, he said, turned out well, and because biologists had determined that Jews are "prepotent" the children of such marriages "will look like Jews only." While Harvard accepted a few Negroes, President A. Lawrence Lowell staunchly refused to allow them to reside in the freshman dormitories with whites.

Oppenheimer was not oblivious to these issues. Indeed, early that autumn of 1922 he joined the Student Liberal Club, founded three years earlier as a forum for students to discuss politics and current events. In its early years, the club attracted large audiences with such speakers as the liberal journalist Lincoln Steffens, Samuel Gompers of the American Federation of Labor and the pacifist A. J. Muste. In March 1923, the club took a formal stand against the university's discriminatory admissions policies. Though it cultivated a reputation for radical views, Robert was not impressed, and wrote to Smith of the "asinine pomposity of the Liberal Club." In this, his first introduction to organized politics, he felt himself "very much a fish out of water." Nevertheless, lunching one day at the club's rooms at 66 Winthrop Street, he was introduced to a senior, John Edsall, who quickly convinced him to help edit a new student journal. Drawing on his Greek, he persuaded Edsall to call the journal *The Gad-fly;* the title page reproduced a quotation in Greek describing Socrates as the gadfly of the Athenians. The first issue of *The Gad-fly* came out in December 1922, and Robert was listed on its masthead as an associate editor. He remembered writing a few unsigned articles, but *The Gad-fly* did not become a permanent fixture on campus and only four issues survive. However, Robert's friendship with Edsall continued.

By the end of his freshman year at Harvard, Robert decided that he had made a mistake in selecting chemistry as his major. "I can't recall how it came over me that what I liked in chemistry was very close to physics," Oppenheimer said. "It's obvious that if you were reading physical chemistry and you began to run into thermodynamical and statistical mechanical ideas you'd want to find out about them. . . . It's a very odd picture; I never had an elementary course in physics." Though committed to a chemistry major, that spring he petitioned the Physics Department for graduate standing, which would allow him to take upper-level physics courses. To demonstrate that he knew something about physics, he listed fifteen books he claimed to have read. Years later, he heard that when the faculty committee met to consider his petition, one professor, George Washington Pierce, quipped, "Obviously, if he [Oppenheimer] says he's read these books, he's a liar, but he should get a Ph.D. for knowing their titles."

His primary tutor in physics became Percy Bridgman (1882–1961), who later won a Nobel Prize. "I found Bridgman a wonderful teacher," Oppenheimer remembered, "because he never really was quite reconciled to things being the way they were and he always thought them out." "A very intelligent student," Bridgman later said of Oppenheimer. "He knew enough to ask questions." But when Bridgman assigned him a lab experiment that required making a copper-nickel alloy in a self-built furnace, Oppenheimer "didn't know one end of the soldering iron from the other." Oppenheimer was so clumsy with the lab's galvanometer that its delicate suspensions had to be replaced every time he used the apparatus. Robert nevertheless showed persistence and Bridgman found the results interesting enough to publish them in a scientific journal. Robert was both precocious and, on occasion, irritatingly brash. One evening Bridgman invited him to his home for tea. In the course of the evening, the professor showed his student a photograph of a temple built, he said, about 400 B.C. in Segesta, Sicily. Oppenheimer quickly disagreed: "I judge from the capitals on the columns that it was built about fifty years earlier."

When the famous Danish physicist Niels Bohr gave two lectures at Harvard in October 1923, Robert made a point of attending both. Bohr had just won the Nobel the previous year for "his investigations of the structure of atoms and of the radiation emanating from them." Oppenheimer would later say that "it would be hard to exaggerate how much I venerate Bohr." Even on this occasion, his first glimpse of the man, he was deeply moved. Afterwards, Professor Bridgman noted that "[t]he impression he [Bohr] made on everyone who met him was a singularly pleasant one personally. I have seldom met a man with such evident singleness of purpose and so apparently free from guile . . . he is now idolized as a scientific god through most of Europe."

Oppenheimer's approach to learning physics was eclectic, even haphazard. He focused on the most interesting, abstract problems in the field, bypassing the dreary basics. Years later, he confessed to feeling insecure about the gaps in his knowledge. "To this day," he told an interviewer in 1963, "I get panicky when I think about a smoke ring or elastic vibrations. There's nothing there—just a little skin over a hole. In the same way my mathematical formation was, even for those days, very primitive. . . . I took a course from [J. E.] Littlewood on number theory—well, that was nice, but that wasn't really how to go about learning mathematics for the professional pursuit of physics."

When Alfred North Whitehead arrived on campus, only Robert and one other undergraduate had the courage to sign up for a course with the philosopher and mathematician. They painstakingly worked their way through the three volumes of *Principia Mathematica,* coauthored by Whitehead and Bertrand Russell. "I had a very exciting time," Oppenheimer recalled, "reading the *Principia* with Whitehead, who had forgotten it, so that he was both teacher and student." Despite this experience, Oppenheimer always thought he was deficient in mathematics. "I never did learn very much. I probably learned a good deal by a method that is never given enough credit, that is, by being with people. . . . I should have learned more mathematics. I think I would have enjoyed it, but it was a part of my impatience that I was careless about it."

But if there were gaps in his education, he could admit to his friend Paul Horgan that Harvard was good for him. In the autumn of 1923, Robert wrote Horgan a satirical letter in which he wrote about himself in the third person: "[Oppenheimer] has grown to be quite a man now you have no idea how Harvard has changed him. I am afraid it is not for the good of his soul to study so hard. He says the most *terrible* things. Only the other night I was arguing with him and I said but you believe in God don't you? And he said I believe in the second law of thermodynamics, in Hamilton's Principle, in Bertrand Russell, and would you believe it Siegfried [*sic*] Freud."

Horgan thought Robert enthralling and charming. Horgan was himself a brilliant young man, and in the course of his long life he would eventually write seventeen novels and twenty works of history, winning the Pulitzer Prize twice. But he would always regard Oppenheimer as a rare and invaluable polymath. "Leonardos and Oppenheimers are scarce," Horgan wrote in 1988, "but their wonderful love and projection of understanding as both private connoisseurs and historical achievers offer us at least an ideal to consider and measure by."

THROUGHOUT HIS YEARS AT HARVARD, Robert kept up a frequent correspondence with his Ethical Culture School teacher and New Mexico guide, Herbert Smith. In the winter of 1923, he tried to convey with elaborate irony what his life was like at Harvard: "Generously, you ask what I do," Oppenheimer wrote Smith. "Aside from the activities exposed in last week's disgusting note, I labor, and write innumerable theses, notes, poems, stories and junk; I go to math lib[rary] and read and to the Phil[osophy] lib and divide my time between Minherr [Bertrand] Russell and the contemplation of a most beautiful and lovely lady who is writing a thesis on Spinoza—charmingly ironic, at that, don't you think?; I make stenches in three different labs, listen to [Professor Louis] Allard gossip about Racine, serve tea and talk learnedly to a few lost souls, go off for the weekend to distill low grade energy into laughter and exhaustion, read Greek, commit faux pas, search my desk for letters, and wish I were dead. Voila."

Dark wit aside, Robert still suffered periodic bouts of depression. Some of these episodes were brought on by his family's visits to Cambridge. Fergusson remembers going out to dinner with Robert and some of his relatives—not his parents—and watching as his friend turned visibly green from the strain of being polite. Afterwards, Robert would drag Fergusson with him as he pounded the pavement for miles, talking all the while in his quiet, even voice about some physics problem. Walking was his only therapy. Fred Bernheim recalled hiking one winter night until 3:00 a.m. On one of these cold winter walks, someone dared the boys to jump in the river. Robert and at least one of his friends stripped and plunged into the icy water.

In retrospect, all of his friends noted that he seemed to be wrestling in these years with inner demons. "My feeling about myself," Oppenheimer later said of this period in his life, "was always one of extreme discontent. I had very little sensitiveness to human beings, very little humility before the realities of this world."

Unfulfilled sexual desires certainly lay behind some of Robert's troubles. At the age of twenty, he was not alone, of course. Few of his friends had a social life that included women. And none of them remember Robert ever taking a woman on a date. Wyman recalled that he and Robert were "too much in love" with intellectual life "to be thinking about girls. . . . We were all going through a series of love affairs [with ideas] . . . but perhaps we lacked some of the more mundane forms of love affairs that make life easier." Robert certainly felt a welter of sensuous desires, as evidenced by some of the decidedly erotic poetry he wrote during this period:

> Tonight she wears a sealskin cape
> glistening black diamonds where the water swathes her thighs
> and noxious glints conspire to surprise
> a pulse condoning eagerness with rape.

In the winter of 1923–24, he wrote what he called "my first love poem"—to honor that "most beautiful and lovely lady who is writing a thesis on Spinoza." He contemplates this mystery woman from afar in the library but apparently never speaks to her.

> No, I know that there have been others who have read Spinoza,
> Even I;
> Others who have crossed their white arms
> Across the umber pages;
> Others too pure to glance, even a second,
> Beyond the sacred sphincter of their erudition.

But what is all that to me?
You must come, I say, and see the sea gulls,
Gold in the late sun;
You must come and talk to me and tell me why
In this same world, little white puffs of cloud-
Like cotton batting, if you will, or lingerie,
I have heard that before—
Little white puffs of cloud should float so quietly across the
Cleanly sky,
And you should sit, pale, in a black dress that would have graced
The stern ascetic conscience of a Benedict,
And read Spinoza, and let the wind blow the clouds by,
And let me drown myself in an ecstasy of dearth . . .

Well, what if I do forget,
Forget Spinoza and your constancy,
Forget everything, till there stays with me
Only a faint half hope and half regret
And the unnumbered stretches of the sea?

Unable to initiate a relationship, he remained aloof, hoping, as the poem says, that the young woman would make the first move: "You must come and talk to me . . ." He feels "a faint half hope and half regret." Such a mix of powerful emotions is not, of course, unusual for a young man coming of age. But Robert had to be told that he was not alone.

Again and again, whenever he was in anguish, Robert turned to his old teacher for help. In the late winter of 1924, he wrote Smith in the great "distress" of some emotional crisis. That letter has not survived, but we have Robert's reply to Smith's letter of reassurance. "What has soothed me most, I think," he told Smith, "is that you perceived in my distress a certain similarity to that from which you had suffered; it had never occurred to me that the situation of anyone who now appeared to me in all respects so impeccable and so enviable could be in any way comparable with my own. . . . Abstractly, I feel that it is a terrible pity that there should be so many good people I shall not know, so many joys missed. But you are right. At least for me the desire is not a need; it is an impertinence."

After Robert finished his first year at Harvard, his father found him a summer job in a New Jersey laboratory. But he was bored. "The job and people are bourgeois and lazy and dead," he wrote Francis Fergusson, who was himself back in lovely Los Pinos. "There is little work and nothing to puzzle at . . . how I envy you! . . . Francis, you choke me with anguish and despair; all I can do is admit to my hierarchy of physico-chemical immutabilities the Chaucerian 'Amour vincit omnia.' " Robert's friends were used to this florid language. "Everything he takes up," Francis later observed, "he exaggerates." Paul Horgan too recalled Robert's "baroque tendency to exaggerate." But it was also true that he quit the lab job and spent the month of August back at Bay Shore, much of the time sailing with Horgan, who had agreed to spend his vacation with him.

IN JUNE 1925, after only three years of study, Robert was graduated summa cum laude with a bachelor's degree in chemistry. He made the dean's list and was one of only thirty students to be selected for membership in Phi Beta Kappa. Tongue in cheek, he wrote Herbert Smith that year: "Even in the last stages of senile aphasia I will not say that education, in an academic sense, was only secondary when I was at college. I plow through about five or ten big scientific books a week, and pretend to research. Even if, in the end, I've got to satisfy myself with testing toothpaste, I don't want to know it till it has happened."

Testing toothpaste was hardly a likely future for a Harvard senior who that year had taken such courses as "Colloid Chemistry," "History of England from 1688 to the Present Time," "Introduction to the Theory of Potential Functions and LaPlace's Equation," "The Analytical Theory of Heat and Problems of Inelastic Vibrations" and "Mathematical Theory of Electricity and Magnetism." But decades later, he would look back on his undergraduate years and confess: "Although I liked to work, I spread myself very thin and got by with murder; I got A's in all these courses which I don't think I should have." He thought he had acquired a "very quick, superficial, eager familiarization with some parts of physics with tremendous lacunae and often with a tremendous lack of practice and discipline."

Skipping the commencement ceremonies, Robert and two friends, William C. Boyd and Frederick Bernheim, celebrated privately with laboratory alcohol in a dorm room. "Boyd and I got plastered," Bernheim recalled. "Robert, I think, only took one drink and retired." Later that weekend, Robert took Boyd to the summer house in Bay Shore and sailed his beloved *Trimethy* to Fire Island. "We took off our clothes," Boyd remembered, "and walked up and down the beach getting a sunburn." Robert could have stayed at Harvard—he was offered a graduate fellowship—but he already had loftier ambitions. He had graduated as a chemistry major, but it was physics that called him, and he knew that in the world of physics Cambridge, England, was "more near the center." Hoping that the eminent English physicist Ernest Rutherford, celebrated as the man who had first developed a model of the nuclear atom in 1911, would take him under his wing, Robert persuaded his physics teacher Percy Bridgman to write a letter of recommendation. In his letter, Bridgman wrote candidly that Oppenheimer had a "perfectly prodigious power of assimilation" but that "his weakness is on the experimental side. His type of mind is analytical, rather than physical, and he is not at home in the manipulations of the laboratory. . . . It appears to me that it is a bit of a gamble as to whether Oppenheimer will ever make any real contributions of an important character, but if he does make good at all, I believe he will be a very unusual success."

Bridgman then closed with remarks—not unusual for that time and place—on Oppenheimer's Jewish background: "As appears from his name, Oppenheimer is a Jew, but entirely without the usual qualifications of his race. He is a tall, well set-up young man, with a rather engaging diffidence of manner, and I think you need have no hesitation whatever for any reason of this sort in considering his application."

With the hope that Bridgman's recommendation would gain him admittance to Rutherford's laboratory, Robert spent the month of August in his beloved New Mexico. Significantly, he took his parents with him and introduced them to his few acres of heaven. The Oppenheimers boarded for a time at Bishop's Lodge on the outskirts of Santa Fe, and then journeyed north to Katherine Page's Los Pinos ranch. "The Parents are really quite pleased with the place," Robert wrote with obvious pride to Herbert Smith, "and are starting to ride a little. Curiously enough they enjoy the frivolous courtesy of the place."

Together with Paul Horgan, who was back from Harvard for the summer, and Robert's brother, Frank, now age thirteen, the boys went for long horseback rides in the mountains. Horgan recalls renting horses in Santa Fe and riding with Robert on the Lake Peak trail across the Sangre de Cristo range and down to the village of Cowles: "We hit the divide at the very top of that mountain in a tremendous thunderstorm . . . immense, huge pounding rain. We sat under our horses for lunch and ate oranges, [and we] were drenched. . . . I was looking at Robert and all of a sudden I noticed his hair was standing straight up, responding to the static. Marvelous." When they finally rode into Los Pinos that night after dark, Katy Page's windows were lit. "It was a very welcome sight," Horgan said. "She received us and we had a beautiful time for several days there. She referred to us always then and afterward as her slaves. 'Here come my slaves.' "

While Mrs. Oppenheimer sat on the shaded, wraparound porch of the Los Pinos ranch house, Page and her "slaves" went out on day-long rides in the surrounding mountains. On one of these expeditions, Robert found a small, uncharted lake on the eastern slopes of the Santa Fe Baldy—which he named Lake Katherine.

It was probably on one of these long rides that he smoked his first tobacco. Page taught the boys to ride light, packing the bare minimum. One night on the trail Robert found himself out of food, and someone offered him a pipe to quell the pangs of hunger. Pipe tobacco and cigarettes quickly became thereafter a lifelong addiction.

Upon his return to New York, Robert opened his mail to learn that Ernest Rutherford had rejected him. "Rutherford wouldn't have me," Oppenheimer recalled. "He didn't think much of Bridgman and my credentials were peculiar." In the event, however, Rutherford passed Robert's application along to J. J. Thomson, Rutherford's celebrated predecessor as director of Cavendish Laboratory. At sixty-nine years of age, Thomson, who had won the 1906 Nobel Prize in physics for his detection of the electron, was well past his prime as a working physicist. In 1919 he had resigned his administrative responsibilities, and by 1925 he came into the laboratory sporadically and tutored only the occasional student. Robert was nonetheless greatly relieved when he learned that Thomson had agreed to supervise his studies. He had chosen physics as his vocation, and he was confident that its future—and his—lay in Europe.

CHAPTER THREE

"*I Am Having a Pretty Bad Time*"

*I am not well, and I am afraid to come to see you now for
fear something melodramatic might happen.*

ROBERT OPPENHEIMER,
January 23, 1926

HARVARD HAD BEEN A MIXED EXPERIENCE for Robert. He
had grown intellectually, but his social experiences had been such
as to leave his emotional life taut and strained. The daily routines
of structured undergraduate existence had provided him with a protective
shield; once again, he had been a superstar in the classroom. Now the shield
was gone, and he was about to undergo a series of nearly disastrous existential crises that would begin that autumn and stretch into the spring of 1926.

In mid-September 1925, he boarded a ship bound for England. He and
Francis Fergusson had agreed that they would meet in the little village of
Swanage in Dorsetshire, in southwest England. Fergusson had spent the
entire summer traveling about Europe with his mother and was now eager
for some male companionship. For ten days they walked along the coastal
cliffs, confiding to each other their latest adventures. Though they had not
seen each other for two years, they had kept in touch through correspondence and remained close.

"When I met him at the station," Fergusson wrote afterwards, "he
seemed more self-confident, strong and upstanding . . . he was far less
embarrassed with mother. This, I afterwards found out, was because he had
nearly managed to fall in love with an attractive gentile in New Mexico."
Still, at the age of twenty-one, Robert, Fergusson sensed, "was completely
at a loss about his sex life." For his part, Fergusson "unfolded to him all the
things that had pleased me, and that I had to keep quiet about." In retrospect, however, Fergusson thought he had unburdened himself too much. "I
was cruel and stupid enough," he wrote, "to go over with Robert [these
things] at length, finally completing what Jean [a friend] would have called
a first-class mental rape."

By then, Fergusson had spent two full years as a Rhodes Scholar at Oxford. Francis had always been more mature than Robert, who now found himself dazzled by his friend's ease and social polish. To begin with, Francis had had a girlfriend for some three years—a young woman Robert had known from the Ethical Culture School named Frances Keeley. He was also impressed that Fergusson had demonstrated the self-confidence to abandon his major in biology for his first passion, literature and poetry. He was moving in elite social circles, visiting upper-class English families in their country houses. Robert found himself envious of his friend's blossoming sophistication. They parted ways—one off to Oxford, the other to Cambridge—promising to meet again over Christmas break.

ROBERT'S ARRIVAL AT CAVENDISH LABORATORY in Cambridge coincided with a time of great excitement in the world of physics. By the early 1920s some European physicists—Niels Bohr and Werner Heisenberg, among others—were building a theory that they called quantum physics (or quantum mechanics). Briefly stated, quantum physics is the study of the laws that apply to the behavior of phenomena on a very small scale, the scale of molecules and atoms. Quantum theory was soon to replace classical physics when dealing with subatomic phenomena such as an electron orbiting around the nucleus of a hydrogen atom.

But if this was a "hot time" for physicists in Europe, Oppenheimer and many more senior American physicists were oblivious. "I was still, in the bad sense of the word, a student," recalled Oppenheimer. "I didn't learn about quantum mechanics until I got to Europe. I didn't learn about electron spin until I got to Europe. I don't believe that they were actually known in '25 in the spring in America; anyway, I didn't know them."

Robert settled into a depressing apartment that he later called a "miserable hole." He took all his meals at the college and spent his days in a corner of J. J. Thomson's basement laboratory, trying to make thin beryllium films for use in the study of electrons. It was a laborious process which required the evaporation of beryllium onto collodion; afterwards, the collodion had to be painstakingly discarded. Clumsy and inept at this meticulous work, Robert soon found himself avoiding the laboratory. Instead, he spent his time attending seminars and reading physics journals. But if his lab work was "quite a sham," it nevertheless provided the occasion for him to meet physicists like Rutherford, Chadwick and C. F. Powell. "I met [Patrick M. S.] Blackett whom I liked very much," Oppenheimer recalled decades later. Patrick Blackett—who would win the Nobel Prize for physics in 1948—soon became one of Robert's tutors. A tall, elegant Englishman with forthright socialist politics, Blackett had completed his physics degree at Cambridge just three years earlier.

In November 1925, Robert wrote Fergusson that "the place is very rich, and has plenty of luscious treasures; and although I am altogether unable to take advantage of them, yet I have a chance to see many people, and a few good ones. There are certainly some good physicists here—the young ones, I mean. . . . I have been taken to all sorts of meetings: High Maths at Trinity, a secret pacifist meeting, a Zionist club, and several rather pallid science clubs. But I have seen no one here who is of any use who is not doing science. . . ." But then he dropped the bravado and confessed: "I am having a pretty bad time. The lab work is a terrible bore, and I am so bad at it that it is impossible to feel that I am learning anything . . . the lectures are vile."

His difficulties in the laboratory were compounded by his deteriorating emotional state. One day Robert caught himself staring at an empty blackboard with a piece of chalk in his hand, muttering over and over, "The point is, the point is . . . the point is." His Harvard friend Jeffries Wyman, who was also at Cambridge that year, detected signs of distress. Walking into his room one day, Wyman found Robert lying on the floor, groaning and rolling from side to side. In another account of this incident, Wyman reported Oppenheimer as telling him "that he felt so miserable in Cambridge, so unhappy, that he used sometimes to get down on the floor and roll from side to side—he told me that." On another occasion, Rutherford witnessed Oppenheimer collapsing in a heap on the floor of the laboratory.

Neither was it any comfort that some of his closest friends were headed toward early domesticity. His Harvard roommate, Fred Bernheim, was also in Cambridge and had met a woman who would soon become his wife. Robert could see that his friendship with Bernheim was, predictably, petering out. "There are some terrible complications with Fred," Oppenheimer explained to Fergusson, "and an awful evening, two weeks ago, in the Moon. I have not seen him since, and blush when I think of him. And a Dostoievskian confession from him."

Robert demanded much of his friends and sometimes his demands were just too much. "In a way," Bernheim recalled, "it was a relief. . . . His intensity and his drive always made me feel slightly uncomfortable." Bernheim felt drained in Robert's presence. Robert stubbornly tried to revive the friendship, but Bernheim finally told him that he was going to marry and that "we couldn't re-establish what we'd had at Harvard." Robert was not so much offended as perplexed that someone he had known so well could decide to spin out of his orbit. Similarly, he was astonished to learn of the early marriage of Jane Didisheim, another classmate from the Ethical Culture School. Robert had always been fond of Jane and seemed taken aback that a woman his own age could already be married (to a Frenchman) and with child.

By the end of that autumn term, Fergusson concluded that Robert had a "first class case of depression." His parents also had some inkling that their son was in crisis. According to Fergusson, Robert's depression "was further increased and made specific by the struggle he was carrying on with his mother." Ella and Julius now insisted on rushing across the Atlantic to be with their troubled son. "He wanted her," Fergusson wrote in his diary, "but felt he ought to discourage her coming. . . . So that when he got on the train for Southampton where he was to meet her, he was exploding with every kind of wild revolution."

Fergusson was a witness to only some of the extraordinary events that followed that winter. But clearly many of the details that Fergusson recorded could only have come from Robert—and it is quite possible, indeed, it is almost certain, that in recounting his experiences, Robert allowed his vivid imagination to color his stories. Fergusson's "Account of the Adventures of Robert Oppenheimer in Europe" is dated simply "February 26," and the context suggests that it was written contemporaneously in February 1926. In any case, Fergusson did not reveal his friend's confidences until many years after Robert's death.

According to Fergusson's account, an episode occurred aboard the train that indicated that Robert was losing control emotionally. "He found himself in a third-class carriage with a man and woman who were making love [kissing and perhaps fondling each other, we assume], and though he tried to read thermodynamics he could not concentrate. When the man left he [Robert] kissed the woman. She did not seem unduly surprised. . . . But he was at once overcome with remorse, fell on his knees, his feet sprawling, and with many tears, begged her pardon." Hastily gathering up his luggage, Robert then fled the compartment. "His reflections were so bitter that, on the way out of the station, when they were going downstairs, and he saw the woman below him, he was inspired to drop his suitcase on her head. Fortunately, he missed." Assuming that Fergusson accurately reported the story he was told, it seems clear that Robert was caught up in a fantasy. He wanted to kiss the woman. He did kiss her? He didn't? What happened exactly in that train compartment is uncertain. But what is reported to have occurred exiting the station surely did not happen, although Robert needed to tell Fergusson that it did. He was in trouble; he was losing control, and his fantastic tale was an expression of his distress.

In this agitated state, Robert proceeded on to the port where he was scheduled to greet his parents. The first person he saw on the gangplank was not his mother or father, but Inez Pollak, a classmate from the Ethical Culture School. Robert had corresponded with Inez while she attended Vassar, and he had seen her on occasion in New York during vacations. In an interview decades later, Fergusson said he thought that Ella "saw to it that there came with them [to England] a young woman that he [Robert] had seen in New York, and she tried to put them together and it didn't work."

In his "diary," Fergusson writes that upon seeing Inez on the gangplank, Robert's first impulse had been to turn and run. "Now it would have been difficult," Fergusson wrote, "to say who was the more terrified, Inez or Robert." For her part, Inez apparently saw in Robert an escape from her life in New York, where her mother had grown intolerable to her. Ella had agreed to escort her to England, thinking that Inez might help to distract Robert from his depression. But at the same time, according to Fergusson, Ella regarded Inez as "ridiculously unworthy" of her son, and as soon as she saw that Robert was actually showing an interest in the girl, she took Robert aside and spoke of how "tiresome it was for Inez to have come over."

Inez nevertheless accompanied the Oppenheimers to Cambridge. Robert busied himself with his physics, but in the afternoons he began taking Inez on long walks about town. According to Fergusson, Robert went through the motions of courting her. He "did a very good, and chiefly rhetorical, imitation of being in love with her. She responded in kind." For a short time, the couple were at least informally engaged. And then one evening they went to Inez's room and crawled into bed together. "There they lay, tremulous with cold, afraid to do anything. And Inez began to sob. Then Robert began to sob." After a time there was a knock at the door, and they heard Mrs. Oppenheimer's voice saying, "Let me in, Inez, why won't you let me in? I know Robert is in there." Ella finally stomped off in a huff, and Robert emerged, miserable and thoroughly humiliated.

Pollak left almost immediately for Italy, taking with her a copy of Dostoyevsky's *The Possessed* as a gift from Robert. Naturally, the collapse of this relationship only deepened Robert's melancholy. Just before classes broke for Christmas, he wrote Herbert Smith a sad, wistful letter. Apologizing for his silence, he explained that "Really I have been engaged in the far more difficult business of making myself for a career. . . . And I have not written, simply because I have lacked the comfortable conviction & assurance which are necessary to an adequately splendid letter." Referring to Francis, he wrote, "He has changed a great deal. Exempli gratia, he is happy. . . . He knows everyone at Oxford; he goes to tea with Lady Ottoline Morrell, the high priestess of civilized society, & the patroness of [T. S.] Eliot & Berty [Bertrand Russell]. . . ."

To the concern of his friends and family, Robert's emotional state continued to deteriorate. He seemed oddly unsure of himself and stubbornly morose. Among other complaints, he talked about his troubled relationship with his head tutor, Patrick Blackett. Robert liked Blackett and eagerly sought his approval, but Blackett, being a hands-on, experimental physicist, hounded Robert to do more of what he wasn't good at—laboratory work. Blackett probably thought nothing of it, but in Oppenheimer's agitated state, the relationship became a source of intense anxiety.

Late that autumn of 1925, Robert did something so stupid that it seemed calculated to prove that his emotional distress was overwhelming him. Consumed by feelings of inadequacy and intense jealousy, he "poisoned" an apple with chemicals from the laboratory and left it on Blackett's desk. Jeffries Wyman later said, "Whether or not this was an imaginary apple, or a real apple, whatever it was, it was an act of jealousy." Fortunately, Blackett did not eat the apple; but university officials somehow were informed of the incident. As Robert himself confessed to Fergusson two months later, "He had kind of poisoned the head steward. It seemed incredible, but that was what he said. And he had actually used cyanide or something somewhere. And fortunately the tutor discovered it. Of course there was hell to pay with Cambridge." If the alleged "poison" was potentially lethal, what Robert had done amounted to attempted murder. But this seems improbable, given what happened next. More likely, Robert had laced the apple with something that merely would have made Blackett sick; but this was still a serious matter—and grounds for expulsion.

As Robert's parents were still visiting Cambridge, the university authorities immediately informed them of what had happened. Julius Oppenheimer frantically—and successfully—lobbied the university not to press criminal charges. After protracted negotiations, it was agreed that Robert would be put on probation and have regular sessions with a prominent Harley Street psychiatrist in London. As Robert's old Ethical Culture School mentor, Herbert Smith, put it, "He was retained at Cambridge for a while only on condition that he had periodic interviews with a psychiatrist."

Robert traveled to London for regularly scheduled sessions, but it was not a good experience. A Freudian psychoanalyst diagnosed dementia praecox, a now archaic label for symptoms associated with schizophrenia. He concluded that Oppenheimer was a hopeless case and that "further analysis would do more harm than good."

Fergusson went to meet Oppenheimer one day just after Robert had finished a session with the psychiatrist. "He looked crazy at that time. . . . I saw him standing on the corner, waiting for me, with his hat on one side of his head, looking absolutely weird. . . . He was sort of standing around, looking like he might run or do something drastic." The two old friends took off together at a more than brisk pace, Robert walking his peculiar walk with his feet turned out at a severe angle. "I asked him how it had been. He said that the guy was too stupid to follow him and that he knew more about his troubles than the doctor did, which was probably true." At the time, Fergusson was still unaware of the "poisoned apple" incident, so he did not understand what had precipitated the psychiatric visits. And though he could see that Robert was in considerable distress, he nevertheless had confidence that his friend had the "ability to bring himself up, to figure out what his trouble was, and to deal with it."

The crisis, however, had not passed. During the Christmas holidays, Robert found himself walking along the Brittany coast near the village of Cancale, where his parents had taken him for the holiday. It was a rainy, dreary winter day and years later Oppenheimer said he had a vivid realization: "I was on the point of bumping myself off. This was chronic."

Sometime shortly after New Year's 1926, Fergusson arranged to meet Oppenheimer in Paris, where Robert's parents had taken him for the remainder of the six-week-long winter break. On one of their long walks through the streets of Paris, Robert finally confided in his friend, explaining what had precipitated his visits to the London psychiatrist. At this point, Robert thought the Cambridge authorities might not even let him return. "My reaction was dismay," Fergusson recalled. "But then, when he talked about it, I thought he had sort of gone beyond it, and that he was having trouble with his father." Robert acknowledged that his parents were very worried, that they were trying to help him, but that "they were not succeeding."

Robert was getting very little sleep and, according to Fergusson, he "began to get very queer." One morning he locked his mother in her hotel room and left. Ella was furious. After this incident, Ella insisted that he see a French psychoanalyst. After several sessions this doctor announced that Robert was suffering a "*crise morale*" associated with sexual frustration. He prescribed "*une femme*" and "a course of aphrodisiacs." Years later, Fergusson observed of that time: "He [Robert] was completely at a loss about his sex life."

Soon, Robert's emotional crisis took another violent turn. Sitting in his Paris hotel room with Robert, Fergusson sensed that his friend was in "one of his ambiguous moods." Perhaps in an attempt to divert him from his depression, Fergusson showed him some poetry written by his girlfriend, Frances Keeley, and then announced that he had proposed to Keeley and she had accepted. Robert was stunned at this news, and he snapped. "I leaned over to pick up a book," Fergusson recalled, "and he jumped on me from behind with a trunk strap and wound it around my neck. I was quite scared for a little while. We must have made some noise. And then I managed to pull aside and he fell on the ground weeping."

Robert may have been provoked by simple jealousy over his friend's love affair. He had already lost one friend, Fred Bernheim, to a woman; perhaps the thought of losing another under the same circumstances was just too much for him at that point. Fergusson himself noted the "deep glares that Robert kept darting theatrically at her [Frances Keeley]. How easy it was for him to act the violent lover; how I know the feeling from experience!"

Despite the choking incident, Fergusson stood by his friend. Indeed, he may even have felt some guilt, since he had been forewarned in a letter by none other than Herbert Smith, who knew Robert's vulnerabilities all too well: "I've a notion, by the way, that your ability to show him [Robert] about should be exercised with great tact, rather than in royal profusion. Your [two] years start and social adaptivity are likely to make him despair. *And instead of flying at your throat—as I remember your being ready to do for George What's-his-name . . . when you were similarly awed by him* (italics added)—I'm afraid he'd merely cease to think his own life worth living." Smith's letter raises the question of whether Fergusson, an aspiring writer, may have conflated his own experience with "George" and Oppenheimer's behavior. But Robert would apologize in a manner that makes Fergusson's story credible.

Fergusson understood that his friend had a "neurotic" streak, but he also thought he could see Robert growing out of it. "He knew that I knew that this was a momentary spasm. . . . I think that I would have been more worried if I hadn't realized how rapidly he was changing. . . . I liked him very much." The two men would remain lifelong friends. All the same, for some months after the assault, Fergusson felt it prudent to be on his guard. He moved out of the hotel, and he hesitated when Robert insisted that he visit him in Cambridge that spring. Robert was no doubt as perplexed by his own behavior as was Francis. He wrote his friend a few weeks after the incident that "You should have, not a letter, but a pilgrimage to Oxford, made in a hair shirt, with much fasting and snow and prayer. But I will keep my remorse and gratitude, and the shame I feel for my inadequacy to you, until I can do something rather less useless for you. I do not understand your forbearance nor your charity, but you must know that I will not forget them."* Through all this turmoil, Robert had become something of his own psychoanalyst, consciously trying to confront his emotional fragility. In a letter to Fergusson on January 23, 1926, he suggested that his mental state had something to do with the *"awful fact of excellence . . . it is that fact now, combined with my inability to solder two copper wires together, which is probably succeeding in getting me crazy."* He then confessed, "I am not well, and I am afraid to come to see you now for fear something melodramatic might happen."

---

*And indeed, he did not forget. Decades later, Oppenheimer arranged an appointment for Fergusson at Princeton's Institute for Advanced Study.

Setting aside his qualms, Fergusson eventually agreed to visit Cambridge early that spring. "He put me in a room next door, and I remember thinking that I'd better make sure that he didn't turn up in the night, so I put a chair up against the door. But nothing happened." By then, Robert seemed to be on the mend. When Fergusson briefly raised the matter, "he said that he needn't worry, that he was over that." Indeed, Robert had been seeing yet another psychoanalyst—his third within four months—in Cambridge. By this time, Robert had read a good deal about psychoanalysis and according to his friend John Edsall, he "took it very seriously." He also thought that his new analyst—a Dr. M—was a "wiser and more sensible man" than either of the doctors he had consulted in London and Paris.

Robert apparently continued to see this analyst throughout the spring of 1926. But over time their relationship broke down. One day in June, Robert dropped by John Edsall's lodgings and told him that "[Dr.] M has decided that there's no use in going on with the analysis any further."

Herbert Smith later ran into one of his psychiatrist friends in New York who knew of the case, and claimed that Robert "gave the psychiatrist in Cambridge an outrageous song and dance. . . . The trouble is, you've got to have a psychiatrist who is abler than the person who's being analyzed. They don't have anybody."

IN MID-MARCH 1926, Robert left Cambridge on a short vacation. Three friends, Jeffries Wyman, Frederick Bernheim and John Edsall, had talked him into accompanying them to Corsica. For ten days they bicycled the length of the island, sleeping in small village inns or camping out in the open. The island's craggy mountains and lightly forested high mesas may well have reminded Robert of New Mexico's rugged beauty. "The scenery was magnificent," recalled Bernheim, "verbal communication with the natives disastrous, and the local fleas abundantly fed each night." Robert's dark moods occasionally overcame him and he sometimes spoke of feeling depressed. He had been reading a great deal of French and Russian literature in recent months, and as they hiked through the mountains, he enjoyed arguing with Edsall over the relative merits of Tolstoy and Dostoyevsky. One evening after being drenched by a sudden thunderstorm, the young men sought shelter in a nearby inn. As they hung their wet clothes by a fire and huddled in blankets, Edsall insisted, "Tolstoy is the writer I most enjoy." "No, no, Dostoyevsky is superior," Oppenheimer said. "He gets to the soul and torment of man."

Later, when the conversation turned to their respective futures, Robert remarked: "The kind of person that I admire most would be one who becomes extraordinarily good at doing a lot of things but still maintains a tear-stained countenance." If Robert seemed burdened by such intensely existential thoughts, his companions nevertheless got a strong impression that he was unburdening himself as they hiked around the island. Clearly relishing the dramatic scenery, the good French food and wines, he wrote his brother Frank: "It's a great place, with every virtue from wine to glaciers, and from langouste to brigantines."

In Corsica, Wyman believed, Robert was "passing through a great emotional crisis." And then something strange happened. "One day," Wyman recalled decades later, "when we'd almost finished our time in Corsica, we were staying in some little inn, the three of us—Edsall, Oppenheimer and I—and we were having dinner together." The waiter approached Oppenheimer and told him when the next boat departed for France. Surprised, Edsall and Wyman asked Oppenheimer why he was rushing back earlier than planned. "I can't bear to speak of it," Robert replied, "but I've got to go." Later in the evening, after they had drunk a little more wine, he relented and said, "Well, perhaps I can tell you why I have to go. I've done a terrible thing. I've put a poisoned apple on Blackett's desk and I've got to go back and see what happened." Edsall and Wyman were stunned. "I never knew," Wyman recalled, "whether it was real or imaginary." Robert didn't elaborate, but he did mention that he had been diagnosed with dementia praecox. Unaware that the "poisoned apple" incident had actually occurred the previous autumn, Wyman and Edsall assumed that Robert, in a fit of "jealousy," had done something to Blackett that spring, just before their trip to Corsica. Clearly, something had happened, but, as Edsall put it later, "he [Robert] spoke of it with a sense of reality that Jeffries and I both felt that this must be some kind of hallucination on his part."

Over the decades, the truth of the poisoned-apple story has been muddied by conflicting accounts. In his 1979 interview with Martin Sherwin, however, Fergusson made it quite plain that the incident occurred in the late autumn of 1925, and not in the spring of 1926: "All this happened during his [Robert's] first term at Cambridge. And just before I met him in London, when he was going to the psychiatrist." When Sherwin asked if he really believed the poisoned-apple story, Fergusson replied, "Yes, I do. I do. His father then had to engineer the authorities of Cambridge about Robert's attempted murder." Talking with Alice Kimball Smith in 1976, Fergusson referred to "the time when he [Robert] tried to poison one of his people. . . . He told me about it at the time, or shortly thereafter, in Paris. I always assumed that it was probably true. But I don't know. He was doing all sorts of crazy things then." Fergusson certainly appeared to Smith to be a reliable source. As she noted after interviewing him, "He doesn't pretend to remember anything he doesn't."

OPPENHEIMER'S PROLONGED ADOLESCENCE was finally coming to an end. Sometime during his short stay in Corsica, something happened to him in the nature of an awakening. Whatever it was, Oppenheimer made sure that it remained a carefully cultivated mystery. Perhaps it was a fleeting love affair—but more likely not. Years later, he would respond to a query from the author Nuel Pharr Davis: "The psychiatrist was a prelude to what began for me in Corsica. You ask whether I will tell you the full story or whether you must dig it out. But it is known to few and they won't tell. You can't dig it out. What you need to know is that it was not a mere love affair, not a love affair at all, but love." The encounter held some kind of mystical, transcendental meaning for Oppenheimer: "Geography was henceforth the only separation I recognized, but for me it was not a real separation." It was, he told Davis, "a great thing in my life, a great and lasting part of it, more to me now, even more as I look back when my life is nearly over."

So what actually happened in Corsica? Probably nothing. Oppenheimer deliberately answered Davis' query about Corsica with an enigma sure to frustrate his biographers. He coyly called it "love" and not a "mere" love affair. Obviously, the distinction was important to him. In the company of his friends, he had no opportunity for a real affair. But he did read a book that appears to have resulted in an epiphany.

The book was Marcel Proust's *A La Recherche du Temps Perdu,* a mystical and existential text that spoke to Oppenheimer's troubled soul. Reading it in the evening by flashlight during his walking tour of Corsica, he later claimed to his Berkeley friend Haakon Chevalier, was one of the great experiences of his life. It snapped him out of his depression. Proust's work is a classic novel of introspection, and it left a deep and lasting impression on Oppenheimer. More than a decade after he first read Proust, Oppenheimer startled Chevalier by quoting from memory a passage in Volume One that discusses cruelty:

> Perhaps she would not have considered evil to be so rare, so extraordinary, so estranging a state, to which it was so restful to emigrate, had she been able to discern in herself, as in everyone, that indifference to the sufferings one causes, an indifference which, whatever other names one may give it, is the terrible and permanent form of cruelty.

As a young man in Corsica, Robert no doubt memorized these words precisely because he saw in himself an indifference to the sufferings he caused others. It was a painful insight. One can only speculate about a man's inner life, but perhaps seeing an expression of his own dark and guilt-laden thoughts in print somehow lightened Robert's psychological burden. It had to be comforting to know that he was not alone, that this was part of the human condition. He no longer need despise himself; he could love. And perhaps it was also reassuring, particularly for an intellectual, that Robert could tell himself that it was a book—and not a psychiatrist—which had helped to wrench him from the black hole of his depression.

OPPENHEIMER RETURNED TO CAMBRIDGE with a lighter, more forgiving attitude about life. "I felt much kinder and more tolerant," he recalled. "I could now relate to others. . . ." By June 1926, he decided to end his sessions with the Cambridge psychiatrist. It also lifted his spirits that spring to leave the "miserable hole" he had till then occupied in Cambridge and move into "less miserable" quarters along the river Cam, halfway to Grantchester, a quaint village one mile south of Cambridge.

As he despised laboratory work, and clearly was inept as an experimental physicist, he now wisely turned to the abstractions of theoretical physics. Even in the midst of his long winter depression, he had managed to read enough to realize that the entire field was in a state of ferment. One day in a Cavendish seminar, Robert watched as James Chadwick, the discoverer of the neutron, opened a copy of *Physical Review* to a new paper by Robert A. Millikan and quipped, "Another cackle. Will there ever be an egg?"

Sometime in early 1926, after reading a paper by the young German physicist Werner Heisenberg, he realized that there was emerging a wholly new way of thinking about how electrons behaved. About the same time, an Austrian physicist, Erwin Schrödinger, published a radical new theory about the structure of the atom. Schrödinger proposed that electrons behaved more precisely as a wave curving around the nucleus of the atom. Like Heisenberg, he crafted a mathematical portrait of his fluid atom and called it quantum mechanics. After reading both papers, Oppenheimer suspected that there had to be a connection between Schrödinger's wave mechanics and Heisenberg's matrix mechanics. They were, in fact, two versions of the same theory. Here was an egg, and not merely another cackle.

Quantum mechanics now became the hot topic at the Kapitza Club, an informal physics discussion group named for its founder, Peter Kapitza, a young Russian physicist. "In a rudimentary way," recalled Oppenheimer, "I began to get pretty interested." That spring he also met another young physicist, Paul Dirac, who would earn his doctorate that May from Cambridge. By then, Dirac had already done some groundbreaking work on quantum mechanics. Robert remarked with considerable understatement that Dirac's work "was not easily understood [and he was] not concerned to be understood. I thought he was absolutely grand." On the other hand, his first impression of Dirac may not have been so favorable. Robert told Jeffries Wyman that "he didn't think he [Dirac] amounted to anything." Dirac was himself an extremely eccentric young man, and notoriously single-minded in his devotion to science. One day some years later, when Oppenheimer offered his friend several books, Dirac politely declined the gift, remarking that "reading books interfered with thought."

It was about this same time that Oppenheimer met the great Danish physicist Niels Bohr, whose lectures he had attended at Harvard. Here was a role model finely attuned to Robert's sensibilities. Nineteen years older than Oppenheimer, Bohr was born—like Oppenheimer—into an upper-class family surrounded by books, music and learning. Bohr's father was a professor of physiology, and his mother came from a Jewish banking family. Bohr obtained his doctorate in physics at the University of Copenhagen in 1911. Two years later, he achieved the key theoretical breakthrough in early quantum mechanics by postulating "quantum jumps" in the orbital momentum of an electron around the nucleus of an atom. In 1922, he won the Nobel Prize for this theoretical model of atomic structure.

Tall and athletic, a warm and gentle soul with a wry sense of humor, Bohr was universally admired. He always spoke in a self-effacing near-whisper. "Not often in life," Albert Einstein wrote to Bohr in the spring of 1920, "has a human being caused me such joy by his mere presence as you did." Einstein was charmed by Bohr's manner of "uttering his opinions like one perpetually groping and never like one who [believed himself to be] in the possession of definite truth." Oppenheimer came to speak of Bohr as "his God."

"At that point I forgot about beryllium and films and decided to try to learn the trade of being a theoretical physicist. By that time I was fully aware that it was an unusual time, that great things were afoot." That spring, with his mental health on the mend, Oppenheimer worked steadily on what would become his first major paper in theoretical physics, a study of the "collision" or "continuous spectrum" problem. It was hard work. One day he walked into Ernest Rutherford's office and saw Bohr sitting in a chair. Rutherford rose from behind his desk and introduced his student to Bohr. The renowned Danish physicist then asked politely, "How is it going?"

Robert replied bluntly, "I'm in difficulties." Bohr asked, "Are the difficulties mathematical or physical?" When Robert replied, "I don't know," Bohr said, "That's bad."

Bohr vividly remembered the encounter—Oppenheimer had looked uncommonly youthful, and after he left the room, Rutherford had turned to Bohr and remarked that he had high expectations for the young man.

Years later, Robert reflected that Bohr's question—"Are the problems mathematical or physical?"—was a very good one. "I thought it put a rather useful glare on the extent to which I became embroiled in formal questions without stepping back to see what they really had to do with the physics of the problem." Later he realized that some physicists rely almost exclusively on mathematical language to describe the reality of nature; any verbal description is "only a concession to intelligibility; it's only pedagogical. I think this is largely true of [Paul] Dirac; I think that his invention is never initially verbal but initially algebraic." By contrast, he realized that a physicist like Bohr "regarded mathematics as Dirac regards words, namely, as a way to make himself intelligible to other people. . . . So there's a very wide spectrum. [At Cambridge] I was simply learning and hadn't learned very much." By temperament and talent, Robert was very much a verbal physicist in the style of Bohr.

Late that spring, Cambridge organized a weeklong visit to the University of Leiden for American physics students. Oppenheimer went on the trip and met a number of German physicists. "It was wonderful," he recalled, "and I realized then that some of the troubles of the winter had been exacerbated by the English customs." Upon his return to Cambridge, he met another German physicist, Max Born, director of the Institute of Theoretical Physics at the University of Göttingen. Born was intrigued by Oppenheimer, partly because the twenty-two-year-old American was grappling with some of the same theoretical problems raised in the recent papers by Heisenberg and Schrödinger. "Oppenheimer seemed to me," Born said, "right from the beginning a very gifted man." By the end of that spring, Oppenheimer had accepted an invitation from Born to study at Göttingen.

CAMBRIDGE HAD BEEN A disastrous year for Robert. He had narrowly escaped expulsion over the "poison apple" incident. For the first time in his life, he had found himself incapable of excelling intellectually. And his closest friends had witnessed more than one episode of emotional instability. But he had overcome a winter of depression, and was now ready to explore an entirely new field of intellectual endeavor. "When I got to Cambridge," Robert said, "I was faced with the problem of looking at a question to which no one knew the answer—but I wasn't willing to face it. When I left Cambridge, I didn't know how to face it very well, but I understood that this was my job; this was the change that occurred that year."

Robert later recalled that he still had "very great misgivings about myself on all fronts, but I clearly was going to do theoretical physics if I could. . . . I felt completely relieved of the responsibility to go back into a laboratory. I hadn't been good; I hadn't done anybody any good, and I hadn't had any fun whatever; and here was something I felt just driven to try."

CHAPTER FOUR

"I Find the Work Hard, Thank God, & Almost Pleasant"

You would like Göttingen, I think. . . . The science is much better than at Cambridge, & on the whole, probably the best to be found. . . . I find the work hard, thank God, & almost pleasant.

ROBERT OPPENHEIMER TO HIS BROTHER, FRANK, *November 14, 1926*

L ATE IN THE SUMMER OF 1926, Robert—in far better spirits and considerably more mature than a year earlier—traveled by train through Lower Saxony to Göttingen, a small medieval town that boasted a city hall and several churches dating back to the fourteenth century. At the corner of Barfüsser Strasse and Jüden Strasse (Barefoot Street and Street of the Jews), he could dine on wienerschnitzel at the four-hundred-year-old Junkers' Hall, sitting beneath a steel engraving of Otto von Bismarck and surrounded by three stories of stained glass. Quaint half-timbered houses were scattered about the town's narrow, winding streets. Nestled on the banks of the Leine Canal, Göttingen had as its chief attraction Georgia Augusta University, founded in the 1730s by a German prince. By tradition, graduates of the university were expected to wade into a fountain that stood before the ancient city hall and kiss the Goose Girl, a bronze maiden that served as the fountain's centerpiece.

If Cambridge could claim to be Europe's great center for experimental physics, Göttingen was undoubtedly the center of theoretical physics. German physicists at the time thought so little of their American counterparts that copies of *Physical Review,* the monthly research journal of the American Physical Society, routinely sat unread for more than a year before the university librarian got around to putting them on the shelf.

It was Oppenheimer's good fortune to arrive shortly before an extraordinary revolution in theoretical physics drew to its close: Max Planck's discovery of quanta (photons); Einstein's magnificent achievement—the special theory of relativity; Niels Bohr's description of the hydrogen atom; Werner Heisenberg's formulation of matrix mechanics; and Erwin Schrödinger's theory of wave mechanics. This truly innovative period began to wind down with Born's 1926 paper on probability and causality. It was completed in 1927 with Heisenberg's uncertainty principle and Bohr's formulation of the theory of complementarity. By the time Robert left Göttingen, the foundations for a post-Newtonian physics had been laid.

As chairman of the physics department, Professor Max Born nurtured the work of Heisenberg, Eugene Wigner, Wolfgang Pauli and Enrico Fermi. It was Born who in 1924 coined the term "quantum mechanics," and it was Born who suggested that the outcome of any interaction in the quantum world is determined by chance. In 1954 he would be awarded the Nobel Prize for physics. A pacifist and a Jew, Born was regarded by his students as an unusually warm and patient teacher. He was the ideal mentor for a young student with Robert's delicate temperament.

That academic year Oppenheimer would find himself in the company of an extraordinary collection of scientists. James Franck, with whom Robert was also studying, was an experimental physicist who had won the Nobel Prize just a year earlier. The German chemist Otto Hahn would in just a few years contribute to the discovery of nuclear fission. Another German physicist, Ernst Pascual Jordan, was collaborating with Born and Heisenberg to formulate the matrix mechanics version of quantum theory. The young English physicist Paul Dirac, whom Oppenheimer had met at Cambridge, was then working on early quantum field theory, and in 1933 he would share a Nobel Prize with Erwin Schrödinger. The Hungarian-born mathematician John Von Neumann would later work for Oppenheimer on the Manhattan Project. George Eugene Uhlenbeck was an Indonesian-born Dutchman who, together with Samuel Abraham Goudsmit, discovered the concept of electron spin in late 1925. Robert quickly drew the attention of these men. He had met Uhlenbeck the previous spring during his weeklong visit to the University of Leiden. "We got along very well immediately," recalled Uhlenbeck. Robert was so deeply immersed in physics that it seemed to Uhlenbeck "as if we were old friends."

Robert found lodging in a private villa owned by a Göttingen physician who had lost his medical license for malpractice. Once very well-to-do, the Cario family now had a spacious granite villa with a walled garden of several acres near the center of Göttingen—and no money. With the family's fortune eaten away by Germany's postwar inflation, they were compelled to take on boarders. Fluent in German, Robert quickly grasped the debilitating political atmosphere of the Weimar Republic. He later speculated that the Carios "had the typical bitterness on which the Nazi movement rested."

That autumn, he wrote his brother that everyone seemed concerned with "trying to make Germany a practically successful & sane country. Neuroticism is very severely frowned upon. So are Jews, Prussians & French."

Outside the university's gate, Robert could see that times were tough for most Germans. "Although this [university] society was extremely rich and warm and helpful to me, it was parked there in a very miserable German mood." He found many Germans "bitter, sullen . . . angry and loaded with all those ingredients which were later to produce a major disaster. And this I felt very much." He had one German friend, a member of the wealthy Ullstein publishing family, who owned a car. He and Robert used to take long drives in the countryside together. But Oppenheimer was struck by the fact that his friend "parked this car in a barn outside Göttingen because he thought it was dangerous to be seen driving it."

Life for the American expatriates—and especially for Robert—was quite a different matter. For one thing, he was never short of money. At twenty-two, he dressed casually in rumpled herringbone suits made of the finest English wool. His fellow students noted that, in contrast to their own cloth luggage, Oppenheimer packed his belongings in gleaming, expensive pigskin suitcases. And when they strolled down to the fifteenth-century Schwartzen Baren—the Black Bear pub—to drink *frisches Bier* or went to sip coffee at the Karon Lanz coffee shop, it was often Robert who picked up the tab. He was transformed; he was now confident, excited and focused. Material possessions were unimportant to him, but the admiration of others was something he sought every day. To that end, he would use his wit, his erudition and his fine things to attract those people he wanted within his orbit of admirers. "He was," said Uhlenbeck, "so to say, clearly a center of all the younger students . . . he was really a kind of oracle. He knew very much. He was very difficult to understand, but very quick." Uhlenbeck thought it remarkable that so young a man already had "a whole group of admirers" trailing around after him.

In contrast to Cambridge, at Göttingen Oppenheimer felt a pleasant camaraderie with his fellow students. "I was part of a little community of people who had some common interests and tastes and many common interests in physics." At Harvard and Cambridge, Robert's intellectual pursuits had been solitary forays into books; at Göttingen, for the first time, he realized that he could learn from others: "Something which for me—more than most people—is important began to take place, namely, I began to have some conversations. Gradually, I guess, they gave me some sense and, perhaps more gradually, some taste in physics, something that I probably would not have ever gotten to if I'd been locked up in a room."

Lodging with him in the Cario family villa was Karl T. Compton, age thirty-nine, a professor of physics at Princeton University. Compton, a future president of MIT, felt intimidated by Oppenheimer's extraordinary versatility. He could hold his own with the young man when the topic was science, but found himself at a loss when Robert began talking about literature, philosophy or even politics. No doubt with Compton in mind, Robert wrote his brother that most of the other American expatriates in Göttingen were "professors at Princeton or California or some such place, married, respectable. They are mostly pretty good at physics, but completely uneducated & unspoiled. They envy the Germans their intellectual adroitness & organization, & want physics to come to America."

In short, Robert thrived in Göttingen. That autumn he wrote enthusiastically to Francis Fergusson, "You would like Göttingen, I think. Like Cambridge, it is almost exclusively scientific, & such philosophers as are here are pretty largely interested in epistemological paradoxes & tricks. The science is much better than at Cambridge, & on the whole, probably the best to be found. They are working very hard here, & combining a fantastically impregnable metaphysical disingenuousness with the go-getting habits of a wall paper manufacturer. The result is that the work done here has an almost demoniac(?) lack of plausibility to it, & is highly successful. . . . I find the work hard, thank God, & almost pleasant."

Most of the time, he felt himself on an even keel emotionally. But there were some momentary relapses. One day, Paul Dirac saw him faint and fall to the floor, just as he had done the previous year in Rutherford's laboratory. "I still was not entirely well," Oppenheimer recalled decades later, "and I had several attacks during the year, but they became much more isolated and interfered less and less with work." Another physics student, Thorfin Hogness, and his wife, Phoebe, also roomed that year in the Cario mansion and found Oppenheimer's behavior sometimes odd. Phoebe often saw him lying in bed, doing nothing. But then these periods of hibernation were invariably followed by episodes of incessant talking. Phoebe thought him "highly neurotic." On occasion, some witnessed Robert trying to overcome an episode of stuttering.

Gradually, with his self-assurance returning, Oppenheimer found that his reputation had preceded him. One of his last acts before leaving Cambridge had been to present two papers before the Cambridge Philosophical Society titled "On the Quantum Theory of Vibration-rotation Bands" and "On the Quantum Theory of the Problem of the Two Bodies." The first paper dealt with molecular energy levels and the second investigated transitions to continuum states in hydrogenic atoms. Both papers represented small but important advances in quantum theory, and Oppenheimer was pleased to learn that the Cambridge Philosophical Society had published them by the time he arrived in Göttingen.

Robert responded to the recognition his publications brought him by throwing himself enthusiastically into seminar discussions—with such abandon that he often annoyed his fellow students. "He was a man of great talent," Professor Max Born later wrote, "and he was conscious of his superiority in a way which was embarrassing and led to trouble." In Born's seminar on quantum mechanics, Robert routinely interrupted whoever was speaking, not excluding Born, and, stepping to the blackboard with chalk in hand, would declare in his American-accented German, "This can be done much better in the following manner. . . ." Though other students complained about these interruptions, Robert was oblivious to his professor's polite, halfhearted attempts to change his behavior. One day, however, Maria Göppert—a future Nobelist—presented Born with a petition written on thick parchment and signed by her and most of the other members of the seminar: Unless the "child prodigy" was reined in, his fellow students would boycott the class. Still unwilling to confront Oppenheimer, Born decided to leave the document on his desk in a place where Robert could not help but see it when he next came to discuss his thesis. "To make this more certain," Born later wrote, "I arranged to be called out of the room for a few minutes. This plot worked. When I returned I found him rather pale and not so voluble as usual." Thereafter his interruptions ceased altogether.

Not that he was by any means completely tamed. Robert could startle even his professors with his bruising candor. Born was a brilliant theoretical physicist, but because he sometimes made small mistakes in his long calculations, he often asked a graduate student to recheck his math. On one occasion, Born recalled, he gave a set of calculations to Oppenheimer. After a few days, Robert returned and said, "I couldn't find any mistake—did you really do this alone?" All of Born's students knew of his propensity for calculation errors, but, as Born later wrote, "Oppenheimer was the only one frank and rude enough to say it without joking. I was not offended; it actually increased my esteem for his remarkable personality."

Born soon began collaborating with Oppenheimer, who wrote one of his Harvard physics professors, Edwin Kemble, a veritable summary of their work: "Almost all of the theorists seem to be working on q-mechanics. Professor Born is publishing a paper on the Adiabatic Theorem, & Heisenberg on 'Schwankungen [fluctuations].' Perhaps the most important idea is one of [Wolfgang] Pauli's, who suggests that the usual Schroedinger ψ [psi] functions are only special cases, and only in special cases—the spectroscopic one—give the physical information we want. . . . I have been working for some time on the quantum theory of aperiodic phenomena. . . . Another problem on which Prof. Born and I are working is the law of deflection of, say, an α-particle by a nucleus. We have not made very much progress with this, but I think we shall soon have it. Certainly the theory will not be so simple, when it is done, as the old one based on corpuscular dynamics." Professor Kemble was impressed; after less than three months in Göttingen, his former student seemed steeped in the excitement of unraveling the mysteries of quantum mechanics.

By February 1927, Robert felt so confident of his mastery of the new quantum mechanics that he was writing his Harvard physics professor, Percy Bridgman, to explain its finer points:

> On the classical quantum theory, an electron in one of two regions of
> low potential which were separated by a region of high potential,
> could not cross to the other without receiving enough energy to clear
> "impediment." On the new theory that is no longer true: the electron
> will spend part of its time in one region, & part in the other. . . . On
> one point the new mechanics suggests a change, however: the
> electrons, which are "free" in the sense defined above, are not "free"
> in the sense that they are carriers of equipartition thermal energy. In
> order to account for the Wiedemann-Franz law one might have to
> adopt the suggestion, due, I think, to Professor Bohr, that when an
> electron jumps from one atom to another the two atoms may
> exchange momentum. With best greetings,
>
> > Yours,
> > J. R. Oppenheimer

Bridgman was no doubt impressed by his former student's command of the new theory. But Robert's lack of tact made others leery. He could be engaging and considerate one moment and in the next rudely cut someone off. At the dinner table, he was polite and formal to an extreme. But he seemed incapable of tolerating banalities. "The trouble is that Oppie is so quick on the trigger intellectually," complained one of his fellow students, Edward U. Condon, "that he puts the other guy at a disadvantage. And, dammit, he is always right, or at least right enough."

Having just earned his Ph.D. from Berkeley in 1926, Condon was struggling to support a wife and an infant child on a small postdoctoral fellowship. It annoyed him that Oppenheimer spent money so casually on food and fine clothes while seeming blissfully unaware of his friend's familial responsibilities. One day, Robert invited Ed and Emilie Condon out for a walk, but Emilie explained that she had to stay with the baby. The Condons were startled when Robert replied, "All right, we'll leave you to your peasant tasks." And yet, despite his occasional cutting remarks, Robert often displayed a sense of humor. Upon seeing Karl Compton's two-year-old daughter pretending to read a small red book—which just happened to be on the topic of birth control—Robert looked over at the very pregnant Mrs. Compton and quipped, "A little late."

PAUL DIRAC ARRIVED IN Göttingen for the winter term of 1927, and he too rented a room in the Cario villa. Robert relished any contact with Dirac. "The most exciting time in my life," Oppenheimer once said, "was when Dirac arrived and gave me the proofs of his paper on the quantum theory of radiation." The young English physicist was perplexed, however, by his friend's determined intellectual versatility. "They tell me you write poetry as well as working at physics," Dirac said to Oppenheimer. "How can you do both? In physics we try to tell people in such a way that they understand something that nobody knew before. In the case of poetry, it's the exact opposite." Flattered, Robert just laughed. He knew that for Dirac life was physics and nothing else; by contrast, his own interests were extravagantly catholic.

He still loved French literature, and while in Göttingen he found time to read Paul Claudel's dramatic comedy *Jeune Fille Violaine,* F. Scott Fitzgerald's short story collections, *The Sensible Thing* and *Winter Dreams,* Anton Chekhov's play *Ivanov* and the works of Johann Hölderlin and Stefan Zweig. When he discovered that two friends were regularly reading Dante in the original Italian, Robert disappeared from Göttingen's cafés for a month and returned with enough Italian to read Dante aloud. Dirac was unimpressed, grumbling, "Why do you waste time on such trash? And I think you're giving too much time to music and that painting collection of yours." But Robert lived comfortably in worlds beyond Dirac's comprehension and so was merely amused by his friend's urgings, during their long walks around Göttingen, to abandon the pursuit of the irrational.

Göttingen was not all physics and poetry. Robert also found himself attracted to Charlotte Riefenstahl, a German physics student, and one of the prettiest women on campus. They had met on a student overnight trip to Hamburg. Riefenstahl was standing on the train platform when she glanced down at the assembled luggage and her eyes were drawn to the one suitcase not made from cheap cardboard or worn brown leather.

"What a beautiful thing," she said to Professor Franck, pointing to the shiny tan leather pigskin grip. "Whose is it?"

"Who else but Oppenheimer's," said Franck with a shrug.

On the train ride back to Göttingen, Riefenstahl asked someone to point out Oppenheimer, and when she sat down beside him he was reading a novel by André Gide, the contemporary French novelist whose works dwelt on the individual's moral responsibility for the affairs of the world. To his astonishment, he discovered that this pretty woman had read Gide and could intelligently discuss his work. Upon arriving in Göttingen, Charlotte casually mentioned how much she admired his pigskin bag. Robert acknowledged the compliment, but seemed perplexed that anyone would bother to admire his luggage.

When Riefenstahl later recounted the conversation to a fellow student, he predicted that Robert would soon try to give her the bag. Among his many eccentricities, everyone knew that Robert felt compelled to give away any possession of his that was admired. Robert was smitten with Charlotte, and courted her as best as he could in his stiff, excessively polite manner.

So too did one of Robert's classmates, Friedrich Georg Houtermans, a young physicist who had made a name for himself with a paper on the energy production of stars. Like Oppenheimer, "Fritz"—or "Fizzl" to some of his friends—had come to Göttingen with a family trust fund. He was the son of a Dutch banker, and his mother was German and half-Jewish, a fact that Houtermans was unafraid to advertise. Contemptuous of authority and armed with a dangerous wit, Houtermans liked to tell his gentile friends, "When your ancestors were still living in the trees, mine were already forging checks!" As a teenager growing up in Vienna, he had been expelled from his gymnasium (high school) for publicly reading the *Communist Manifesto* on May Day. He and Oppenheimer were virtual contemporaries, and both would receive their doctorates in 1927. They shared a passion for literature—and Charlotte. As fate would have it, Oppenheimer and Houtermans would later both work on developing an atomic bomb—but Houtermans would do so in Germany.

PHYSICISTS HAD BEEN IMPROVISING quantum theory for nearly a quarter of a century when suddenly, in the years 1925–27, a series of dramatic breakthroughs made it possible to construct a radical and cohesive theory of quantum mechanics. New discoveries were then made so rapidly that it was hard to keep up with the literature. "Great ideas were coming out so fast during that period," Edward Condon recalled, "that one got an altogether wrong impression of the normal rate of progress in theoretical physics. One had intellectual indigestion most of the time that year, and it was most discouraging." In the highly competitive race to publish the new findings, more papers on quantum theory were written from Göttingen than from Copenhagen, Cavendish or anywhere else in the world. Oppenheimer himself published seven papers out of Göttingen, a phenomenal output for a twenty-three-year-old graduate student. Wolfgang Pauli began to refer to quantum mechanics as *Knabenphysik*—"boys' physics"—because the authors of so many of these papers were so young. In 1926, Heisenberg and Dirac were only twenty-four years old, Pauli was twenty-six and Jordan was twenty-three.

The new physics was, to be sure, highly controversial. When Max Born sent Albert Einstein a copy of Heisenberg's 1925 paper on matrix mechanics, an intensely mathematical description of the quantum phenomenon, he explained somewhat defensively to the great man that it "looks very mystical, but is certainly correct and profound." But after reading the paper that autumn, Einstein wrote Paul Ehrenfest that "Heisenberg has laid a big quantum egg. In Göttingen they believe in it. (I don't.)" Ironically, the author of the theory of relativity would forever believe the *Knabenphysik* incomplete if not profoundly flawed. Einstein's doubts were only heightened when in 1927 Heisenberg published his paper on the central role of *uncertainty* in the quantum world. What he meant was that it is impossible to determine at any given moment both an entity's precise position *and* its precise momentum: "We cannot know, as a matter of principle, the present in all its details." Born agreed, and argued that the outcome of any quantum experiment depended on chance. In 1927, Einstein wrote Born: "An inner voice tells me that this is not the true Jacob. The theory accomplishes a lot, but it does not bring us closer to the secrets of the Old One. In any case, I am convinced that He does not play dice."

Obviously, quantum physics was a young man's science. The young physicists in turn regarded Einstein's stubborn refusal to embrace the new physics as a sign that his time had passed. A few years down the road, Oppenheimer would visit Einstein in Princeton—and he came away distinctly unimpressed, writing his brother with cocky irreverence that "Einstein is completely cuckoo." But in the late 1920s, the boys from Göttingen (and Bohr's Copenhagen) still had hopes of recruiting Einstein to their quantum vision.

The first of Oppenheimer's papers written at Göttingen demonstrated that quantum theory made it possible to measure the frequencies and intensities of the molecular band spectrum. He had become obsessed with what he called the "miracle" of quantum mechanics precisely because the new theory explained so much about observable phenomena in a "harmonious, consistent and intelligible way." By February 1927, Born was so impressed with Oppenheimer's work on the application of quantum theory to transitions in the continuous spectrum that he found himself writing S. W. Stratton, the president of the Massachusetts Institute of Technology: "We have here a number of Americans. . . . One man is quite excellent, Mr. Oppenheimer." For sheer brilliance, Robert's peers ranked him with Dirac and Jordan: "There are three young geniuses in theory here," reported one young American student, "each less intelligible to me than the others."

Robert got into the habit of working all night and then sleeping through a good part of the day. Göttingen's damp weather and poorly heated buildings wreaked havoc on his delicate constitution. He walked around with a chronic cough which friends attributed to either his frequent colds or his chain-smoking. But in other respects, life in Göttingen was pleasantly bucolic. As Hans Bethe later observed of this golden age in theoretical physics, ". . . life at the centers of the development of quantum theory, Copenhagen and Göttingen, was idyllic and leisurely, in spite of the enormous amount of work accomplished."

Oppenheimer invariably sought out those young men with growing reputations. Others could not help but feel they had been snubbed. "He [Oppenheimer] and Born became very close friends," Edward Condon said rather peevishly years later, "and saw a great deal of each other, so much so, that Born did not see much of the other theoretical physics students who had come there to work with him."

Heisenberg passed through Göttingen that year and Robert made a point of meeting the brightest of Germany's young physicists. Just three years older than Oppenheimer, Heisenberg was articulate, charming and tenacious in argument with his peers. Both men possessed original intellects and knew it. The son of a professor of Greek, Heisenberg had studied with Wolfgang Pauli at the University of Munich, and later he had done postdoctoral work with both Bohr and Born. Like Oppenheimer, Heisenberg had a way of using his intuition to cut to the root of a problem. He was an oddly charismatic young man, whose sparkling intellect commanded attention. By all accounts, Oppenheimer admired Heisenberg and respected his work. He could not have known then that in the years ahead they would become shadowy rivals. Oppenheimer would one day find himself contemplating Heisenberg's loyalty to wartime Germany and wondering whether the man was capable of building an atomic bomb for Adolf Hitler. But in 1927, he was building on Heisenberg's discoveries in quantum mechanics.

That spring, prompted by a remark from Heisenberg, Robert became interested in using the new quantum theory to explain, as he put it, "why molecules were molecules." In very short order, he found a simple solution to the problem. When he showed Professor Born his notes, the older man was startled and very pleased. They then agreed to collaborate on a paper, and Robert promised that while he was in Paris for Easter, he would write up his notes into a first draft. But Born was "horrified" when he received from Paris a very spare, four- or five-page paper. "I thought that this was about right," Oppenheimer recalled. "It was very light of touch and it seemed to me all that was necessary." Born eventually lengthened the paper to thirty pages, padding it, Robert thought, with unnecessary or obvious theorems. "I didn't like it, but it was obviously not possible for me to protest to a senior author." For Oppenheimer, the central new idea was everything; the context and the academic window dressing were clutter that disturbed his acute aesthetic sense.

On the Quantum Theory of Molecules was published later that year. This joint paper containing the "Born-Oppenheimer approximation"—in reality, just the "Oppenheimer approximation"—is still regarded as a significant breakthrough in using quantum mechanics to understand the behavior of molecules. Oppenheimer had recognized that the lighter electrons in molecules travel with a much greater speed than the heavier nuclei. By integrating out the higher frequency electron motions, he and Born were then able to calculate the "effective wave-mechanical" phenomena of nuclear vibrations. The paper laid the foundation for developments more than seven decades later in high-energy physics.

Late that spring, Robert submitted his doctoral thesis, the heart of which contained a complicated calculation for the photoelectric effect in hydrogen and X rays. Born recommended that it be accepted "with distinction." The one fault he noted was that the paper was "difficult to read." Nevertheless, Born recorded that Oppenheimer had written "a complicated paper and he did it very well." Years later, Hans Bethe, another Nobel laureate, observed that "[i]n 1926 Oppenheimer had to develop all the methods himself, including the normalization of wave functions in the continuum. Naturally, his calculations were later improved upon, but he correctly obtained the absorption coefficient at the K edge and the frequency dependence in its neighborhood." Bethe concluded: "Even today this is a complicated calculation, beyond the scope of most quantum mechanics textbooks." A year later, in a related field, Oppenheimer published the first paper to describe the phenomenon of quantum mechanical "tunneling," whereby particles literally are able to "tunnel" through a barrier. Both papers were formidable achievements.

On May 11, 1927, Robert sat down for his oral examination and emerged a few hours later with excellent grades. Afterwards one of his examiners, the physicist James Franck, told a colleague, "I got out of there just in time. He was beginning to ask *me* questions." At the last moment, the university's authorities discovered to their indignation that Oppenheimer had failed to register formally as a student—and so they threatened to withhold his degree. He was finally awarded his doctorate only after Born interceded and falsely told the Prussian Ministry of Education that "economic circumstances render it impossible for Herr Oppenheimer to remain in Göttingen after the end of the summer term."

That June, Professor Edwin Kemble happened to be visiting Göttingen and soon wrote a colleague: "Oppenheimer is turning out to be even more brilliant than we thought when we had him at Harvard. He is turning out new work very rapidly and is able to hold his own with any of the galaxy of young mathematical physicists here." Curiously, the professor added, "Unfortunately, Born tells me that he has the same difficulty about expressing himself clearly in writing which we observed at Harvard." Oppenheimer had long since become an extremely expressive writer. But it was also true that his physics papers were usually brief to the point of being cursory. Kemble thought Robert's command of language was indeed remarkable, but that he became "two different people" when talking about physics and about any other general topic.

Born was disheartened to see Oppenheimer depart. "It's all right for you to leave, but I cannot," he told him. "You have left me too much homework." A parting gift from Robert to his mentor was a valuable edition of LaGrange's classic text *Mécanique Analytique*. Decades later, long after he had been forced to flee Germany, Born wrote Oppenheimer: "This [book] has survived all upheavals: revolution, war, emigration and return, and I am glad that it is still in my library, for it represents very well your attitude to science which comprehends it as a part of the general intellectual development in the course of human history." By then Oppenheimer had long eclipsed Born in notoriety—although not in scientific achievement.

Göttingen was the scene of Oppenheimer's first real triumph as a young man coming of age. Becoming a scientist, Oppenheimer later remarked, is "like climbing a mountain in a tunnel: you wouldn't know whether you were coming out above the valley or whether you were ever coming out at all." This was particularly so for a young scientist on the cusp of the quantum revolution. More of a witness to this upheaval than a participant, he nevertheless demonstrated that he had the raw intellect and motivation to make physics his life's work. In nine short months he had combined real academic success with a renewal of his personality and his own sense of worth. The profound emotional inadequacies that only a year before had threatened his very survival had been trumped by serious achievements, and the confidence that flowed from them. The world now beckoned.

CHAPTER FIVE

"I Am Oppenheimer"

*God knows I'm not the simplest person, but compared to
Oppenheimer, I'm very, very simple.*

I. I. RABI

BY THE END OF HIS YEAR IN GÖTTINGEN, Oppenheimer was showing unmistakable signs of homesickness. In his casual remarks about things German he sounded like a chauvinistic American. Nothing in Germany could compare to the desert landscapes of New Mexico. "He's too much," complained a Dutch student. "According to Oppenheimer, even the flowers seem to smell better in America." He threw a party at his apartment the night before leaving, and among many others, the lovely, dark-haired Charlotte Riefenstahl came to say good-bye. Robert made a point of giving her the pigskin satchel she had admired when they had first met. She kept it for the next three decades, calling it "The Oppenheimer."

After a quick side trip with Paul Dirac to Leiden, Robert sailed for New York from Liverpool in mid-July 1927. It felt good to be home. He had not only survived, he had triumphed, bringing back a hard-earned doctorate. Among theoretical physicists, it was known that young Oppenheimer had firsthand knowledge of the latest European breakthroughs in quantum mechanics. Barely two years after graduating from Harvard, Robert was a rising star in his field.

Earlier that spring, he had been encouraged to take a Rockefeller Foundation–funded postdoctoral fellowship awarded by the National Research Council to promising young scientists. He had accepted, and decided to spend the fall term at Harvard before moving to Pasadena, California, where he had been offered a teaching post at the California Institute of Technology (Caltech), a leading center of scientific research. So, even as he unpacked his bags at the Oppenheimer home on Riverside Drive, Robert knew that his immediate future was set. In the meantime, he had six weeks to become reacquainted with his fifteen-year-old brother, Frank, and to visit with his parents.

To his regret, Julius and Ella had decided to sell the Bay Shore house the previous winter. But as his sailboat, the *Trimethy,* was still temporarily moored near the house, Robert took Frank out, as he had so many times in the past, for a wild sail along the Long Island coast. In August the brothers joined their parents for a short vacation on Nantucket. "My brother and I," Frank recalled, "spent most of the days painting with oils on canvas the dunes and grassy hills." Frank worshipped his brother. Unlike Robert, he was good with his hands and loved tinkering with things, taking apart electric motors and watches and putting them back together. Now, at the Ethical Culture School, he too was gravitating toward physics. When Robert had left for Harvard, he had given Frank his microscope, and Frank had used it one day to look at his own sperm. "Never having heard of sperm," Frank said, "it was really a marvelous discovery."

At the end of that summer, Robert was pleased to hear that Charlotte Riefenstahl had accepted a teaching post at Vassar College. When, in September, her boat arrived in New York harbor, he was at dockside to meet her. Traveling with her were two other triumphant Göttingen alumni—Samuel Goudsmit and George Uhlenbeck—with Uhlenbeck's new wife, Else. Oppenheimer knew both men as accomplished physicists. Together, Goudsmit and Uhlenbeck had discovered the existence of electron spin in 1925. Robert spared no expense in serving as their host in New York.

"We all got the real Oppenheimer treatment," Goudsmit recalled, "but it was for Charlotte's benefit really. He met us in this great chauffeur-driven limousine, and took us downtown to a hotel he had selected in Greenwich Village." Over the next few weeks, he escorted Charlotte all over New York, taking her to all his old haunts, from the city's great art galleries to the most expensive restaurants he could find. Charlotte protested, "Is the Ritz really the only hotel you know?" And as an indication of how serious were his intentions, he introduced Charlotte to his parents at the spacious apartment on Riverside Drive. But though Charlotte admired Robert and was flattered by his attentions, she sensed that he was emotionally unavailable. He evaded all her attempts to get him to talk about his past. She found the Oppenheimer home stifling and overprotective, and the couple began to drift apart. Charlotte's teaching position at Vassar kept her out of New York—and Oppenheimer's fellowship required his presence at Harvard. Charlotte eventually returned to Germany; in 1931 she married Robert's Göttingen classmate Fritz Houtermans.

BACK AT HARVARD THAT AUTUMN, he renewed his friendship with William Boyd, who was in Cambridge finishing his doctorate in biochemistry. Robert confided in him about his troubled year at Cambridge. Boyd was not surprised; he had always thought of Robert as an emotionally taut young man who could nevertheless handle his troubles. Poetry was still a passion with Robert, and when he showed Boyd a poem he had written, his friend encouraged him to submit it to Harvard's literary magazine, *Hound and Horn*. It appeared in the June 1928 issue:

CROSSING

It was evening when we came to the river
with a low moon over the desert
that we had lost in the mountains, forgotten,
what with the cold and the sweating
and the ranges barring the sky.
And when we found it again,
In the dry hills down by the river,
half withered, we had
the hot winds against us.

There were two palms by the landing;
The yuccas were flowering; there was
a light on the far shore, and tamarisks.
We waited a long time, in silence.
Then we heard the oars creaking
and afterwards, I remember,
the boatman called to us.
We did not look back at the mountains.

J. R. Oppenheimer

New Mexico was calling to Robert. He desperately missed that "low moon over the desert" and the sheer physical sensations—"the cold and the sweating"—that had made him feel so alive during his two summers out West. He could not plausibly do cutting-edge physics in New Mexico—but he had accepted a position at Caltech in Pasadena at least partly because it was near the desert he loved. At the same time, he also wanted to be free of Harvard and that "separate prison" that had confined him for so long. Part of his recovery from the crisis of the previous year had come from the recognition that he needed a new beginning. Corsica, Proust and Göttingen had afforded him that new beginning; remaining at Harvard now would seem too much like a step backwards. So, shortly after Christmas 1927, Robert packed his bags and moved to Pasadena.

California suited him. After only a few months he was writing Frank: "I have had trouble getting time to work, for Pasadena is a pleasant place, and hundreds of pleasant people are continually suggesting pleasant things to do. I am trying to decide whether to take a professorship at the University of California next year, or go abroad."

Despite his teaching duties at Caltech and Pasadena's distractions, Oppenheimer published six papers in 1928, all of them on various aspects of quantum theory. His productivity was all the more remarkable in that late the same spring his doctor decided that his persistent cough might be a symptom of tuberculosis. After attending a seminar on theoretical physics at Ann Arbor, Michigan, in June, Robert headed for the dry mountain air of New Mexico. Earlier that spring he had written his brother Frank, now nearly sixteen years old, suggesting that the two of them "might knock around for a fortnight on the desert" sometime that summer.

Robert had begun to take an almost paternal interest in helping his younger brother navigate the rough shoals of adolescence—a difficult voyage, as he knew only too well. That March, in response to Frank's confession that he had been distracted from his studies by a member of the opposite sex, Robert had written a letter filled with advice that bordered on self-conscious analysis. It was, he suggested, the young woman's "profession to make you waste your time with her; it is your profession to keep clear." No doubt drawing on his own checkered experience, Robert remarked that dating was "only important for people who have time to waste. For you, and for me, it isn't." His bottom line was "Don't worry about girls, and don't make love to girls, unless you have to: DON'T DO IT AS A DUTY. Try to find out, by watching yourself, what you really want; if you approve of it, try to get it; if you disapprove of it, try to get over it." Robert admitted that he was being dogmatic, but he told Frank that he hoped his words would be of some use "as the fruit and outcome of my erotic labours. You are very young, but much more mature than I was."

ROBERT WAS QUITE RIGHT; young Frank was far more mature than his brother had been at the same age. He had the same icy blue eyes and shock of bushy black hair. Born with the Oppenheimer lankiness, he would soon stand six feet but weigh a mere 135 pounds. He was in many ways as gifted intellectually as his brother, but seemed unburdened by Robert's intense nervous energy. If Robert could sometimes seem manic in his obsessions, Frank was a calming presence and ever congenial. As an adolescent, Frank had known his brother at a distance, mainly through his letters, and during vacations when they had gone sailing together. It was during this trip to New Mexico—without their parents—that Frank bonded with his sibling as an adult.

When the brothers arrived in Los Pinos, they bunked at Katherine Page's ranch, and despite his persistent coughing, Robert insisted on mounting a series of extended expeditions on horseback into the surrounding hills. They'd make do with a little peanut butter, some canned artichokes, Vienna sausages and Kirschwasser and whiskey. As they rode, Frank would listen as Robert talked excitedly about physics and literature. At night, the older brother would pull out a worn copy of Baudelaire and read aloud by the light of a campfire. That summer of 1928, Robert was also reading the 1922 novel *The Enormous Room,* an account by e. e. cummings of his four-month incarceration in a French wartime prison camp. He loved cummings' notion that a man stripped of all his possessions can nevertheless find personal freedom in the most spartan of surroundings. The story would take on a new meaning for him after 1954.

Frank Oppenheimer noticed that his brother's passions were always mercurial. Robert seemed to divide the world into people who were worth his time and those who were not. "For the former group," Frank said, "it was wonderful. . . . Robert wanted everything and everyone to be special, and his enthusiasms communicated themselves and made these people feel special. . . . Once he had accepted someone as worthy of attention or friendship, he would always be ringing or writing them, doing them small favors, giving them presents. He couldn't be humdrum. He would even work up those enthusiasms for a brand of cigarettes, even elevating them to something special. His sunsets were always the best." Frank observed that his brother could like all manner of people—they could be famous or not—but in liking them he had a way of making these people into heroes: "Anybody who struck him with their wisdom, talent, skill, decency or devotion became, at least temporarily, a hero to him, to themselves and to his friends."

One day that July, Katherine Page took the Oppenheimer brothers on a ride about a mile up into the mountains above Los Pinos. After riding through a pass at 10,000 feet, they came upon a meadow perched on Grass Mountain and covered with thick clover and blue and purple alpine flowers. Ponderosa and white pine trees framed a magnificent view of the Sangre de Cristo Mountains and the Pecos River. Nestled in the meadow at an altitude of 9,500 feet was a rustic cabin built from half-trunks and adobe mortar. A hardened clay fireplace dominated one wall of the cabin and a narrow wooden staircase led upstairs to two small bedrooms. The kitchen had a sink and wood stove, but there was no running water, and the only bathroom was a windy outhouse built at the end of a covered porch.

"Like it?" Katherine asked Robert.

When Robert nodded, she explained that the cabin and 154 acres of pasture and brook were for rent.

"Hot dog!" Robert exclaimed.

"No, perro caliente!" quipped Katherine, translating Robert's exclamation into Spanish.

Later that winter, Robert and Frank persuaded their father to sign a four-year lease on the ranch; they named it Perro Caliente. They continued to lease it until 1947 when Oppenheimer purchased it for $10,000. The ranch would be Robert's private haven for years to come.

After two weeks in New Mexico, the brothers left in the early fall of 1928 to join their parents at the luxurious Broadmoor Hotel in Colorado Springs. Both Robert and Frank took some rudimentary driving lessons and then bought a used six-cylinder Chrysler roadster. Their plan was to drive to Pasadena. "We had a variety of mishaps," Frank said with understatement, "but finally got there." Outside of Cortez, Colorado, with Frank at the wheel, the car skidded on some loose gravel and landed upside-down in a gulley. The windshield was shattered and the car's cloth top was ruined. Robert fractured his right arm and two bones in his right wrist. After getting a tow to Cortez they got the roadster running again—but the very next evening Frank managed to run the car up onto a slab of rock. Unable to move, they spent the night lying on the desert floor, "sipping from a bottle of spirits . . . and sucking on some lemons we had with us."

When they finally arrived in Pasadena, Robert went straight away to Caltech's Bridge Laboratory. With one arm in a bright red sling, he walked in, disheveled and unshaven, and announced, "I am Oppenheimer."

"Oh, are you Oppenheimer?" replied a physics professor, Charles Christian Lauritsen, who thought he "looked more like a tramp than a college professor." "Then you can help. Why am I getting the wrong results from this confounded cascade voltage generator?"

Oppenheimer was back in Pasadena only to pack his belongings and prepare for a return to Europe. Earlier that spring of 1928 he had received job offers from ten American universities, including Harvard, and two from abroad. All of them were attractive positions with competitive salaries. Robert decided to accept a double appointment in the physics departments at the University of California, Berkeley, and Caltech. The plan was for him to teach one semester at each school. He chose Berkeley precisely because its physics program lacked any theoretical component. Berkeley was in that sense "a desert," so for that reason he "thought it would be nice to try to start something."

He did not intend to "start something" immediately, however. For at the same time, Robert asked for, and shortly received, a fellowship so that he could return to Europe for another year. He felt that he still needed the seasoning, particularly in mathematics, that would come with an additional year of postdoctoral studies. He wanted to study under Paul Ehrenfest, a greatly admired physicist at the University of Leiden in the Netherlands. As he embarked for Leiden, his plan was that after a term with Ehrenfest he might move on to Copenhagen, where he hoped to get to know Niels Bohr.

In the event, Ehrenfest was out of sorts and distracted, suffering from one of his recurrent bouts of depression. "I don't think that I was of great interest to him then," Oppenheimer recalled. "I have a recollection of quiet and gloom." In retrospect, Robert thought he wasted his term in Leiden and that this was his own fault. Ehrenfest insisted on simplicity and clarity, traits that Robert had not yet embraced. "I probably still had a fascination with formalism and complication," he said, "so that the large part of what had me stuck or engaged was not his dish. And some of the things that were his dish I didn't appreciate how really valuable it would be to have them in clear, good order." Ehrenfest thought Robert was too quick with his answers to any question—and sometimes hidden behind his quickness were errors.

Ehrenfest in fact found it emotionally draining to work with the young man. "Oppenheimer is now with you," Max Born wrote his Leiden colleague. "I should like to know what you think of him. Your judgment will not be influenced by the fact that I have never suffered as much with anybody as with him. He is doubtless very gifted but completely without mental discipline. He's outwardly very modest, but inwardly very arrogant." Ehrenfest's reply is lost, but Born's next letter is indicative: "Your information about Oppenheimer was very valuable to me. I know that he is a very fine and decent man, but you can't help it if someone gets on your nerves."

Only six weeks after his arrival, Oppenheimer astonished his peers by giving a lecture in Dutch, yet another language he had taught himself. His Dutch friends were so impressed by this spirited delivery that they began calling him "Opje"—an affectionate contraction of his last name—and he would bear the new nickname for life. His facility with this new language may have been assisted by a woman. According to the physicist Abraham Pais, Oppenheimer had an affair with a young Dutch woman named Suus (Susan).

This Dutch affair must have been brief, because Robert soon decided to leave Leiden. Though he had intended to go to Copenhagen, Ehrenfest convinced him he would be better off studying under Wolfgang Pauli in Switzerland. Ehrenfest wrote Pauli: "For the development of his great scientific talents, Oppenheimer needs right now to be lovingly spanked in shape! He really deserves that treatment . . . since he is an especially lovable chap." Ehrenfest usually sent his students to Bohr. But in this case Ehrenfest was certain, Oppenheimer recalled, "that Bohr with his largeness and vagueness was not the medicine I needed but that I needed someone who was a professional calculating physicist and that Pauli would be right for me. I think he used the phrase *herausprugeln* [to thrash out]. . . . It was clear that he was sending me there to be fixed up."

Robert also thought that Switzerland's mountain air might do him good. He had ignored Ehrenfest's nagging admonishments on the evils of smoking, but now his persistent cough suggested to him that he might still have a lingering case of tuberculosis. When concerned friends urged him to rest, Oppenheimer shrugged and said that rather than take care of the cough, he "prefers to live while he is alive."

On his way to Zurich, he stopped in Leipzig and heard Werner Heisenberg give a talk on ferromagnetism. Robert had, of course, met the future head of the German atomic bomb program at Göttingen a year earlier, and while no great friendship had ensued, they had developed a mutual if reserved respect. Upon arriving in Zurich, Wolfgang Pauli told him about his own work with Heisenberg. By then, Robert was very much interested in what he called the "electron problem and relativistic theory." That spring he nearly collaborated on a paper with Pauli and Heisenberg. "At first [we] thought the three of us should publish together; then Pauli thought he might publish it with me and then it seemed better to make some reference to it in their paper and let [my paper] be a separate publication. But Pauli said, 'You really made a terrible mess of the continuous spectra and you have a duty to clean it up, and besides, if you clean it up you may please the astronomers.' So that's how I got into that." Robert's paper was published the following year under the title "Notes on Theory of Interaction of Field and Matter."

Oppenheimer grew to be very fond of Pauli. "He was such a good physicist," Robert joked, "that things broke down or blew up when he merely walked into a laboratory." Only four years older than Oppenheimer, the precocious Pauli had established his reputation in 1920, the year before he obtained his Ph.D. at the University of Munich, when he published a two-hundred-page article on both the special and general theories of relativity. Einstein himself praised the essay for its clear exposition. After studying under Max Born and Niels Bohr, Pauli taught first at Hamburg and then, in 1928, at the Swiss Federal Institute of Technology in Zurich. By then he had published what became known as the "Pauli exclusion principle," which explained why each "orbital" in an atom may be occupied by only two electrons at a time.

Pauli was a pugnacious young man with a biting wit; like Oppenheimer, he was always quick to jump to his feet and aggressively question a lecturer if he perceived the slightest flaw in an argument. He frequently disparaged other physicists by saying that they were "not even wrong." And he once said of another scholar that he was "so young and already so unknown."

Pauli appreciated Oppenheimer's ability to discern the heart of a problem, but he found himself frustrated by Robert's inattentiveness to detail. "His ideas are always very interesting," Pauli said, "but his calculations are always wrong." After listening to Robert lecture one day, and hearing him pause, groping for words and murmuring little "nim-nim-nim" sounds, Pauli took to calling him the "nim-nim-nim man." Yet Pauli was fascinated by this complicated young American. "His strength," Pauli soon wrote Ehrenfest, "is that he has many and good ideas, and has much imagination. His weakness is that he is much too quickly satisfied with poorly based statements, that he does not answer his own often quite interesting questions for lack of perseverance and thoroughness. . . . Unfortunately, he has a very bad trait: he confronts me with a rather unconditional belief in authority and considers all I say as final and definitive truth. . . . I do not know how to make him give that up."

Another student, Isidor I. Rabi, spent a lot of time with Robert that spring. Having met in Leipzig, they traveled together to Zurich. "We got along very well," Rabi recalled. "We were friends until his last day. I enjoyed the things about him that some people disliked." Six years older than Oppenheimer, Rabi had spent his childhood, like Robert, in New York City. But his was a far different New York than Robert's gilded life on Riverside Drive. Rabi's family lived in a two-room flat on the Lower East Side. His father was a manual laborer and the family was poor. And unlike Oppenheimer, Rabi grew up with no ambiguity about his identity. The Rabis were Orthodox Jews and God was a part of daily life. "Even in casual conversation," Rabi remembered, "God entered, not every paragraph, more like every sentence." As he grew older, the formal religion fell away: "This was the church I failed," he quipped.

But Rabi remained comfortable as a Jew. Even in Germany in those years of festering anti-Semitism, Rabi insisted on introducing himself as an Austrian Jew precisely because he knew Austrian Jews were stereotypically the most disliked. Oppenheimer, by contrast, never advertised his Jewish identity. Decades later, Rabi thought he knew why: "Oppenheimer was Jewish, but he wished he weren't and tried to pretend he wasn't. . . . The Jewish tradition, even if you don't know it in detail, is so strong that you renounce it at your own peril. [This] doesn't mean you have to be Orthodox, or even practice it, but if you turn your back on it, having been born into it, you're in trouble. So that poor Robert, an expert in Sanskrit and French literature . . . [Rabi's voice here trailed off into silent thought.]"

Rabi later speculated that Robert "never got to be an integrated personality. It happens sometimes, with many people, but more frequently, perhaps, because of their situation, with brilliant Jewish people. With enormous capacities in every direction, it is hard to choose. He wanted everything. He reminded me very much of a boyhood friend of mine, who's a lawyer, about whom someone said, 'He'd like to be president of the Knights of Columbus and B'nai B'rith.' God knows I'm not the simplest person, but compared to Oppenheimer, I'm very, very simple."

Rabi loved Robert, but he could also proclaim to a friend for outrageous effect, "Oppenheimer? A rich spoiled Jewish brat from New York." Rabi thought he knew the type. "He was East German Jewish, and what happened to them was that they began to value the German culture above their own. You can see very easily why—with those immigrant Polish Jews and their very crude form of worship." The remarkable thing, Rabi thought, was that so many of these highly assimilated German Jews nevertheless couldn't in the end bring themselves to renounce their identity. The doors would open for them, but many refused to pass through. "I think in the Bible," Rabi said, "it says God complains that they're such obstinate people." In Rabi's eyes, Oppenheimer was similarly conflicted, but the difference may have been that he was unconsciously obstinate. "I don't know if he thought of himself as being Jewish," Rabi recalled many years later. "I think he had fantasies thinking he was not Jewish. I remember once saying to him how I found the Christian religion so puzzling, such a combination of blood and gentleness. He said that is what attracted him to it."

Rabi never told Oppenheimer what he thought of this ambivalence: "I didn't think it would be worthwhile telling him these things. . . . Can't change a man, that comes from inside." Rabi just felt he knew better than Oppenheimer himself who he was. "Whatever you want to say about Oppenheimer, he certainly wasn't a WASP."

Despite their differences, a close bond developed between Rabi and Oppenheimer. "I was never in the same class with him," Rabi later said. "I never ran into anyone who was brighter than he was." Still, Rabi's own brilliance was never in doubt. In just a few years, his experiments in a molecular-beam laboratory at Columbia University would produce seminal results for a wide range of fields in both physics and chemistry. Like Oppenheimer, he did not have the hands of an experimentalist; because he was clumsy, he often let others handle the equipment. But he had an uncanny ability to design experiments that produced results. And perhaps this was explained by the fact that during his stint in Zurich Rabi acquired, unlike most experimentalists, a very firm grasp of the theoretical. "Rabi was a great experimentalist," recalled Oppenheimer's student Wendell Furry, "and he was no slouch as a theorist." In the rarefied world of physics, Rabi would come to be regarded as the deep thinker and Oppenheimer as the great synthesizer. Together, they were formidable.

Their friendship transcended physics. Rabi shared Oppenheimer's interest in philosophy, religion and art. "We felt a certain kinship," Rabi said. It was that rare brand of friendship, forged in youth, that survives long separations. "You start off," Rabi recalled, "just where you left off." Robert particularly valued Rabi's candor. "I was not, as it were, put off by his manner," Rabi recalled. "I never flattered him, I was always honest with him." He always found Oppenheimer "stimulating, very stimulating." Over the years, and particularly at those times when most people felt intimidated by Oppenheimer, Rabi was perhaps the only man who could tell him in his straightforward fashion when he was being stupid. Near the end of his life, Rabi confessed, "Oppenheimer meant a great deal to me. I miss him."

In Zurich, Rabi knew his friend was working very hard on the quite difficult task of calculating the opacity of the surfaces of stars to their internal radiation—but Robert concealed his efforts under a calculated "air of easy nonchalance." Indeed, among friends, he avoided talking physics and became animated only when the topic turned to America. When the young Swiss physicist Felix Bloch stopped by Robert's apartment in Zurich, he happened to admire the beautiful Navajo rug Robert had slung over his sofa. This led Robert into a long and excited discourse on the merits of America. "There was no mistaking the intensity of Oppenheimer's affection for his country," remarked Bloch. "His attachment was most apparent." Robert could also talk at length about literature, "especially the Hindu classics and the more esoteric Western writers." Pauli joked with Rabi that Oppenheimer "seemed to treat physics as an avocation and psychoanalysis as a vocation."

To his friends, Robert seemed physically fragile and mentally robust. He smoked incessantly and nervously bit his fingernails. "The time with Pauli," he recalled later, "seemed just very, very good indeed. But I did get quite sick and had to go away for a while. I was told not to do any physics." After a six weeks' rest, an apparently mild case of tuberculosis was in remission. Oppenheimer returned to Zurich and resumed his frantic pace.

By the time Robert left Zurich in June 1929 to return to America, he had established an international reputation for his work in theoretical physics. Between 1926 and 1929 he published sixteen papers, an astonishing output for any scientist. If he had been a little too young to participate in the initial flowering of quantum physics in 1925–26, under Wolfgang Pauli's supervision he had clearly caught the second wave. He was the first physicist to master the nature of continuum wave functions. His most original contribution, in the opinion of the physicist Robert Serber, was his theory of field emission, an approach that permitted him to study the emission of electrons from metals, induced by a very strong field. In these early years he was also able to achieve breakthroughs in the calculation of the absorption coefficient of X rays and the elastic and inelastic scattering of electrons.

And what could any of this mean, in a practical sense, for humanity? However weirdly unintelligible—today as much as then—to the average citizen, quantum physics nevertheless explains our physical world. As the physicist Richard Feynman once observed, "[Quantum mechanics] describes nature as absurd from the point of view of common sense. And it fully agrees with experiment. So I hope you can accept nature as She is— absurd." Quantum mechanics seems to study that which doesn't exist—but nevertheless proves true. It works. In the decades to come, quantum physics would open the door to a host of practical inventions that now define the digital age, including the modern personal computer, nuclear power, genetic engineering, and laser technology (from which we get such consumer products as the CD player and the bar-code reader commonly used in supermarkets). If the youthful Oppenheimer loved quantum mechanics for the sheer beauty of its abstractions, it was nevertheless a theory that would soon spawn a revolution in how human beings relate to the world.

CHAPTER SIX

"Oppie"

> *I think that the world in which we shall live these next thirty years will be a pretty restless and tormented place; I do not think that there will be much of a compromise possible between being of it, and being not of it.*

ROBERT OPPENHEIMER
August 10, 1931

R OBERT'S TIME IN ZURICH had been productive and stimulating, but as always, with the coming of summer, he craved the exhilaration and the invigorating calmness induced by Perro Caliente. There was a rhythm now to his life: intense intellectual work, at times to the point of near exhaustion, followed by a month or more of renewal on horseback in the Sangre de Cristo Mountains of New Mexico.

In the spring of 1929, Robert wrote to brother Frank, urging him to bring their parents out West in June. He suggested further that, once the sixteen-year-old Frank had gotten Julius and Ella settled into a comfortable lodge in Santa Fe, he should take a friend up to their ranch above Los Pinos and "open up the place, get horses, learn to cook, make the hacienda as nearly habitable as you can, and see the country." He would join Frank in mid-July.

Frank needed no further prodding, and in June he arrived in Los Pinos with two friends from the Ethical Culture School, Ian Martin and Roger Lewis. Lewis was to become a regular visitor to Perro Caliente. Frank found a Sears, Roebuck catalogue and mail-ordered everything: beds, furniture, a stove, pots and pans, sheets and rugs. "It was a great spree," Frank recalled. "The stuff arrived shortly before my brother did that first summer. Old Mr. Windsor hauled it up to Perro Caliente with a horse and wagon." Robert arrived with two gallons of bootleg whiskey, a large quantity of peanut butter and a bag of Vienna sausages and chocolate. He arranged to borrow from Katherine Page a saddle horse named Crisis.

Aptly named, Crisis was a large, half-castrated stallion that no one but Robert could ride.

For the next three weeks, he and the boys spent their days hiking and riding through the mountains. After a particularly grueling day on horseback, Robert wrote a friend wistfully, "My two great loves are physics and New Mexico. It's a pity they can't be combined." At night, Robert sat by the light of a Coleman lantern, reading his physics books and preparing his lectures. On one trip, fully eight days long, they rode all the way to Colorado and back, a distance of more than 200 miles. When they weren't surviving on plain peanut butter, Robert introduced them to *nasi goreng,* an exceedingly spicy Indo-Dutch dish that Else Uhlenbeck had taught him to cook in the Netherlands. These were Prohibition years, but Robert always had plenty of whiskey on hand. "We'd get sort of drunk," Frank recalled, "when we were high up [in the mountains], and we'd all act kind of silly. . . . Everything my brother did would sort of be special. If he went off into the woods to take a leak, he'd come back with a flower. Not to disguise the fact that he'd made a leak, but just to make it an occasion, I guess." If he picked wild strawberries, Robert would serve them with Cointreau.

The Oppenheimer brothers spent hours in the saddle together, talking. "I think we probably rode about a thousand miles a summer," Frank Oppenheimer recalled. "We'd start off very early in the morning, and saddle up a horse, sometimes a packhorse, and start riding. Usually we'd have some new place that we wanted to go, often where there was no trail, and we really knew the mountains, the Upper Pecos, the surface of the whole mountain range. . . . There were wonderful flowers all the time. The place was very lush."

During one memorable ride up the Valle Grande, they were attacked by deer flies, which sting like bees. "So we set the horses to a full wild run up the length of the Valle (two miles), overtaking each other over and over again to pass the welcome flask after slowing enough to take a swig."

Robert showered his brother with gifts—a fine watch at the end of that summer, and two years later a secondhand Packard roadster—but he also invested time in tutoring Frank on love, music, art, physics—and his own philosophy of life: "The reason why a bad philosophy leads to such hell is that it is what you think and want and treasure and foster in times of preparation that determine what you do in the pinch, and that it takes an error to father a sin." Their times together at Perro Caliente were an intense part of Frank's education. When, later that summer, Frank wrote his brother a letter describing his encounter with a burro, Robert replied, "Your tales of a burro were immensely entertaining—so entertaining in fact that I showed them to one or two friends." Robert then went on to critique Frank's prose: "What you said, for instance, about Truchas and Ojo Caliente [in New Mexico] at night was much more convincing and honest and in the end communicative of emotion than your bits of purple writing about miscellaneous sunsets of the past."

In mid-August, Robert ambivalently packed his bag and drove to Berkeley, where he moved into a sparsely furnished room in the Faculty Club. Frank remained in New Mexico until early September, when Robert wrote him that he already missed the "gay times at Perro Caliente." He was nevertheless busy preparing his lectures and getting to know his colleagues. "The undergraduate college here," he wrote Frank, "seems not to be worth much, or I should suggest that you come here next year. For it is a beautiful place and the people are pleasant. I think that I am going to keep my room at the Faculty Club. . . . Tomorrow I have promised to cook Nasi Goreng on a camp fire. . . ." Soon, Robert's new friends in Berkeley would be calling his exotic dish "nasty gory" and trying to avoid it whenever possible.

THE UNIVERSITY OF CALIFORNIA, Berkeley, had hired Oppenheimer to introduce the new physics to graduate students. It did not occur to anyone, least of all to Robert, that he might teach undergraduates. In his first course, a graduate-level class on quantum mechanics, Robert jumped right in and tried to explain Heisenberg's uncertainty principle, the Schrödinger equation, Dirac's synthesis, field theory and Pauli's latest thinking on quantum electrodynamics. "I had for non-relativistic quantum mechanics a pretty good feeling, a pretty good understanding of what it was about," he later recalled. He started with wave-particle duality, the notion that quantum entities may behave as either particles or waves, depending on the circumstances of the experiment. "I would just make the paradox as bald and inescapable as possible." Initially, his lectures were largely incomprehensible to most students. When told that he was moving too fast, he only reluctantly tried to slow the pace and soon complained to his department chairman, "I'm going so slowly that I'm not getting anywhere."

Oppenheimer nevertheless always delivered a performance in the classroom—although during his first year or two of teaching, his presentations sounded more like a liturgy than a physics lecture. He tended to mumble in a soft, almost inaudible voice that got even lower when he was trying to emphasize a point. In the beginning, also, he stammered a good deal. Though he spoke without notes, he invariably laced his lectures with quotes from famous scientists and the occasional poet. "I was a very difficult lecturer," Oppenheimer recalled. His friend Linus Pauling, then an assistant professor of theoretical chemistry at Caltech, gave him this unfortunate advice in 1928: "If you want to give a seminar or lecture, decide what it is you want to talk about and then find some agreeable subject of contemplation not remotely related to your lecture and then interrupt that from time to time to say a few words." Years later, Oppenheimer commented, "So you can see how bad it must have been."

He played with his words, inventing complicated puns. There were no broken phrases in Robert's speech. He had the extraordinary ability to speak in complete, grammatically correct English sentences, without notes, pausing on occasion, as if between paragraphs, to stutter his oddly lilting hum that sounded like "nim-nim-nim." The relentless patter of his voice was interrupted only by puffs on his cigarette. Every so often, he would twirl toward the blackboard and write out an equation. "We were always expecting him," recalled one early graduate student, James Brady, "to write on the board with it [the cigarette] and smoke the chalk, but I don't think he ever did." As his students filed out of the classroom one day, Robert spotted a Caltech friend, Professor Richard Tolman, sitting in the back. When he asked Tolman what he thought of the lecture, he replied, "Well, Robert, that was beautiful, but I didn't understand a damn word."

Robert eventually transformed himself into a skilled and charismatic lecturer, but during his first years at Berkeley he seemed oblivious to the basic principles of communication. "Robert's blackboard manners were inexcusable," said Leo Nedelsky, one of his earliest graduate students. Once, when questioned about a particular equation on the blackboard, Oppenheimer replied, "No, not that one; the one underneath." But when perplexed students pointed out that there was no equation underneath, Robert said, "Not below, underneath. I have written over it."

Glenn Seaborg, later a chairman of the United States Atomic Energy Commission, complained of Professor Oppenheimer's "tendency to answer your question even before you had fully stated it." Frequently he interrupted guest speakers with comments like, "Oh, come now! We all know that. Let's get on with it." He refused to suffer fools—or even ordinary physicists—and he never hesitated to impose his own exceedingly high standards on others. In these early years at Berkeley, some thought he "terrorized" his students with sarcasm. "He could . . . be very cruel in his remarks," recalled one colleague. But as he matured as a teacher, he grew more tolerant of his students. "He was always very kind and considerate to anybody below him," recalled Harold Cherniss. "But not at all to people who might be considered his intellectual equals. And this, of course, irritated people, made people very angry, and made him enemies."

Wendell Furry, who studied at Berkeley from 1932 to 1934, complained that Oppenheimer expressed himself "somewhat obscurely and very quickly with flashes of insight which we couldn't follow." But even so, Furry recalled, "He praised all of our efforts even when we weren't so hot." One day in class, after a particularly difficult lecture, Oppenheimer quipped, "I can make it clearer; I can't make it simpler."

As difficult as he was, or perhaps because he was so difficult, most of his students took his courses more than once; indeed, one student, a young Russian woman recalled only as Miss Kacharova, took the course three times, and when she tried to enroll again, Oppenheimer refused to allow it. "She went on a hunger strike," recalled Robert Serber, "and forced her way in that way." For those who stuck it out, Oppenheimer found numerous ways to reward their hard work. "One learned from him through conversation and personal contact," Leo Nedelsky said. "When you took a question to him, he would spend hours—until midnight perhaps—exploring every angle with you." He invited a good number of his doctoral students to collaborate with him on papers, and he made sure they were listed as coauthors. "It is easy for a famous scientist to have lots of students doing the dirty work for him," said one colleague. "But Opje helps people with their problems and then gives them the credit." He encouraged his students to call him "Opje," the Dutch nickname he had acquired in Leiden. Robert himself began signing his letters with "Opje." Gradually, his Berkeley students anglicized "Opje" into "Oppie."

Over time, Oppenheimer developed a uniquely open teaching style in which he encouraged all of his students to interact with each other. Instead of holding office hours and seeing each student individually, he required his eight to ten graduate students and half-dozen postdoctoral fellows to meet together in his office in Room 219, LeConte Hall. Each student had a small desk and chair where he or she sat and watched as Oppenheimer paced the room. Oppie himself had no desk, only a table in the middle of the room piled high with stacks of papers. A blackboard covered with formulae dominated one wall. Shortly before the appointed hour, these young men (and the occasional woman) would straggle in and wait for Oppie as they casually sat on the edge of a table or leaned against the wall. When he arrived, he zeroed in on each student's particular research problem in turn and solicited comments from everyone. "Oppenheimer was interested in everything," Serber recalled, "and one subject after another was introduced and coexisted with all the others. In an afternoon, we might discuss electrodynamics, cosmic rays and nuclear physics." By focusing on the unsolved problems in physics, Oppenheimer gave his students a restless sense of standing on the edge of the unknown.

Very soon it was clear that Oppie had become a "Pied Piper" of theoretical physics. Word spread around the country that if you wished to enter this field, Berkeley was the place to do it. "I didn't start to make a school," Oppenheimer later said, "I didn't start to look for students. I started really as a propagator of the theory which I loved, about which I continued to learn more, and which was not well understood but which was very rich." In 1934, three of the five students awarded National Research Council fellowships in physics that year chose to study under Oppenheimer. And yet, while they came for Oppenheimer, they came as well for an experimental physicist named Ernest Orlando Lawrence.

Lawrence was everything that Robert Oppenheimer was not. Reared in South Dakota and educated at the universities of South Dakota, Minnesota, Chicago and Yale, Lawrence was a young man supremely confident of his talents. Of Norwegian Lutheran stock, Lawrence had an untroubled all-American demeanor. As a college student, he had paid his tuition peddling aluminum pots and pans to his farmer neighbors. An extrovert, he would use his natural affinity for salesmanship to promote his academic career. Some of his friends thought him a bit of a social climber, but unlike Robert, he possessed not a shred of existential angst or introspection. By the early 1930s, Lawrence was the premier experimental physicist of his generation.

At the time Oppenheimer arrived at Berkeley in the autumn of 1929, Lawrence, twenty-eight years old, was lodging in a room at the Faculty Club. The two very boyish physicists quickly became best friends. They talked almost daily and socialized in the evenings. On weekends they occasionally went horseback-riding. Robert, of course, rode in a Western saddle, but Ernest insisted on distinguishing himself from his farm background by affecting jodhpurs and an English saddle. Robert admired his new friend for his "unbelievable vitality and love of life." Here was a man, he saw, who could "work all day, run off for tennis, and work half the night." But he could also see that Ernest's interests were "primarily active [and] instrumental" while his own were "just the opposite."

Even after Lawrence married, Oppie was a frequent dinner guest, invariably bringing orchids for Ernest's wife, Molly. When Molly gave birth to their second son, Ernest insisted on naming the boy Robert. Molly acquiesced, but over the years she grew to think of Oppenheimer as somewhat faux, a man whose elaborate affectations betrayed a certain shallowness of character. Early in her marriage, she did not come between the two friends; but later, when circumstances changed, Molly would push her husband to see Oppie in a different light.

Lawrence was a builder—and he had the fundraising skills to realize his ambitions. In the months before meeting Oppenheimer, he had conceived the notion of building a machine capable of penetrating the so far unassailable nucleus of the atom, which existed, he quipped, "like a fly inside a cathedral." And not only was the nucleus tiny and elusive, it was also protected by a skin called the Coulomb barrier. Physicists estimated that it would take a stream of hydrogen ions, propelled with the potential of perhaps a million volts, to penetrate it. Generating such levels of high energy seemed an impossibility in 1929; but Lawrence conceived of a way around the impossible. He suggested that a machine could be built that used relatively small 25,000-volt potential to accelerate protons back and forth in an alternating electric field. By means of vacuum tubes and an electromagnet, the ion particles might then be accelerated by the electric field to greater and greater speeds along a spiral path. He was not sure how big an accelerator had to be to penetrate an atom's nucleus—but he was convinced that with a large enough magnet and a big enough circular chamber, he could break the million-volt mark.

By early 1931, Lawrence had built his first crude accelerator, a machine with a small 4.5-inch chamber within which he generated 80,000-volt protons. A year later, he had an eleven-inch machine that produced million-volt protons. Lawrence now dreamed of building ever bigger accelerators, machines weighing many hundreds of tons and costing tens of thousands of dollars. He coined a new name for his invention, the "cyclotron," and persuaded the president of the University of California, Robert Gordon Sproul, to give him an old wooden building adjacent to LeConte Hall, the physics building that sat high on the upper end of Berkeley's beautiful campus. Lawrence named it the Berkeley Radiation Laboratory. Theoretical physicists around the world soon realized that what Lawrence had created in his "Rad Lab" would allow them to explore the innermost reaches of the atom. In 1939, Lawrence won the Nobel Prize for physics.

Lawrence's relentless drive for ever larger and more powerful cyclotrons epitomized the trend toward the kind of "big science" associated with the rise of corporate America in the early twentieth century. Only four industrial laboratories existed in the country in 1890; forty years later there were nearly one thousand such facilities. In most of these labs a culture of technology, not science, was supreme. Over the years, theoretical physicists like Oppenheimer, devoted to pure "small" science, would find themselves alienated from the culture of these big labs, which were often devoted to "military science." Even in the 1930s, however, some young physicists couldn't stand the atmosphere. Robert Wilson, a student of both Oppenheimer and Lawrence, decided to leave Berkeley for Princeton, having concluded that the science associated with these big machines was "an activity that epitomized team research at its worst."

Building cyclotrons with eighty-ton magnets required large sums of money. But Lawrence was adept at enlisting financial support from such

Berkeley regents as oil entrepreneur Edwin Pauley, banker William H. Crocker and John Francis Neylan, a national power broker who happened to be William Randolph Hearst's chief counsel. In 1932, President Sproul sponsored Lawrence for membership in San Francisco's elite Bohemian Club, a fraternity of California's most influential businessmen and politicians. The members of the Bohemian Club would never have thought to welcome Robert Oppenheimer; he was Jewish and too otherworldly. But the Midwestern farm boy Lawrence slipped effortlessly into this elite society. (Later, Neylan got Lawrence into the even more exclusive Pacific Union Club.) Gradually, as Lawrence repeatedly took the money of these powerful men, he found himself also sharing their conservative, anti–New Deal politics.

By contrast, Oppenheimer had a laissez-faire attitude toward the role of money in his own research. When one of his graduate students wrote him for help in raising money for a particular project, Oppie replied whimsically that such research, "like marriage and poetry, should be discouraged and should occur only despite such discouragement."

On February 14, 1930, Oppenheimer finished writing a seminal paper, "On the Theory of Electrons and Protons." Drawing on Paul Dirac's equation on the electron, Oppenheimer argued that there had to be a positively charged counterpart to the electron—and that this mysterious counterpart should have the same mass as the electron itself. It could not, as Dirac had suggested, be a proton. Instead, Oppenheimer predicted the existence of an "anti-electron—the positron." Ironically, Dirac had failed to pick up on this implication in his own equation, and he willingly gave Oppenheimer the credit for this insight—which soon impelled him, Dirac, to propose that perhaps there existed "a new kind of particle, unknown to experimental physics, having the same mass and opposite charge to an electron." What he was very tentatively proposing was the existence of antimatter. Dirac suggested naming this elusive particle an "anti-electron."

Initially, Dirac himself was not at all comfortable with his own hypothesis. Wolfgang Pauli and even Niels Bohr emphatically rejected it. "Pauli thought it was nonsense," Oppenheimer later said. "Bohr not only thought it was nonsense but was completely incredulous." It took someone like Oppenheimer to push Dirac into predicting the existence of antimatter. This was Oppenheimer's penchant for original thinking at its best. In 1932 the experimental physicist Carl Anderson proved the existence of the positron, the positively charged antimatter counterpart to the electron. Anderson's discovery came fully two years after Oppenheimer's calculations suggested its theoretical existence. A year later, Dirac won his Nobel Prize.

Physicists around the globe were racing to solve the same set of problems, and the competition to be first was fierce. Oppenheimer proved to be a productive dilettante in this race. Working with a small number of students, he still managed to skip from one critical problem to another just in time to publish a short letter on a particular topic a month or two ahead of the competition. "It was amazing," recalled one Berkeley colleague, "that Oppenheimer and his group essentially got something on all these problems, about the same time as the competition." The result might not be elegant or even particularly accurate in all the details—others would have to come along and clean up his work. But Oppenheimer invariably had the essence. "Oppie was extremely good at seeing the physics and doing the calculation on the back of the envelope and getting all the main factors. . . . As far as finishing and doing an elegant job like Dirac would do, that wasn't Oppie's style." He worked "fast and dirty, like the American way of building a machine."

In 1932, Ralph Fowler, one of Oppie's former teachers from Cambridge, England, visited Berkeley and had a chance to observe his old student. In the evenings, Oppie persuaded Fowler to play his particularly complicated version of tiddlywinks for hours on end. Some months later, at a time when Harvard was trying to recruit Oppenheimer away from Berkeley, Fowler wrote that "his work is apt to be full of mistakes due to lack of care, but it is work of the highest originality and he had an extremely stimulating influence in a theoretical school as I had ample opportunity of learning last fall." Robert Serber agreed: "His physics was good, but his arithmetic awful."

Robert did not have the patience to stick with any one problem very long. As a result, it was frequently he who opened the door through which others then walked to make major discoveries. In 1930 he wrote what would become a well-known paper on the infinite nature of spectral lines using direct theory. A splitting of the line in a spectrum of hydrogen suggested a small difference in the energy levels of two possible states of the hydrogen atom. Dirac had argued that these two states of hydrogen should have precisely the same energy. In his paper, Oppenheimer disagreed, but his results were inconclusive. Years later, however, an experimental physicist, Willis E. Lamb, Jr., one of Oppenheimer's doctoral students, resolved the issue. The so-called "Lamb shift" correctly attributed the difference between the two energy levels to the process of self-interaction—whereby charged particles interact with electromagnetic fields. Lamb won a Nobel Prize in 1955, in part for his precise measurement of the Lamb shift, a key step in the development of quantum electrodynamics.

During these years, Oppenheimer wrote important, even seminal, papers on cosmic rays, gamma rays, electrodynamics and electron-positron showers. In the field of nuclear physics, he and Melba Phillips calculated the yield of protons in deuteron reactions. Phillips, an Indiana farm girl, born in 1907, was Oppenheimer's first doctoral student. Their calculations on proton yields became widely known as the "Oppenheimer-Phillips process." "He was an idea man," recalled Phillips. "He never did any great physics, but look at all the lovely ideas that he worked out with his students."

Physicists today agree that Oppenheimer's most stunning and original work was done in the late 1930s on neutron stars—a phenomenon astronomers would not actually be able to observe until 1967. His interest in astrophysics was initially sparked by his friendship with Richard Tolman, who introduced him to astronomers working at Pasadena's Mt. Wilson Observatory. In 1938, Oppenheimer wrote a paper with Robert Serber titled "The Stability of Stellar Neutron Cores," which explored certain properties of highly compressed stars called "white dwarfs." A few months later, he collaborated with another student, George Volkoff, on a paper titled "On Massive Neutron Cores." Laboriously deriving their calculations from slide rules, Oppenheimer and Volkoff suggested there was an upper limit—now called the "Oppenheimer-Volkoff limit"—to the mass of these neutron stars. Beyond this limit they would become unstable.

Nine months later, on September 1, 1939, Oppenheimer and a different collaborator—yet another student, Hartland Snyder—published a paper titled "On Continued Gravitational Contraction." Historically, of course, the date is best known for Hitler's invasion of Poland and the start of World War II. But in its quiet way, this publication was also a momentous event. The physicist and science historian Jeremy Bernstein calls it "one of the great papers in twentieth-century physics." At the time, it attracted little attention. Only decades later would physicists understand that in 1939 Oppenheimer and Snyder had opened the door to twenty-first-century physics.

They began their paper by asking what would happen to a massive star that has begun to burn itself out, having exhausted its fuel. Their calculations suggested that instead of collapsing into a white dwarf star, a star with a core beyond a certain mass—now believed to be two to three solar masses—would continue to contract indefinitely under the force of its own gravity. Relying on Einstein's theory of general relativity, they argued that such a star would be crushed with such "singularity" that not even light waves would be able to escape the pull of its all-encompassing gravity. Seen from afar, such a star would literally disappear, closing itself off from the rest of the universe. "Only its gravitation field persists," Oppenheimer and Snyder wrote. That is, though they themselves did not use the term, it would become a black hole. It was an intriguing but bizarre notion—and the paper was ignored, with its calculations long regarded as a mere mathematical curiosity.

Only since the early 1970s, when the technology of astronomical observation caught up with theory, have numerous such black holes been detected by astronomers. At that time, computers and technical advances in radio telescopes made black-hole theory the centerpiece of astrophysics. "Oppenheimer's work with Snyder is, in retrospect, remarkably complete and an accurate mathematical description of the collapse of a black hole," observed Kip Thorne, a Caltech theoretical physicist. "It was hard for people of that era to understand the paper because the things that were being smoked out of the mathematics were so different from any mental picture of how things should behave in the universe."

Characteristically, however, Oppenheimer never took the time to develop anything so elegant as a theory of the phenomenon, leaving this achievement to others decades later. And the question remains: Why? Personality and temperament appear to be critical. Robert instantly saw the flaws in any idea almost as soon as he had conceived it. Whereas some physicists—Edward Teller immediately comes to mind—boldly and optimistically promoted all of their new ideas, regardless of their flaws, Oppenheimer's rigorous critical faculties made him profoundly skeptical. "Oppie was always pessimistic about all the ideas," recalled Serber. Turned on himself, his brilliance denied him the dogged conviction that is sometimes necessary for pursuing and developing original theoretical insights. Instead, his skepticism invariably propelled him on to the next problem.* Having made the initial creative leap, in this case to black-hole theory, Oppenheimer quickly moved on to another new topic, meson theory.

Years later, Robert's friends and peers in the world of physics, who generally agreed that he was brilliant, would ruminate on why he never won a Nobel Prize. "Robert's own knowledge of physics was profound," recalled Leo Nedelsky. "Perhaps only Pauli knew more physics and knew it more profoundly than Robert." And yet, winning a Nobel, like much in life, is a matter of commitment, strategy, ability, timing, and, of course, chance. Robert had a commitment to doing cutting-edge physics, to attacking problems that interested him; and he certainly had the ability. But he did not have the right strategy—and his timing was off. Finally, the Nobel Prize is a distinction awarded to scientists who achieve something specific. By contrast, Oppenheimer's genius lay in his ability to synthesize the entire field of study. "Oppenheimer was a very imaginative person," recalled Edwin Uehling, a postdoctoral student who studied under him during the years 1934–36. "His knowledge of physics was extremely comprehensive. I am not sure that one should say that he didn't do Nobel Prize–quality work; but

*More than two decades later, another physicist, John Wheeler, tried to talk with Oppenheimer about his old work on spent neutron stars. But by then, he expressed no interest in what was rapidly becoming the hottest topic in physics.

it just didn't happen to lead to that kind of result which the Nobel Prize committee regarded as exciting."

"The work is fine," Oppenheimer wrote to Frank in the autumn of 1932. "Not fine in the fruits but the doing. . . . We have been running a nuclear seminar, in addition to the usual ones, trying to make some order out of the great chaos. . . ." While Oppenheimer was a theorist who knew how incompetent he was in the laboratory, he nevertheless stayed close to experimentalists like Lawrence. Unlike many European theorists, he appreciated the potential benefit from close collaboration with those who were involved in testing the validity of the new physics. Even in high school, his teachers had noted his gift for explaining technical things in plain language. As a theorist who understood what the experimentalists were doing in the laboratory, he had that rare quality of being able to synthesize a great mass of information from disparate fields of research. An articulate synthesizer was exactly the kind of person needed for building a world-class school of physics. Some physicists have suggested that Oppenheimer possessed the knowledge and resources to publish a comprehensive "bible" of quantum physics. By 1935, he certainly had the material for such a book at hand. His basic lectures explaining quantum mechanics were so popular on campus that his secretary, Miss Rebecca Young, had his lecture notes mimeographed and sold them to students. The proceeds were used for the physics department's petty cash fund. "Had Oppenheimer gone one step further and compiled his lectures and papers," argues one colleague, "his work would have made one of the finest textbooks on quantum physics ever written."

ROBERT HAD PRECIOUS LITTLE TIME for diversions. "I need physics more than friends," he confessed to Frank in the autumn of 1929. He managed to go horseback-riding once a week in the hills overlooking San Francisco Bay. "And from time to time," he wrote Frank, "I take out the Chrysler, and scare one of my friends out of all sanity by wheeling corners at seventy. The car will do seventy-five without a tremor. I am and shall be a vile driver." One day he crashed his car while recklessly racing the coast train near Los Angeles; Robert escaped unscathed but for a moment he thought his passenger, a young woman named Natalie Raymond, was dead. Actually, Raymond had only been knocked unconscious. When Julius found out about the accident, he gave her a Cézanne drawing and a small Vlaminck painting.

Raymond was a beautiful woman in her late twenties when she met Oppenheimer at a Pasadena party. "Natalie was a dare-devil, an adventurer, as was Robert to some extent," wrote a mutual friend. "This may have been the common ground of their natures. Robert grew up (or did he?), Natalie less so." Robert called her Nat, and they saw quite a bit of each other in the early 1930s. Frank Oppenheimer described her as "quite a lady," and Robert himself wrote Frank after seeing her at a New Year's Eve party: "Nat has learned to dress. She wears long graceful things in gold and blue and black, and delicate long earrings, and likes orchids, and even has a hat. To the vicissitudes and anguishes of fortune which have brought this change to her I need say nothing." After spending an evening with her at Radio City Music Hall listening to a "most marvelous" Bach concert, he wrote Frank, "The last days were impregnated with Nat; her always new & always moving miseries." She even spent part of the summer of 1934 with Robert and others at Perro Caliente. But the relationship ended when she moved to New York to work as a free-lance book editor.

Nat wasn't the only woman in Oppenheimer's life. In the spring of 1928, he had met Helen Campbell at a Pasadena party. Though she was already engaged to a Berkeley physics instructor, Samuel K. Allison, Helen found herself strongly attracted to Oppenheimer. He took her out to dinner and they had a few walks together. When Oppenheimer returned to Berkeley in 1929, they resumed their friendship. By then Helen was a married woman, and she watched with some amusement as she observed "young wives falling for Robert, charmed by his conversation, gifts of flowers, etc." She realized that he "had an eye for women and that his attentions to her should not be taken too seriously." She thought he "liked to talk to slightly discontented women and seemed specially sensitive to lesbianism." He possessed plenty of charisma.

"Everyone wants rather to be pleasing to women," Robert wrote to his brother in 1929, "and that desire is not altogether, though it is very largely, a manifestation of vanity. But one cannot aim to be pleasing to women, any more than one can aim to have taste, or beauty of expression, or happiness; for these things are not specific aims which one may learn to attain; they are descriptions of the adequacy of one's living. To try to be happy is to try to build a machine with no other specification than that it shall run noiselessly."

When Frank wrote him to complain about his problems with "the jeunes filles Newyorkaises," Robert replied, "I should say that you were wrong to let the creatures worry you . . . you should not associate with them unless it is for you a genuine pleasure; and that you should have truck only with those girls who not only pleased you, but who were pleased, and who put you at your ease. The obligation is always on the girl for making a go of conversation: if she does not accept the obligation, nothing you can do will

make the negotiations pleasant." Obviously, relations with the opposite
sex were still a matter of uneasy negotiations for Robert, let alone his
seventeen-year-old brother.

To most of his friends, Robert was a maddening bundle of contradictions. Harold F. Cherniss was getting his doctorate in Berkeley's Classical
Greek Department when he first met Oppenheimer in 1929. Cherniss had
just married a childhood friend of Oppenheimer's, Ruth Meyer, who also
had known Robert at the Ethical Culture School. Cherniss was immediately
taken by Oppenheimer: "His mere physical appearance, his voice, and his
manners made people fall in love with him—male, female. Almost everybody." But he admitted that "the longer I was acquainted with him, the more
intimately I was acquainted with him, the less I knew about him." A keen
observer of people, Cherniss sensed a disconnect in Robert. Here was a
man, he thought, who was "very sharp intellectually." People thought him
complicated simply because he was interested in so many things, and knew
so much. But on an emotional level, "he wanted to be a simple person, simple in the good sense of the word." Robert "wanted friends very much,"
Cherniss said. And yet, despite his tremendous personal charm, "he didn't
quite know how to make friends."

CHAPTER SEVEN

"The Nim Nim Boys"

*Tell me, what has politics
to do with truth,
goodness and beauty.*

ROBERT OPPENHEIMER

IN THE SPRING OF 1930, Julius and Ella Oppenheimer came out to visit their son in Pasadena. The stock market crash of the previous autumn had plunged the nation into a deep economic depression, but Julius had fortuitously decided to retire in 1928, selling his interest in Rothfeld, Stern and Co. He had also sold both the Riverside Drive apartment and their Bay Shore summer home, moving with Ella into a smaller apartment on Park Avenue. The Oppenheimer family fortune was unscathed. Robert immediately introduced his parents to his closest friends, Richard and Ruth Tolman. The elder Oppenheimers had what Julius called a "delightful" dinner, and several teas, with the Tolmans, and Ruth later took them to Los Angeles to hear a Tchaikovsky concert. Observing that "[Robert's] reconstructed Chrysler emitted all sorts of groans," Julius decided to buy him a new Chrysler despite his son's "severe protests." "Now, that brother has it," Julius subsequently wrote his son Frank, "he is most delighted with it, and he has reduced his speed about 50% from what he used to drive, so we hope no further accidents will occur." Robert named his new car the *Gamaliel*, the Hebrew name of a number of prominent ancient rabbis. As an adolescent, he had tried to hide his Jewish ancestry; it was a measure of his newly developed confidence and maturity that he now felt comfortable advertising it.

Around this time, Frank wrote him to complain that the brother he had known had "completely vanished." Robert had written back in protest that this could not be so. Nevertheless, Robert realized that during his two-year absence in Europe, Frank—eight years his junior—must have done some growing up. "For purposes of recognition it will suffice for you to know that

I am six feet tall, have black hair, blue eyes and at present a split lip, and that I answer to the call of Robert."

He then went on to try to answer a question posed by his younger brother: "How far is it wise to respond to a mood?" Robert's answer suggests that his fascination with the psychological was still acute: ". . . my own conviction is that one should use moods, but not be greatly deflected by them; thus one should try to use the gay times to do those things one wants to do which require gaiety, and the sober moods for the work one wants, and the low moods for giving oneself hell."

MORE THAN MOST PROFESSORS, Oppenheimer included his students in his social life. "We did everything together," said Edwin Uehling. On Sunday mornings, Oppenheimer frequently dropped by the Uehlings' apartment to have breakfast and listen to a broadcast of the New York Symphony. Every Monday evening, Oppenheimer and Lawrence led a colloquium on physics open to all graduate students from both Berkeley and Stanford. They dubbed it the "Monday Evening Journal Club," in part because the focus of discussion was usually a recently published article in *Nature* magazine or *Physical Review.*

For a short time, Robert dated his doctoral student Melba Phillips, and one evening he drove her out to Grizzly Peak, in the Berkeley hills, with a fine view of San Francisco Bay in the distance. After wrapping a blanket around Phillips, Oppenheimer announced, "I'll be back presently. I'm going for a walk." He came back shortly and briefly leaned in toward the car window and said, "Melba, I think I'll walk on down to the house, why don't you bring the car down?" Melba, however, had dozed off and didn't hear him. When she awoke, she waited patiently for Oppie to return, but finally, after two hours had gone by with no sign of him, she hailed a passing policeman and said, "My escort went for a walk hours ago and he hasn't returned." Fearing the worst, police combed the bushes for Oppenheimer's body. Phillips eventually drove herself home in Oppie's car, and the police went to his quarters in the Faculty Club—where they roused a sleepy Oppenheimer from his bed. Apologizing, he explained to the police that he had forgotten all about Miss Phillips: "I'm awfully erratic, you know. I just walked and walked—and I was home and I went to bed. I'm so sorry." A reporter on the police beat heard the story, and the next day's *San Francisco Chronicle* ran a short story on the front page headlined "Forgetful Prof Parks Girl, Takes Self Home." It was Oppenheimer's first exposure to the press. Newspapers around the world picked up on the story. Frank Oppenheimer happened to read it in a paper in Cambridge, England. Naturally, both Oppie and Melba were embarrassed, and, somewhat defensively, he explained to friends that he had told Melba he was going to walk home, but that she must have dozed off and hadn't heard him.

In 1934, Oppenheimer moved into an apartment on the lower level of a small house at 2665 Shasta Road, perched on one of the steep switchbacks in the Berkeley Hills. Often he invited students over for a simple dinner of "eggs à la Oppie," invariably laced with Mexican chilies and washed down with red wine. On occasion he would subject his guests to his potent martini, shaken with elaborate ceremony and poured into chilled glasses. Sometimes he dipped the rims of the martini glasses in lime juice and honey. Winter or summer, he always kept the windows wide open, which meant that in winter his guests would crowd around the large fireplace that dominated the dark-paneled living-room draped with Indian rugs from New Mexico. His father had given him a small Picasso lithograph which he hung on the wall. If everyone seemed tired of physics, the conversation might turn to art or literature—or he would suggest a movie. The little redwood house enjoyed a view of San Francisco and the Golden Gate Bridge. Oppie called it "the most beautiful harbor in the world." From the road above, the house was almost entirely concealed by a grove of eucalyptus, pine and acacia. He told his brother, Frank, that he usually slept on the porch "under the Yaqui and the stars, and imagine I am on the porch at *Perro Caliente.*"

In these years, Oppie's professional garb was always a gray suit, a blue denim shirt and clunky, round-toed black shoes, worn but well-polished. But away from the university he changed out of this academic uniform into a blue workshirt and faded blue jeans held up by a broad leather belt with a Mexican silver buckle. His long bony fingers were now stained a deep yellow from nicotine.

Consciously or not, some of Oppie's students began imitating his quirks and eccentricities. They came to be called the "nim nim boys," because they mimicked his "nim nim" humming. Almost all of these budding young physicists began chain-smoking Chesterfields, Oppie's brand, and, like Oppie, flicked their lighters whenever anyone took out a cigarette. "They copied his gestures, his mannerisms, his intonations," recalled Robert Serber. Isidor Rabi observed, "He [Oppenheimer] was like a spider with this communication web all around him. I was once in Berkeley and said to a couple of his students, 'I see you have your genius costumes on.' By the next day, Oppenheimer knew that I had said that." It was a cult or mystique that some found annoying. "We weren't supposed to like Tchaikovsky," Edwin Uehling reported, "because Oppenheimer never liked Tchaikovsky."

His students were constantly reminded that, unlike most physicists, he read books far outside his field. "He read a good deal of French poetry," recalled Harold Cherniss. "He read almost everything [novels and poetry] that came out." Cherniss saw him reading the classical Greek poets, but also such contemporary novelists as Ernest Hemingway. He particularly liked Hemingway's *The Sun Also Rises.*

Even during the Depression, Oppie's circumstances were decidedly flush. To begin with, by October 1931, when he was promoted to associate professor, he had an annual salary of $3,000 and his father continued to provide him with additional funds. Although Julius hadn't had enough money from the sale of the firm to set up the independent foundation he wanted to establish, there was enough for a trust fund, "so that Robert will never have to give up his research."

Like his father, Robert was instinctively generous, and he never hesitated to share with his students his fine taste in food and wine. At Berkeley, after leading a late-afternoon seminar, he often invited a roomful of students to join him for dinner at Jack's Restaurant, one of San Francisco's most pleasant eating establishments. Prior to 1933, Prohibition was still the law of the land, but Oppenheimer, said one old friend, "knew all the best restaurants and speakeasies in San Francisco." In those years, one still had to take the ferry from Berkeley to San Francisco, and often (after 1933), while waiting at the ferry terminal, everyone would have a quick drink at one of the bars that now lined the wharf. Once they had made their way to Jack's at 615 Sacramento Street, Oppie chose the wines and guided his students in their selections from the menu. He always picked up the check. "The world of good food and good wines and gracious living was far from the experience of many," said one of his students. "Oppenheimer introduced us to an unfamiliar way of life. . . . We acquired something of his tastes." Once a week or so Oppie dropped by Leo Nedelsky's house, where a number of his students rented rooms, including J. Franklin Carlson and Melba Phillips. Almost every night at 10:00, tea and cake were served and everyone sat around playing tiddlywinks and discussing anything and everything. Most people left by midnight, but sometimes the conversation would last until two or three in the morning.

One night late in the spring semester, 1932, Oppie announced that Frank Carlson—who suffered from occasional bouts of depression—needed help in finishing his thesis. "Frank has done this work," Oppenheimer said, "and now it's got to be written up." In response, Oppie's other students pitched in and formed what amounted to a sort of little factory: "Frank [Carlson] wrote," Phillips recalled, "Leo [Nedelsky] edited . . . I proofread and wrote all the equations in the thesis." Carlson got his thesis accepted that June and served as Oppenheimer's research associate for the academic year 1932–33.

Each spring, after Berkeley's semester ended in April, Oppie's students would follow him 375 miles south to Caltech in Pasadena, where he taught the spring quarter. They thought nothing of giving up the leases on their Berkeley area apartments and moving into garden cottages in Pasadena for $25 a month. Additionally, in the summer some of them even followed him for a few weeks to the University of Michigan's summer physics seminar in Ann Arbor.

In the summer of 1931, Oppie's former teacher in Zurich, Wolfgang Pauli, showed up at the Ann Arbor seminar. On one occasion, Pauli kept interrupting Oppie's presentation until finally another eminent physicist, H. A. Kramers, shouted, "Shut up, Pauli, and let us hear what Oppenheimer has to say. You can explain how wrong it is afterward." Such sharp-tongued banter only enhanced the aura of free-wheeling brilliance that surrounded Oppenheimer.

DURING THE SUMMER OF 1931 Ella Oppenheimer fell ill and was diagnosed with leukemia. On October 6, 1931, Julius cabled Robert: "Mother critically ill. Not expected to live . . ." Robert rushed home and sat vigil at his mother's bedside. He found her "terribly low, almost beyond hope." He wrote Ernest Lawrence: "I have been able to talk to her a little; she is tired and sad, but without desperation; she is unbelievably sweet." Ten days later, he was reporting that the end was approaching: "She is comatose, now; and death is very near. We cannot help feeling now a little grateful that she should not have to suffer more. . . . The last thing she said to me was 'Yes—California.' "

Toward the end, Herbert Smith came to the Oppenheimer home to com- fort his former student. After several hours of desultory conversation, Robert looked up and said, "I'm the loneliest man in the world." Ella died on October 17, 1931, age sixty-two. Robert was twenty-seven years old. When a family friend tried to console him by saying, "You know, Robert, your mother loved you very much," he muttered softly in reply, "Yes, I know. Maybe she loved me too much."

A grief-stricken Julius continued to reside in New York City, but soon he was visiting his son in California on a regular basis. Father and son grew even closer. Indeed, Robert's students and colleagues at Berkeley were rather taken by the manner in which he made room in his life for his father. During the winter of 1932, father and son shared a cottage in Pasadena, where Robert was teaching that term. Robert had lunch with his father every day and took him one evening a week to an elite dinner club that met at Caltech; Robert used the German word *Stammtisch* (a table reserved for regular guests) for these dinners, where a designated speaker gave a presen- tation, followed by vigorous discussion. Julius was enormously pleased to be included in these events and wrote Frank: "They are very good fun. . . . I am meeting lots of Robert's friends and yet I believe that I have not interfered with his activities. He is always busy and has had a couple of short talks with Einstein." Twice a week, Julius played bridge with Ruth Uehling, and they became good friends. "Nobody could make a woman feel more important than the way he [Julius] could," recalled Ruth later. "He was terribly proud of his son. . . . He couldn't understand how he had produced Robert." Julius also talked passionately about the art world, and when Ruth visited him in New York in the summer of 1936, he proudly showed her his collection of paintings. "He made me sit all day before the beautiful Van Gogh with a blazing sun, to see," she recalled, "how the light changed it."

Among other friends, Robert introduced his father to Arthur W. Ryder, a professor of Sanskrit at Berkeley. Ryder was a Hoover Republican and a sharp-tongued iconoclast. He was "fascinated" by Oppenheimer, and Robert, for his part, thought Ryder the quintessential intellectual. His father agreed: "He is an astounding person," Julius said, "a remarkable combination of austereness thru which peeps the gentlest kind of soul." Robert later credited Ryder with giving him a renewed "feeling for the place of ethics." Here was a scholar, he said, who "felt and thought and talked as a stoic." He regarded Ryder as one of those rare people who "have a tragic sense of life, in that they attribute to human actions the completely decisive role in the difference between salvation and damnation. Ryder knew that a man could commit irretrievable error, and that in the face of this fact, all others were secondary."

Robert felt himself drawn to both Ryder and the ancient language that was his friend's vocation. Soon Ryder was giving Oppenheimer private tutorials in Sanskrit each Thursday evening. "I am learning Sanskrit," Robert wrote Frank, "enjoying it very much, and enjoying again the sweet luxury of being taught." While most of his friends saw this new obsession as slightly odd, Harold Cherniss—who had introduced Oppie to Ryder— thought it made perfect sense. "He liked things that were difficult," Cherniss said. "And since almost everything was easy for him, the things that really would attract his attention were essentially the difficult." Besides, Oppie had a "taste for the mystical, the cryptic."

With his facility for languages, it wasn't long before Robert was reading the Bhagavad-Gita. "It is very easy and quite marvelous," he wrote Frank. He told friends that this ancient Hindu text—"The Lord's Song"—was "the most beautiful philosophical song existing in any known tongue." Ryder gave him a pink-covered copy of the book which found its way onto the bookshelf closest to his desk. Oppie took to passing out copies of the Gita as gifts to his friends.

Robert was so enraptured by his Sanskrit studies that when, in the autumn of 1933, his father bought him yet another Chrysler, he named it the *Garuda,* after the giant bird god in Hindu mythology that ferries Vishnu across the sky. The Gita—which constitutes the heart of the Sanskrit epic Mahabharata—is told in the form of a dialogue between the incarnate god Krishna and a human hero, Prince Arjuna. About to lead his troops into mortal combat, Arjuna refuses to engage in a war against friends and relatives. Lord Krishna replies, in essence, that Arjuna must fulfill his destiny as a warrior to fight and kill.*

Ever since his emotional crisis of 1926, Robert had been trying to achieve some kind of inner equilibrium. Discipline and work had always been his guiding principles, but now he self-consciously elevated these traits to a philosophy of life. In the spring of 1932, Robert wrote his brother a long letter explaining why. The fact that discipline, he argued, "is good for the soul is more fundamental than any of the grounds given for its goodness. I believe that through discipline, though not through discipline alone, we can achieve serenity, and a certain small but precious measure of freedom from the accidents of incarnation . . . and that detachment which preserves the world which it renounces. I believe that through discipline we learn to preserve what is essential to our happiness in more and more adverse circumstances, and to abandon with simplicity what would else have seemed to us indispensable." And only through discipline is it possible "to see the world without the gross distortion of personal desire, and in seeing it so, accept more easily our earthly privation and its earthly horror."

Like many Western intellectuals enthralled with Eastern philosophies, Oppenheimer the scientist found solace in their mysticism. He knew, moreover, that he was not alone; he knew that some of the poets he admired most, like W. B. Yeats and T. S. Eliot, had themselves dipped into the Mahabharata. "Therefore," he concluded in his letter to the twenty-year-old Frank, "I think that all things which evoke discipline: study, and our duties to men and to the commonwealth, and war, and personal hardship, and even the need for subsistence, ought to be greeted by us with profound gratitude; for only through them can we attain to the least detachment; and only so can we know peace."

In his late twenties, Oppenheimer already seemed to be searching for an earthly detachment; he wished, in other words, to be engaged as a scientist with the physical world, and yet detached from it. He was not seeking to escape to a purely spiritual realm. He was not seeking religion. What he sought was peace of mind. The Gita seemed to provide precisely the right

*Oppenheimer was quite obviously moved by this ancient existential epic. But when his old friend from Zurich days, Isidor Rabi, passed through Berkeley and learned that Oppie was studying Sanskrit, he wondered, "Why not the Talmud?"

philosophy for an intellectual keenly attuned to the affairs of men and the pleasures of the senses. One of his favorite Sanskrit texts was the Meghaduta, a poem that discusses the geography of love from the laps of naked women to the soaring mountains of the Himalayas. "The Meghaduta I read with Ryder," he wrote Frank, "with delight, some ease, and great enchantment. . . ." Yet another of his favorite parts of the Gita, the Satakatrayam, contains these fatalistic lines:

> *Vanquish enemies at arms . . .*
> *Gain mastery of the sciences*
> *And varied arts . . .*
> *You may do all this, but karma's force*
> *Alone prevents what is not destined*
> *And compels what is to be.*

Unlike the Upanishads, the Gita celebrates a life of action and engagement with the world. As such, it was compatible with Oppenheimer's Ethical Culture upbringing; but there also were important differences. The Gita's notions of karma, destiny and earthly duty would seem to be at odds with the humanitarianism of the Ethical Culture Society. Dr. Adler had disparaged the teaching of any inexorable "laws of history." Ethical Culture stressed instead the role of individual human will. There was nothing fatalistic about John Lovejoy Elliott's social work in the immigrant ghettos of lower Manhattan. So perhaps the attraction Oppenheimer felt to the fatalism of the Gita was at least partly stimulated by a late-blooming rebellion against what he had been taught as a youth. Isidor Rabi thought so. Rabi's wife, Helen Newmark, had been a classmate of Robert's at the Ethical Culture School, and Rabi later recalled, "From conversations with him I have the impression that his own regard for the school was not affectionate. Too great a dose of ethical culture can often sour the budding intellectual who would prefer a more profound approach to human relations and man's place in the universe."

Rabi speculated that young Oppenheimer's Ethical Culture heritage may have become an immobilizing burden. It is impossible to know the full results of one's actions, and sometimes even good intentions lead to horrific outcomes. Robert was acutely attuned to the ethical, and yet endowed with ambition and an expansive, curious intelligence. Like many intellectuals aware of the complexities of life, perhaps he sometimes felt paralyzed to the point of inaction. Oppenheimer later reflected upon precisely this dilemma: "I may, as we all have to, make a decision and act or I may think about my motives and my peculiarities and my virtues and my faults and try to decide why I am doing what I am. Each of these has its place in our life, but clearly the one forecloses the other." At the Ethical Culture School, Felix Adler had subjected himself to "constant self-analysis and self-evaluation by the same high standards and objectives that he set for others." But as Oppenheimer approached his thirties, he became increasingly uncomfortable with this relentless introspection. As the historian James Hijiya has suggested, the Gita provided an answer to this psychological dilemma: celebrate work, duty and discipline—and worry little as to the consequences. Oppenheimer was acutely attuned to the consequences of his actions, but, like Arjuna, he was also driven to do his duty. So duty (and ambition) overrode his doubts—though doubt remained, in the form of an ever-present awareness of human fallibility.

In June 1934, Oppenheimer returned to the University of Michigan summer school session on physics and lectured on his latest critique of the Dirac equation. The lecture so impressed Robert Serber, then a young post-doctoral fellow, that he decided on the spot to switch his research fellow-ship from Princeton to Berkeley. A week or two after Serber drove into Berkeley, Oppie invited him to a movie house, where they saw *Night Must Fall,* a thriller starring Robert Montgomery. It was the beginning of a life-long friendship.

The son of a politically well-connected Philadelphia lawyer, Serber grew up in a decidedly left-wing political culture. His father was Russian-born, and both parents were Jewish. When Serber was twelve, his mother died. Not long afterward, his father remarried; his new wife was Frances Leof, a muralist and potter who later, according to FBI documents, joined the Communist Party. Robert Serber quickly became a part of the extended Leof family, centered around the household of his stepmother's uncle, a charismatic Philadelphia doctor, Morris V. Leof, and his wife, Jenny. The Leof household was run as a political and artistic salon; regular visitors included the playwright Clifford Odets, the left-wing journalist I. F. Stone, and the poet Jean Roisman, who later married the left-liberal trial lawyer Leonard Boudin. Young Robert Serber soon became captivated by the charms of Charlotte Leof, the younger of Morris and Jenny's two daughters. In 1933, he and Charlotte married in a civil ceremony shortly after her grad-uation from the University of Pennsylvania. Charlotte took her politics straight from her radical father, and throughout the 1930s she was a fervent activist on behalf of a variety of left-wing causes. Not surprisingly, given all these family associations, Serber's own political leanings were certainly to the left, although the FBI concluded years later that "no definite evidence is known of Robert Serber's Communist membership."

At Berkeley, Serber studied theoretical physics with Oppenheimer, and in the course of a few years he published a dozen papers, including seven that he coauthored with his mentor. The papers dealt with such topics as cosmic ray particles, the disintegration of high-energy protons, nuclear photoeffects at high energy levels and stellar nuclear cores. Oppie told Lawrence that Serber was "one of the few really first rate theoretical men that he worked with."

They were also the closest of friends. In the summer of 1935, Oppie invited the Serbers to visit him in New Mexico. But Serber was completely unprepared for the conditions at Perro Caliente. When they arrived, after driving on unpaved roads for hours, the Serbers found Frank Oppenheimer, Melba Phillips and Ed McMillan already there. Oppie greeted them nonchalantly and suggested that because the cabin was already full, perhaps they ought to take two horses and ride north eighty miles to Taos. That meant a three-day ride across the Jicoria Pass at 12,500 feet. Serber had never been on a horse! Following Oppie's instructions, the Serbers saddled up, packing only a change of socks and underwear, a toothbrush, a box of chocolate graham crackers, a pint of whiskey and a bag of oats to feed the horses. Three days later, with muscles aching and leg skin rubbed raw by so many hours in the saddle, the Serbers arrived in Taos. After a night in the inn at Ranchos de Taos, they rode back to meet Oppenheimer. Along the way, Charlotte twice fell off her horse and arrived with her jacket splattered with blood.

Life at Perro Caliente was rough. At nearly 9,000 feet, the thin air left many visitors wheezing. "For the first few days there," Serber later wrote, "any physical task left one gasping for breath." Five years after the Oppenheimer brothers had first taken a lease out on the ranch, the cabin was still sparsely furnished, with simple wooden chairs, a sofa in front of the fireplace, a Navajo rug on the floor. Frank had run a pipe from a spring above the cabin, so now there was running water. But that was it. Serber soon realized that for Oppie the ranch was merely a place to sleep in between long, grueling rides into the wilderness. He recounts that once, on a night ride with his host in a thunderstorm, they came to a fork in the trail. Oppie said, "That way it's seven miles home, but this way it's only a little longer, and it's much more beautiful!"

Despite the hardships, the Serbers spent a part of each summer from 1935 to 1941 at Perro Caliente. Oppenheimer had many other visitors to the ranch. Once he ran into the German-born physicist Hans Bethe hiking in the region and persuaded him to stop by. Other physicists, among them Ernest Lawrence, George Placzek, Walter Elsasser and Victor Weisskopf, all spent a few days there. All his visitors were surprised by how much their seemingly fragile friend clearly relished the spartan conditions.

On occasion, Robert's expeditions verged on the truly calamitous. Once he and three friends—George and Else Uhlenbeck and Roger Lewis— camped overnight at Lake Katherine below the east side of a peak called Santa Fe Baldy. Owing to the high altitude, Robert and the two other men suddenly came down with symptoms of altitude sickness. They made it through a freezing night in sleeping bags and woke up the next morning to discover that two of the horses had run off. Robert nevertheless persuaded the men to climb North Truchas Peak, the highest peak, at 13,024 feet, in the southern Sangre de Cristo range. They scaled the summit in a thunder-storm and then had to walk back sopping wet, all the way to Los Pinos, where Katherine Page served them all stiff drinks. The next morning, the two horses that had deserted them reappeared and Else laughed at the sight of Oppenheimer, clad in pink pajamas, chasing them back into the corral.

UNTIL ABOUT 1934, Oppenheimer displayed little interest in current events or politics. He was not so much ignorant as he was indifferent, and he certainly was not politically active. But later—at a time when he wished to highlight his political naïveté—he cultivated the myth that he was obliv-ious to politics and practical affairs: he claimed that he owned neither a radio nor a telephone and that he never read a newspaper or magazine. And he liked to tell the story that he first heard about the stock market crash of October 29, 1929, months after the event. He said he never cast a vote until the 1936 presidential election. "To many of my friends," he testified in 1954, "my indifference to contemporary affairs seemed bizarre, and they often chided me for being too much of a highbrow. I was interested in man and his experience; I was deeply interested in my science; but I had no understanding of the relations of man to his society." Years later, Robert Serber observed that this self-portrait of Oppenheimer as "an unworldly, withdrawn un-esthetic person who didn't know what was going on—all [this was] exactly the opposite of what he was really like."

At Berkeley, Oppenheimer surrounded himself with friends and col-leagues who took an intense interest in political and social issues. From the autumn of 1931, his landlady at 2665 Shasta Road was Mary Ellen Wash-burn, a tall, commanding woman who wore colorful, full-length batik dresses and loved to socialize. Her husband, John Washburn, was an accountant who may also have taught economics at the university. Their home was a longstanding social hub for Berkeley's intellectuals—and, like Mary Ellen herself, many of these people had strong sympathies with the political left. The FBI would later conclude that Mary Ellen was an "active member of the Communist Party in Alameda County."

A young professor of French literature named Haakon Chevalier had been attending parties hosted by the Washburns since the 1920s. The Serbers came to these parties, as did a beautiful young medical student named Jean Tatlock. It was only natural that Oppie, a bachelor living downstairs, dropped by for these social occasions. He was always gracious and usually charmed everyone. But one evening, while he was discoursing at length about a particular poem, the guests heard John Washburn, by now deep in his cups, mutter, "Never since the Greek tragedies has there been heard the unrelieved pomposity of a Robert Oppenheimer."

"We were not political at all in any overt way," recalled Melba Phillips. Oppie once remarked to Leo Nedelsky, "I know three people who are interested in politics. Tell me, what has politics to do with truth, goodness and beauty?" But after January 1933, when Adolf Hitler came to power in Germany, politics began to intrude into Oppenheimer's life. By April of that year, German Jewish professors were being summarily dismissed from their jobs. A year later, in the spring of 1934, Oppenheimer received a circular letter soliciting funds to support German physicists as they attempted to emigrate from Nazi Germany. He immediately agreed to earmark for this purpose three percent of his salary (about $100 a year) for two years. Ironically, one of the refugees who may have been assisted by this fund was Robert's former professor in Göttingen, Dr. James Franck. When Hitler first came to power, Franck, who had won two Iron Crosses during World War I, was one of the few Jewish physicists permitted to keep his post. But a year later he was forced into exile when he refused to dismiss other Jews from their jobs. By 1935, he was teaching physics at Johns Hopkins University in Baltimore. Similarly, Max Born was forced to flee Göttingen in 1933 and ended up teaching in England.

The news from Germany was certainly grim. But by 1934, it would have been difficult for anyone to ignore the political turmoil right in Berkeley's backyard. Almost five years of depression had impoverished millions of ordinary citizens. Early that year, labor strife turned violent. In late January, 3,000 lettuce pickers in the Imperial Valley went on strike. Acting on behalf of employers, police arrested hundreds of workers. The strike was quickly broken, and wages fell from 20 cents to 15 cents an hour. Then, on May 9, 1934, more than 12,000 longshoremen set up picket lines at ports up and down the West Coast. By the end of June, the dock strike had virtually strangled the economies of California, Oregon and Washington. Early in July, authorities attempted to open the port of San Francisco; police lobbed tear gas bombs at thousands of longshoremen and a riot ensued. After four days of running skirmishes, several policemen fired into a crowd; three men were wounded and two of them died. July 5, 1934, became known as

"Bloody Thursday." That same day, the Republican governor ordered the California National Guard to seize control of the streets.

Eleven days later, on July 16, San Francisco labor unions called a general strike. For four days the city was paralyzed. Federal mediators at last stepped in, and by July 30 the largest strike in the history of the West Coast ended. The longshoremen returned to work having achieved almost none of their wage demands, but it was clear to all that the unions had achieved a major political victory. The strike had garnered popular sympathy for the longshoremen's plight and greatly strengthened the union movement. On August 28, 1934, in a sign that the political atmosphere had shifted significantly to the left, the radical writer Upton Sinclair stunned the California establishment by decisively winning the Democratic gubernatorial nomination. Although Sinclair lost the general election—partly as a result of intense slander and fear-mongering on the part of the Republicans—California politics would never be the same.

Such dramatic events could not go unnoticed by Oppenheimer or his students. Berkeley itself was split between critics and supporters of the strike. When the longshoremen initially walked out on May 9, 1934, a conservative member of the physics faculty, Leonard Loeb, recruited "Cal" (University of California, Berkeley) football players to act as strikebreakers. Significantly, Oppenheimer later invited some of his students, including Melba Phillips and Bob Serber, to come along with him to a longshoremen's rally in a large San Francisco auditorium. "We were sitting up high in a balcony," recalled Serber, "and by the end we were caught up in the enthusiasm of the strikers, shouting with them, 'Strike! Strike! Strike!' " Afterwards, Oppie went to the apartment of a friend, Estelle Caen, where he was introduced to Harry Bridges, the charismatic longshoreman union leader.

IN THE AUTUMN OF 1935, Frank Oppenheimer returned from two years of study at Cavendish Laboratory in Cambridge, England, and accepted a tuition scholarship to complete his graduate work at Caltech. Robert's old friend Charles Lauritsen agreed to serve as Frank's thesis adviser. Frank immediately plunged into research on beta ray spectroscopy, a topic he had already studied at Cavendish. "It was very nice to be a beginning graduate student knowing what you wanted to do," Frank recalled.

Robert was still dividing his time between Berkeley and Caltech, spending the late spring every year in Pasadena, where he stayed with his good friends Richard and Ruth Tolman. The Tolmans had built a whitewashed Spanish-style house near the campus, and in the backyard were a lush garden and a one-bedroom guest house which Robert occupied whenever he was in town. Robert had met the Tolmans in the spring of 1929, and that summer the couple had visited the Oppenheimer ranch in New Mexico. Robert would later describe the friendship as "very close." He admired Tolman's "wisdom and broad interests, broad in physics and broad throughout." But he also admired Tolman's "extremely intelligent and quite lovely wife." Ruth was then a clinical psychologist completing her graduate training. For Oppenheimer, the Tolmans "made a sweet island in the Southern California horror." In the evenings, Tolman often hosted informal dinners attended by Frank and other Oppenheimer friends like Linus Pauling, Charlie Lauritsen, Robert and Charlotte Serber, and Edwin and Ruth Uehling. Often Frank and Ruth would play the flute.

In 1936, Oppenheimer lobbied vigorously to obtain Serber an appointment in the Berkeley physics department as his research assistant. The department chairman, Raymond Birge, only very reluctantly agreed to allocate Serber a salary of $1,200 a year. Over the next two years, Oppie tried repeatedly to get Serber appointed to a tenure-track position as an assistant professor. But Birge stubbornly refused, writing another colleague that "one Jew in the department was enough."

Oppenheimer was unaware of this remark at the time, but he was not unfamiliar with the sentiment. If anything, anti-Semitism in polite society was on the rise in America during the 1920s and '30s. Many universities had followed Harvard's lead in the early twenties and imposed restrictive quotas on the number of Jewish students. Elite law firms and social clubs in major cities like New York, Washington, D.C., and San Francisco were segregated by both race and religion. The California establishment was no different on this score from the East Coast establishment. Still, if Oppenheimer could not aspire to become, like his friend Ernest Lawrence, a part of California's establishment, he was happy where he was. "I had decided where to make my bed," he recalled. And it was a bed he was "content" to be in.

Indeed, never once in the 1930s did he revisit Europe, or even, aside from his summers in New Mexico and trips to the Ann Arbor summer seminar, leave California. When Harvard proposed to double his salary if he moved east, he brushed the offer aside. Twice in 1934, the newly formed Institute for Advanced Study in Princeton tried to lure him away from Berkeley, but Oppenheimer was resolute: "I could be of absolutely no use at such a place. . . ." He wrote his brother: "I turned down these seductions, thinking more highly of my present jobs, where it is a little less difficult for me to believe in my usefulness, and where the good California wine consoles for the hardness of physics and the poor powers of the human mind." He thought he "had not grown up, but had grown up a little." His theoretical work was flourishing, in part because classes took up but five hours a week and that left him "a lot of time for physics and for a lot of other things. . . ." And then he met a woman who would change his life.

PART TWO

CHAPTER EIGHT

"In 1936 My Interests Began to Change"

> *Jean was Robert's truest love. He loved her the most. He was devoted to her.*
>
> ROBERT SERBER

J EAN TATLOCK WAS ONLY TWENTY-TWO years old when Robert met her in the spring of 1936. They were introduced at a party hosted by Oppie's landlady, Mary Ellen Washburn, in the house on Shasta Road. Jean was finishing her first year at Stanford University School of Medicine, which was then located in San Francisco. That autumn, Oppenheimer recalled, he "began to court her, and we grew close to each other."

Jean was a shapely woman with thick, dark curly hair, hazel-blue eyes with heavy black lashes and naturally red lips; some thought she looked "like an old Irish princess." Five feet seven inches tall, she never weighed more than 128 pounds. She had but one tiny physical imperfection, a "sleeping" eyelid that drooped slightly as a result of a childhood accident. But even this barely perceptible flaw added to her allure. Her beauty captivated Robert, but so too did her shy melancholy. "Jean was very private about her despair," a friend, Edith A. Jenkins, later wrote.

Robert knew her as the daughter of Berkeley's eminent Chaucer scholar Professor John S. P. Tatlock, one of the few faculty members outside the physics department with whom he had a more than casual acquaintance. Over lunch at the Faculty Club, Tatlock was often dazzled by the knowledge of English literature displayed by this young physics professor. In turn, when Oppenheimer met Jean, he quickly realized that she had soaked up her father's literary sensibilities. Jean favored the dark, morose verse of Gerard Manley Hopkins. She also loved the poems of John Donne—a passion that she passed on to Robert, who, years later, turned to Donne's sonnet "Batter my heart, three-person'd God . . ." for inspiration in assigning the code name "Trinity" to the first test of an atomic bomb.

Jean owned a roadster that she often drove with the top down, singing in her fine contralto voice lyrics from *Twelfth Night.* A free-spirited woman with a hungry, poetic mind, she was always the one person in the room, whatever the circumstances, who remained unforgettable. A college classmate at Vassar remembered her as "the most promising girl I ever knew, the only one of all that I saw around me in college that even then seemed touched with greatness." Jean was born in Ann Arbor, Michigan, on February 21, 1914, and she and her older brother, Hugh, grew up in Cambridge, Massachusetts, and later in Berkeley. Her father had spent most of his career at Harvard, but after retiring, he began teaching at Berkeley. When Jean was ten, she began spending her summers on a Colorado dude ranch. A childhood friend and college classmate, Priscilla Robertson, would write in a "letter" addressed to Jean after her death, "You had a wise mother, who gentled you and never tried to break you, and yet who kept you from the dangers of your passionate kind of adolescence."

Before she went to Vassar College in 1931, her parents allowed her a year off to travel in Europe. She stayed with a friend of her mother's in Switzerland who was a devoted follower of Carl Jung. This family friend introduced Jean to the close-knit community of psychoanalysts centered around Freud's former friend and rival. The Jungian school—with its emphasis on the idea of the collective human psyche—strongly appealed to the young Tatlock. By the time she left Switzerland, she was seriously interested in psychology.

At Vassar, she studied English literature and wrote for the college's *Literary Review.* This daughter of an English scholar had spent much of her childhood listening to her parents reading aloud the works of Shakespeare and Chaucer. As a teenager, she had spent two full weeks at Stratford-on-Avon, seeing a performance of Shakespeare each night. Both her intellect and her stunning good looks intimidated her classmates; Jean always seemed mature beyond her years, "having gotten by nature and experience a depth that most girls don't get until after graduation."

She was also what would later be called, in irony, a "premature antifascist"—an early opponent of Mussolini and Hitler. When a professor gave her Max Eastman's *Artists in Uniform,* hoping that it might serve as a sobering antidote to her woolly-headed admiration of Russian communism, Jean confided to a friend, "I just wouldn't want to go on living if I didn't believe that in Russia everything is better."

She spent 1933–34 at the University of California, Berkeley, taking premed courses, before graduating from Vassar in June 1935. A friend later wrote Tatlock: "It was this social conscience, added to your earlier contact with Jung, that made you want to be a doctor. . . ." While at Berkeley, she also found time to report and write for the *Western Worker,* the Pacific Coast

organ of the Communist Party. A dues-paying Party member, Jean regularly
attended two CP meetings a week. A year before she met Robert, Tatlock
wrote Priscilla Robertson: "I find I am a complete Red when anything at
all." Her anger and passion were easily aroused by the stories she encountered of social injustice and inequity. Her reporting for the *Western Worker*
reinforced her outrage as she covered such incidents as the trial of three
children arrested for selling copies of the *Western Worker* on the streets of
San Francisco, and the trial of twenty-five lumber-mill workers accused of
staging a riot in Eureka, California.

Still, like many American communists, Jean was not a very good ideologue. "I find it impossible to be an ardent Communist," she wrote Robertson, "which means breathing, talking and acting it, all day and all night."
She aspired, moreover, to become a Freudian psychoanalyst, and at the time
the Communist Party insisted that Freud and Marx were irreconcilable.
This intellectual schism seems not to have fazed Tatlock, but probably had
much to do with her on-again, off-again ardor for the Party. (As an adolescent, she had rebelled against the religious dogma she had been taught by
the Episcopal Church; she told a girlfriend that every day she scrubbed her
forehead to wipe away the spot where she had been christened. She hated
any form of religious "claptrap.") Unlike many of her Party comrades, Jean
still had "a feeling for the sanctity and sense of the individual soul," even as
she expressed exasperation with those of her friends who shared an interest
in psychology but scorned political action: ". . . their interest in psychoanalysis amounts to a disbelief in any other positive form of social action."
For her, psychological theory was like expert surgery, "a therapeutic method
for specific disorders."

Jean Tatlock, in sum, was a complicated woman certain to hold the
interest of a physicist with an acute sense of the psychological. She was,
according to a mutual friend, "worthy of Robert in every way. They had
much in common."

AFTER JEAN AND OPPIE began dating that autumn, it quickly became
clear to everyone that this was a very intense relationship. "All of us were a
bit envious," one of Jean's closest friends, Edith Arnstein Jenkins, later
wrote. "I for one had admired him [Oppenheimer] from a distance. His pre-
cocity and brilliance already legend, he walked his jerky walk, feet turned
out, a Jewish Pan with his blue eyes and his wild Einstein hair. And when
we came to know him at the parties for Loyalist Spain, we knew how those
eyes would hold one's own, how he would listen as few others listen and
punctuate his attentiveness with 'Yes! Yes! Yes!' and how when he was

deep in thought he would pace so that all the young physicist-apostles who surrounded him walked the same jerky, pronated walk and punctuated their listening with 'Yes! Yes! Yes!' "

Jean Tatlock was well aware of Oppenheimer's eccentricities. Perhaps because she herself felt life to the bone, she could empathize with a man whose own passions were so odd. "You must remember," she told a friend, "that he was lecturing to learned societies when he was seven, that he never had a childhood, and so is different from the rest of us." Like Oppenheimer, she was decidedly introspective. She had, as noted, already decided to become a psychoanalyst and psychiatrist.

Prior to meeting Tatlock, Oppenheimer's students noticed that he had been seeing many women. "There were a half dozen at least," recalled Bob Serber. But with Tatlock, things were different. Oppie kept her to himself and rarely brought her into his circle of friends in the physics department. His friends only saw them together at the irregular parties hosted by Mary Ellen Washburn. Serber recalled Tatlock as "very good-looking and quite composed in any social gathering." Politically, Serber recognized that she was decidedly "left-wing—more so than the rest of us." And though she was obviously "a very intelligent girl," he could see that she had a dark side. "I don't know whether it was a manic-depressive case or what, but she did have these terrible depressions." And when Jean was down, so was Oppie. "He'd be depressed some days," Serber said, "because he was having troubles with Jean."

The relationship nevertheless survived these episodes for more than three years. "Jean was Robert's truest love," a friend would later say. "He loved her the most. He was devoted to her." And so perhaps it was only natural that Jean's activism and social conscience awakened in Robert the sense of social responsibility that had been so often discussed at the Ethical Culture School. He soon became active in numerous Popular Front causes.

"Beginning in late 1936," Oppenheimer would explain to his interrogators in 1954, "my interests began to change. . . . I had had a continuing, smoldering fury about the treatment of Jews in Germany. I had relatives there [an aunt and several cousins], and was later to help in extricating them and bringing them to this country. I saw what the Depression was doing to my students. Often they could get no jobs, or jobs which were wholly inadequate. And through them, I began to understand how deeply political and economic events could affect men's lives. I began to feel the need to participate more fully in the life of the community."

For a time, he became particularly interested in the plight of migrant farm workers. Avram Yedidia, a neighbor of one of Oppenheimer's students, was working for the California State Relief Administration in 1937–38, when he became acquainted with the Berkeley physicist. "He manifested

deep interest in the plight of the unemployed," Yedidia recalled, "and showered us with questions on work with migrants who came to this area from the
dust bowl of Oklahoma and Arkansas. . . . Our perception then—which I
feel was shared by Oppenheimer—had been that our work was vital and, in
the language of today, 'relevant' while his was esoteric and remote."

The Depression had caused many Americans to reconsider their political outlook. Nowhere was this truer than in California. In 1930, three out of
every four California voters were registered Republicans; eight years later,
Democrats outnumbered Republicans by a margin of two to one. In 1934,
the muckraking writer Upton Sinclair nearly won the governorship with his
radical platform to End Poverty in California (EPIC). That year *The Nation*
editorialized: "If ever a revolution was due, it was due in California.
Nowhere else has the battle between labor and capital been so widespread
and bitter, and the casualties so large; nowhere else has there been such
a flagrant denial of the personal liberties guaranteed by the Bill of
Rights. . . ." In 1938 another reformer, Culbert L. Olson, a Democrat, was
elected governor with the open support of the state Communist Party. Olson
had campaigned under the slogan of a "united front against fascism."

Although the political left as a whole in California was momentarily
mainstream, the California Communist Party was still a tiny minority, even
on the various campuses of the University of California. In Alameda
County, where Berkeley was located, the Party claimed between five hundred and six hundred members, including a hundred longshoremen working
in the Oakland shipyards. California communists were generally thought to
be a voice for moderation in the national Party. With only 2,500 members in
1936, the state Party grew to more than 6,000 by 1938. Nationwide the
Communist Party (USA) had approximately 75,000 members in 1938, but
many of these new recruits remained less than a year. All told, during the
1930s, about 250,000 Americans affiliated themselves with the CPUSA for
at least a short time.

For many New Deal Democrats, no stigma was attached to those who
were involved in the CPUSA and its numerous cultural and educational
activities. Indeed, in some circles the Popular Front carried a certain cachet.
Numerous intellectuals who never joined the Party nevertheless were willing to attend a writers' congress sponsored by the CP, or volunteer to teach
workers at a "People's Educational Center." So it was not particularly
unusual for a young Berkeley academic like Oppenheimer to savor in this
way a bit of the intellectual and political life of Depression-era California.
"I liked the new sense of companionship," he later testified, "and at the time
felt that I was coming to be part of the life of my time and country."

It was Tatlock who "opened the door" for Robert into this world of politics. Her friends became his friends. These included Communist Party

members Kenneth May (a graduate student at Berkeley), John Pitman (a reporter for *People's World*), Aubrey Grossman (a lawyer), Rudy Lambert and Edith Arnstein. One of Tatlock's best friends was Hannah Peters, a German-born medical doctor whom she had met at Stanford medical school. Dr. Peters, who soon became Oppenheimer's physician, was married to Bernard Peters (formerly, Pietrkowski), another refugee from Nazi Germany.

Born in Posen in 1910, Bernard studied electrical engineering in Munich until Hitler came to power in 1933. Though he later denied being a Communist Party member, he did attend several communist rallies as a spectator, and on one occasion he was present at an anti-Nazi demonstration in which two people were injured. Soon he was arrested and imprisoned in Dachau, an early Nazi concentration camp. After three terrifying months, he was transferred to a Munich prison—and then, without explanation, released. (In another version of this story, Peters managed to escape from the prison.) He then spent several months traveling at night on a bicycle through southern Germany and across the Alps to Italy. There he found his Berlin-born girlfriend, Hannah Lilien, age twenty-two, who had fled to Padua to study medicine. In April 1934, the couple immigrated to the United States. They were married in New York on November 20, 1934, and, after Hannah received her medical degree in 1937 at Long Island Medical School in New York, they moved to the San Francisco Bay area. During a stint at Stanford University School of Medicine, Hannah worked on research projects with Dr. Thomas Addis, a friend and mentor of Jean Tatlock's. By the time Oppenheimer met the Peterses through Jean, Bernard was working as a longshoreman.

In 1934, Peters had written a 3,000-word account of the horrors he had witnessed in Dachau. He described in sickening detail the torture and summary execution of individual prisoners. One prisoner, he reported, "died in my hands a few hours after the beating. All skin was removed from his back, his muscles were hanging down in shreds." Peters no doubt shared his graphic account of Nazi atrocities with his friends when he arrived on the West Coast. Whether Oppenheimer read Peters' report on Dachau or merely heard him talk about it, he must have been deeply moved by these stories. There was a note of authenticity and worldliness in Peters' extraordinary life. Another of Oppenheimer's graduate students, Philip Morrison, always thought Peters was "a little different from most of us, more mature, marked with a special seriousness and intensity . . . his experience went far beyond ours. . . . He had seen and felt the barbarous darkness that mantled Nazi Germany, [and] had worked among the longshoremen in San Francisco Bay."

When Peters displayed an interest in physics, Oppie encouraged him to take a course in the subject at Berkeley. He proved to be a talented student and, despite his lack of an undergraduate degree, Robert got him enrolled in Berkeley's physics graduate program. Peters soon became Oppenheimer's designated note-taker in his course on quantum mechanics and wrote his thesis under Oppie's supervision. Not surprisingly, Oppie and Jean Tatlock frequently socialized with Hannah and Bernard Peters. Although the couple always insisted that they never joined the Communist Party, their politics were clearly left-wing. By 1940 Hannah had a private practice in a poverty-stricken district of downtown Oakland, and this experience "strengthened a conviction that had been growing for some years, namely that adequate medical care can only be provided by a comprehensive health insurance scheme with federal backing." Hannah also insisted on racial integration in her practice, accepting black patients at a time when few other white physicians did so. Both views stamped her as a radical—and the FBI concluded that she was a member of the CP.

All these new friends drew Oppenheimer into their world of political activism. On the other hand, it would be wrong to suggest that Tatlock and her circle were solely responsible for his political awakening. Sometime around 1935, Oppenheimer's father lent him a copy of *Soviet Communism: A New Civilization?*, a rosy description of the Soviet state written by the well-known British socialists Sidney and Beatrice Webb. He was favorably impressed by what it said about the Soviet experiment.

In the summer of 1936, Oppenheimer is said to have taken all three volumes of the German-language edition of *Das Kapital* with him on a three-day train trip to New York City. As his friends tell the story, by the time he arrived in New York, he had read the three volumes cover to cover. In fact, his exposure to Marx occurred several years earlier, probably in the spring of 1932. His friend Harold Cherniss remembered Oppie visiting him in Ithaca, New York, that spring and boasting that he had read *Das Kapital.* Cherniss just laughed; he didn't think of Oppie as political, but he knew his friend read widely: "I suppose somewhere someone said to him, 'You don't know about this? You haven't seen it?' So he got this wretched book and read it!"

Though they had yet to be introduced, Haakon Chevalier knew of Oppenheimer by reputation—and it was not for his work in physics. In July 1937 Chevalier noted in his diary a remark by a mutual friend that Oppenheimer had bought and read the complete works of Lenin. Chevalier, impressed, commented that this would make Oppenheimer "better read than most party members." Although Chevalier considered himself a relatively sophisticated Marxist, he had never plowed through *Das Kapital.*

Born in 1901 in Lakewood, New Jersey, Haakon Chevalier might never-theless easily have been mistaken for an expatriate. His father was French and his mother had been born in Norway. "Hoke," as his friends called him, spent parts of his early childhood in Paris and Oslo; consequently, he spoke fluent French and Norwegian. But his parents brought him back to America in 1913, and he finished high school in Santa Barbara, California. He stud-ied at both Stanford and Berkeley, but interrupted his college studies in 1920 to spend eleven months working as a seaman aboard a merchant ship sailing between San Francisco and Cape Town. After this adventure, Chevalier returned to Berkeley and received his doctorate in Romance lan-guages in 1929, specializing in French literature.

Six feet one inch tall, with blue eyes and wavy brown hair, Hoke cut a debonair figure as a young man. In 1922, he married Ruth Walsworth Bosley—but divorced her on grounds of desertion in 1930, and a year later married Barbara Ethel Lansburgh, twenty-four, one of his Berkeley stu-dents. The blond, green-eyed Lansburgh came from a wealthy family and owned a stunning redwood seaside home at Stinson Beach, twenty miles north of San Francisco. "He was a terribly charismatic teacher," recalled their daughter Suzanne Chevalier-Skolnikoff. "That drew her to him."

In 1932, Chevalier published his first book, a biography of Anatole France. That same year, he began writing book reviews and essays for the left-leaning *New Republic* and *Nation* magazines. By the mid-1930s, he had become a fixture on the Berkeley campus, teaching French and opening his rambling redwood home on Chabot Road in Oakland to an eclectic collec-tion of students, artists, political activists and visiting writers such as Edmund Wilson, Lillian Hellman and Lincoln Steffens. Frequently partying late into the night, Chevalier was so often tardy to his morning classes that his department finally barred him from teaching in the morning.

An ambitious intellectual, Chevalier was also politically active. He joined the American Civil Liberties Union, the Teachers' Union, the Inter-Professional Association and the Consumer's Union. He became a friend and supporter of Caroline Decker, a leader of the California Cannery and Agricultural Workers, a radical union representing Mexican-American farm laborers. In the spring of 1935, the Berkeley campus mobilized to protest the expulsion of a student who had offended university authorities by adver-tising his communist affiliations. The meeting held to protest this expulsion was then broken up by the football team, egged on by the coach. According to one account, only one faculty member—Haakon Chevalier—"gave shel-ter and moral support to the trailed [*sic*] and terrorized students."

In 1933, Chevalier had visited France, where he managed to meet such left-wing literary figures as André Gide, André Malraux and Henri Barbusse. He returned to California convinced that he was destined "to witness the transition from a society based on the pursuit of profit and the exploitation of man by man to a society based on production for use and on human cooperation."

By 1934, he had translated André Malraux's acclaimed novel of the Chinese uprising of 1927, *La Condition Humaine* (*Man's Fate*) and his *Le Temps du Mépris* (*The Time of Contempt*), novels inspired by what Chevalier thought of as "the new vision of man."

As for so many on the left, the outbreak of the Spanish Civil War was a turning point for Chevalier. In July 1936, right-wing factions in the Spanish army rose against the democratically elected left-wing government in Madrid. Led by General Francisco Franco, the fascist rebels expected to overthrow the Republic within weeks. But popular resistance was tenacious, and a brutal civil war ensued. The United States and the European democracies, suspicious of communist influence in the Spanish government, and encouraged by the Catholic Church, declared an arms embargo against both sides. This handed a distinct advantage to the fascists, who received generous aid from Hitler's Germany and Mussolini's Italy. Only the Soviet Union aided the besieged Republican government. In addition, volunteers from around the world, mostly communists but also other leftists, joined international brigades to defend the Republic. During the years 1936–39, the defense of the Spanish Republic was the cause célèbre in liberal circles everywhere. Over these years, some 2,800 Americans volunteered to fight the fascists by joining the communist-sponsored Abraham Lincoln Brigade.

In the spring of 1937, Chevalier accompanied Malraux on a tour through California. Recently wounded in the Spanish Civil War, Malraux was promoting his novels and fundraising on behalf of the Spanish Medical Bureau, a group that sent medical aid to the Republic. For Chevalier, Malraux personified the serious intellectual who was also politically committed.

By 1937, all the evidence indicates, Chevalier was committed to the Communist Party. His 1965 memoir, *Oppenheimer: The Story of a Friendship,* is remarkably forthcoming in describing his political outlook in the 1930s. But even then, writing eleven years after the high point of McCarthyism had passed, he thought it prudent to be vague about the critical question of Party membership. The late 1930s, he wrote, were "a time of innocence. . . . We were animated by a candid faith in the efficiency of reason and persuasion, in the operation of democratic processes and in the ultimate triumph of justice." Like-minded men such as Oppenheimer, he wrote, believed that abroad the Spanish Republic would triumph over the winds of fascist Europe, and at home, the reforms of the New Deal were clearing the way for a new social compact based on racial and class equality. Many intellectuals had such hopes—but some also joined the Communist Party.

By the time Oppenheimer met him, Chevalier was a committed Marxist intellectual, probably a Party member and, quite likely, a respected though informal adviser to Party officials in San Francisco. Over the years, he had seen Oppenheimer from a distance, spotting him at the Faculty Club and elsewhere on campus. He had heard, however, through the Berkeley grapevine, that this brilliant young physicist was now "anxious to do something more than *read* about the problems that beset the world. He wanted to *do* something."

Chevalier and Oppenheimer were finally introduced at an early meeting of a newly formed teachers' union. Chevalier later dated this first encounter as taking place in the autumn of 1937. But if they met, as both men later said, at this union meeting, that would place the event a full two years earlier, in the autumn of 1935. That was when Local 349 of the Teachers' Union, an affiliate of the American Federation of Labor (AFL), expanded to admit university professors. "A group of people from the faculty talked about it," Oppenheimer later testified, "and met, and we had lunch at the Faculty Club or some place and decided to do it." Oppenheimer was elected recording secretary. Chevalier later served as the local's president. Within a few months, Local 349 had about a hundred members, forty of whom were professors or teaching assistants at the university.

Neither Oppenheimer nor Chevalier could remember the exact circumstances of their first encounter, only that they liked each other immediately. Chevalier recalled a "hallucinatory feeling . . . that I had always known him." He felt both dazzled by Oppenheimer's intellect and charmed by his "naturalness and simplicity." That very day, according to Chevalier, they agreed to create a regular discussion group of six to ten people who would meet every week or two to discuss politics. These salons met regularly from the autumn of 1937 through the late fall of 1942. During these years, Chevalier regarded Oppenheimer "as my most intimate and steadfast friend." Initially, their friendship arose from shared political commitments. But, as Chevalier explained later, "our intimacy, however, even at the beginning was by no means purely ideological, but full of personal overtones, of warmth, curiosity, reciprocity, of intellectual give and take, rapidly developing into affection." Chevalier quickly learned to call his new friend by his nickname, Oppie, and Oppenheimer in turn found himself dropping by the Chevalier household for dinner. From time to time, they went out to a movie or concert. "Drinking was for him a social function that called for a certain ritual," Chevalier wrote in his memoirs. Oppie made the "best martinis in the world," invariably drunk with his trademark toast, "To the confusion of our enemies." It was, Chevalier thought, quite clear who their enemies were.

FOR JEAN TATLOCK, it was the causes, not the Party or its ideology, that were important. "She told me about her Communist Party memberships," Oppenheimer later testified. "They were on-again, off-again affairs, and never seemed to provide for her what she was seeking. I do not believe that her interests were really political. She was a person of deep religious feeling. She loved this country and its people and its life." By the autumn of 1936, the single cause that captivated her most was the plight of Republican Spain.

It was Tatlock's passionate nature to push Oppenheimer to move from theory to action. One day he commented that while he was certainly an "underdogger," he would have to settle for being on the periphery of these political struggles. "Oh for God's sake," protested Jean, "don't *settle* for anything." She and Oppenheimer soon began organizing fund-raisers for a variety of Spanish relief groups. In the winter of 1937–38, Jean introduced Robert to Dr. Thomas Addis, the chairman of the Spanish Refugee Appeal. A distinguished professor of medicine at Stanford University, Dr. Addis had encouraged Tatlock in her studies at the Stanford University School of Medicine; he was both a friend and a mentor. He also happened to be a friend of Haakon Chevalier, Linus Pauling (Oppie's Caltech colleague), Louise Bransten and many other people in Oppie's circle of Berkeley acquaintances. Addis himself quickly became "a good friend" of Oppenheimer's.

Tom Addis was an extraordinarily cultivated Scotsman. Born in 1881, he was raised in a strict Calvinist household in Edinburgh. (Even as a young doctor, he still carried a small Bible in his pocket.) He received his medical degree from the University of Edinburgh in 1905 and did postdoctoral research in Berlin and Heidelberg as a Carnegie Scholar. He was the first medical researcher to demonstrate that normal plasma could be used to treat hemophilia. In 1911 he became chief of the Clinical Laboratory at Stanford University School of Medicine in San Francisco. At Stanford he commenced a long and distinguished career as a physician-scientist, becoming a pioneer in the treatment of kidney disease. He wrote two books on nephritis and more than 130 scientific papers, becoming America's leading expert on the disease. In 1944 he was elected to membership in the prestigious National Academy of Sciences.

Even as he was building his reputation as a physician-scientist, Addis was always politically active. When war broke out in Europe in 1914, Addis violated U.S. neutrality laws by raising funds for the British war effort. Indicted in 1915, he was formally pardoned by President Woodrow Wilson in 1917. The following year, Addis became an American citizen. Though he came from a privileged background—his uncle, Sir Charles Addis, was a director of the Bank of England—he had a pronounced distaste for money. In California, he became a well-known advocate of civil rights for Negroes, Jews and union members, signing numerous petitions and lending his name to scores of civic organizations. He was a friend of the radical longshoreman union leader Harry Bridges.

In 1935, Addis attended an academic conference of the International Physiological Congress in Leningrad and he returned from his visit to the Soviet Union with glowing accounts of the socialist state's progress in public health. He was particularly impressed that Soviet doctors had experimented with human cadaveric kidney transplants as early as 1933. Thereafter he lobbied vigorously for national health insurance, which eventually prompted the American Medical Association to expel him. But his Stanford colleagues regarded his admiration for the Soviet system as "an act of faith," a tolerable foible on the part of a respected scientist. Pauling thought him "a great man, of a rare sort—a combination of scientist and clinician. . . ." Others called him a genius. "He was not one of those who have an inner need to play it safe, to appear sound and rational," recalled Dr. Horace Gray, a colleague. "He was an explorer, a liberal open mind, a non-conformist without being rebellious."

By the late 1930s, the FBI was reporting that Addis was one of the Communist Party's major recruiters of white-collar professionals. Oppenheimer himself later thought Addis was either a Communist or "close to one." "Injustice or oppression in the next street," wrote a medical colleague at Stanford, "or in the city, or in South Africa or Europe or Java or in any spot inhabited by men was a personal affront to Tom Addis, and his name, from its early alphabetical place, was conspicuous on lists of sponsors of scores of organizations fighting for democracy and against fascism."

For a dozen years, Addis served off and on as chairman or vice-chairman of the United American Spanish Aid Committee, and it was in this capacity that he first approached Oppenheimer for financial contributions. By 1940, Addis was claiming that his committee had been "instrumental" in rescuing many thousands of refugees, including many European Jews, from concentration camps in France. Already sympathetic to the cause of the Spanish Republic, Oppenheimer found himself charmed and deeply impressed by Addis' sophisticated blend of utilitarian commitment and intellectual rigor. Dr. Addis was an intellectual much like himself, a man of broad interests whose knowledge of poetry, music, economics and science "reached into his work. . . . There was no division to all these things."

One day Oppenheimer received a phone call from Addis, inviting him to come to his Stanford laboratory. They met in private and Addis told him, "You are giving all this money [for the Spanish Republic cause] through these relief organizations. If you want it to do good, let it go through Communist channels . . . and it will really help." Thereafter Oppenheimer regularly gave cash payments in person to Dr. Addis, usually in Addis' lab or at his home. "He made it clear," Oppenheimer later said, "that this money . . . would go straight to the fighting effort." After a while, however, Addis suggested that it would be more convenient to give these regular contributions to Isaac "Pop" Folkoff, a veteran member of the San Francisco Communist Party. Oppenheimer donated in cash because he thought it might not be entirely legal to contribute money for military equipment, as opposed to medical aid. His annual donations for Spanish relief work given through the Communist Party amounted to about $1,000—a hefty sum in the 1930s. But after the fascist victory in 1939, Addis and then Folkoff solicited money for such other causes as the Party's efforts to organize migratory farm workers in California. Robert's last such contribution was apparently made in April 1942.

Folkoff, a former garment worker then in his late seventies, was paralyzed in one hand. At the time he met Oppenheimer, he was head of the Party's finance committee in the Bay Area. "He was a respected old left-winger," recalled Steve Nelson, a political commissar in the Abraham Lincoln Brigade who became the Party chairman in San Francisco in 1940. "I don't mean to denigrate him, but the guy dabbled as a worker and became interested in philosophy. He became quite versed in Marxist philosophy. So he had a kind of prestige and dignity and trustworthiness. He used to meet with the professionals around the movement and collected money from them." Nelson confirmed that Folkoff collected money from both Oppenheimer brothers.

When Oppenheimer was asked in 1954 about these donations to the Communist Party, he explained, "I doubt that it occurred to me that the contributions might be directed to other purposes than those I had intended, or that such purposes might be evil. I did not then regard Communists as dangerous; and some of their declared objectives seemed to me desirable."

The Communist Party was often in the forefront of such progressive causes as desegregation, better working conditions for migratory farm workers, and the fight against fascism in the Spanish Civil War, and Oppenheimer gradually became active in a number of these causes. Early in 1938, he subscribed to the *People's World,* the Party's new West Coast newspaper. He read the paper regularly, taking an interest, he later explained, in its "formulation of issues." Late in January 1938, his name found its way into the *People's World,* when the paper reported that Oppenheimer, Haakon Chevalier and several other Berkeley professors had raised $1,500 to purchase an ambulance to be shipped to the Spanish Republic.

That spring, Robert and 197 other Pacific Coast academics signed a petition urging President Roosevelt to lift the arms embargo on the Spanish Republic. Later that year, he joined the Western Council of the Consumer's Union. In January 1939, Robert was appointed to the executive committee of the California chapter of the American Civil Liberties Union. In 1940, he was listed as a sponsor of Friends of the Chinese People, and became a member of the national executive committee of the American Committee for Democracy and Intellectual Freedom, a group that publicized the plight of German intellectuals. With the exception of the ACLU, all of these organizations were labeled "Communist front organizations" in 1942 and 1944 by the House Committee on Un-American Activities.

Oppenheimer was particularly active in Local 349 of the East Bay Teachers' Union. "It was a time of great tension in the faculty," recalled Chevalier. "The few of us who were more or less left-wingers were very conscious of the fact that we were frowned upon by the elders." In meetings of the faculty council, the conservatives "always won." Most Berkeley academics refused to have anything to do with a union. The exceptions included Jean Tatlock's psychology professor, Edward Tolman, the brother of Oppenheimer's Caltech friend Richard Tolman. Over the next four years, Robert worked hard to increase the union's membership. According to Chevalier, he rarely missed a union meeting and could be counted on for the most menial of tasks. Chevalier recalled staying up with him until two in the morning on one occasion, addressing envelopes for a mailing to the union's several hundred members. It was tedious work for an unpopular cause. One evening Oppenheimer appeared as the featured speaker at the Oakland High School auditorium. The event had been widely publicized, and the Teachers' Union fully expected hundreds of public school teachers to show up to hear Oppenheimer expound on the promise of the union cause. Fewer than a dozen people came. He nevertheless stood up and made his union pitch in a voice characteristically so soft that he could hardly be heard.

Some sensed that Oppenheimer's politics were always driven by the personal. "Somehow one always knew he felt guilty about his gifts, about his inherited wealth, about the distance that separated him from others," observed Edith Arnstein, a friend of Tatlock's and a Party member. Even in the early 1930s, when he was not yet politically active, he had always been aware of what was going on in Germany. Only a year after Hitler came to power in 1933, Oppenheimer was contributing sizable sums to assist German Jewish physicists to escape Nazi Germany. These were men he knew and admired. Similarly, he talked often with anguish about the plight of his relatives in Germany. In the autumn of 1937, Robert's aunt Hedwig Oppenheimer Stern (Julius' youngest sister) and her son Alfred Stern and his family landed in New York as refugees from Nazi Germany. Robert had sponsored them legally and paid their expenses, and soon he persuaded them to settle in Berkeley. Robert's generosity toward the Sterns was not fleeting. He always regarded them as family; decades later, when Hedwig Stern died, her son wrote Oppenheimer, "As long as she could think and feel, she was all for you."

That autumn, Robert was introduced to another refugee from Europe, Dr. Siegfried Bernfeld, a highly respected Viennese disciple of Sigmund Freud. Fleeing the Nazi contagion, Bernfeld had first gone to London, where another Freudian, Dr. Ernest Jones, advised him, "Go West, don't settle here." By September 1937, Bernfeld had settled in San Francisco, a city he knew had then only one practicing analyst. His wife, Suzanne, was also a psychoanalyst. Her father had been a major art gallery impresario in Berlin who had helped to introduce artists like Cézanne and Picasso to the German public. When they arrived in San Francisco, the Bernfelds sold one of the last paintings left in their once impressive art collection to pay their living expenses. An eloquent teacher and passionate idealist, Dr. Bernfeld was one of a handful of Freudian analysts who was trying to integrate psychoanalysis with Marxism. As a young man in Austria, Bernfeld had become politically active, first as a Zionist, and later as a socialist. Tall and gaunt, he wore a distinctive porkpie hat, a felt hat with a low, flat top. Oppenheimer was deeply impressed—and soon took to wearing a porkpie hat like Bernfeld's.

Within weeks of landing in San Francisco, Dr. Bernfeld organized an ecumenical group of the city's leading intellectuals to discuss psychoanalysis on a regular basis. In addition to Oppenheimer, Bernfeld invited Dr. Edward Tolman, Dr. Ernest Hilgard, Drs. Donald and Jean Macfarlane (friends of Frank Oppenheimer's), Erik Erikson (a German-born psychoanalyst trained by Anna Freud), the pediatrician Dr. Ernst Wolff (who was to become Jean Tatlock's boss at Mt. Zion Hospital's Child Guidance Clinic), Dr. Stephen Pepper, a philosophy professor at Berkeley, and the well-known anthropologist Dr. Robert Lowie to be regular members of this interdisciplinary study group. They met in private homes, drank good wine, smoked cigarettes and talked about such psychoanalytic issues as "fear of castration" and the "psychology of war."

Oppenheimer, of course, had painful memories of his youthful encounters with psychiatrists. But that was no doubt part of his attraction to the topic. He must have been particularly interested in Erikson's work on the problem of "identity formation" in young adults. A prolonged adolescence, Erikson argued, accompanied by "chronic malignant disturbance," was sometimes an indication that an individual was having trouble shedding fragments of his personality that he finds undesirable. Seeking "wholeness," and yet fearing a threatened loss of identity, some young adults experience such a sense of rage that they strike out at others in arbitrary acts of destruction. Oppenheimer's behavior and problems back in 1925–26 had conformed in significant ways to this thesis. He had thrown himself into theoretical physics, carving out for himself a robust identity. But the scars remained. As the physicist and science historian Gerald Holton has observed, "Some psychological damage remained, however, not least a vulnerability that ran through his personality like a geological fault, to be revealed at the next earthquake."

Bernfeld would sometimes talk about individual therapy cases. Like his mentor Freud, he lectured without notes, smoking one cigarette after another. "Bernfeld was one of the most eloquent speakers I've ever heard," recalled another psychoanalyst, Dr. Nathan Adler. "I sat at the edge of my seat listening not only to what he said, but to the way he spoke. It was an aesthetic experience." Oppenheimer, the only physicist in the group, was remembered as someone who was "intensely interested" in psychoanalysis. In any case, Robert's curiosity about the psychological complemented his interest in physics. Recall Wolfgang Pauli's complaint to Isidor Rabi in Zurich that Oppenheimer "seemed to treat physics as an avocation and psychoanalysis as a vocation." Things metaphysical still took priority. And so during the years 1938 to 1941, he found the time to attend Bernfeld's seminars, a study group that in 1942 gave rise to the formation of the San Francisco Psychoanalytic Institute and Society.

Oppenheimer's exploration of the psychological was encouraged by his intense, often mercurial, relationship with Jean Tatlock—who was, after all, training to become a psychiatrist. Though not a member of Bernfeld's monthly group, Jean knew some of these men and later was analyzed by Dr. Bernfeld as part of her training. Moody and introspective, Tatlock shared Robert's obsession with the unconscious. Furthermore, it made sense that Oppenheimer the political activist would choose to study psychoanalysis under the tutelage of a Marxist Freudian analyst like Dr. Bernfeld.

Some of Oppenheimer's oldest friends found his sudden political activism distasteful—most particularly Ernest Lawrence, who could easily sympathize with the plight of his friend's persecuted relatives, but on a more personal level thought that what was happening in Europe was not our affair. He separately told both Oppie and his brother Frank, "You're too good a physicist to get mixed up in politics and causes." Such things, he said, were best left to the experts. One day Lawrence walked into the Rad Lab and saw that Oppie had written on the blackboard, "Cocktail Party Benefit for Spanish Loyalists at Brode's, everyone at Lab invited." Seething, Lawrence stared at the message and then erased it. To Lawrence, Oppie's politics were a nuisance.

CHAPTER NINE

"[Frank] Clipped It Out
and Sent It In"

*We [Chevalier and Oppenheimer] both were and were not
[members of the Communist Party]. Any way you want to
look at it.*

HAAKON CHEVALIER

O N SEPTEMBER 20, 1937, Julius Oppenheimer died of a heart
attack at the age of sixty-seven. Robert had known that his father
was no longer robust, but his sudden death came as a shock. In the
nearly six years since Ella's death in 1931, Julius had developed a close and
tender relationship with his sons. He visited them both frequently, and it
was often the case that Robert's friends became his father's friends.

Julius' fortune had diminished somewhat after eight years of depression. Even so, by the time of his death, his estate, divided equally between
Robert and Frank, amounted to the still quite substantial sum of $392,602.
The annual income from this inheritance gave each of the brothers an average of $10,000 to supplement his earnings. But, as if to underscore a certain
ambivalence about his wealth, Robert immediately wrote out a will leaving
his entire estate to the University of California, earmarked for graduate
fellowships.

The Oppenheimer brothers had always been extremely close. Robert
formed remarkably intense relationships with a number of people, but none
was either as deep or as durable as the bond he forged with his brother.
Their correspondence in the 1930s reflected an intensity of emotion unusual
for siblings, and particularly so for brothers eight years apart in age.
Robert's letters often read more like a father's than an older brother's. At
times, he wrote with what must have seemed a maddening condescension to
Frank, who so obviously wished to emulate him. Frank patiently tolerated
whatever his strong-willed brother said or did; only years later did he admit

that Robert's "youthful cockiness . . . stayed with my brother a little longer than it should have."

They were alike—and yet not. No one disliked Frank Oppenheimer. He was Oppie without an edge, endowed with much of the Oppenheimer brilliance and none of the abrasiveness. "Frank himself is a sweet, lovable person," observed the physicist Leona Marshall Libby, a friend of both brothers. She called him a "delta function," a mathematical device used by physicists in which delta is defined as zero—except at a specified place or time, at which point it becomes infinity. When called upon, Frank always possessed an infinite reservoir of goodwill and cheer. Years later, Robert himself said of his brother, "He is a much finer person than I am."

At one time, Robert had tried to talk Frank out of choosing physics as his profession. When Frank was only thirteen, and clearly set on following in his brother's footsteps, Robert wrote: "I don't think you would enjoy reading about relativity very much until you have studied a little geometry, a little mechanics, a little electrodynamics. But if you want to try, Eddington's book is the best to start on. . . . And now a final word of advice: try to understand really, to your own satisfaction, thoroughly and honestly, the few things in which you are most interested; because it is only when you have learnt to do that, when you realize how hard and how very satisfying it is, that you will appreciate fully the more spectacular things like relativity and mechanistic biology. If you think I'm wrong please don't hesitate to tell me so. I'm only talking from my own very small experience."

By the time he got to Baltimore and Johns Hopkins University, Frank was determined to demonstrate that he was made of the same stuff as his brother. Like Robert, he was a polymath; he loved music, and, unlike his brother, he actually played an instrument, the flute, extremely well. At Hopkins, he regularly played in a quartet. But he was committed to physics. During his second year, Frank met Robert in New Orleans, where they attended the annual meeting of the American Physical Society. Afterwards, Robert wrote Ernest Lawrence that "we had a fine holiday together; and I think that it settled definitely Frank's vocation for physics." After rubbing shoulders with a good number of physicists, all of them bubbling over with enthusiasm for their work, Robert observed that "it is impossible not to conceive for them a great liking and respect, and for their work a great attraction." On the second day of the conference, Robert took Frank to a joint session on biochemistry and psychology, and, while "it was enormously rowdy and very funny," it also "discouraged an excessive faith in either of these sciences."

But then, only a few months later, Robert cautioned Frank not to commit himself to physics without first exploring the alternatives. He thought

Frank's intellectual appetite might be whetted by some course work in the biological sciences. While declaring that "I know very well surely that physics has a beauty which no other science can match, a rigor and austerity and depth," he urged Frank to take an advanced course in physiology: "Genetics certainly involves a rigorous technique, and a constructive and complicated theory.... By all means, and with my whole blessing, learn physics, all there is of it, so that you understand it, and can use it and contemplate it, and, if you should want, teach it; but do not plan yet to 'do' it: to adopt physical research as a vocation. For that decision you should know something more of the other sciences, and a good deal more of physics."

Frank ignored this bit of sibling counsel. After earning his undergraduate degree in physics in only three years, he spent 1933–35 studying at Cavendish Laboratory in England under some of the same physicists who had taught Robert, and he met such friends of his brother's as Paul Dirac and Max Born. By then, however, Robert was more than reconciled to his brother's chosen course: "You know how happy I was," he wrote Frank in 1933, "with your decision to go to Cambridge...." But now he longed to see his brother. "There has seldom been a time," he wrote Frank in early 1934, "when I have missed you so as in these last days.... I take it that Cambridge has been right for you, and that physics has gotten now very much under your skin, physics and the obvious excellences of the life it brings. I take it that you have been working very hard, getting your hand in the laboratory, and learning mathematics at close hand, and finding in this, and in the natural austerity of life in Cambridge, at last an adequate field for your unremitting need of discipline and order." If at times Robert sounded patronizing in his role as elder brother, his letters to Frank make it clear that he was as dependent on the closeness of this brotherly bond as was Frank.

Unlike Robert, Frank excelled at experimental physics; he liked getting his hands dirty in the laboratory. He loved tinkering with machines and once built his brother a custom phonograph. As Robert observed, Frank had a way of "reducing a specific and rather complex situation to its central irreducible *Fragestellung* [formulation of a question]." After studying two years in England and several months in Italy—where he observed and acquired a loathing for Mussolini's fascism—Frank applied to several universities to complete his Ph.D. in experimental physics. He was conflicted about whether to go to Caltech, but Robert "did something," and suddenly Caltech offered him a tuition scholarship based on merit, and his decision was made.

In the laboratory he worked under Robert's old friend Charlie Lauritsen, experimenting with a beta-ray spectograph. Whereas Robert had taken only two years to complete his doctorate, Frank spent a leisurely four earning his. In part, this was because experimental work was often simply more time-consuming than theoretical physics. But it was also Frank's choice, by temperament and inclination, to fill his life with more than physics. He loved music and was accomplished enough as a flutist that his brother and many friends thought he could have played professionally. Drawing on his mother's artistic sensibilities, he loved painting and read a great deal of poetry. In contrast to Robert's assiduously correct European manners, friends thought Frank rather sloppy in dress and "bohemian" in manner.

During his first year at Caltech, Frank met Jacquenette "Jackie" Quann, a twenty-four-year-old French-Canadian woman who was studying economics at Berkeley. They met in Berkeley in the spring of 1936, when Robert took his brother to visit a friend, Wenonah Nedelsky, and Jackie happened to be there baby-sitting. To pay the bills, she worked as a waitress. Plain and outspoken, she possessed a down-to-earth demeanor that rebuffed pretentiousness. "Jackie prided herself on being working-class," said Bob Serber, "and she had no use for intellectuals." Her ambition was to be a social worker. She wore her hair in a simple page-boy cut and never bothered with lipstick or other makeup. She was not the kind of woman Robert Oppenheimer would have chosen for his brother. But later that spring, Robert, Frank, Jackie and Wenonah (recently separated from her husband, Leo) went out together two or three times. In June, Frank invited Jackie to come up to Perro Caliente that summer. They arrived in a brand-new $750 Ford pickup truck, a gift from Robert.

When, later that summer, Frank informed Robert that he intended to marry Jackie, Robert tried to talk him out of it. Jackie and Robert did not get along. She recalled that "he was always saying things like, 'Of course, you're much older than Frank'—I'm eight months older, actually—and saying that Frank wasn't ready for it."

This time, however, Frank ignored his brother's advice, and married Jackie on September 15, 1936. "It was an act of emancipation and rebellion on his part," wrote Robert, "against his dependence on me." Robert continued to disparage Jackie by referring to her as "the waitress my brother has married." On the other hand, he continued to "arrange things" for his brother and his new wife. "The three of us saw each other a great deal in Pasadena, Berkeley and Perro Caliente," recalled Frank, "and between my brother and me there was the continuing sharing of ideas, enterprises and friends."

Jackie had always been a political firebrand. "She could drive you crazy with her political rants," recalled a relative. As an undergraduate at Berkeley, she had joined the Young Communist League, and later she worked for a year in Los Angeles for the Communist Party newspaper. Frank was quite comfortable with her politics. "I had been close to sort of slightly left-wing things starting in high school," he recalled. "I remember once I went with some friends to hear a concert at Carnegie Hall that didn't have a conductor. It was a kind of 'down with the bosses' movement."

Like Robert, Frank was a product of the Ethical Culture School, where he had learned to debate moral and ethical issues. At sixteen, he had worked, together with some of his school friends, on Al Smith's 1928 presidential campaign. At Johns Hopkins, many of his peers were to the left of the Democratic Party. But at the time, Frank disliked long-winded political discussions. "I used to tell people," he recalled, "unless I meant to do something about it, I didn't want to talk about it." He recalled being "dismayed" in 1935 by what he heard at a Communist Party meeting in Cambridge, England. "It sounded to me sort of empty," Frank recalled. During a visit to Germany, however, he quickly acquired an appreciation of the fascist menace: "The whole society seemed corrupt." His father's relatives had told him "some of the terrible things" that were happening in Hitler's Germany, and he was inclined to support any group determined to "do something about it."

Upon his return to California that autumn, he was deeply moved by the deplorable condition of local farm laborers and Negroes. The Depression was taking a terrible toll on millions of people. Another graduate student in physics at Caltech, William "Willie" Fowler, used to say that the reason he was a physicist was that he didn't want to have to worry about people—and now he was upset because he was being forced by the Depression to do just that. Frank felt the same way. He began reading up on labor history and eventually read a great deal of Marx, Engels and Lenin.

One day early in 1937, Jackie and Frank saw a membership coupon in the local Communist newspaper, *People's World.* "I clipped it out and sent it in," recalled Frank. "We were really quite overt about it—completely overt about it." But it was some months before anybody from the Party responded. Like many professionals, Frank was asked to join the Party under an alias, and he chose the name Frank Folsom. "When I joined the Communist Party," he later testified, "for some reason which I did not understand at the time and have never understood since, they requested that my right name and another name be written down. This seemed to me ludicrous. I never used any name but my own, and at the same time, because of the fact that it seemed so ludicrous, I wrote down the name of a California jail [Folsom]." In 1937 his Communist Party "book number" was 56385. One day he absent-mindedly left his green-colored Party card in his shirt pocket when sending it to the laundry. The shirt came back with the Party card neatly preserved in an envelope.

By 1935, it was not at all unusual for Americans who were concerned with economic justice—including many New Deal liberals—to identify with the Communist movement. Many laborers, as well as writers, journalists and teachers, supported the most radical features of Franklin Roosevelt's New Deal. And even if most intellectuals didn't actually join the Communist Party, their hearts lay with a populist movement that promised a just world steeped in a culture of egalitarianism.

Frank's attachment to communism had deep American roots. As he later explained: "The intellectuals who were drawn toward the left by the horror, the injustices and fears of the thirties did, in varying degrees, identify with the history of protest in America.... John Brown, Susan B. Anthony, Clarence Darrow, Jack London, and even with movements such as the abolitionists, the early AFL and the IWW."

Initially, the Party assigned Frank and Jackie to what was called a "street unit" in Pasadena; most of their comrades were local neighborhood residents, and quite a few were poor, unemployed Negroes. Their Party cell membership fluctuated between ten and as many as thirty people. They had regular, open meetings attended by both communists and members of various organizations connected with the New Deal, such as the Workers' Alliance, an organization of unemployed laborers. There was a lot of talk and not much action, which frustrated Frank. "We tried to integrate the city swimming pool," he said. "They just allowed blacks in Wednesday afternoon and evening, and then they drained the pool Thursday morning." But despite their efforts, the pool remained segregated.

A little later, Frank agreed to try to organize a Party unit at Caltech. Jackie remained with the street unit for a while, but she too eventually joined the Caltech group. She and Frank recruited about ten members, including fellow graduate students Frank K. Malina, Sidney Weinbaum and Hsue-Shen Tsien. Unlike the Pasadena street unit, this Caltech group "was essentially a secret group." Frank was the only member who remained open about his political affiliation. Most of the others, he explained, "were scared of losing their jobs."

Frank understood that his association with the Party offended some people. "I remember a friend of my father's, an old man, saying he wouldn't send his son to a college at which I was teaching." The Stanford physicist Felix Bloch once tried to persuade him to quit the party, but Frank wouldn't hear of it. Most of his friends, however, cared little one way or the other. Party membership was just one aspect of his life. By then, Frank was devoted to his studies in beta-ray spectroscopy at Caltech. Like his brother, he stood on the edge of a promising career. But his politics—if not necessarily his Party membership—were both an open book and an extracurricular activity. One day Ernest Lawrence ran into Frank, whom he liked very much, and asked him why he wasted so much time with "causes." It baffled Lawrence, who saw himself as a man of science above politics, even though he spent much of his own time ingratiating himself with the businessmen and financiers on the Board of Regents who directed the policies of the University of California. In his own way, Lawrence was as much of a political animal as Frank; he just owed his allegiance to different "causes."

Frank and Jackie opened their home to regular Tuesday evening CP meetings. According to one "reliable confidential" FBI informant, Frank continued to host these meetings until about June 1941. Robert attended at least once—which he later claimed was the only time he participated in a "recognizable" Communist Party meeting. The topic was the ongoing concern of racial segregation at Pasadena's municipal swimming pool. Robert later testified that the meeting "made a rather pathetic impression on me."

Like his brother, Frank was active in the East Bay Teachers' Union, the Consumer's Union and the cause of migratory farm workers in California. One evening he gave a flute recital in Pasadena, with Ruth Tolman on the piano, in a local auditorium; proceeds from the event went to the Spanish Republic. "We spent a lot of time at meetings, political meetings," Frank later said. "There were many issues." "He frequently spoke," a Stanford colleague told the FBI, "of instances of economic oppression which he seemed to resent." Another informant claimed that Frank "continually showed a great admiration for the Soviet Union in its internal and external policies." On occasion, Frank could be strident. He assailed one colleague—who reported the conversation to the FBI later—as a "hopeless Bourgeois not in sympathy with the Proletariat."

Robert later made light of his brother's communist associations. Although a Party member, Frank did a lot of other things: "He was passionately fond of music. He had many wholly non-Communist friends. . . . He spent his summers at the ranch. He couldn't have been," Robert summed up, "a very hard working Communist during those years."

Soon after Frank joined the Party, he made a point of driving up to Berkeley, where he spent the night with his brother and told him the news. "I was quite upset about it," Robert testified in 1954, without explaining just why he was unhappy over Frank's taking this step. Party membership, to be sure, was not without its risks. But in 1937 there was little stigma attached to it among Berkeley liberals. "It wasn't regarded," Robert testified, "perhaps foolishly, as a great state crime to be a member of the Communist Party or as a matter of dishonor or shame." Still, it was clear that the University of California administration was hostile to anyone affiliated with the CP, and Frank was in the process of trying to build an academic career. And, unlike Robert, Frank didn't have tenure. If Robert was upset with Frank's decision, perhaps he thought his younger brother was being unwisely headstrong in making such a commitment, or was too much under the influence of his radical wife. Despite Robert's own political awakening, he felt no compulsion to join the Communist Party as a matter of principle. Frank, on the other hand, evidently felt an emotional need to make a formal commitment. The brothers may have shared common political instincts, but Frank was proving himself to be far more impetuous. He still very much idolized Robert, but with his marriage, and his politics, he was trying to stake out his own persona and step out of Robert's shadow.

In 1943 a colleague of Frank's during his two years at Stanford University told an FBI agent that "in his opinion Frank Oppenheimer had followed the lead and dictates of his brother, J. Robert Oppenheimer, on all of his political attitudes and affiliations." This anonymous source had it mostly wrong—Frank had joined the Party independently, against his brother's advice. The informant had one thing right, however: he assured the FBI that he believed both Oppenheimers were "basically loyal to this country. . . ." In the eyes of their friends (and of the FBI), the Oppenheimer brothers were extraordinarily close. What Frank did would always reflect on Robert. And, try as he might to arrange things for his brother, Robert would never quite be able to protect Frank from the glare of his own fame.

COMPARED WITH HIS GUILELESS BROTHER, Robert was an enigma. All of their friends knew where his political sympathies lay—but the exact nature of his relationship to the Communist Party remains to this day hazy and vague. He later described his friend Haakon Chevalier as "a parlor pink. He had very wide connections with all kinds of front organizations; he was interested in left-wing writers . . . he talked quite freely of his opinions." The description might easily have been applied to Oppenheimer himself.

Without question, Robert was surrounded by relatives, friends and colleagues who at some point or other were members of the Communist Party. As a left-wing New Dealer, he gave considerable sums of money to causes championed by the Party. But he always insisted that he was never a card-carrying member of the CP. Instead, he said, his associations with the Party were "very brief and very intense." He was referring to the Spanish Civil War period, but afterwards he continued to participate in meetings in which dues-paying Communist Party members discussed current events. These meetings, encouraged by the Party, were specifically designed to involve independent intellectuals like Oppenheimer and blur the boundaries of Communist Party identity. But never having been a formal, card-carrying member left Oppenheimer the option of deciding for himself how he wished to define his relationship to the Party. For a brief time, he may well have thought of himself as an unaffiliated comrade. There is no doubt that in later years he minimized the extent of his associations with the Party. Quite bluntly, any attempt to label Robert Oppenheimer a Party member is a futile exercise—as the FBI learned to its frustration over many years.

In fact, his associations with Communists were a natural and socially seamless outgrowth of his sympathies and his station in life. As a professor at the University of California in the late 1930s, Oppenheimer lived in a politically charged environment. Moving in such circles, he inevitably left the impression with many of his friends who were formal Party members that he was one of them. Robert, after all, wanted to be liked and he certainly believed in the social justice goals the Party espoused and worked for. His friends could think what they wanted. Not surprisingly, some in the Party did think he was a comrade. And naturally, when the FBI used wiretaps to monitor the conversations of these people talking about Oppenheimer, they occasionally heard bona fide Party members discuss him as one of their own. And yet again, other FBI wiretaps record Party members complaining about Oppenheimer's aloofness and unreliability. Most importantly, there is no evidence that he ever submitted himself to Party discipline. Given his strong personal alignment with much if not most of the Party program, where he did disagree he never trimmed his views to conform to the Party line. Tellingly, he expressed qualms about the totalitarian nature of the Soviet regime. He openly admired Franklin Roosevelt and defended the New Deal. And while he was a member of various Popular Front organizations dominated by the Communist Party, he was also a staunch civil libertarian and a prominent member of the American Civil Liberties Union. In short, he was a classic fellow-traveling New Deal progressive who admired the Communist Party's opposition to fascism in Europe, and its championing of labor rights at home. It is neither surprising nor revealing that he worked with Party members in support of those goals.

All this ambiguity is compounded by the fact that during these years of the Popular Front, the Communist Party's very organizational structure, particularly in California, led to a blurring of the distinction between casual affiliation and actual membership. As Jessica Mitford wrote in her irreverent memoir of her experiences in the San Francisco branch of the Party, "In those days . . . the Party was a strange mixture of openness and secrecy." The conspiratorial-sounding "cell" of three to five members had been replaced by "branches" or "clubs"—"a nomenclature deemed more consistent with American political tradition." Hundreds of people might belong to these "clubs," in which Party business was conducted in a fairly open and informal manner; everyone was welcome and people, often including FBI informers, attended weekly meetings in rented halls without too much attention paid as to whether their party dues were up-to-date. On the other hand, Mitford reports that she and her husband "were at first assigned to the Southside Club, one of the few 'closed' or secret branches, reserved for government workers, doctors, lawyers, and others whose occupations could have been jeopardized by open affiliation with the Party."

Many left-of-center, pro-union, anti-fascist intellectuals in the late 1930s never affiliated with the Communist Party. And yet, many who did join the Party chose to hide their affiliation even if, like Oppenheimer, they were politically active on behalf of causes supported by the Party. So numerous were the Party's secret members that Communist Party chief Earl Browder griped in June 1936 about too many prominent figures in American society hiding their Party identity. "How shall we dissipate the Red Scare from among the Reds?" he asked. "Some of these comrades hide as a shameful secret their Communist opinions and affiliations; they hysterically beg the Party to keep as far away from their work as possible."

Years later, Haakon Chevalier insisted that Oppenheimer was one such secret Party member. But when closely questioned about the unit Robert allegedly belonged to, Chevalier described an innocuous gathering of friends more akin to the "discussion group" he reported in his 1965 memoir than the sort of official "closed unit" described by Mitford. "We/he initiated it," Chevalier, referring to Oppenheimer, told Martin Sherwin. "It was a closed unit and unofficial. There's no record of it. . . . It was not known to anyone except one person. I don't know who he was, but [he was] in the top echelon of the party in San Francisco." This "unofficial" group known only to "one person" initially contained just six or seven members, though at one point as many as twelve were participating in its discussions. "We discussed things that were going on locally and in the state and in the country and in the world," recalled Chevalier.

It is Chevalier's version of this story that is reflected in the FBI files. The FBI first opened a file on Oppenheimer in March 1941. His name had come to the Bureau's attention quite by accident the previous December. For almost a year the FBI had been wiretapping the conversations of William Schneiderman, the California Communist Party's state secretary, and Isaac "Pops" Folkoff, the state treasurer. The wiretaps were not authorized by any court or by the Attorney General, and were therefore illegal. But in December 1940, when one of the Bureau's agents in San Francisco overheard Folkoff referring to a 3:00 p.m. appointment at Chevalier's house as a meeting of "the big boys," an agent was sent to jot down license plate numbers. One of the cars found to be parked outside Chevalier's home was Oppenheimer's Chrysler roadster. By the spring of 1941, the FBI was identifying Oppenheimer as a professor "reported from other sources as having Communistic sympathies." The FBI noted that he served on the Executive Committee of the American Civil Liberties Union—which the Bureau labeled "a Communist Party front group." Inevitably, an investigative file was opened on Oppenheimer which would eventually grow to some 7,000 pages. That same month, Oppenheimer's name was put on a list of "persons to be considered for custodial detention pending investigation in the event of a national emergency."

Another FBI document, citing the investigative documents of "T-2, another Government agency," claimed that Oppenheimer was a member of a "professional section" of the Communist Party. One of these "T-2" documents found in Oppenheimer's FBI file included a two-page excerpt from a longer unidentified report listing the membership of various branches of the Communist Party. Names and addresses are provided for the "Longshoreman's Branch," the "Seaman's Branch" and the "Professional Section." Nine members are listed for this "Professional Section": Helen Pell, Dr. Thomas Addis, J. Robert Oppenheimer, Haakon Chevalier, Alexander Kaun, Aubrey Grossman, Herbert Resner, George R. Andersen and I. Richard Gladstein. Oppenheimer clearly knew some of these individuals (Pell, Addis, Chevalier and Kaun), and it is equally clear that at least some of them were in fact Communist Party members. But it is impossible to evaluate the credibility of this undated document.

According to Chevalier, who spoke with Martin Sherwin at length and in detail, each member of this alleged "closed unit" paid dues to the Communist Party—except for Oppenheimer. "Oppenheimer paid his separately," Chevalier speculated, "because he probably paid a lot more than he was supposed to." Or, as Robert always insisted, he made contributions to causes, but never paid dues at all. "But the rest of us paid to one member who was also a known member, an open member [of the Party]," Chevalier continued. "I'm not supposed to say, but it was Philip Morrison." Otherwise, according to Chevalier, the group took no "orders" from the Party and functioned simply as a group of academics who met to share ideas about international affairs and politics. Morrison, of course, has long acknowledged that he joined the Young Communist League in 1938 and the CP itself in 1939 or 1940. When asked about Chevalier's recollection, Morrison flatly denied that he had been in the same Party unit as Oppenheimer. As a student, he pointed out, he would never have been assigned to a unit with faculty members.

When asked by Sherwin in 1982, "What made you a member of the Communist Party as opposed to just a group of people who were Left?" Chevalier replied, "I don't know. We paid dues." When Sherwin pressed him again, "Did you receive any orders from the Party?" Chevalier said, "No. In a sense we weren't [regular Party members]." At the time, he explained, it was possible for men like himself and Oppenheimer to think of themselves as politically committed intellectuals who were nevertheless free from Party discipline. Members of this group contributed money to the Party's causes; they gave speeches at Party-sponsored events; and they drafted articles and pamphlets for Party publications. And yet, explained Chevalier, *"We both were and were not. Any way you want to look at it."* Pressed further to explain this ambiguity, Chevalier said, "It had a kind of shadowy existence. It existed, but it wasn't identified, and that had some influence because we had our views about certain things that were happening which were transmitted to the center, and we were consulted about certain things. . . . Apparently, the same thing happened in many other parts of the United States, closed units for professionals or people who didn't want to be identified in any way."

The ambiguous nature of Oppenheimer's relationship to the CP, as described by Chevalier, is corroborated by Steve Nelson, a charismatic Communist Party leader in San Francisco and a friend of Oppenheimer's in the years 1940–43. Nelson saw Oppenheimer socially, but it was also his job to serve as one of the Party's liaisons to the university community. "I met socially with this group," Nelson explained in a 1981 interview, "that included some Party members and some non-Party people where they discussed freely what's ahead of us. . . . This group was discussing questions of foreign policy. The general mood, which included Oppenheimer's mood, was that it would be tragic if the United States, England and France do not form some kind of alliance against Italy; it would be tragic. I don't remember now whether it was Chevalier or Bob [Oppenheimer] or any other member who expressed himself along these lines. But this was the tone of the meeting."

Nelson reinforced Chevalier's ambiguous description of Oppenheimer's party membership. "I don't know that I could prove or disprove the point," Nelson said. "So I'll just leave it at that—that he was a close sympathizer. I know that to be a fact because we had a number of discussions of policies of the left. . . . Now that doesn't mean that he was a member of the Party. I think he was a close friend of a number of Party members on campus."

Nelson himself left the Communist Party in 1957. In 1981 he published a memoir in which he briefly discussed his relationship to Oppenheimer. When he showed the manuscript to one of his old California comrades, still a Party member, this old communist thought he had been "too easy" on Oppenheimer—Nelson should have attacked Oppenheimer for having denied his affiliation with the Party. "My own estimate of Oppenheimer," Nelson remarked, "was that he had this association with the left. Whether one had a Party card or not didn't matter. He was associated with the causes of the left, and that was enough to murder him politically. . . ."

All the members of this allegedly closed Party unit are dead. But one of them left behind an unpublished memoir. Gordon Griffiths (1915–2001) joined the Communist Party in Berkeley in June 1936, just before leaving for Oxford. Upon his return in the summer of 1939, Griffiths quietly renewed his Party membership. But because his wife, Mary, had become disillusioned with the Party, Griffiths asked for a low-profile assignment. Eventually, he was given the job of "liaison with the Faculty group at the University of California." Griffiths took the assignment in the autumn of 1940 and left it in the spring of 1942. In his memoir, he writes that out of the several hundred faculty members in Berkeley, only three were members of this "faculty Communist group": Arthur Brodeur (an authority on Icelandic sagas and Beowulf in the English department), Haakon Chevalier—and Robert Oppenheimer.

Griffiths acknowledges Oppenheimer's denial of ever having been a Party member. Oppenheimer's defenders, Griffiths points out, have always explained Oppenheimer's fellow-traveling with the assertion that he was politically naïve. "A great deal of energy was spent by well-intentioned liberals who felt that this was the only way to defend his case. Perhaps at the time—at the height of the McCarthyite period—it was. . . . But the time has come to set the record straight, and to put the question as it should have been put: not whether or not he had or had not been a member of the Communist Party, but whether such membership should, in itself, constitute an impediment to his service in a position of trust."

Griffiths' memoir adds few details to Chevalier's description of what he called the "closed unit." Understandably, Griffiths clearly believes that the mere fact of Oppenheimer's attendance in these meetings qualified him as a Communist. He writes that the group met regularly, twice a month, either at Chevalier's house or at Oppenheimer's house. Griffiths usually brought along recent Party literature to distribute, and he collected dues from Brodeur and Chevalier, *but not from Oppenheimer.* "I was given to understand that Oppenheimer, as a man of independent wealth, made his contribution through some special channel. Nobody carried a party card. If payment of dues was the only test of membership, I could not testify that Oppenheimer was a member, but I can say, without any qualification, that all three men considered themselves to be Communists."

The faculty group, Griffiths recalls, didn't actually do very much "that could not have been done as a group of liberals or Democrats." They encouraged each other to devote their energies to such good causes as the Teachers' Union and the plight of refugees from the Spanish Civil War. "There was never any discussion of the exciting developments in theoretical physics, classified or otherwise, let alone any suggestion of passing information to the Russians. In short, there was nothing subversive or treasonable about our activity. . . . The meetings were devoted mainly to the discussion of events on the world and national scene, and to their interpretation. In these discussions, Oppenheimer was always the one who gave the fullest and most profound explanations, in the light of his understanding of Marxist theory. To describe his attachment to left-wing causes as the result of political naïveté, as many have done, is absurd, and diminishes the intellectual stature of a man who saw the implications of what was happening in the political world more deeply than most."

Kenneth O. May, the Berkeley CP functionary who assigned Griffiths to this group, later told the FBI that Haakon Chevalier and other Berkeley professors attended its meetings, but that he "did not consider the people who participated in these gatherings to consist of a CP group."

Once a graduate student in Berkeley's math department, Ken May was a friend of Oppenheimer's. May joined the Communist Party in 1936, and he visited Russia for five weeks in 1937 and again in 1939 for two weeks. He returned enamored of the Soviet political and economic model. During local elections in Berkeley in 1940, May gave a speech before the school board defending the right of local Communist Party candidates to hold a meeting on the grounds of a public school. When the speech drew coverage in the local press, his father, a conservative UC Berkeley political scientist, publicly disinherited him and the university canceled his teaching assistantship. The next year, May campaigned as a Communist for a seat on the Berkeley city council while still a graduate student in the math department. His affiliation with the Communist Party was thus no secret when he first met Oppenheimer. May was a friend of Jean Tatlock's, and the two men were probably introduced at a meeting of the Teachers' Union sometime in 1939.

Years later, after he had left the Party, May told the FBI he had visited Oppenheimer's home on several occasions to talk politics and recalled seeing him at "informal gatherings . . . which were held for the purpose of discussing theoretical questions concerning Socialism." He added that he did not consider Oppenheimer to be either a Party member or someone "under CP discipline." Oppenheimer was an independent intellectual, and, May explained to the FBI, "the CP tended to distrust intellectuals as a group in the management of CP affairs, but at the same time, the Party was anxious to influence the thinking of such people along CP lines and to gain the prestige and support of Communist objectives which they would lend to the Party. For this reason, May would keep in touch with the subject [Oppenheimer] and other professional people; he would discuss Communism with them and would provide them with CP literature."

Oppenheimer, May explained to the FBI agents, was the kind of man

who was quite willing to "agree with CP aims and objectives at any particular time if he had decided in his own mind that they had merit. He would
not, however, condone those objectives with which he did not agree." May
observed that the "subject openly associated with whomever he pleased,
Communists or not."

The FBI would never resolve the question of whether or not Robert was
a CP member—which is to say that there was scant evidence that he was.
Much of the evidence in the FBI files on this issue is circumstantial and
contradictory. If a few of the FBI's informants claimed that Oppenheimer
was a Communist, most of its informants merely painted a portrait of a fellow traveler. And some emphatically denied that he was ever a Party member. The Bureau had only its suspicions, and the conjectures of others. Only
Oppenheimer himself knew—and he always insisted that he had never been
a member of the Communist Party.

CHAPTER TEN

"More and More Surely"

This was a very decisive week in his life, and he told me so. . . . That weekend started Oppenheimer's turning away from the Communist Party.

VICTOR WEISSKOPF

O N AUGUST 24, 1939, the Soviet Union stunned the world by announcing that on the previous day it had signed a nonaggression pact with Nazi Germany. One week later, World War II commenced when Germany and the Soviet Union simultaneously invaded Poland. Commenting on these momentous events, Oppenheimer wrote his fellow physicist Willie Fowler: "I know Charlie [Lauritsen] will say a melancholy I told you so over the Nazi Soviet pact, but I am not paying any bets yet on any aspect of the hocuspocus except maybe that the Germans are pretty well into Poland. Ca stink."

No issue of the day was more vigorously debated within left-intellectual circles than the August 1939 Nazi-Soviet Non-Aggression Pact. Many American Communists resigned from the Party. As Chevalier put it with marked understatement, the Soviet-German pact "confused and upset many people." But Chevalier remained loyal to the CP and defended the pact as a necessary strategic decision. In August 1939, he and four hundred others signed an open letter, published in the September 1939 issue of *Soviet Russia Today,* which attacked the "fantastic falsehood that the USSR and the totalitarian states are basically alike." Oppenheimer's name did not appear on the letter. According to Chevalier, it was in the fall of 1939 "that Opje proved himself to be such an impressive and effective analyst. . . . Opje had a simple, lucid way of presenting facts and arguments that allayed misgivings and carried conviction." Chevalier claimed that, at a time when communists were suddenly extremely unpopular even among Californian intellectuals, Oppenheimer patiently explained that the Nazi-Soviet pact was not so much an alliance as a treaty of necessity motivated by the West's appeasement of Hitler at Munich.

Chevalier was deeply alarmed by the wave of war hysteria that seemed to be turning "seasoned liberals into reactionaries and peace-lovers into war-mongers." One evening after midnight, on his way home from a meeting of the League of American Writers, Chevalier dropped by Oppenheimer's home. Robert was still awake, working on a physics lecture. After Robert offered him a drink, Hoke explained that he needed his help in editing an antiwar pamphlet, sponsored by the League. Obliging his friend, Robert sat down and read the manuscript. When he had finished, he stood up and said, "It's no good." He told Hoke to sit before his typewriter and then he proceeded to dictate new language. An hour later, Hoke left with "a completely new text."

Robert was not himself a member of the League of American Writers, so his editing of the pamphlet was simply a favor to his friend. As redrafted, the pamphlet made an impassioned argument for keeping America out of the European war. Robert may have similarly helped to write or edit two other pamphlets in February and April 1940, respectively. Both were titled *Report to Our Colleagues,* and were signed "College Faculties Committee, Communist Party of California." Their purpose was to explain the consequences of the war in Europe. More than a thousand copies were mailed to individuals at various universities on the West Coast.

According to Chevalier, Oppenheimer not only drafted the reports but also paid for their printing and distribution. Not surprisingly, their discovery—combined with Chevalier's claim—has made them part of the debate as to whether or not Robert was a member of the CP.* Gordon Griffiths corroborates Chevalier's assertion of Oppenheimer's involvement with the production of these pamphlets. "They were printed on expensive bond, no doubt paid for by Oppie. He was not their sole author, but he took special pride in them. . . . Free of jargon, these letters were stylistically elegant and intellectually cogent."

"The outbreak of war in Europe," the pamphlet dated February 20, 1940, states, "has changed profoundly the course of our own political development. In the last month strange things have happened to the New Deal. We have seen it attacked, and *more and more surely* we have seen it abandoned. There is a growing discouragement of liberals with the movement for a democratic front and red-baiting has grown to a national sport. Reaction is mobilized."

Chevalier, in an interview, insisted that the language here is distinctively Oppenheimer's. "You can recognize his style. He has certain little manner-

---

*Phil Morrison recalled helping Oppenheimer mail a pamphlet he had written analyzing the Soviet attack on Finland in the autumn of 1939. That pamphlet has not been found.

isms, using certain words. 'More and more surely.' That's very characteristic of him. You wouldn't ordinarily find use of 'surely' in such a context." Chevalier's claim is too thin a reed on which to rely for a positive identification of Oppenheimer as *the* author of the pamphlet, but it does suggest that Robert might have had a hand in editing a draft of it. While "more and more surely" does sound like Oppenheimer, much else in the pamphlet decidedly does not.

But what do these "reports" propose? More than anything else, a defense of the New Deal and its domestic social programs:

> The Communist Party is being attacked for its support of the Soviet policy. But the total extermination of the Party here cannot reverse that policy: it can only silence some of the voices, some of the clearest voices, that oppose a war between the United States and Russia. What the attack can do directly, what it is meant to do, is to disrupt the democratic forces, to destroy unions in general and CIO unions in particular, to make possible the cutting of relief, to force abandonment of the great program of peace, security and work that is the basis of the movement toward a democratic front.

On April 6, 1940, the College Faculties Committee of the Communist Party of California issued another *Report to Our Colleagues*. Like the first pamphlet, this report carried no by-line. But again Chevalier insisted that Oppenheimer was among the pamphlet's anonymous authors.

> The elementary test of a good society is its ability to keep its members alive. It must make it possible for them to feed themselves and it must protect their persons from violent death. Today unemployment and war constitutes [sic] so serious a threat to the well being and security of the members of our society that many are asking whether that society is capable of meeting its most essential obligations. Communists ask much more of society than this: they ask for all men that opportunity, discipline, and freedom which have characterized the high cultures of the past. But we know that today, with the knowledge and power that are ours, no culture which ignores the elementary needs, no culture based on the denial of opportunity, on indifference to human want, can be either honest or fruitful.

As in February, domestic issues are the focus of this report. The plight of the nation's millions of unemployed is examined, and the decision of

California and national Democrats to cut the budget for welfare relief is attacked. "The cutting of relief and the simultaneous increase in the budget for armaments are connected not only by arithmetic considerations. Roosevelt's abandonment of the program for social reform, the attack, where once there had been support, on the labor movement, and the preparation for war, these are related and parallel developments." From 1933 to 1939, the pamphlet observes, the Roosevelt Administration had "followed a policy of social reform." But since August 1939, "not a single new measure of progressive purpose has been proposed . . . and the measures of the past have not even been defended against reactionary attack." Where once the Roosevelt Administration had voiced its "disgust" for the antics of the House Un-American Activities Committee under Martin Dies, now it was "coddling" these reactionaries. Where once it had defended organized labor, civil liberties and the unemployed, now it was attacking labor leaders like John L. Lewis and pouring money into armaments.

Roosevelt himself, a man whom the pamphleteers once thought "something of a progressive," had now become a "reactionary" and even a "warmonger." This transformation has happened because of the war in Europe. "It is a common thought, and a likely one, that when the war is over Europe will be socialist, and the British Empire gone. We think that Roosevelt is assuming the role of preserving the old order in Europe and that he plans, if need be, to use the wealth and the lives of this country to carry it out."

If Oppenheimer had anything to do with this second pamphlet, his rational style had abandoned him. Is it possible that he really thought of Roosevelt as a "war-monger"? The one reference to the president in Oppenheimer's correspondence during this period suggests that he was disappointed with FDR, but hardly ready to denounce him.* If Oppenheimer had something to do with drafting these pamphlets, his words reveal someone primarily worried about the impact on domestic politics of a world teetering on the brink of a great disaster.

B Y   T H E   L A T E   1930 S, Oppenheimer was a senior professor with a fairly prominent public persona. He was giving speeches on political issues and

---

*More than a year after the April 1940 pamphlet was published, he wrote his old friends Ed and Ruth Uehling: "My own views could, im Kleinem, hardly be gloomier, either for what will happen locally & nationally, or in the world. I think we'll go to war—that the Roosevelt faction will win over the Lindbergh. I don't think we'll get anywhere near the Nazis. Later I think the Hearst-Lindbergh side will kick the administration 'humanitarians' out. I see no good for a long time; & the only cheerful thing in these parts is the strength & toughness & political growth of organized labor."

signing public petitions. His name appeared on occasion in the local newspapers. San Francisco was then a fiercely polarized city; the longshoremen's strikes in particular had hardened the political extremes on both the left and the right. And when the conservative backlash began, Oppenheimer was sensitive to the effect, or potential effect, of his political activities on the reputation of the university. Indeed, in the spring of 1941 he confided to his Caltech colleague Willie Fowler that "I may be out of a job . . . because UC is going to be investigated next week for radicalism and the story is that the committee members are no gentlemen and that they don't like me."

"The University of California was an obvious target," observed Martin D. Kamen, a former graduate. "And Oppenheimer was very prominent because he was quite vocal and active. He would occasionally get somewhat alarmed about what was happening, and maybe he'd have to draw his horns in and he would become quiet. Then when something happened to provoke him . . . he became active. So he wasn't consistent."

In contrast to Chevalier's assertions about Oppenheimer's communist sympathies in 1940, other friends saw Oppie becoming disillusioned with the Soviet Union. By 1938, American newspapers regularly reported on the wave of political terror orchestrated by Stalin against thousands of alleged traitors within the Soviet Communist Party. "I read about the purge trials, though not in full detail," Robert wrote in 1954, "and could never find a view of them which was not damning to the Soviet system." While his friend Chevalier gladly signed a statement in the April 28, 1938, *Daily Worker* commending the Moscow trial verdicts against Trotskyite and Bukharinite "traitors," Oppenheimer never defended Stalin's deadly purges.

In the summer of 1938 two physicists who had spent several months in the Soviet Union—George Placzek and Victor Weisskopf—visited Oppie at his ranch in New Mexico. Over the next week, they had several long conversations about what was taking place there. "Russia is not what you think it is," they told an initially "skeptical" Oppenheimer. They talked about the case of Alex Weissberg, an Austrian engineer and communist who was suddenly arrested merely for associating with Placzek and Weisskopf. "It was an absolutely scary experience," Weisskopf said, "We called up our friends, and they said they didn't know us." Weisskopf told his friend, "It's worse than you can imagine. It's a morass." Oppie asked probing questions that showed how disturbed he was by their reports.

Sixteen years later, in 1954, Oppenheimer explained to his interrogators, "What they reported seemed to me so solid, so unfanatical, so true, that it made a great impression; and it presented Russia, even when seen from their limited experience, as a land of purge and terror, of ludicrously bad management and of a long-suffering people."

There seemed no reason, however, why news of Stalin's abuses should cause him to alter his principles or renounce his sympathies with the American left. It was clear, as Weisskopf remembered, that Oppie "still believed to a great extent in communism." Oppie trusted Weisskopf. "He really had a deep attachment to me," the latter recalled, "which I found very touching." Robert knew that Weisskopf, an Austrian social democrat, was not saying these things out of any antipathy for the left. "We were very much convinced—both sides—that socialism was the desirable development."

Nevertheless, Weisskopf thought that this was the first time Oppenheimer was really shaken. "I know that these conversations had a very deep influence on Robert," he said. "This was a very decisive week in his life, and he told me so. . . . That weekend started Oppenheimer's turning away from the Communist Party." Weisskopf insists that Oppie "saw the Hitler danger very clearly. . . . And in 1939, Oppenheimer was already very far from the Communist group."

Shortly after hearing from Weisskopf and Placzek, Oppenheimer expressed his concerns to Edith Arnstein, Jean Tatlock's old friend: "Opje said he came to me because he knew I would not be shaken in my political loyalties, and he needed to talk." He explained that he had heard from Weisskopf about the arrest of various Soviet physicists. He said he was reluctant to believe the report but neither could he dismiss it. "He was depressed and agitated," Arnstein later wrote, "and I suppose now I know how he was feeling, but then I was scornful of what I saw as his gullibility."

That autumn some friends noticed that he was no longer as loquacious about his political views, although privately he engaged his close friends in political discussions. "Opje is fine and sends you his greetings," wrote Felix Bloch to I. I. Rabi in November 1938. "[H]onestly, I don't think you wore him out but at least he does not praise Russia too loudly any more which is already some progress."

WHATEVER THE STATE OF his associations with Communist Party members, Oppenheimer had always been enamored of Franklin Roosevelt and the New Deal. His friends saw him as an ardent Roosevelt supporter. Ernest Lawrence recalled being lobbied vigorously by his friend in the days just prior to the 1940 presidential election. Oppie was incredulous that his old friend was undecided. That evening he presented such a passionate defense of Roosevelt's campaign for a third term that Lawrence finally promised to cast another vote for FDR.

Oppenheimer's political views continued to evolve, in reaction chiefly to the disastrous war news. In the late spring and early summer of 1940,

Oppie was clearly distressed by the collapse of France. That summer, Hans Bethe encountered him at a conference of the American Physical Society in Seattle. Bethe had an inkling of Oppenheimer's political loyalties, so he was struck one evening when his friend gave a "beautifully eloquent speech" about how the fall of Paris to the Nazis threatened all of Western civilization. "We have to defend Western values against the Nazis," Bethe recalled Oppenheimer saying. "And because of the Molotov–von Ribbentrop pact we can have no truck with the Communists." Years later, Bethe told the physicist-historian Jeremy Bernstein: "He had sympathies to the far left, mostly, I believe, on humanitarian grounds. The Hitler-Stalin pact had confused most people with Communist sympathies into staying completely aloof from the war against Germany until the Nazis invaded Russia in 1941. But Oppenheimer was so deeply impressed by the fall of France [a year before the invasion of Russia] that this displaced everything else in his mind."

ON SUNDAY, JUNE 22, 1941, the Chevaliers were driving back from a picnic at the beach with Oppenheimer when they heard the news over the radio that the Nazis had invaded the Soviet Union. That evening, everyone stayed up late listening to the latest news bulletins, trying to make sense of what had happened. Chevalier recalled Oppie saying that Hitler had committed a major blunder. By turning against the Soviet Union, Oppenheimer argued, Hitler had "destroyed at one stroke the dangerous fiction, so prevalent in liberal and political circles, that fascism and communism were but two different versions of the same totalitarian philosophy." Now communists everywhere would be welcomed as allies of the Western democracies. And that was a development both men thought was long overdue.

After the Japanese attack on Pearl Harbor on December 7, 1941, the country was suddenly at war. "Our little group in Berkeley," Chevalier recalled, "inevitably reflected the country's changing mood." Chevalier said the group "continued to meet irregularly"—though Oppenheimer himself rarely attended, owing to his busy travel schedule. "When we did meet," Chevalier wrote, "our business was largely confined to discussing the progress of the war and events on the home front."

Chevalier always insisted that Oppenheimer, the man he considered his closest friend, shared his own leftist political views right up to the moment Oppenheimer left Berkeley in the spring of 1943: "[W]e shared the ideal of a socialist society . . . there was never any wavering, any weakening of his position. He was firm as a rock." But Chevalier was clear that Oppenheimer was not an ideologue. "There was no blindness in him, no narrow partisanship, no automatic hewing to a line."

CHEVALIER'S DESCRIPTION of Oppenheimer essentially presented a left-wing intellectual not under Party discipline. But over the years, as he turned to writing about his friendship with Oppie, Chevalier tried to suggest something else. In 1948, he produced the outline of a novel in which the protagonist, a brilliant physicist working to build an atomic bomb, is also the de facto leader of a "closed unit" of the Communist Party. In 1950, Chevalier set aside the partially written manuscript after he was unable to find a publisher for it. But in 1954, after the Oppenheimer security hearing, he returned to the novel, and in 1959 G. P. Putnam's Sons published it under the ponderous title *The Man Who Would Be God.*

In the novel, the Oppenheimer character, one Sebastian Bloch, decides to join the Communist Party—but to his surprise, the local CP leader refuses to let him formally join. "Sebastian would meet with the unit regularly, and in every way act as though he were a bona fide member, and the other members would so regard him; but he would pay no dues—he could make his own financial arrangements with the party, but outside the unit." Later in the novel, Chevalier describes the weekly meetings of this closed party unit as "informal seminars of the kind that were constantly being held on all sorts of subjects among professors and students on the campus." The members discuss "ideas and theory," current events, the "activity of this or that member of the Teachers' Union," and "support to be given to a labor union campaign, a strike, an individual or group under attack on a civil liberties issue." In response to the Soviet Union's invasion of Finland in November 1939, Chevalier has the Oppenheimer alter ego propose that the party unit publish essays that would explain the international situation "in language palatable to cultivated, critical minds." The Oppenheimer character pays for the printing and postage costs and does most of the writing himself. "It was his baby," writes the novelist. "A number of these 'Reports to the Faculty' appeared over the next few months."

This thinly disguised roman à clef did not sell well, and Chevalier was unhappy with the reviews. *Time* magazine's reviewer, for instance, thought the "novel's underlying tone suggests an ex-worshipper stomping on a fallen idol." But Chevalier could not let the matter drop. In the summer of 1964, he wrote Oppenheimer to say that he had nearly finished writing a memoir about their friendship. He explained, "I tried to tell the essential story in my novel. But readers in America were disturbed by the blend of truth and fiction, and it has become clear to me that for the record I have to tell the story straight. . . . an important part of the story concerns your and my membership in the same unit of the CP from 1938 to 1942. I should like to deal with this in its proper perspective, telling the facts as I remember them. As this is one of the things in your life which, in my opinion, you have least to be ashamed of, and as your commitment, attested among other things by your 'Reports to Our Colleagues,' which today make impressive reading, was a deep and genuine one, I consider that it would be a grave omission not to give it its due prominence." Chevalier then asked whether Oppenheimer would have any objections to the telling of this story.

Two weeks later, Oppenheimer wrote back a terse note:

> Your letter asks whether I would have any objections. Indeed I do.
> What you say of yourself I find surprising. Surely in one respect
> what you say of me is not true. I have never been a member of the
> Communist Party, and thus have never been a member of a
> Communist Party unit. I, of course, have always known this. I
> thought you did too. I have said so officially time and time again. I
> said so publicly in response to what Crouch said in 1950. I said so in
> the AEC hearings ten years ago.
>
> As ever,
Robert Oppenheimer

Chevalier reasonably concluded that Oppenheimer's denial also meant to warn him that he could face a libel suit if he wrote that Oppenheimer had joined the Communist Party. So the following year, he published *Oppenheimer: The Story of a Friendship* without the bald allegation. Instead, throughout the book, the alleged CP "closed unit" is described merely as a "discussion group."

Chevalier told Oppenheimer that he had been compelled to write this book because "history, though coy, needs truth to be her handmaiden." But in this case, the "truth" lies in each man's perception. Were all members of the Berkeley "discussion group" also members of the CP? Apparently, Chevalier believed they were; Oppenheimer insisted that he, at least, was not. He would fund specific causes through the CP—the Spanish Republic, farm workers, civil rights and consumer protection. He would attend meetings, offer his advice and even help the Party's intellectuals write position papers. But he did not have a Party card, he did not pay dues, he was totally free of Party discipline. His friends might have had reasons to think he was a comrade, but it was clear to him that he was not.

John Earl Haynes and Harvey Klehr—two historians of American Communism—have written that "to be a Communist was to be part of a rigid mental world tightly sealed from outside influences. . . ." This certainly does not describe Robert Oppenheimer at any time. He was reading Marx, but he was also reading the Bhagavad-Gita, Ernest Hemingway and Sigmund Freud—and, in those years, the last was grounds for expulsion from the Party. In short, Oppenheimer never entered into that peculiar social contract expected of Party members.

Robert was probably closer to the party in the 1930s than he later admitted, or even remembered, but he was not nearly as close to it as his friend Haakon believed. This is neither surprising nor deceitful. The so-called "secret units" of the Party—the sort of association that Oppenheimer is alleged to have had—were organizations without formal rosters or established rules and with little, if any, regimentation, as Chevalier explained to Martin Sherwin. For obvious organizational reasons, the Party chose to see those individuals associated with "secret units" as having made substantial personal commitments. On the other hand, each "committed" member could set the limits of his commitment, and that commitment could change over time, even very short periods of time, as it did, for instance, with Jean Tatlock.

Chevalier seemed always to be committed to the Party, and in those days when he and Robert were close friends, it is not surprising that he considered Robert equally committed. Perhaps for a time he was, but we do not, and we cannot, know the extent of his commitment. But what we can say with confidence is that Robert's period of high commitment was short and did not last.

The bottom line is that Robert always wished to be, and was, free to think for himself and to make his own political choices. Commitments have to be put in perspective to be understood, and the failure to do that was the most damaging characteristic of the McCarthy period. The most relevant political fact about Robert Oppenheimer was that in the 1930s he was devoted to working for social and economic justice in America, and to achieve this goal he chose to stand with the left.

CHAPTER ELEVEN

"I'm Going to Marry a Friend of Yours, Steve"

Her career was advancing Robert's...

ROBERT SERBER

B Y THE END OF 1939, Oppenheimer's often stormy relationship with Jean Tatlock had disintegrated. Robert loved her and wanted to marry her despite their problems. "We were at least twice close enough to marriage to consider ourselves engaged," he later recalled. But he often brought out the worst in Jean. He annoyed her with his old habit of showering friends with gifts. Jean didn't want to be catered to in this way. "No more flowers, please, Robert," she told him one day. But inevitably, the next time he came to pick her up at a friend's house, he came armed with the usual bouquet of gardenias. When Jean saw the flowers, she threw them to the floor and told her friend, "Tell him to go away, tell him I am not here." Bob Serber claimed that Jean went through phases when "she disappeared for weeks, months sometimes, and then would taunt Robert mercilessly. She would taunt him about whom she had been with and what they had been doing. She seemed determined to hurt him, perhaps because she knew Robert loved her so much."

In the end, it was Tatlock who made the final break. Jean could be as strong-willed as Oppenheimer himself. Confused and highly distraught, she now rejected his latest offer of marriage. By then she had spent three years in medical school. Not many women became doctors in the 1930s. Her determination to pursue a career as a psychiatrist surprised some of her friends, who explained it as characteristic of a sometimes bold and impetuous woman. And yet they knew it also made sense. From her politics to her interest in the psychological, Tatlock had always been motivated by the desire to help others in a practical, hard-headed manner. Becoming a psychiatrist suited her temperament and intelligence, and by June 1941 Tatlock had a medical degree from the Stanford University School of Medicine. She spent the year 1941–42 as an intern at St. Elizabeth's psychiatric hospital in Washington, D.C., and the following year she was a resident physician at Mount Zion Hospital in San Francisco.

ON THE REBOUND, Robert was seen dating a number of "mostly very attractive youngish girls." Among others, he had relationships with Haakon Chevalier's sister-in-law, Ann Hoffman, and Estelle Caen, the sister of the *San Francisco Chronicle*'s columnist Herbert Caen. Bob Serber recalled a half-dozen girlfriends, including a British émigré named Sandra Dyer-Bennett. He broke several hearts. Still, whenever Tatlock phoned him in a low mood, he came to her and talked her out of her depression. They remained the closest of friends, and occasional lovers.

And then, in August 1939, he attended a garden party in Pasadena hosted by Charles Lauritsen, and in the course of the afternoon he was introduced to a twenty-nine-year-old married woman named Kitty Harrison. Bob Serber happened to witness the encounter. Kitty, he could see, was immediately mesmerized. "I fell in love with Robert that day," Kitty later wrote, "but hoped to conceal it." Soon afterward, Robert surprised his friends by turning up at a party in San Francisco unannounced with Kitty Harrison on his arm. That evening Kitty was wearing a corsage of flaming orchids. Everyone was rather uncomfortable, since the hostess of the party was Estelle Caen, Oppie's most recent lover. Chevalier called it "a not altogether happy occasion." Some of Oppie's friends—who very much liked Tatlock and had assumed they would reconcile—snubbed his new lady. Kitty seemed altogether too flirtatious and manipulative. Years later, Robert recalled that "there was among our friends much concern. . . ." But when it became clear that Kitty was not a passing fancy, his friends resigned themselves. "Oh, let's face it," said one woman. "It may be scandalous, but at least Kitty has humanized him."

A petite brunette, Katherine "Kitty" Puening Harrison was as attractive as Tatlock but worlds apart in temperament. The orchids she wore the evening she met Oppie's friends were no accident; she cultivated these flamboyant flowers in her apartment and wore them to make a statement. No one would ever find in the vivacious Kitty a touch of the morose. If she'd had some hard knocks in life, she had nevertheless always responded by making swift decisions to move on. If Tatlock looked like an Irish princess, Puening sometimes claimed to be the real thing, only of German royalty. "Kitty was related on her mother's side to all the crowned heads of Europe," recalled Robert Serber. "When she was a girl, she used to spend her summers visiting her uncle, the king of the Belgians." Kitty had been born on August 8, 1910, in Recklinghausen, a small town in North Rhine–Westphalia, Germany. She had come to America two years later, when her parents, Franz Puening, thirty-one, and Kaethe Vissering Puening, thirty, immigrated to Pittsburgh, Pennsylvania. Trained as a metallurgical engineer, Franz Puening had landed an engineering position with a steel company.

An only child, Kitty led a privileged childhood, growing up in the wealthy Pittsburgh suburb of Aspinwall. She later told friends that her father was "a prince of a small principality in Westphalia" and her mother was related to Queen Victoria. Her grandfather Bodewin Vissering was a royal Hanoverian crown-land lessee, and an elected member of the city council of Hanover. The ancestors of her grandmother Johanna Blonay had been, from the time of the eleventh-century Crusades, royal vassals to the House of Savoy, one of Europe's oldest surviving dynasties. The Blonays served as administrators and court advisers in various Savoy principalities in parts of Italy, Switzerland and France and occupied a magnificent château south of Lake Geneva.

Kaethe Vissering was beautiful and imposing. For a short time, she was engaged to a cousin, Wilhelm Keitel—who later served as Hitler's field marshal and in 1946 was tried and hanged at Nuremberg as a war criminal. While Kitty's mother made a point of taking her back as a child to visit her "princely" relatives in Europe, her father made her promise never to speak about her blue-blooded ancestry. As a young woman, however, Kitty occasionally let it be known that she came from a noble family. Friends of the family recall her receiving letters from her German relatives addressed to "Her Highness, Katherine."

As German immigrants, the Puenings sometimes had a difficult time in Pittsburgh during World War I. As an enemy alien, Franz Puening was placed under surveillance by local authorities, and even young Kitty had a hard time with neighborhood kids. Kitty's first language was not English, and even later in life she could speak a beautiful High German. As an adolescent, she found her mother "imperious." They didn't get along. She was a spunky, exuberant girl who paid little attention to social convention. "She was wild as hell in high school," recalled Pat Sherr, a friend who knew her later.

Kitty began what became a checkered college career. She enrolled at the University of Pittsburgh, but within a year, she left for Germany and France. Over the next couple of years she studied at the University of Munich, the Sorbonne and the University of Grenoble. She spent most of her time, however, in Paris cafés, hanging out with musicians. "I spent little time on school work," Kitty recalled. On the day after Christmas 1932, she impulsively married one of these young men, a Boston-born musician named Frank Ramseyer. Several months into the marriage, Kitty found her husband's diary—he kept it in mirror writing—and learned that he was both a drug addict and a homosexual. Retreating to America, she enrolled at the University of Wisconsin and began studying biology. On December 20, 1933, a Wisconsin court awarded her an annulment—and impounded the court testimony on grounds of obscenity.

Ten days later, Kitty was invited by a friend in Pittsburgh to a New Year's Eve party. Her friend, Selma Baker, said she had met a communist, and asked Kitty if she would like to meet the guy. "The consensus was that none of us had met a real live communist," recalled Kitty, "and that it would be interesting to see one." That evening she met Joe Dallet, the twenty-six-year-old son of a wealthy Long Island businessman. "Joe was three years older than I," Kitty remembered. "I fell in love with him at this party and I never stopped loving him." Less than six weeks later, she left Wisconsin to marry Dallet and join him in Youngstown, Ohio.

"He was a handsome son-of-a-bitch," recalled a friend. "Just a gorgeous guy." A tall, gaunt young man with a thick mop of dark curly hair, Dallet seemed capable of almost anything. Born in 1907, he spoke fluent French, played the classical piano with ease and knew his dialectical materialism. Both his parents were first-generation Americans of German-Jewish origin, and by the time Joe was an adolescent his father had made a small fortune in the silk trade. Although he and his sisters attended a temple in the middle-class Jewish community of Woodmere, Long Island, when he turned thirteen Joe refused a bar-mitzvah. For a time he went to private school before enrolling in Dartmouth College in the autumn of 1923. By then he was already politically radical and went out of his way to champion, in a belligerent fashion, what he called "proletarian ideals." His Dartmouth classmates regarded him as an eccentric, "an utter misfit in college." After failing most of his courses, he dropped out halfway through his sophomore year and took a job with an insurance company in New York. Successful, he nevertheless quit his job in disgust one day and literally assumed a new life as a laboring man. His transformation seems to have been precipitated by the execution, in August 1927, of the Italian-born anarchists Nicola Sacco and Bartolomeo Vanzetti. "It is difficult to tell what would have become of me," Dallet wrote his sister, "had not a couple of 'wops' been burned to death in the electric chair of the state of Mass. on August 22, 1927."

Determined to "throttle the evidence of his earlier sheltered life," Dallet went to work first as a social worker and then as a longshoreman and coal miner. After joining the Communist Party in 1929, he wrote his worried family, "Certainly now you must see that I am doing what I believe in, want to do, do best, and most enjoy doing. . . . you must see that I am really happy." He spent a few months in Chicago, where, after speaking before a crowd of thousands, he was beaten by the notorious "Red Squad" of the city police.

By 1932, Dallet was a union organizer in Youngstown, Ohio, where he served on the front lines of the rough-and-tumble CIO campaign to bring steelworkers into the fold of organized labor. He bristled with physical courage in the often violent confrontations with the steel companies' thugs. On several occasions, local police clapped him into jail to keep him from speaking at labor rallies. At one point he ran for mayor on the Communist Party ticket. Kitty, despite being his wife, was only allowed to join the Young Communist League after proving her commitment by hawking the *Daily Worker* on the streets and handing out leaflets to steelworkers. "I used to wear tennis shoes," she recalled, "when I handed out Communist Party leaflets at factory gates so that I could get a fast running start when the police arrived."

Her party dues were ten cents a week. The couple lived in a dilapidated boardinghouse for five dollars' rent a month and, ironically, survived on government relief checks of $12.50 every two weeks. Down the hallway for a time lived two other Communist Party stalwarts, John Gates and Arvo Kusta Halberg—who later changed his name to Gus Hall and rose to become chairman of the Communist Party USA. "The house had a kitchen," Kitty later said, "but the stove leaked and it was impossible to cook. Our food consisted of two meals a day, which we got at a grimy restaurant." During the summer of 1935 she served the Party as its "literary agent," which meant that she tried to encourage members to buy and read Marxist classics.

Kitty stuck it out until 1936, when she told Joe that she could no longer live under such conditions. Joe's whole life was the Party, and while Kitty hadn't abandoned her political beliefs, they began arguing. According to a mutual friend, Steve Nelson, Joe "was a bit dogmatic about her reluctance to accept party loyalty as strongly as he did." In Joe's eyes, Kitty was just acting like a young "middle-class intellectual who couldn't quite see the working class attitude." Kitty resented his condescension. After two and a half years of living in extreme poverty, she announced that they had to separate. "The poverty became more and more depressing to me," she recalled. Finally, in June 1936, she fled to London, where her father had taken an assignment to build an industrial furnace. For a while she heard nothing from Dallet—until one day she discovered that her mother had been intercepting his letters. Now eager for a reconciliation, she was pleased to learn that her husband was coming to Europe.

Early in 1937, Dallet had decided to volunteer to fight with a communist-sponsored brigade in the Spanish Civil War on behalf of the Republic and against the fascists. He and his old comrade Steve Nelson shipped out aboard the same cruise liner, the *Queen Mary,* in March 1937. Joe, clearly still in love, told Nelson that he had hopes that he and Kitty would soon work things out.

Kitty was waiting for them at dockside when their ship arrived in Cherbourg, France. She and Joe spent a week together in Paris—with Nelson tagging along. "I was like a third wheel," Nelson recalled. "Kitty impressed me as a very cute young woman; not very tall, short, blonde [*sic*] and the very friendly type." She had brought enough money with her from London so that the three of them could stay in a decent hotel and eat out in good French restaurants. Nelson remembered eating exotic French cheeses and sipping wine over lunch as he listened to Kitty scheme about how much she wanted to accompany Joe to the battlefields in Spain. The problem was that the Communist Party had decided that wives could not join their husbands in Spain. "Joe raised holy hell," Nelson recalled of these luncheons. "He'd say, 'This is bureaucratic; she could do a lot of work, she could drive an ambulance.' Kitty was determined to go." But all their efforts to bend the rules were in vain; by the end of the week Dallet was forced to leave Kitty behind as he and Nelson departed for Spain. On their last day together, Kitty took Dallet and Nelson shopping for warm flannel shirts, wool-lined gloves and wool socks. She then returned to London to await an opportunity to rejoin her husband. They corresponded often and Kitty got in the habit of sending him a snapshot of herself each week.

On their way into Spain, Dallet and Nelson were arrested by French authorities; after a trial in April, they served a sentence of twenty days in prison and were then released. When Dallet finally smuggled himself into Spain in late April, he wrote Kitty, "I adore you and can't wait to reach A. [Albacete] and get your letter." By July, he was still writing her upbeat, glowing accounts of his experiences: "It's a bloody interesting country, a bloody interesting war and the most bloody interesting job of all the bloody interesting jobs I've ever had, to give the fascists a real bloody licking."

Kitty had genuinely liked her husband's friend and took the trouble to write Nelson's wife, Margaret—a woman she had yet to meet—about their week in Paris together. "We had a nice few days," she wrote. "I don't suppose they were too good a preparation for the tough journey ahead, but they were fun." She reported that they had attended a splendid mass meeting of 30,000 people protesting the West's stance of strict neutrality in the Spanish Civil War. "The most thrilling part to us since we couldn't understand the speeches at the meeting was the subway ride there. Hundreds of young communist leaders held up the subway until they got on, singing the Internationale and shouting anti-fascist slogans. Everyone joined in and by the time we arrived at Grenelle (the meeting station), it seemed as if the whole of Paris was roaring out the Internationale. I may be the emotional type (though I doubt it), but it made me feel as though I'd suddenly grown triple my size, brought tears to my eyes, and made me want to shout a big belly-roar." Kitty signed the letter, "Comradely yours, Kitty Dallet."

In Spain, Joe Dallet was soon assigned as "political commissar" to the 1,500-man McKenzie-Papineau Battalion, a largely Canadian unit that by then had absorbed many American volunteers from the Abraham Lincoln Brigade. That summer, he and his men began their combat training. "Man, what a feeling of power you have when entrenched behind a heavy machine gun!" he wrote Kitty. "You know how I always enjoyed gangster movies for the mere sound of the machine guns. Then you can imagine my joy at finally being on the business end of one."

The war was not going well for the Republican cause. Dallet and his men were outmanned and outgunned by the Spanish fascists, who were being supplied with aircraft and artillery by Germany and Italy. And, as Dallet soon discovered, the Spanish left was further weakened by fierce, sometimes deadly, sectarian politics. In a letter to Kitty dated May 12, 1937, Dallet ominously wrote that his Spanish communist superiors had promised a "cleaning out" of the anarchists among the troops. By that autumn, Dallet was supervising "trials" of deserters; by one account, a handful of these men may have been executed. Dallet himself became extremely unpopular with his own troops. These feelings, according to a friend of Dallet's, amounted to "near hatred." Some thought him an ideological zealot. According to a Comintern report dated October 9, 1937, "A percentage of the men openly declare their dissatisfaction with Joe and there is some talk of removal. . . ."

Four days later, he went into battle for the first time, leading his battalion in an offensive against the fascist-held town of Fuentes del Ebro. A few days earlier, an old friend had found him at night sitting alone in a small hut by the faint light of a kerosene lamp. Dallet confided that he felt lonely and knew that he was extremely unpopular. He said he was determined to prove to them that he was not one of those "safe behind the line" political officers; he would demonstrate his courage by being the first man over the parapet. When his friend argued that this might be a foolish way to lead an entire battalion, Dallet was adamant.

On the day of the battle, Dallet kept his word. He was the first man out of the trenches and had advanced only a few yards toward the fascist lines when he was hit in the groin by machine-gun fire. The battalion's machine-gun commander later reported: "The attack started at 1:40 PM. Joe Dallet, battalion commissar, went over with the First Company on the left flank, where the fire was heaviest. He was leading the advance when he fell, mortally wounded. He behaved heroically until the very end, refusing to permit the first-aid men to approach him in his exposed position." Suffering dreadfully, he was trying to crawl back to the trenches when a second round of machine-gun fire killed him. He was barely thirty years old.

Steve Nelson—who himself had been wounded in August—heard of Dallet's death shortly afterwards while on a trip to Paris. Before his death, Dallet had written Kitty, telling her that Nelson would be passing through Paris, so Kitty had decided to make the trip from London to meet him. She planned to go on from Paris to Spain. Knowing that he had to tell her the tragic news, Nelson arranged to meet her in the lobby of her hotel. "She was crushed," Nelson recalled. "She literally collapsed and hung on to me. I became a substitute for Joe, in a sense. She hugged me and cried, and I couldn't maintain my composure." When Kitty plaintively cried, "What am I going to do now?" Nelson on impulse invited her to move in with him and his wife, Margaret, back in New York. Kitty agreed, but not until Nelson talked her out of going on to Spain, where she thought she could volunteer as a hospital worker.

Kitty returned to America the twenty-seven-year-old widow of a CP war hero. The American Communist Party made sure that his sacrifice would be remembered. Party chief Earl Browder wrote that Dallet had joined those who had given "themselves completely to the task of stopping fascism." One of the Party's few genuine Ivy League communists, Dallet had become a martyr of the working class. With Kitty's permission, in 1938 the Party published *Letters from Spain,* a collection of Joe's letters to his wife.

Kitty spent a couple of months with the Nelsons in their cramped apartment in New York City. She saw some of Joe's old friends, all of whom were Party members. Kitty herself later told government investigators that she had at some point met as acquaintances such well-known Communist Party officials as Earl Browder, John Gates, Gus Hall, John Steuben and John Williamson. But she said she had ceased to be a member of the Party when she left Youngstown in June 1936 and stopped paying Party dues. "She seemed to be in a very unsettled state," Margaret Nelson recalled. "I was under the impression that she was under a great emotional strain." Other friends testify that Kitty remained deeply affected by Dallet's death for a long time.

And then, in early 1938, she visited a friend in Philadelphia and decided to stay, enrolling in the University of Pennsylvania for the spring semester. She studied chemistry, math and biology and seemed ready, finally, to get her college degree. Sometime that spring or summer, she ran into a British-born doctor, Richard Stewart Harrison, whom she had known as a teenager. Harrison, a tall, handsome man with piercing blue eyes, had practiced medicine in England, and was then finishing an internship to become licensed in the United States. Older and apolitical, Harrison seemed to offer Kitty something she now desperately wanted: stability. Making another of her impetuous decisions, Kitty married Harrison on November 23, 1938. This marriage, she later said, was "singularly unsuccessful from the start." She told a friend that it was "an impossible marriage" and that she "was ready to leave him long before she did." Harrison soon left for Pasadena, where he had a residency lined up. Kitty stayed in Philadelphia and in June 1939 obtained her Bachelor of Arts degree, with honors in botany. Two weeks later, she agreed to follow Harrison to California and maintain the pretense of a stable marriage because, she said, "of his conviction that a divorce might ruin a rising young doctor."

At twenty-nine, Kitty finally seemed ready to take charge of her own life. Although seemingly locked into a dead-end marriage, she now was determined to get on with her own career. Her main interest was botany, and that summer she won a research fellowship to begin graduate studies at the University of California's Los Angeles campus. Her ambition was to earn a doctorate and, perhaps, a professorship in botanical studies.

In August 1939, she and Harrison attended the garden party in Pasadena where she met Oppenheimer. Kitty began her graduate studies at UCLA that autumn, but she did not forget the tall young man with such bright blue eyes. Sometime over the next few months they met again and began to date—and, though Kitty was still married, they made no effort to conceal the affair. They were frequently seen driving in Robert's Chrysler coupe. "He would ride up [near my office] with this cute young girl," recalled Dr. Louis Hempelman, a physician who taught at Berkeley. "She was very attractive. She was tiny, skinny as a rail, just like he was. They'd give each other a fond kiss and go their separate ways. Robert always had that porkpie hat on."

In the spring of 1940, Oppenheimer—rather audaciously—invited Richard Harrison and Kitty to spend some time that summer at Perro Caliente. At the last moment, Dr. Harrison later told the FBI, he decided he could not go, but encouraged Kitty to go anyway. As it happened, Bob and Charlotte Serber had been invited by Oppie to come to the ranch at the same time and when they drove into Berkeley from Urbana, Illinois—where Serber had been teaching—Oppie explained that he had invited the Harrisons but Richard couldn't make it. "Kitty might come alone," he said. "You could bring her with you. I'll leave it up to you. But if you do it might have serious consequences." Kitty went with the Serbers eagerly—and stayed a full two months on the ranch.

Just a day or two after her arrival, Kitty and Robert—she always insisted on calling him Robert—rode horses to Katherine Page's dude ranch at Los Pinos. They spent the night and then rode back the next morning. They were followed a few hours later by Page—the woman whom young Oppenheimer had been so infatuated with in the summer of 1922—who mischievously presented Kitty with her nightgown, which she explained had been found under Robert's pillow at Los Pinos.

At the end of the summer, Oppenheimer phoned Dr. Harrison to tell him that his wife was pregnant. The two men agreed that the thing to do was for Harrison to divorce Kitty so that Oppenheimer could marry her. It was all very civilized. Harrison told the FBI that "he and the Oppenheimers were still on good terms and that he realized that they all had modern views concerning sex."

Even though Bob Serber was a witness to the passionate affair of that summer of 1940, he was still astonished in October when he heard from Oppie that he was marrying. When told the news, he wasn't sure if Oppenheimer had said his prospective bride was Jean or Kitty. It could have been either. Oppenheimer had walked off with another man's wife—and some of his friends were genuinely scandalized. Oppie was not a womanizer, but he was the kind of man who was strongly attracted to women who were attracted to him. Kitty had been irresistible.

One evening that autumn of 1940, Robert happened to share a platform with Steve Nelson at a Berkeley fundraiser on behalf of refugees from the Spanish Civil War. Newly arrived in San Francisco, Nelson had never heard of Oppenheimer. As the featured speaker, Oppenheimer said the fascist victory in Spain had led directly to the outbreak of general war in Europe. He argued that those like Nelson who had served in Spain had fought a delaying action.

Afterwards, Oppenheimer approached Nelson, and with a broad smile said, "I'm going to marry a friend of yours, Steve." Nelson couldn't think who that could be. So Robert explained, "I'm going to marry Kitty."

"Kitty Dallet!" Nelson exclaimed. He had lost touch with her since her stay with him and Margaret in New York. "She's back there, sitting in the hall," Oppenheimer said, and he motioned to her to come up. The two old friends hugged and agreed to get together. Soon afterwards, the Nelsons came to the Oppenheimers' for a picnic dinner. Sometime that autumn, Kitty moved to Reno, Nevada, for the required residency of six weeks, and there, on November 1, 1940, she obtained a divorce decree. That very day she married Robert in Virginia City, Nevada. A court janitor and a local clerk signed the marriage certificate as witnesses. By the time the newly-weds returned to Berkeley, Kitty was wearing a maternity dress.

At the end of November, Margaret Nelson phoned Kitty to say that she had just given birth to a daughter, and that they had named the child Josie, in honor of Joe Dallet. Kitty immediately invited the Nelsons to visit and use the spare bedroom in their new house. Over the next couple of years, the Nelsons visited the Oppenheimer household on numerous occasions, although the visits gradually grew less frequent. In later years, their children would play together. "I also saw Robert at Berkeley now and then," Nelson wrote in his memoirs, "because I was responsible for working with people from the university, getting them to conduct classes and discussions." They also had one-on-one meetings. An FBI wiretap, for instance, shows that Oppenheimer met with Nelson on Sunday, October 5, 1941, apparently to pass him a check for $100, earmarked as a donation for striking farm workers. But the relationship went far beyond political transactions. When Josie Nelson turned two in November 1942, Oppenheimer surprised her mother by turning up on their doorstep, bearing a gift for the child. Margaret was "astounded" and touched by this typical act of kindness. "With all of his brilliance," she thought, "there were very strong human qualities."

Though pregnant, Kitty continued her biology studies and insisted to her friends that she still intended to make a professional career for herself as a botanist. "Kitty was very excited about the fact that she was going back to school," Maggie Nelson said. "She was very much taken up with that." But despite their common interest in science, Kitty and Robert were temperamentally poles apart. "He was gentle, mild," recalled one friend who knew them both. "She was strident, assertive, aggressive. But that's often what makes a good marriage, the opposites."

Most of Robert's relatives were put off by Kitty. Plain-spoken Jackie Oppenheimer always thought she was "a bitch" and resented the way she thought Kitty cut Robert off from his friends. Decades later she vented her animosity: "She could not stand sharing Robert with anyone," recalled Jackie. "Kitty was a schemer. If Kitty wanted anything, she would always get it. . . . She was a phony. All her political convictions were phony, all her ideas were borrowed. Honestly, she's one of the few really evil people I've known in my life."

Kitty certainly had a sharp tongue and easily antagonized some of Robert's friends, but some thought her "very smart." Chevalier considered her intelligence more intuitive than astute or profound. And as their friend Bob Serber recalled, "Everybody was talking about Kitty being a communist." But it was also true that she had a stabilizing influence on Robert's life. "Her career," Serber said, "was advancing Robert's career, which was the overwhelming, controlling influence on her from then on."

SOON AFTER THEIR HASTY WEDDING, Oppie and Kitty rented a large house at 10 Kenilworth Court, north of the campus. After selling his aging Chrysler coupe, he presented his bride with a new Cadillac; they nicknamed it "Bombsight." Kitty persuaded her husband to dress in a style more suited to his station in life. And so he began for the first time to wear tweed jackets and more expensive suits. But he kept his brown porkpie hat. "A certain stuffiness overcame me," he later confessed of married life. At this point in their marriage, Kitty was an excellent cook, and so they entertained frequently, inviting close friends like the Serbers, the Chevaliers and other Berkeley colleagues. Their liquor cabinet was always well stocked. One evening Maggie Nelson recalled a discussion in which Kitty confessed that "their bill for liquor was even higher than their bill for food."

One evening early in 1941, John Edsall, Robert's friend from his Harvard and Cambridge years, dropped by for dinner. Now a professor of chemistry, Edsall hadn't seen Robert in over a decade. He was startled by the change. The introspective boy he had known in Cambridge and Corsica was now a figure of commanding personality. "I felt that he obviously was a far stronger person," Edsall recalled, "that the inner crises that he had been through in those earlier years he had obviously worked out and achieved a great deal of inner resolution of them. I felt a sense of confidence and authority, although still tension and [a] lack of inner ease in some respects . . . he could reach and see intuitively things that most people would be able to follow only very slowly and hesitatingly, if at all. This was not only in physics, but in other things as well."

By then, Robert was about to become a father. Their child was born on May 12, 1941, in Pasadena, where Oppenheimer was on his regular spring teaching schedule at Caltech. They christened the boy Peter—but Robert impishly nicknamed him "Pronto." Kitty told some of her friends, tongue in cheek, that the eight-pound baby was premature. It had been a difficult pregnancy for Kitty, and that spring Oppenheimer himself was suffering from a case of infectious mononucleosis. By June, however, they had both recovered their health enough to invite the Chevaliers to visit them. They arrived in mid-June and spent a week catching up with their old friends. Haakon had recently befriended the surrealist artist Salvador Dalí and spent the days sitting in Oppie's garden working on a translation of Dalí's book *The Secret Life of Salvador Dalí.*

A few weeks later, Oppie and Kitty approached the Chevaliers to ask an enormous favor. Kitty badly needed a rest, Robert explained. Would the Chevaliers take two-month-old Peter, along with his German nanny, while he and Kitty escaped to Perro Caliente for a month? Haakon saw the request as a confirmation of his own feeling that Oppie was his closest, most intimate friend. "Deeply flattered," the Chevaliers promptly agreed and kept Peter for not one but two full months, until Kitty and Oppenheimer returned for the fall semester. This rather unusual arrangement, however, may have had long-term consequences for mother and child. Kitty never properly bonded with Peter. Even a year later, friends noticed that it was always Robert who took them into the baby's room and showed him off with obvious pride and delight. "Kitty seemed quite uninterested," said this old friend.

Robert felt reinvigorated almost as soon as he arrived at Perro Caliente. That first week he and Kitty found the energy to nail new shingles on the cabin's roof. They went for long rides in the mountains. One day Kitty showed her spunk by cantering her horse in a meadow while standing up in the saddle. Robert was pleased when in late July he ran into his old friend Hans Bethe, the Cornell physicist he had first met in Göttingen, and persuaded him to visit them at the ranch. Unfortunately, soon afterwards Robert was trampled by a horse he was trying to corral for Bethe to ride and had to have X rays taken at the hospital in Santa Fe. In more ways than one, it was a memorable visit.

Upon their return, the Oppenheimers retrieved baby Peter and moved into a newly purchased home at Number One, Eagle Hill, in the hills overlooking Berkeley. Earlier that summer, Robert had briskly toured the house once and then immediately agreed to pay the full asking price of $22,500—plus another $5,300 for two adjoining lots. A Spanish-style, one-story villa with whitewashed walls and a red-tiled roof, their new home stood on a knoll surrounded on three sides by a steep wooded canyon. They had a stunning view of the sunset over the Golden Gate Bridge. The large living room had redwood floors, twelve-foot-high beamed ceilings and windows on three sides. An image of a ferocious lion was carved into a massive stone fireplace. Floor-to-ceiling bookcases lined each end of the living room. French doors opened onto a lovely garden framed by live oak trees. The house came with a well-equipped kitchen and a separate apartment over the garage for guests. It was already partially furnished, and Barbara Chevalier helped Kitty with some of the interior decorating. Everyone thought it a charming, well-designed structure. Oppenheimer called it home for nearly a decade.

CHAPTER TWELVE

"We Were Pulling the New Deal to the Left"

I had had about enough of the Spanish cause, and there were other and more pressing crises in the world.

ROBERT OPPENHEIMER

O N SUNDAY, JANUARY 29, 1939, Luis W. Alvarez—a promising young physicist who worked closely with Ernest Lawrence—was sitting in a barber's chair, reading the *San Francisco Chronicle*. Suddenly, he read a wire service story reporting that two German chemists, Otto Hahn and Fritz Strassmann, had successfully demonstrated that the uranium nucleus could be split into two or more parts. They had achieved fission by bombarding uranium, one of the heaviest of the elements, with neutrons. Stunned by this development, Alvarez "stopped the barber in mid-snip, and ran all the way to the Radiation Laboratory to spread the word." When he told Oppenheimer the news, his reply was, "That's impossible." Oppenheimer then went to the blackboard and proceeded to prove mathematically that fission couldn't happen. Someone must have made a mistake.

But the next day, Alvarez successfully repeated the experiment in his laboratory. "I invited Robert over to see the very small natural alpha-particle pulses on our oscilloscope and the tall, spiking fission pulses, twenty-five times larger. In less than fifteen minutes he not only agreed that the reaction was authentic but also speculated that in the process extra neutrons would boil off that could be used to split more uranium atoms and thereby generate power or make bombs. It was amazing to see how rapidly his mind worked. . . ."

Writing his Caltech colleague Willie Fowler a few days later, Oppie remarked, "The U business is unbelievable. We first saw it in the papers, wired for more dope, and have had a lot of reports since. . . . Many points are still unclear: where are the short lived high energy betas one would expect? . . . In how many ways does the U come apart. At random, as one might guess, or only in certain ways? . . . It is I think exciting, not in the rare way of positrons and mesotrons, but in a good honest practical way." Here was a significant discovery, and he could hardly contain his excitement. At the same time, he also saw its deadly implications. "So I think it really not too improbable that a ten cm [centimeter] cube of uranium deuteride (one should have something to slow the neutrons without capturing them) might very well blow itself to hell," he wrote his old friend George Uhlenbeck.

Coincidentally, that same week, a twenty-one-year-old graduate student named Joseph Weinberg found his way to Room 219 in LeConte Hall and knocked on the door. Cocky and opinionated, Weinberg had been sent packing in mid-year by his physics professor at Wisconsin, Gregory Breit, who told him that Berkeley was one of the few places in the world where "a person as crazy as you could be acceptable." He belonged with Oppenheimer, Breit had said, ignoring Weinberg's protests that Oppenheimer's papers in *Physical Review* were the only articles that he couldn't understand.

"There was a tremendous hubbub behind the door," Weinberg recalled, "so I knocked very loudly and after a moment somebody sprang out with a great puff of smoke and noise as the door opened and closed again."

"What the hell do you want?" the man asked Weinberg.

"I'm seeking Professor J. Robert Oppenheimer," young Weinberg said.

"Well, you found him," replied Oppenheimer.

Behind the door, Weinberg could hear excited men shouting and arguing. "What are you doing here?" Oppenheimer asked.

He had just come from Wisconsin, Weinberg explained.

"And what did you do there?"

"I worked with Professor Gregory Breit," replied Weinberg.

"That's a lie," snapped Oppenheimer, "that's your first lie."

"Sir?"

"You're here," explained Oppenheimer. "You worked away from Breit, you worked loose from Breit."

"That would be a more accurate statement," conceded Weinberg.

"Very well," Oppenheimer said, "congratulations! Come in and join the madness."

Oppenheimer introduced Weinberg to Ernest Lawrence, Linus Pauling, and several of Oppenheimer's graduate students: Hartland Snyder, Philip Morrison and Sydney M. Dancoff. Weinberg was astonished to meet these luminaries of physics. "It was first names all around, which was ridiculous," he later recalled. Afterwards, Weinberg went out to lunch with Morrison and Dancoff and while sitting at a table in the student union restaurant, the Heartland, they discussed the significance of a telegram from Niels Bohr about the discovery of fission. Someone got out a napkin and began sketching a bomb based on the notion of a chain reaction. "On the basis of the data," said Weinberg, "we designed a bomb." Phil Morrison did some preliminary calculations and came to the conclusion that it wouldn't work, that the chain reaction would fizzle before exploding. "You see," Weinberg recalled, "at that time we didn't know that the uranium could eventually be purified and isolated in much greater concentrations—which of course could lead to fission." Within a week, Morrison recalls walking into Oppie's office and seeing on the blackboard "a drawing—a very bad, an execrable drawing—of a bomb."

The very next day, Oppenheimer sat down with Weinberg to define his course of study. "You think you're going to be a physicist," Oppie teased him, "so what have you done?" Flustered, Weinberg replied, "Do you mean lately?" Oppenheimer leaned back and roared with laughter. He didn't really expect a new graduate student to have done anything original. But Weinberg volunteered that he had worked on a theoretical problem and when he explained it, Oppenheimer interrupted to say, "You have this written up, of course?" Weinberg didn't, but he rashly promised to have a paper ready the next morning. "He looked at me," Weinberg recalled, "and said coldly, 'How about 8:30 a.m.?' " Trapped by his own cockiness, Weinberg spent the rest of the day and all night writing up that paper. He got it back from Oppenheimer a day later with one unpronounceable word scribbled across the flyleaf, "Snoessigenheellollig."

"I looked at him," Weinberg recalled, "and he said, 'Of course, you know what that means?'" Weinberg knew the word was Dutch slang, but he could decipher only just enough of it to know that it was a favorable comment. Oppie grinned and explained that, roughly translated, it meant "ducky."

"But why Dutch?" Weinberg asked.

"That I cannot tell you—I dare not tell you," replied Oppie. He then spun around and left the room, closing the door behind him. A moment later, however, the door cracked open; Oppenheimer poked his head in the room and said, "I really shouldn't tell you but then maybe I owe it to you—because the paper reminded me of [Paul] Ehrenfest."

Weinberg was stunned. He knew enough about Ehrenfest's reputation to grasp what Oppie was saying. "That was the only compliment he ever paid me. . . . He loved Ehrenfest, [who] had the knack of making things luminously clear and witty and pregnant in the simplest terms." That same week, Oppenheimer flattered Weinberg by having him present this paper in place of a previously scheduled seminar. But afterwards, as if to compensate for the flattery, Oppenheimer told him with a sneer that what he had presented was "kid stuff." There was, he said, a "grown-up way to do this kind of problem," and he suggested that Weinberg should get onto it right away. Weinberg duly spent the next three months laboring to produce an elaborate calculation. In the end, he had to report back that he could find no trace of the empirical relationship that he had predicted from his initial and very simple-minded argument. "Now you have learned a lesson," Oppenheimer told him. "Sometimes the elaborate, the learned method, the grown-up method is not as good as the simple and childishly naïve method."

Weinberg was a devoted disciple of Bohr's even prior to his arrival in Berkeley. Like many physicists, he found himself attracted to the discipline chiefly because it promised to open the door to fundamental philosophical insights. "I was interested in the fun of tampering with the laws of nature," Weinberg said. And indeed, when for a period he considered dropping physics, he only continued with it after a friend encouraged him to read Niels Bohr's classic work *Atomic Theory and the Description of Nature.* "I read Bohr and I was reconciled with physics," Weinberg said. "It really reconverted me." In Bohr's hands, quantum theory became a joyous celebration of life. The day Weinberg arrived at Berkeley, he happened to mention to Phil Morrison that Bohr's book was one of the few volumes he had thought worth bringing along. Phil burst out laughing, because at Berkeley, among those in Oppenheimer's tight-knit circle, Bohr's little book was considered the Bible. Weinberg happily realized that at Berkeley, "Bohr was God and Oppie was his prophet."

WHEN A STUDENT WAS STUMPED and just couldn't finish a paper, it was not unheard of for Oppie to just do it himself. One night in 1939, he invited Joe Weinberg and Hartland Snyder up to his home on Shasta Road. The two young graduate students had collaborated on a paper but felt unable to write a satisfactory conclusion. "He gave us the usual obligatory tumbler of whiskey," Weinberg recalled, "and he put on some music to keep me busy. Hartland drifted around looking at books while Oppie sat down at the typewriter. After a half-hour he had hammered out the last paragraph. A beautiful paragraph." The paper, "Stationary States of Scalar and Vector Fields," was published in *Physical Review* in 1940.

Oppenheimer's lectures were invariably accompanied by a slew of formulas written on the blackboard. But like most theoreticians, he had no respect for mere formulas. Weinberg, whom Oppenheimer had come to regard as one of his brightest students, observed that mathematical formulas were like temporary hand-holds for a rock-climber. Each hand-hold more or less dictates the position of the next hand-hold. "A record of that," Weinberg said, "is a record of a particular climb. It gives you very little of the shape of the rock." For Weinberg and others, "being in a course with Oppie was like experiencing lightning flashes five or ten times in an hour, so brief that you might've missed them. If you were scrounging formulas off a blackboard, you might very well not have known they were there at all. Very often these flashes were basic philosophical insights that placed physics in a human context."

Oppenheimer thought that no one could be expected to learn quantum mechanics from books alone; the verbal wrestling inherent in the process of explanation is what opens the door to understanding. He never gave the same lecture twice. "He was very keenly aware," Weinberg recalled, "of the people in his class." He could look into the faces of his audience and suddenly decide to change his entire approach because he had sensed what their particular difficulties were with the subject at hand. Once he gave an entire lecture on a problem he knew would pique the interest of just one student. Afterwards, that student rushed up to him and said he wanted permission to tackle the problem. Oppenheimer replied, "Good, that's why I gave the seminar."

Oppenheimer gave no final exams, but he handed out plenty of homework assignments. During each class hour he presented a non-Socratic lecture, "delivered at high speed," recalled Ed Geurjoy, a graduate student from 1938 to 1942. Students felt free to interrupt Oppie with a question. "He generally would answer patiently," Geurjoy said, "unless the question was manifestly stupid, in which event his response was likely to be quite caustic."

Oppenheimer was brusque with some students, but he treated those who were vulnerable with a gentle hand. One day when Weinberg was in Oppenheimer's office, he began rummaging through papers stacked on the trestle table in the center of the room. Picking out one paper, he began reading the first paragraph, oblivious to Oppie's irritated look. "This is an excellent proposal," Weinberg exclaimed, "I'd sure as hell like to work on it." To his astonishment, Oppenheimer replied curtly, "Put that down; put it back where you found it." When Weinberg asked what he had done wrong, Oppenheimer said, "That was not for you to find."

A few weeks later, Weinberg heard that another student who was struggling to find a thesis topic had begun work on the proposal he had read that day. "[The student] was a very genial, decent man," Weinberg recalled. "But, unlike a few of us who enjoyed the kind of challenge that Oppie threw out like sparks, he was often baffled and nonplussed and ill at ease. Nobody had the courage to tell him, 'Look, you're out of your depth.' " Weinberg now realized that Oppie had planted this thesis problem for this very student. It was a distinctly easy problem, "But it was perfect for him," Weinberg said, "and it got him his Ph.D. It would have been difficult for him to get it with Oppie if Oppie had treated him the way he treated me or Phil Morrison or Sid Dancoff." Instead, Weinberg insisted years later, Oppie nurtured this student as a father would have treated a baby learning to walk. "He waited for him to discover that proposal accidentally, on his own terms, to pick it up and to express his interest, to find his way to it. . . . He needed special treatment, and by God, Oppie was going to give it to him. It showed a great deal of love, sympathy and human understanding." The student in question, Weinberg reported, went on to do distinguished work as an applied physicist.

Weinberg quickly became a devoted member of Oppenheimer's inner circle. "He knew that I adored him," Weinberg said, "as we all did." Philip Morrison, Giovanni Rossi Lomanitz, David Bohm and Max Friedman were some of the other graduate students who regarded Oppenheimer as their mentor and role model during these years. These were unconventional young men who, in the words of Morrison, prided themselves on being "self-conscious and daring intellectuals." All of them were studying theoretical physics. And all of them were active in one or another Popular Front cause. Some, like Philip Morrison and David Bohm, have acknowledged that they joined the Communist Party. Others were merely on the fringe. Joe Weinberg was probably in the Party for at least a brief time.

Morrison, born in 1915 in Pittsburgh, grew up not far from Kitty Oppenheimer's childhood home. After a public school education, he received a B.S. in physics from Carnegie Institute of Technology in 1936. That autumn, he went to Berkeley to study theoretical physics under Oppenheimer. A victim of childhood polio, he arrived on campus wearing a brace on one leg. As a child convalescing from the disease, he had spent a great deal of time in bed, and learned to speed-read at five pages a minute. As a graduate student, Morrison impressed everyone with his wide range of knowledge about almost anything, from military history to physics. In 1936, he joined the Communist Party. But though he didn't hide his left-wing political views, neither did he advertise his Party membership. His office-mate at Berkeley in the late 1930s, Dale Corson, was unaware that Morrison was a CP member.

"We were all close to communism at the time," Bohm recalled. Actually, until 1940–41, Bohm didn't have much sympathy for the Communist Party. But then, with the collapse of France, it seemed to him that no one but the communists had the will to resist the Nazis. Indeed, many Europeans appeared to prefer the Nazis to the Russians. "And I felt," Bohm said, "that there was such a trend in America too. I thought the Nazis were a total threat to civilization. . . . It seemed that the Russians were the only ones that were really fighting them. Then I began to listen to what they said more sympathetically."

Late in the autumn of 1942, the newspapers were filled with accounts of the battle for Stalingrad; for a time that fall, it seemed as if the entire outcome of the war depended on the sacrifices being made by the Russian people. Weinberg later said that he and his friends suffered every day along with the Russian people. "No one can feel the way we felt," he recalled. "Even when we saw the sham of what was going on in the Soviet Union, of the show trials, we turned our eyes away from them."

In November 1942, just as the Russians opened up an offensive to push the Nazis back from the outskirts of Stalingrad, Bohm began attending regular meetings at a Berkeley chapter of the Communist Party. Typically, fifteen people might show up. After a while, Bohm found the meetings "interminable," and decided that the group's various plans to "stir up things on the campus" didn't amount to much. "I had the feeling that they were really rather ineffective." Gradually, Bohm just stopped attending. But he remained a passionate and enthusiastic intellectual Marxist, reading Marxist texts together with his closest friends at the time, Weinberg, Lomanitz and Bernard Peters.

Phil Morrison recalled that his Party unit meetings were attended by "many people who were not communist. It would be very hard to say which members were communists." The meetings were often like college bull sessions. They discussed, Morrison recalled, "everything under the sun." As a cash-starved graduate student, Morrison was assessed Party dues of a mere twenty-five cents a month. Morrison remained a Party member through the Nazi-Soviet Pact, but, like many of his American comrades, he drifted out of the Party soon after Pearl Harbor. By then he was teaching at the University of Illinois, and his small Party unit simply decided that their priority should be to assist the war effort, and that left no time for "discussing politics."

David Hawkins came to Berkeley in 1936 to study philosophy. Almost immediately, he fell in with a number of Oppenheimer's students, including Phil Morrison, David Bohm and Joe Weinberg. Hawkins encountered Oppenheimer one day at a meeting of the Teachers' Union; they were discussing the plight of underpaid teaching assistants and Hawkins recalled being struck by Oppenheimer's eloquence and sympathetic demeanor: "He was very persuasive, very cogent, elegant in language and able to listen to what other people said and incorporate it in what he would say. I had the impression that he was a good politician in the sense that if several people spoke he could summarize what they said and they would discover that they had agreed with each other as a result of his summary. A great talent."

Hawkins had met Frank Oppenheimer at Stanford and, like Frank, he joined the Communist Party in late 1937. Like the Oppenheimer brothers and many other academics, he was incensed by the antilabor vigilantism sweeping California's farm factories. Even so, his political activities were very much part-time; he didn't meet a full-time Party functionary like Steve Nelson until sometime in 1940. Like many in the academy, Hawkins felt it necessary to hide his affiliation with the Party. "We were pretty secretive," he said, "we would have lost our jobs. You could be on the left, you could engage in some of these activities, but you couldn't say, 'I'm a Communist Party member.' " Hawkins didn't think about revolution either. "The centralization of a technological society," he later said, "made it very hard to think about barricades in the streets . . . we were self-consciously a left-wing component of the New Deal. We were pulling the New Deal to the left. That was our mission in life." It was an accurate description of Robert Oppenheimer's political objectives as well as his own.

By 1941, Hawkins was active in local campus politics as a junior faculty member in the philosophy department. He participated in the same study groups attended by Weinberg, Morrison and others in private homes around Berkeley. "We were all very much interested in historical materialism and the theory of history," Hawkins recalled. "I was very much impressed with Phil, and he and I became close friends."

Some of these meetings occurred in Oppenheimer's home. When asked years later whether he thought Oppenheimer had been a member of the Party, Hawkins replied, "Not that I know of. But you know, again, I would say it wouldn't have mattered very much. In a sense, it's not an important question. He was clearly identified with many of these left-wing activities."

MARTIN D. KAMEN was another of Oppie's acolytes. A chemist by training, he had written his doctoral dissertation in Chicago on a problem in nuclear physics. In just a few years, he and another chemist, Sam Ruben, would use Lawrence's cyclotron to discover the radioactive isotope carbon-14. In early 1937, he followed a girlfriend to Berkeley, where Ernest Lawrence hired him for $1,000 a year to work in the Rad Lab. "It was like Mecca," Kamen recalled of Berkeley. Oppenheimer quickly learned that Kamen was a serious musician—he played the violin with Frank Oppenheimer—and enjoyed talking about literature and music. "I think he took a shine to me," Kamen said, "because I could talk to him about things other than physics." They spent a lot of time together from 1937 until the war broke out.

Like everyone else who entered Oppenheimer's circle, Kamen admired the charismatic physicist. "Everyone sort of regarded him very affectionately as being sort of nuts," Kamen said. "He was very brilliant, but somehow superficial. He had the approach of a dilettante." At times, Kamen thought Oppie's eccentricities were calculated performances. Kamen recalled going with him to a New Year's Eve party at Estelle Caen's home. On the drive over, Oppie said he knew Estelle lived on a particular street, but he had forgotten the number of the house. He remembered only that it was a multiple of seven. "So we drove up and down the street," Kamen recalled, "and finally found Number 3528, a multiple of seven, all right. Thinking about it now, I wonder sometimes whether he wasn't pulling everybody's leg a little bit. . . . He had this overwhelming temptation just to snow you."

Kamen was no left-wing activist, and he certainly was never a communist. But he joined Oppenheimer on the Berkeley cocktail circuit, attending various fundraising affairs for the Joint Anti-Fascist Refugee Committee and Russian War Relief. Oppenheimer also involved him in an ill-fated attempt to organize a union at the Radiation Laboratory. It all began with a labor union election fight inside the Shell Development Company's plant in nearby Emeryville. Shell had a large number of white-collar workers, engineers and chemists who had Ph.D.s, many from Berkeley. A Congress of Industrial Organizations (CIO)–sponsored union, the Federation of Architects, Engineers, Chemists and Technicians (FAECT-CIO), launched an organizing drive in the plant. In response, Shell management was encouraging its employees to join a company union. At one point, a Shell chemist named David Adelson appealed to Oppenheimer to lend his prestige to the FAECT organizing drive. Adelson belonged to a professional unit of the Alameda County (California) Communist Party, and he thought Oppenheimer would be sympathetic. He was right. One evening, Oppenheimer gave a union-sponsored talk at the Berkeley home of one of his former graduate students, Herve Voge, who was then employed by Shell. More than fifteen people attended and listened respectfully as Oppenheimer talked about the likelihood of America getting into the war. "When he spoke," recalled Voge, "everyone listened."

In the autumn of 1941, Oppenheimer agreed to hold an organizing meeting in his Eagle Hill home, and, among others, he invited Martin Kamen to attend. "I was not happy about that," recalled Kamen, "but I said, 'Yes, I'll come.' " Kamen was worried about the notion of recruiting employees of the Radiation Lab—who were now essentially working for the U.S. Army and had signed security pledges—into a controversial union like FAECT. But he came to the meeting and listened to Oppenheimer's union pitch. Fifteen people were present, including Oppenheimer's psychologist friend Ernest Hilgard, Joel Hildebrand of the Berkeley chemistry department, and a young British chemical engineer, George C. Eltenton, employed by the Shell Development Company. "We all sat in a circle in Oppenheimer's living room," Kamen recalled. "Everybody said, 'Yeah, it's great, it's marvelous.' " When it came Kamen's turn to speak, he said, "Wait. Has anybody told [Ernest] Lawrence about this? We're working in the Radiation Lab and we have no independence in this matter. We have to get the permission of Lawrence on this."

Oppenheimer had not anticipated this consideration and Kamen thought he seemed shaken by his interruption. The two-hour meeting broke up without the unanimous support Oppenheimer had expected. A couple of days later, he found Kamen and said, "Gee, I don't know. Maybe I did the wrong thing." He then explained, "I went to see Lawrence, and Lawrence blew a gasket." Lawrence—whose politics had become increasingly conservative over the years—was incensed that a communist-backed union was trying to organize the people in his laboratory. When he demanded to know who was behind this, Oppenheimer insisted, "I can't tell you who they are. They'll have to come tell you themselves." Lawrence was furious, not only because he was violently opposed to his physicists and chemists joining a union, but because the incident demonstrated that his old friend was still wasting his precious time on left-wing politics. Lawrence had repeatedly scolded Oppenheimer about his "leftwandering activities" but once again Oppie argued with his usual eloquence that scientists had a responsibility to help society's "underdogs."

No wonder Lawrence was annoyed. That autumn Lawrence was trying, unsuccessfully, to bring Oppenheimer aboard the bomb project. "If he would just stop these nonsensical things," he complained to Kamen, "we could get him on the project, but it's impossible to get the Army to accept him."

OPPENHEIMER BACKED OFF from the union in the autumn of 1941, but the notion of organizing the scientists in the Rad Lab did not die. A little more than a year later, in early 1943, Rossi Lomanitz, Irving David Fox, David Bohm, Bernard Peters and Max Friedman, all Oppenheimer students, did join the union (FAECT Local 25). The usual motivations for forming a union were conspicuously absent. Lomanitz, for one, was making $150 a month at the Rad Lab—more than double his previous salary. No one had complaints about the working conditions; everyone in the lab was eager to put in as many hours as they could. "It seemed like a dramatic thing to do," recalled Lomanitz. "It was kind of a thing of youth. . . . It was a ridiculous reason for forming a union."

Friedman was persuaded by Lomanitz and Weinberg to be the organizer within the Radiation Laboratory. "It was a title, I never did anything," he recalled. But he thought in principle it was a fine idea to form a union. "Partly we were afraid of what the atom bomb might be used for. That was part of it. And part of it was that we thought the scientists shouldn't just be [working on the bomb project] without any voice in what happens to their effort."

The union rapidly drew the attention of Army intelligence officers who had the Radiation Laboratory under surveillance, and in August 1943 the War Department was warned that several people inside the Radiation Laboratory were "active communists." Joe Weinberg's name was mentioned. An attached intelligence report stated that Local 25 of FAECT was "an organization known to be dominated and controlled by Communist Party members or Communist Party sympathizers." Secretary of War Henry L. Stimson weighed in with a note to the president: "Unless this can be at once stopped, I think the situation is very alarming." Soon afterwards, the CIO was formally asked by the Roosevelt Administration to stop its organizing drive at the Berkeley lab.

By 1943, however, Oppenheimer had long since turned his back on union organizing. He did so not because he had changed his political views but because he had come to the realization that unless he followed Lawrence's advice he would not be allowed to work on a project that he believed might be necessary to defeat Nazi Germany. During their arguments in the autumn of 1941 over his union-organizing activities, Lawrence had told him that James B. Conant, the president of Harvard University, had rebuked him for having discussed fission calculations with Oppenheimer, who was not then officially in the bomb project.

In point of fact, Oppenheimer had been collaborating with Lawrence since early 1941, when Lawrence began using his cyclotron to develop an electromagnetic process for separating uranium isotope 235 (U-235), which might be necessary to create a nuclear explosion. Oppenheimer and many other scientists around the country were aware that a Uranium Committee had been authorized by President Roosevelt in October 1939 to coordinate research on fission. But by June 1941 many physicists began to fear that the German scientific community might easily be far more advanced in fission research. That autumn Lawrence, worried by the lack of progress toward a practical bomb project, wrote to Compton and insisted that Oppenheimer be included in a secret meeting scheduled for October 21, 1941, at General Electric's laboratory in Schenectady, New York. "Oppenheimer has important new ideas," Lawrence wrote. Knowing that Oppenheimer's name was widely associated with radical politics, Lawrence wrote Compton an additional note, reassuring him: "I have a great deal of confidence in Oppenheimer."

Oppie attended the meeting in Schenectady on October 21, and his calculations on the amount of U-235 necessary for an effective weapon were an essential part of the meeting's final report to Washington. A hundred kilograms, he calculated, would be sufficient to produce an explosive chain reaction. The meeting, attended by Conant, Compton, Lawrence and a handful of others, had a profound effect on Oppenheimer. Disheartened by the war news—the Nazis were at that moment advancing on Moscow—Oppenheimer was anxious to help prepare America for the coming war. He envied those of his colleagues who had gone off to work on radar; "but it was not until my first connection with the rudimentary atomic-energy enterprise," he later testified, "that I began to see any way in which I could be of direct use."

A month later, Oppenheimer wrote a note to Lawrence assuring him that his union activities were over: ". . . there will be no further difficulties at any time with [the union]. . . . I have not spoken to everyone involved, but all those to whom I have spoken agree with us; so you can forget it."

But though Oppenheimer ceased his union activities, that same autumn he couldn't refrain from taking a strong public stand on a question of civil liberties. Across the continent, a New York politician, State Senator F. R. Coudert, Jr., was using his position as co-chair of New York's Joint Legislative Committee to Investigate the Public Educational System to orchestrate a highly publicized witch-hunt against alleged subversives in New York City's public universities. By September 1941, City College alone had dismissed twenty-eight staff members, some of whom were members of the New York branch of the Teachers' Union—the same union to which Oppenheimer belonged in Berkeley. The American Committee for Democracy and Intellectual Freedom (ACDIF), to which Oppenheimer also belonged, published a statement condemning the dismissals. In response, Senator Coudert accused the ACDIF of having ties to communists, and a *New York Times* editorial lent support to Coudert's attack.

Into this political thicket waded Oppenheimer with a strongly worded protest. His letter of October 13, 1941, was by stages polite in tone, witty, ironical and then cuttingly sarcastic. Oppenheimer reminded the senator that the Bill of Rights guaranteed not merely the right to hold any belief, however radical, but the right to express that belief in speech or in writing with "anonymity." The activities, he wrote, of "teachers who were communists or communist sympathizers consisted precisely in meeting, speaking their views, and publishing them (often anonymously), in engaging, that is, in practices specifically protected by the Bill of Rights." Concluding on a note of defiance, he observed that "it took your own statement, with its sanctimonious equivocations and its red baiting, to get me to believe that the stories of mixed cajolery, intimidation and arrogance on the part of the committee of which you are chairman, are in fact true."

IN THE LATE 1930s, Robert Oppenheimer found himself in the center of things. And that's where he wanted to be. "Everything that happened," said Kamen, "you'd go to Oppenheimer, and tell him what it was and he would think about it and come up with an explanation. He was the official explainer." And then, beginning in 1941, Oppenheimer had some reason to think that he was being kept out of the loop. "All of a sudden," Kamen said, "nobody's talking to him. He's out of it. There's something big going on over there, but he doesn't know what it is. And so he was getting more and more frustrated and Lawrence is very worried because he feels that, after all, Oppenheimer can certainly figure out what's going on, so the security is nonsense to keep him out of it. Better to have him in. And I imagine that's what finally happened; they said it's easier to monitor him if he's inside the project than outside."

On Saturday evening, December 6, 1941, Oppenheimer attended a fundraiser for veterans of the Spanish Civil War. He later testified that the next day, after hearing of the surprise Japanese attack on Pearl Harbor, he decided "that I had had about enough of the Spanish cause, and that there were other and more pressing crises in the world."

CHAPTER THIRTEEN

"The Coordinator of Rapid Rupture"

> *Now I could see at firsthand the tremendous intellectual power of Oppenheimer who was the unquestioned leader of our group.... The intellectual experience was unforgettable.*

HANS BETHE

OPPENHEIMER'S STEADY AND OFTEN BRILLIANT contributions at the "uranium problem" meetings he was invited to attend were impressive. He was rapidly becoming indispensable. His politics aside, he was the perfect recruit for this scientific team. His comprehension of the issues was profound, his interpersonal skills were now finely honed, and his enthusiasm for the problems at hand was infectious. In less than a decade and a half, Oppenheimer had transformed himself through his work and his social life from an awkward scientific prodigy into a sophisticated and charismatic intellectual leader. It did not take long for those he worked with to be convinced that if the problems associated with building an atomic bomb were to be solved quickly, Oppie had to play an important role in the process.

Oppenheimer and many other physicists around the country had known as early as February 1939 that an atomic bomb was a real possibility. But arousing the government's interest in the matter would take time. A month before war broke out in Europe (September 1, 1939), Leo Szilard had persuaded Albert Einstein to sign his name to a letter (written by Szilard) addressed to President Franklin Roosevelt. The letter warned the president "that extremely powerful bombs of a new type may be constructed." He pointed out that "a single bomb of this type, carried by boat and exploded in port, might very well destroy the whole port together with some of the surrounding territory." Ominously, he suggested that the Germans might already be working on such a bomb: "I understand Germany has actually stopped the sale of uranium from the Czechoslovakian mines which she has taken over...."

Upon receipt of Einstein's letter, President Roosevelt established an ad hoc "Uranium Committee" headed by a physicist, Lyman C. Briggs. And then, for nearly two years, very little happened. But across the Atlantic Ocean, two German physicists living as refugees in Britain, Otto Frisch and Rudolph Peierls, persuaded the British wartime government that an atomic bomb project was a matter of real urgency. In the spring of 1941, a top-secret British group code-named the MAUD Committee produced a report on "The Use of Uranium for a Bomb." It suggested that a bomb made from plutonium or uranium might be small enough to carry in existing aircraft— and that such a bomb might be constructed within two years. About the same time, in June 1941, the Roosevelt Administration created an Office of Scientific Research and Development (OSRD) to marshal science for military purposes. The OSRD was chaired by Vannevar Bush, an engineer and MIT professor who was then president of the Carnegie Institution in Washington, D.C. Initially, Bush told President Roosevelt that the possibility of making an atomic bomb was "very remote." But after reading the MAUD report, Bush changed his mind. Although the matter was still "highly abstruse," he wrote Roosevelt on July 16, 1941, "one thing is certain: if such an explosion were made it would be thousands of times more powerful than existing explosives, and its use might be determining."

Suddenly, things began to happen. Bush's July memorandum persuaded Roosevelt to replace Briggs' Uranium Committee with a high-powered group that would report directly to the White House. Code-named the S-1 Committee, this group included Bush, Harvard's James Conant, War Secretary Henry Stimson, Chief of Staff George C. Marshall and Vice President Henry Wallace. These men believed that they were in a race against the Germans, a race that might easily determine the outcome of the war. Conant served as chairman of S-1, and together with Bush he now began to organize the government's enormous resources to recruit scientists around the country to work on the bomb project.

In January 1942, Robert was elated to learn that he might be put in charge of fast-neutron research in Berkeley—work that he considered critical to the project. Oppenheimer "would be a tremendous asset in every way," Lawrence had told Conant. "He combines a penetrating insight of the theoretical aspects of the whole program with solid common sense, which sometimes in certain directions seems to be lacking...." So in May, Oppenheimer was formally appointed S-1's director of fast-neutron research with the curious title Coordinator of Rapid Rupture. Almost immediately, he began to organize a highly secret summer seminar of top theoretical physicists whose job it was to outline a bare-bones design of an atomic bomb. Hans Bethe was first on his list of invitees. Now thirty-six years old, the German-born Bethe had fled Europe in 1935 and moved to Cornell University, where he became a professor of physics in 1937. So concerned was Oppenheimer to assure Bethe's attendance that he enlisted Harvard's senior theoretical physicist, John H. Van Vleck, to help recruit him. He told Van Vleck that the "essential point is to enlist Bethe's interest, to impress on him the magnitude of the job we have to do." Bethe was then working on military applications of radar, a project he viewed as far more practical than anything associated with nuclear physics. But he was eventually persuaded to spend the summer at Berkeley. So too was Edward Teller, a Hungarian-born physicist then teaching at George Washington University in Washington, D.C. Also recruited were Oppenheimer's Swiss physicist friends, Felix Bloch of Stanford University and Emil Konopinski of Indiana University. Oppenheimer also invited Robert Serber and several other former students. He called this outstanding group of physicists his "luminaries."

Soon after his appointment as Coordinator of Rapid Rupture, Oppenheimer asked Serber to be his assistant, and by early May 1942 he and Charlotte were ensconced in a room above Oppie's garage at One Eagle Hill. He considered Serber one of his closest friends. Since 1938, when Serber moved to the University of Illinois at Urbana, they had written each other almost every Sunday.* Over the next few months, Serber became Oppie's shadow, his note-taker and facilitator. "We were together almost all the time," recalled Serber. "He had two people to talk to, that was Kitty or me."

The summer seminar of 1942 met in the northwest corner of the fourth-floor attic of LeConte Hall, above Oppenheimer's office on the second floor. The two rooms had French doors opening out onto a balcony, and so for security reasons a thick wire netting was securely fastened over the entire balcony. Oppenheimer had the only key to the room. One day, Joe Weinberg was sitting in the attic office with Oppenheimer and several other physicists when there was a pause in the conversation and Oppie said, "Oh geez, look." And he pointed to the sunlight streaming through the French doors, which cast a shadow across the papers on the table and clearly outlined the wire netting. "It was as if for a moment," Weinberg said, "all of us were dappled with the shadow of the wire netting." It was eerie, Weinberg thought; they were trapped in a symbolic cage.

As the weeks went by, Oppie's "luminaries" began to appreciate his talents as their instigator and rapporteur. "As Chairman," Edward Teller later wrote, "Oppenheimer showed a refined, sure, informal touch. I don't know

---

*When Serber later ran into difficulties in retaining his security clearance, he found it prudent to destroy this correspondence.

how he had acquired this facility for handling people. Those who knew him well were really surprised." Bethe agreed: "His grasp of problems was immediate—he could often understand an entire problem after he had heard a single sentence. Incidentally, one of the difficulties that he had in dealing with people was that he expected them to have the same faculty."

They began their deliberations by studying a previous man-made explosion: the detonation in 1917 of a fully loaded ammunition ship in Halifax, Nova Scotia. In this tragic accident, an estimated 5,000 tons of TNT had decimated 2½ square miles of downtown Halifax and killed 4,000 people. They quickly estimated that any fission weapon might easily explode with a force two to three times that of the Halifax explosion.

Oppenheimer then directed his colleagues' attention to the development of the basic design of a fission device that could be small enough to be militarily deliverable. They quickly determined that a chain reaction could probably be achieved with a uranium core placed inside a metal sphere only eight inches in diameter. Other design specifications required extremely precise calculations. "We were forever inventing new tricks," Bethe recalled, "finding ways to calculate, and rejecting most of the tricks on the basis of the calculations. Now I could see at firsthand the tremendous intellectual power of Oppenheimer who was the unquestioned leader of our group. . . . The intellectual experience was unforgettable."

While Oppenheimer soon concluded that there were no major theoretical gaps to fill in designing a fast-neutron reaction device, the seminar's calculations on the actual amount of fissionable material needed were necessarily vague. They simply lacked hard experimental data. But what they did know suggested that the amount of fissionable material necessary for a weapon might easily be twice the estimated amount indicated to the president just four months earlier. The discrepancy implied that the fissionable materials could not be refined in small amounts in a mere laboratory but would have to be manufactured in a large industrial plant. The bomb would be very expensive.

At times, Robert despaired of being able to solve so many imponderables. He so feared that they were already in a losing race against the Germans that he impatiently dismissed any research efforts that seemed too time-consuming. When one scientist proposed a laborious scheme for measuring fast-neutron scattering, Oppenheimer argued that "we would do better to have a rapid and qualitative survey of scattering. . . . Landenburg's method [is] so tedious and uncertain that we may well have lost the war before he has found an answer."

In July, their deliberations were temporarily sidetracked when Edward Teller informed the group of calculations he had completed on the feasibility of a hydrogen or "super" bomb. Teller had come to Berkeley that summer convinced that a fission bomb was a sure thing. But bored with discussions of a mere fission weapon, he had entertained himself with calculations on another problem, suggested to him by Enrico Fermi over lunch a year earlier. Fermi had observed that a fission weapon might be used to ignite a quantity of deuterium—a heavy form of hydrogen—thus producing a far more powerful *fusion* explosion, a super bomb. Teller stunned Oppenheimer's group in July with calculations suggesting that a mere twenty-six pounds of liquid heavy hydrogen, ignited by a fission weapon, could produce an explosion equivalent to one million tons of TNT. Magnitudes of this scale raised the possibility, Teller suggested, that even a fission bomb might inadvertently ignite the earth's atmosphere, seventy-eight percent of which was hydrogen. "I didn't believe it from the first minute," Bethe said later. But Oppenheimer thought it advisable to hop a train East and personally report to Compton on both the super bomb and Teller's apocalyptic calculations. He tracked Compton down at his summer cottage on a lake in northern Michigan.

"I'll never forget that morning," Compton later wrote in a tone of high drama. "I drove Oppenheimer from the railroad station down to the beach looking out over the peaceful lake. There I listened to his story. . . . Was there really any chance that an atomic bomb would trigger the explosion of the nitrogen in the atmosphere or the hydrogen in the ocean? . . . Better to accept the slavery of the Nazis than to run a chance of drawing the final curtain on mankind."

In the event, Bethe soon ran further calculations that convinced both Teller and Oppenheimer of the *near*-zero impossibility of igniting the atmosphere. Oppenheimer spent the rest of the summer writing up the group's summary report. In late August 1942, Conant sat reading it and scribbled notes to himself headed "Status of the Bomb." According to Oppenheimer and his colleagues, an atomic device would explode with "150 times energy of previous calculation"—but it would need a critical mass of fissionable material six times the previous estimate. An atomic bomb was entirely feasible, but it would require the marshaling of massive technical, scientific and industrial resources.

One evening before the summer seminar ended, Oppenheimer invited the Tellers to dinner at his home on Eagle Hill. Teller vividly recalled Oppenheimer saying with absolute conviction that "only an atomic bomb could dislodge Hitler from Europe."

By September 1942, Oppenheimer's name was being floated within the bureaucracy as the obvious candidate to direct a secret weapons lab that would be dedicated to the development of an atomic bomb. Bush and

Conant certainly thought Oppenheimer was the right man for the job; everything he had done over the summer had borne out their confidence. But there was a problem: The Army was still refusing to issue him a security clearance.

Oppenheimer himself was aware that one of his problems was his many communist friends. "I'm cutting off every communist connection," he said in a phone conversation with Compton, "for if I don't, the government will find it difficult to use me. I don't want to let anything interfere with my usefulness to the nation." Nevertheless, in August 1942, Compton was informed that the War Department had "turned thumbs down on O." His security file contained numerous reports of his allegedly "questionable" and "Communistic" associations. Oppie himself had filled out a security questionnaire in early 1942, listing the many organizations he had joined, including some considered by the FBI to be communist front groups.

Despite all this, Conant and Bush began pushing the War Department to approve clearances for Oppenheimer and other scientists with left-wing backgrounds. In September, they took him with them to Bohemian Grove. In this beautiful setting, amid giant redwood trees, Oppenheimer attended his first meeting of the highly secret S-1 Committee. In early October, Bush told Secretary of War Stimson's executive assistant, Harvey Bundy, that even though Oppenheimer was "decidedly left-wing politically," he had "contributed substantially" to the project and ought to be cleared for further work.

By then, Bush and Conant had taken steps to bring the military into the project. Bush took his case to Gen. Brehon B. Somervell, the senior officer in charge of Army logistics. Somervell, already familiar with the S-1 project, informed Bush that he already had a man picked to supervise S-1 and lend it new urgency. On September 17, 1942, Somervell met with a forty-six-year-old career Army officer, Col. Leslie R. Groves, in the corridor outside a congressional hearing room. Groves had been the Army Corps of Engineers' key man on the construction of the newly completed Pentagon. Now he wanted an overseas combat assignment. But Somervell told him to forget it: He was staying in Washington.

"I don't want to stay in Washington," Groves said evenly.

"If you do the job right," Somervell replied, "it will win the war."

"Oh, that thing," said Groves, who was familiar with S-1. He was not impressed. He was already dispensing far more money on Army construction projects than S-1's expected $100 million budget. But Somervell had made up his mind and Groves had to accept his fate, which included a promotion to the rank of general.

Leslie Groves was used to getting others to do his bidding, a talent he shared with Oppenheimer. Otherwise, the two men were opposites. Nearly six feet tall and weighing over 250 pounds, Groves had muscled his way through life. Gruff and plainspoken, he had no time for the subtleties of diplomacy. "Oh yes," Oppenheimer once remarked, "Groves is a bastard, but he's a straightforward one!" By temperament and training, he was an authoritarian. Politically, he was a conservative who barely concealed his contempt for the New Deal.

The son of a Presbyterian army chaplain, Groves had studied engineering at the University of Washington in Seattle and later at the Massachusetts Institute of Technology. He graduated fourth in his class at West Point. Men serving under him grudgingly admired his ability to get things done. "General Groves is the biggest S.O.B. I have ever worked for," wrote Col. Kenneth D. Nichols, his aide throughout the war. "He is most demanding. He is most critical. He is always a driver, never a praiser. He is abrasive and sarcastic. He disregards all normal organizational channels. He is extremely intelligent. He has the guts to make timely, difficult decisions. He is the most egotistical man I know. . . . I hated his guts and so did everybody else, but we had our form of understanding."

On September 18, 1942, Groves formally took charge of the bomb project—officially designated the Manhattan Engineer District, but most often referred to as the Manhattan Project. That very day, he arranged to buy 1,200 tons of high-grade uranium ore. The next day, he ordered the acquisition of a site in Oak Ridge, Tennessee, where the uranium could be processed. Later that month, he began a tour across the country of all the laboratories engaged in experimental work on uranium isotope separation. On October 8, 1942, he met Oppenheimer at a Berkeley luncheon hosted by the president of the university. Soon afterwards, Robert Serber saw Groves walk into Oppenheimer's office, accompanied by Colonel Nichols. Groves took off his Army jacket and handed it to Nichols, saying, "Take this and find a dry cleaner and get it cleaned." Serber was astounded by this treatment of a colonel as a mere errand boy: "That was Groves' way."

Oppenheimer understood that Groves guarded the entrance to the Manhattan Project, and he therefore turned on all his charm and brilliance. It was an irresistible performance, yet Groves was most struck by Oppie's "overweening ambition," a quality he thought would make him a reliable and perhaps even pliable partner. He was also intrigued by Robert's suggestion that the new lab should be located in some isolated rural site rather than in a large city—a notion that fit nicely with Groves' concerns for security. But more than anything else, he just liked the man. "He's a genius," Groves later told a reporter. "A real genius. While Lawrence is very bright, he's not a genius, just a good hard worker. Why, Oppenheimer knows about everything. He can talk to you about anything you bring up. Well, not exactly. I guess there are a few things he doesn't know about. He doesn't know anything about sports."

Oppenheimer was the first scientist Groves had met on his tour who grasped that building an atomic bomb required finding practical solutions to a variety of cross-disciplinary problems. Oppenheimer pointed out that the various groups working on fast-neutron fission at Princeton, Chicago and Berkeley were sometimes just duplicating each other's work. These scientists needed to collaborate in a central location. This, too, appealed to the engineer in Groves, who found himself nodding in agreement when Oppenheimer pitched the notion of a central laboratory devoted to this purpose, where, as he later testified, "we could begin to come to grips with chemical, metallurgical, engineering, and ordnance problems that had so far received no consideration."

A week after their first meeting, Groves had Oppenheimer flown to Chicago, where he could join him on the Twentieth Century Limited, a luxury passenger train bound for New York. They continued their discussions aboard the train. By then, Groves already had Oppenheimer in mind as a candidate for the directorship of the proposed central laboratory. He perceived three drawbacks to Oppenheimer's selection. First, the physicist lacked a Nobel Prize, and Groves thought that fact might make it difficult for him to direct the activities of so many of his colleagues who had won that prestigious award. Second, he had no administrative experience. And third, "[his political] background included much that was not to our liking by any means."

"It was not obvious that Oppenheimer would be director," Hans Bethe noted. "He had, after all, no experience in directing a large group of people." No one to whom Groves broached the idea showed any enthusiasm for Oppenheimer's appointment. "I had no support, only opposition," Groves later wrote, "from those who were the scientific leaders of that era." For one thing, Oppenheimer was a theorist, and building an atomic bomb at this point required the talents of an experimentalist and engineer. As much as he admired Oppie, Ernest Lawrence, among others, was astonished that Groves had selected him. Another great friend and admirer, I. I. Rabi, simply thought him a most unlikely choice: "He was a very impractical fellow. He walked about with scuffed shoes and a funny hat, and more important, he didn't know anything about equipment." One Berkeley scientist remarked, "He couldn't run a hamburger stand."

When Groves proposed Oppenheimer's name to the Military Policy Committee, there was, again, considerable opposition. "After much discussion I asked each member to give me the name of a man who would be a better choice. In a few weeks it became clear that we were not going to find a better man." By the end of October, the job was Oppenheimer's. Rabi, who didn't like Groves, grudgingly observed, after the war, that the appointment "was a real stroke of genius on the part of General Groves, who was not generally considered to be a genius. . . . I was astonished."

IMMEDIATELY AFTER his appointment, Oppenheimer began to explain his new mission to a few key figures in the scientific community. On October 19, 1942, he wrote Bethe: "It is about time that I wrote to you and explained some of my wires and actions. I came east this time to get our future straight. It is turning out to be a very big order and I am not at liberty to tell all that is going on. We are going to have a laboratory for the military applications, probably in a remote spot and ready for use, I hope, within the next few months. The essential problems have to do with taking reasonable precautions about secrecy and nevertheless making the situation effective, flexible, and attractive enough so that we can get the job done."

By the autumn of 1942, it was more or less an open secret around Berkeley that Oppenheimer and his students were exploring the feasibility of a powerful new weapon associated with the atom. He had sometimes talked about his work, even to casual acquaintances. John McTernan, an attorney for the National Labor Relations Board, and a friend of Jean Tatlock's, ran into Oppenheimer one evening at a party and vividly recalled the encounter: "He talked very fast, trying to explain his work on this explosive device. I didn't understand a word he was saying. . . . And then, the next time I saw him he made it clear that he was no longer free to talk about it." Almost anyone who had friends in the physics department might have heard speculation about such work. David Bohm thought that "many people all around knew what was going on at Berkeley. . . . It didn't take much to piece it together."

A young graduate student in the psychology department, Betty Goldstein, arrived on campus fresh from Smith in the autumn of 1942 and befriended several of Oppenheimer's graduate students. The future Betty Friedan began dating David Bohm, who was writing his doctoral dissertation in physics under Oppie's supervision. Bohm—who decades later became a world-famous physicist and philosopher of science—fell in love with Betty, and introduced her to his friends, Rossi Lomanitz, Joe Weinberg and Max Friedman. They all socialized on weekends and sometimes saw each other in what Friedan characterized as "various radical study groups."

"They were all working on some mysterious project they couldn't talk about," Friedan recalled, "because it had something to do with the war." By the end of 1942, when Oppenheimer began recruiting some of his students, it was pretty clear to everyone that a very big weapon was going to be built.

"Many of us thought," said Lomanitz, " 'My God, what kind of a situation it's going to be to bring a weapon like that [into the world]; it might end up by blowing up the world.' Some of us brought this up to Oppenheimer; and basically his answer was, 'Look, what if the Nazis get it first?' "

STEVE NELSON—whose job it was to serve as the Communist Party's liaison to the Berkeley university community—had also heard the rumors about a new weapon. Some of these rumors were actually published when local newspapers quoted a congressman boasting about the weapons research being conducted at Berkeley. Rossi Lomanitz heard Nelson say in a public speech: "I've heard some of these congressmen talk about how there's some big weapon being developed here. I'll tell you, people's wars aren't being won by big weapons." And then Nelson went on to argue that this war would be won when a second front was opened up in Europe. The Soviets were fighting four-fifths of the Nazi armies and desperately needed relief. "It's going to take the American people making that sacrifice—that's how this war is going to be won."

Lomanitz had met Nelson at various public meetings of the Communist Party and, he said, "respected him a great deal." He regarded Nelson as a hero of the Spanish Republic, a veteran labor organizer and a courageous critic of racial segregation. By his own account, Lomanitz, while strongly sympathetic to the Party in many ways, never formally became a member. "I attended some Communist Party meetings," he said, "because at that time meetings were much more open. There wasn't any great distinction. . . . Who was officially a member or what it took to be officially a member, I can't tell you to this day. It just wasn't all that conspiratorial."

In his memoirs, Nelson described his relationship to Oppenheimer's students like Lomanitz, Weinberg and others: "I was responsible for working with people from the university, getting them to conduct classes and discussions. A number of Oppenheimer's graduate students in the field of physics were quite active. Our contacts were more on their terms than ours. They lived in a more rarefied intellectual and cultural atmosphere, although they were friendly and not at all pretentious."

BY THE EARLY SPRING of 1943, the FBI had installed a microphone in Nelson's home. In the early morning hours of March 30, 1943, Bureau agents overheard a man they could identify only as "Joe" talking about his work at the Radiation Lab. "Joe" had arrived at Nelson's home at 1:30 a.m. and was obviously anxious to speak with him. The two men talked in whispers. Nelson began by saying that he was looking for a "comrade who was absolutely trustworthy." "Joe" insisted that he was that man. "Joe" then explained that "certain portions of the project were to be moved to some remote section of the country, hundreds of miles away," where highly secret experimental explosions could be carried out.

The conversation then turned to discussing "the professor." Nelson commented that "he's very much worried now and we make him feel uncomfortable."

"Joe" agreed, saying that the professor (the transcript makes clear that the reference is to Oppenheimer) had "kept me off the project because he's afraid of two things. First of all, that my being there will attract more attention. . . . That's one excuse. The other is, he fears that I will propagandize . . . a strange thing for him to fear. But he's changed a bit."

Nelson: "I know that."

Joe: "You won't hardly believe the change that has taken place."

Nelson then explained that he "used to be very intimate with the guy, not only from a Party relationship, but also for a personal relationship." Oppenheimer's wife, he said, used to be the wife of his (Nelson's) best friend, who was killed in Spain. Nelson said he had always tried to keep Oppenheimer "politically up to date, but that he is not so sound as he would have people believe. . . . Well, you know, he probably impresses you fellows as brilliant in his field and I don't doubt that. But in other way[s] he had to admit a couple of times that he was off—when he tried to teach Marx, you know, and when he tried to teach Lenin to somebody else. You know what I mean. He's just not a Marxist."

Joe: "Yes, it's interesting. He rather resents the fact that I don't have deviations."

At this, Nelson and "Joe" laughed.

Nelson then observed that Oppenheimer "would like to be on the right track but I think now he's gone a little further away from whatever associations he had with us. . . . Now, he's got the one thing in the world, and that's this project and that project is going to wean him from his friends."

Clearly, Nelson was annoyed with his old friend's attitude. He knew Oppenheimer wasn't interested in money—"No," interjected Joe, "he's quite wealthy"—but he sensed that it was ambition that was now driving Oppenheimer's actions. "[He] wants to make a name for himself, unquestionably."

Joe disagreed: "No that's not necessarily it, Steve. He's internationally very well known."

Nelson: "Well, I'll tell you, to my sorrow, his wife is influencing him in the wrong direction."

Joe: "It's something we all suspected. . . ."

Having established that Oppenheimer was not going to be forthcoming with information about the project, Nelson now focused on "Joe" and tried to coax him into revealing information about the project that might be useful to the Soviets.

The FBI's twenty-seven-page transcript—based on an illegal bug—then has Joe cautiously, even anxiously, discussing details of the project that might be helpful to America's wartime ally. Speaking in a whisper, Nelson asked how soon such a weapon would become available. Joe's guess was that it would take at least one year to produce enough of this separated material for an experimental trial. "Oppie, for instance," Joe volunteered, "thinks that it might take as long as a year and a half." "So," Nelson said, "as far as the question of turning the material over. I don't know whether he'd come through but I think it's done every day." At this point in the transcript, an FBI or Army Counter-Intelligence official analyzing the transcript, writes, "Said in such a fashion as to indicate that Oppenheimer was overly cautious in withholding such information from Steve."

If the transcript implicates Joe in passing information to Nelson, it also demonstrates that Oppenheimer had become security-conscious, and Nelson concluded that he had become uncooperative and overly cautious.*

AN FBI TRANSCRIPT of Nelson's conversation with the then still unidentified "Joe" was soon delivered to Lt. Col. Boris T. Pash at G-2 Army intelligence in San Francisco. Pash, Chief of Counter-Intelligence for the Ninth Army Corps on the West Coast, was stunned. He had spent much of

---

*The few documents available from Soviet archives suggest that NKVD officials knew that Oppenheimer was working on "Enormoz"—their code-name for the Manhattan Project. They thought of him as a possibly sympathetic fellow-traveler or even a secret member of the American Communist Party—and so they were particularly frustrated that he seemed to be so unapproachable.

The notion, however, that Oppenheimer could have been recruited as a spy is simply far-fetched. There is no credible evidence linking him to espionage. Two Soviet-era intelligence documents mention Oppenheimer's name. An October 2, 1944, memorandum written in Moscow by NKVD deputy chief Vselovod Merkulov and addressed to his boss, Lavrenty Beria, seems to implicate Oppenheimer as a source of information about "the state of work on the problem of uranium and its development abroad." Merkulov claims, "In 1942 one of the leaders of scientific work on uranium in the USA, professor Oppenheimer unlisted member of the apparat of Comrade Browder informed us about the beginning of work. At the request of Comrade Kheifets . . . he provided cooperation in access to the research for several tested sources including a relative of Comrade Browder." (See Jerrold L. & Leona P. Schecter, *Sacred Secrets: How Soviet Intelligence Operations Changed American History*, Washington, DC: Brassey's, 2002.) But there is no evidence to support any of these claims and no evidence that Grigory Kheifets, the NKVD agent stationed in San Francisco, ever met Oppenheimer. On close examination, however, it quickly becomes clear that Merkulov was making this claim only to inflate the credentials of his San Francisco agent and save Kheifets'

his career hunting communists. A native-born San Franciscan, he had as a young man accompanied his father, a Russian Orthodox bishop, to Moscow during World War I. When the Bolsheviks seized power, Pash joined the counterrevolutionary White Army and fought in the 1918–20 civil war. He returned to America after marrying a Russian aristocrat. During the 1920s and '30s, while employed as a high school football coach, Pash spent his summers as a reserve U.S. Army intelligence officer. After America entered World War II, he assisted in the internment of Japanese-Americans on the West Coast and then was assigned as the Manhattan Project's chief counter-intelligence officer. Pash had little patience for bureaucracy; he considered himself a man of action. While his admirers described him as "cunning and shrewd," others regarded him as a "crazy Russian." Pash considered the Soviet Union America's mortal enemy—and not just a temporary wartime ally.

Pash quickly leaped to the conclusion that the Nelson-"Joe" transcript was not only evidence of espionage but also confirmation that his suspicions about Oppenheimer were well founded. The next day he flew to Washington, where he briefed General Groves on the transcript. Because the wiretap on Nelson was illegal, the authorities couldn't press charges against him or the mysterious "Joe." But they could use the information to trace the full extent of Nelson's activities and contacts inside the Radiation Lab. Lieutenant Colonel Pash was soon authorized to investigate whether the Berkeley lab was the target of espionage.

Pash later testified that he and his colleagues "knew" that "Joe" had furnished technical information and "timetables" pertaining to the bomb project

life. In the summer of 1944 Kheifets had been suddenly "recalled for inactivity" back to Moscow. Facing allegations that he was a double agent, Kheifets understood that his life was in danger. By floating the claim that he had developed Oppenheimer as a source of information on the American bomb project, Kheifets saved his position and his life.

Furthermore, another Soviet-era document directly contradicts the October 1944 Merkulov memo. Notes taken in the Soviet archives by a former KGB agent, Alexander Vassiliev, report that in February 1944 Merkulov received a message describing Oppenheimer. "According to data we have, [Oppenheimer] has been cultivated by the 'neighbors' (GRU-Soviet military intelligence) since June 1942. In case Oppenheimer is recruited by them, it is necessary to have him passed to us. If the recruitment is not realized, we must get from the 'neighbors' all the materials on [Oppenheimer] and begin his active cultivation through channels we have . . . brother, 'Ray' [Frank Oppenheimer], also a professor at the University of California and a member of the compatriot organization but politically closer to us than [Robert Oppenheimer]."

This document demonstrates that by early 1944 Robert Oppenheimer had not been recruited by the NKVD to serve as a source, an agent or a spy of any sort. And, of course, by 1944 Oppenheimer was living behind barbed wire in Los Alamos and it was well-nigh impossible for him to be recruited while Groves and the U.S. Army's Counter-Intelligence had him under twenty-four-hour surveillance.

to Steve Nelson. Initially, Pash's investigation focused on Lomanitz, merely because Pash had information that Lomanitz was a Communist Party member. A tail was put on Lomanitz, and one day in June 1943 he was observed standing just outside U.C. Berkeley's Sather Gate with several friends. They were posing, with their arms draped over each other's shoulders, for a photographer who routinely sold his services to students on campus. After the photo was taken and Lomanitz and his friends walked away, a government agent walked up to the photographer and bought the negative. Lomanitz' friends were quickly identified as Joe Weinberg, David Bohm and Max Friedman—all of them Oppie's students. From that moment on, these young men were marked as subversives.

Lieutenant Colonel Pash testified that his investigators "determined in the first place that these four men I mentioned were very frequently together." Without divulging "investigative techniques or operational procedures," Pash explained that "we had an unidentified man and we had this photograph. As a result of our study we determined and were sure that Joe was Joseph Weinberg." He also claimed that he had "sufficient information" to name both Weinberg and Bohm as Communist Party members.

Pash was convinced that he had stumbled upon a sophisticated ring of wily Soviet agents, and he felt that any means necessary should be used to break the suspects. In July 1943, the FBI field office in San Francisco reported that Pash wanted to kidnap Lomanitz, Weinberg, Bohm and Friedman, take them out to sea in a boat and interrogate them "after the Russian manner." The FBI noted that any information gathered in such a fashion could not be used in court, "but apparently Pash did not intend to have anyone available for prosecution after questioning." This was too much for the FBI: "Pressure was brought to bear to discourage this particular activity."

Pash nevertheless stepped up his surveillance of Steve Nelson. The FBI had placed a microphone in Nelson's office even before they bugged his home, and the conversations they overheard suggested that he had methodically gathered information on the Berkeley Radiation Lab from a number of young physicists whom he knew to be sympathetic to the Soviet war effort. As early as October 1942, the FBI bug picked up a conversation between Nelson and Lloyd Lehmann, an organizer for the Young Communist League who also worked at the Rad Lab: "Lehmann advised Nelson that a very important weapon was being developed and that he was in on the research end of this development. Nelson then asked Lehmann if Opp. [Oppenheimer] knew he was a 'YCLer' and added that Opp. was 'too jittery.' Nelson went on to state that Opp. at one time was active in the Party but was then inactive and further stated that the reason the Government left Opp. alone was because of his ability in the scientific field." After noting that Oppenheimer had worked on the "Teachers' Committee"—a reference to the Teachers' Union—and the Spanish Aid Committee, Nelson wryly commented that "he can't cover his past."

IN THE SPRING OF 1943, just as David Bohm was trying to write up his thesis research on the collisions of protons and deuterons, he was suddenly told that such work was classified. Since he lacked the necessary security clearance, his own notes on scattering calculations were seized and he was informed that he was barred from writing up his own research. He appealed to Oppenheimer, who then wrote a letter certifying that his student had nevertheless met the requirements for a thesis. On this basis, Bohm was awarded his Ph.D. by Berkeley in June 1943. Although Oppenheimer personally requested the transfer of Bohm to Los Alamos, Army security officers flatly refused to give him clearance. Instead, a disbelieving Oppenheimer was told that because Bohm still had relatives in Germany, he couldn't be cleared for special work. This was a lie; in fact Bohm was banned from Los Alamos because of his association with Weinberg. He spent the war years working in the Radiation Lab, where he studied the behavior of plasmas.

Although barred from working on the Manhattan Project, Bohm was able to continue his work as a physicist. Lomanitz and several others were not so fortunate. Shortly after Ernest Lawrence appointed him to serve as the liaison between the Rad Lab and the Manhattan Project's plant at Oak Ridge, Lomanitz received a draft notice from the Army. Both Lawrence and Oppenheimer interceded for him, but to no avail. Lomanitz spent the remainder of the war years in various stateside Army camps.

Max Friedman was called in and fired from his job in the Radiation Lab. He taught physics for a while at the University of Wyoming, and late in the war, Phil Morrison got him a job at the Met Lab in Chicago. But security officers caught up to him after six months there, and he was fired. After the war, when his name surfaced in the HUAC investigations into atomic spying, the only job he could get was at the University of Puerto Rico. Like Lomanitz, Friedman had been associated with union organizing within the Rad Lab for Local 25 of FAECT. Army intelligence officers equated such activities with subversive tendencies and they easily jumped to the conclusion that they should get rid of Lomanitz and Friedman.

As for Weinberg, he was put under close surveillance, and when no other evidence emerged to connect him to espionage, he too was drafted and sent to an Army post in Alaska.

Shortly before leaving for Los Alamos, Oppenheimer phoned Steve

Nelson and asked his friend to meet him at a local restaurant. They met for lunch in an eatery on Berkeley's main strip. "He appeared excited to the point of nervousness," Nelson later wrote. Over a big mug of coffee, Robert told him, "I just want to say good-bye to you . . . and I hope to see you when the war is over." He explained that he couldn't say where he was going, but that it had something to do with the war effort. Nelson merely asked if Kitty was going with him, and then the two friends chatted about the war news. As they parted, Robert commented that it was too bad the Spanish Loyalists hadn't managed to hold out a little longer "so that we could have buried Franco and Hitler in the same grave." Writing later, in his memoirs, Nelson noted that this was the last time he ever saw Oppenheimer, "for Robert's connection with the Party had been tenuous at best, anyway."

CHAPTER FOURTEEN

"The Chevalier Affair"

I talked to Chevalier and Chevalier talked to Oppenheimer,
and Oppenheimer said he didn't want to have anything to do
with this.

GEORGE ELTENTON

A MAN'S LIFE CAN TURN on a small event, and for Robert Oppenheimer such an incident occurred in the winter of 1942–43 in the kitchen of his Eagle Hill home. It was merely a brief conversation with a friend. But what was said, and how Oppie chose to deal with it, so shaped the remainder of his life that one is drawn to comparisons with the tragedies of classical Greece and Shakespeare. It became known as "the Chevalier affair," and over time it took on some of the qualities of *Rashomon,* the 1951 film by Akira Kurosawa in which descriptions of an event vary according to the perspective of each participant.

Knowing that they would soon be leaving Berkeley, the Oppenheimers invited the Chevaliers to their home for a quiet dinner. They counted Haakon and Barbara among their closest friends and wanted to share with them a special farewell. When the Chevaliers arrived, Oppie went into the kitchen to prepare a tray of martinis. Hoke followed, and relayed a recent conversation he had had with their mutual acquaintance George C. Eltenton, a British-born physicist educated at Cambridge employed by the Shell Oil Company.

Exactly what each man said is lost to history; neither made contemporaneous notes of the conversation. At the time, neither appears to have considered it a momentous exchange, even though the topic was an outrageous proposal. Eltenton, Chevalier reported, had solicited him to ask his friend Oppenheimer to pass information about his scientific work to a diplomat Eltenton knew in the Soviet consulate in San Francisco.

By all accounts—Chevalier's, Oppenheimer's and Eltenton's—Oppie angrily told Hoke that he was talking about "treason" and that he should have nothing to do with Eltenton's scheme. He was unmoved by Eltenton's argument, prevalent in Berkeley's left-wing circles, that America's Soviet allies were fighting for survival while reactionaries in Washington were sabotaging the assistance that the Soviets were entitled to receive.

Chevalier always insisted that he was merely alerting Oppie to Eltenton's proposal rather than acting as his conduit. In either case, that is the interpretation that Oppenheimer put on what his friend told him. Viewing it thus—as a dead end that he had buried—allowed him to brush it aside for the time being as yet another manifestation of Hoke's overwrought concern for Soviet survival. Should he have informed the authorities immediately? His life would have been very different if he had. But, at the time, he could not have done so without implicating his best friend, whom he believed to be, at worst, an overenthusiastic idealist.

The martinis mixed, the conversation over, the two friends rejoined their wives.

IN HIS MEMOIR, *The Story of a Friendship,* Chevalier recounts that he and Oppenheimer talked only briefly about Eltenton's proposition. He insisted that he was not soliciting information from Oppie, but was merely passing on to his friend the fact that Eltenton had proposed a means of sharing information with Soviet scientists. He thought it important that Oppie know of it. "He was visibly disturbed," wrote Chevalier, "we exchanged a remark or two, and that was all." Then they returned to the living room with their martinis to join their wives. Chevalier remembered that Kitty had just bought an early-nineteenth-century French edition of a book on mycology with hand-drawn, painted illustrations of orchids—her favorite flower. Sipping their drinks, the two couples perused the beautiful book before sitting down to dinner. Thereafter, Chevalier "dismissed the whole thing from my mind."

In 1954, at his security hearing, Oppenheimer testified that Chevalier had followed him into the kitchen and said something like, "I saw George Eltenton recently." Chevalier then added that Eltenton had a "means of getting technical information to Soviet scientists." Oppenheimer continued: "I thought I said [to Chevalier], 'But that is treason,' but I am not sure. I said anyway something. 'This is a terrible thing to do.' Chevalier said or expressed complete agreement. That was the end of it. It was a brief conversation."

After Robert's death, Kitty reported yet another version of the story. While in London visiting Verna Hobson (Oppie's former secretary and Kitty's friend), she said that "the minute Chevalier came into the house she could see that something was up." She made a point of not leaving the men alone together, and finally, when Chevalier realized that he could not get Robert off by himself, he related his conversation with Eltenton in her presence. Kitty said it was she who then blurted out, "But that would be treason!" According to this version, Oppenheimer was so determined to keep Kitty out of it that he took her words in his mouth and always claimed that he and Chevalier were alone in the kitchen when they discussed Eltenton. On the other hand, Chevalier always insisted that Kitty never entered the kitchen while he and Robert discussed Eltenton's proposition, and Barbara Chevalier's recollection of the incident does not include Kitty.

Decades later, Barbara, by then an embittered ex-wife, wrote a "diary" that adds a somewhat different perspective. "I was not, of course, in the kitchen when Haakon spoke to Oppie, but I knew what he was going to tell him. I also know that Haakon was one hundred percent in favor of finding out what Oppie was doing and reporting it back to Eltenton. I believe Haakon also believed that Oppie would be in favor of cooperating with the Russians. I know because we had a big fight over it beforehand."

At the time Barbara wrote this—some forty years later—she had a low opinion of her ex-husband. She thought him foolish, "a man of limited horizons, fixed ideas and immutable habits." Soon after Eltenton's approach, Haakon had told her, "The Russians want to know." As she remembered things, she had tried to persuade her husband not to pursue the matter with Oppenheimer. "The absurd ridiculousness of the situation never occurred to him," she wrote in her unpublished memoir in 1983. "This innocent teacher of modern French literature to be the conduit to Russians of what Oppie was doing."

OPPENHEIMER KNEW ELTENTON only because the two of them had attended union organizing meetings on behalf of the Federation of Architects, Engineers, Chemists and Technicians (FAECT). Eltenton had attended one of these union meetings in Oppenheimer's home. All told, he had seen Eltenton on four or five occasions.

Eltenton, a thin, Nordic-featured man, and his wife, Dorothea (Dolly), were English. Although Dolly was a first cousin of the British aristocrat Sir Hartley Shawcross, the Eltentons were decidedly left-wing in their politics. In the mid-1930s, they had observed the Soviet experiment firsthand, in Leningrad, where George had been employed by a British firm.

Chevalier had first met Dolly Eltenton in 1938, when she walked into the office of the League of American Writers in San Francisco and volunteered her secretarial services. Dolly, whose politics were, if anything, more radical than her husband's, worked as a secretary for the pro-Soviet American Russian Institute in San Francisco. Moving to Berkeley, the couple naturally gravitated into its left-wing social circuit. Chevalier had seen them at many of the same fundraising parties attended by Oppenheimer.

So when Eltenton phoned him one day to say that he wanted to have a chat, Chevalier drove over to his Berkeley home at 986 Cragmont Avenue a day or two later. Eltenton talked earnestly about the war and its still uncertain outcome. The Soviets, he pointed out, were bearing the brunt of the Nazi onslaught—four-fifths of the Wehrmacht was fighting on the Eastern Front—and much might depend on how effectively the Americans aided their Russian allies with arms and new technology. It was very important that there be close collaboration between Soviet and American scientists.

Eltenton had been approached by Peter Ivanov, he said, whom he believed to be a secretary in the Soviet Consulate General in San Francisco. (Actually, Ivanov was a Soviet intelligence officer.) Ivanov had remarked that "in many ways the Soviet Government did not feel it was getting the scientific and technical cooperation which it felt it deserved." He had then asked Eltenton whether he knew anything about what was going on "up on the Hill," meaning the Berkeley laboratory.

In 1946, the FBI interrogated Eltenton about the Chevalier incident, and he reconstructed his conversation with Ivanov as follows: "I told him [Ivanov] that I, personally, knew very little of what was going on, whereupon he asked me whether I knew Professor E. O. Lawrence, Dr. J. R. Oppenheimer or a third party whose name I do not recall." (Eltenton later thought the third scientist named by Ivanov was Luis Alvarez.) Eltenton replied that he knew only Oppenheimer, but not well enough to discuss the issue. Ivanov had pressed him, asking if he knew anyone else who could approach Oppenheimer. "On thinking the matter over I said that the only mutual acquaintance whom I could think of was Haakon Chevalier. He asked me whether I would be willing to discuss the matter with [Chevalier]. After assuring myself that Mr. Ivanov was genuinely convinced that there were no authorized channels through which such information could be obtained and having convinced myself that the situation was of such a critical nature that I would be in my own mind free in conscience to approach Haakon Chevalier I agreed to contact the latter."

According to Eltenton, he and Chevalier agreed "with considerable reluctance" that Oppenheimer should be approached. Eltenton assured Chevalier that if Oppenheimer had any useful information, Ivanov could get it "safely transmitted." From Eltenton's account, the two men clearly understood what they were contemplating. "The question of remuneration was raised by Mr. Ivanov, but no sum was mentioned since I did not wish to accept payment for what I was doing."

A few days later—Eltenton told the FBI in 1946—Chevalier informed him that he had seen Oppenheimer, but that "there was no chance whatsoever of obtaining any data and Dr. Oppenheimer did not approve." Ivanov later came by Eltenton's house and was likewise told that Oppenheimer would not cooperate. That was the end of it, although somewhat later Ivanov asked Eltenton if he had any information about a new drug called penicillin. Eltenton had no idea what this was—though he said he later called Ivanov's attention to an article about it in *Nature* magazine.

The accuracy of Eltenton's account of the affair is confirmed by another FBI interview. At the same time that FBI agents were interrogating Eltenton, another team picked up Chevalier and asked him similar questions. As their interviews proceeded, the two teams of agents coordinated their questions through phone calls, checking each man's recollections against the other's and probing any inconsistencies. In the end, there were only minor differences in their statements. Chevalier said that to the best of his recollection he had not mentioned Eltenton's name to Oppenheimer (although in his memoirs he recalled that he had). And he did not mention to his interrogators that Eltenton had made reference to Lawrence and Alvarez: "I wish to state that to my present knowledge and recollection I approached no one except Oppenheimer to request information concerning the work of the Radiation Laboratory. I may have mentioned the desirability of obtaining this information with any number of people in passing. I am certain that I never made another specific proposal in this connection." Oppenheimer, he said, had "dismissed my approach without discussion."

In other words, the two men confessed that they had talked about funneling scientific information to the Soviets, but each confirmed that Oppenheimer had rejected the idea out of hand.

OVER THE YEARS, historians have surmised that Eltenton was a Soviet agent who had worked as a recruiter throughout the war. In 1947, when the details of his interrogation began to leak from FBI sources, he fled to England, and for the rest of his life he refused to talk about the incident. *Was* Eltenton a Soviet spy? Certainly, no one can dispute that he proposed funneling scientific information about a war project to the Soviets. But an investigation of his behavior in 1942–43 suggests that he was more likely a misguided idealist than a serious Soviet agent.

For nine years—1938 to 1947—Eltenton car-pooled to work at Shell every day with a neighbor, Herve Voge. Voge, a physical chemist who had once taken a class from Oppenheimer, was also employed at Shell's facility in Emeryville, an eight-mile drive from Berkeley. Four other men car-pooled with them in 1943: Hugh Harvey, an Englishman whose politics were pretty middle-of-the-road; Lee Thurston Carlton, whose political views were leftist; and Harold Luck and Daniel Luten. They called their car pool the "red-herring ride club" because Luten was always bringing up red herrings in their lively discussions. Voge vividly recalled these "ride club" conversations: "I remember this very well, everybody knew that there were important things going on at the radiation laboratory in Berkeley; it was obvious. People were coming there and there was a lot of hush hush talk. . . ."

One day as they drove to work, Eltenton got exercised about the war news and said, "I would like Russia to win this war, rather than the Nazis, and I would like to do anything I can to help them." Voge claims that Eltenton then said, "I'm going to try to talk to Chevalier or Oppenheimer and tell them that I would be very happy to forward any information that they feel is useful to the Russians."

Voge thought Eltenton's political views, which he wore on his sleeve, were at best simpleminded and immature; at worst, he was "a dupe of the Russian consulate." Eltenton openly talked about his friends in the Soviet Consulate in San Francisco, and boasted that he could get this information sent to Russia through his contacts at the consulate. (Indeed, FBI agents observed him meeting on several occasions in 1942 with Ivanov.) Eltenton brought up the subject more than once, Voge recalled: "He would continu-ally say, 'You know, we're fighting on the same side as the Russians, why don't we help them?' " When some of his car-pool buddies questioned whether this "isn't the kind of thing that should go through official chan-nels?," Eltenton responded, "Well, I'll do what I can."

A few weeks later, however, he told Voge and the others, "I talked to Chevalier and Chevalier talked to Oppenheimer, and Oppenheimer said he didn't want to have anything to do with this." Eltenton seemed disap-pointed, but Voge was pretty sure that that was the end of his little scheme.

This story, which Voge related to Martin Sherwin in 1983, is buttressed by what he told the FBI in the late 1940s. After the war, Voge almost lost his job because of his association with Eltenton; when the FBI said they could clear his name if he agreed to act as an informant, Voge refused. But the FBI did persuade him to sign a statement about Eltenton, which read in part: "George and Dolly Eltenton are admittedly suspicious characters. They had lived in the Soviet Union and were openly sympathetic to the regime. George made apparently open efforts to aid the Russians during World War II." Describing his conversations with Eltenton in the "red-herring ride club," Voge wrote, "We were never able to convince George of the evils of communism and he never converted any of us to his views."

Years later, when Eltenton's name surfaced in the 1954 Oppenheimer hearings, Voge thought the government had it all wrong about Eltenton: "If he'd really been a genuine spy, he wouldn't have talked that openly at all. He would have pretended to be a much different type of person."

PART THREE

CHAPTER FIFTEEN

"He'd Become Very Patriotic"

When I was with him, I was a larger person. . . . I became
very much of an Oppenheimer person and just idolized him.

ROBERT WILSON

ROBERT WAS BEGINNING A NEW LIFE. As the director of a
weapons laboratory that would integrate the diverse efforts of the
far-flung sites of the Manhattan Project and mold them quickly into
a usable atomic weapon, he would have to conjure up skills he did not yet
have, deal with problems he had never imagined, develop work habits
entirely at odds with his previous lifestyle, and adjust to attitudes and
modes of behavior (such as security considerations) that were emotionally
awkward and alien to his experience. It is not too much of an exaggeration
to suggest that in order to succeed, at age thirty-nine, Robert Oppenheimer
would have to remake a significant part of his personality if not his intellect,
and he was going to have to do all this in short order. Every aspect of his
new job was on a fast-track schedule. Very few things—including Oppenheimer's transformation—could meet that impossible schedule; yet it is a
measure of his commitment and willpower that he came very close.

Robert had often mused about combining his passion for physics with
his fierce attraction to the desert high country of New Mexico. Now he had
his chance. On November 16, 1942, he and Edwin McMillan, another
Berkeley physicist, accompanied an Army officer, Maj. John H. Dudley, to
Jemez Springs, a deep canyon forty miles northwest of Santa Fe. After
inspecting dozens of potential sites across the American Southwest, Dudley
had finally settled on Jemez Springs as a suitable home for the proposed
new weapons laboratory. Oppenheimer remembered it from his horseback
trips as a "lovely spot and in every way satisfactory."

But when the three men arrived at Jemez Springs, he and McMillan
began arguing with Dudley that the snake of land at the bottom of the
canyon was too narrow and confined for the town they envisioned building.

Oppenheimer complained that it had no view of the magnificent mountain scenery, and that the site's steep canyons would make it nearly impossible to fence in. "We were arguing about this when General Groves showed up," recalled McMillan. Groves took one look at the site and said, "This will never do." When he turned to Oppenheimer and asked if there was something else around that had prospects, "Oppie proposed Los Alamos as though it was a brand new idea."

"If you go on up the canyon," Oppenheimer told him, "you come out on top of the mesa and there's a boys' school there which might be a usable site." Reluctantly, the men piled back into their cars and drove northwest about thirty miles across a lava mesa called the Pajarito (Little Bird) Plateau. It was already late afternoon when they pulled up to the Los Alamos Ranch School. Through the haze of drizzly snowfall, Oppenheimer, Groves and McMillan saw a group of schoolboys out on a playing field running around in shorts. The school's 800-acre grounds included the "Big House," its main building; Fuller Lodge, a beautiful manor house built in 1928 from 800 huge ponderosa logs; a rustic dormitory; and a few other, smaller buildings. Behind the lodge there was a pond that the boys used for ice skating in the winter and canoeing during the summer. The school stood at an elevation of 7,200 feet, just about at timberline. To the west, the snow-capped Jemez Mountains rose to 11,000 feet. From the spacious porch of Fuller Lodge, one could look forty miles east across the Rio Grande Valley to Oppenheimer's beloved Sangre de Cristo mountain range, rising to a height of 13,000 feet. By one account, as Groves surveyed the scene he suddenly announced, "This is the place."

Within two days, the Army initiated the paperwork to buy the school, and four days later, after a quick trip to Washington, D.C., Oppenheimer returned with McMillan and Ernest Lawrence to inspect what had been designated "Site Y." Wearing cowboy boots, Oppenheimer took Lawrence on a tour of the school buildings. For security purposes, they had introduced themselves under assumed names. But a Los Alamos student, Sterling Colgate, recognized the scientists. "Suddenly we knew the war had arrived here," Colgate recalled. "These two characters showed up, Mr. Smith and Mr. Jones, one wearing a porkpie hat and the other a normal hat, and these two guys went around as if they owned the place." Colgate, a high school senior, had studied physics and he had seen photographs of Oppenheimer and Lawrence in a textbook. Soon afterwards, an armada of bulldozers and construction crews invaded the school grounds. Oppenheimer, of course, knew Los Alamos well. Perro Caliente was a forty-mile horseback ride across the plateau. He and his brother had explored the Jemez Mountains on horseback over many summers.

Oppenheimer got what he wanted—a spectacular view of the Sangre de Cristo Mountains—and General Groves got a site so isolated there was only a winding gravel road and one phone line into the place. Over the next three months, construction crews built cheap barracks with shingled or tin roofs. Similar buildings were constructed to serve as crude chemistry and physics laboratories. Everything was painted Army green.

Oppenheimer seemed unaware of the utter chaos that had descended on Los Alamos—although years later, he confessed, "I am responsible for ruining a beautiful place." Focused on recruiting the scientists he needed for the project, he had no time for the administrative tasks associated with building a small town. John Manley, an experimental physicist whom Oppie had tapped as one of his assistants, had serious qualms about the site. Manley had just come from Chicago, where on December 2, 1942, the Italian émigré physicist Enrico Fermi had led a team that conducted the world's first controlled nuclear chain reaction. Chicago was a big city, home to an eminent university, world-class libraries and a large pool of experienced machinists, glass-blowers, engineers and other technicians. Los Alamos had nothing. "What we were trying to do," wrote Manley, "was build a new laboratory in the wilds of New Mexico with no initial equipment except the library of Horatio Alger books or whatever it was that those boys in the Ranch School read, and the pack equipment that they used to go horseback riding, none of which helped us very much in getting neutron-producing accelerators." Manley thought that if Oppenheimer had been an experimental physicist, he would have understood that "experimental physics is really 90 percent plumbing," and he never would have agreed to having a laboratory built in such a setting.

The logistics were horrendously complicated. Oppenheimer and the initial group of scientists planned to arrive at Los Alamos by mid-March 1943. By then, Robert assured Hans Bethe, there would be a viable community run by a city engineer. There would be bachelor quarters and homes for families with one, two and three bedrooms. These furnished quarters would all come with electricity—but for security reasons there would be no phones. The kitchens would be equipped with wood-fired stoves and hot-water heaters. There would be fireplaces and a refrigerator. Servants would be available on occasion for any heavy housework. There would be a school for young children, a library, a laundry, a hospital and garbage collection. An Army post exchange would serve as the community's grocery store and mail-order house. A recreation officer would arrange for regular movies and hiking trips in the nearby mountains. And Oppie promised there would be a cantina for beer, Cokes and light lunches, a regular mess hall for unmarried people and a "fancy" café where married couples could eat out in the evening.

. . .

FOR THE LABORATORIES, they ordered the shipment of two Van de Graaff generators from Michigan, a cyclotron from Harvard and a Cockcroft-Walton machine from the University of Illinois. All were essential. The Van de Graaff generators would be used to run basic physics measurements. The Cockcroft-Walton machine, the first particle accelerator, was necessary for experiments in which various elements could be artificially transmuted into other elements.

The construction of Los Alamos, the recruitment of scientists and the assembling of all the equipment necessary for the world's first nuclear weapons laboratory required a meticulous and patient administrator. In early 1943, Oppenheimer was neither. He had never supervised anything larger than his graduate seminars. In 1938, he had been responsible for fifteen graduate students; now he was directing the work of hundreds, soon to be thousands, of scientists and technicians. Nor did his peers believe he was temperamentally suited for the job. "He was something of an eccentric—almost a professional eccentric when I knew him before 1940," recalled Robert Wilson, a young experimental physicist who was then studying under Ernest Lawrence. "He just wasn't the kind of person that you would think would be an administrator." As late as December 1942, James Conant wrote Groves that he and Vannevar Bush were "wondering whether we have found the right man to be the leader."

Even John Manley had serious misgivings about serving as Oppie's deputy. "I was somewhat frightened of his evident erudition," Manley recalled, "and his lack of interest in mundane affairs." Manley was particularly worried about the laboratory's organization. "I bugged Oppie for I don't know how many months about an organization chart—who was going to be responsible for this and who was going to be responsible for that." Oppenheimer ignored his pleas until finally, one day in March 1943, Manley climbed to the top floor of LeConte Hall and pushed open the door of Oppenheimer's office. When Oppenheimer glanced up to see him standing there, he knew exactly what Manley wanted. Grabbing a piece of paper, he threw it down on his desk and said, "Here's your damned organization chart." Oppenheimer envisioned four broad divisions within the laboratory: experimental physics, theoretical physics, chemistry and metallurgy, and, finally, ordnance. Group leaders within each of these divisions would report to the division chiefs, and the division chiefs would report to Oppenheimer. It was a beginning.

In early 1943, Oppenheimer sent the twenty-eight-year-old Robert Wilson to Harvard to arrange for the safe shipment of Harvard's cyclotron to

Los Alamos. On March 4, Wilson arrived at Los Alamos to inspect the building that would house the cyclotron. He found utter chaos; there seemed to be no schedule, no planning and no line of responsibility. Wilson complained about the situation to Manley, and the two men agreed they should confront Oppenheimer. Their meeting in Berkeley was a disaster: Oppenheimer became angry and swore at them. Stunned, Wilson and Manley left wondering if he was up to the challenge.

A Quaker by ancestry, Wilson was a pacifist when the European war erupted: "So it was quite a change for me to find in fact that I would be working on this horrible project." But, like everyone else he knew at Los Alamos, Wilson feared above all the prospect of the Nazis' winning the war with an atomic weapon. And while privately he still hoped that they might someday prove that an atomic bomb was not possible, he was eager to build it if it could be built. Hardworking and serious-minded by temperament, Wilson initially found himself annoyed by Oppenheimer's arrogant demeanor. "I sort of disliked him," he later said. "He was such a smart-aleck and didn't suffer fools gladly. And maybe I was one of the fools he hadn't suffered."

In the end, however disconnected from his responsibilities Oppenheimer may have seemed before he moved to Los Alamos, he quickly demonstrated his capacity for change. Wilson was surprised after several months at Los Alamos to see his boss metamorphose into a charismatic and efficient administrator. The once eccentric theoretical physicist, a long-haired, left-wing intellectual, was now becoming a first-rate, highly organized leader. "He had style and he had class," Wilson said. "He was a very clever man. And whatever we felt about his deficiencies, in a few months he had corrected those deficiencies, and obviously knew a lot more than we did about administrative procedures. Whatever our qualms were, why, they were soon allayed." By the summer of 1943, Wilson noticed that "when I was with him, I was a larger person. . . . I became very much of an Oppenheimer person and just idolized him. . . . I changed around completely."

EVEN SO, through these early planning stages, Oppenheimer was often incredibly naïve. On the organization chart he gave Manley, he had listed himself as both director of the lab and chief of the theoretical division. But it soon became clear to his colleagues, and finally to Robert, that he hadn't the time to do both jobs, so he appointed Hans Bethe to head the theoretical division. He also told General Groves that he thought he would need only a handful of scientists. Major Dudley claims that when they were first scouting the site, Oppenheimer remarked that he thought six scientists, joined by a number of engineers and technicians, could do the job. While this is probably an exaggeration, the point is clear: Oppenheimer at first greatly underestimated the magnitude of the job. The initial construction contract budgeted $300,000—but within a year $7.5 million had been spent.

When Los Alamos opened in March 1943, a hundred scientists, engineers and support staff converged on the new community; within six months there were a thousand and a year later there were 3,500 people living on the mesa. By the summer of 1945, Oppenheimer's wilderness outpost had grown into a small town of at least 4,000 civilians and 2,000 men in uniform. They lived in 300 apartment buildings, fifty-two dormitories and some 200 trailers. The "Technical Area" alone enclosed thirty-seven buildings, including a plutonium purification plant, a foundry, a library, an auditorium and dozens of laboratories, warehouses, and offices.

To the dismay of nearly all his colleagues, Oppenheimer had originally accepted General Groves' suggestion that all the scientists in the new lab should become commissioned Army officers. In mid-January 1943, Oppenheimer visited the Presidio, an Army base in San Francisco, to arrange for his commission as a lieutenant colonel. He actually took the Army physical—and failed it. Army doctors reported that at 128 pounds Oppenheimer was 11 pounds under the minimum weight and 27 pounds under the ideal weight for a man his age and height. They noted he had a "chronic cough" dating back to 1927, when X rays of his chest had confirmed a case of tuberculosis. He also reported a history of "lumbo-sacral strain": Every ten days or so, he said, he felt moderate pains shooting down his left leg. For all these reasons, the Army doctors deemed him "permanently incapacitated for active service." But because Groves had already instructed the doctors that Oppenheimer had to be cleared for duty, he was asked to sign a note acknowledging the existence of "the above physical defects," and requesting that he nevertheless be placed on extended active duty.

After the physical, Oppenheimer had an officer's uniform tailored for him. His motivations were complex. Perhaps donning a colonel's uniform was a visible sign of acceptance important to a man who was self-conscious about his Jewish heritage. But wearing a uniform was also the patriotic thing to do in 1942. Across the country, men and women were donning military uniforms in a symbolic, primordial ritual of defending the tribe, the country—and the uniform was a visible statement of this commitment. There was a lot of apple pie in Robert's psyche. "Oppie would get a faraway look in his eyes," recalled Robert Wilson, "and tell me that this war was different from any war ever fought before; it was a war about the principles of freedom. . . . He was convinced that the war effort was a mass effort to overthrow the Nazis and upset Fascism and he talked of a people's army and a people's war. . . . The language had changed so little. It's the same kind of [political] language, except that now it has a patriotic flavor, whereas before it had just a radical flavor."

Soon after Oppenheimer began making his rounds to recruit physicists to Los Alamos, however, he discovered that his peers flatly opposed the notion of having to work under military discipline. By February 1943, his old friend Isidor Rabi and several other physicists had persuaded him that the "laboratory must demilitarize." Rabi was one of the few among Oppie's friends who could tell him when he was being foolish. "He thought it would be fine to go in uniform because we were at war; it would bring us closer to the American people, that sort of crap. I know he wanted seriously to win the war, but we couldn't make a bomb that way." In addition to being "very wise, he was very foolish."

By the end of that month, Groves agreed to a compromise: During the lab's experimental work, the scientists would remain civilians, but when the time came to test the weapon, everyone would don a uniform. Los Alamos would be fenced and designated an Army post—but within the "Technical Area" of the lab itself, the scientists would report to Oppenheimer as "Scientific Director." The Army would control access to the community, but it would not control the exchange of information among the scientists; that was Oppenheimer's responsibility. Hans Bethe congratulated Oppie on his negotiations with the Army, writing him that "I think that you have now earned a degree in High Diplomacy."

Rabi played a critical role in this and other organizational issues. "Without Rabi," Bethe later said, "it would have been a mess because Oppie did not want to have an organization. Rabi and [Lee] Dubridge [then head of MIT's Radiation Laboratory] came to Oppie and said, 'You have to have an organization. The laboratory has to be organized in divisions and the divisions into groups. Otherwise, nothing will ever come of it.' And Oppie, well, that was all new to him. Rabi made Oppie more practical. He talked Oppie out of putting on a uniform."

One of Oppenheimer's great disappointments was his failure to persuade Isidor Rabi to relocate to Los Alamos. He so wanted Rabi aboard that he offered him the associate directorship of the laboratory—but to no avail. Rabi had fundamental doubts about the whole notion of building a bomb. "I was strongly opposed to bombing ever since 1931, when I saw those pictures of the Japanese bombing that suburb of Shanghai. You drop a bomb and it falls on the just and the unjust. There is no escape from it. The prudent man can't escape, [nor] the honest man. . . . During the war with Germany, we [in the Rad Lab] certainly helped to develop devices for bombing . . . but this was a real enemy and a serious matter. But atomic bombing just carried the principle one step further and I didn't like it then and I don't now. I think it's terrible." To Rabi's way of thinking, this war would be won with a far less exotic technology—radar. "I thought it over," Rabi recalled, "and turned him down. I said, 'I'm very serious about this war. We could lose it with insufficient radar.' "

Rabi also gave a less practical but more profound reason for not joining: He did not, he told Oppenheimer, wish to make "the culmination of three centuries of physics" a weapon of mass destruction. This was an extraordinary statement, one that Rabi knew might well resonate with a man of Oppenheimer's philosophical bent. But if Rabi was already thinking about the moral consequences of an atomic bomb, Oppenheimer, in the midst of this war, for once had no patience for the metaphysical. He now brushed aside his friend's objection. "I think if I believed with you that this project was 'the culmination of three centuries of physics,' " he wrote Rabi, "I should take a different stand. To me it is primarily the development in time of war of a military weapon of some consequence. I do not think that the Nazis allow us the option of [not] carrying out that development." Only one thing mattered now to Oppenheimer: building the weapon before the Nazis did.

If Rabi refused to move to Los Alamos, Oppenheimer nevertheless prevailed upon him to come to the first colloquium, and thereafter to serve as one of the project's rare visiting consultants. Rabi became, as Hans Bethe put it, "the fatherly adviser to Oppie." "I never went on the payroll at Los Alamos," Rabi said. "I refused to. I wanted to have my lines of communication clear. I was not a member of any of their important committees, or anything of the sort, but just Oppenheimer's adviser."

Moreover, Rabi was instrumental in persuading both Hans Bethe and many others to move to Los Alamos. He also urged Oppenheimer to appoint Bethe as chief of the theoretical division, which he called "the nerve center of the project." Oppenheimer trusted Rabi's judgments in all these matters and acted quickly upon his suggestions.

When Rabi warned him that "morale is sinking" among the group of physicists working in Princeton, Oppenheimer decided to import the entire Princeton team of twenty scientists to Los Alamos. This turned out to be a particularly serendipitous decision, as the Princeton group included not only Robert Wilson but a brilliant and cheerfully mischievous twenty-four-year-old physicist named Richard Feynman. Oppenheimer had immediately recognized the genius in Feynman and knew he wanted him at Los Alamos. However, Feynman's wife, Arline, was battling tuberculosis and Feynman made it clear he could not move to Los Alamos without her. Feynman thought that had ended the matter, but one day in the winter of early 1943 he received a long-distance phone call from Chicago. It was Oppenheimer, calling to say that he had located a tuberculosis sanatorium for Arline in Albuquerque. Feynman, he assured him, could work in Los Alamos and visit Arline on the weekends. Feynman was touched, and persuaded.

Oppenheimer was relentless in his pursuit of men to work on the mesa— "The Hill," as it was soon nicknamed. He had begun in the autumn of 1942, even before Los Alamos had been selected as "Site Y." "We should start now," he wrote Manley, "on a policy of absolutely unscrupulous recruiting of anyone we can lay hands on." Among his early targets was Robert Bacher, an MIT administrator and experimental physicist. Only after months of persistent lobbying did Bacher finally agree to move to Los Alamos in June 1943 and direct the project's division of experimental physics. Oppenheimer wrote Bacher earlier that spring that his qualifications made him "very nearly unique, and that is why I have pursued you with such diligence for so many months." He believed strongly, Oppenheimer wrote him, "in your stability and judgment, qualities on which this stormy enterprise puts a very high premium." Bacher came—but warned that he would resign if he were ever asked to put on an Army uniform.

ON MARCH 16, 1943, Oppie and Kitty boarded a train bound for Santa Fe, a sleepy town of 20,000 people. They checked into La Fonda, the best hotel in town, where Oppenheimer spent a few days recruiting people to run a Santa Fe liaison office for the laboratory. One day, Dorothy Scarritt McKibbin, a forty-five-year-old Smith College graduate, was standing in the lobby of La Fonda, waiting to be interviewed for a job she had been told nothing about. "I saw a man walking on the balls of his feet and garbed in a trench coat and porkpie hat," McKibbin said. Oppenheimer introduced himself as "Mr. Bradley" and asked about her background. Widowed twelve years earlier, McKibbin had first moved to New Mexico to cure a mild case of tuberculosis and, like Oppenheimer, had fallen in love with the stark beauty of the place. By 1943, McKibbin knew everyone there was to know in Santa Fe society, including such artists and writers as the poet Peggy Pond Church, the watercolorist Cady Wells and the architect John Gaw Meem. She was also a friend of the dancer and choreographer Martha Graham, who spent her summers in New Mexico during the late 1930s. Oppenheimer could see that this sophisticated, well-connected and self-confident woman would not be easily intimidated, and when he realized that McKibbin knew Santa Fe and its environs better than he did, he hired her to run a discreet office at 109 East Palace Avenue in the downtown area.

McKibbin was immediately smitten by Oppenheimer's easy grace and charming manners. "I knew that anything he was connected with would be alive," she recalled, "and I made my decision. I thought to be associated with that person, whoever he was, would be simply great! I never met a person with a magnetism that hit you so fast and so completely as his did. I didn't know what he did. I thought maybe if he were digging trenches to put in a new road, I would love to do that. . . . I just wanted to be allied and have something to do with a person of such vitality and radiant force. That was for me."

McKibbin may have had no idea what Oppenheimer was doing, but she nevertheless soon became the "gatekeeper to Los Alamos." From her unmarked office she greeted hundreds of scientists and their families bound for The Hill. Some days she fielded a hundred phone calls and issued dozens of passes. She would come to know everyone and everything about the new community—but it took her a year to figure out that they were building an atomic bomb. McKibbin and Oppenheimer were to become lifelong friends. Robert called her by her nickname, "Dink," and quickly learned to rely on her good judgment and her ability to get things done.

At thirty-nine, Oppenheimer seemed not to have aged in twenty years. He still had long, very black and crinkly hair that stood nearly straight up. "He had the bluest eyes I've ever seen," McKibbin said, "very clear blue." They reminded her of the pale, icy blue color of gentians, a wildflower that grew on the slopes of the Sangre de Cristo Mountains. The eyes were mesmerizing. They were large and round and guarded by heavy eyelashes and thick, black eyebrows. "He always looked at the person he was talking to; he always gave everything he could to the person he was talking to." He still spoke very softly, and though he could talk with great erudition about almost anything, he could still seem charmingly boyish. "When he was impressed with something," McKibbin later recalled, "he'd say 'Gee' and it was just lovely to hear him say 'Gee.' " Robert's collection of admirers was growing exponentially at Los Alamos.

BY THE END of the month, Robert, Kitty and Peter moved up to The Hill and settled into their new home—a rustic one-story log-and-stone house built in 1929 for May Connell, the sister of the Ranch School's director and an artist who served as a matron for the Ranch School boys. "Master's Cottage #2" sat at the end of "Bathtub Row"—named with impeccable logic because it and five other log homes from the Ranch School period were the only houses on the mesa equipped with bathtubs. Located on a quiet unpaved street in the middle of the new community, the Oppenheimer home was partially shielded by shrubbery and boasted a small garden. With two tiny bedrooms and a study, the house was modest compared to One Eagle Hill. Because the schoolmasters had taken all their meals in the school cafeteria, the house lacked a kitchen, a drawback soon rectified at Kitty's insistence. But its living room was pleasant, with high ceilings, a stone fireplace and an enormous plate-glass window overlooking the garden. It would be their home until the end of 1945.

That first spring in 1943 was something of an unexpected nightmare for most of the new residents. With the melting of the snows, mud was everywhere and everyone's shoes were constantly caked with it. On some days the mud engulfed car tires in a quicksand-like grip. By April, the population of scientists had risen to thirty. Most of the new arrivals were boarded in tin-roofed plywood barracks. In the one concession to aesthetics, Oppenheimer persuaded the Army's engineers to lay out the housing so as to follow the natural contours of the land.

Hans Bethe was disheartened by what he saw. "I was rather shocked," he said. "I was shocked by the isolation, and I was shocked by the shoddy buildings . . . everybody was always afraid that a fire might break out and the whole project might burn down." Still, Bethe had to admit that the setting was "absolutely beautiful. . . . Mountains behind us, desert in front of us, mountains again on the other side. It was late winter, and in April there's still snow on the mountains, so it was lovely to look at. But clearly, we were very far from anything, very far from anybody. We learned to live with it."

The breathtaking scenery compensated in part for the utilitarian ugliness of the town. "We could gaze beyond the town, fenced in by steel wire," wrote Bernice Brode, the wife of physicist Robert Brode, "and watch the seasons come and go—the aspens turning gold in the fall against the dark evergreens; blizzards piling up snow in winter; the pale green of spring buds; and the dry desert wind whistling through the pines in summer. It was surely a touch of genius to establish our strange town on the mesa top, although many sensible people sensibly said that Los Alamos was a city that never should have been." When Oppenheimer spoke of the mesa's beauty during a recruiting trip to the University of Chicago, an urbane Leo Szilard was heard to exclaim, "Nobody could think straight in a place like that. Everybody who goes there will go crazy."

Everyone had to change lifelong habits. At Berkeley, Oppenheimer had refused to schedule a class before 11:00 a.m., so he could socialize late into the evening; at Los Alamos, he was invariably on his way to the Technical Area by 7:30 a.m. The Tech Area—known simply as the "T"—was surrounded by a 9½-foot-high woven wire fence, topped by two strands of barbed wire. Military police guarding the gate inspected everyone's colored badges. A white badge designated a physicist or other scientist who had the right to roam freely throughout the "T." On occasion, Oppenheimer absent-mindedly forgot about the all-too-visible armed guards stationed everywhere. One day he drove up to Los Alamos' main gate and, without even slowing down, whizzed through. The astonished MP shouted a warning and then fired a shot at the car's tires. Oppenheimer stopped, backed up the car and, after murmuring an apology, drove off. Understandably worried about Oppenheimer's safety, Groves wrote to him in July 1943 requesting that he refrain from driving an automobile for more than a few miles—and, for good measure, "refrain from flying in airplanes."

Like everyone else, Oppenheimer worked six days a week, taking off Sunday. But even on workdays he usually wore casual clothes, reverting to his New Mexico wardrobe of jeans or khaki pants with a blue tieless workshirt. His colleagues followed suit. "I don't recall seeing a shined pair of shoes during working hours," wrote Bernice Brode. As Oppie walked to the "T," his colleagues often fell in behind him and listened quietly as he softly murmured his thoughts of the morning. "There goes the mother hen and all the little chickens," observed one Los Alamos resident. "His porkpie hat, his pipe, and something about his eyes gave him a certain aura," recalled a twenty-three-year-old WAC who worked the telephone switchboard. "He never needed to show off or shout. . . . He could have demanded Priority One with his telephone calls but never did. He never really needed to be as kind as he was."

The director's studied informality endeared him to many who might otherwise have felt intimidated in his presence. Ed Doty, a young technician with the Army's Special Engineer Detachment (SED), wrote his parents after the war about how "several times Dr. Oppenheimer has called for something or other . . . and every time, when I would answer the phone with 'Doty,' the voice at the other end would say, 'This is Oppy.' " His informality contrasted sharply with the manner of General Groves, who "demanded attention, demanded respect." Oppie, on the other hand, got attention and respect naturally.

From the beginning, Oppenheimer and Groves had agreed that everyone's salaries were to be pegged according to each recruit's previous job. This resulted in wide disparities since a relatively young man recruited from private industry might well be paid much more than an older, tenured professor. To compensate for this inequality, Oppenheimer decreed that rents would be pro-rated according to salary. When the young physicist Harold Agnew challenged Oppenheimer to explain why a plumber could earn nearly three times the pay of a college graduate, Oppie replied that the plumbers had no idea of the laboratory's importance to the war effort, whereas the scientists did—and that, explained Oppenheimer, justified the pay difference. The scientists, at least, were not working for the money. Oppenheimer himself had been six months in Los Alamos when his secretary reminded him one day that he had not yet received a salary check.

Everyone put in long hours. The laboratory was open day and night and Oppenheimer encouraged people to set their own schedules. He refused to allow time clocks to be installed, and a siren was introduced only in October 1944, when one of General Groves' efficiency experts complained about the laxness in regular work hours. "The work was terribly demanding," Bethe recalled. The leader of the Theoretical Division thought that as science his work was "much less difficult than many things I have done at other times." But the deadlines were highly stressful. "I had the feeling, and this came in my dreams," Bethe said, "that I was behind a terribly heavy cart which I had to push up a hill." Scientists accustomed to working with limited resources and virtually no deadlines now had to adjust to a world of unlimited resources and exacting deadlines.

Bethe worked in Oppenheimer's headquarters, the T-Building ("T" for "Theoretical"), a drab two-story green structure that quickly became the spiritual center of The Hill. Nearby sat Dick Feynman, who was as gregarious as Bethe was serious. "For me," Bethe recalled, "Feynman sort of materialized from Princeton. I hadn't known about him, but Oppenheimer had. He was very lively from the beginning, but he didn't start insulting me until about two months after he came." The thirty-seven-year-old Bethe liked to have someone around who was willing to argue with him, and the twenty-five-year-old Feynman loved to argue. When the two of them were together, everyone in their building could hear Feynman yelling, "No, no, you're crazy," or "That's nuts!" Bethe would then quietly explain why he was right. Feynman would calm down for a few minutes and then erupt again with "That's impossible, you're mad!" Their colleagues soon nicknamed Feynman "The Mosquito" and Bethe "The Battleship."

"OPPENHEIMER AT LOS ALAMOS," Bethe said, "was very different from the Oppenheimer I had known. For one thing, the Oppenheimer before the war was somewhat hesitant, diffident. The Oppenheimer at Los Alamos was a decisive executive." Bethe was hard-pressed to explain the transformation. The man of "pure science" he knew at Berkeley had been entirely focused on exploring the "deep secrets of nature." Oppenheimer had not been remotely interested in anything like an industrial enterprise—and yet at Los Alamos he was directing an industrial enterprise. "It was a different problem, a different attitude," Bethe said, "and he completely changed to fit the new role."

He rarely gave orders, and instead managed to communicate his desires, as the physicist Eugene Wigner recalled, "very easily and naturally, with just his eyes, his two hands, and a half-lighted pipe." Bethe remembered that Oppie "never dictated what should be done. He brought out the best in all of us, like a good host with his guests." Robert Wilson felt similarly: "In his presence, I became more intelligent, more vocal, more intense, more prescient, more poetic myself. Although normally a slow reader, when he handed me a letter I would glance at it and hand it back prepared to discuss the nuances of it minutely." He also admitted that in retrospect there was a certain amount of "self-delusion" in these feelings. "Once out of his presence the bright things that had been said were difficult to reconstruct or remember. No matter, the tone had been established. I would know how to invent what it was that had to be done."

Oppenheimer's frail, ascetic physique only accentuated his charismatic authority. "The power of his personality is the stronger because of the fragility of his person," John Mason Brown observed some years later. "When he speaks he seems to grow, since the largeness of his mind so affirms itself that the smallness of his body is forgotten."

He had always had a knack for anticipating the next question to be faced in solving any theoretical physics problem. But now he surprised his colleagues with his seemingly instantaneous comprehension of any facet of engineering. "He could read a paper—I saw this many times," recalled Lee Dubridge, "and you know, it'd be fifteen or twenty typed pages, and he'd say, 'Well, let's look this over and we'll talk about it.' Oppie would then flip through the pages in about five minutes and then he'd brief everybody on exactly the important points. . . . He had a remarkable ability to absorb things so rapidly. . . . I don't think there was anything around the lab of any significance that Oppie wasn't fully familiar with and knew what was going on." Even when there was disagreement, Oppenheimer had an instinct for preempting arguments. David Hawkins, the Berkeley philosophy student Oppenheimer had recruited to serve as his personal assistant, had many opportunities to observe his boss in action: "One would listen patiently to an argument beginning, and finally Oppenheimer would summarize, and he would do it in such a way that there was no disagreement. It was a kind of magical trick that brought respect from all those people, some of them superiors in terms of their scientific record. . . ."

It helped that Oppenheimer could turn on—and off—his personal charm. Those who knew him from Berkeley understood that this was a man with a remarkable flair for drawing others into his orbit. And those, like Dorothy McKibbin, who met him for the first time in New Mexico invariably found themselves eager to please him. "He made you do the impossible," McKibbin recalled. One day, she was called from Santa Fe to the Site and asked if she would help to alleviate the ongoing housing crisis by taking over a lodge ten miles up the road and turning it into housing for a hundred employees. McKibbin resisted. "Well," she protested, "I've never run a hotel before." At that moment the door of Oppenheimer's office opened and he stuck his head out and said, "Dorothy, I wish you would." He then withdrew his head and closed the door. McKibbin said, "I will."

"I think he had no great reluctance about using people," recalled John Manley. "If he found that people were useful to him, why it was just natural to him to use them." But Manley thought many people, himself included, enjoyed being used by Robert because he did it so adroitly. "I think that he really realized that the other person knew that this was going on; it was like a ballet, each one knowing the part and the role he's playing, and there wasn't any subterfuge in it."

He listened to and often accepted the advice of others. When Hans Bethe suggested everyone would benefit from a weekly open-ended colloquium, Oppenheimer immediately agreed. When Groves first learned of it, he tried to stop it, but Oppenheimer insisted that such a free exchange of ideas among the "white badge" scientists was essential. "The background of our work is so complicated," Oppie wrote Enrico Fermi, "and information in the past has been so highly compartmentalized, that it seems that we shall have a good deal to gain from a leisurely and thorough discussion."

The first colloquium was convened on April 15, 1943, in the now empty schoolboys' library. Standing before a small blackboard, Oppenheimer offered some perfunctory words of welcome and then introduced Bob Serber, his former student. Serber, he explained, would brief the assembled scientists, numbering no more than forty, on the task at hand. Speaking from notes with his habitual stutter, the shy and awkward Serber took center stage. "Security was terrible," Serber later wrote. "We could hear carpenters banging down the hall and at one point a leg appeared through the beaverboard ceiling, presumably belonging to an electrician working up above." After only a few minutes, Oppenheimer sent John Manley up to whisper in Serber's ear that he should stop using the word "bomb" in favor of something more neutral like "gadget."

"The object of the project," Serber said, "is to produce a practical military weapon in the form of a bomb in which the energy is released by a fast-neutron chain reaction in one or more of the materials known to show nuclear fission." Summarizing what Oppenheimer's team had learned from their Berkeley summer sessions, Serber reported that by their calculations an atomic bomb might conceivably produce an explosion equivalent to 20,000 tons of TNT. Any such "gadget," however, would need highly enriched uranium. This core of enriched uranium, approximately the size of a cantaloupe, would weigh about thirty-three pounds. They could also construct a weapon from the even heavier element of plutonium—produced via a neutron-capture process using U-238. A plutonium bomb would need far less critical mass, and the plutonium core might therefore weigh only eleven pounds and appear no larger than an orange. Either core would need to be packed within a thick shell of ordinary uranium the size of a basketball. This would bring the weight of either device to about a ton—still something deliverable by airplane.*

Most of the scientists in Serber's audience already understood the theoretical possibilities inherent in the new physics—but compartmentalization had kept many of them in the dark about the details. Few had realized how many of the basic questions had already been answered, at least in broad outline. The obstacles remaining to building a practical military weapon were large but not insurmountable. Some of the physics of building an atomic bomb was still uncertain, but the real imponderables lay in the field of engineering and ordnance design. Producing sufficient amounts of either U-235 or plutonium would require a massive industrial effort. And even if sufficient bomb-grade materials could be produced, no one was quite sure how to design an atomic bomb that would detonate efficiently. But even a onetime skeptic like Bethe understood, as he later put it, "That once plutonium was made, it was almost certain that a nuclear bomb could be made as well." Thus, the real news to Serber's audience was that they had a mission that could contribute enormously to the war effort. This fact alone lifted morale. Serber's first talk conveyed what Oppenheimer wanted: a sense of mission and a realization that they had the means to change history. But could they solve the technical problems before the Germans? Could they indeed help win the war?

Over the next two weeks, Serber gave four more hour-long lectures, stimulating the kind of creative dialogue that Oppenheimer wanted. Among many other issues, Serber briefly summarized the actual mechanics of what he called "shooting"—the problem of how to bring together the critical masses of the uranium or plutonium so as to initiate a chain reaction. Serber dwelled on the most obvious method—the gun assembly—whereby criticality would be achieved by firing a slug of uranium into another mass of U-235, leading to an explosion. But he also suggested that "the pieces might be mounted on a ring as in the [accompanying] sketch. If explosive material were distributed around the ring and fired, the pieces would be

*Little Boy, the world's first combat atomic bomb, weighed 9,700 pounds when it was dropped on Hiroshima from a B-29 bomber named the *Enola Gay*.

blown inward to form a sphere." The idea of imploding fissionable material had first been suggested by Oppenheimer's old friend Richard Tolman during the summer of 1942, and he and Serber had thereupon written a memorandum on the subject for Oppenheimer. Tolman later wrote two other memos on implosion, and in March 1943 Vannevar Bush and James Conant urged Oppenheimer to explore the implosion design. Oppenheimer reportedly replied, "Serber is looking into it." Although Tolman's proposal had not included the notion of actually compressing solid material so as to increase its density, the idea was sufficiently well formulated to warrant inclusion in Serber's lecture notes, if only as an aside. But this was enough to spark the interest of another physicist, Seth Neddermeyer, who asked Oppenheimer's permission to investigate its potential. Soon, Neddermeyer and a small team of scientists could be found in a canyon near Los Alamos, testing implosion explosives.

Serber's lectures would have a long life. Using Serber's notes, Ed Condon typed up the lectures as a twenty-four-page summary. This became a mimeographed booklet, titled *The Los Alamos Primer,* which was passed out to newly arriving scientists. Among others, Enrico Fermi attended some of Serber's lectures, and he then remarked to Oppenheimer, "I believe your people actually want to make a bomb." Oppenheimer was struck by the note of surprise in Fermi's voice as he said this. Fermi had just come from Chicago, where he found the atmosphere among the scientists oddly subdued in comparison to the exhilaration he often encountered among the men in Oppie's mesa laboratory. Everyone, whether in Chicago or Los Alamos or elsewhere, held the sobering thought that if an atomic bomb was possible, the Germans might be ahead in the race to build one. But whereas at Chicago, many of the senior scientists were troubled and even depressed by this realization, at Los Alamos, under Oppenheimer's charismatic leadership, this awareness seemed only to inspire the men to forge ahead with their work.

Fermi took Oppenheimer aside one day and suggested another way to kill large numbers of Germans. Perhaps, he said, radioactive fission products could be used to poison Germany's food supply. Oppenheimer seems to have taken the proposal seriously. After urging Fermi not to mention the matter to anyone else, Oppenheimer reported the idea to General Groves and later discussed it with Edward Teller. Teller reportedly told him that separating out strontium-90 from a chain-reacting pile was feasible. But by May 1943, Oppenheimer had decided to recommend a delay in action on the proposal—for a gruesome reason: "In this connection," he wrote Fermi, "I think that we should not attempt a plan unless we can poison food sufficient to kill a half a million men, since there is no doubt that the actual number affected will, because of non-uniform distribution, be much smaller than this." The idea was dropped, but only because there seemed no efficient way to poison large numbers of the enemy population.

Wartime compelled some mild-mannered men to contemplate what was once unthinkable. In late October 1942, Oppenheimer received a letter marked "secret" from his old friend and colleague Victor Weisskopf, who wrote to report alarming news in a letter he had just received from the physicist Wolfgang Pauli, then residing in Princeton. Pauli had written that their former German colleague, the Nobel Prize–winning physicist Werner Heisenberg, had just been appointed director of the Kaiser-Wilhelm Institute, a nuclear research facility in Berlin. Moreover, Pauli had learned that Heisenberg was scheduled to give a lecture in Switzerland. Weisskopf reported further that he had discussed this news with Hans Bethe, and the two men had agreed that something should be done immediately: "I believe," Weisskopf wrote Oppenheimer, "that by far the best thing to do in this situation would be to organize a kidnapping of Heisenberg in Switzerland. That's what the Germans would do if, say you or Bethe would appear in Switzerland." Weisskopf even volunteered himself for the job.

Oppenheimer immediately wrote back, thanking Weisskopf for his "interesting" letter. He said he had already learned of Heisenberg's scheduled visit to Switzerland and had discussed the issue with the "proper authorities" in Washington. "I doubt that you will hear further of the matter, but [I] wanted to thank you and assure you that it is receiving the attention it deserves." The "proper authorities" with whom Oppenheimer had indeed already talked of this matter were Vannevar Bush and Leslie Groves, and he now passed on Weisskopf's letter to them. But he did not endorse the proposal—even a successful kidnapping of Heisenberg would alert the Nazis to the high priority the Allies assigned to nuclear research. On the other hand, Oppenheimer could not refrain from remarking to Bush "that Heisenberg's proposed visit to Switzerland would seem to afford us an unusual opportunity."

Much later, Groves seriously pursued the notion of kidnapping or assassinating Heisenberg; in 1944 he dispatched OSS agent Moe Berg to Switzerland, where the former baseball player stalked the German physicist in December 1944—but ultimately decided not to attempt an assassination.

CHAPTER SIXTEEN

"Too Much Secrecy"

. . . this policy puts you in the position of trying to do an extremely difficult job with three hands tied behind your back. . . .

DR. EDWARD CONDON *to Oppenheimer*

T HE DIRECTOR'S FIRST REAL ADMINISTRATIVE CRISIS occurred early that first spring. With General Groves' approval, Oppenheimer had appointed his former Göttingen classmate Edward U. Condon as associate director. Condon's job was to relieve Oppenheimer of some of his administrative burdens and to serve as liaison with the Army's military commander at Los Alamos. Two years older than Oppenheimer, Condon was both a brilliant physicist and a seasoned laboratory administrator. After earning his doctorate at Berkeley in 1926, Condon had won postdoctoral appointments in Göttingen and Munich. For the next decade he taught at several universities, including Princeton, and published the first English-language textbook on quantum mechanics. In 1937, he left Princeton to become associate director of research at Westinghouse Electric Company, a major industrial research center. Over the next few years, he supervised the company's research in nuclear physics and microwave radar. By the autumn of 1940, he was working full-time on war-related projects, primarily radar, at MIT's Radiation Laboratory. In short, Condon was, in terms of experience at least, significantly more qualified than Oppenheimer to lead the new laboratory in Los Alamos.

Condon had not been as politically active as Oppenheimer in the 1930s, and he certainly was not affiliated with the Communist Party. He thought of himself as a "liberal" New Dealer, a loyal Democrat who voted for Franklin Roosevelt. Raised as a Quaker, Condon once told a friend, "I join every organization that seems to have noble goals. I don't ask whether it contains Communists." An idealist with strong civil-libertarian instincts, Condon believed that good science could not come without a free exchange of ideas, and he lobbied vigorously for regular contacts between physicists at Los Alamos and the other labs around the country. Inevitably, he quickly attracted the ire of General Groves, who heard repeated reports of security infractions from his military representatives in Los Alamos. "Compartmentalization of knowledge, to me," Groves insisted, "was the very heart of security."

In late April 1943, Groves was angered to learn that Oppenheimer had traveled to the University of Chicago, where he had discussed the production schedule for plutonium with the director of the Manhattan Project's Metallurgical Laboratory (Met Lab), the physicist Arthur Compton. The general blamed Condon for this ostensible infringement of security. Descending on Los Alamos, Groves stormed into Oppenheimer's office and confronted the two men. Condon stood his ground against the general, but, to his astonishment, he realized that Oppenheimer was not backing him up. Within a week, Condon decided to tender his resignation. He had intended to stay for the project's duration, but had lasted just six weeks.

"The thing which upsets me most is the extraordinarily close security policy," he wrote Oppenheimer in his resignation letter. "I do not feel qualified to question the wisdom of this since I am totally unaware of the extent of enemy espionage and sabotage activities. I only want to say that in my case I found that the extreme concern with security was morbidly depressing—especially the discussion about censoring mail and telephone calls." Condon explained that he was "so shocked that I could hardly believe my ears when General Groves undertook to reprove us. . . . I feel so strongly that this policy puts you in the position of trying to do an extremely difficult job with three hands tied behind your back. . . ." If he and Oppenheimer truly could not meet with a man like Compton without violating security, then "I would say the scientific position of the project is hopeless."

Condon concluded that he could better contribute to the war effort by returning to Westinghouse and working on radar technology. He left saddened and perplexed by Oppie's apparent unwillingness to defy Groves. Condon was unaware that Oppenheimer had yet to receive his own security clearance. The Army's security bureaucracy was still trying to block Oppenheimer's clearance and Oppie knew he could not press Groves about security—not if he wanted to keep his job.

Oppenheimer had much invested in his relationship with Groves. The previous autumn, each man had taken the measure of the other and arrogantly calculated that he could dominate their relationship. Groves believed the charismatic physicist was essential to the success of the project. And precisely because Oppenheimer came with left-wing political baggage, Groves thought he could use Oppie's past to control him. Robert's calculation was equally straightforward. He understood that he could keep his job only if Groves continued to consider him far and away the best director available. He realized that his communist associations gave Groves a certain hold over him, but by demonstrating his unique competence, he believed, he would convince the general to allow him to run the laboratory as he saw fit. Oppenheimer didn't disagree with Condon; he too was convinced that onerous security regulations could smother the scientists. But he was confident that over time he would prevail. After all, in the end, Groves needed Oppenheimer's skills as much as Oppenheimer needed Groves' approval.

In retrospect, they were a perfect team to lead the effort to beat Germans in the race to build a nuclear weapon. If Robert's style of charismatic authority tended to breed consensus, Groves exercised his authority through intimidation. "Basically his way of running projects," observed Harvard chemist George Kistiakowsky, "was to scare his subordinates to a point of blind obedience." Robert Serber thought that with Groves it was a "matter of policy to be as nasty as possible to his subordinates." Oppie's secretary, Priscilla Green Duffield, always remembered how the general would stride past her desk and, without even a hello, say something rude such as, "Your face is dirty." This crude behavior made Groves the object of most of the complaints on the mesa, and this deflected criticism from Oppenheimer. But Groves refrained from such behavior around Oppenheimer, and it was a measure of Oppenheimer's leverage in their relationship that he usually got his way.

Robert did what was necessary to appease Groves. He became what the general wanted, a deft and efficient administrator. At Berkeley, his office desk had typically been stacked with foot-high piles of paper. Dr. Louis Hempelmann, the Berkeley physician who came to Los Alamos and became the Oppenheimers' close friend, observed that on the mesa, Robert "was a clean-desk man. Never any paper there." There was also a physical transformation: Oppie cut his long, curly hair. "He had his hair [so] closely clipped," remarked Hempelmann, "I almost didn't recognize him."

In point of fact, even as Condon was quitting Los Alamos, Groves' compartmentalization policy was breaking down. Oppenheimer may have avoided a confrontation over the issue, but the policy was becoming a sham. As the work progressed, it became increasingly important to have all "white badge" scientists free to discuss their ideas and problems with each other. Even Edward Teller understood that compartmentalization was an impediment to efficiency. Early in March 1943, he explained to Oppenheimer that he had written an official letter to him discussing "my old anxiety: too much secrecy." But then he confided, "I did not do so to annoy you but to give you a possibility to use the statement at any time in case you see any advantage in doing so." Groves soon realized what he was up against. Try as he might, he could not even get the most responsible and senior scientists to cooperate. On one occasion when Ernest Lawrence was visiting Los Alamos and due to lecture there to a small group of scientists, Groves took the physicist aside and carefully briefed him on what he was not allowed to say to his audience. To his dismay, just a few moments later, Groves heard Lawrence up at the blackboard saying, "I know General Groves doesn't want me to say this, but . . ." Officially nothing changed, but in practice compartmentalization among the scientists grew more and more lax.

Groves often blamed the collapse of compartmentalization on Condon's influence over Oppenheimer. "He [Condon] did a tremendous amount of damage at Los Alamos in the initial setup," Groves testified in 1954. "I could never make up my own mind as to whether Dr. Oppenheimer was the one who was primarily at fault in breaking up the compartmentalization, or whether it was Dr. Condon." It was one thing, he thought, to have the top twenty to thirty scientists freely talking to each other. But when hundreds of men ignored the policy, compartmentalization became a joke.

Groves eventually came to recognize that at Los Alamos the rules of science had trumped the principles of military security. "While I may have dominated the situation in general," he testified, "I didn't have my own way in a lot of things. So when I say that Dr. Oppenheimer did not always keep the faith with respect to the strict interpretation of the security rules, if I could say that he was no worse than any of my other leading scientists, I think that would be a fair statement."

In May 1943, Oppenheimer presided over a meeting in which it was decided that a General Colloquium would be held every other Tuesday evening. He persuaded Teller to organize the meetings. When Groves said he was "disturbed" by the wide-ranging scope of these discussions, Oppie replied quite firmly that he was "committed" to the colloquia. His only concession was to agree to restrict attendance to scientists. He also argued adamantly that his people had to be able to exchange information with their counterparts at other Manhattan Project sites. That June, for instance, he insisted on Enrico Fermi being permitted to visit Los Alamos from the Met Lab in Chicago. He told Groves that because Fermi's trip was of the "highest importance," he simply would not take responsibility for its cancellation. Groves relented and Fermi was allowed to visit.

Late in the summer of 1943, Oppenheimer explained his views on security to a Manhattan Project security officer: "My view about the whole damn thing, of course, is that the [basic] information we are working on is probably known to all the governments that care to find out. The information about what we're doing is probably of no use because it is so damn complicated." The danger, he said, was not that technical information about the bomb might leak to another country. The real secret was "the intensity of our effort" and the scale of the "international investment involved." If other governments understood the resources America was throwing into the bomb effort, they might attempt to duplicate the bomb project. Oppenheimer didn't think even this knowledge would "have any effect on Russia," but "it might have a very big effect on Germany, and I am as convinced about that . . . as everyone else is."

Even as Oppenheimer was distracted by the demands of Groves' security officers, some of his younger protégés were complaining that the Army's clumsy management of the Manhattan Project was wasting precious time. By the time Los Alamos opened in March 1943, four years had passed since the discovery of fission and most of the physicists working on the project assumed that their German counterparts had at least a two-year lead. Feeling a desperate sense of urgency, they were angered by the Army's security precautions, the plodding bureaucracy—and anything that seemed to cause delays. That summer, Phil Morrison reported in a "Dear Opje" letter from the Met Lab that "the drive which accompanied last winter's work seems nearly gone. Relations between our people and the contractor's are impossibly bad . . . the result is intolerable and incompatible with speedy success." A dozen of the Chicago lab's younger scientists were so alarmed that they had signed a letter addressed to President Roosevelt reporting that it was their "sober judgment that this project is losing time. The Army direction is conventional and routine. . . ." Speed was essential. And yet, the Army was not consulting the "few scientific leaders who alone are competent in this new field. The life of our nation is endangered by such a policy."

Three weeks later, on August 21, 1943, Hans Bethe and Edward Teller wrote Oppenheimer of their own frustrations with the pace of the project. "Recent reports both through the newspapers and through the secret service, have given indications that the Germans may be in possession of a powerful new weapon which is expected to be ready between November and Jan." The new weapon, they warned, was probably "Tube-Alloys"—the British code name for an atomic bomb. "It is not necessary," they wrote, "to describe the probable consequences which would result if this proves to be the case." They then complained that the private companies responsible for the production of bomb-grade uranium were retarding the program. The solution, they argued, was to "make available adequate funds for the additional program, directly to those scientists who are most experienced in the various phases of the problem."

Oppenheimer shared their concerns. He too was worried that they might be falling behind the Germans, and so he worked harder and exhorted his people to do the same.

WITH THE TITLE of scientific director, Oppenheimer's authority inside Los Alamos was nearly absolute. Though he ostensibly shared power with a military post commander, Oppie reported directly to General Groves. The first post commander, Lt. Col. John M. Harmon, had numerous arguments with the scientists and as a result he was replaced in April 1943, after only four months on the job. His successor, Lt. Col. Whitney Ashbridge, understood that his job was to minimize friction and keep the scientists happy. Ashbridge, coincidentally a graduate of the Los Alamos Ranch School, lasted until the autumn of 1944, when, overworked and exhausted, he suffered a mild heart attack. He was replaced by Col. Gerald R. Tyler. Thus, Oppenheimer literally worked through three army colonels.

Security was always a headache. At one point, Army security stationed armed military police outside Oppenheimer's "Bathtub Row" house. The MPs inspected everyone's pass, including Kitty's, before allowing them to enter the house. Kitty frequently forgot to take her pass when she left and always made a scene when they wouldn't let her back in. Still, she was not entirely unhappy about their presence: Always ready to seize an opportunity, she occasionally used the MPs as baby-sitters for Peter. When the sergeant in charge of the detail realized what was happening, he had the MPs withdrawn.

As part of his understanding with General Groves, Oppenheimer had agreed to name a three-man committee to be responsible for internal security. He appointed his assistants David Hawkins and John Manley, and a chemist, Joe Kennedy. They were responsible for security inside the laboratory (the T-Section), which was enclosed within a second, inner barbed-wire fence that was off-limits to MPs and soldiers. The internal security committee dealt with such prosaic matters as checking to make sure that scientists locked their file cabinets when they left their offices. If someone was caught leaving a secret document on his desk overnight, then that scientist was required to patrol the lab the next night and try to catch someone else. One day, Serber saw Hawkins and Emilio Segrè having an argument. "Emilio, you left a secret paper out last night," Hawkins said, "and you have to go around tonight." Segrè retorted, "That paper, it was all wrong. It would only have confused the enemy."

Oppenheimer struggled constantly to protect his people from The Hill's security apparatus. He and Serber had numerous discussions about how to "save" various people from being dismissed. "If they had had their way," Serber said of the security division, "there wouldn't have been anybody left." Indeed, in October 1943 the army's security investigators recommended that Robert and Charlotte Serber both be removed from Los Alamos. The FBI charged, with typical hyperbole, that the Serbers were "entirely saturated with Communist beliefs and all of their associates were known radicals."

While Robert Serber's views were certainly leftist, he had never been as politically active as his wife. Charlotte had poured her energies in the late 1930s into such projects as raising funds for the Spanish Republicans. But, of course, Oppenheimer himself had been more politically active than Charlotte. It is unclear from the documentary record how the Army was overruled, but Oppie probably vouched personally for the Serbers' loyalty. One day Capt. Peer de Silva, the chief resident security officer, confronted Oppenheimer with Serber's political background, only to have Oppenheimer dismiss it all as unimportant: "Oppenheimer volunteered information that he had known Serber was formerly active in Communist activities and stated that, in fact, Serber had told him so." Oppenheimer explained that he had told Serber, prior to bringing him to Los Alamos, that he would have to drop his political activities. "Serber promised me he would, therefore, I believe him." Incredulous, De Silva thought this evidence of Oppenheimer's naïveté, or worse.

Like many Hill wives, Charlotte Serber worked in the Tech Area. And though G-2's security file on the Serbers noted her family's left-wing background, Charlotte's job as scientific librarian literally made her the gatekeeper for The Hill's most important secrets. Oppenheimer placed enormous trust in her. Casually dressed in jeans or slacks, Charlotte presided over the library as a social hangout and a "center for all gossip."

One day, Oppenheimer called Charlotte into his office. Oppie explained that rumors were beginning to circulate in Santa Fe about the secret facility on the mesa. He had suggested to Groves that it might be wise to plant their own rumors as a diversion. "Therefore," said Oppie, "for Santa Fe purposes, we are making an electric rocket." He then explained that he wanted the Serbers and another couple to frequent some of the bars in Santa Fe. "Talk. Talk too much," Oppie said. "Talk as if you had too many drinks. . . . I don't care how you manage it, say we are building an electric rocket." Accompanied by John Manley and Priscilla Greene, Bob and Charlotte Serber soon drove down to Santa Fe and tried to spread the rumor. But no one was interested, and G-2 never picked up any talk about electric rockets.

Richard Feynman, an incorrigible practical joker, had his own way of dealing with security regulations. When the censors complained that his wife, Arline, now a patient at a tuberculosis sanatorium in Albuquerque, was sending him letters in code and asked for the code, Feynman explained that he didn't have the key to it—it was a game he played with his wife to practice his code-breaking. Feynman also drove security personnel to distraction when he went on a nighttime safecracking spree, opening the combination locks for secret file cabinets all over the laboratory. On another occasion, he noticed a hole in the fence surrounding Los Alamos—so he walked out the main gate, waved to the guard, and then crawled back through the hole and walked out the main gate again. He repeated this several times. Feynman was almost arrested. His antics became part of Los Alamos lore.

The Army's relations with the scientists and their families were always shaky. General Groves set the tone. In private with his own men, Groves routinely labeled Los Alamos civilians "the children." He instructed one of his commanders: "Try to satisfy these temperamental people. Don't allow living conditions, family problems, or anything else to take their minds off their work." Most of the civilians made it clear that they found Groves "distasteful"—and he made it clear that he didn't care what they thought.

Oppenheimer got along with Groves—but he found most of the Army's counterintelligence officers obtuse and offensive. One day Captain de Silva barged into one of Oppenheimer's regular Friday afternoon meetings of all the group leaders, and announced, "I have a complaint." De Silva explained that a scientist had come into his office to talk and, without asking his permission, had sat on the corner of his desk. "I didn't appreciate it," fumed the captain. To the amusement of everyone else in the room, Oppenheimer replied, "In this laboratory, Captain, anybody can sit on anybody's desk."

Captain de Silva, the only West Point graduate resident at Los Alamos, could not laugh at himself. "He was profoundly suspicious of everyone," recalled David Hawkins. That Oppenheimer had appointed Hawkins, a former Communist Party member, to the lab's security committee, only fueled De Silva's suspicions. Oppenheimer liked Hawkins and thought highly of his abilities. He also knew that Hawkins was a loyal American, whose left-wing politics—like his own—were reformist rather than revolutionary.

Some of the security restrictions were deeply annoying to everyone. When Edward Teller said that his people were complaining about their mail being opened, Oppie replied bitterly, "What are they griping about? I am not allowed to talk to my own brother." He chafed at the notion that he was being watched. "He complained constantly," Robert Wilson recalled, "that his telephone calls were monitored." At the time, Wilson thought this "somewhat paranoiac"; only much later did he realize that Oppie had indeed been under near-total surveillance.

Even before Los Alamos opened in March 1943, Army counterintelligence instructed J. Edgar Hoover to suspend FBI surveillance of Oppenheimer. As of March 22, Hoover complied, but he instructed his agents in San Francisco to continue their surveillance of individuals who might have been connected with Oppenheimer in the Communist Party. On that date, the Army informed the FBI that it had arranged for full-time technical and physical surveillance of Oppenheimer. A large number of Army Counter-Intelligence Corps (CIC) officers had already been placed in undercover assignments even before Oppenheimer arrived in Los Alamos. One such agent, Andrew Walker, was assigned to serve as Oppenheimer's personal driver and bodyguard. Walker later confirmed that CIC officers monitored Oppenheimer's mail and his home telephone. Oppie's office was wiretapped.

OPPENHEIMER, meanwhile, was himself becoming highly security-conscious. The once casual college professor now could be seen carefully pinning a classified memo inside his hip pocket so as not to lose it. He even tried to placate the Army security officers, giving them his valuable time and complying with virtually all of their requests. But the pressure of the work, the sensation of being constantly watched, the fear of failure—all of this and more—began to take its toll. At one point in the summer of 1943, Oppenheimer confessed to Robert Bacher that he was thinking of quitting. He felt hounded by the investigations into his past. Besides, he told Bacher, the strain of the job was just too much. After listening to Oppie list his inadequacies, Bacher told him simply, "There isn't anybody else who can do it."

So Oppie persevered. But once, in June 1943, he did something which he should have known would surely heighten the concerns of CIC officers. Despite his marriage to Kitty, Robert had continued to see Jean Tatlock about twice each year between 1939 and 1943. He later explained, "we had been very much involved with one another, and there was still very deep feeling when we saw each other." He and Jean had met around New Year's Eve in 1941 and occasionally ran into each other at parties in Berkeley. But Oppie also visited Jean at her apartment and in her office at the children's hospital where she was employed as a psychiatrist. Once he went to see Jean at her father's house around the corner from his own home on Eagle Hill Drive, and on another occasion they had drinks at the Top of the Mark, an elegant restaurant with one of the best views in San Francisco.

Oppenheimer may or may not have resumed his love affair with Jean during these years; we know only that he continued to see her and that the emotional bonds between them were unbroken. Sometime after Robert married Kitty in 1940, Jean was visiting their old friend Edith Arnstein, now married, in her San Francisco apartment. Jean was standing at the window, holding Edith's baby girl, Margaret Ludmilla, when Edith asked her if she regretted refusing to marry Oppie. She replied "Yes," and that she probably would have married him "had she not been so mixed up."

By the time Oppenheimer left Berkeley in the spring of 1943, Jean was Dr. Jean Tatlock, a woman on the threshold of a rewarding medical career. She was a pediatric psychiatrist at Mount Zion Hospital, where most of her patients were mentally troubled children. She seemed to have found a career that suited her temperament and intellect.

Jean had told Oppie that she "had a great desire" to see him before he and Kitty left that spring for Los Alamos. But for some reason, Oppie refused. Security could not have been the issue, since he had made a point of saying good-bye to Steve Nelson. Perhaps Kitty objected. Whatever the case, he left for Los Alamos without saying good-bye to Jean, and he felt guilty about it. They corresponded, but Jean told her friends that she found his letters mystifying. She implored him in several anguished letters to return. Robert knew she was seeing a psychologist, his good friend Dr. Siegfried Bernfeld, Freud's disciple and the leader of the study group he had attended regularly for several years. Oppenheimer knew that Dr. Bernfeld was Jean's training analyst—and he also knew that "she was extremely unhappy."

So when he had occasion in June 1943 to return to Berkeley, Oppie made a point of calling Jean and taking her to dinner. Military intelligence agents stalked him throughout his visit, and later reported to the FBI what they had observed: "On June 14, 1943, Oppenheimer traveled via Key Railway from Berkeley to San Francisco . . . where he was met by Jean Tatlock who kissed him." They then walked arm in arm to her car, a 1935 green Plymouth coupe; she drove him to the Xochimilco Café, a cheap combination bar, café and dance hall. They had a few drinks with dinner and then at about 10:50 p.m. Jean drove them back to her top-floor apartment at 1405 Montgomery Street in San Francisco. At 11:30 p.m. the lights were extinguished, and Oppenheimer was not observed until 8:30 a.m. next day, when he and Jean Tatlock left the building together. The FBI report noted that "the relationship of Oppenheimer and Tatlock appears to be very affectionate and intimate." Again that evening, the agents watched as Tatlock met Oppenheimer at the United Airlines office in downtown San Francisco: "Tatlock arrived on foot and Oppenheimer rushed to meet her. They greeted each other affectionately and walked to her car nearby; thence to dinner at Kit Carson's Grill." After dinner, Jean drove him to the airport, where he caught a flight back to New Mexico. Oppie never saw her again. Eleven years later, he was asked by his interrogators, "Did you find out why she had to see you?" He replied, "Because she was still in love with me."

Reports of Oppenheimer's visit with Tatlock, a known Communist Party member, made their way to Washington, and soon she was being described as a possible conduit for passing atomic secrets to Soviet intelligence. On August 27, 1943, in a memo justifying a wiretap on Tatlock's phone, the FBI suggested that Oppenheimer himself "might either use her as a go-between or use her telephone from which to place important calls affecting the Comintern Apparatus. . . ."

On September 1, 1943, FBI chief J. Edgar Hoover wrote the Attorney General that in connection with the Bureau's investigation of Soviet Comintern espionage agents "it has been determined that Jean Tatlock . . . has become the paramour of an individual possessed of vital secret information regarding this nation's war effort." Hoover asserted that Tatlock was "a contact of members of the Comintern Apparatus in the San Francisco area and it is reported that she is not only in a position to solicit secret information from the man with whom she associates, but is also in a position to pass the information on to espionage agents within the Apparatus." Hoover recommended tapping her telephone "for the purpose of determining the identities of espionage agents within the Comintern Apparatus," and late that summer one tap was installed by either Army intelligence or the FBI.

On June 29, 1943, just two weeks after Oppenheimer spent the night with Tatlock, Col. Boris Pash, Chief of Counter-Intelligence on the West Coast, wrote a memo to the Pentagon recommending that Oppenheimer be denied a security clearance and fired. Pash reported that he had information that Oppenheimer "may still be connected with the Communist Party." All of his evidence was circumstantial. He cited Oppenheimer's visit to Tatlock, and a phone call Oppenheimer made to David Hawkins, "a party member who has contacts with both Bernadette Doyle and Steve Nelson."

Pash believed that if Oppenheimer himself was not prepared to transmit scientific information directly to the Party, "he may be making that information available to his other contacts, who, in turn, may be furnishing" knowledge about the Manhattan Project to the Soviet Union. Pash naturally wondered whether Tatlock could be the conduit. He would also have learned from his FBI colleagues that, as late as August 1943, Tatlock was politically active in Communist Party affairs.

In Pash's mind, Tatlock was a prime spy suspect and he was hoping that a wiretap on her phone would prove it. Short of that, Pash intended to use the fact of Oppenheimer's relationship with Tatlock as a weapon against him. In late June, he marshaled his thoughts along these lines in a long memo to Groves' new security aide, Lt. Col. John Lansdale, a smart thirty-one-year-old lawyer from Cleveland. Pash told Lansdale that if Oppenheimer could not be fired outright, he should be called to Washington and threatened in person with the "Espionage Act and all its ramifications." He should be informed that military intelligence knew all about his Communist Party affiliations and that the government would not tolerate leaks of any kind to his friends in the Party. Like General Groves, Pash thought Oppenheimer's ambition and pride could be used to keep him in check: "It is the opinion of this office," Pash wrote, "that subject's personal inclinations would be to protect his own future and reputation and the high degree of honor which would be his if his present work is successful, and, consequently, it is felt that he would lend every effort to cooperating with the Government in any plan which would leave him in charge."

By then, however, Lansdale had met Oppenheimer and, unlike Pash, he liked and trusted him. But he also understood that while Oppie was a key man in the project, his political associations were troubling. Shortly after receiving Pash's recommendations, Lansdale wrote Groves a concisely worded, two-page memo summarizing the evidence. Lansdale listed all the "front" groups (as defined by the FBI) Oppie had joined over the years, from the American Civil Liberties Union [sic] to the American Committee for Democracy and Intellectual Freedom. He cited his association and friendship with such known or suspected communists as William Schneiderman, Steve Nelson, Dr. Hannah L. Peters—identified by Lansdale as "organizer of the Doctors Branch, Professional Section, Communist Party, Alameda County, California"—Isaac Folkoff and such personal friends as Jean Tatlock, "with whom Oppenheimer is alleged to have an illicit association," and Haakon Chevalier, "believed to be a Communist Party member." Most damaging of all, Lansdale noted that Steve Nelson's assistant, Bernadette Doyle, "is reported by a very reliable informant [i.e., a telephone intercept] to have referred to J. R. Oppenheimer and his brother, Frank, as being regularly registered within the Communist Party."

Yet Lansdale did not recommend firing Oppenheimer. Instead, he advised Groves in July 1943, "you should tell Oppenheimer substantially that we know that the Communist Party . . . is attempting to discover information" about the Manhattan Project. Tell him, Lansdale wrote, that "we know who some of the traitors engaged in this activity are. . . ." Others, he noted, remained concealed, and for that reason the Army was going to methodically remove any individuals from the project who seemed to be followers of the Communist Party line. There would be no mass discharges, only careful investigations based on substantial evidence. To this end, Lansdale wanted to use Oppenheimer: "He should be told that we have hesitated to take him into our confidence in this matter . . . because of his known interest in the Communist Party and his association with and friendship for certain members of the Communist Party." Lansdale seemed to think that this approach would encourage Oppenheimer to name names. In short, Lansdale was telling Groves that if he intended to keep Oppenheimer as his scientific director, he should press him to become an informer.

OVER THE ENSUING MONTHS and years, indeed, as long as Oppenheimer was in the employ of the government, he was harassed by variations of the Pash-Lansdale strategy. At Los Alamos, he was assigned assistants who in reality were "specially trained Counter Intelligence Corps agents who will not only serve as bodyguards for subject but also as undercover agents for this office." His driver and bodyguard, Andrew Walker, was a CIC agent who reported directly to Colonel Pash; his mail was monitored, his phone tapped, his office wired. Even after the war, he was subjected to close physical and electronic surveillance. His past associations were raised repeatedly by congressional committees and the FBI, and he was made to understand—repeatedly—that he was himself suspected of being a Communist Party member.

CHAPTER SEVENTEEN

"Oppenheimer Is Telling the Truth..."

I would be perfectly willing to be shot if I had done anything wrong.

ROBERT OPPENHEIMER *to Lt. Col. Boris Pash*

ENERAL GROVES AGREED WITH Lieutenant Colonel Lans-dale's recommendations. They would keep Oppenheimer as scientific director of the project, but Lansdale would set out to reel Oppenheimer into his security web. Not surprisingly, Pash vigorously objected to this subtle strategy, but on July 20, 1943, Groves instructed the Manhattan Project security division to issue Oppenheimer his security clearance. This was to be done "irrespective of the information which you have concerning Mr. Oppenheimer. He is absolutely essential to the project." Pash was not the only security officer who seethed at this decision. When Groves' aide, Lt. Col. Kenneth Nichols, informed Oppenheimer that his clearance had been issued, Nichols warned him, "In the future, please avoid seeing your questionable friends, and remember, whenever you leave Los Alamos, we will be tailing you." Nichols already strongly distrusted Oppenheimer, not merely because of his past associations with communists but because he believed Oppenheimer was endangering security by recruiting "questionable people" at Los Alamos. The more he saw of Oppenheimer, the more Nichols grew to despise him. That Groves didn't share this sentiment, and was actually coming to trust the physicist, irritated Nichols and only accentuated his resentment of Oppenheimer.

If Oppenheimer couldn't be eliminated, there were others more vulnerable—Oppenheimer's protégé Rossi Lomanitz, for example. On July 27, 1943, the twenty-one-year-old physicist was called into Ernest Lawrence's office and told that he was being promoted to group leader in the Radiation Lab. But three days later, as the result of an investigative report by Pash, Lomanitz received a special-delivery letter from his draft board ordering him to appear for a physical examination the very next day. He immediately called Oppenheimer in Los Alamos and told him what had happened. Oppie fired off a cable that afternoon to the Pentagon, saying that a "very serious mistake is being made. Lomanitz now only man at Berkeley who can take this responsibility." Despite this intervention, Lomanitz was shortly inducted into the Army.

A few days later, Lansdale dropped by Oppenheimer's Los Alamos office for a long chat. Lansdale warned Oppenheimer against any further efforts to help Lomanitz, saying that the young physicist had been guilty of "indiscretions which could not be overlooked or condoned." Lansdale avowed that even after joining the Radiation Lab, Lomanitz had continued his political activities. "That makes me mad," Oppenheimer said. Lomanitz had promised him, he explained, that if he came aboard the bomb project he would abstain from political work.

Lansdale and Oppenheimer then had a general discussion about the Communist Party. Lansdale declared that, as a military intelligence officer, he was not concerned with a man's political beliefs. His only concern was with preventing the transmission of classified information to unauthorized persons. To Lansdale's surprise, Oppenheimer vigorously disagreed, saying that he did not want anyone working for him on the project who was a current member of the Communist Party. According to Lansdale's memorandum of the conversation, Oppenheimer explained that "one always had a question of divided loyalty." Discipline inside the Communist Party "was very severe and was not compatible with complete loyalty to the project." He made it clear to Lansdale that he was speaking only of those who were current members of the Party. Former members were another matter—he knew several former Party members who were now working in Los Alamos.

Before Lansdale could ask him for the names of these former members, their conversation was interrupted by someone walking into the room. Afterwards, Lansdale had the distinct impression that Oppenheimer was "trying to indicate that he had been a member of the party, and had definitely severed his connections upon engaging in this work." Lansdale's overall impression was that Oppenheimer "gave every appearance of sincerity." The scientist was "extremely subtle in his allusions" but also "anxious" to explain his position. In the months to come, the two men would occasionally spar over security issues, but Lansdale would always believe that Oppenheimer was loyal and devoted to America.

Oppenheimer himself, however, came away worried from this conversation with Lansdale. The fact that Lomanitz had been dismissed from the Rad Lab despite his intercession was troublesome. Unaware of the exact "indiscretions" that had provoked this action, Oppenheimer surmised that the cause was union organizing on behalf of FAECT. In this context, he recalled that George Eltenton, the Shell engineer who had asked Chevalier to approach him about passing project information to the Soviets, had also been active in FAECT. The conversation in his kitchen some six months earlier with Chevalier about Eltenton's scheme—which he had dismissed as ridiculous—now appeared serious. Oppie's meeting with Lansdale thus triggered a fateful decision: He decided he had to tell the authorities about Eltenton's activities.

General Groves later told the FBI that Oppenheimer first came to him with Eltenton's name sometime in early or mid-August. But Oppenheimer didn't stop there. On August 25, 1943, during a visit to Berkeley on project business, Robert walked into the office of Lt. Lyall Johnson, the army security officer for the Rad Lab. After a brief discussion about Lomanitz, he told Johnson that there was a man in town who worked at the Shell Development Corporation and was active in FAECT. His name, he said, was Eltenton, and he ought to be watched. He intimated that Eltenton may have been trying to obtain information about the Rad Lab's work. Oppenheimer left without saying much more. Lieutenant Johnson immediately called his superior, Colonel Pash, who instructed him to have Oppenheimer return the next day for an interview. Overnight, they placed a small microphone in the base of the phone on Johnson's desk and ran a connection to a recording device in the adjoining room.

The next day, Oppenheimer appeared for what would be a fateful interrogation. When he walked into Johnson's office, he was startled to be introduced to Pash, still a stranger, but nevertheless a man whose reputation had preceded him. As the three men sat down, it was clear that Pash himself would conduct the interview.

Pash began with transparent obsequiousness: "This is a pleasure. . . . General Groves has, more or less, I feel, placed a certain responsibility in me and it's like having a child, that you can't see, by remote control. I don't mean to take much of your time."

"That's perfectly all right," Oppenheimer replied. "Whatever time you choose."

When Pash then began to ask him about his conversation of the day before with Lieutenant Johnson, Oppenheimer interrupted and began talking about the subject he had expected to discuss, Rossi Lomanitz. He explained that he didn't know whether he should talk to Rossi, but he wanted to tell him that he had been indiscreet.

Pash interrupted and said he had more serious concerns. Were there "other groups" interested in the Rad Lab?

"Oh, I think that is true," Oppenheimer replied, "but I have no firsthand knowledge." But then he went on to say, "I think it is true that a man, whose name I never heard, who was attached to the Soviet consul, has indicated indirectly through intermediary people concerned in this project that he was in a position to transmit, without danger of leak, or scandal, or anything of that kind, information which they might supply." He then indicated that he was concerned about possible "indiscretions" on the part of people who might move in the same circles. Having revealed as "fact" an effort by someone in the Soviet consulate to collect information on the Rad Lab's activities, Oppenheimer plunged ahead and, without interruption from Pash, explained his personal position: "To put it quite frankly—I would feel friendly to the idea of the Commander in Chief informing the Russians that we were working on this problem. At least, I can see that there might be some arguments for doing that, but I do not feel friendly to the idea of having it moved out the back door. I think that it might not hurt to be on the lookout for it."

Pash—a man reared to loathe the Bolsheviks—responded evenly, "Could you give me a little more specific information as to exactly what information you have? You can readily realize that phase [the transmittal of secret information] would be, to me, as interesting, pretty near, as the whole project is to you."

"Well, I might say," replied Oppenheimer, "that the *approaches* were always to other people, who were troubled by them, and sometimes came and discussed *them* with me."

Oppenheimer had used the plural, and he began to elaborate about more than one such approach. He had not come to this interview prepared. Indeed, he had expected to be asked to expand on his conversation with Lieutenant Johnson about Lomanitz. Suddenly he was facing Pash, and a line of questioning that was making him anxious—and all too loquacious.

The memory of his brief conversation with Chevalier six months ago in his Berkeley kitchen was now hazy. Perhaps Chevalier had mentioned to him (as Eltenton later told the FBI) that Eltenton had suggested approaching three scientists: Lawrence and Alvarez in addition to himself. But perhaps he had in mind several other conversations about the notion that the Soviets ought to have access to new weapons technology. And why not? Many of his friends, students and colleagues worried daily about a fascist victory in Europe. They understood, quite correctly, that only the Soviet army could prevent such a calamity. Many of the physicists then working in the Rad Lab were not joining the Army only because they had been convinced—in quite a few cases by Oppenheimer himself—that their special project would materially contribute to the war effort. These men often discussed whether their government was doing everything it could to help those bearing the brunt of the fascist onslaught. Surely, Oppenheimer had heard many of his colleagues and students giving voice to the desire to help the beleaguered Russians—at a time when, after all, the Soviets were being promoted in the American press as heroic allies.

So Oppenheimer now tried to explain to Pash that the people who approached him about assisting the Soviets all came to him with an attitude of "bewilderment rather than one of cooperation." They were sympathetic to the notion of helping our ally, but troubled by the idea of providing information, as Oppenheimer put it, "out the back door." Oppenheimer now reported what he had already told Groves and Lieutenant Johnson: that George Eltenton, who worked at the Shell Development Corporation, should be watched. "He has probably been asked," Oppenheimer said, "to do what he can to provide information." Eltenton, he said, had talked to a friend who was also an acquaintance of one of the men on the project.

When Pash pressed him to name who had been approached, Oppenheimer politely refused, on the grounds that the individuals were entirely innocent. "I'll tell you one thing," Oppenheimer said, "I have known of two or three cases, and I think two of the men were with me at Los Alamos— they are men who are very closely associated with me." These two Los Alamos men were approached separately but within a week of each other. A third man, an employee of the Rad Lab, had already left or was scheduled to be transferred to "Site X"—the Oak Ridge facility of the Manhattan Project in Tennessee. These approaches came not from Eltenton but from a third party, a man Oppenheimer refused to name because, he said, "I think it would be a mistake." He explained that it was his "honest opinion" that the man was himself innocent. He conjectured that this individual had bumped into Eltenton at a party and Eltenton had said, "Do you suppose you could help me? This is a very serious thing because we know that important work is going on here, and we think this ought to be made available to our allies, and would you see if any of those guys are willing to help us with it."

Other than identifying this "third party" as a member of the Berkeley faculty, Oppenheimer stubbornly refused to say more, insisting, "I think I have told you where the initiative came from [Eltenton] and that the other things were almost purely accident. . . ." Oppenheimer had identified Eltenton because he considered him as "dangerous to this country." He would not, in the same breath, name his friend Hoke, whom he believed to be an innocent. "The intermediary between Eltenton and the project," Oppenheimer told Pash, "thought it was the wrong idea, but said that this was the situation. I don't think he supported it. In fact, I know it."

While refusing to name Chevalier or any names other than Eltenton's, Oppie talked freely and in considerable detail about the nature of the approach to his friends. In an effort to place all of this in a benign context, he told Pash, "Let me give you the background. The background was—well, you know how difficult it is with the relations between these two allies, and there are a lot of people who don't feel very friendly to Russia, so that the information—a lot of our secret information, our radar and so on, doesn't get to them, and they are battling for their lives and they would like to have an idea of what is going on and this is just to make up, in other words, for the defects of our official communication. That is the form in which it was presented."

"Oh, I see," Pash responded.

"Of course," Oppenheimer rushed to acknowledge, "the actual fact is that since it is not a communication which ought to be taking place, it is treasonable." But the spirit of the approach was not treason at all, Oppie continued. Aiding our Soviet allies was "more or less a policy of the Government. . . ." The men involved were merely being asked to compensate for the bureaucracy's "defects" in official communications with the Russians. Oppenheimer even spelled out how the information would be transmitted to the Russians. As he understood it from his friends who had been approached by Eltenton's contact, an interview would be arranged with Eltenton. They were told that "this man Eltenton . . . had very good contacts with a man from the [Soviet] embassy attached to the consulate who was a very reliable guy (that's his story) and who had a lot of experience in microfilm work, or whatever the hell."

"SECRET INFORMATION." "Treasonable." "Microfilm." Oppenheimer had used all of these words, surely alarming Pash who already was convinced that Oppenheimer was a dangerous security risk, if not a hardened communist agent. Pash would never understand the man who sat before him. Although he and Oppenheimer lived in adjacent cities, they came from different worlds. The former high school football coach and intelligence officer must have been astonished that Oppie could sound so self-assured as he spoke of treasonable activities and in the same breath confidently explained why he could not, as a matter of principle, name the names of men he knew to be innocent.

In some respects, Oppenheimer had become a changed man in the six months since his conversation with Chevalier. Los Alamos had transformed him; he was now the bomb laboratory's director, the science administrator upon whose shoulders the ultimate success of the project rested. But in other respects, he was the same self-assured, brilliant professor of physics who demonstrated every day that he had an informed opinion about an astonishingly broad array of topics. He understood Pash had a job to do, but

Oppie was confident that he could decide on his own who was a security risk (Eltenton) and who was not (Chevalier). He even explained to Pash his belief that "association with the Communist movement is not compatible with the job on a secret war project, it is just that the two loyalties cannot go [together]." Furthermore, he told Pash, "I think that a lot of brilliant and thoughtful people have seen something in the Communist movement, and that they maybe belong there, maybe it is a good thing for the country. I hope it doesn't belong on the war project. . . ."

As he had told Lansdale just a few weeks earlier, Party discipline subjected members to the pressures of dual loyalties. As an example he cited Lomanitz, to whom he still felt "a sense of responsibility." Lomanitz, he said, "may have been indiscreet in circles [meaning the Communist Party] which would lead to trouble." He had no doubt that people often approached Lomanitz and they "might feel it their duty if they got word of something to let it go further. . . ." For this reason, it would simplify things for everyone if it were agreed that communists should stay away from secret war projects.

Incredibly—in retrospect—Oppenheimer repeatedly tried to convince Pash that pretty much all the people involved in these contacts were well-meaning innocents. "I'm pretty sure that none of the guys here, with the possible exception of the Russian, who is doing probably his duty by his country—but the other guys really were just feeling they didn't do anything but they were considering the step, which they would have regarded as thoroughly in line with the policy of this Government, just making up for the fact that there were a couple of guys in the State Department who might block such communications." He pointed out that State was sharing some information with the British, and so many people thought there wasn't a great deal of difference between that and sharing similar information with the Soviets. "A thing like this going on, let us say, with the Nazis would have a somewhat different color," he told Pash.

From Pash's perspective, all of this was outrageous and, moreover, quite beside the point. Eltenton and at least one other individual—the unnamed faculty member—were trying to get information about the Manhattan Project, and that was espionage. Pash nevertheless patiently listened to Oppenheimer lecture him on his view of the security problem, and then he returned the focus of the conversation back to Eltenton and the unnamed intermediary. Pash explained that it might be necessary for him to come back to Oppenheimer and press him again for more names. Oppenheimer again explained that he was only trying to "act reasonably" and "draw the line" between those, like Eltenton, who took the initiative and those who reacted negatively to such approaches.

They continued to spar a little longer. Pash tried to use a bit of irony, saying, "I am not persistent (ha ha) but—"

"You are persistent," interrupted Oppenheimer, "and it is your duty."

Toward the end of the interrogation, Oppenheimer returned to his earlier concerns about the FAECT union: The main thing Pash needed to know was that "there are some things there which would bear watching." He even suggested that "it wouldn't hurt to have a man in the local of this union FAECT—to see what may happen and what he can pick up." Pash immediately picked up on this suggestion and asked if Oppenheimer knew anyone in the union who might be willing to serve as an informant. He replied, no, that he had only heard that "a boy called [David] Fox is president of it."

Oppenheimer then made it clear to Pash that as director at Los Alamos, he was certain that "everything is 100 percent in order. . . . I think that's the truth," he said, and added for emphasis, "I would be perfectly willing to be shot if I had done anything wrong."

When Pash indicated that he might be visiting Los Alamos, Oppenheimer quipped, "My motto is God bless you." As Oppenheimer rose to leave, the tape recorder captured Pash saying, "the best of luck." Oppenheimer replied, "Thank you very much."

It was a bizarre—and ultimately disastrous—performance. Oppenheimer had raised the red flag of espionage, identified Eltenton as the culprit, described an unnamed "innocent" intermediary and reported that this innocent person had contacted several other scientists who likewise were innocent. He was certain of his judgments, he had assured Pash, so there was no need to name names.

Recall that, unbeknownst to Oppenheimer, this conversation was recorded and transcribed. It became a part of Oppenheimer's security file, and because he would later claim that his report of approach*es* (whether it was two or three is not clear) was inaccurate—a "cock and bull" story whose origins he himself could not explain—he could never prove whether he had lied to Pash, or had told Pash the truth and lied later. It was as if he unknowingly had swallowed a time bomb; a decade would pass before it exploded.

IN THE AFTERMATH of Oppenheimer's encounter with Pash, Lansdale and Groves realized they had a serious problem on their hands. On September 12, 1943, Lansdale sat down with Robert for yet another long and frank conversation. Having read the transcript of Oppenheimer's interrogation, he was determined to get to the bottom of the alleged espionage approach. Surreptitiously, he, too, recorded the conversation.

Lansdale began with an obvious attempt to flatter Oppenheimer. "I want to say this without any intent of flattery . . . you're probably the most intelligent man I ever met." He then confessed that he had not been entirely straight with him during their previous conversations, but now he wanted to be "perfectly frank." Lansdale then explained that "we have known since February that several people were transmitting information about this project to the Soviet Government." He claimed that the Soviets knew the scale of the project, knew about the facilities at Los Alamos, Chicago and Oak Ridge—and had a general sense of the project's timetable.

Oppenheimer seemed genuinely shocked by this news. "I might say that I have not known that," he told Lansdale. "I knew of this *one attempt* to obtain information which was earlier, or I don't, I can't remember the date, though I've tried."

The conversation soon turned to the role of the Communist Party, and both men agreed on having heard that it was Party policy that anyone doing confidential war work should resign their Party membership. Robert volunteered that his own brother, Frank, had severed his ties to the Party. Moreover, eighteen months before, when they had started work on the project, Robert said he had told Frank's wife, Jackie, that she should stop socializing with CP members. "Whether they have, in fact, done that, I don't know." He confessed that it still worried him that his brother's friends were "very left wing, and I think it is not always necessary to call a unit meeting for it to be a pretty good contact."

Lansdale in turn explained his approach to the whole problem of security. "You know as well as I do," Lansdale told Oppenheimer, "how difficult it is to prove communism." Besides, their goal was to build the "gadget," and Lansdale suggested that a man's politics really didn't matter so long as he was contributing to the project. After all, everyone was risking their lives to get the job done, and "we don't want to protect the thing [the project] to death." But if they thought a man was engaged in espionage, they had to make a decision on whether to prosecute him or just weed him out of the project.

At this point, Lansdale brought up what Oppenheimer had told Pash about Eltenton—and Oppenheimer once again said he didn't think it would be right to name the individual who had approached him. Lansdale pointed out that Oppenheimer had spoken of "three persons on the project" who had been contacted and all three told this intermediary "to go to hell in substance." Oppenheimer agreed. So Lansdale asked him how he could be sure that Eltenton hadn't approached other scientists. "I don't," Oppenheimer replied. "I can't know that." He understood why Lansdale thought it important to discover the channel through which this initial approach had been made, but he still felt it would be wrong to involve these other people.

"I hesitate to mention any more names because of the fact that the other names I have do not seem to be people who were guilty of anything. . . . They are not people who are going to get tied up in it in any other way. That is, I have a feeling that this is an extremely erratic and unsystematic thing." He therefore felt "justified" in withholding the name of the intermediary "because of a sense of duty."

Changing direction, Lansdale asked Oppenheimer for the names of those individuals working on the project in Berkeley who he thought were Party members or had once been Party members. Oppenheimer named some names. He said he had learned on his last visit to Berkeley that both Rossi Lomanitz and Joe Weinberg were Party members. He thought a secretary named Jane Muir was a member. At Los Alamos, he said, he knew that Charlotte Serber had at one time been a Party member. As to his good friend, Bob Serber, "I think it is possible, but I don't know."

"How about Dave Hawkins?" Lansdale asked.

"I don't think he was, I would not say so."

"Now," said Lansdale, "have you yourself ever been a member of the Communist Party?"

"No," replied Oppenheimer.

"You've probably belonged to every front organization on the Coast," Lansdale suggested.*

"Just about," Oppenheimer replied casually.

"Would you in fact have considered yourself at one time a fellow traveler?"

"I think so," replied Oppenheimer. "My association with these things was very brief and very intense."

At a later point, Lansdale got Oppenheimer to explain why he might have gone through a relatively brief period of intense association with the Party—yet never joined. Oppenheimer remarked that a lot of these people they had been discussing had joined the Party out of "a very deep sense of right and wrong." Some of these people, Oppenheimer said, "have a very deep fervor," something akin to a religious commitment.

"But I can't understand;" interrupted Lansdale, "here's the particular thing about it. They are not adhering to any constant ideals. . . . They may be adhering to Marxism, but they follow the twistings and turnings of a line designed to assist the foreign policy of another country."

Oppenheimer agreed, saying, "This conviction makes it not only hysterical. . . . I think absolutely unthinkable[.] My membership in the Communist Party. [Quite clearly, what he means here is that actually joining the Communist Party was for him "unthinkable."] At the period in which I was

*During the 1954 security hearing, these words were attributed to Oppenheimer.

involved there were so many positions in which I did fervently believe, in correctments [*sic*] and aims of the party. . . ."

Lansdale: "Can I ask what period that was?"

Oppenheimer: "That was the time of the Spanish War, up to the [Nazi-Soviet] pact."

Lansdale: "Up to the pact. That is the time you broke, you might say?"

Oppenheimer: "*I never broke. I never had anything to break.* I gradually disappeared from one after another of the organizations." (Emphasis added.)

When Lansdale once again pressed him for names, Oppenheimer replied, "I would regard it as a low trick to involve someone where I would be[t] dollars to doughnuts he wasn't involved."

Lansdale ended the interview with a sigh and said, "O.K., sir."

TWO DAYS LATER, on September 14, 1943, Groves and Lansdale had another conversation with Oppenheimer about Eltenton. They were on a train ride between Cheyenne and Chicago, and Lansdale wrote up a memorandum of the conversation. Groves brought up the Eltenton affair, but Oppenheimer said he would only name the intermediary if ordered to do so. A month later, Oppenheimer again refused to name the intermediary. But curiously, Groves accepted Robert's position. He attributed it to Oppenheimer's "typical American schoolboy attitude that there is something wicked about telling on a friend." Pressed by the FBI for more information about the whole affair, Lansdale informed the Bureau that both he and Groves "believed that Oppenheimer is telling the truth. . . ."

MOST OF GROVES' subordinates did not share his trust in Oppenheimer. Early in September 1943, Groves had a conversation with another of the Manhattan Project's security officers, James Murray. Frustrated that Oppenheimer had finally been awarded a security clearance, Murray posed a hypothetical question for Groves: Suppose twenty individuals in Los Alamos were found to be definite communists and this evidence was laid before Oppenheimer. How would Oppenheimer react? Groves replied that Dr. Oppenheimer would say that all scientists are liberals and that this was nothing to be alarmed about. Groves then told Murray a story. Some months earlier, he said, Oppenheimer was asked to sign a secrecy pledge that among other things stated that he would "always be loyal to the United States." Oppenheimer signed the pledge, but he first struck out those words and wrote, "I stake my reputation as a scientist." If a "loyalty" oath was personally distasteful, Oppenheimer was nevertheless pledging his absolute trustworthiness as a scientist. It was an arrogant act—but one calculated to make it clear to Groves that science was the altar at which Oppenheimer worshipped and that he had pledged his unreserved commitment to the success of the project.

Groves went on to explain to Murray that he believed Oppenheimer would regard any subversive activity at Los Alamos as a personal betrayal. "In other words," Groves said, "it is not a question of the country's safety, but rather whether a person might be working against OPP [Oppenheimer] in stopping him from obtaining the reputation which will be his, with the complete development of the project." In Groves' eyes, Oppenheimer's personal ambitions guaranteed his loyalty. According to Murray's notes of the conversation, Groves explained that Oppenheimer's "wife is pressing him for fame and that his wife's attitude is that [Ernest] Lawrence has received all the limelight and honors in this matter so far, and she would rather that Dr. OPP have these honors because she thinks her own husband is more deserving. . . . this is the Doctor's one big chance to gain a name for himself in the history of the world." For this reason, Groves concluded, "it is believed that he will continue to be loyal to the United States. . . ."

Fierce ambition was a character trait Groves respected and trusted. It was a trait he shared with Oppie, and together they had a single transcendent goal—to build this primordial weapon that would defeat fascism and win the war.

GROVES CONSIDERED himself a good judge of character, and in Oppenheimer he believed he had found a man of unswerving integrity. Still, he also knew that the Army-FBI investigation of the Eltenton affair would go nowhere without further names. So finally, in early December 1943, Groves ordered Oppenheimer to name the intermediary who had approached him with Eltenton's request. Oppenheimer, having committed himself to respond frankly if ordered, reluctantly named Chevalier, insisting that his friend was harmless and innocent of espionage. Putting together what Robert had told Pash on August 26 with this new information, Colonel Lansdale wrote the FBI on December 13, "Professor J. R. Oppenheimer stated that three members of the DSM project [an early designation for the bomb program] had advised him that they were approached by an unnamed professor at the University of California to commit espionage." When ordered to name the professor, Lansdale said, Oppenheimer had identified Chevalier as the intermediary. Lansdale's letter mentioned no other names, either because Oppenheimer was still refusing to identify the three men approached by Chevalier, or more likely, because Groves had asked him only for the name of the intermediary. This so rankled the FBI that two months later, on February 25, 1944, the Bureau pressed Groves to get Oppenheimer to reveal the names of the "other scientists." Groves apparently did not even bother to reply to this request, for the Bureau was never able to find a reply in its records.

And yet, in *Rashomon* fashion, there is yet another version of this story. On March 5, 1944, FBI agent William Harvey wrote a summary memorandum titled "Cinrad." "In March 1944,"* Harvey reported, "General Leslie R. Groves conferred with Oppenheimer. . . . Oppenheimer finally stated that only one person had been approached by Chevalier, that one person being his brother, Frank Oppenheimer." In this version, Chevalier is supposed to have approached Frank—not Robert—in the fall of 1941. Frank is reported to have immediately informed his brother—who promptly phoned Chevalier and "gave him hell."

If Frank was involved, this of course would put the story in a quite different light. But the story is not only problematic, it is certainly incorrect. Why would Chevalier approach Frank, whom he hardly knew, rather than Robert, his closest friend? And it seems quite ridiculous that anyone would ask Frank for information in the autumn of 1941 about a project that didn't get started until the summer of 1942, at the earliest. Moreover, both Chevalier and Eltenton, in simultaneous interviews with the FBI, confirmed that the Eagle Hill kitchen conversation was between Oppenheimer and Chevalier and it occurred in the winter of 1942–43. Furthermore, Harvey's March 5 memo is the only roughly contemporaneous document that mentions Frank Oppenheimer, and after searching its files the FBI reported that "the original source of the story involving Frank Oppenheimer has not been located in Bureau files." Nevertheless, because Harvey's report was now part of Oppenheimer's FBI dossier, this part of the story would acquire a robust life of its own.†

*Harvey probably had the date wrong.

†Over the years, such thoughtful historians as Richard Rhodes, Gregg Herken and Richard G. Hewlett and Jack M. Holl have suggested that Frank Oppenheimer was somehow involved in the Eltenton scheme.

CHAPTER EIGHTEEN

"Suicide, Motive Unknown"

I am disgusted with everything. . . .

JEAN TATLOCK
January 1944

L T. COL. BORIS PASH had spent two frustrating months in the autumn of 1943 trying to discover who had talked to Oppenheimer about passing information to the Soviet consulate. To no avail, he and his agents had repeatedly interviewed various Berkeley students and faculty members. Pash had been dogged and stubborn in his investigation—and so antagonistic toward Oppenheimer as to finally lead Groves to conclude that Pash was wasting the Army's time and resources on an investigation that was going nowhere. This was what had finally prompted Groves, in early December 1943, to order Oppenheimer to name the contact—Chevalier. At the same time, Groves decided that Pash's talents could be put to better use elsewhere. In November, he was made military commander of a secret mission, code-named Alsos, to determine the status of the Nazi regime's bomb program by capturing German scientists. Pash was transferred to London, where he would spend the next six months preparing a top-secret team of scientists and soldiers to follow the Allied troops into Europe. But even after Pash's departure, his friends at the FBI office in San Francisco continued monitoring Jean Tatlock's phone conversations from her apartment on Telegraph Hill. Months had gone by, and they had learned nothing to confirm their suspicions that the young psychiatrist was Oppenheimer's (or anyone's) conduit for passing information to the Soviets. But no one at Bureau headquarters in Washington told them to stop the surveillance.

Early in 1944—just after the holiday season—Tatlock was coping with one of her black moods. When she visited her father in his Berkeley home on Monday, January 3, he found her "despondent." Upon leaving him that day, she promised to phone him the next evening. When she failed to call on

Tuesday night, John Tatlock tried phoning her, but Jean never answered. Wednesday morning he tried again, and then went to her apartment on Telegraph Hill. Arriving at about 1:00 p.m., he rang the doorbell and after getting no response, Professor Tatlock, age sixty-seven, climbed through a window.

Inside the flat, he discovered Jean's body "lying on a pile of pillows at the end of the bathtub, with her head submerged in the partly filled tub." For whatever reason, Professor Tatlock did not call the police. Instead, he picked his daughter up and laid her on the sofa in the living room. On the dining room table, he found an unsigned suicide note, scribbled in pencil on the back of an envelope. It read in part, "I am disgusted with everything. . . . To those who loved me and helped me, all love and courage. I wanted to live and to give and I got paralyzed somehow. I tried like hell to understand and couldn't. . . . I think I would have been a liability all my life—at least I could take away the burden of a paralyzed soul from a fighting world." From there the words ran into a jagged, illegible line.

Stunned, Tatlock began rummaging about the apartment. Eventually, he found a stack of Jean's private correspondence and some photographs. Whatever he read in this correspondence inspired him to light a fire in the fireplace. With his dead daughter stretched out on the sofa beside him, he methodically burned her correspondence and a number of photographs. Hours passed. The first phone call he made was to a funeral parlor. Someone at the funeral parlor finally called the police. When they arrived at 5:30 p.m., accompanied by the city's deputy coroner, papers were still smoldering in the fireplace. Tatlock told the police that the letters and photos had belonged to his daughter. Four and a half hours had passed since he had discovered her body.

Professor Tatlock's behavior was, to say the least, unusual. But relatives who stumble upon the suicide of a loved one often behave oddly. That he methodically searched the apartment, however, suggests that he may have known what he was looking for. Clearly, what he saw in Jean's papers motivated him to destroy them. It wasn't politics: Tatlock sympathized with many of his daughter's political causes. His motive can only have been something more personal.

The coroner's report stated that death had occurred at least twelve hours earlier. Jean had died sometime during the evening of Tuesday, January 4, 1944. Her stomach contained "considerable recently ingested, semi-solid food"—and an undetermined quantity of drugs. One bottle labeled "Abbott's Nembutal C" was found in the apartment. It still contained two tablets of the sleeping pills. There was also an envelope marked "Codeine ½ gr" that contained only traces of white powder. Police also found a tin box labeled

"Upjohn Racephedrine Hydrochloride, ⅜ grain." The tin still contained eleven capsules. The coroner's toxicological department conducted an analysis of her stomach and found "barbituric acid derivative, a derivative of salicylic acid and a faint trace of chloral hydrate (uncorroborated)." The actual cause of death was "acute edema of the lungs with pulmonary congestion." Jean had drowned in her bathtub.

At a formal inquest in February 1944, a jury determined Jean Tatlock's death to be "Suicide, motive unknown." The newspapers reported that a $732.50 bill from her analyst, Dr. Siegfried Bernfeld, was found in the apartment, evidence that she had "taken her own troubles to a psychologist." Actually, as a psychiatrist in training, Jean was required to undergo analysis and pay for it herself. If recurring episodes of manic depression drove her to suicide, it was tragic. By all accounts, her friends thought she had reached a new plateau in her life. Her achievements were considerable. Her colleagues at Mount Zion Hospital—the foremost center in Northern California for training analytic psychiatrists—thought her an "outstanding success" and were shocked that she had taken her own life.

When Jean's childhood friend Priscilla Robertson learned of her death, she wrote her a posthumous letter, trying to understand what had happened. Robertson did not think a "personal heartbreak" would have pushed Jean to suicide: "For you were never starved for affection—your insatiable hunger was for creativity. And you longed to find perfection in yourself, not out of pride but in order to have a good instrument to serve the world. When you found that your medical training, completed, did not give you all the power for good that you had hoped for, when you found yourself entangled in the small routine of hospital conventions and in the huge messes which the war made in the lives of your patients, far beyond any doctor's power to patch up—then you turned, in your eleventh hour, again to psychoanalysis." Robertson speculated that perhaps it was this experience, "which always brings introspective despair in mid-course," that had stirred up agonies "too deep to be assuaged."

Robertson and many other friends were unaware that Tatlock was struggling to cope with issues surrounding her sexual orientation. Jackie Oppenheimer later reported Jean as telling her that her psychoanalysis had revealed latent homosexual tendencies. At the time, Freudian analysts regarded homosexuality as a pathological condition to be overcome.

Some time after Jean's death, one of her friends, Edith Arnstein Jenkins, went for a walk with Mason Roberson, an editor of *People's World.* Roberson had known Jean well and he said that Jean had confided to him that she was a lesbian; she told Roberson that in an effort to overcome her attraction to women she "had slept with every 'bull' she could find." This prompted

Jenkins to recall one occasion when she had entered the Shasta Road house on a weekend morning and seen Mary Ellen Washburn and Jean Tatlock "sitting up and smoking over the newspaper in Mary Ellen's double bed." In remarks suggesting her perception of a lesbian relationship, Jenkins later wrote in her memoir that "Jean seemed to need Mary Ellen," and she quoted Washburn as saying, "When I first met Jean, I was put off by her [large] breasts and her thick ankles."

Mary Ellen Washburn had a particular reason to be devastated when she heard the news of Tatlock's death; she confided to a friend that Jean had called her the night before she died and had asked her to come over. Jean had said she was "very depressed." Unable to come that night, Mary Ellen was understandably filled with remorse and guilt afterwards.

The taking of one's own life invariably becomes an imponderable, a mystery to the living. For Oppenheimer, Jean Tatlock's suicide was a profound loss. He had invested much of himself in this young woman. He had wanted to marry her, and even after his marriage to Kitty, he had remained a loyal friend to her in her need—and an occasional lover. He had spent many hours walking and talking her out of her depressions. And now she was gone. He had failed.

The day after the suicide was discovered, Washburn cabled the Serbers in Los Alamos. When Robert Serber went to tell Oppenheimer the sad news, he could see that Oppie had already heard. "He was deeply grieved," Serber recalled. Oppenheimer then left the house and went for one of his long, lonely walks high into the pines surrounding Los Alamos. Given what he knew about Jean's psychological state over the years, Oppenheimer must have felt a basketful of painfully conflicting emotions. Together with regret, anger, frustration and deep sadness, he surely also felt a sense of remorse and even guilt. For if Jean had become a "paralyzed soul," his looming presence in her life must somehow have contributed to this paralysis.

For reasons of love and compassion, he had become a key member of Jean's psychological support structure—and then he had vanished, mysteriously. He had tried to maintain the connection, but after June 1943 it was made very clear to him that he could not continue his relationship with Jean without jeopardizing his work in Los Alamos. He was trapped by circumstances. He had obligations to a wife he loved and a child. He had responsibilities to his colleagues in Los Alamos. From this perspective, he had acted reasonably. But in Jean's eyes, it may have seemed as if ambition had trumped love. In this sense, Jean Tatlock might be considered the first casualty of Oppenheimer's directorship of Los Alamos.

Tatlock's suicide was front-page news in San Francisco newspapers. That morning, the FBI office in San Francisco cabled J. Edgar Hoover a summary of what had been reported in the papers. The cable concluded: "No direct action will be taken by this office due possible unfavorable publicity. Direct inquiries will be made discreetly in view of elapse of time and Bureau will be advised."

In the years since, a number of historians and journalists have speculated about Tatlock's suicide. According to the coroner, Tatlock had eaten a full meal shortly before her death. If it was her intention to drug and then drown herself, as a doctor she had to have known that undigested food slows the metabolizing of drugs into the system. The autopsy report contains no evidence that the barbiturates had reached her liver or other vital organs. Neither does the report indicate whether she had taken a sufficiently large dose of barbiturates to cause death. To the contrary, as previously noted, the autopsy determined that the cause of death was asphyxiation by drowning. These curious circumstances are suspicious enough—but the disturbing information contained in the autopsy report is the assertion that the coroner found "a faint trace of chloral hydrate" in her system. If administered with alcohol, chloral hydrate is the active ingredient of what was then commonly called a "Mickey Finn"—knockout drops. In short, several investigators have speculated, Jean may have been "slipped a Mickey," and then forcibly drowned in her bathtub.

The coroner's report indicated that no alcohol was found in her blood. (The coroner, however, did find some pancreatic damage, indicating that Tatlock had been a heavy drinker.) Medical doctors who have studied suicides—and read the Tatlock autopsy report—say that it is possible she drowned herself. In this scenario, Tatlock could have eaten a last meal with some barbiturates to make herself sleepy and then self-administered chloral hydrate to knock herself out while kneeling over the bathtub. If the dose of chloral hydrate was large enough, Tatlock could have plunged her head into the bathtub water and never revived. She then would have died from asphyxiation. Tatlock's "psychological autopsy" fits the profile of a high-functioning individual suffering from "retarded depression." As a psychiatrist working in a hospital, Jean had easy access to potent sedatives, including chloral hydrate. On the other hand, said one doctor shown the Tatlock records, "If you were clever and wanted to kill someone, this is the way to do it."

Some investigators, as well as Jean's brother, Dr. Hugh Tatlock, have continued to question the bizarre nature of Jean's death. In 1975 they became increasingly suspicious of the conclusion that she had committed suicide after the U.S. Senate's Church Committee hearings on CIA assassination plots were made public. One of the star witnesses was none other than the irrepressible Boris Pash, who had not only directed the wiretapping of Jean's phone but had also proposed to interrogate Weinberg, Lomanitz, Bohm and Friedman "in the Russian manner" and then dispose of their bodies at sea.

Pash served from 1949 through 1952 as the CIA's Chief of Program Branch 7 (PB/7), a special operations unit within the Office of Policy Coordination, the original CIA clandestine service. Pash's boss, the Director of Operations Planning for OPC, told the Senate investigators that Colonel Pash's Program Branch 7 unit was responsible for assassinations and kidnapping as well as other "special operations." Pash denied that he had been delegated responsibility for assassinations, but acknowledged that it was "understandable" that others in the CIA "could have had the impression that my unit would undertake such planning." Former CIA officer E. Howard Hunt, Jr., told the *New York Times* on December 26, 1975, that in the mid-1950s he had been informed by his superiors that Boris T. Pash was in charge of a special operations unit responsible for the "assassination of suspected double agents and similar low-ranking officials. . . ."

Despite the CIA's claim that it had no records dealing with assassinations, the Senate Committee staff investigation concluded that Pash's unit was indeed assigned "responsibility for assassinations and kidnappings." It was documented, for example, that while working in the CIA's Technical Services Division in the early 1960s, Pash was involved in the attempt to design poisoned cigars destined for Fidel Castro.

Clearly, Col. Boris Pash, a veteran anti-Bolshevik turned counterintelligence officer, had all the credentials requisite for an assassin in a Cold War spy novel. But despite his colorful résumé, no one has produced evidence linking him to Tatlock's death. Indeed, by January 1944, Pash had been transferred to London. Jean's unsigned suicide note suggests that she died by her own hand—a "paralyzed soul"—and this is certainly what Oppenheimer always believed.

CHAPTER NINETEEN

"Would You Like to Adopt Her?"

Here at Los Alamos, I found a spirit of Athens, of Plato, of an ideal republic.

JAMES TUCK

L os ALAMOS WAS ALWAYS AN ANOMALY. Hardly anyone was over fifty, and the average age was a mere twenty-five. "We had no invalids, no in-laws, no unemployed, no idle rich and no poor," wrote Bernice Brode in a memoir. Everyone's driver's license had numbers and no name; their address was simply P.O. Box 1663. Surrounded by barbed wire, on the inside Los Alamos was transforming itself into a self-contained community of scientists, sponsored and protected by the U.S. Army. Ruth Marshak recalled arriving at Los Alamos and feeling "as if we shut a great door behind us. The world I had known of friends and family would no longer be real to me."

That first winter of 1943–44, the snows came early and stayed late. "Only the oldest men in the Pueblo," wrote a longtime resident, "remember so much snow on the ground for so many weeks." On some mornings the temperature fell to well below zero, draping the valley below in a thick fog. But the harshness of the winter served only to enhance the natural beauty of the mesa, and to connect the transplanted urbanites to this strange new mystical landscape. Some Los Alamos residents skied until May. When the snows finally melted, the drenched highlands blossomed with lavender mariposas and other wildflowers. Almost every day in the spring and summer, dramatic thunderstorms rolled in over the mountains for an hour or two in the late afternoon, cooling the terrain. Flocks of bluebirds, juncos and towhees perched in the spring-green cottonwoods around Los Alamos. "We learned to watch the snow on the Sangres, and to look for deer in Water Canyon," Phil Morrison later wrote, with a lyricism that reflected the emotional attachment to the land that seized many residents. "We found that on the mesas and in the valley there was an old and strange culture; there were our neighbors, the people of the pueblos, and there were the caves in Otowi canyon to remind us that other men had sought water in the dry land."

LOS ALAMOS was an army camp—but it also had many of the characteristics of a mountain resort. Just before arriving, Robert Wilson had finished reading Thomas Mann's *The Magic Mountain,* and sometimes he now felt as if he had been transported to that magical dominion. It was a "golden time," said the English physicist James Tuck: "Here at Los Alamos, I found a spirit of Athens, of Plato, of an ideal republic." It was an "island in the sky," or, as some new arrivals dubbed it, "Shangri-La."

Within a very few months, Los Alamos' residents forged a sense of community—and many of the wives credited Oppenheimer. Early on, in a nod to participatory democracy, he appointed a Town Council; later it became an elected body and, though it had no formal power, it met regularly and helped Oppie keep in touch with the community's needs. Here the mundane complaints of life—the quality of PX food, housing conditions and parking tickets—could be vented. By the end of 1943, Los Alamos had a low-power radio station that broadcast news, community announcements and music, the last drawn in part from Oppenheimer's large personal collection of classical records. In small ways, he made it known that he understood and appreciated the sacrifices everyone was making. Despite the lack of privacy, the spartan conditions and the recurring shortages in water, milk and even electricity, he infected people with his own special sense of jocular élan. "Everyone in your house is quite mad," Oppie told Bernice Brode one day. "You should get on fine together." (The Brodes lived in an apartment above Cyril and Alice Kimball Smith and Edward and Mici Teller.) When the local theater group put on a production of *Arsenic and Old Lace,* the audience was stunned and delighted to see Oppenheimer, powdered white with flour and looking stiff as a corpse, carried on stage and laid out on the floor with the other victims in Joseph Kesselring's comedy. And when, in the autumn of 1943, a young woman, the wife of a group leader, suddenly died of a mysterious paralysis—and the community feared a polio contagion—Oppenheimer was the first to visit the grieving husband.

At home, Oppie was the cook. He was still partial to exotic hot dishes like *nasi goreng,* but one of his stock dinners included steak, fresh asparagus and potatoes, prefaced by a gin sour or martini. On April 22, 1943, he hosted the first big party on The Hill—to celebrate his thirty-ninth birthday. He plied his guests with the driest of dry martinis and gourmet food, though the food was always on the scanty side. "The alcohol hits you harder at 8,000 feet," recalled Dr. Louis Hempelmann, "so everybody, even the most sober people, like Rabi, was just feeling no pain at all. Everyone was dancing." Oppie danced the fox-trot in his usual Old World style, holding his arm stiffly in front of him. Rabi amused everyone that night when he took out his comb and played it like a harmonica.

Kitty refused to play the social role of a director's wife. "Kitty was strictly a blue jeans and Brooks Brothers shirt kind of gal," recalled one Los Alamos friend. Initially, she worked part-time as a lab technician under the supervision of Dr. Hempelmann, whose job it was to study the health hazards of radiation. "She was awful bossy," he recalled. Only occasionally did she invite old Berkeley friends over for dinner, and she seldom hosted open house parties. However, Deke and Martha Parsons, the Oppenheimers' next-door neighbors, did like to entertain, and held many such events. Oppie encouraged everyone to work hard and play hard. "On Saturdays we raised whoopee," wrote Bernice Brode, "on Sundays we took trips, the rest of the week we worked."

On Saturday evenings, the lodge was often packed with square dancers, the men dressed in jeans, cowboy boots and colorful shirts, the women wearing long dresses bulging with petticoats. Not surprisingly, the resident bachelors hosted the rowdiest parties. These dorm parties were fueled by a concoction of half lab alcohol and half grapefruit juice mixed in a thirty-two gallon G.I. can and chilled with a chunk of smoking dry ice. One of the younger scientists, Mike Michnoviicz, sometimes played his accordion while everyone danced.

Occasionally, some of the physicists gave piano and violin recitals. Oppenheimer dressed up for these Saturday evening affairs, wearing one of his tweedy suits. Invariably, he was the center of attraction. "If you were in a large hall," Dorothy McKibbin recalled, "the largest group of people would be hovering around what, if you could get your way through, would be Oppenheimer. He was great at a party and women simply loved him." On one occasion, someone threw a theme party: "Come As Your Suppressed Desire." Oppie came dressed in his ordinary suit, with a napkin draped over his arm—as if to imply that he wished merely to be a waiter. It was a pose no doubt designed to reflect a studied humility rather than any real inner longing for anonymity. As the scientific director of the most important project in the war, Oppenheimer was actually living his "suppressed" desire.

On Sundays, many residents went for hikes or picnics in the nearby mountains, or rented the horses boarded at the Los Alamos Ranch School's former stables. Oppenheimer rode his own horse, Chico, a beautiful fourteen-year-old chestnut, on a regular route from the east side of town west toward the mountain trails. Oppie could make Chico "single-foot"—trot by placing each of his hooves down at a different time—over the roughest trails. Along the way, he greeted everyone he encountered with a wave of his mud-colored porkpie hat and a passing remark. Kitty was also a "very good horsewoman, really European trained"; initially, she rode Dixie, a full standardbred pacer who had once run the races in Albuquerque. Later she switched to a thoroughbred. An armed guard always accompanied them.

Oppenheimer's physical stamina atop a horse or hiking in the mountains invariably surprised his companions. "He always looked so frail," recalled Dr. Hempelmann. "He was always so painfully thin, of course, but he was amazingly strong." During the summer of 1944, he and Hempelmann rode together over the Sangre de Cristo Mountains to his Perro Caliente ranch. "It nearly killed me," said Hempelmann. "He was on his horse with the 'single-foot' gait, perfectly comfortable, and my horse had to go into a hard trot to keep up with him. I think the first day we must have ridden thirty to thirty-five miles, and I was nearly dead." Though rarely sick, Oppie suffered from smoker's cough, the result of a four- or five-pack-a-day habit. "I think he only picked up a pipe," said one of his secretaries, "as an interlude from the chain-smoking." He was given to uncontrolled, protracted spasms of coughing, and his face would sometimes flush purple as he persisted in talking through his cough. Just as he made a ceremony of mixing his martinis, Oppie smoked his cigarettes with singular style. Where most men used their index finger to tap ashes off the end of their cigarettes, he had the peculiar mannerism of brushing the ash from the tip by using the end of his little finger. The habit had so callused the tip of his finger that it appeared almost charred.

Gradually, life on the mesa became comfortable, if hardly luxurious. Soldiers chopped firewood and stacked it for use in each apartment's kitchen and fireplace. The Army also collected the garbage and stoked the furnaces with coal. Every day the Army bused in Pueblo Indian women from the nearby settlement of San Ildefonso to work as housekeepers. Dressed in deerskin-wrapped boots and colorful Pueblo shawls and wearing abundant turquoise and silver jewelry, the Pueblo women quickly became a familiar sight around town. Early each morning, after checking in with the Army's Maid Service Office near the town water tower, they could be seen walking along the dirt roads toward their assigned Los Alamos households for half a day—which was why the residents began calling them their "half-days." The idea, endorsed by Oppenheimer and administered by the Army, was that such maid service would allow the wives of project scientists to work as secretaries, lab assistants, schoolteachers or "computing-machine operators" in the Tech Area. This in turn would help the Army keep the population of Los Alamos to a minimum and support the morale of so many intelligent and energetic women. Maid service was assigned largely on the basis of need, depending on the importance and hours of a housewife's job and the number of young children, as well as on occasions of illness. Not always perfect, this bit of army socialism greatly eased life on the mesa and helped to turn the isolated laboratory into a fully employed, effective community.

Los Alamos always had an unusually high percentage of single men and women, and naturally, the Army had little success in keeping the sexes apart. Robert Wilson, the youngest of the lab's group leaders, was chairman of the Town Council when the military police ordered the closing of one of the women's dormitories and the dismissal of its female residents. A tearful group of young women, supported by a determined group of bachelors, appeared before the Council to appeal the decision. Wilson later recalled what happened: "It seems that the girls had been doing a flourishing business of requiting the basic needs of our young men, and at a price. All understandable to the Army until disease reared its ugly head, hence their interference." In the event, the Town Council decided that the number of girls plying their trade was few; health measures were taken and the dormitory was kept open.

EVERY FEW WEEKS, residents of The Hill were permitted to spend an afternoon in Santa Fe, shopping. Some would also take the occasion to drop by the bar at La Fonda for a drink. Oppenheimer frequently spent the night in Dorothy McKibbin's beautiful, thick-walled adobe home on the Old Santa Fe Trail. In 1936, McKibbin had spent $10,000 to build a classic Hispanic ranch house on an acre and a half of land just south of Santa Fe. With its carved Spanish doors and wraparound porch, the house looked as if it had been there for many decades. Dorothy filled it with local antique furniture and Navajo rugs. As the project's "gatekeeper," she held a "Q" (top-level) security badge, and so Oppenheimer frequently used her home to hold sensitive meetings in Santa Fe. McKibbin loved playing "den mother" on these occasions—but she also treasured the many quiet evenings she spent alone with Oppenheimer, cooking his favorite dinner of steak and asparagus, while he mixed "the best dry martinis you ever had." For Oppenheimer, McKibbin's home was a refuge from the constant surveillance he lived with on The Hill. "Dorothy loved Robert Oppenheimer," David Hawkins later said. "He was her special one, and she, his."

WHILE MOST Los Alamos spouses adapted reasonably well to the stark climate, isolation and rhythms of the mesa, Kitty increasingly felt trapped.

She wanted desperately what Los Alamos could give her husband—but as a bright woman with ambitions to be a botanist, she felt stymied professionally. After a year of doing blood counts for Dr. Hempelmann, she quit. She also felt isolated socially. If she was in a good mood, she could be charming and warm with friends or strangers. But everyone sensed she had a sharp edge. Often she seemed tense and unhappy. At social gatherings she could make small talk, but, as one friend put it, "She wanted to make big talk." Joseph Rotblat, a young Polish physicist, saw her occasionally at parties or in the Oppenheimer home for dinner. "She seemed to be very much aloof," Rotblat said, "a haughty person."

Oppenheimer's secretary, Priscilla Greene Duffield, had an ideal perch from which to observe Kitty. "She was a very intense, very intelligent, very vital kind of person," Duffield recalled. But she also thought Kitty was "very difficult to handle." Pat Sherr, a neighbor and the wife of another physicist, felt overwhelmed by Kitty's meteoric personality. "She was outwardly very gay and exuded some warmth," recalled Sherr. "I later realized that it wasn't any real warmth for people, but it was part of her terrible need for attention, for affection."

Like Robert, Kitty tended to shower people with gifts. When Sherr complained one day about the kerosene stove in her cabin, Kitty gave her an old electric stove. "She would give me gifts and envelop me totally," Sherr said. Other women found her abrupt manner to verge on insulting. But so, too, did many men, even though Kitty seemed to prefer the company of men. "She's also one of the very few people I've heard men—and very nice men—call a bitch," recalled Duffield. But it was also clear to Duffield that her boss trusted Kitty and turned to her for advice about all manner of issues. "He would give her judgment as much weight as that of anyone whose advice he chose to ask," she said. Kitty never hesitated to interrupt her husband, but, recalled one close friend, "It never seemed to bother him."

EARLY IN 1945, Priscilla Greene Duffield had a baby and Oppenheimer suddenly needed a new secretary. Groves offered him in turn several seasoned secretaries, but Oppenheimer rejected them all, until one day he told Groves that he wanted Anne T. Wilson, a pretty blond, blue-eyed twenty-year-old whom he had met in Groves' office in Washington. "He [Oppenheimer] stopped at my desk—which was right outside the general's door—and we made conversation," Wilson said of Oppenheimer. "I was just practically dumbstruck because here was this legendary character and part of his legend was that all women fell on their faces in front of him."

Flattered, Wilson agreed to move out to Los Alamos. Before she went, however, John Lansdale, Groves' counterintelligence chief, approached her with an offer: He would pay her $200 a month if she sent him just one letter each month reporting on what she saw in Oppenheimer's office. Shocked, Wilson flatly refused. "I told him," she later said, " 'Lansdale, I want you just to pretend you never even mentioned such a thing to me.' " Groves had assured her, she said, that once she moved out to Los Alamos, her loyalties were to be to Oppenheimer. But, perhaps not surprisingly, she learned after the war that Groves had ordered that she be placed under surveillance whenever she left Los Alamos—after working in his office, he believed, Anne Wilson knew too much to be left unwatched.

Upon arriving in Los Alamos, Wilson learned that Oppenheimer was sick in bed with chickenpox, accompanied by a 104-degree fever. "Our thin, ascetic Director," wrote the wife of another physicist, "looked like a 15th century portrait of a saint with his fever-stricken eyes peering out from a face checkered with red patches and covered by a straggling beard." Soon after he recovered, Wilson was invited over to the Oppenheimer home for drinks. Her host served her one, and then another, of his famous martinis, and since she was not yet acclimated to the altitude, the powerful concoction quickly went to her head. Wilson remembered having to be escorted back to her room in the nurses' quarters.

Anne Wilson was fascinated by her charismatic new boss and deeply admired him. But at twenty, she was not attracted romantically to Oppenheimer, a married man twice her age in 1945. Still, Anne was a beautiful young woman, smart and sassy—and people began to talk on The Hill about the director's new secretary. Several weeks after her arrival, Anne began receiving a single rose in a vase, delivered every three days from a florist in Santa Fe. The mysterious roses came without a card. "I was totally baffled, so I went around in my childlike way, saying, 'I've got a secret lover. Who is sending all these gorgeous roses?' I never found out. But finally, one person said to me, 'There is only one person who would do that, and that's Robert.' Well, I said it's ridiculous."

As might happen in any small town, rumors soon began circulating that Oppenheimer was having an affair with Wilson. She said it never happened: "I have to tell you that I was too young to appreciate him. Maybe I thought a forty-year-old man was ancient." Inevitably, Kitty heard the rumors, and one day she confronted Wilson and asked her point-blank if she had designs on Robert. Annie was thunderstruck. "She could not have misread my astonishment," Wilson recalled.

In the years that followed Anne got married, Kitty relaxed, and an enduring friendship developed between the two women. If Robert *was* attracted to Anne, the anonymous single red rose was a subtle gesture not out of character. He was not the kind of man who initiated sexual conquests. As Wilson herself observed, women "gravitated" to Oppenheimer: "He really was a man of women," Wilson said. "I could see that and I heard plenty of that." But at the same time, the man himself was still painfully shy and even unworldly. "He was enormously empathetic," Wilson said. "This was, I think, the secret of his attraction for women. I mean, it felt almost that he could read their minds—many women have said this to me. Women at Los Alamos who were pregnant could say, 'The only one who would understand was Robert.' He had a really almost saintly empathy for people." And if he was attracted to other women, he was still devoted to his marriage. "They were terribly close," Hempelmann said of Kitty and Robert. "He would come home in the evenings whenever he could. I think she was proud of him, but I think she would have liked to have been more in the center of things."

THE SECURITY NET that enveloped Robert naturally also included his wife. Soon Kitty found herself being gently interrogated by Colonel Lansdale. A skillful and empathetic interviewer, Lansdale quickly decided that Kitty could provide him key insights into her husband. "Her background was not good," he later testified. "For that reason I took as many occasions as I could to talk to Mrs. Oppenheimer." When she served him a martini, he wryly noted that she was not the kind to serve tea. "Mrs. Oppenheimer impressed me as a strong woman with strong convictions. She impressed me as the type of person who could have been, and I could see she certainly was, a communist. It requires a very strong person to be a real communist." And yet, in the course of their meandering conversations, Lansdale realized that Kitty's ultimate loyalty was to her husband. He also sensed that while she was politely playing her part, she "hated me and everything I stood for."

The rambling interrogation turned into a dance. "As we say in the lingo," Lansdale later said, "she was trying to rope me, just as I was trying to rope her. . . . I felt she'd go to any lengths for what she believed in. The tactic I fell back on was to try to show her I was a person of balance, honestly wanting to evaluate Oppenheimer's position. That's why our talks ran on so long.

"I was sure she'd been a communist and not sure her abstract opinions had ever changed much. . . . She didn't care how much I knew of what she'd done before she met Oppenheimer or how it looked to me. Gradually I began to see that nothing in her past and nothing in her other husband's meant anything to her compared with him. I became convinced that in him she had an attachment stronger than communism, that his future meant more to her than communism. She was trying to sell me on the idea he was her life, and she did sell me." Later, Lansdale reported his conclusions to Groves: "Dr. Oppenheimer was the most important thing in her life. . . . her strength of will was a powerful influence in keeping Dr. Oppenheimer away from what we would regard as dangerous associations."

INSIDE THE BARBED WIRE, Kitty sometimes felt as if she were living under a microscope. The Army commissary often had foods and goods available on the outside only with a ration card. The theater showed two movies a week for only 15 cents a show. Medical care was free. So many young couples had babies—some eighty births were recorded the first year, and about ten a month thereafter—that the small seven-room hospital was labeled "RFD," for "rural free delivery." When General Groves complained about all the new babies, Oppenheimer wryly observed that the duties of a scientific director did not include birth control. And that was also true for the Oppenheimers. By then, Kitty was pregnant again. On December 7, 1944, she gave birth in the Los Alamos barracks hospital to a daughter, Katherine, whom they nicknamed "Tyke." A sign was posted over the crib saying "Oppenheimer," and for several days people filed by to take a peek at the boss's baby girl.

Four months later, Kitty announced she "just had to go home to see her parents." Whether because of postpartum depression, or the excess of martinis in the Oppenheimer home, or the state of her marriage, Kitty was on the verge of an emotional collapse. "Kitty had begun to break down, drinking a lot," recalled Pat Sherr. Kitty and Robert were also having problems with their two-year-old son. Like any toddler, Peter was a handful. And according to Sherr, Kitty "was very, very impatient with him." Sherr, a trained psychologist, thought Kitty "had absolutely no intuitive understanding of the children." Kitty had always been mercurial. Her sister-in-law, Jackie Oppenheimer, observed that Kitty "would go off on a shopping trip for days to Albuquerque or even to the West Coast and leave the children in the hands of the maid." Upon her return, Kitty would bring an enormous present for Peter. "She must have felt so guilty and unhappy," said Jackie, "the poor woman."

IN APRIL 1945, Kitty left for Pittsburgh, taking Peter with her. But she decided to leave her four-month-old baby girl in the care of her friend Pat Sherr, who had recently had a miscarriage. The Los Alamos pediatrician Dr. Henry Barnett suggested that it would be good for Sherr to care for a child.

Thus "Tyke"—or Toni, as they later called her—was moved into Sherr's home. Kitty and young Peter were gone for three and a half months, until July 1945. Robert, of course, was working long hours, so he came by only twice a week to visit his baby daughter.

The strain on Robert over these incredibly intense two years was taking its toll. Physically, that toll was obvious: His coughing was incessant and his weight was down to 115 pounds, skin-and-bones for a man 5 feet 10 inches tall. His energy level never flagged, but he seemed to be literally disappearing little by little, day after day. The psychological toll was, if anything, harsher—albeit less obvious. Robert had spent a lifetime dealing with and managing his mental stresses. Nevertheless, "Tyke's" birth and Kitty's departure left him unusually vulnerable.

"It was all very strange," remembered Sherr. "He would come and sit and chat with me, but he wouldn't ask to see the baby. She might as well have been God knows where, but he never asked to see her."

"Finally, one day I said, 'Wouldn't you like to see your daughter, she's growing beautifully?' And he said, 'Yeah, yeah.' "

Two months went by, and then during one of Robert's visits he said to Sherr, "You seem to have grown to love Tyke very much." Sherr responded matter-of-factly, "Well, I love children, and when you take care of a baby, whether it's yours or someone else's, it becomes a part of your life."

Sherr was stunned when Oppenheimer then asked, "*Would you like to adopt her?*"

"Of course not," she replied, "she has two perfectly good parents." When she asked why he would say such a thing, Robert replied, "Because I can't love her."

Sherr reassured him, saying that such feelings were not unusual for a parent who has been separated from a child, and that over time he would become "attached" to the baby.

"No, I'm not an attached kind of person," Oppenheimer said. When Sherr asked if he had discussed this with Kitty, Robert said, "No, no, no. I was feeling you out first because I thought it was important for this child to have a loving home. And you have given her this."

Sherr was embarrassed and upset by the conversation. It struck her that, however outlandish the suggestion, it was nevertheless motivated by genuine emotion. "It seemed to me that he was a man of great conscience; for him to be able to say this to me. . . . Now here was a person who was conscious of his feelings—and at the same time feeling guilt about the feeling—and wanting somehow or other to give his child the fair deal that he felt he couldn't give her."

When Kitty finally returned to Los Alamos in July 1945, she characteristically showered Sherr with gifts. Kitty found Los Alamos in a state of high tension; the men were working longer hours, and their wives felt more isolated than ever. Kitty took to inviting small groups of women over for daily cocktails. Jackie Oppenheimer, who visited Los Alamos in 1945, remembered one such event. "It was known that we didn't get on too well," Jackie said, "and she seemed determined that we should be seen together. On one occasion, she asked me to cocktails—this was four o'clock in the afternoon. When I arrived, there was Kitty and just four or five other women—drinking companions—and we just sat there with little conversation, drinking. It was awful and I never went again."

At the time, Pat Sherr did not think Kitty was an alcoholic. "She drank somewhat," Sherr recalled. "Come four o'clock, she would have her drink and continue on, but she didn't have slurred speech." Kitty's drinking would definitely become an issue later in her life, but according to another close friend, Dr. Hempelmann, "She certainly didn't drink more than anybody else did out at Los Alamos." Alcohol flowed freely on the mesa, and as the months rolled by, some people felt oppressed by the small town's isolation. "At first, it was lots of fun," Hempelmann recalled, "but as things wore on and everybody got tired and tense and irritable, it wasn't so good. Everybody was living in each other's pockets. You'd play with the same people that you worked with. And a friend would ask you out to dinner, and you didn't have anything else to do, but you just didn't want to go. So they would know. If they drove by your house, they would see that your car was still there. Everybody knew everything about everybody else."

ASIDE FROM the periodic afternoon excursions in Santa Fe, one of the few permitted escapes from Los Alamos was dinner at Miss Edith Warner's adobe house at Otowi—the "place where the water makes noise"—on the Rio Grande, about twenty miles down the winding road. Oppie first met Miss Warner while on a pack trip from Frijoles Canyon with Frank and Jackie; one of their horses had run off and Oppie had given chase. He ended up at Miss Warner's "tea house." "We had tea and chocolate cake and talk," Oppenheimer later wrote; "it was my first unforgettable meeting." Wearing blue jeans and cowboy boots with spurs, Robert looked, thought Miss Warner, like the "slim and wiry hero of a Western movie."

Miss Warner, the daughter of a Philadelphia clergyman, had first come to the Pajarito Plateau in 1922, after suffering a nervous breakdown at the age of thirty. Together with her companion, an elderly Native American, Atilano Montoya—known about the pueblo as Tilano—she ran what she called a tea room for tourists out of her home. Her life was simple in the extreme.

One evening soon after Oppie moved to the mesa, he took General

Groves to the house at Otowi Bridge for tea. With the closing of the Ranch School and the imposition of wartime gas rationing, which discouraged tourist traffic, Edith gently confessed that she was wondering how she would make ends meet. As they sipped their tea, Groves offered to put her in charge of all the food services on The Hill. It was a big job with good pay. Edith said she would consider the idea. When they left, Robert escorted Groves to their car, but then returned and knocked on Edith's door. Standing with hat in hand and the moonlight full on his face, he told her, "Don't do it." Then he abruptly turned and walked back to his car.

A few days later, Oppenheimer reappeared on Miss Warner's doorstep and proposed that she host three small dinners each week for parties of no more than ten. By providing the scientists a brief diversion from life on The Hill, Oppie explained, she would be making a real contribution to the war effort. General Groves had given his consent to the idea—and Edith herself regarded it as a godsend.

"Along about April," wrote Miss Warner at the end of that year, "the X's began coming down from Los Alamos for dinner once a week, and they were followed by others." After cooking all day, Miss Warner presided, wearing a simple shirtwaist dress and Indian moccasins. Everyone sat at one long, hand-carved wooden table set in the center of a dining room with whitewashed adobe walls and low-slung, hand-hewn beams. Miss Warner, aged fifty-one, served her "hungry scientists" generous portions of home-cooked food. They ate ragoût of lamb by candlelight off traditional Indian black ceramic plates and bowls, hand-coiled by the local potter, Maria Martinez. Afterwards, her guests huddled briefly together by the fireplace for warmth before making the long drive back up to the mesa. In return for this evening of candle-lit adobe ambience, Miss Warner charged her guests the token sum of $2 per head. She knew only that these mysterious people were working "for some very secret project. . . . Santa Fe calls it a submarine base—as good a guess as any!"

Dinner at Miss Warner's became such a sought-after pleasure that teams of five couples had permanent reservations for the same night each week. Oppenheimer made sure that he and Kitty had first choice on Edith's calendar, but soon the Parsonses, Wilsons, Bethes, Tellers, Serbers and others became regulars, while many other Los Alamos couples vied for the prestige of an invitation. Oddly enough, the calm, quiet Miss Warner had a special rapport with Oppenheimer's vivacious, sharp-tongued wife. "Kitty and I understood each other," Warner later said. "She was very close to me, and I to her."

One day in early 1944, Oppie brought along the Danish Nobelist Niels Bohr, and introduced him to Miss Warner as "Mr. Nicholas Baker"—an alias Bohr was assigned at Oppenheimer's initiative. Everyone called the gentle, unassuming Dane "Uncle Nick." The softspoken, mumbling Bohr conversed in stumbling half-sentences—but then, Miss Warner wasn't much of a talker either. Years later, Bohr attested to this most unlikely friendship by writing Miss Warner's sister a note "in gratitude for the friendship of your sister." Miss Warner had a near-mystical regard for both Bohr and Oppenheimer: "He [Bohr] has a great stillness in him, a calm inexhaustible source. . . . Robert has the same thing in him."

Bohr was not, of course, the only memorable personality to dine at Miss Warner's table. James Conant (chairman of S-1 or Section One of the Office of Scientific Research and Development), Arthur Compton (a Nobelist and director of the Metallurgical Laboratory at the University of Chicago) and the Nobelist Enrico Fermi visited the house at Otowi Bridge. But it was only Oppie's framed photograph that Miss Warner kept on her Philadelphia dresser. Phil Morrison could easily have been speaking for Oppenheimer when, late in 1945, he wrote Miss Warner a long letter of thanks for his many evenings in her company: "Not the smallest part of the life we came to lead, Miss Warner, was you. Evenings in your place by the river, by the table so neatly set, before the fireplaces so carefully contrived, gave us a little of your assurance, allowed us to belong, took us from the green temporary houses and the bulldozed roads. We shall not forget. . . . I am glad that at the foot of our canyons there is a house where the spirit of Bohr is so well understood."

CHAPTER TWENTY

"Bohr Was God, and Oppie Was His Prophet"

They didn't need my help in making the atom bomb.

NIELS BOHR

THE "RACE" FOR THE ATOMIC BOMB had begun more or less as a straggle. A few scientists, almost all European émigrés, were panicked in 1939 over the possibility that their former colleagues in Germany might take the lead in putting the discovery of fission to military use. They alerted the U.S. government to this danger, and the government supported conferences and small nuclear research projects. Committees of scientists did studies and wrote reports. But it was not until the spring of 1941, more than two years after the discovery of nuclear fission in Germany, that Otto Frisch and Rudolph Peierls, German émigré physicists working in Britain, figured out how a usable atomic bomb could be produced quickly in time for use during the war. From that time forward, everyone involved with the combined American-British-Canadian atomic bomb project was totally focused on winning this deadly race. Thoughts about the postwar implications of a nuclear-armed world remained dormant until December 1943, when Niels Bohr arrived at Los Alamos.

Oppenheimer was enormously gratified to have Bohr at his side. The fifty-seven-year-old Danish physicist had been smuggled out of Copenhagen aboard a motor launch on the night of September 29, 1943. Arriving safely on the Swedish coast, he was taken to Stockholm—where German agents plotted his assassination. On October 5, British airmen sent to his rescue helped Bohr into the bomb bay of an unmarked British Mosquito bomber. When the plywood aircraft approached an altitude of 20,000 feet, the pilot instructed Bohr to don the oxygen mask built into his leather helmet. But Bohr failed to hear the instructions—he later said the helmet was too small for his large head—and soon he fainted from lack of oxygen. He nevertheless survived the air journey and upon landing in Scotland, he remarked that he had had a pleasant nap.

Greeting him on the tarmac was his friend and colleague James Chadwick, who took him to London and began briefing him on the British-American bomb project. Bohr had understood since 1939 that the discovery of nuclear fission made an atomic bomb feasible, but he believed that the engineering necessary for separating out U-235 would require an immense, and therefore impractical, industrial effort. Now he was told that the Americans were turning their great industrial resources to exactly this purpose. "To Bohr," wrote Oppenheimer later, "[it] seemed completely fantastic."

A week after his arrival in London, Bohr was joined by his twenty-one-year-old son Aage (pronounced "Awa"), a promising young physicist who later would earn his own Nobel Prize. Over the next seven weeks, father and son were thoroughly briefed about "Tube Alloys"—the British code name for the bomb project. Bohr agreed to become a consultant to the British, who then agreed to send him to America. In early December, he and his son boarded a ship for New York. General Groves was not happy about the idea of Bohr's participation, but, given the Dane's prestige in the world of physics, he reluctantly granted him permission to visit the mysterious "Site Y" in the New Mexico desert.

Groves' displeasure had been sparked by intelligence reports suggesting that Bohr was a loose cannon. On October 9, 1943, the *New York Times* reported that the Danish physicist had arrived in London bearing "plans for a new invention involving atomic explosions." Groves was incensed, but there was nothing he could do beyond trying to contain Bohr. This proved to be a hopeless task: Bohr was irrepressible. In Denmark, he had simply walked up to the palace door and knocked if he wished to see the king. And he did pretty much the same thing in Washington, D.C., where he visited Lord Halifax, the British ambassador, and Supreme Court Justice Felix Frankfurter, an intimate of President Roosevelt's. His message to these men was clear: The making of the atomic bomb was a foregone conclusion, but it was not too soon to consider what would happen after its development. His deepest fear was that its invention would inspire a deadly nuclear arms race between the West and the Soviet Union. To prevent this, he insisted, it was imperative that the Russians be told about the existence of the bomb project, and be assured that it was no threat to them.

Such views, of course, horrified Groves, who was desperate to get Bohr out to Los Alamos, where the loquacious physicist could be isolated. To ensure that Bohr got there without breaking security, Groves personally joined him and his son on the train from Chicago. Caltech's Richard Tolman, Groves' science adviser, also came along. Groves and Tolman had agreed to take turns watching over the Danish visitor, to make sure he didn't wander out of the compartment. After an hour with Bohr, however, Tolman came out exhausted and told Groves, "General, I can't stand it any more. I am reneging, you are in the Army, you have to do it."

So as Groves listened to Bohr's characteristic "whispering mumble," every so often he would try to interrupt and explain to him the importance of compartmentalization. It was an effort foredoomed to failure. Bohr had a broad overview of the Manhattan Project and an insatiable concern for the social and international implications of science. Not only that, more than two years earlier, in September 1941, Bohr had met with his former student Werner Heisenberg, the German physicist who led the German atomic bomb program. Groves had debriefed Bohr about what he knew about the German project—but he certainly didn't want him to talk to others about it. "I think I talked to him about twelve hours straight on what he was not to say."

They arrived in Los Alamos late on the evening of December 30, 1943, and immediately went to a small reception in Bohr's honor hosted by Oppenheimer. Groves complained later that "within five minutes after his [Bohr's] arrival he was saying everything he promised not to say." Bohr's first question to Oppenheimer was, "Is it really big enough?" In other words, would the new weapon be so powerful as to make future wars inconceivable? Oppenheimer immediately understood the import of the question. For more than a year, he had concentrated his energies entirely on the administrative details related to setting up and running the new lab; but over the next few days and weeks, Bohr sharply focused Oppie's mind on the bomb's postwar consequences. "That is why I went to America," Bohr later said. "They didn't need my help in making the atom bomb."

That evening, Bohr told Oppenheimer that Heisenberg was working quite vigorously on a uranium reactor that could produce a runaway chain reaction, and thereby create an immense explosion. Oppenheimer convened a meeting the next day, the last day of 1943, to discuss Bohr's concerns. Attending were Bohr, Aage and some of the best minds at Los Alamos, including Edward Teller, Richard Tolman, Robert Serber, Robert Bacher, Victor Weisskopf and Hans Bethe. Bohr then tried to convey to these men the quite extraordinary nature of his encounter with Heisenberg in September 1941.

Bohr recounted how his brilliant German protégé had received special permission from the Nazi regime to attend a conference in German-occupied Copenhagen. Though not himself a Nazi, Heisenberg was certainly a German patriot who had chosen to remain in Nazi Germany. He was undoubtedly Germany's most eminent physicist; if the Germans had an atomic bomb project, Heisenberg was the obvious candidate to direct it. When he arrived in Copenhagen, he sought out Bohr, and what the two old friends said to each other has become an enduring enigma. Heisenberg later maintained that he had guardedly mentioned the uranium problem and tried to suggest to his old friend that while a fission weapon was quite possible in principle, it would "require a terrific technical effort, which, one can only hope, cannot be realized in this war." He claimed that he was implying—but, worried about German surveillance, and fearful for his own life, could not explicitly say—that he and other German physicists wanted to persuade the Nazi regime that it would not be feasible to build such a weapon in time for use in this war.

But if this was Heisenberg's message, Bohr was not listening. All the Danish physicist heard was Germany's leading physicist telling him that a fission weapon was indeed possible and that, if developed, it would be decisive in this war. Alarmed and angry, Bohr cut short their conversation.

Later, Bohr himself reported that he was not quite sure what Heisenberg had meant to say. Years later, he would compose numerous drafts—as was his habit—of a letter to Heisenberg that in the end he never sent. In all versions of the letter, it is quite clear that Heisenberg had shocked Bohr by merely mentioning atomic weapons. In one draft, for instance, Bohr wrote:

> On the other hand, I remember quite clearly the impression it made on me when, at the beginning of the conversation, you told me without preparation that you were certain that the war, if it lasted long enough, would be decided with atomic weapons. I did not respond to this at all, but as you perhaps regarded this as an expression of doubt, you related how in the preceding years you had devoted yourself almost exclusively to the question and were quite certain that it could be done, but you gave no hint about efforts on the part of German scientists to prevent such a development.

What was said or not said between Bohr and Heisenberg remains a source of considerable controversy. Oppenheimer himself later wrote, cryptically: "Bohr had the impression that they [Heisenberg and his colleague Carl Friedrich von Weizsäcker] came less to tell what they knew than to see if Bohr knew anything that they did not. I believe that it was a standoff."

One thing is clear, however: Bohr came away from the encounter with a great fear that the Germans might end the war with an atomic weapon. In New Mexico, he conveyed this fear to Oppenheimer and his team of scientists. Not only did he tell them that Heisenberg had confirmed the existence of a German bomb project, Bohr also displayed a drawing of what he said was a bomb, allegedly sketched by Heisenberg himself. One glance, however, persuaded everyone that the sketch depicted, not a bomb, but a uranium reactor. "My God," Bethe said when he saw the drawing, "the Germans are trying to throw a reactor at London." If it was disquieting to learn that the Germans were indeed working on a bomb project, it was reassuring that they seemed to be pursuing a highly impractical design. After discussion of the issue, even Bohr was persuaded that such a "bomb" would fizzle. The next day, Oppenheimer wrote Groves to explain that an exploding uranium pile would actually "be a quite useless military weapon."

OPPENHEIMER ONCE observed that "it is easy, as history has shown, for even wise men not to know what Bohr was talking about." Like Bohr, Oppenheimer was never simple or straightforward. At Los Alamos, the two men sometimes seemed to be mimicking each other. "Bohr at Los Alamos was marvelous," Oppenheimer later wrote. "He took a very lively technical interest. But his real function, I think, for almost all of us, was not the technical one." Instead, Bohr came "most secretly of all," as Oppenheimer explained, to advance a political cause—the case for openness in science as well as international relations, the only hope to forestall a postwar nuclear arms race. This was a message Oppenheimer was ready to hear. For nearly two years, he had been preoccupied with complex administrative responsibilities. As the months rolled by, he was becoming less and less a theoretical physicist and more and more a science administrator. This transformation had to be intellectually stifling for him. So when Bohr showed up on the mesa speaking in deeply philosophical terms about the project's implications for humanity, Oppenheimer felt rejuvenated. He assured Groves that Bohr's presence had greatly helped morale. Up until then, Oppenheimer later wrote, the work "often looked so macabre." Bohr soon "made the enterprise seem hopeful, when many were not free of misgiving." He spoke contemptuously of Hitler and underscored the role scientists could play in his defeat. "His own high hope that the outcome would be good, that the objectivity, the cooperation of the sciences would play a helpful part, we all wanted to believe."

Victor Weisskopf recalled Bohr telling him that "this bomb may be a terrible thing, but it might also be the 'Great Hope.' " Early that spring, Bohr attempted to put his concerns on paper, drafting and redrafting a memorandum and then sharing it with Oppenheimer. By April 2, 1944, he had a draft that contained several basic insights. No matter how things ultimately worked out, Bohr argued, "it is already evident that we are presented with one of the greatest triumphs of science and technique, destined deeply to influence the future of mankind." In the very near term, "a weapon of an unparalleled power is being created which will completely change all future conditions of warfare." That was the good news. The bad news was equally clear, and prophetic: "Unless, indeed, some agreement about the control of the use of the new active materials can be obtained in due time, any temporary advantage, however great, may be outweighed by a perpetual menace to human security."

In Bohr's mind, the atomic bomb was already a fact—and control over this menace to humanity required "a new approach to the problem of international relationship. . . ." In the coming atomic era, humanity would not be safe unless secrecy was banished. The "open world" Bohr imagined was not a utopian dream. This new world already existed in the multinational communities of science. In a very pragmatic sense, Bohr believed the laboratories in Copenhagen, Cavendish and elsewhere were practical models for this new world. International control of atomic energy was only possible in an "open world" based on the values of science. For Bohr, it was the communitarian culture of scientific inquiry that produced progress, rationality and even peace. "Knowledge is itself the basis of civilization," he wrote, "[but] any widening of the borders of our knowledge imposes an increased responsibility on individuals and nations through the possibilities it gives for shaping the conditions of human life." It followed that in the postwar world each nation had to feel confident that no potential enemy was stockpiling atomic weapons. That would only be possible in an "open world" where international inspectors had full access to any military and industrial complexes and full information about new scientific discoveries.

Finally, Bohr concluded that such a sweeping new regime of international control could be inaugurated after the war only by promptly inviting the Soviet Union's participation in postwar atomic energy planning—before the bomb was a reality and before the war was over. A postwar nuclear arms race could be prevented, Bohr believed, if Stalin was informed of the existence of the Manhattan Project and assured that it posed no threat to the Soviet Union. An early agreement among the wartime allies for the postwar international control of atomic energy was the only alternative to a nuclear-armed world. Oppenheimer agreed—indeed, he had shocked his security officers the previous August when he told Colonel Pash that he would "feel friendly" to the idea of the president informing the Russians about the bomb project.

It was easy to see the effect Bohr had on Oppenheimer. "[He] knew Bohr from way back and they were pretty close personally," said Weisskopf. "Bohr was the one who really discussed these political and ethical problems with Oppenheimer, and probably that was the time [early 1944] when he began to think seriously about it." One afternoon that winter, Oppenheimer and David Hawkins were walking Bohr back to his guest quarters in Fuller Lodge when Bohr playfully insisted on testing the thickness of the ice on Ashley Pond. The usually daring Oppenheimer afterwards turned to Hawkins and exclaimed, "My God, suppose he should slip? Suppose he should fall through? What would we all do then?"

The very next day, Oppenheimer beckoned Hawkins into his office, pulled a folder from his secure file cabinet, and let him read a letter Bohr had written to Franklin Roosevelt. Oppie obviously set great store by the precious document. According to Hawkins, "the implication was that Roosevelt had fully understood. And this was a great source of joy and optimism. . . . It's interesting. We all lived under this illusion, you see, for the rest of the time at Los Alamos, that Roosevelt had understood."

BOHR HAD long since converted his particular "Copenhagen" interpretation of quantum physics into a philosophical world view that he named "Complementarity." Bohr was forever trying to take his insights into the physical nature of the world and apply them to human relations. As the historian of science Jeremy Bernstein later wrote: "Bohr was not satisfied to limit the idea of complementarity to physics. He saw it everywhere: instinct and reason, free will, love and justice, and on and on." He quite understandably saw it in the work at Los Alamos too. Everything about the project was fraught with contradictions. They were building a weapon of mass destruction that would defeat fascism and end all wars—but also make it possible to end all civilization. Oppenheimer quite naturally found it comforting to be told by Bohr that the contradictions in life were nevertheless all of a piece—and therefore complementary.

Oppenheimer so admired Bohr that in years to come he often took it upon himself to translate him to the rest of humanity. Not many understood what Bohr meant by an "open world." And those who did were sometimes positively alarmed by the audacity of what Bohr was proposing. In the early spring of 1944, Bohr received a letter, long delayed in the mails, from one of his former students, the Russian physicist Peter Kapitza. Writing from Moscow, Kapitza warmly invited Bohr to come settle there, "where everything will be done to give you and your family a shelter and where we now have all the necessary conditions for carrying on scientific work." Kapitza then passed on the greetings of a number of Russian physicists whom Bohr knew—broadly suggesting that they would all be delighted to have him join them in their "scientific work." Bohr thought this a splendid opportunity, and he actually hoped that Roosevelt and Churchill would authorize him to

The Oppenheimers.
Julius Oppen-
heimer (above, left)
arrived in New York
City from Germany
in 1888. In 1903 he
married Ella Friedman
(above, right), a German-
American painter born in
Baltimore. Robert, born in
1904, sits (right) in his father's
lap.

As a young child, Robert (seated on the right with a friend) had a passion for blocks and collecting rock specimens.

Ella and Robert.

"I was an unctuous, repulsively good little boy," Oppenheimer later said. "My life as a child did not prepare me for the fact that the world is full of cruel and bitter things."

Oppenheimer (right) riding in Central Park.

Robert and his younger brother Frank.

Robert attended the Ethical Culture School where he was taught to develop his "ethical imagination," to see "things not as they are, but as they might be."

Oppenheimer studied at Göttingen University, where he received his doctorate in quantum physics under Max Born (right). There he was befriended by physicists Paul Dirac (center, right) and the Dutch physicist Hendrik Kramers (below, left). Later he studied briefly in Zurich with I. I. Rabi, H. M. Mott-Smith and Wolfgang Pauli (bottom, right, sailing with Robert on Lake Zurich).

Professor Oppenheimer (above, left) in 1929 at Caltech, where he had accepted a dual appointment with the University of California, Berkeley, and where he quickly became an apostle for the new quantum physics. "I need physics more than friends," Robert confessed. Oppenheimer (above, right) between physicists William A. Fowler and Luis Alvarez. "I started really as a propagator of the theory which I loved, about which I continued to learn more, and which was not well understood but which was very rich." Robert Serber (below, right) was one of his students and then a lifelong friend.

"My two great loves are physics and New Mexico," Oppenheimer wrote. "It's a pity they can't be combined." Oppenheimer spent his summers at Perro Caliente, his 154 acre ranch (above) with a view of the Sangre de Christo mountains. Robert and his horse, Crisis (right), went for long rides with his brother Frank and other friends, including Berkeley physicist Ernest Lawrence (below).

Oppenheimer with the Italian physicist Enrico Fermi and Ernest Lawrence.

Joe Weinberg, Rossi Lomanitz, David Bohm and Max Friedman were some of Oppie's acolytes at Berkeley. "They copied his gestures, his mannerisms, his intonations," recalled Bob Serber.

In the world of quantum physics, Weinberg said, "Niels Bohr (left) was God, and Oppie was his prophet."

Jean Tatlock was Oppie's fiancée for four years—and a Communist Party member, although with reservations. "I find it impossible to be an ardent Communist," she wrote.

Tatlock's mentor at Stanford's School of Medicine was Dr. Thomas Addis (above, right). Dr. Addis persuaded Oppenheimer to donate money for the Spanish cause through the Communist Party.

.deral Bureau of Investigatio
United States Department of Justice

San Francisco, California
March 28, 1941

Director
Federal Bureau of Investigation
Washington, D. C.

Re: J. ROBERT OPPENHEIMER,
with alias;
INTERNAL SECURITY (C)

Dear Sir:

The above individual was included in a list recently furnished to the Bureau by this office of persons to be considered for custodial detention pending investigation in the event of a national emergency.

The Bureau recently furnished this office with custodial detention cards on certain individuals. From an examination of these cards, it is noted that there has been no such card made up on this individual. Therefore, a card should be made on him based on the information contained in the report of Special Agent R. B. HOOD, dated March 28, 1941, at San Francisco, California.

Very truly yours,

R. J. L. PIEPER
Special Agent in Charge

RJLB
100-3132

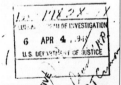

By 1941, Oppenheimer was on a FBI list of suspected radicals to be detained in the event of a national emergency.

In 1943 Haakon Chevalier (above, left), a Berkeley professor of French literature, told Oppie of a scheme George Eltenton (above, right) had to provide scientific information for the Soviet war effort. Oppie eventually reported the event to an Army counterintelligence officer, Col. Boris Pash (left).

Below, Martin Sherwin with Chevalier after interviewing him in Paris in 1982.

Kitty Puening grew up in
Pittsburgh. Here she is (above)
in jodhpurs at 21; in a 1936
passport photograph (top, right);
and in the mycology lab at
Berkeley (right). In 1939 she met
and fell in love with Oppenheimer,
who is pictured on his Radiation
Laboratory security badge
(opposite, top).

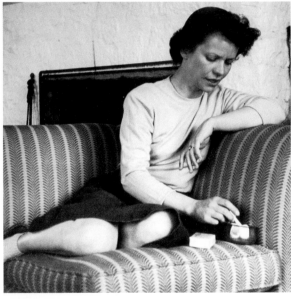

Kitty, here seated in their Los Alamos cottage, was a mercurial personality. "She was a very intense, very intelligent, very vital kind of person…very difficult to handle."

Kitty felt stymied professionally at Los Alamos. She worked in the medical clinic, conducting blood counts, but after a year she quit. At social gatherings she could make small talk, but as one friend put it, "She wanted to make big talk."

Peter Oppenheimer was born in May 1941. Above, being fed by Robert, and below laughing with Kitty.

"He (Robert) was great at a party and women simply loved him," said Dorothy McKibbin.

Oppenheimer entertains McKibbin (on his right) and Victor Weisskopf (kneeling) in his Los Alamos cottage.

Below, Hans Bethe, chief of the theoretical division.

Above, a scientific colloquium at Los Alamos with (left to right) Norris Bradbury, John Manley, Enrico Fermi and J.M.B. Kellogg seated in the front row. Oppenheimer, Richard Feynman and Phillip Porter sit behind them.

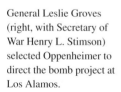

Robert brought his brother Frank (center, inspecting an alpha calutron) to Los Alamos in 1945 to work on the Trinity test of the first atomic bomb.

General Leslie Groves (right, with Secretary of War Henry L. Stimson) selected Oppenheimer to direct the bomb project at Los Alamos.

Oppenheimer pours coffee while touring southern New Mexico, in late 1944, scouting out a site for the Trinity explosion.

Wearing his porkpie hat, Oppenheimer leans over the "Gadget" atop the Trinity site tower, just hours before the test. Below, the Trinity explosion.

Hiroshima after the bomb. More than 95 percent of the roughly 225,000 people killed in Hiroshima and Nagasaki were civilians, mostly women and children. At least half the victims died of radiation poisoning in the months following the initial blast. This photograph by Yosuke Yamahata of a mother and child (right) was taken less then twenty-four hours after the bombing of Nagasaki.

accept Kapitza's invitation. As Oppenheimer later explained it to his colleagues, Bohr wished "to propose to the rulers of Russia, who were then our Allies, via these scientists, that the United States and the United Kingdom 'trade' their atomic knowledge for an open world . . . that we propose to the Russians that atomic knowledge would be shared with them if they would agree to open Russia and make it an open country and part of an open world."

To Bohr's thinking, secrecy was dangerous. Knowing Kapitza and other Russian physicists, Bohr thought them perfectly capable of grasping the military implications of fission. In fact, he surmised from Kapitza's letter that the Soviets already did know something about the British-American atomic program—and he thought it would only sow dangerous suspicions if the Russians concluded that the new weapon was being developed without them. Other physicists at Los Alamos agreed. Robert Wilson later recalled "bugging" Oppenheimer about why British scientists, but no Russian scientists, were working at Los Alamos. "It seemed to me that down the line," Wilson said, "that was going to make for some very hard feelings." By the war's end, it is clear that Oppenheimer agreed, but during the war, he was circumspect, knowing that he was under constant surveillance, and so he always refused to be drawn into such conversations. Either he wouldn't answer at all or he'd mutter that it wasn't the business of the scientists to determine such things. "I don't know," Wilson later said, "I felt that perhaps he thought that I was testing him."

Not surprisingly, Bohr's attitude was not shared by the generals and politicians who employed the scientists. General Groves, for instance, never really thought of the Russians as allies. In 1954, he told the Atomic Energy Commission's hearing board that "there was never from about two weeks from the time I took charge of this project any illusion on my part but that Russia was our enemy and that the project was conducted on that basis. I didn't go along with the attitude of the country as a whole that Russia was a gallant ally." Winston Churchill had a similar view of the Soviets, and he was outraged to learn of the Kapitza-Bohr correspondence from British intelligence. "How did he [Bohr] come into this business?" Churchill exclaimed to his science adviser, Lord Cherwell. "It seems to me Bohr ought to be confined or at any rate made to see that he is very near the edge of mortal crimes."

Despite personal meetings with both Roosevelt and Churchill in the spring and summer of 1944, Bohr failed to persuade either leader that the Anglo-American monopoly in atomic matters was shortsighted. Groves later told Oppenheimer that he thought Bohr "was at times a thorn in the sides of everyone dealing with him, possibly because of his great mental capacity." Ironically, as his influence with such political leaders waned, Bohr's stature among the physicists at Los Alamos rose to new heights. Once again, Bohr was God and Oppie was his prophet.

BOHR HAD come to Los Alamos in December 1943 alarmed by what he had learned from his encounter with Heisenberg about the potential for a German bomb. He left Los Alamos that spring persuaded by intelligence reports that the Germans probably did not have a viable bomb program: ". . . from leakage regarding the activities of German scientists," he noted, "it is practically certain that no substantial progress has been achieved by the Axis Powers." If Bohr was convinced, then Oppenheimer too must have realized that the German physicists were in all likelihood far behind in the race to build a bomb. According to David Hawkins, Oppenheimer was told by General Groves at the end of 1943 that a German source had recently claimed that the Germans had abandoned their early bomb program. Groves suggested that it was hard to evaluate such a report; the German source might be passing on disinformation. Oppenheimer just shrugged. Hawkins recalled thinking to himself that it was too late—the men at Los Alamos "were committed to building a bomb regardless of German progress."

"The Impact of the Gadget on Civilization"

My feeling about Oppenheimer was, at that time, that this was a man who is angelic, true and honest and he could do no wrong. . . . I believed in him.

ROBERT WILSON

EVERYONE SENSED OPPIE'S PRESENCE. He drove himself around The Hill in an Army jeep or in his own large black Buick, dropping in unannounced on one of the laboratory's scattered offices. Usually he'd sit in the back of the room, chain-smoking and listening quietly to the discussion. His mere presence seemed to galvanize people to greater efforts. "Vicki" Weisskopf marveled at how often Oppie seemed to be physically present at each new breakthrough in the project. "He was present in the laboratory or in the seminar room when a new effect was measured, when a new idea was conceived. It was not that he contributed so many ideas or suggestions; he did so sometimes, but his main influence came from his continuous and intense presence, which produced a sense of direct participation in all of us." Hans Bethe recalled the day Oppie dropped in to a session on metallurgy and listened to an inconclusive debate over what type of refractory container should be used for melting plutonium. After listening to the argument, Oppie summed up the discussion. He didn't directly propose a solution, but by the time he left the room the right answer was clear to all.

By contrast, General Groves' visits were always interruptions—and sometimes comically disruptive. One day, Oppie was showing Groves around a lab when the general put his considerable weight on one of three rubber tubes funneling hot water into a casing. As McAllister Hull recalled for the historian Charles Thorpe, "It [the rubber tube] pops off the wall and a stream of water just below the boiling point shoots across the room. And if you've ever seen a picture of Groves, you know what it hit." Oppenheimer looked over his soaking-wet general and quipped, "Well, just goes to show the incompressibility of water."

Oppie's interventions sometimes proved to be absolutely essential to the success of the project. He understood that the single major impediment to building a usable weapon quickly was the meager supply of fissionable material. And so he was constantly looking for ways to accelerate the production of these materials. Early in 1943, Groves and his S-1 Executive Committee had settled on gaseous diffusion and electromagnetic technologies to separate out enriched fissionable uranium for the Los Alamos bomb lab. At the time, another possible technology, based on liquid thermal diffusion, had been rejected as unfeasible. But in the spring of 1944, Oppenheimer read some year-old reports about liquid thermal diffusion and decided that this had been a mistake. He thought this technology represented a relatively cheap path to providing partially enriched uranium for the electromagnetic process. So in April 1944, he wrote Groves that a liquid thermal diffusion plant might serve as a stopgap measure; its production of even slightly enriched uranium could then be fed to the electromagnetic diffusion plant and thereby accelerate production of fissionable material. It was his hope, he wrote, "that the production of the Y-12 [electromagnetic] plant could be increased by some 30 to 40 percent, and its enhancement somewhat improved, many months earlier than the scheduled date for K-25 [gaseous diffusion] production."

After sitting on Oppie's recommendation for a month, Groves agreed to explore it. A plant was rushed into production, and by the spring of 1945 it was producing just enough extra partially enriched uranium to guarantee a sufficient amount of fissionable material for one bomb by the end of July 1945.

Oppenheimer had always possessed a high degree of confidence in the uranium gun-design program—whereby a "slug" of fissionable material would be fired into a target of additional fissionable matter, creating "criticality" and a nuclear explosion. But in the spring of 1944, he suddenly faced a crisis that threatened to derail the entire effort to design a plutonium bomb. While Oppenheimer had authorized Seth Neddermeyer to conduct explosive experiments aimed at creating an implosion design bomb—a loosely packed sphere of fissionable material that could be instantly compressed to reach criticality—he had always hoped that a straightforward gun assembly would prove viable for the plutonium bomb. In July 1944, however, it became clear from tests performed on the first small supplies of plutonium that an efficient plutonium bomb could not be triggered within the "gun-barrel" design. Indeed, any such attempt would undoubtedly lead to a catastrophic pre-detonation inside the plutonium "gun."

One solution might have been to separate further the plutonium materials in an attempt to make a more stable element. "One could have separated out those bad plutonium isotopes from the good ones," John Manley explained, "but that would have meant duplicating everything that had been done for uranium isotope separation—all those big plants—and there was just no time to do that. The choice was to junk the whole discovery of the chain reaction that produced plutonium, and all of the investment in time and effort of the Hanford [Washington] plant, unless somebody could come up with a way of assembling the plutonium material into a weapon that would explode."

On July 17, 1944, Oppenheimer convened a meeting in Chicago with Groves, Conant, Fermi and others, to resolve the crisis. Conant urged that they aim merely to build a low-efficiency implosion bomb based on a mixture of uranium and plutonium. Such a weapon would have had an explosive equivalent of only several hundred tons of TNT. Only after successfully testing such a low-efficiency bomb, Conant said, would the lab have the confidence to proceed with a larger weapon.

Oppenheimer rejected this notion on the grounds that it would lead to unacceptable delays. Despite having been skeptical about the implosion idea when it was first broached by Serber, Oppenheimer now marshaled all his persuasive powers to argue that they gamble everything on an implosion-design plutonium bomb. It was an audacious and brilliant gamble. Since the spring of 1943, when Seth Neddermeyer had volunteered to experiment with the concept, little progress had been made. But in the autumn of 1943, Oppenheimer brought the Princeton mathematician John von Neumann to Los Alamos, and von Neumann calculated that implosion was possible, at least theoretically. Oppenheimer was willing to bet on it.

The next day, July 18, Oppenheimer summarized his conclusions for Groves: "We have investigated briefly the possibility of an electromagnetic separation. . . . It is our opinion that this method is in principle a possible one but that the necessary developments involved are in no way compatible with present ideas of schedule. . . . In the light of the above facts, it appears reasonable to discontinue the intensive effort to achieve higher purity for plutonium and to concentrate attention on methods of assembly which do not require a low neutron background for their success. At the present time the method to which an overriding priority must be assigned is the method of implosion."

Oppenheimer's assistant, David Hawkins, later explained, "The implosion was the only real hope [for a plutonium bomb], and from current evidence not a very good one." Neddermeyer and his men in the Ordnance Division were making very little progress on the implosion design. Neddermeyer, shy and retiring, liked to work alone, and methodically. He later admitted that Oppenheimer "became terribly impatient with me in the spring of 1944. . . . I think he felt very badly because I seemed not to push things as for war research but acted as though it were just a normal research situation." Neddermeyer was also one of the few men on the mesa who seemed immune to Oppie's charms. In his frustration, Oppie uncharacteristically began to lose his temper. "Oppenheimer lit into me," Neddermeyer recalled. "A lot of people looked up to him as a source of wisdom and inspiration. I respected him as a scientist, but I just didn't look up to him that way. . . . He could cut you cold and humiliate you right down to the ground. On the other hand, I could irritate him." Stoked by this personality conflict, the crisis over the implosion design came to a head late that summer when Oppenheimer announced a major reorganization of the lab.

Early in 1944, Oppenheimer had persuaded an explosives expert from Harvard, George "Kisty" Kistiakowsky, to move to Los Alamos. Kistiakowsky was opinionated and strong-willed. Inevitably, he had numerous run-ins with his ostensible superior, Captain "Deke" Parsons. Neither did Kistiakowsky get along with Neddermeyer, who seemed to him far too lackadaisical in his approach. Early in June 1944, Kistiakowsky wrote Oppenheimer a memo threatening to resign. In response, Oppenheimer swiftly called Neddermeyer in and told him that Kistiakowsky was replacing him. Angry and hurt, Neddermeyer walked out. Although he would feel an "enduring bitterness," he nevertheless was persuaded to remain in Los Alamos as a senior technical adviser. Acting decisively, Oppenheimer had announced this change without first consulting Captain Parsons. "Parsons was furious," recalled Kistiakowsky. "He felt that I had bypassed him and that was outrageous. I can understand perfectly how he felt, but I was a civilian, so was Oppie, and I didn't have to go through him."

Parsons chafed at what he considered a loss of control over his Ordnance Division, and in September he sent Oppie a memorandum proposing to give himself broad decision-making powers over all aspects of the implosion bomb project. Oppenheimer gently but firmly refused: "The kind of authority which you appear to request from me is something I cannot delegate to you because I do not possess it. I do not, in fact, whatever protocol may suggest, have the authority to make decisions which are not understood and approved by the qualified scientists of the laboratory who must execute them." As a military man, Navy Captain Parsons wanted the authority in order to short-circuit the debates among his scientists. "You have pointed out," Oppenheimer wrote him, "that you are afraid your position in the laboratory might make it necessary for you to engage in prolonged argument and discussion in order to obtain agreement upon which the progress of the work would depend. Nothing that I can put in writing can eliminate this necessity." The scientists had to be free to argue—and Oppenheimer would arbitrate disputes only for the purpose of reaching some kind of collegial consensus. "I am not arguing that the laboratory should be so constituted," he told Parsons. "It is in fact so constituted."

In the midst of this ongoing crisis associated with the design of the plutonium bomb, Isidor Rabi paid one of his periodic visits to Los Alamos. He later remembered a gloomy session with a number of top scientists on the project as they talked of the urgency they felt about finding a way to make the plutonium bomb work. The conversation soon turned to the enemy: "Who were the German scientists? We knew them all," Rabi recalled. "What were they doing? We went over the whole thing again and looked at the history of our own development and tried to see where they could have been cleverer, where they might have had better judgment and avoided this error or that error. . . . We finally arrived at the conclusion that they could be exactly up to us, or perhaps further. We felt very solemn. One didn't know what the enemy had. One didn't want to lose a single day, a single week. And certainly, a month would be a calamity." As Philip Morrison summed up their attitude in mid-1944, "The only way we could lose the war was if we failed in our jobs."

Despite the reorganization, by late 1944 Kistiakowsky's group had still not managed to manufacture shaped explosives (called lenses) that would precisely crush a loosely packed, grapefruit-sized sphere of plutonium symmetrically into a sphere the size of a golf ball. Without such lenses, an implosion bomb seemed impractical. Captain Parsons was so pessimistic that he went to Oppenheimer and proposed that they abandon the lenses and try instead to create a non-lens type of implosion. In January 1945, the issue was hotly debated between Parsons and Kistiakowsky in the presence of both Groves and Oppenheimer. Kistiakowsky insisted that implosion could not be achieved without the lenses, and he promised that his men would soon be able to make them. In a decision critical to the success of the plutonium bomb, Oppenheimer backed him. During the next few months, Kistiakowsky and his team managed to perfect the implosion design. By May 1945, Oppenheimer felt fairly confident that the plutonium gadget would work.

Bomb-building was more engineering than theoretical physics. But Oppenheimer was as singularly adept at marshaling his scientists to overcome technical and engineering obstacles as he had been at stimulating his students to new insights at Berkeley. "Los Alamos might have succeeded without him," Hans Bethe later said, "but certainly only with much greater strain, less enthusiasm, and less speed. As it was, it was an unforgettable experience for all the members of the laboratory. There were other wartime laboratories of high achievement. . . . But I have never observed in any one of these other groups quite the spirit of belonging together, quite the urge to reminisce about the days of the laboratory, quite the feeling that this was really the great time of their lives. That this was true of Los Alamos was mainly due to Oppenheimer. He was a leader."

IN FEBRUARY 1944, a team of British scientists led by the German-born Rudolf E. Peierls arrived in Los Alamos. Oppenheimer had first met this brilliant but unassuming theoretical physicist in 1929, when both men were studying under Wolfgang Pauli. Peierls had emigrated from Germany to England in the early 1930s, and in 1940 he and Otto R. Frisch had written the seminal paper "On the Construction of a Superbomb," which had persuaded both the British and American governments that a nuclear weapon was feasible. During the next several years, Peierls worked on all aspects of Tube Alloys, the British bomb program. In 1942 and again in September 1943, Prime Minister Winston Churchill sent Peierls to America to help expedite work on the bomb. Peierls visited Oppenheimer in Berkeley and was "very impressed with his command of things. . . . He was the first person I met on that trip who had thought about the weapon itself and the implications of the physics of what would be going on."

Dr. Peierls spent only two and a half days on his first visit to Los Alamos. But Oppenheimer reported to Groves that they had agreed the British team could contribute substantially to studying the hydrodynamics of implosion. A month later, Peierls moved back to Los Alamos for the duration of the war. He admired how articulate and quick Oppenheimer was to understand anyone—but he particularly admired the way "he could stand up to General Groves."

As Peierls and his team settled into Los Alamos in the spring of 1944, Oppenheimer decided to give Peierls the job ostensibly held by Edward Teller. The mercurial Hungarian physicist was supposed to be working on a complicated set of calculations necessary for the implosion bomb. But Teller wasn't performing. Obsessed with the theoretical challenges posed by a "Super" thermonuclear bomb, Teller had no interest in a fission bomb. After Oppenheimer decided in June 1943 that wartime exigencies dictated a low priority for the Super, Teller became increasingly uncooperative. He seemed oblivious to any responsibility to contribute to the war effort. Always loquacious, he talked incessantly about a hydrogen bomb. Neither could he contain his resentment at having to work under Bethe. "I was not happy about having him as my boss," Teller recalled. To be sure, his resentment was fueled by Bethe's criticisms. Every morning Teller would have a bright new idea about how to make an H-bomb work—and overnight Bethe would prove it cockeyed. After one particularly trying encounter with Teller, Oppie quipped to Charles Critchfield, "God protect us from the enemy without and the Hungarians within."

Oppenheimer, understandably, became increasingly annoyed by Teller's behavior. One day that spring, Teller walked out of a meeting of section leaders and refused to do some calculations Bethe needed for his work on the implosion project. Extremely angry, Bethe complained to Oppie. "Edward essentially went on strike," Bethe recalled. When Oppenheimer confronted him about the incident, Teller finally asked to be relieved of all responsibility for work on the fission bomb. Oppenheimer agreed, and wrote General Groves that he wished to replace Teller with Peierls: "These calculations were originally under the supervision of Teller, who is, in my opinion and Bethe's, quite unsuited for this responsibility. Bethe feels that he needs a man under him to handle the implosion program."

Feeling slighted, Teller let it be known that he was thinking of leaving Los Alamos altogether. No one would have been surprised if Oppenheimer had let him go. Everyone thought of Teller as a "prima donna"; Bob Serber called him "a disaster to any organization." But instead of firing him, Oppenheimer gave Teller what he wanted, freedom to explore the feasibility of a thermonuclear bomb. Oppenheimer even agreed to give him a precious hour of his time once a week just to talk about whatever was on Teller's mind.

Not even this extraordinary gesture satisfied Teller, who thought that his friend had become a "politician." Oppie's colleagues wondered why he bothered with Teller. Peierls considered Teller "somewhat wild; he can back an idea for a time and then it turns out to be nonsense." Oppenheimer could be impatient with fools; but he was aware that Teller was no fool. He tolerated him because, in the end, he might contribute something to the project. When, later that summer, he hosted a reception for Churchill's special representative, Lord Cherwell (Frederic A. Lindemann), Oppenheimer realized afterwards that he had inadvertently left Rudolf Peierls off the invitation list. The next day he apologized to Peierls and then quipped, "It could have been worse—it could have been Teller."

IN DECEMBER 1944, Oppenheimer urged Rabi to make another visit to Los Alamos. "Dear Rab," he wrote, "We have been wondering for some time when you could come out again. The crises here are so continuous that it is hard to find one time which would be better or worse than another from our point of view." Rabi had just been awarded the Nobel Prize in physics in

recognition for "his resonance method for recording the magnetic properties of atomic nuclei." Oppie congratulated him: "It is nice to have the prize
go to a man who is out of his adolescence rather than just entering it."

Swamped with administrative work, Oppenheimer still found time to
write the occasional personal letter. In the spring of 1944, he wrote to a family of German refugees whose escape from Europe he had facilitated. They
were utter strangers, but in 1940 he had given the Meyers family—a mother
and four daughters—a sum of money to pay their expenses to the United
States. Four years later, the Meyers repaid Oppenheimer and proudly
informed him that they had become American citizens. He wrote back that
he understood the "pride" they felt, and he thanked them for the money: "I
hope it has not been a hardship for you. . . ." He then offered to return the
money if they had any further need for it. (Years later, one of the Meyers
daughters wrote in gratitude: "[I]n 1940 you brought us all over and we
could save our lives.") For Oppenheimer, the rescue of the Meyerses
from the Nazi contagion was important in several respects. It was in the
first instance a politically noncontroversial extension of his antifascist
activism—and that felt good. Secondly, while a small act of generosity, it
was nevertheless a profound and welcome reminder of why he was racing
to build a horrific weapon.

And racing he was. Restlessness was part of his character—or so
thought Freeman Dyson, a young physicist who came to know and admire
Oppenheimer after the war. But Dyson also saw restlessness as Oppie's
tragic flaw: "Restlessness drove him to his supreme achievement, the fulfillment of the mission of Los Alamos, without pause for rest or reflection."

"Only one man paused," Dyson wrote. "The one who paused was
Joseph Rotblat from Liverpool. . . ." A Polish physicist, Rotblat had been
stranded in England when the war broke out. He was recruited by James
Chadwick into the British bomb project and by early 1944 found himself in
Los Alamos. One evening in March 1944, Rotblat experienced a "disagree-
able shock." General Groves came for dinner at the Chadwicks' and in the
course of casual banter over the dinner table, he said, "You realize of course
that the main purpose of this project is to subdue the Russians." Rotblat was
shocked. He had no illusions about Stalin—the Soviet dictator had, after all,
invaded his beloved Poland. But thousands of Russians were dying every
day on the Eastern Front and Rotblat felt a sense of betrayal. "Until then I
had thought that our work was to prevent a Nazi victory," he later wrote,
"and now I was told that the weapon we were preparing was intended for
use against the people who were making extreme sacrifices for that very
aim." By the end of 1944, six months after the Allies had landed on the
beaches of Normandy, it was clear that the war in Europe would soon be

over. Rotblat saw no point in continuing to work on a weapon that was no longer needed to defeat the Germans.* After saying good-bye to Oppenheimer at a going-away party, he left Los Alamos on December 8, 1944.

IN THE AUTUMN of 1944, the Soviets received the first of many intelligence reports directly from Los Alamos. The spies overlooked by Army counterintelligence included Klaus Fuchs, a German physicist with British citizenship, and Ted Hall, a precociously brilliant nineteen-year-old with a Harvard B.S. in physics. Hall arrived in Los Alamos in late January 1944, while Fuchs came in August as part of the British team led by Rudolf Peierls.

Fuchs, born in 1911, was raised in a German Quaker family. Studious and idealistic, he joined the German socialist party, the SPD, while studying at the University of Leipzig in 1931—the same year his mother committed suicide. In 1932, alarmed by the growing political strength of the Nazis, Fuchs broke with the socialists and joined the Communist Party, which was more actively resisting Hitler. In July 1933, he fled Hitler's Germany and became a political refugee in England. Over the next few years, his family was decimated by the Nazi regime. His brother escaped to Switzerland, leaving behind a wife and child who later died in a concentration camp. His father was sent to prison for "anti-government agitation," and in 1936 his sister Elizabeth killed herself after her husband was arrested and sent to a concentration camp. Fuchs had every reason to hate the Nazis.

In 1937, after earning a doctorate in physics in Bristol, Fuchs won a postgraduate fellowship to work with Oppenheimer's former professor Max Born, who by then was teaching at Edinburgh. After the war began, Fuchs was interned in Canada as an enemy alien, and Professor Born helped to obtain his release by attesting that Fuchs was "among the two or three most gifted theoretical physicists of the young generation." He and thousands of other anti-Nazi German refugees were released at the end of 1940; Fuchs was given permission to return to his work in England. Although the British Home Office knew all about his communist past, by the spring of 1941 Fuchs was working with Peierls and other British scientists on the highly classified Tube Alloys project. In June 1942, Fuchs received British citizenship—by then, he was already passing information to the Soviets about the British bomb program.

When Fuchs arrived in Los Alamos, neither Oppenheimer nor anyone

---

*In 1995, Joseph Rotblat was awarded the Nobel Peace Prize for his work on nuclear disarmament.

else had any suspicion that he was a Soviet spy. After he was arrested in 1950, Oppie told the FBI that he had thought Fuchs was a Christian Democrat, and certainly not a "political fanatic." Bethe considered Fuchs one of the best men in his division. "If he was a spy," Bethe told the FBI, "he played his role beautifully. He worked days and nights. He was a bachelor and had nothing better to do, and he contributed very greatly to the success of the Los Alamos project." Over the next year, Fuchs passed detailed written information to the Soviets about the problems and advantages of the implosion-type bomb design over the gun method. He was unaware that the Soviets were getting confirmation of his information from another Los Alamos resident.

By September 1944, Ted Hall was working on the calibration tests needed for the implosion-design bomb. Oppenheimer heard that Hall was one of the best young technicians on the mesa when it came to creating a test implosion. An extremely bright man, Hall that autumn was sitting on the edge of an intellectual precipice. He was a socialist in outlook, an admirer of the Soviet Union, but not yet a formal communist, and neither was he disgruntled or unhappy with his work or his station in life. No one recruited him. But all that year he had listened to "older" scientists—in their late twenties and early thirties—talk about their fear of a postwar arms race. On one occasion, sitting at the same Fuller Lodge dinner table with Niels Bohr, he heard Bohr's concerns for an "open world." Prompted by his conclusion that a postwar U.S. nuclear monopoly could lead to another war, in October 1944 Hall decided to act: ". . . it seemed to me that an American monopoly was dangerous and should be prevented. I was not the only scientist to take that view."

While on a fourteen-day leave from Los Alamos, Hall boarded a train to New York City and simply walked into a Soviet trade office and gave a Soviet official a handwritten report on Los Alamos. It described the laboratory's purpose and listed the names of the leading scientists working on the bomb project. In the months that followed, Hall managed to pass the Soviets much additional information, including critical information on the design for the implosion bomb. Hall was the perfect "walk-in" spy; he knew what the Russians needed to know about the atomic bomb project; he needed nothing himself and expected nothing. His sole purpose was to "save the world" from a nuclear war that he believed was inevitable if the United States emerged from the war with an atomic monopoly.

Oppenheimer knew nothing about Hall's espionage activities. But he did know that a group of twenty or so scientists, some of them group leaders, had begun meeting informally once a month to talk about the war, politics and the future. "It used to be in the evenings," recalled Rotblat, "usually at somebody's house like the Tellers', someone who had fairly large rooms. People would meet to discuss the future of Europe, the future of the world." Among other issues, they talked about the exclusion of Soviet scientists from the project. According to Rotblat, Oppenheimer came to at least one of their meetings and Rotblat said later, "I always thought he was a soul mate in the sense that we had the same humanitarian approach to problems."

BY LATE 1944, a number of scientists at Los Alamos began to voice their growing ethical qualms about the continued development of the "gadget." Robert Wilson, now chief of the lab's experimental physics division, had "quite long discussions with Oppie about how it might be used." Snow was still on the ground when Wilson went to Oppenheimer and proposed holding a formal meeting to discuss the matter more fully. "He tried to talk me out of it," Wilson later recalled, "saying I would get into trouble with the G-2, the security people."

Despite his respect, even reverence, for Oppie, Wilson thought little of this argument. He told himself, "All right. So what? I mean, if you're a good pacifist, then clearly you are not going to be worried about being thrown in jail or whatever they would do—have your salary reduced or horrible things like that." So Wilson told Oppenheimer that he hadn't talked him out of at least having an open discussion about an issue that was obviously of great importance. Wilson then put up notices all over the lab announcing a public meeting to discuss "The Impact of the Gadget on Civilization." He chose this title because earlier, at Princeton, "just before we'd come out, there'd been many sanctimonious talks about the 'impact' of something else, with all very scholarly kinds of discussions."

To his surprise, Oppie showed up on the appointed evening and listened to the discussion. Wilson later thought about twenty people attended, including such senior physicists as Vicki Weisskopf. The meeting was held in the same building that housed the cyclotron. "I can remember," Wilson said, "it being very cold in our building. . . . We did have a pretty intense discussion of why it was that we were continuing to make a bomb after the war had been [virtually] won."

This may not have been the only occasion when the morality and politics of the atomic bomb were discussed. A young physicist working on implosion techniques, Louis Rosen, remembered a packed daytime colloquium held in the old theater. Oppenheimer was the speaker and, according to Rosen, the topic was "whether the country is doing the right thing in using this weapon on real live human beings." Oppenheimer apparently argued that as scientists they had no right to a louder voice in determining the gadget's fate than any other citizen. "He was a very eloquent and persuasive guy," Rosen said. The chemist Joseph O. Hirschfelder recalled a similar discussion held in Los Alamos' small wooden chapel in the midst of a thunderstorm on a cold Sunday evening in early 1945. On this occasion, Oppenheimer argued with his usual eloquence that, although they were all destined to live in perpetual fear, the bomb might also end all war. Such a hope, echoing Bohr's words, was persuasive to many of the assembled scientists.

No official records were kept of these sensitive discussions. So memories prevail. Robert Wilson's account is the most vivid—and those who knew Wilson always thought him a man of singular integrity. Victor Weisskopf later recalled having political discussions about the bomb at various times with Willy Higinbotham, Robert Wilson, Hans Bethe, David Hawkins, Phil Morrison and William Woodward, among others. Weisskopf recalled that the expected end of the war in Europe "caused us to think more about the future of the world after the war." At first, they simply met in their apartments, and pondered questions such as "What will this terrible weapon do to this world? Are we doing something good, something bad? Should we not worry about how it will be applied?" Gradually, these informal discussions became formal meetings. "We tried to organize meetings in some of the lecture rooms," Weisskopf said, "and then we ran into opposition. Oppenheimer was against that. He said that's not our task, and this is politics, and we should not do this." Weisskopf recalled a meeting in March 1945, attended by forty scientists, to discuss "the atomic bomb in world politics." Oppenheimer again tried to discourage people from attending. "He thought we should not get involved in questions about the use of the bomb. . . ." But, contrary to Wilson's memory, Weisskopf later wrote that "the thought of quitting did not even cross my mind."

Wilson believed it would have reflected badly on Oppenheimer if he had chosen not to appear. "You know, you're the director, a little bit like a general. Sometimes you have got to be in front of your troops, sometimes you've got to be in back of them. Anyway, he came and he had very cogent arguments that convinced me." Wilson wanted to be convinced. Now that it seemed so clear that the gadget would not be used on the Germans, he and many others in the room had doubts but no answers. "I thought we were fighting the Nazis," Wilson said, "not the Japanese particularly." No one thought the Japanese had a bomb program.

When Oppenheimer took the floor and began speaking in his soft voice, everyone listened in absolute silence. Wilson recalled that Oppenheimer "dominated" the discussion. His main argument essentially drew on Niels Bohr's vision of "openness." The war, he argued, should not end without the world knowing about this primordial new weapon. The worst outcome would be if the gadget remained a military secret. If that happened, then the next war would almost certainly be fought with atomic weapons. They had to forge ahead, he explained, to the point where the gadget could be tested. He pointed out that the new United Nations was scheduled to hold its inaugural meeting in April 1945—and that it was important that the delegates begin their deliberations on the postwar world with the knowledge that mankind had invented these weapons of mass destruction.

"I thought that was a very good argument," said Wilson. For some time now, Bohr and Oppenheimer himself had talked about how the gadget was going to change the world. The scientists knew that the gadget was going to force a redefinition of the whole notion of national sovereignty. They had faith in Franklin Roosevelt and believed that he was setting up the United Nations precisely to address this conundrum. As Wilson put it, "There would be areas in which there would be no sovereignty, the sovereignty would exist in the United Nations. It was to be the end of war as we knew it, and this was a promise that was made. That is why I could continue on that project."

Oppenheimer had prevailed, to no one's surprise, by articulating the argument that the war could not end without the world knowing the terrible secret of Los Alamos. It was a defining moment for everyone. The logic— Bohr's logic—was particularly compelling to Oppenheimer's fellow scientists. But so too was the charismatic man who stood before them. As Wilson recalled that moment, "My feeling about Oppenheimer was, at that time, that this was a man who is angelic, true and honest and he could do no wrong. . . . I believed in him."

CHAPTER TWENTY-TWO

"Now We're All Sons-of-Bitches"

Well, Roosevelt was a great architect, perhaps Truman will be a good carpenter.

ROBERT OPPENHEIMER

O N THURSDAY AFTERNOON, April 12, 1945—just two years after the lab's opening—word suddenly spread of Franklin Roosevelt's death. Work was suspended and Oppenheimer notified everyone to assemble at the flagpole near the administrative building for a formal announcement. He then scheduled a memorial service for that Sunday. "Sunday morning found the mesa deep in snow," Phil Morrison later wrote. "A night's fall had covered the rude textures of the town, silenced its business, and unified the view in a soft whiteness, over which the bright sun shone, casting deep blue shadows behind every wall. It was no costume for mourning, but it seemed recognition of something we needed, a gesture of consolation. Everybody came to the theater, where Opje spoke very quietly for two or three minutes out of his heart and ours."

Oppenheimer had drafted a eulogy of three short paragraphs. "We have been living through years of great evil," he said, "and great terror." And during this time Franklin Roosevelt had been, "in an old and unperverted sense, our leader." Characteristically, Oppenheimer turned to the Bhagavad-Gita: "Man is a creature whose substance is faith. What his faith is, he is." Roosevelt had inspired millions around the globe to have faith that the terrible sacrifices of this war would result in "a world more fit for human habitation." For this reason, Oppenheimer concluded, "we should dedicate ourselves to the hope, that his good works will not have ended with his death."

Oppenheimer still nurtured the hope that Roosevelt and his men had learned from Bohr that the formidable new weapon they were building would require a radical new openness. "Well," he told David Hawkins afterwards, "Roosevelt was a great architect, perhaps Truman will be a good carpenter."

. . .

AS HARRY TRUMAN moved into the White House, the war in Europe was nearly won. But the war in the Pacific was coming to its bloodiest climax. On the evening of March 9–10, 1945, 334 B-29 aircraft dropped tons of jellied gasoline—napalm—and high explosives on Tokyo. The resulting firestorm killed an estimated 100,000 people and completely burned out 15.8 square miles of the city. The fire-bombing raids continued and by July 1945, all but five of Japan's major cities had been razed and hundreds of thousands of Japanese civilians had been killed. This was total warfare, an attack aimed at the destruction of a nation, not just its military targets.

The fire bombings were no secret. Ordinary Americans read about the raids in their newspapers. Thoughtful people understood that strategic bombing of cities raised profound ethical questions. "I remember Mr. Stimson [the secretary of war] saying to me," Oppenheimer later remarked, "that he thought it appalling that there should be no protest over the air raids which we were conducting against Japan, which in the case of Tokyo led to such extraordinarily heavy loss of life. He didn't say that the air strikes shouldn't be carried on, but he did think there was something wrong with a country where no one questioned that. . . ."

On April 30, 1945, Adolf Hitler committed suicide, and eight days later Germany surrendered. When Emilio Segrè heard the news, his first reaction was, "We have been too late." Like almost everyone at Los Alamos, Segrè thought that defeating Hitler was the sole justification for working on the "gadget." "Now that the bomb could not be used against the Nazis, doubts arose," he wrote in his memoirs. "Those doubts, even if they do not appear in official reports, were discussed in many private discussions."

AT THE UNIVERSITY of Chicago's Met Lab, Leo Szilard was frantic. The peripatetic physicist knew time was running out. Atomic bombs would soon be ready, and he expected that they would be used on Japanese cities. Having been the first to urge President Roosevelt to initiate a program to build atomic weapons, he now made repeated attempts to prevent their use. First, he drafted a memorandum to President Roosevelt—introduced by another letter from Einstein—in which he warned the president that "our 'demonstration' of atomic bombs will precipitate" an arms race with the Soviets. But when Roosevelt died before Szilard could see him, he managed to get an appointment to see the new president, Harry Truman, on May 25. In the meantime, he decided to write Oppenheimer, warning him "that if a race in the production of atomic bombs should become unavoidable, the prospects of this country cannot be expected to be good." In the absence of a clear policy to avoid such an arms race, Szilard wrote, "I doubt whether it is wise to show our hand by using atomic bombs against Japan." He had listened to the proponents of using the bomb, and he felt their arguments "were not strong enough to dispel my doubts." Oppie did not reply.

On May 25, Szilard and two colleagues—Walter Bartky of the University of Chicago and Harold Urey of Columbia University—appeared at the White House, only to be told that Truman had referred them to James F. Byrnes, soon to be designated secretary of state. Dutifully, they traveled to Byrnes' home in Spartanburg, South Carolina, for a meeting that concluded, to say the least, unproductively. When Szilard explained that the use of the atomic bomb against Japan risked turning the Soviet Union into an atomic power, Byrnes interrupted, "General Groves tells me there is no uranium in Russia." No, Szilard replied, the Soviet Union has plenty of uranium.

Byrnes then suggested that the use of the atomic bomb on Japan would help persuade Russia to withdraw its troops from Eastern Europe after the war. Szilard was "flabbergasted by the assumption that rattling the bomb might make Russia more manageable." "Well," Byrnes said, "you come from Hungary—you would not want Russia to stay in Hungary indefinitely." This only incensed Szilard, who later wrote, "I was concerned at this point that . . . we might start an arms race between America and Russia which might end with the destruction of both countries. I was not disposed at this point to worry about what would happen to Hungary." Szilard left in a somber mood. "I was rarely," he wrote, "as depressed as when we left Byrnes' house and walked toward the station."

Back in Washington, Szilard made another attempt to block the use of the bomb. On May 30, hearing that Oppenheimer was in the capital for a meeting with Secretary of War Stimson, Szilard phoned General Groves' office and made an appointment to see Oppenheimer that morning. Oppenheimer considered Szilard a meddler, but decided he had to hear him out.

"The atomic bomb is shit," Oppenheimer said after listening to Szilard's arguments.

"What do you mean by that?" Szilard asked.

"Well," Oppenheimer replied, "this is a weapon which has no military significance. It will make a big bang—a very big bang—but it is not a weapon which is useful in war." At the same time, Oppie told Szilard that if the weapon was used, he thought it important that the Russians be informed of this in advance. Szilard argued that merely telling Stalin about the new weapon would not by itself prevent an arms race after the war.

"Well," Oppenheimer insisted, "don't you think that if we tell the Russians what we intend to do and then use the bomb on Japan, the Russians will understand it?"

"They'll understand it only too well," Szilard replied.

Szilard left the meeting once again disheartened, knowing that this, his third attempt to stop the bomb, had failed. Over the next few weeks, he worked feverishly to establish a public record that would show that at least a vocal minority of the scientists involved in the Manhattan Project had opposed the use of the bomb on a civilian target.

The next day, May 31, Oppenheimer attended a critical meeting of Stimson's so-called Interim Committee, an ad hoc group of government officials brought together to advise the secretary of war on the future of atomic policy. Members of the Committee included Stimson, Assistant Secretary of the Navy Ralph A. Bard, Dr. Vannevar Bush, James F. Byrnes, William L. Clayton, Dr. Karl T. Compton, Dr. James B. Conant and George L. Harrison, an aide to Stimson. Four scientists were present, having been invited to serve the Committee as a panel of scientific consultants: Oppenheimer, Enrico Fermi, Arthur Compton and Ernest Lawrence. Also in attendance that day were Gen. George C. Marshall, General Groves and two Stimson assistants, Harvey H. Bundy and Arthur Page.

Stimson controlled the agenda—and it did not include a decision on whether the bomb should be used against Japan. That was more or less a foregone conclusion. As if to emphasize this point, Stimson began the meeting with a general explanation of his responsibilities to the president on military matters. No one could escape the implication that decisions on the military use of the bomb would be controlled exclusively by the White House, with no input from the scientists who over the past two years had been building the bomb. But Stimson was a wise man who had paid careful attention to all discussions regarding the implications of nuclear weapons. Oppenheimer and the other scientists thus were reassured to hear him say that he and the other members of the Interim Committee did not regard the bomb "as a new weapon merely but as a revolutionary change in the relations of man to the universe." The atomic bomb might become "a Frankenstein which would eat us up," or it could secure the global peace. Its import, in either case, "went far beyond the needs of the present war."

Stimson then quickly turned the discussion to the future development of atomic weapons. Oppenheimer reported that within three years it might be possible to produce a bomb with an explosive force of 10 million to 100 million tons of TNT. Lawrence jumped in with the recommendation that "a sizable stockpile of bombs and material should be built up"; more money had to be spent on nuclear plant expansion if Washington wanted the country to "stay out in front." Initially, the official minutes of the meeting have

Stimson declaring that everyone agrees with Lawrence's proposal to build up stockpiles of both weapons and industrial plants. But then the minutes begin to reflect Oppenheimer's seeming ambivalence. He observed that the Manhattan Project had merely "plucked the fruits of earlier research." He strongly urged Stimson to allow most scientists, once the war was over, to go back to their universities and research laboratories, "to avoid the sterility" of wartime work.

Unlike Lawrence, Oppenheimer did not want the Manhattan Project to continue to dominate scientific inquiry after the war. As he addressed the meeting in his characteristically hushed tones, Oppie's words were persuasive to many in the room. Vannevar Bush interrupted to say that he "agreed with Dr. Oppenheimer that only a nucleus of the present staff should be retained and that as many as possible should be released for broader and freer inquiry." Compton and Fermi—but not Lawrence—chimed in with their approval. Although he had not made the point explicitly, Oppenheimer had staked out an argument for refocusing the work of the weapons labs after the war.

When Stimson asked about the nonmilitary potential of the project, Oppenheimer again dominated the discussion. He pointed out that up until then their "immediate concern had been to shorten the war." But it should be understood, he said, that "fundamental knowledge" about atomic physics was "so widespread throughout the world" that he thought it wise for the United States to offer a "free interchange of information" on the development of peacetime uses of the atom. Echoing his discussion of the previous day with Szilard, Oppenheimer said, "If we were to offer to exchange information before the bomb was actually used, our moral position would be greatly strengthened."

Picking up on this cue, Stimson began discussing the prospects for "a policy of self-restraint." He referred to the possibility that an international organization should be established to guarantee "complete scientific freedom." Perhaps the bomb could be controlled in the postwar world by an "international control body" armed with the right of inspection. While the scientists in the room nodded their heads, a heretofore silent General Marshall suddenly cautioned against putting too much faith in the effectiveness of any inspection mechanism. Russia was obviously the "paramount concern."

Marshall's stature was such that not many men challenged his judgment. But Oppenheimer had an agenda—Bohr's—and he now quietly and forcefully brought the revered general around to his point of view. Who knew, he admitted, what the Russians were doing in this field of atomic weapons? But he nevertheless "expressed the hope that the fraternity of interest among scientists would aid in the solution." He pointed out that "Russia had always been friendly to science." Perhaps, he suggested, we should open discussions with them in a tentative fashion, and explain what we had developed "without giving them any details of our productive effort."

"We might say that a great national effort had been put into this project," he said, "and express a hope for cooperation with them in this field." Oppenheimer finished by saying that he "felt strongly that we should not prejudge the Russian attitude in this matter."

Somewhat surprisingly, Oppenheimer's statement now roused Marshall into a detailed defense of the Russians. Relations between Moscow and Washington had been marked, he said, by a long history of charges and countercharges. But "most of these allegations have proven unfounded." On the question of the atomic bomb, Marshall said he was "certain that we need have no fear that the Russians, if they had knowledge of the project, would disclose this information to the Japanese." Far from trying to keep the bomb secret from the Russians, Marshall "raised the question whether it might be desirable to invite two prominent Russian scientists to witness the test."

Oppenheimer must have been pleased to hear such words coming from the country's top military officer. And he must have been quickly disheartened to hear James Byrnes, Truman's personal representative to the Interim Committee, protest vigorously that if such a thing happened, he feared Stalin would then ask to be brought into the atomic project. Between the lines of the dry and unemotional official record, a careful reader can discern a debate. Vannevar Bush pointed out that even the British "do not have any of our blueprints on plants," and clearly, the Russians could be told a lot more about the project without giving them the engineering designs for the bomb. Indeed, Oppenheimer and all of the scientists in the room understood that such information could not remain secret for very long. Inevitably, the physics of the bomb was soon to be known to most physicists.

But Byrnes was already beginning to think of the bomb as an American diplomatic weapon. Running roughshod over Oppenheimer's and Marshall's arguments, the secretary of state–designate reinforced Lawrence by insisting that they had to "push ahead as fast as possible in [atomic] production and research to make certain that we stay ahead and at the same time make every effort to better our political relations with Russia." The minutes record that Byrnes' view was "generally agreed to by all present." And yet Oppenheimer—and surely many others in the room—understood that they could not rush to "stay ahead" in atomic weapons without pushing the Russians into an arms race with the United States. This gaping contradiction was papered over by Arthur Compton, who stressed the importance of maintaining American superiority through "freedom of research" while also reaching a "cooperative understanding" with Russia. On this ambiguous conclusion, the committee adjourned at 1:15 p.m. for a one-hour lunch.

Over lunch, someone raised the question of the use of the bomb on Japan. No notes were taken, but when the formal meeting resumed, the discussion continued to focus on the effect of the impending bombing. Stimson, always alert to the political implications of any decision, altered the agenda to allow the discussion to carry on. Someone commented that one atomic bomb would have no more effect than some of the massive bomber strikes launched against Japanese cities that spring. Oppenheimer seemed to agree, but added that "the visual effect of an atomic bombing would be tremendous. It would be accompanied by a brilliant luminescence which would rise to a height of 10,000 to 20,000 feet. The neutron effect of the explosion would be dangerous to life for a radius of at least two-thirds of a mile."

"Various types of targets and the effects to be produced" were discussed, and then Secretary Stimson summarized what seemed to be a general agreement: ". . . that we could not give the Japanese any warning; that we could not concentrate on a civilian area; but that we should seek to make a profound psychological impression on as many of the inhabitants as possible." Stimson said he agreed with James Conant's suggestion "that the most desirable target would be a vital war plant employing a large number of workers and closely surrounded by workers' houses." Thus, with such delicate euphemisms, did the president of Harvard University select civilians as the target of the world's first atomic bomb.

Oppenheimer voiced no disagreement with the choice of the defined target. Instead, he seems to have initiated a discussion of whether several such strikes could be mounted simultaneously. He thought multiple atomic bombing "would be feasible." General Groves vetoed this idea, and then went on to complain that the program had been "plagued since its inception by the presence of certain scientists of doubtful discretion and uncertain loyalty." Groves had in mind Leo Szilard, who he had just learned had attempted to see Truman in his effort to persuade the president not to use the bomb. After Groves' comments, the minutes record that it was "agreed" that after the bomb was used, steps would be taken to sever these scientists from the program. Oppenheimer seems to have given his assent, if only silently, to this purge.

Last, someone—most likely, one of the scientists—asked what the scientists might tell their colleagues about the Interim Committee's deliberations. It was agreed that the four scientists in attendance should "feel free to tell their people" that they had met with a committee chaired by the secretary of war and had been given "complete freedom to present their views on any phase of the subject." On this note, the meeting was adjourned at 4:15 p.m.

Oppenheimer had played an ambiguous role in this critical discussion. He had vigorously advanced Bohr's notion that the Russians should soon be briefed about the impending new weapon. He had even persuaded General Marshall, until Byrnes had effectively derailed the idea. On the other hand, he had evidently felt it prudent to remain silent as General Groves made clear his intention to dismiss dissident scientists like Szilard. Neither had Oppenheimer offered an alternative to, let alone criticism of, Conant's euphemistic definition of the proposed "military" target—"a vital war plant employing a large number of workers and closely surrounded by workers' houses." Though he had clearly argued for some of Bohr's ideas about openness, in the end he had won nothing and acquiesced to everything. The Soviets would not be adequately informed about the Manhattan Project, and the bomb would be used on a Japanese city without warning.

MEANWHILE, a group of scientists in Chicago, spurred on by Szilard, organized an informal committee on the social and political implications of the bomb. In early June 1945, several members of the committee produced a twelve-page document that came to be known as the Franck Report, after its chairman, the Nobelist James Franck. It concluded that a surprise atomic attack on Japan was inadvisable from any point of view: "It may be very difficult to persuade the world that a nation which was capable of secretly preparing and suddenly releasing a weapon as indiscriminate as the [German] rocket bomb and a million times more destructive, is to be trusted in its proclaimed desire of having such weapons abolished by international agreement." The signatories recommended a demonstration of the new weapon before representatives of the United Nations, perhaps in a desert site or on a barren island. Franck was dispatched with the Report to Washington, D.C., where he was informed, falsely, that Stimson was out of town. Truman never saw the Franck Report; it was seized by the Army and classified.

By contrast to the people in Chicago, the scientists in Los Alamos, working feverishly to test the plutonium implosion bomb model as soon as possible, had little time to think about how or whether their "gadget" should be used on Japan. But they also felt that they could rely on Oppenheimer. As the Met Lab biophysicist Eugene Rabinowitch, one of the seven signatories of the Franck Report, observed, the Los Alamos scientists shared a widespread "feeling that we can trust Oppenheimer to do the right thing."

One day, Oppenheimer called Robert Wilson into his office and explained that he was a consultant to the Interim Committee that was advising Stimson on how the bomb should be used. He asked Wilson for his views. "He gave me some time to think about it. . . . And so I came back and said I felt that it should not be used, and that the Japanese should be alerted to it in some manner." Wilson pointed out that in just a few weeks they would be conducting a test of the bomb. Why not invite the Japanese to send a delegation of observers to witness the test?

"Well," Oppenheimer replied, "supposing it didn't go off?"

"And I turned to him, coldly," Wilson recalled, "and said, 'Well, we could kill 'em all.' " Within seconds, Wilson—a pacifist—regretted having said "such a bloodthirsty thing."

Wilson was flattered to have been asked, but disappointed that his views had not changed Oppie's thinking. "He should have had no business talking to me about it in the first place," Wilson said. "But he clearly wanted some advice from somebody and he liked me, and I was very fond of him."

Oppenheimer also talked with Phil Morrison, his former student and, since his transfer from the Met Lab in Chicago, one of his closest friends in Los Alamos. Morrison remembers participating in a meeting of Groves' Target Committee in the spring of 1945. Two such meetings took place in Oppenheimer's office on May 10 and 11, and the official minutes record the participants' agreement that the target for the bomb should be located "in a large urban area of more than three miles diameter." They even discussed targeting the emperor's palace in downtown Tokyo. Morrison, sitting in as a technical expert, remembers speaking up in favor of some kind of formal warning to the Japanese, if a demonstration seemed impractical: "I thought even leaflet warning would have been enough." But when he suggested this, the notion of a warning was quickly dismissed by an unidentified Army officer. "If we give a warning they'll follow us and shoot us down," said the officer dismissively. "It's very easy for you to say and it's not easy for me to accept." And Morrison got no support for his position from Oppenheimer.

"Essentially," he recalled much later, "I was given rather a hard time. I was excluded from having any real comment. . . . I came away with the realization that we had little influence on what was going to happen." Morrison's recollection was confirmed by David Hawkins, who also was in the room. "Morrison represented the concerns of many of us," Hawkins wrote. "He said that he proposed that a warning be sent to the Japanese . . . giving them a chance to evacuate. The officer sitting across from him—name not known, or remembered—spoke vehemently against the proposal, saying something like, 'They'd send up everything they have against us, and I'd be in that plane.' "

. . .

IN MID-JUNE, Oppenheimer convened a meeting in Los Alamos of the Scientific Panel—himself, Lawrence, Arthur Compton and Enrico Fermi—to discuss their final recommendations to the Interim Committee. The four scientists had a freewheeling discussion about the Franck Report, which Compton summarized for them. Of special interest was its call for a non-lethal, but dramatic, demonstration of the power of the atomic bomb. Oppenheimer was ambivalent: "I set forth my anxieties and the arguments . . . against dropping [the bomb] . . . but I did not endorse them," he later reported.

On June 16, 1945, Oppenheimer signed a short memorandum summarizing the Scientific Panel's recommendations "on the immediate use of nuclear weapons." Addressed to Secretary Stimson, it was a diffident document. The panel members recommended, first, that prior to the use of the bomb, Washington should inform Britain, Russia, France and China of the existence of atomic weapons and "welcome suggestions as to how we can cooperate in making this development contribute to improved international relations." Secondly, the panel reported that there was no unanimity among their scientific colleagues on the initial use of these weapons. Some of the men who were building them proposed a demonstration of the "gadget" as an alternative. "Those who advocate a purely technical demonstration would wish to outlaw the use of atomic weapons, and have feared that if we use the weapons now our position in future negotiations will be prejudiced." Although Oppenheimer surely sensed that most of his colleagues at Los Alamos and at Chicago's Met Lab favored such a demonstration, he now weighed in on the side of those who "emphasize the opportunity of saving American lives by immediate military use. . . ."

Why? Oddly enough, his reasoning was essentially as Bohrian as that of the men who favored a demonstration. He had become convinced that the military use of the bomb in this war might eliminate all wars. Oppenheimer explained that some of his colleagues actually believed that the use of the bomb in this war might "improve the international prospects, in that they are more concerned with the prevention of war than with the elimination of this specific weapon. We find ourselves closer to these latter views; we can propose no technical demonstration likely to bring an end to the war; we see no acceptable alternative to direct military use."

Having offered such a clear, unambiguous endorsement of "military use," the panel could come to no conclusion as to how to define "military use." As Compton later informed Groves, "There was not sufficient agreement among the members of the panel to unite upon a statement as to how or under what conditions such use was to be made." Oppenheimer ended his memo with a curious disclaimer: "[I]t is clear that we, as scientific men, have no proprietary rights . . . no claim to special competence in solving the political, social, and military problems which are presented by the advent of atomic power." It was an odd conclusion—and one that Oppenheimer would soon abandon.

There was much that Oppenheimer did not know. As he later recalled, "We didn't know beans about the military situation in Japan. We didn't know whether they could be caused to surrender by other means or whether the invasion was really inevitable. But in the backs of our minds was the notion that the invasion was inevitable because we had been told that." Among other things, he was unaware that military intelligence in Washington had intercepted and decoded messages from Japan indicating that the Japanese government understood the war was lost and was seeking acceptable surrender terms.

On May 28, for instance, Assistant Secretary of War John J. McCloy urged Stimson to recommend that the term "unconditional surrender" be dropped from America's demands on the Japanese. Based on their reading of intercepted Japanese cable traffic (code-named "Magic"), McCloy and many other ranking officials could see that key members of the Tokyo government were trying to find a way to terminate the war, largely on Washington's terms. On the same day, Acting Secretary of State Joseph C. Grew had a long meeting with President Truman and told him the very same thing. Whatever their other objectives, Japanese government officials had one immutable condition, as Allen Dulles, then an OSS agent in Switzerland, reported to McCloy: "They wanted to keep their emperor and the constitution, fearing that otherwise a military surrender would only mean the collapse of all order and of all discipline."

On June 18, Truman's chief of staff, Adm. William D. Leahy, wrote in his diary: "It is my opinion at the present time that a surrender of Japan can be arranged with terms that can be accepted by Japan. . . ." The same day, McCloy told President Truman that he believed the Japanese military position to be so dire as to raise the "question of whether we needed to get Russia in to help us defeat Japan." He went on to tell Truman that before a final decision was taken to invade the Japanese home islands, or to use the atomic bomb, political steps should be taken that might well secure a full Japanese surrender. The Japanese, he said, should be told that they "would be permitted to retain the Emperor and a form of government of their own choosing." In addition, he said, "the Japs should be told, furthermore, that we had another and terrifyingly destructive weapon which we would have to use if they did not surrender."

According to McCloy, Truman seemed receptive to these suggestions. American military superiority was such that by July 17 McCloy was writing in his diary: "The delivery of a warning now would hit them at *the* moment. It would probably bring what we are after—the successful termination of the war."

According to Gen. Dwight D. Eisenhower, when he was informed of the existence of the bomb at the Potsdam Conference in July, he told Stimson he thought an atomic bombing was unnecessary because "the Japanese were ready to surrender and it wasn't necessary to hit them with that awful thing." Finally, President Truman himself seemed to think that the Japanese were very close to capitulation. Writing in his private, handwritten diary on July 18, 1945, the president referred to a recently intercepted cable quoting the emperor to the Japanese envoy in Moscow as a "telegram from Jap Emperor asking for peace." The cable said: "Unconditional surrender is the only obstacle to peace. . . ." Truman had extracted a promise from Stalin that the Soviet Union would declare war on Japan by August 15—an event that he and many of his military planners thought would be decisive. "He'll [Stalin] be in the Jap war on August 15," Truman wrote in his diary on July 17. "Fini Japs when that comes about."

Truman and the men around him knew that the initial invasion of the Japanese home islands was not scheduled to take place until November 1, 1945—at the earliest. And nearly all the president's advisers believed the war would be over prior to that date. It would surely end with the shock of a Soviet declaration of war—or it might end with the kind of political overture to the Japanese that Grew, McCloy, Leahy and many others envisioned: a clarification of the terms of surrender to specify that the Japanese could keep their emperor. But Truman—and his closest adviser, Secretary of State James F. Byrnes—had decided that the advent of the atomic bomb gave them yet another option. As Byrnes later explained, ". . . it was ever present in my mind that it was important that we should have an end to the war before the Russians came in."

Short of a clarification of the terms of surrender—a move Byrnes opposed on domestic political grounds—the war could end prior to August 15 only with the use of the new weapon. Thus, on July 18, Truman noted in his diary, "Believe Japs will fold up before Russia comes in." Finally, on August 3, Walter Brown, a special assistant to Secretary Byrnes, wrote in his diary, "President, Leahy, JFB [Byrnes] agreed Japs looking for peace. (Leahy had another report from the Pacific.) President afraid they will sue for peace through Russia instead of some country like Sweden."

Isolated in Los Alamos, Oppenheimer had no knowledge of the "Magic" intelligence intercepts, no knowledge of the vigorous debate going on among Washington insiders over the surrender terms, and no idea that the president and his secretary of state were hoping that the atomic bomb would allow them to end the war without a clarification of the terms of unconditional surrender, and without Soviet intervention.

No one can be certain of Oppenheimer's reaction had he learned that on the eve of the Hiroshima bombing, the president *knew* the Japanese were "looking for peace," and that the military use of atomic bombs on cities was an option rather than a necessity for ending the war in August. But we do know that after the war he came to believe that he had been misled, and that this knowledge served as a constant reminder that it was henceforth his obligation to be skeptical of what he was told by government officials.

TWO WEEKS after Oppenheimer wrote his June 16 memo summarizing the views of the science panel, Edward Teller came to him with a copy of a petition that was circulating throughout the Manhattan Project's facilities. Drafted by Leo Szilard, the petition urged President Truman not to use atomic weapons on Japan without a public statement of the terms of surrender: ". . . the United States shall not resort to the use of atomic bombs in this war unless the terms which will be imposed upon Japan have been made public in detail and Japan knowing these terms has refused to surrender. . . ." Over the next few weeks, Szilard's petition garnered the signatures of 155 Manhattan Project scientists. A counter-petition mustered only two signatures. In a separate July 12, 1945, Army poll of 150 scientists in the project, seventy-two percent favored a demonstration of the bomb's power as against its military use without prior warning. Even so, Oppenheimer expressed real anger when Teller showed him Szilard's petition. According to Teller, Oppie began disparaging Szilard and his cohorts: "What do they know about Japanese psychology? How can they judge the way to end the war?" These were judgments better left in the hands of men like Stimson and General Marshall. "Our conversation was brief," Teller wrote in his memoirs. "His talking so harshly about my close friends and his impatience and vehemence greatly distressed me. But I readily accepted his decision. . . ."

Teller claims in his memoirs to have thought in 1945 that use of the bomb without a demonstration and a warning "would be of uncertain expediency and of deplorable morality." But his actual reply to Szilard, dated July 2, 1945, shows that he came to quite the contrary conclusion. "I am not really convinced of your objections [to immediate military use of the weapon]," Teller wrote. The gadget was indeed a "terrible" weapon, but Teller thought the only hope for humanity was to "convince everybody that the next war would be fatal. For this purpose actual combat use might even be the best thing." At no point did Teller even hint that he thought a demonstration practical, or a warning necessary. "The accident that we worked out this dreadful thing," Teller wrote Szilard, "should not give us the responsibility of having a voice in how it is to be used."

This, of course, was one of the arguments Oppenheimer had advanced in his June 16 memo to Stimson. He was convinced that nothing more need be done by the scientific community. He told Ralph Lapp and Edward Creutz, two physicists at Los Alamos who had agreed to circulate Szilard's petition, that, "since an opportunity has been given to people here to express, through him, their opinions on the matters concerned, the proposed method [the petition] was somewhat redundant and probably not very satisfactory." Oppie could be persuasive. Creutz explained to Szilard, somewhat apologetically, "Because of his [Oppenheimer's] very frank and non-peremptory treatment of the situation, I should like to abide by his suggestions." Oppie would not expedite the petition to Washington; instead, it would be sent through normal Army channels—and it would arrive too late.

Oppie informed Groves of the Szilard petition—and did so in a disparaging tone: "The enclosed note [from Szilard to Creutz] is a further incident in the developments which I know you have watched with interest." Groves' aide, Colonel Nichols, called Groves that same day and in the course of their discussion of the Szilard petition, "Nichols asked why not get rid of the lion [Szilard] and general stated can't do that at this time." Groves understood that firing or arresting Szilard would inspire a revolt among the other scientists. But with Oppenheimer equally annoyed by Szilard's actions, Groves felt confident that the problem could be safely contained until the bomb was ready.

THE SUMMER of 1945 was unusually hot and dry on the mesa. Oppenheimer pushed the men in the Tech Area to work longer hours; everyone seemed on edge. Even Miss Warner, isolated as she was down in the valley, noticed a change: "There was tension and accelerated activity on the Hill. . . . Explosions on the Plateau seemed to increase and then to cease." She observed much more traffic on the road headed south—toward Alamogordo.

Initially, General Groves had opposed the idea of a test of the implosion bomb, on the grounds that plutonium was so scarce that not an ounce should be wasted. Oppenheimer convinced him that a full-scale test was absolutely necessary because of the "incompleteness of our knowledge." Without a test, he told Groves, "the planning of the use of the gadget over enemy territory will have to be done substantially blindly."

More than a year earlier, in the spring of 1944, Oppenheimer had spent three days and nights bouncing around the barren, dry valleys of southern New Mexico in a three-quarter-ton Army truck, searching for a suitably isolated stretch of wilderness where the bomb could be safely tested. Accompanying him were Kenneth Bainbridge, an experimental physicist from Harvard, and several Army officers, including the Los Alamos security officer, Capt. Peer de Silva. At night, the men slept in the truck's flatbed to avoid rattlesnakes. De Silva later remembered Oppenheimer lying in a sleeping bag, gazing up at the stars and reminiscing about his student days at Göttingen. For Oppenheimer, it was a rare opportunity to savor the spartan desert he so loved. Several expeditions later, Bainbridge finally selected a desert site sixty miles northwest of Alamogordo. The Spanish had called the area the Jornada del Muerto—the "Journey of Death."

Here the Army staked out an area eighteen by twenty-four miles in size, evicted a few ranchers by eminent domain and began building a field laboratory and hardened bunkers from which to observe the first explosion of an atomic bomb. Oppenheimer dubbed the test site "Trinity"—though years later, he wasn't quite sure why he chose such a name. He remembered vaguely having in mind a John Donne poem that opens with the line "Batter my heart, three-person'd God . . ." But this suggests that he may also have once again been drawing from the Bhagavad-Gita; Hinduism, after all, has its trinity in Brahma the creator, Vishnu the preserver, and Shiva the destroyer.

EVERYONE WAS exhausted from working such long hours. Groves called for speed, not perfection. Phil Morrison was told that "a date near August tenth was a mysterious final date which we, who had the technical job of readying the bomb, had to meet at whatever cost in risk or money or good development policy." (Stalin was expected to enter the Pacific War no later than August 15.) Oppenheimer recalled, "I did suggest to General Groves some changes in the bomb design which would have made more efficient use of the material. . . . He turned them down as jeopardizing the promptness of availability of these bombs." Groves' timetable was driven by President Truman's scheduled meeting with Stalin and Churchill in Potsdam in mid-July. Oppenheimer later testified at his security hearing, "I believe we were under incredible pressure to get it done before the Potsdam meeting and Groves and I bickered for a couple of days." Groves wanted a tested and usable bomb in Truman's hands before that conference ended. Earlier that spring, Oppenheimer had agreed to a target date of July 4—but this soon proved to be unrealistic. By the end of June, after further pressure from Groves, Oppenheimer told his people that they were now aiming for Monday, July 16.

Oppenheimer had delegated Ken Bainbridge to supervise preparations at the Trinity site, but he also sent his brother, Frank, to serve as Bainbridge's chief administrative assistant. To Robert's delight, Frank had arrived in Los Alamos in late May, leaving Jackie and their five-year-old daughter, Judith, and three-year-old son, Michael, in Berkeley. Frank had spent the early war years working with Lawrence in the Radiation Lab. The FBI and Army intelligence kept a close watch on him, but he seems to have followed Lawrence's advice and abandoned all political activity.

Frank began camping out at the Trinity site in late May 1945. Conditions were spartan, to say the least. The men slept in tents and toiled in hundred-degree weather. As the target date approached, Frank felt it only prudent to prepare for disaster. "We spent several days finding escape routes through the desert," he recalled, "and making little maps so everybody could be evacuated."

On the evening of July 11, 1945, Robert Oppenheimer walked home and said goodbye to Kitty. He told her that if the test was successful, he would get a message to her saying, "You can change the sheets." For good luck, she gave him a four-leaf clover from their garden.

Two days prior to the scheduled test, Oppenheimer checked into the Hilton Hotel in nearby Albuquerque. Joining him were Vannevar Bush, James Conant and other S-1 officials who flew in from Washington to observe the test. "He was very nervous," recalled Joseph O. Hirschfelder, a chemist. As if people were not already anxious enough, a last-minute test-firing of the implosion explosives (without the plutonium core) had just indicated that the bomb was likely to be a dud. Everyone began quizzing Kistiakowsky. "Oppenheimer became so emotional," Kistiakowsky recalled, "that I offered him a month's salary against ten dollars that our implosion charge would work." That evening, in an effort to relieve the tension, Oppie recited for Bush a stanza from the Gita that he had translated from the Sanskrit:

> *In battle, in forest, at the precipice in the mountains*
> *On the dark great sea, in the midst of javelins and arrows,*
> *In sleep, in confusion, in the depths of shame,*
> *The good deeds a man has done before defend him.*

That night Robert slept only four hours; Gen. Thomas Farrell, Groves' executive officer, who was trying to sleep on a bunk in the next room, heard him coughing miserably half the night. Robert awoke that Sunday, July 15, exhausted and still depressed by the news of the previous day. But as he ate breakfast in the Base Camp mess hall, he received a phone call from Bethe informing him that the dummy implosion test had failed only because of blown circuits in the wiring. There was no reason, Bethe said, why Kistia-kowsky's design on the actual device shouldn't work. Relieved, Oppen-heimer now turned his attention to the weather. That morning the skies over Trinity were clear, but his meteorologist, Jack Hubbard, told him that the winds around the site were picking up. Speaking on the phone to Groves shortly before the general flew in from California for the test, Oppie warned him, "The weather is whimsical."

In the late afternoon, as thunderclouds moved in, Oppie drove to the Trinity tower for one last look at his "gadget." Alone, he climbed the tower and inspected his creation, an ugly metal globe studded with detonator plugs. Everything seemed in order, and after surveying the landscape he climbed down, got back into his vehicle and drove over to the McDonald Ranch, where the last of the men who had assembled the gadget were pack-ing up their gear. A violent storm was brewing. Back at Base Camp, Oppie talked with Cyril Smith, one of his senior metallurgists. Oppenheimer did most of the talking, chatting aimlessly about family and life on the mesa. At one point, the conversation turned briefly philosophical. Scanning the dark-ening horizon, Oppie muttered, "Funny how the mountains always inspire our work." Smith thought it a moment of calm—quite literally before the gathering storm.

To relieve the tension, some of the scientists organized a betting pool—with a dollar a bet to predict the size of the explosion. Teller characteristi-cally bet high, putting his dollar on 45,000 tons of TNT; Oppenheimer bet low, a very modest 3,000 tons. Rabi staked his money on 20,000 tons. And Fermi alarmed some of the Army guards by taking side bets on whether the bomb would ignite the atmosphere.

That night, those few scientists who managed to sleep a bit were awak-ened by an extraordinary noise. As Frank Oppenheimer recalled, "All the frogs in that area had gathered in a little pond by the camp and copulated and squawked all night long." Oppenheimer hung out in the Base Camp mess hall, alternately gulping down black coffee and rolling one cigarette after another, and smoking them nervously down to the butt. For a time, he pulled out a copy of Baudelaire and sat quietly reading poetry. By then, the storm was pelting the tin roof with a strong downpour. As lightning flashes pierced the darkness outside, Fermi, fearing that the storm's winds might drench them with radioactive rain, said he favored a postponement. "There could be a catastrophe," he warned Oppenheimer.

On the other hand, Oppie's chief weatherman, Hubbard, assured him that the storm would pass before sunrise. Hubbard recommended postpon-ing the hour of detonation, moving it from 4:00 to 5:00 a.m. An agitated Groves paced the mess hall. Groves disliked Hubbard and thought him "obviously confused and badly rattled"; he had gone so far as to bring along his own Army Air Force meteorologist. Not trusting Hubbard's assurances, the general was, even so, vigorously opposed to any postponement. At one point, he pulled Oppenheimer aside and listed all the reasons why the test should proceed. Both men knew that everyone was so exhausted that any postponement would have meant delaying the test for at least two or three days. Worried that some of the more cautious scientists might convince Oppie to postpone the test, Groves took him to the control center at South Shelter–10,000 yards. This was less than six miles from the Trinity site.

At 2:30 a.m., the whole test site was being raked by thirty-mile-an-hour winds and severe thundershowers. Still, Jack Hubbard and his small team of forecasters predicted that the storm would clear at dawn. Outside the bunker at South–10,000 yards, Oppenheimer and Groves paced the ground, glancing to the skies every few minutes to see if they could discern a change in the weather. At around 3:00 a.m., they went back inside the bunker and talked. Neither man could stomach a delay. "If we postpone," Oppenheimer said, "I'll never get my people up to pitch again." Groves was even more adamant that the test should proceed. Finally, they announced their decision: They would schedule the shot for 5:30 a.m. and hope for the best. An hour later, the skies began to clear and the wind abated. At 5:10 a.m., the voice of Sam Allison, the Chicago physicist, boomed across a loudspeaker outside the control center, "It is now zero minus twenty minutes."

RICHARD FEYNMAN was standing twenty miles from the Trinity site when he was handed dark glasses. He decided he wouldn't see anything through the dark glasses, so instead he climbed into the cab of a truck facing Alamogordo. The truck windshield would protect his eyes from harmful ultraviolet rays, and he'd be able actually to see the flash. Even so, he reflexively ducked when the horizon lit up with a tremendous flash. When he looked up again, he saw a white light changing into yellow and then orange: "A big ball of orange, the center that was so bright, becomes a ball of orange that starts to rise and billow a little bit and get a little black around the edges, and then you see it's a big ball of smoke with flashes on the inside of the fire going out, the heat." A full minute and a half after the explosion, Feynman finally heard an enormous bang, followed by the rumble of man-made thunder.

James Conant had expected a relatively quick flash of light. But the white light so filled the sky that for a moment he thought "something had gone wrong" and the "whole world has gone up in flames."

Bob Serber was also twenty miles away, lying face down and holding a piece of welder's glass to his eyes. "Of course," he wrote later, "just at the moment my arm got tired and I lowered the glass for a second, the bomb went off. I was completely blinded by the flash." When his vision returned thirty seconds later, he saw a bright violet column rising to 20,000 or 30,000 feet. "I could feel the heat on my face a full twenty miles away."

Joe Hirschfelder, the chemist assigned to measure the radioactive fallout from the explosion, later described the moment: "All of a sudden, the night turned into day, and it was tremendously bright, the chill turned into warmth; the fireball gradually turned from white to yellow to red as it grew in size and climbed into the sky; after about five seconds the darkness returned but with the sky and the air filled with a purple glow, just as though we were surrounded by an aurora borealis. . . . We stood there in awe as the blast wave picked up chunks of dirt from the desert soil and soon passed us by."

Frank Oppenheimer was next to his brother when the gadget exploded. Though he was lying on the ground, "the light of the first flash penetrated and came up from the ground through one's [eye]lids. When one first looked up, one saw the fireball, and then almost immediately afterwards, this unearthly hovering cloud. It was very bright and very purple." Frank thought, "Maybe it's going to drift over the area and engulf us." He hadn't expected the heat from the flash to be nearly that intense. In a few moments, the thunder of the blast was bouncing back and forth on the distant mountains. "But I think the most terrifying thing," Frank recalled, "was this really brilliant purple cloud, black with radioactive dust, that hung there, and you had no feeling of whether it would go up or would drift towards you."

Oppenheimer himself was lying facedown, just outside the control bunker, situated 10,000 yards south of ground zero. As the countdown reached the two-minute mark, he muttered, "Lord, these affairs are hard on the heart." An Army general watched him closely as the final countdown commenced: "Dr. Oppenheimer . . . grew tenser as the last seconds ticked off. He scarcely breathed. . . . For the last few seconds he stared directly ahead and then when the announcer shouted 'Now!' and there came this tremendous burst of light followed shortly thereafter by the deep growling roar of the explosion, his face relaxed into an expression of tremendous relief."

We don't know, of course, what flashed through Oppie's mind at this seminal moment. His brother recalled, "I think we just said 'It worked.' "

Afterwards, Rabi caught sight of Robert from a distance. Something about his gait, the easy bearing of a man in command of his destiny, made Rabi's skin tingle: "I'll never forget his walk; I'll never forget the way he stepped out of the car. . . . his walk was like High Noon . . . this kind of strut. He had done it."

Later that morning, when William L. Laurence, the *New York Times* reporter selected by Groves to chronicle the event, approached him for comment, Oppenheimer reportedly described his emotions in pedestrian terms. The effect of the blast, he told Laurence, was "terrifying" and "not entirely undepressing." After pausing a moment, he added, "Lots of boys not grown up yet will owe their life to it."

Oppenheimer later said that at the sight of the unearthly mushroom cloud soaring into the heavens above Point Zero, he recalled lines from the Gita. In a 1965 NBC television documentary, he remembered: "We knew the world would not be the same. A few people laughed, a few people cried. Most people were silent. I remembered the line from the Hindu scripture, the Bhagavad-Gita; Vishnu is trying to persuade the prince that he should do his duty, and to impress him, takes on his multi-armed form and says, 'Now I am become death, the destroyer of worlds.' I suppose we all thought that, one way or another." One of Robert's friends, Abraham Pais, once suggested that the quote sounded like one of Oppie's "priestly exaggerations."*

Whatever flashed through Oppenheimer's mind, it is certain that the men around him felt unvarnished euphoria. Laurence described their mood in his dispatch: "The big boom came about 100 seconds after the Great Flash—the first cry of a new-born world. It brought the silent, motionless silhouettes to life, gave them a voice. A loud cry filled the air. The little groups that hitherto had stood rooted to the earth like desert plants broke into dance." The dancing lasted but a few seconds and then the men began shaking hands, Laurence reported, "slapping each other on the back, laughing like happy children." Kistiakowsky, who had been thrown to the ground by the blast, threw his arms around Oppie and gleefully demanded his ten dollars. Oppie pulled out his empty wallet, and told Kisty he'd have to wait. (Later, back in Los Alamos, Oppie made a ceremony of presenting Kistiakowsky with an autographed ten-dollar bill.)

As Oppenheimer left the control center, he turned to shake hands with Ken Bainbridge, who looked him in the eye and muttered, "Now we're all sons-of-bitches." Back at Base Camp, Oppie shared a brandy with his brother and General Farrell. Then, according to one historian, he phoned Los Alamos and asked his secretary to pass a message to Kitty: "Tell her she can change the sheets."

---

*Laurence, the *New York Times* reporter, later said that he would never forget the "shattering impact" of Oppenheimer's words. But curiously, he didn't use the Gita quotes in his 1945 *Times* stories—or in his 1947 book, *Dawn over Zero: The Story of the Atomic Bomb.* A 1948 *Time* magazine article used the quote, and Laurence himself published it in his 1959 book *Men and Atoms.* But Laurence might have picked it up from Robert Jungk's 1958 history *Brighter Than a Thousand Suns.*

PART FOUR

CHAPTER TWENTY-THREE

"Those Poor Little People"

A stone's throw from despair.

ROBERT OPPENHEIMER

A FTER THE RETURN to Los Alamos, everybody seemed to be partying. With his usual exuberance, Richard Feynman was sitting on the hood of a jeep beating his bongo drums. "But one man, I remember, Bob Wilson, was just sitting there moping," Feynman wrote later.

"What are you moping about?" asked Feynman.

"It's a terrible thing that we made," replied Wilson.

"But you started it," Feynman said, remembering that it had been Wilson who had recruited him to Los Alamos from Princeton. "You got us into it."

Wilson aside, euphoria was only to be expected. Everyone who had come to Los Alamos had come for a good reason. Everyone had worked hard to accomplish a difficult task. The work itself became satisfying, and the stunning accomplishment at Alamogordo infected everyone with an overwhelming feeling of excitement. In the process, even someone with as lively a mind as Feynman's was elated. But later he said of that moment, "You stop thinking, you know; you just stop." Bob Wilson seemed to Feynman "[the] only one who was still thinking about it, at that moment."

But Feynman was wrong. Oppenheimer was thinking about it too. In the days after the Trinity test, his mood began to change. Everyone at Los Alamos eased off on the long hours spent in the lab. They knew that after Trinity, the gadget had become a weapon, and weapons were controlled by the military. Anne Wilson, Oppenheimer's secretary, remembered a series of meetings with Army Air Force officers: "They were picking targets." Oppenheimer knew the names of the Japanese cities on the list of potential targets—and the knowledge was clearly sobering. "Robert got very still and ruminative, during that two-week period," Wilson recalled, "partly because he knew what was about to happen, and partly because he knew what it meant."

One day soon after the Trinity test, Oppenheimer startled Wilson with a sad, even morose remark. "He was beginning to feel very down," Wilson said. "I didn't know of other people who were quite in the mood he was in, but he used to come from his house walking over to the Technical Area, and I used to come from the nurses' quarters and somewhere along the way we often bumped into each other. That morning, he's puffing on his pipe and he's saying, 'Those poor little people, those poor little people'—referring to the Japanese." He said it with an air of resignation. And deadly knowledge.

That very week, however, Oppenheimer was working hard to make sure that the bomb exploded efficiently over those "poor little people." On the evening of July 23, 1945, he met with Gen. Thomas Farrell and his aide, Lt. Col. John F. Moynahan, two senior officers designated to supervise the bombing run over Hiroshima from the island of Tinian. It was a clear, cool, starry night. Pacing nervously in his office, chain-smoking, Oppenheimer wanted to make sure that they understood his precise instructions for delivering the weapon on target. Lieutenant Colonel Moynahan, a former newspaperman, published a vivid account of the evening in a 1946 pamphlet: " 'Don't let them bomb through clouds or through an overcast,' [Oppenheimer said.] He was emphatic, tense, his nerves talking. 'Got to see the target. No radar bombing; it must be dropped visually.' Long strides, feet turned out, another cigarette. 'Of course, it doesn't matter if they check the drop with radar, but it must be a visual drop.' More strides. 'If they drop it at night there should be a moon; that would be best. Of course, they must not drop it in rain or fog. . . . Don't let them detonate it too high. The figure fixed on is just right. Don't let it go up [higher] or the target won't get as much damage.' "

The atomic bombs that Oppenheimer had organized into existence were going to be used. But he told himself that they were going to be used in a manner that would not spark a postwar arms race with the Soviets. Shortly after the Trinity test, he had been relieved to hear from Vannevar Bush that the Interim Committee had unanimously accepted his recommendation that the Russians be clearly informed of the bomb and its impending use against Japan. He assumed that such forthright discussions were taking place at that very moment in Potsdam, where President Truman was meeting with Churchill and Stalin. He was later appalled to learn what actually happened at that final Big Three conference. Instead of an open and frank discussion of the nature of the weapon, Truman coyly confined himself to a cryptic reference: "On July 24," Truman wrote in his memoirs, "I casually mentioned to Stalin that we had a new weapon of unusual destructive force. The Russian premier showed no special interest. All he said was that he was glad to hear it and hoped we would make 'good use of it against the Japanese.' " This fell far short of what Oppenheimer had expected. As the historian Alice Kimball Smith later wrote, "what actually occurred at Potsdam was a sheer travesty. . . ."

ON AUGUST 6, 1945, at exactly 8:14 a.m., a B-29 aircraft, the *Enola Gay,* named after pilot Paul Tibbets' mother, dropped the untested, gun-type uranium bomb over Hiroshima. John Manley was in Washington that day, waiting anxiously to hear the news. Oppenheimer had sent him there with one assignment—to report to him on the bombing. After a five-hour delay in communications from the aircraft, Manley finally received a teletype from Captain Parsons—who was the "arming" officer on the *Enola Gay*—that "the visible effects were greater than the New Mexico test." But just as Manley was about to call Oppenheimer in Los Alamos, Groves stopped him. No one was to disseminate any information about the atomic bombing until the president himself announced it. Frustrated, Manley went for a midnight walk in Lafayette Park, across from the White House. Early the next morning, he was told that Truman would make an announcement at 11:00 a.m. Manley finally got Oppie on the phone just as the president's statement was released on nationwide radio. Although they had agreed to use a pre-arranged code for conveying the news over the phone, Oppenheimer's first words to Manley were: "Why the hell did you think I sent you to Washington in the first place?"

That same day, at 2:00 p.m., General Groves picked up the phone in Washington and called Oppenheimer in Los Alamos. Groves was in a congratulatory mood. "I'm proud of you and all of your people," Groves said.

"It went all right?" Oppie asked.

"Apparently it went with a tremendous bang. . . ."

"Everybody is feeling reasonably good about it," Oppie said, "and I extend my heartiest congratulations. It's been a long road."

"Yes," Groves replied, "it has been a long road and I think one of the wisest things I ever did was when I selected the director of Los Alamos."

"Well," replied Oppenheimer diffidently, "I have my doubts, General Groves."

Groves replied, "Well, you know I've never concurred with those doubts at any time."

Later in the day, the news was announced over the Los Alamos public address system: "Attention please, attention please. One of our units has just been successfully dropped on Japan." Frank Oppenheimer was standing in the hallway right outside his brother's office when he heard the news. His first reaction was "Thank God, it wasn't a dud." But within seconds, he recalled, "One suddenly got this horror of all the people that had been killed."

A soldier, Ed Doty, described the scene for his parents in a letter he wrote the next day: "This last 24 hours has been quite exciting. Everyone has been keyed up to a pitch higher than anything I have ever seen on such a mass scale before. . . . People came out into the hallways of the building and milled around like a Times Square New Year's crowd. Everyone was looking for a radio." That evening a crowd gathered in an auditorium. One of the younger physicists, Sam Cohen, remembers a cheering, foot-stamping audience waiting for Oppenheimer to appear. Everyone expected him to come onstage from the auditorium wings, as was his custom. But Oppenheimer chose to make a more dramatic entrance from the rear, making his way up the center aisle. Once onstage, according to Cohen, he clasped his hands together and pumped them over his head like a prize-fighter. Cohen remembers Oppie telling the cheering crowd that it was "too early to determine what the results of the bombing might have been, but he was sure that the Japanese didn't like it." The crowd cheered and then roared its approval when Oppie said he was "proud" of what they had accomplished. By Cohen's account, "his [Oppenheimer's] only regret was that we hadn't developed the bomb in time to have used it against the Germans. This practically raised the roof."

It was as if he had been called upon to act out a stage role, one to which he was truly not suited. Scientists are not meant to be conquering generals. And yet, he was only human and so must have felt the thrill of pure success; he had grabbed a metaphorical gold ring and he was happily waving it aloft. Besides, the audience expected him to appear flushed and triumphant. But the moment was short-lived.

For some who had just seen and felt the blinding light and blasting wind of the explosion at Alamogordo, the expected news from the Pacific was something of an anticlimax. It was almost as if Alamogordo had drained their capacity for astonishment. Others were merely sobered by the news. Phil Morrison heard the news on Tinian, where he had helped to prepare the bomb and load it aboard the *Enola Gay*. "That night we from Los Alamos had a party," Morrison recalled. "It was war and victory in war, and we had a right to our celebration. But I remember sitting . . . on the edge of a cot . . . wondering what it was like on the other side, what was going on in Hiroshima that night."

Alice Kimball Smith later insisted that "certainly no one [at Los Alamos] celebrated Hiroshima." But then she admitted that "a few people" tried to assemble a party in the men's dormitories. It turned into a "memorable fiasco. People either stayed away or beat a hasty retreat." Smith, to be sure, was referring only to the scientists, who appear to have had a decidedly muted—and different—reaction than the military enlisted men. Doty wrote home: "There were parties galore. Invited to three of them, I managed to get to only one. . . . It lasted until three." He reported that people were "happy, very happy. We listened to the radio and danced and listened to the radio again . . . and laughed and laughed at all that was said." Oppenheimer attended one party, but upon leaving he saw a clearly distraught physicist retching his guts out in the bushes. The sight made him realize that an accounting had begun.

Robert Wilson had been horrified by the news from Hiroshima. He had never wanted the weapon to be used, and thought he had grounds for believing it would not be. In January, Oppenheimer had persuaded him to continue his work—but only so that the bomb could be demonstrated. And Oppenheimer, he knew, had participated in the Interim Committee's deliberations. Rationally, he understood that Oppie had been in no position to make him any firm promises—that this was a decision for the generals, Secretary of War Stimson and, ultimately, the president. But he nevertheless felt his trust had been abused. "I felt betrayed," Wilson wrote in 1958, "when the bomb was exploded over Japan without discussion or some peaceful demonstration of its power to the Japanese."

Wilson's wife, Jane, happened to be visiting San Francisco when she heard the news about Hiroshima. Rushing back to Los Alamos, she greeted her husband with congratulatory smiles, only to find him "very depressed," she said. And then, three days later, another bomb devastated Nagasaki. "People were going around banging garbage can covers and so on," Jane Wilson recalled, "and he wouldn't join in, he was sulking and unhappy." Bob Wilson recalled, "I remember being just ill . . . sick . . . to the point that I thought I would be—you know, vomit."

Wilson was not alone. "As the days passed," wrote Alice Kimball Smith, the wife of the Los Alamos metallurgist Cyril Smith, "the revulsion grew, bringing with it—even for those who believed that the end of the war justified the bombing—an intensely personal experience of the reality of evil." After Hiroshima, most people on the mesa understandably felt at least a moment of exhilaration. But after the news from Nagasaki, Charlotte Serber observed, a palpable sense of gloom settled over the laboratory. Word soon spread that "Oppie says that the atomic bomb is so terrible a weapon that war is now impossible." An FBI informant reported on August 9 that Oppie was a "nervous wreck."

On August 8, 1945, as Stalin had promised Roosevelt at the Yalta Conference and confirmed to Truman at Potsdam, the Soviet Union declared war on Japan. It was a devastating event for the emperor's hawkish advisers, who had argued that the Soviet Union could be induced to help Japan obtain more lenient surrender terms than the American doctrine of "unconditional surrender implied." Two days later—a day after Nagasaki was devastated by the plutonium bomb—the Japanese government sent an offer of surrender, with one condition: that the status of Japan's emperor be guaranteed. The next day, the Allies agreed to alter the terms of unconditional surrender: The authority of the emperor to rule would be "subject to the Supreme Commander of the Allied powers. . . ." On August 14, Radio Tokyo announced the government's acceptance of this clarification and, therewith, its surrender. The war was over—and within weeks, journalists and historians began to debate whether it might have ended on similar terms and around the same time without the bomb.

The weekend after the Nagasaki bombing, Ernest Lawrence arrived in Los Alamos. He found Oppenheimer weary, morose and consumed with qualms about what had happened. The two old friends fell to arguing about the bomb. Reminded that it had been Lawrence who had argued for a demonstration and Oppie who had blocked it, Oppie stung Lawrence with a biting remark about how Lawrence cared only for the rich and powerful. Lawrence tried to reassure his old friend that precisely because the bomb was so terrible, it would never be used again.

Hardly reassured, Oppie spent much of his time that weekend drafting a final report on behalf of the Scientific Panel to Secretary Stimson. His conclusions were pessimistic: ". . . it is our firm opinion that no military countermeasures will be found which will be adequately effective in preventing the delivery of atomic weapons." In the future these devices, already vastly destructive, would become only bigger and more lethal. Only three days after America's victory, Oppenheimer was telling Stimson and the president that the nation had no defense against these new weapons: "We are not only unable to outline a program that would assure to this nation for the next decades hegemony in the field of atomic weapons; we are equally unable to insure that such hegemony, if achieved, could protect us from the most terrible destruction. . . . We believe that the safety of this nation—as opposed to its ability to inflict damage on an enemy power—cannot lie wholly or even primarily in its scientific or technical prowess. It can be based only on making future wars impossible."

That week he personally hand-carried the letter to Washington, D.C., where he met with Vannevar Bush and George Harrison, Stimson's aide in the War Department. "It was a bad time," he reported to Lawrence at the end of August, "too early for clarity." He had tried to explain the futility scientists felt concerning any further work on the atomic bomb. He implied that the bomb should be made illegal, "just like poison gases after the last war." But he found no encouragement from the people he saw in Washington. "I had the fairly clear impression from the talks that things had gone most badly at Potsdam, and that little or no progress had been made in interesting the Russians in collaboration or control."

In fact, he doubted that any serious effort in this direction had been made. Before Oppie left Washington, he gloomily noted that the president had issued a gag order on any further disclosures about the atomic bomb—and Secretary of State Byrnes sent word, after reading Oppie's letter to Truman, that in the present international situation there was "no alternative to pushing the Med [Manhattan Engineer District] program full steam ahead." Oppie returned to New Mexico even more depressed than when he had left.

A few days later, Robert and Kitty went alone to Perro Caliente, their cabin near Los Pinos, and spent a week trying to sort out the consequences of the incredibly intense last two years. It was the first time they had spent any real time alone in three years. Robert took the opportunity to catch up on some of his personal correspondence, replying to letters from old friends, many of whom had only recently learned from the newspapers what he had been doing during the war. He wrote his former teacher Herbert Smith: "You will believe that this undertaking has not been without its misgivings; they are heavy on us today, when the future, which has so many elements of high promise, is yet only a stone's throw from despair." Similarly, he wrote his Harvard roommate Frederick Bernheim: "We are at the ranch now, in an earnest but not too sanguine search for sanity. . . . There would seem to be some great headaches ahead."

On August 7, Haakon Chevalier had written him a note of congratulations: "Dear Opje, You are probably the most famous man in the world today. . . ." Oppie replied on August 27 with a three-page handwritten letter. Chevalier later described it as filled with the "affection and the informal intimacy that had always existed between us." Regarding the bomb, Oppie wrote Chevalier: "the thing had to be done, Haakon. It had to be brought to an open public fruition at a time when all over the world men craved peace as never before, were committed as never before both to technology as a way of life and thought and to the idea that no man is an island." But he was by no means comfortable with this defense. "Circumstances are heavy with misgiving, and far, far more difficult than they should be, had we power to re-make the world to be as we think it."

Oppenheimer had long since decided to resign his job as scientific director. By the end of August, he knew that Harvard, Princeton and Columbia

University were offering him jobs—but his instinct was to return to California. "I have a sense of belonging there which I will probably not get over," he wrote his friend James Conant, Harvard's president. His old friends at Caltech, Dick Tolman and Charlie Lauritsen, were encouraging him to come full-time to Pasadena. Incredibly, a formal offer from Caltech was delayed when its president, Robert Millikan, raised objections. Oppenheimer, he wrote Tolman, was not a good teacher, his original contributions to theoretical physics were probably behind him—and perhaps Caltech had enough Jews on its faculty. But Tolman and others persuaded Millikan to change his mind, and an offer was extended to Oppenheimer on August 31.

By then, Oppenheimer also had been invited to return to Berkeley, which he felt was his real home. Still, he hesitated. He told Lawrence that he had "got in bad" with President Robert G. Sproul and Monroe Deutsch, the university provost. Furthermore, his relations with the physics department chairman, Raymond Birge, were so strained that Oppie said to Lawrence that he thought Birge should be replaced. Lawrence, angered by what he saw as a cavalier display of arrogance, retorted that if Oppie felt this way perhaps he shouldn't come back to Berkeley.

Oppenheimer wrote Lawrence a note of explanation: "I have very mixed and sad feelings about our discussions on Berkeley." Oppie reminded his old friend "how much more of an underdogger I have always been than you. That is a part of me that is unlikely to change, for I am not ashamed of it." He had not decided what to do, but Lawrence's "very strong, very negative reactions" gave him pause.

Even as "Oppenheimer" was becoming a household name around the globe, the man who defined himself as an "underdogger" was plunging into depression. When they returned to Los Alamos, Kitty told her friend Jean Bacher, "You just can't imagine how terrible it's been for me; Robert was just definitely beside himself." Bacher was struck by Kitty's emotional state. "She was just afraid for what was going to happen [given] the terrible reaction that he [Robert] had."

The enormity of what had happened in Hiroshima and Nagasaki had affected him profoundly. "Kitty didn't often share her feelings," Bacher said. "But she just said she didn't know how she would stand it." Robert had shared his distress with others as well. According to his Ethical Culture School classmate, Jane Didisheim, Robert wrote her a letter soon after the war ended "that shows so clearly and so sadly his disappointment and his grief."

On The Hill, many people had similar emotional responses—particularly after Bob Serber and Phil Morrison returned from Hiroshima and Nagasaki in October with the first group of scientific observers. Until then, people sometimes gathered in their homes to try to grasp what had happened. "But Phil was the only one who really made me understand it," recalled Jean Bacher. "He's got quite a wizard tongue and descriptive power. I was just absolutely undone. I went home and I couldn't go to sleep; I just shook all night, it was such a shock."

Morrison had landed in Hiroshima just thirty-one days after the *Enola Gay* dropped its deadly load. "Virtually everyone in the street for nearly a mile around was instantly and seriously burned by the heat of the bomb," Morrison said. "The hot flash burned suddenly and strangely. They [the Japanese] told us of people who wore striped clothing upon whom the skin was burned in stripes. . . . There were many who thought themselves lucky, who crawled out of the ruins of their homes only slightly injured. But they died anyway. They died days or weeks later from the radium-like rays emitted in great numbers at the moment of the explosion."

Serber described how in Nagasaki he noticed that the sides of all the telephone poles facing the explosion were charred. He followed a line of such charred poles out beyond two miles from ground zero. "At one point," Serber recounted, "I saw a horse grazing. On one side all its hair was burnt off, the other side was perfectly normal." When Serber somewhat flippantly remarked that the horse nevertheless seemed to be "happily grazing," Oppenheimer "scolded me for giving the impression that the bomb was a benevolent weapon."

Morrison gave a formal briefing in Los Alamos on what he had seen, but he also summarized his report for a local Albuquerque radio station: "We circled finally low over Hiroshima and stared in disbelief. There below was the flat level ground of what had been a city, scorched red. . . . But no hundreds of planes had visited this town during a long night. One bomber and one bomb, had, in the time it takes a rifle bullet to cross the city, turned a city of three hundred thousand into a burning pyre. That was the new thing."

Miss Edith Warner first heard the news of Hiroshima from Kitty, who came one day to fetch fresh vegetables: "Much was now explained," Warner noted afterwards. More than one physicist felt compelled to visit the house at Otowi Bridge and explain themselves to the gentle Miss Warner. Morrison himself wrote her of his hope that "people of intelligence and goodwill everywhere can understand and share our sense of crisis." Having helped to build the weapon, Morrison and many other like-minded physicists now believed the only wise course of action left was to place international controls over all things nuclear. "The scientists know," Miss Warner wrote approvingly in her Christmas letter of 1945, "that they cannot go back to the laboratories leaving atomic energy in the hands of the armed forces or the statesmen."

Oppenheimer knew that in some fundamental sense the Manhattan Project had achieved exactly what Rabi had feared it would achieve—it had made a weapon of mass destruction "the culmination of three centuries of physics." And in doing so, he thought, the project had impoverished physics, and not just in a metaphysical sense; and soon he began to disparage it as a scientific achievement. "We took this tree with a lot of ripe fruit on it," Oppenheimer told a Senate committee in late 1945, "and shook it hard and out came radar and atomic bombs. [The] whole [wartime] spirit was one of frantic and rather ruthless exploitation of the known." The war had "a notable effect on physics," he said. "It practically stopped it." He soon came to believe that during the war we "perhaps witnessed a more total cessation of true professional activity in the field of physics, even in its training, than [in] any other country." But the war also had focused attention on science. As Victor Weisskopf later wrote: "The war had made it obvious by the most cruel of all arguments, that science is of the most immediate and direct importance to everybody. This had changed the character of physics."

At noon on Friday, September 21, 1945, Oppenheimer went to say farewell to Henry Stimson. It was both Stimson's last day in office as secretary of war and his seventy-eighth birthday. Oppenheimer knew that Stimson was scheduled to give a parting presentation at the White House that afternoon in which he would advocate, "very belatedly," thought Oppenheimer, the case for "an open approach on the atom. . . ." By Stimson's diary account, he would bluntly tell President Truman that "we should approach Russia at once with an opportunity to share on a proper quid pro quo the bomb."

Robert genuinely liked and trusted the old man. He was sorry to see him leaving at such a critical juncture in the emerging debate over how to handle the atomic bomb in the postwar era. On this occasion, Oppenheimer briefed him one more time about some technical aspects of the bomb, and then Stimson asked him to accompany him to the Pentagon barbershop, where he had his thin gray hair trimmed. When it was time to go, Stimson rose from the barber's chair, shook Oppenheimer's hand and said, "Now it is in your hands."

CHAPTER TWENTY-FOUR

"I Feel I Have Blood on My Hands"

If atomic bombs are to be added as new weapons to the arsenals of a warring world, or to the arsenals of nations preparing for war, then the time will come when mankind will curse the names of Los Alamos and Hiroshima.

ROBERT OPPENHEIMER
October 16, 1945

ROBERT OPPENHEIMER WAS NOW a celebrity, his name familiar to millions of Americans. Photographs of his chiseled features stared out from magazine covers and newsprint across the nation. His achievements had become synonymous with the achievements of all science. "Hats off to the men of research," editorialized the *Milwaukee Journal*. Never again, chimed in the *St. Louis Post-Dispatch*, should America's "science-explorers . . . be denied anything needful for their adventures." We must admire their "glorious achievement," opined *Scientific Monthly*. "Modern Prometheans have raided Mount Olympus again and have brought back for man the very thunderbolts of Zeus." *Life* magazine observed that physicists now seemed to wear "the tunic of Superman."

Oppenheimer grew comfortable with the adulation. It was as if he had spent the previous two and a half years atop the mesa training for this new role. It had transformed him into a scientist-statesman—and an icon. Even his affectations, the pipe-smoking and the ever-present porkpie hat, soon became internationally recognizable.

He soon began to make his private broodings public. "We have made a thing, a most terrible weapon," he told an audience of the American Philosophical Society, "that has altered abruptly and profoundly the nature of the world . . . a thing that by all the standards of the world we grew up in is an evil thing. And by so doing . . . we have raised again the question of whether science is good for man. . . ." The "father" of the atomic bomb explained that it was by definition a weapon of terror and aggression. And it was cheap. The combination might someday prove deadly to whole civilizations. "Atomic weapons, even with what we know today," he said, "can be cheap . . . atomic armament will not break the economic back of any people that want it. The pattern of the use of atomic weapons was set at Hiroshima." The Hiroshima bomb, he said, was used "against an essentially defeated enemy. . . . it is a weapon for aggressors, and the elements of surprise and of terror are as intrinsic to it as are the fissionable nuclei."

Some of his friends were astounded by his ability to speak, often extemporaneously, with such eloquence and poise. Harold Cherniss was present one day when he addressed an assembly of students at U.C. Berkeley. Thousands packed into the men's gymnasium to hear the famous scientist. Cherniss, however, was apprehensive, because "I thought that he was no public speaker." After being introduced by President Sproul, Oppenheimer got up and spoke without notes for three-quarters of an hour. Cherniss was stunned by his hold on the audience: "From the moment he began to speak until the end, not a whisper in the whole place. This was the kind of magic that he exercised." Cherniss, indeed, thought his friend perhaps spoke too well for his own good. "The ability to speak in public like that is a poison—it's very dangerous for the person who has it." Such a talent might lead a man to think his velvet tongue was an effective political armor.

THROUGHOUT THAT AUTUMN, Oppenheimer shuttled between Los Alamos and Washington, trying to use his sudden celebrity to influence high-ranking government officials. He spoke on behalf of virtually all the civilian scientists at Los Alamos. On August 30, 1945, some 500 of them had squeezed into the auditorium and agreed to form a new organization, the Association of Los Alamos Scientists (ALAS). Within days, Hans Bethe, Edward Teller, Frank Oppenheimer, Robert Christy and others had drafted a strongly worded statement on the dangers of an arms race, the impossibility of any defense against the atomic bomb in future wars, and the need for international control. Oppenheimer was asked to forward "The Document," as it became known, to the War Department. Everyone fully expected that the statement would shortly be released to the press.

On September 9, Oppenheimer sent the report to Stimson's assistant, George Harrison. In his cover letter, he noted that "The Document" had been circulated to more than 300 scientists, and only three had declined to sign it. Oppie wrote that while he had had nothing to do with its formulation, "The Document" certainly reflected his personal views, and he hoped that the War Department would approve its publication. Harrison soon phoned Oppie to say that Stimson wanted more copies for circulation within the government. But Harrison added that the War Department did not wish to release it—at least not just yet.

Unhappy at this delay, the ALAS scientists pressed Oppenheimer to do something. While admitting that he too was disturbed, Oppie argued that the Administration must have a good reason, and he urged his friends to be patient. On September 18, he flew to Washington and two days later phoned to say that "the situation looked real good." "The Document" was being passed around, and he thought the Truman Administration wanted to do the right thing. However, by the end of the month, the Administration had classified it. The ALAS scientists were also stunned to learn that their own trusted emissary had reversed himself and now concurred in the decision to suppress it. To some of his colleagues, it appeared that the more time Oppie spent in Washington, D.C., the more compliant he became.

Oppenheimer insisted that he had a good reason for his change of heart: The Truman Administration was about to propose legislation on atomic energy, and he explained to scientists at Los Alamos that public debate of the sort reflected in the "famous memo" was very desirable—but that they should wait, as a matter of courtesy, until President Truman released his own message on atomic energy to Congress. Oppenheimer's appeal was hotly debated back in Los Alamos, but ALAS' leader, William "Willy" Higinbotham, argued that "the suppression of the document is a matter of political expediency, the reasons for which we are not in a position to know or evaluate." ALAS, however, had "one representative who does know what is going on and knows personally the people involved, that is, Oppie." A motion was then carried unanimously "that Willy tell Oppie that we are strongly behind him."

Oppenheimer was, in fact, doing his best to reflect the deep concern his fellow scientists held for the future. Late in September, he told Under Secretary of State Dean Acheson that most Manhattan Project scientists were strongly disinclined to work any longer on weapons—and "not merely a super bomb, but any bomb." After Hiroshima and the end of the war, such work, he said, was felt to be "against the dictates of their hearts and spirits." He was a scientist, he told a reporter disdainfully, not an "armaments manufacturer." Not every scientist, of course, felt this way. Edward Teller was still promoting the "Super" to anyone with the patience to listen. When Teller asked Oppenheimer to urge that research on the Super continue, Oppie cut him short: "I neither can nor will do so." It was a reaction that Teller would never forget—or forgive.

WHEN PRESIDENT TRUMAN issued his message to Congress on October 3, 1945, many scientists initially thought it reassuring. Drafted by Herbert Marks, a young lawyer working for Acheson, the message urged Congress to establish an atomic energy commission with power to regulate the entire industry. Unbeknownst even to Washington insiders, Oppenheimer had helped Marks write the message. Not surprisingly, it reflected Oppie's own sense of urgency about both the dangers and the potential benefits of atomic energy. The release of atomic energy, Truman pronounced, "constitutes a new force too revolutionary to consider in the framework of old ideas." Time was of the essence. "The hope of civilization," Truman warned, "lies in international arrangements looking, if possible, to the renunciation of the use and development of the atomic bomb. . . ." Oppenheimer thought he had won the president's commitment to seek the abolition of atomic weapons.

But if Oppie had managed to shape the larger message, he had no control over the legislation introduced the following day by Senator Edwin C. Johnson of Colorado and Representative Andrew J. May of Kentucky. The May-Johnson bill embodied a policy that contrasted sharply with the tenor of the president's speech. Most scientists read it as a victory for the military. For one thing, the bill proposed harsh prison terms and hefty fines for any violations of security. Inexplicably to his colleagues, Oppenheimer announced his support for the May-Johnson legislation. On October 7, he returned to Los Alamos and urged the members of ALAS' executive committee to support the bill. As a measure of his still formidable powers of persuasion, he succeeded. His rationale was simple. Time was of the essence, and any bill that quickly set up legislation to oversee the domestic aspects of atomic energy would pave the way for the next step: an international agreement to ban nuclear weapons. Oppie had rapidly become a Washington insider—a cooperative and focused supporter of the Administration, guided by hope and sustained by naïveté.

But as scientists read the bill's fine print, they became alarmed. May-Johnson proposed to centralize all power over atomic energy in the hands of a nine-member commission appointed by the president. Military officers would be allowed to sit on the commission. Scientists were subject to prison terms of up to ten years for even minor security violations. But, as in 1943, when he initially endorsed the notion of drafting Los Alamos scientists into the Army, the details and implications that troubled his colleagues didn't alarm Oppenheimer. Based on his wartime experience, he felt he could work with Groves and the War Department. Others were not so sure. Leo Szilard was outraged, and vowed to work to defeat the bill. A Chicago physicist, Herbert L. Anderson, wrote a colleague in Los Alamos to confess that his confidence in Oppenheimer, Lawrence and Fermi had been shaken. "I believe that these worthy men were duped—that they never had a chance to see this bill." Indeed, Oppie had persuaded Lawrence and Fermi to endorse May-Johnson before they had read the bill's particulars. Both men soon withdrew their support.

In his own Senate testimony on October 17, 1945, Oppenheimer confessed that his prepared statement had been written "considerably before" he actually read the bill: "The Johnson bill, I don't know much about . . . you could do almost anything under that bill." He just knew that good men like Henry Stimson, James Conant and Vannevar Bush had helped to draft the legislation, and "if they like the philosophy of this bill," well, that was good enough for him. It was all a matter of finding nine good men who could be trusted to execute the proposed commission's powers "wisely." When questioned about the wisdom of allowing military officers to sit on the commission, Oppenheimer responded, "I think it is a matter not of what uniform a man wears but what kind of man he is. I cannot think of an administrator in whom I would have more confidence than General [George C.] Marshall."

Szilard, watching from the sidelines, thought Oppenheimer's testimony "a masterpiece. . . . He talked in such a manner that the congressmen present thought he was for the bill but the physicists present all thought that he was against the bill." The left-wing New York City newspaper *PM* reported that Oppenheimer had launched an "oblique attack" on the bill.

Frank Oppenheimer argued with his brother. An activist in ALAS, Frank believed it was time to go public and try to educate citizens about the need for international controls. "He said there wasn't time for this," recalled Frank, "he'd been in the Washington scene, he saw that everything was moving—he felt that he had to change things from within." Perhaps Robert was making a calculated gamble that he could use his prestige and contacts to persuade the Truman Administration to take a quantum leap toward international controls—and he really didn't care if this was done under a civilian or a military atomic regime. Or, perhaps he simply could not bring himself to press for a policy that might lead the Administration to define him as an outsider, a "troublemaker." He wanted to sit center stage during the first act of the Atomic Age.

ALL THIS WAS too much for Robert Wilson, who rewrote the suppressed ALAS "Document" and mailed it to the *New York Times,* which promptly published the statement on its front page. "Mailing it was a serious violation of security," Wilson later wrote. "For me, it was a declaration of independence from our leaders at Los Alamos, not that I did not continue to admire and cherish them. But the lesson we learned early on was that the Best and the Brightest, if in a position of power, were frequently constrained by other considerations and were not necessarily to be relied upon."

As opposition to May-Johnson grew from scientists outside Los Alamos, ALAS members began to have second thoughts. Victor Weisskopf told his colleagues on the ALAS executive committee that "Oppie's suggestions [should] be studied more critically." Within the month, ALAS broke with Oppenheimer and began to mobilize against the legislation. Willy Higinbotham was dispatched to Washington, D.C., with instructions to mount a campaign against the bill. Szilard and other scientists testified against the legislation; this extraordinary lobbying soon commanded the front pages of newspapers and magazines around the country. It was a rebellion—and it succeeded.

To the surprise of many in Washington, the energetic lobbying of the scientists defeated the May-Johnson bill. In its place a new bill was introduced by a freshman senator from Connecticut, Brien McMahon, which proposed to give control over nuclear energy policy to an exclusively civilian Atomic Energy Commission, the AEC. But by the time the Atomic Energy Act was signed by President Truman on August 1, 1946, it had been so altered that many in the "atomic scientists'" movement wondered whether theirs had been a pyrrhic victory. The law included, for example, provisions that subjected scientists working in the field of nuclear physics to a security regime far more draconian than anything they had experienced at Los Alamos. So while many of his peers, including his own brother, were baffled by Oppie's initial support for the May-Johnson bill, no one held it against him for very long. His ambivalence about the whole issue had been justified. If he had failed to challenge the Pentagon's agenda, he had nevertheless understood that the truly important problem was achieving effective international controls against the manufacture of atomic bombs.

IN THE MIDST of this congressional debate, Oppenheimer formally resigned his directorship of Los Alamos. On October 16, 1945, at an award ceremony marking the occasion, thousands of people, virtually the entire population of the mesa, turned out to say good-bye to their forty-one-year-old leader. Dorothy McKibbin briefly greeted Oppie just before he rose to give his farewell address. He had no prepared remarks and McKibbin noted that "his eyes were glazed over, the way they were when he was deep in thought. Afterwards, I realized that in those few moments Robert had been preparing his acceptance speech." A few minutes later, sitting on a dais under a blazing New Mexico sun, Oppenheimer rose to accept a scrolled Certificate of Appreciation from General Groves. Speaking in a low, quiet voice, he expressed his hope that in the years ahead everyone associated with the lab's work would be able to look back on their achievements with pride. But on a sober note, he warned, *"Today that pride must be tempered with a profound concern. If atomic bombs are to be added as new weapons to the arsenals of a warring world, or to the arsenals of nations preparing for war, then the time will come when mankind will curse the names of Los Alamos and Hiroshima."*

He went on: "The peoples of this world must unite or they will perish. This war, that has ravaged so much of the earth, has written these words. The atomic bomb has spelled them out for all men to understand. Other men have spoken them, in other times, of other wars, of other weapons. They have not prevailed. There are some, misled by a false sense of history, who hold that they will not prevail today. It is not for us to believe that. By our works we are committed, committed to a world united, before this common peril, in law and in humanity."

His words reassured many on The Hill that despite his curious support for the May-Johnson bill, he was still one of them. "That day he was us," wrote one Los Alamos resident. "He spoke to us, and for us."

Sitting on the dais with him that morning was Robert G. Sproul, president of the University of California, Berkeley. Stunned by Oppenheimer's stark language, Sproul was further unsettled by the private words they exchanged between speeches. Sproul had come with the intention of winning Oppenheimer back to Berkeley. He knew Oppie was disaffected. On September 29, the physicist had written him to say that he was undecided about his future. Several other institutions had offered him tenured faculty positions with salaries ranging from two to three times what he was paid at Berkeley. And, despite his long years in Berkeley, Oppie said he was aware "of a certain lack of confidence on the part of the University for what it must inevitably have regarded as my indiscretions of the past." By "indiscretions," Oppenheimer was referring to Sproul's annoyance with his political activities on behalf of the Teachers' Union. It would be wrong, he wrote Sproul, to return to Berkeley if the university and the physics department didn't really want him. And "[i]t would seem wrong to me to return at a salary so out of proportion to those of other institutions."

Sproul, a rigid and conservative man, had always thought Oppenheimer troublesome, so he had hesitated when Ernest Lawrence proposed that they offer to double Oppie's salary. Lawrence argued that "how much we pay Professor Oppenheimer really means nothing because the Government will place such large sums at our disposal if Oppenheimer is here, that his salary will be insignificant." Reluctantly, Sproul acquiesced. But now, as the two men sat on the dais and discussed the matter, Oppenheimer brushed aside Sproul's offer, repeating in substance what he had said in his letter: He was aware that his colleagues in the physics department and Sproul himself were not enthusiastic about having him return "because of his difficult temperament and poor judgment." He then abruptly informed Sproul that he had decided to teach at Caltech, but even so, he asked Sproul for a formal extension of his leave of absence—thus leaving the door open to a return to Berkeley at a later date. Though understandably miffed by the tenor of this conversation, Sproul felt compelled to agree to Oppie's request.

Oppenheimer's behavior suggests he was unsure about his next step, but certain that it had to be a significant one. Part of him wanted to re-create the good years he had lived in Berkeley. And yet, increasingly comfortable with his postwar stature, he was also drawn by new ambitions. He temporarily resolved this conundrum by rejecting the offers from Harvard and Columbia in favor of Caltech's. He could remain in California, while keeping open the option to return to Berkeley. In the meantime, he would spend many exhausting days shuttling aboard propeller-driven airplanes back and forth to Washington, D.C.

Indeed, on October 18, just a day after the awards ceremony in Los Alamos, Oppenheimer was back in Washington for a conference at the Statler Hotel. In the presence of a half dozen senators, Oppenheimer outlined in stark terms the perils to the country posed by the atomic bomb. Also in attendance was Henry A. Wallace, vice president during Roosevelt's third term (1941–45), now serving as Truman's commerce secretary. Seizing the occasion, Oppenheimer walked up to Wallace and said he very much wanted to talk with him privately. Wallace invited him to take a walk the following morning.

Walking with the former vice president through downtown Washington toward the Commerce Department, Oppie revealed his deepest anxieties about the bomb. He rapidly outlined the dangers inherent in the Administration's policies. Afterwards, Wallace wrote in his diary that "I never saw a man in such an extremely nervous state as Oppenheimer. He seemed to feel that the destruction of the entire human race was imminent." Oppie complained bitterly that Secretary of State Byrnes "felt that we could use the bomb as a pistol to get what we wanted in international diplomacy." Oppenheimer insisted that this would not work. "He says the Russians are a proud people and have good physicists and abundant resources. They may have to lower their standard of living to do it but they will put everything they have got into getting plenty of atomic bombs as soon as possible. He thinks the mishandling of the situation at Potsdam has prepared the way for the eventual slaughter of tens of millions or perhaps hundreds of millions of innocent people."

Oppenheimer admitted to Wallace that even the previous spring, well before the Trinity test, many of his scientists were "enormously concerned" about a possible war with Russia. He had thought that the Roosevelt Administration had worked out a plan to communicate with the Soviets about the bomb. This hadn't happened, he suspected, because the British had objected. Still, he thought Stimson had a very "statesmanlike" view of the whole matter, and he referred approvingly to the secretary of war's September 11 memo to President Truman which, he said, had "advocated turning over to Russia . . . the industrial know-how as well as the scientific information." At this point, Wallace interrupted to say that Stimson's views on this point had never even been introduced at a Cabinet meeting. Obviously disturbed to hear this news, Oppenheimer said that his scientists back in New Mexico were completely disheartened: ". . . all they think about now are the social and economic implications of the bomb."

At one point, Oppie asked Wallace if he thought it would do any good for him to see the president. Wallace encouraged him to try to get an appointment through the new secretary of war, Robert P. Patterson. On this note, the two men parted. Wallace subsequently noted in his diary: "The guilt consciousness of the atomic bomb scientists is one of the most astounding things I have ever seen."

Six days later, at 10:30 a.m. on October 25, 1945, Oppenheimer was ushered into the Oval Office. President Truman was naturally curious to meet the celebrated physicist, whom he knew by reputation to be an eloquent and charismatic figure. After being introduced by Secretary Patterson, the only other individual in the room, the three men sat down. By one account, Truman opened the conversation by asking for Oppenheimer's help in getting Congress to pass the May-Johnson bill, giving the Army permanent control over atomic energy. "The first thing is to define the national problem," Truman said, "then the international." Oppenheimer let an uncomfortably long silence pass and then said, haltingly, "Perhaps it would be best first to define the international problem." He meant, of course, that the first imperative was to stop the spread of these weapons by placing international controls over all atomic technology. At one point in their conversation, Truman suddenly asked him to guess when the Russians would develop their own atomic bomb. When Oppie replied that he did not know, Truman confidently said he knew the answer: "Never."

For Oppenheimer, such foolishness was proof of Truman's limitations. The "incomprehension it showed just knocked the heart out of him," recalled Willie Higinbotham. As for Truman, a man who compensated for his insecurities with calculated displays of decisiveness, Oppenheimer seemed maddeningly tentative, obscure—and cheerless. Finally, sensing that the president was not comprehending the deadly urgency of his message, Oppenheimer nervously wrung his hands and uttered another of those regrettable remarks that he characteristically made under pressure. "Mr. President," he said quietly, "I feel I have blood on my hands."

The comment angered Truman. He later informed David Lilienthal, "I told him the blood was on my hands—to let me worry about that." But over the years, Truman embellished the story. By one account, he replied, "Never mind, it'll all come out in the wash." In yet another version, he pulled his handkerchief from his breast pocket and offered it to Oppenheimer, saying, "Well, here, would you like to wipe your hands?"

An awkward silence followed this exchange, and then Truman stood up to signal that the meeting was over. The two men shook hands, and Truman reportedly said, "Don't worry, we're going to work something out, and you're going to help us."

Afterwards, the President was heard to mutter, "Blood on his hands, dammit, he hasn't half as much blood on his hands as I have. You just don't go around bellyaching about it." He later told Dean Acheson, "I don't want to see that son-of-a-bitch in this office ever again." Even in May 1946, the encounter still vivid in his mind, he wrote Acheson and described Oppenheimer as a "cry-baby scientist" who had come to "my office some five or six months ago and spent most of his time wringing his hands and telling me they had blood on them because of the discovery of atomic energy."

On this important occasion, the composure and powers of persuasion of the usually charming and self-possessed Oppenheimer had abandoned him. His habit of relying on spontaneity worked well when he was at ease, but, time and again, under pressure he would say things that he would regret profoundly, and that would do him serious harm. On this occasion he had had the opportunity to impress the one man who possessed the power to help him return the nuclear genie to the bottle—and he utterly failed to take advantage of the opportunity. As Harold Cherniss had observed, his facile articulateness was dangerous—a lethal two-edged sword. It was often a sharp instrument of persuasion, but it could also be used to undercut the hard work of research and preparation. It was a form of intellectual arrogance that periodically led him to behave foolishly or badly, an Achilles' heel of sorts that would have devastating consequences. Indeed, it would eventually provide his political enemies with the opportunity to destroy him.

Curiously, this was neither the first nor the last time that Oppenheimer antagonized somebody in a position of authority. Again and again in his life, he showed himself capable of the greatest consideration; he could be patient, gracious and tender with his students—unless they asked him a patently foolish question. But with those in authority, he was often impatient and candid to the point of rudeness. On this occasion, Truman's gross misunderstanding and ignorance of the implications of atomic weapons had prompted Oppenheimer to say something that he should have realized might antagonize the president.

Truman's interactions with scientists were never elevated. The president struck many of them as a small-minded man who was in way over his head. "He was not a man of imagination," said Isidor Rabi. And scientists were hardly alone in this view. Even a seasoned Wall Street lawyer like John J. McCloy, who served Truman briefly as assistant secretary of war, wrote in his diary that the president was "a simple man, prone to make up his mind quickly and decisively, perhaps too quickly—a thorough American." This was not a great president, "not distinguished at all . . . not Lincolnesque, but an instinctive, common, hearty-natured man." Men as different as McCloy, Rabi and Oppenheimer all thought Truman's instincts, particularly in the field of atomic diplomacy, were neither measured nor sound—and sadly, certainly were not up to the challenge the country and the world now faced.

BACK ON THE MESA, no one thought of Oppenheimer as a "cry-baby scientist." On November 2, 1945, a wet and cold evening, the former director returned to The Hill. The Los Alamos theater was again packed to its capacity to hear Oppie talk about what he called "the fix we are in." He began by confessing, "I don't know very much about practical politics." But that wasn't important, because there were issues to be faced that spoke directly to scientists. What has happened, he said, has forced us "to reconsider the relations between science and common sense."

He spoke for an hour—much of it extemporaneously—and his audience was mesmerized; years later, people were still saying, "I remember Oppie's speech. . . ." They remembered this night in part because he explained so well the welter of confused emotions they all felt about the bomb. What they had done was no less than an "organic necessity." If you were a scientist, he said, "you believe that it is good to find out how the world works . . . that it is good to turn over to mankind at large the greatest possible power to control the world and to deal with it according to its lights and values." Besides, there was a "feeling that there was probably no place in the world where the development of atomic weapons would have a better chance of leading to a reasonable solution, and a smaller chance of leading to disaster, than within the United States." Nevertheless, as scientists, Oppenheimer told them, they could not escape responsibility for "the grave crisis." Many people, he said, will "try to wiggle out of this." They will argue that "this is just another weapon." Scientists knew better. "I think it is for us to accept it as a very grave crisis, to realize that these atomic weapons which we have started to make are very terrible, that they involve a change, they are not just a slight modification. . . .

"It is clear to me that wars have changed. It is clear to me that if these first bombs—the bomb that was dropped on Nagasaki—that if these can destroy ten square miles, then that is really quite something. It is clear to me that they are going to be very cheap if anyone wants to make them." As a result of this quantitative change, the very nature of war had changed: Now the advantage rested with the aggressor, not the defender. But if war had become intolerable, then very "radical" changes were required in the relations between nations, "not only in spirit, not only in law, but also in conception and feeling." The one thing he wished to "hammer home," he said, was "what an enormous change in spirit is involved."

The crisis called for a historical transformation of international attitudes and behavior, and he was looking to the experiences of modern science for guidance. He thought he had what he called an "interim solution." First, the major powers should create a "joint atomic energy commission," armed with powers "not subject to review by the heads of State," to pursue the peaceful applications of atomic energy. Second, concrete machinery should be set up to force the exchange of scientists, "so that we would be quite sure that the fraternity of scientists would be strengthened." And finally, "I would say that no bombs be made." He didn't know if these were good proposals, but they were a start. "I know that many of my friends here see pretty much eye to eye. I would speak especially of Bohr. . . ."

But if Bohr and most other scientists approved, everyone knew they were a distinct minority in the country at large. Later in his remarks, Oppie admitted that he was "troubled" by numerous "official statements" characterized by an "insistent note of unilateral responsibility for the handling of atomic weapons." Earlier that week, President Truman had given a bellicose Navy Day speech in New York's Central Park that seemed to revel in America's military power. The atomic bomb, Truman had said, would be held by the United States as a "sacred trust" for the rest of the world, and "we shall not give our approval to any compromises with evil." Oppenheimer said he disliked Truman's triumphalist tone: "If you approach the problem and say, 'We know what is right and we would like to use the atomic bomb to persuade you to agree with us,' then you are in a very weak position and you will not succeed . . . you will find yourselves attempting by force of arms to prevent a disaster." Oppie told his audience that he was not going to argue with the president's motives and aims—but "we are 140 million people, and there are two billion people living on earth." However confident Americans might be that their views and ideas will prevail, the absolute "denial of the views and ideas of other people, cannot be the basis of any kind of agreement."

No one left the auditorium that night unmoved. Oppie had spoken to them on intimate terms, articulating many of their doubts, fears and hopes. For decades afterwards, his words would resonate. The world he had described was as subtle and complicated as the quantum world of the atom itself. He had begun humbly, and yet, like the best of politicians, he had spoken a simple truth that cut to the core of the issue. The world had changed; Americans would behave unilaterally at their peril.

A FEW DAYS LATER, Robert, Kitty and their two young children, Peter and Toni, climbed into the family Cadillac and drove to Pasadena. Kitty was particularly relieved to leave Los Alamos behind. But so, too, was Robert. Here on his beloved mesa he had achieved something unique in the annals of science. He had transformed the world and he had been transformed. But he could not shake a sense of brooding ambivalence.

Soon after his arrival at Caltech, Robert received a letter from the occupant of the small house by the Otowi Bridge. Edith Warner wrote him with the salutation "Dear Mr. Opp." Someone had given her a copy of his farewell speech. "It seemed almost as though you were pacing my kitchen, talking half to yourself and half to me," she wrote. "And from it came the conviction of what I've felt a number of times—you have, in lesser degree, that quality which radiates from Mr. Baker [Niels Bohr's alias]. It has seemed to me in these past few months that it is a power as little known as atomic energy. . . . I think of you both, hopefully, as the song of the river comes from the canyon and the need of the world reaches even this quiet spot."

CHAPTER TWENTY-FIVE

"People Could Destroy New York"

I find that physics and the teaching of physics, which is my life, now seems irrelevant.

ROBERT OPPENHEIMER

OPPENHEIMER WAS NOW AN INFLUENTIAL VOICE in Washington—and the fact of his influence attracted the scrutiny of J. Edgar Hoover. That autumn, the FBI director began circulating derogatory information about the physicist's ties to communists. On November 15, 1945, Hoover sent a three-page summary of Oppenheimer's FBI file to both the White House and the secretary of state. Hoover reported that Communist Party officials in San Francisco had been overheard referring to Oppenheimer as a "regularly registered" member of the Party. "Since the use of the atomic bomb," Hoover wrote, "individual Communists in California who knew Oppenheimer prior to his assignment to the atomic bomb project have expressed interest in re-establishing their old contacts."

Hoover's information was problematic. It was certainly true that FBI wiretaps had overheard some California communists referring to Oppenheimer as a Party member. But this wasn't surprising, since there were many Party members who, before the war, had assumed that Robert was similarly committed—and all who had known Oppenheimer before the war naturally wanted to claim the famous "atomic bomb" physicist as one of their own. Thus, just four days after the atomic bombing of Hiroshima, an FBI wiretap recorded a CP organizer, David Adelson, remarking, "Isn't it nice that Oppenheimer is getting the credit he is?" Another Party activist, Paul Pinsky, replied, "Yes, shall we claim him as a member?" Adelson laughed and said, "Oppenheimer is the guy who originally gave me the push. Remember that session?" Pinsky replied, "Yes," and then Adelson said, "As soon as they get the gestapo from around him I am going to get hold of him and put the bee on him. The guy is so big now that no one can touch him, but he has got to come out and express some ideas."

Clearly, Adelson and Pinsky thought Oppenheimer was sympathetic to their political agenda. But was he a comrade? Even the FBI recognized that Pinsky's question—"Shall we claim him as a member?"—"appears to leave some doubt as to the Subject's [Oppenheimer's] actual membership in the Party."

Similarly, on November 1, 1945, the FBI listened to a conversation among members of the Executive Committee of the North Oakland Club, a branch of the Alameda County Communist Party. One Party functionary, Katrina Sandow, stated that Oppenheimer was a Communist Party member. Another CP official, Jack Manley, boasted that he and Steve Nelson were "close to Oppenheimer," whom he called "one of our men." Manley said the Soviet Union had its own very large uranium deposits and that it was "foolish" to think America could keep a monopoly over the new weapon. Significantly, he claimed that Oppenheimer had "talked it over in great detail with us" two or three years before. Manley also said that he knew other scientists at the Rad Lab who were working on an even more powerful bomb than the one dropped on Japan. He innocently claimed that he intended to get "a simplified diagram of the bomb and print it in all the local papers . . . in order that the public would understand it."

The White House and the State Department did nothing with Hoover's wiretaps. But Hoover pushed his agents to continue. By the end of 1945, the FBI had a wiretap inside Frank Oppenheimer's home outside Berkeley. At a New Year's Day party on January 1, 1946, the FBI wiretap overheard Oppie, who had come to visit his brother, talking with Pinsky and Adelson. They tried to persuade him to make a speech about the atomic bomb at a rally they were organizing, but Oppie politely declined (though Frank agreed to do it). Adelson and Pinsky were not surprised. They had talked about the physicist with another Party official, Barney Young, who said that the Party had tried to communicate with Oppenheimer, but the physicist had "done nothing towards maintaining contact." Oppie's old friend Steve Nelson, the head of the Oakland CP, had tried repeatedly to resume their friendship—but Oppie had failed to respond.

Steve Nelson never did meet Oppenheimer again. Other Party functionaries may have thought of him as someone once on the fringes of the Party. But even Haakon Chevalier knew that Oppenheimer had never subjected himself to Party discipline. Then and now, he had always taken an "individualistic course." This made it difficult for anyone other than Oppenheimer himself to know exactly what his relationship had been to the Communist Party—and what that meant to him. The FBI would never be able to prove Oppenheimer's Party membership. But over the next eight years, Hoover and his agents would generate some 1,000 pages each year of memos, surveillance reports and wiretap transcripts on Oppenheimer, all directed toward the goal of discrediting this "individualistic" thinker. A wiretap on Oppenheimer's home phone at One Eagle Hill was installed on May 8, 1946.

Hoover personally directed the investigation—and he had few scruples. Early in March 1946, the FBI used a Catholic priest in an attempt to turn Oppie's former secretary at Los Alamos, Anne Wilson, into an informant. Father John O'Brien, a Baltimore priest, claimed he knew Wilson as "a Catholic girl" and thought he could persuade her to cooperate with the FBI "for the purpose of developing information concerning Oppenheimer's contacts and activities particularly with regard to possible disclosure of atomic bomb secrets by him." Hoover agreed to the attempt, scribbling on the action memo, "OK if Father will keep quiet about it."

Father O'Brien then requested "derogatory information concerning Oppenheimer that could be used in giving a 'pep talk' to the girl." His FBI handler told him this would not be a safe tactic—not, at least, until they had sounded out Wilson. The priest met with Wilson on the evening of March 26, 1946; the next morning he phoned the FBI to report that "the girl could not be persuaded to cooperate on the basis of her religious convictions and patriotism. . . ." Loyal and feisty, Wilson told the priest that she had "complete faith in Oppenheimer's integrity." Though she knew the tall, blond, handsome priest as a former high school teacher and close family friend, Wilson refused to give Father O'Brien any information. She "expressed resentment over the fact that security agencies" were watching Oppenheimer. Wilson said Oppenheimer had told her that the FBI had him under surveillance, and she thought this outrageous.

Oppie was angry about the surveillance. One day in Berkeley, he was talking to his former student Joe Weinberg when he suddenly pointed to a brass plate on the wall and said, "What the hell is that?" Weinberg tried to explain that the university had ripped out an old intercom system and covered the hole in the wall with this brass spacer. But Oppie interrupted him and said, "That was and always has been a concealed microphone." He then stomped out of the room, slamming the door behind him.

Oppenheimer, to be sure, was not Hoover's only target. In the spring of 1946, the FBI chief was investigating scores of high-ranking Truman Administration officials and disseminating outlandish allegations. Based on so-called "reliable informants," he questioned the loyalty of numerous officials associated with atomic energy policy, including John J. McCloy, Herbert Marks, Edward U. Condon and even Dean Acheson.

Hoover's investigations of Oppenheimer and other members of the Truman Administration in 1946 were a prelude to the politics of anticommunism—the use of the charge of "communist," "communist sympathizer" or "fellow traveler" to silence or destroy a political opponent. It was not in fact a new tactic: Such charges had proven lethal at the state level in the late 1930s. But with the growing rift between the United States and the Soviet Union, it was easy to focus attention on the need to protect our "atomic secrets," and from this need emerged the justification to put anyone associated with nuclear research under close surveillance. Hoover was suspicious of anyone who deviated from the most conservative positions on nuclear issues; and no one working on atomic energy policy was more suspect to him than Robert.

ON A LATE afternoon during the bitterly cold Christmas week of 1945, Oppenheimer visited Isidor Rabi in his New York City apartment on Riverside Drive. Watching the sun set from Rabi's living room window, the two old friends could see ice floes bathed in yellow and pink floating down the Hudson River. Afterwards, the two men sat alone in the spreading darkness, smoking their pipes and talking about the dangers of an atomic arms race. Rabi later claimed that he "originated" the idea of international control—and that Oppie became its "salesman." Oppenheimer, of course, had been thinking along these lines ever since his talks with Bohr at Los Alamos. But perhaps their conversation that evening inspired Oppie to refine those ideas into a concrete plan. "So it came to me," Rabi recalled, "there must be two things: It [the bomb] must be under international control, because if it was under national control there was bound to be rivalry; [second,] we also believed in nuclear energy, that the continuation of this industrial age would depend on it." Rabi and Oppenheimer thus proposed an international atomic authority that would have real clout because it would control both the bomb and peaceful uses of atomic energy. Potential proliferators would face the certain penalty of a punitive closing down of their energy plants if they were found to be acquiring atomic weapons.

Four weeks later, in late January 1946, Oppenheimer was heartened to learn that negotiations begun several months earlier had resulted in an agreement among the Soviet Union, the United States and other countries to establish a United Nations Atomic Energy Commission. In response, President Truman appointed a special committee to draw up a concrete proposal for international control of nuclear weapons. Dean Acheson was to chair the committee, and other members included such leading lights of the American foreign policy establishment as former Assistant Secretary of War John J. McCloy, Vannevar Bush, James Conant and Gen. Leslie Groves. When Acheson complained to his personal assistant, Herbert Marks, that he knew nothing about atomic energy, Marks suggested he create a Board of Consultants. A brilliant and gregarious young lawyer, Marks had once worked for David Lilienthal, the chairman of the Tennessee Valley Authority—and now he suggested that Lilienthal could help to devise a coherent plan. Though not a scientist, Lilienthal, a liberal New Dealer, was an experienced administrator who had worked with hundreds of engineers and technicians. He would bring gravitas to their deliberations. He quickly agreed to chair the Board of Consultants, and four other men were appointed to join him: Chester I. Barnard, president of New Jersey Bell Telephone Company; Dr. Charles A. Thomas, vice president of Monsanto Chemical Company; Harry A. Winne, vice president of General Electric Company—and Oppenheimer.

Oppenheimer was delighted with this development. Here, at last, was the opportunity he had been awaiting to address the major problems associated with controlling the atomic bomb. Acheson's committee and his board of consultants began meeting intermittently that winter to sketch out a preliminary plan. As the only physicist, Oppenheimer naturally dominated the discussions and impressed these strong-minded men with his clarity and his vision. He needed unanimity and he was determined to get it. Right from the start, he enthralled Lilienthal.

They met for the first time in Oppenheimer's hotel room at Washington's Shoreham Hotel. "He walked back and forth," Lilienthal noted in his diary, "making funny 'hugh' sounds between sentences or phrases as he paced the room, looking at the floor—a mannerism quite strange. Very articulate. . . . I left liking him, greatly impressed with his flash of mind, but rather disturbed by the flow of words." Later, after spending more time in his company, Lilienthal gushed, "He [Oppenheimer] is worth living a lifetime just to know that mankind has been able to produce such a being. . . ."

General Groves had seen Oppie work his charms on people, but this time he thought the physicist overdid it: "Everybody genuflected. Lilienthal got so bad he would consult Oppie on what tie to wear in the morning." "Jack" McCloy was almost equally entranced. McCloy had met Oppie early in the war years, and he still thought of him as a man of wide culture, possessed of an "almost musically delicate mind," an intellectual of "great charm."

"All the participants, I think," Acheson later wrote in his memoirs, "agree that the most stimulating and creative mind among us was Robert Oppenheimer's. On this task he was also at his most constructive and accommodating. Robert could be argumentative, sharp, and, on occasion, pedantic, but no such problem intruded here."

Acheson admired Oppenheimer's quick wit, his clear vision—and even his sharp tongue. Early in their deliberations, Oppie was a guest in Acheson's Georgetown house. After cocktails and dinner, he stood by a small blackboard, chalk in hand, and lectured his host and McCloy on the intricacies of the atom. As a visual aid, he drew little stick figures to represent electrons, neutrons and protons chasing one another about and generally carrying on in unpredictable ways. "Our bewildered questions seemed to distress him," Acheson later wrote. "At last he put down the chalk in gentle despair, saying, 'It's hopeless! I really think you two believe neutrons and electrons *are* little men!' "

By early March 1946, the board of consultants had a draft report of some 34,000 words, written by Oppenheimer and reworked by Marks and Lilienthal. Over a period of ten days in mid-March, they held four all-day meetings in Washington, D.C., at Dumbarton Oaks, a stately Georgetown mansion furnished with Byzantine artworks. From the walls, which towered nearly three stories, hung magnificent tapestries; a shaft of sunlight bathed El Greco's painting *The Visitation* in one corner. A Byzantine cat sculpted in ebony sat encased in glass. Near the end of their deliberations, Acheson, Oppenheimer and the other men took turns reading aloud sections of the draft report. When they had finished, Acheson glanced up, removed his reading glasses, and said, "This is a brilliant and profound document."

Oppenheimer had persuaded his fellow panel members to endorse a dramatic and comprehensive plan. Half-measures, he had argued, were not sufficient. A simple international covenant banning atomic weapons was not enough unless people everywhere could be assured that it would be enforced. Neither was a regime of international inspectors sufficient. It would take more than 300 such inspectors just to monitor one diffusion plant at Oak Ridge. And what would an inspection regime do about those countries that professed to be exploiting the peaceful applications of atomic energy? As Oppenheimer had explained, it would be very hard for inspectors to detect a diversion of enriched uranium or plutonium from civilian nuclear energy plants to military purposes. The peaceful exploitation of atomic energy was inextricably linked to the technical ability to produce a bomb.

Having defined the dilemma, Oppenheimer turned again to the internationalism of modern science for a solution. He proposed an international agency that would monopolize all aspects of atomic energy, and apportion its benefits as an incentive to individual countries. Such an agency would both control the technology and develop it for strictly civilian purposes. Oppenheimer believed that in the long run, "without world government there could be no permanent peace, that without peace there would be atomic warfare." World government was obviously not an immediate prospect, so Oppenheimer argued that in the field of atomic energy all countries should agree to a "partial renunciation" of sovereignty. Under his plan, the proposed Atomic Development Authority would have sovereign ownership of all uranium mines, atomic power plants and laboratories. No nation would be permitted to build bombs—but scientists everywhere would still be allowed to exploit the atom for peaceful purposes. As he explained the concept in a speech in early April, "What is here proposed is such a partial renunciation, sufficient, but not more than sufficient, for an Atomic Development Authority to come into being, to exercise its functions of development, exploitation and control, to enable it to protect the world against the use of atomic weapons and provide it with the benefits of atomic energy."

Complete and total transparency would make it impossible for any nation to marshal the enormous industrial, technical and material resources necessary to build an atomic weapon in secrecy. Oppenheimer understood that one couldn't uninvent the weapon; the secret was out. But one could construct a system so transparent that the civilized world would at least have ample warning if a rogue regime set about making such a weapon.

But on one point, Oppenheimer's political vision clouded his scientific judgment. He also suggested that fissionable materials might be permanently "denatured," or contaminated, and thus made useless for bomb-making. But as eventually became clear, any process that denatured uranium and plutonium could be reversed. "Oppenheimer screwed it up later," Rabi said, "by suggesting that the uranium could be poisoned, or denatured, which was crazy. . . . It was such a blunder that I never even chided him for it."

The sense of urgency that everyone came to share was reflected in the plan's endorsement by businessmen like Monsanto's Charles Thomas and the Republican Wall Street lawyer John J. McCloy. Herbert Marks later remarked, "Only something as drastic as the atomic bomb could have got Thomas to suggest that the mines be internationalized. Don't forget he's the vice president of a hundred-and-twenty-million-dollar firm."

Soon afterwards, Oppenheimer's report—which became known as the Acheson-Lilienthal Report—was submitted to the White House. Oppenheimer was pleased; surely, the president would now understand the urgent need to control the atom.

But his optimism was misplaced. While Secretary of State Byrnes made a pretense of saying that he was "favorably impressed," he was, in fact, shocked by the sweeping scope of the report's recommendations. A day later, he persuaded Truman to appoint his (Byrnes') longtime business partner, the Wall Street financier Bernard Baruch, to "translate" the Administration's proposals to the United Nations. Acheson was appalled. Lilienthal wrote in his diary, "When I read the news last night, I was quite sick. . . . We need a man who is young, vigorous, not vain, and who the Russians would feel isn't out simply to put them in a hole, not really caring about international cooperation. Baruch has none of these qualities." When Oppenheimer learned of this appointment, he told his Los Alamos friend Willie Higinbotham, by then president of the newly created Federation of Atomic Scientists, "We're lost."

In private, Baruch was already expressing "great reservations" about the Acheson-Lilienthal Report's recommendations. For advice, he turned to two conservative bankers, Ferdinand Eberstadt and John Hancock (a senior partner at Lehman Brothers), and Fred Searls, Jr., a mining engineer and close personal friend. Both Baruch and Secretary of State Byrnes happened to be board members and investors in Newmont Mining Corporation, a major company with a large stake in uranium mines. Searls was Newmont's chief executive officer. Not surprisingly, they were alarmed by the idea that privately owned mines might be taken over by an international Atomic Development Authority. None of these men seriously contemplated internationalizing the newly emerging nuclear industry. And, as far as atomic weapons were concerned, Baruch thought of the American bomb as the "winning weapon."

Oppenheimer's prestige was so pervasive that even as Baruch prepared to gut the Acheson-Lilienthal Report, he made an effort to recruit Robert as his scientific adviser. Early in April 1946, they met in New York to discuss the possibility of working together. From Oppie's point of view, the meeting was an unmitigated disaster. When pressed, he had to admit that his plan was not exactly compatible with the current Soviet system of government. He insisted, however, that the American position "should be to make an honorable proposal and thus find out whether they have the will to cooperate." Baruch and his advisers argued that the Acheson-Lilienthal proposals needed to be amended in several basic ways: The United Nations should authorize the United States to maintain a stockpile of atomic weapons to serve as a deterrent; the proposed Atomic Development Authority should not control the uranium mines; and finally, the Authority should not have a veto power over the development of atomic energy. The exchange led Oppenheimer to conclude that Baruch thought his job was to prepare "the American people for a refusal by Russia."

Afterwards, Baruch escorted Oppenheimer to the elevator and tried to reassure him: "Don't let these associates of mine worry you. Hancock is pretty 'Right' but [with a wink] I'll watch him. Searls is smart as a whip, but he sees Reds under every bed."

Needless to say, this encounter with Baruch was not reassuring. Oppenheimer left convinced that the old man was a fool. He told Rabi that he "despised Baruch." Soon afterwards, he told Baruch he had decided not to join him as his scientific adviser. Rabi thought this was a mistake: "He did something hard to forgive; he refused to be on the staff. So they got poor old Richard Tolman instead." Tolman, in ill health, had neither the stamina nor the force of personality to stand up to someone like Baruch. As for Oppenheimer, Baruch told Lilienthal, "It is too bad about that young man [Oppenheimer]. Such great promise. But he won't cooperate. He will regret his attitude."

Baruch was right, and Oppenheimer had second thoughts about his decision. Just hours after he turned the job down, he phoned Jim Conant and confessed that he thought he had been foolish. Should he change his mind? Conant told him it was too late, that Baruch had lost confidence in him.

In the weeks ahead, Oppenheimer, Acheson and Lilienthal did their best to keep the Acheson-Lilienthal plan alive, lobbying the bureaucracy and the media. In response, Baruch complained to Acheson that he was "embarrassed" that he was being undercut. Hoping that he could still influence Baruch, Acheson agreed to bring everyone together at Blair House on Pennsylvania Avenue on Friday afternoon, May 17, 1946.

But as Acheson worked to contain the atomic genie, others were working to contain, if not destroy, Oppenheimer. That same week, J. Edgar Hoover was urging his agents to step up their surveillance of Oppenheimer. Though he hadn't a shred of evidence, Hoover now floated the possibility that Oppenheimer intended to defect to the Soviet Union. Having decided that Oppenheimer was a Soviet sympathizer, the FBI director reasoned that "he would be far more valuable there as an advisor in the construction of atomic plants than he would be as a casual informant in the United States." He instructed his agents to "follow Oppenheimer's activities and contacts closely. . . ."

A week before this summit, Oppenheimer, in a phone call to Kitty, told her that the meeting was "an attempt to box the old guy [Baruch] in. . . . It is not a very happy situation." He then added, "I don't want anything from them and if I can work on his [Baruch's] conscience, that is the best angle I have. It just isn't worth anything otherwise." Kitty urged him to be clear with himself about "what the old man wants." Oppie agreed, and then, upon hearing the clicking sound of an operator cutting her key in and out, he asked Kitty, "Are you still there? I wonder who is listening to us?" Kitty replied, "The FBI, dear." Oppie said, "They are—the FBI?" He then quipped, "The FBI must just have hung up." Kitty giggled, and then they resumed their conversation.

Kitty had guessed correctly. Two days earlier, the FBI had wiretapped the Oppenheimer home in Berkeley (and Hoover forwarded a transcript of this conversation to Secretary of State Byrnes, "as of possible interest to you and the President"). Hoover also ordered his agents to tail Oppenheimer on his travels around the country.

Whether Oppenheimer's disparaging remarks reached Baruch is not known, but the meeting at Blair House did not go well. Baruch made it clear that he and his people were moving away from the whole notion of international ownership of uranium mines. Then the discussion broke apart completely over the question of "penalties." Why, Baruch asked, was there no provision for the punishment of violators of the agreement? What would happen to a country found to be building nuclear weapons? Baruch thought a stockpile of nuclear weapons should be set aside and automatically used against any country found in violation. He called this "condign punishment." Herb Marks said such a provision was completely inconsistent with the spirit of the Acheson-Lilienthal plan. Besides, Marks pointed out, it would take a renegade nation at least a year to prepare atomic weapons, and that would provide the international community time to respond. Acheson himself tried to explain in judicious tones that they had indeed grappled with this question, and had concluded that "if a major power violated a treaty, or wanted a trial of strength, then no matter what words or provisions were set forth in the treaty, it was obvious that the international organization had broken down. . . ."

Baruch nevertheless insisted that a law without a penalty was useless. Disregarding the opinion of most scientists, he decided that the Soviets would not be able to build their own atomic weapons for at least two decades. If so, he reasoned, there was no pressing reason to relinquish the American monopoly anytime soon. Consequently, the plan he intended to submit to the United Nations would substantially amend—indeed, fundamentally alter—the Acheson-Lilienthal proposals: The Soviets would have to give up their right to a veto in the Security Council over any actions by the new atomic authority; any nation violating the agreement would immediately be subjected to an attack with atomic weapons; and, before being given access to any of the secrets relating to the peaceful uses of atomic energy, the Soviets would have to submit to a survey of their uranium resources.

Acheson and McCloy vigorously objected to such an early emphasis on punitive provisions. This, and the fact that Baruch clearly intended, at least for some years, to preserve the American monopoly on atomic weapons, would doom the plan. The Soviets would never agree to such conditions, particularly at a time when the United States was continuing to build and test atomic weapons. What Baruch was proposing was not cooperative control over nuclear energy but an atomic pact designed to prolong the U.S. monopoly. McCloy angrily insisted that there was no such condition as complete security, and that it would be "presumptuous" to suggest such harsh and automatic penalty provisions. The next day, Justice Felix Frankfurter wrote McCloy: "I am told that it was a real bullfight—and that you were so disgusted with the gentleman on the other side that you just sputtered 'dust in the air.' "

While Republican John McCloy was merely angry, Oppenheimer's anger led to depression. He wrote Lilienthal after it was all over to say that he was "still very heavy of heart." Once again demonstrating his political perspicacity, Oppenheimer predicted, accurately as it happened, how the whole process would unfold: "The American disposition will be to take plenty of time and not force the issue in a hurry; that then a 10–2 report will go to the [Security Council] and Russia will exercise her veto and decline to go along. This will be construed by us as a demonstration of Russia's warlike intentions. And this will fit perfectly into the plans of that growing number who want to put the country on a war footing, first psychologically, then actually. The Army directing the country's research; Red-baiting; treating all labor organizations, CIO first, as Communist and therefore traitorous, etc. . . ." As he talked, Oppenheimer paced back and forth in his frenetic style, speaking, Lilienthal later noted in his diary, in a "really heart-breaking tone."

Oppie told Lilienthal that he had talked in San Francisco with a Soviet scientist, a technical adviser to the Soviet foreign minister, Andrei Gromyko, who had stressed that Baruch's proposal was meant to preserve America's atomic monopoly. "The American proposal," he had said, "was designed to permit the United States to maintain its own bombs and plants almost indefinitely—30 years, 50 years, as long as we thought necessary—whereas it wants Russia's uranium, and therefore her chance of producing materials, to be taken over and controlled by the ADA [Atomic Development Authority] at once."

On June 11, 1946, the FBI overheard Oppenheimer talking with Lilienthal about Baruch's proposals for "condign punishment." "They worry me like hell," he told Lilienthal.

"Yes, it is very bad," Lilienthal replied. "Even in the short run point of view, it will take all the—"

"Take all the fun out of it," Oppenheimer interrupted. "But they don't see that and they never will. They just haven't lived in the right world."

"They have lived in an unreal world," Lilienthal agreed, "and it is populated by figures and statistics and bonds, and I can't understand them and they can't understand us."

Two days earlier, Oppenheimer had taken his case to the public by publishing a long essay in the *New York Times Magazine* that explained the plan for an international Atomic Development Authority in layman's language.

> It proposes that *in the field of atomic energy* there be set up a world government. That *in this field* there be a renunciation of sovereignty. That in this field there be no legal veto power. That in this field there be international law. How is this possible in a world of sovereign nations? There are only two ways in which this ever can be possible: One is conquest. That destroys sovereignty. And the other is the partial renunciation of that sovereignty. What is here proposed is such a partial renunciation, sufficient, but not more than sufficient, for an atomic development authority to come into being; to exercise its functions of development, exploitation and control; to enable it to live and grow, and to protect the world against the use of atomic weapons and provide it with the benefits of atomic energy.

Early that summer, Oppenheimer ran into his former student Joe Weinberg, who was still teaching physics at Berkeley. When Weinberg asked him, "What do we do if this effort in international control fails?" Oppie pointed out the window and replied, "Well, we can enjoy the view—as long as it lasts."

ON JUNE 14, 1946, Baruch presented his plan to the United Nations, dramatically proclaiming in biblical language that he offered the world a choice between "the quick and the dead." As Oppenheimer and everyone else associated with the original Acheson-Lilienthal plan predicted, Baruch's proposal was promptly rejected by the Soviets. Moscow's diplomats instead proposed a simple treaty to ban the production or use of atomic weapons. This proposal, Oppenheimer told Kitty in a phone call the next day, was "Not too bad." No one could be surprised by Soviet objections to the veto provisions of the Baruch proposal. And yet, Oppie observed to his wife that Baruch was declaiming loudly how sorely disappointed he was, all the while "knowing it was a damn fool performance."

Nevertheless, as Oppie predicted, the Truman Administration rejected the Soviet response out of hand. Negotiations continued in a desultory fashion for many months, but without result. An early opportunity for a good-faith effort to prevent an uncontrolled nuclear arms race between the two major powers had been lost. It would take the terrors of the 1962 Cuban missile crisis, and the massive Soviet buildup that followed it, before an

American administration would propose, in the 1970s, a serious and acceptable arms control agreement. But by then tens of thousands of nuclear warheads had been built. Oppenheimer and many of his colleagues always blamed Baruch for this missed opportunity. Acheson angrily observed later, "It was his [Baruch's] ball and he balled it up. . . . He pretty well ruined the thing." Rabi was equally blunt: "It's simply real madness what has happened."

Over the years, critics of Oppenheimer's 1946 proposals for international control have charged him with political naïveté. Stalin, they argue, would never have accepted inspections. Oppenheimer himself understood this point. "I cannot tell," he wrote years later, "and I think that no one can tell, whether early actions along the lines suggested by Bohr would have changed the course of history. There is not anything that I know of Stalin's behavior that gives one any shred of hope on that score. But Bohr understood that this action was to create a change in the situation. He did not say, except once in jest, 'another experimental arrangement,' but this is the model he had in mind. I think that if we had acted in accordance, wisely and clearly and discreetly in accordance with his views, we might have been freed of our rather sleazy sense of omnipotence, and our delusions about the effectiveness of secrecy, and turned our society toward a healthier vision of a future worth living for."

Later that summer, Lilienthal visited Oppenheimer in his Washington hotel room and the two men talked late into the night about what had happened. "He is really a tragic figure," Lilienthal wrote in his diary, "with all his great attractiveness, brilliance of mind. As I left him he looked so sad: 'I am ready to go anywhere and do anything [Oppie said], but I am bankrupt of further ideas. And I find that physics and the teaching of physics, which is my life, now seems irrelevant.' It was this last that really wrung my heart."

Oppenheimer's anguish was real and deep. He felt a personal responsibility for the consequences of his work at Los Alamos. Every day the newspaper headlines gave him evidence that the world might once again be on the road to war. "Every American knows that if there is another major war," he wrote in the *Bulletin of the Atomic Scientists* on June 1, 1946, "atomic weapons will be used. . . ." This meant, he argued, that the real task at hand was the elimination of war itself. "We know this because in the last war, the two nations which we like to think are the most enlightened and humane in the world—Great Britain and the United States—used atomic weapons against an enemy which was essentially defeated."

He had made this observation earlier in a speech at Los Alamos, but to publish it in 1946 was an extraordinary admission. Less than a year after the events of August 1945, the man who had instructed the bombardiers exactly how to drop their atomic bombs on the center of two Japanese cities had come to the conclusion that he had supported the use of atomic weapons against *"an enemy which was essentially defeated."* This realization weighed heavily on him.

A major war was not Oppie's only worry; he was concerned too about nuclear terrorism. Asked in a closed Senate hearing room "whether three or four men couldn't smuggle units of an [atomic] bomb into New York and blow up the whole city." Oppenheimer responded, "Of course it could be done, and people could destroy New York." When a startled senator then followed by asking, "What instrument would you use to detect an atomic bomb hidden somewhere in a city?" Oppenheimer quipped, "A screwdriver [to open each and every crate or suitcase]." There was no defense against nuclear terrorism—and he felt there never would be.

International control of the bomb, he later told an audience of Foreign Service and military officers, is "the only way in which this country can have security comparable to that which it had in the years before the war. It is the only way in which we will be able to live with bad governments, with new discoveries, with irresponsible governments such as are likely to arise in the next hundred years, without living in fairly constant fear of the surprise use of these weapons."

AT THIRTY-FOUR seconds after 9:00 a.m. on July 1, 1946, the world's fourth atomic bomb exploded above the lagoon of Bikini Atoll, a part of the Marshall Islands in the Pacific Ocean. A fleet of abandoned Navy vessels of all shapes and sizes were either sunk or exposed to murderous radiation. A large crowd of congressmen, journalists and diplomats from numerous countries, including the Soviet Union, witnessed this demonstration. Oppenheimer had been one of many scientists invited to see the show, but he was conspicuously absent.

Two months earlier, his frustrations mounting, Oppenheimer had decided he would not attend the Bikini tests. On May 3, 1946, he wrote President Truman, ostensibly to explain his decision. His real intent, however, was to challenge Truman's entire posture. He began by outlining his "misgivings," which he asserted were shared "not unanimously, but very widely" by other scientists. Then, with devastating logic, he decimated the whole exercise. If the purpose of the tests, as stated, was to determine the effectiveness of atomic weapons in naval warfare, the answer was quite simple: "If an atomic bomb comes close enough to a ship, even a capital ship, it will sink it." One need only determine how close the bomb had to be to the ship—and this could be deduced from mathematical calculations.

The cost of the tests as planned might easily reach $100 million. "For less than one percent of this," Oppenheimer explained, "one could obtain more useful information."

Likewise, if the tests hoped to obtain scientific data on radiation's effects on naval equipment, rations and animals, this information too could be obtained more cheaply and more accurately "by simple laboratory methods." Proponents of testing argued, Oppenheimer wrote, that "we must be prepared for the possibility of atomic warfare." If this was the true purpose behind the tests, then surely everyone understood that "the overwhelming effectiveness of atomic weapons lies in their use for the bombardment of cities." By comparison, "the detailed determination of the destruction of atomic weapons against naval craft would appear trivial." Finally—and this was undoubtedly Oppenheimer's fiercest objection—he questioned "the appropriateness of a purely military test of atomic weapons, at a time when our plans for effectively eliminating them from national armaments are in their earliest beginnings." (The Bikini tests were being conducted virtually simultaneously with Baruch's presentation at the United Nations.)

Oppenheimer concluded that he could have remained on the presidential commission to observe the Bikini tests—but that perhaps the president might think it "most undesirable for me to turn in, after the tests are completed, a report" critical of the whole exercise. Under the circumstances, he wrote, perhaps he could better serve the president elsewhere.

If Oppenheimer thought his letter might persuade Truman to postpone or cancel the Bikini tests, he was mistaken. Instead of focusing on the substance of Oppenheimer's dissent, the president remembered his first encounter with him. Affronted by the letter, Truman now forwarded it to Acting Secretary of State Dean Acheson with a short note in which he described Oppenheimer as that "cry-baby scientist" who had earlier claimed to have blood on his hands. "I think he has concocted himself an alibi in this letter." Truman misunderstood. Oppie's letter was actually a declaration of personal independence, and through it, once again, he further alienated the president of the United States.

"Oppie Had a Rash and Is Now Immune"

He [Oppenheimer] thinks he's God.

PHILIP MORRISON

OPPENHEIMER WENT ABOUT TEACHING physics at Caltech, but his heart wasn't in it. "I did actually give a course," he later said, "but it is obscure to me how I gave it now. . . . The charm went out of teaching after the great change in the war. . . . I was always called away and distracted because I was thinking about other things." Indeed, he and Kitty never set up house in Pasadena. Kitty remained in the Berkeley house on Eagle Hill and Robert commuted, staying one or two nights a week in the guest cottage behind the home of his old friends Richard and Ruth Tolman. But the phone calls from Washington never stopped, and as the months passed, this arrangement proved to be awkward. Late in the spring of 1946, in the midst of his peripatetic negotiations in Washington, New York and Los Alamos, Oppenheimer announced his intention to resume his teaching post at Berkeley in the autumn.

Though truly disheartened by the moral and intellectual fiasco of the "Baruch Plan," Oppenheimer and Lilienthal continued to work together. On October 23, the FBI overheard the two men discussing who should be named to the Atomic Energy Commission (AEC), which had been created by the August 1 passage of the McMahon Act. Oppenheimer told his new friend, "I owe you a statement which I haven't thought it discreet to make until tonight, and that is, in a very grim world since I last saw you, I have not been a despondent man. I just can't tell you, Dave, how I admire what you are doing and how it has changed the whole world for me."

Lilienthal thanked him and remarked, "I think we're going to get a hold of this damn thing yet."

That autumn, President Truman appointed Lilienthal chairman of the

Atomic Energy Commission, and, as required by Congress, he created a General Advisory Committee (GAC) to assist the AEC commissioners. Despite Truman's dislike of Oppenheimer, the "father of the atomic bomb" could hardly be kept off such a committee. So, following the recommendations of a variety of advisers, Truman appointed him together with I. I. Rabi, Glenn Seaborg, Enrico Fermi, James Conant, Cyril S. Smith, Hartley Rowe (a Los Alamos consultant), Hood Worthington (a Du Pont company official), and Lee DuBridge, who had recently been appointed president of Caltech. Truman left it up to these men to choose their own chairman. But when a news report wrongly implied that Conant would chair the committee, Kitty Oppenheimer huffily asked Robert why *he* hadn't been named chairman. Robert assured his wife that "it is not a major issue." In fact, DuBridge and Rabi were quietly lobbying behind the scenes for Oppenheimer. By the time the GAC gathered for its first formal meeting in early January 1947, the fix was in. Delayed by a snowstorm, Oppenheimer arrived late, to learn that his colleagues had unanimously elected him their chairman.

By then, Oppie was disillusioned with both the Soviet and American positions. Neither country seemed prepared to do what was necessary to avoid a nuclear arms race. As a result of both the broadening of his despair and his new responsibilities, his views began to change. That January, Hans Bethe came to visit him in Berkeley, and Oppie confessed in several long conversations that he had "given up all hope that the Russians would agree to a plan." The Soviet attitude appeared inflexible; their proposal to ban the bomb seemed designed to "deprive us immediately of the one weapon which would stop the Russians from going into Western Europe." Bethe agreed.

Later that spring, Oppenheimer used his influence as chairman of the GAC to toughen the American negotiating position. In March 1947, he flew to Washington, where Acheson gave him a preview of the president's soon-to-be-announced Truman Doctrine. "He wanted me to be quite clear," Oppenheimer later testified, "that we were entering an adversary relationship with the Soviets, and whatever we did in the atomic talk we should bear that in mind." Oppenheimer acted on this advice almost immediately; soon afterwards, he met with Frederick Osborn, Bernard Baruch's successor at the United Nations atomic energy negotiations. To Osborn's surprise, Oppenheimer told him that the United States should withdraw from the UN talks. The Soviets, he said, would never agree to a workable plan.

Oppenheimer's attitude toward the Soviet Union was now following the general trajectory of the emerging Cold War. By his own account, during the war he had already begun to turn away from his left-wing internationalist enthusiasms. He was also troubled by a speech Stalin gave on February 9, 1946; Oppenheimer—like most observers in the West—characterized it as a reflection of Soviet fears of "encirclement and their need to keep their guard up and to rearm." In addition, he was disheartened by what he was learning about Soviet wartime espionage. According to an FBI informant—identified as "T-1," an administrator on the Berkeley campus—Oppenheimer returned from receiving a 1946 briefing in Washington "terribly depressed." "T-1 reported that an unnamed government official had "given Oppenheimer 'the facts of life' concerning the Communist conspiracy, and as a result Oppenheimer had become thoroughly disillusioned with Communism."

The briefing Oppenheimer received pertained to a Canadian spy scandal, precipitated by the defection of Soviet code clerk Igor Gouzenko which led to the arrest of Alan Nunn May, a British physicist working in Montreal who had spied for the Soviets. Oppenheimer was genuinely shaken by this evidence of "treachery" on the part of a fellow scientist, and later that year, when the FBI came to interview him about the Chevalier affair, he "commented on the fact that often Communists in various countries outside of the Soviet Union could be led into situations where they would be acting, either knowingly or unknowingly, as spies for the Soviet Union." He could not "reconcile the treachery employed by them [the Soviets] in their international relationships with the high purposes and the democratic aims ascribed to the Soviets by the local [American] Communists."

The failure of the Baruch Plan had made things worse. The dream of international control would have to await a change in geopolitical circumstances. He understood now that the ideological differences between the United States and the Soviet Union were unlikely soon to be reconciled. "It is clear," he told an audience of Foreign Service and Army officers in September 1947, "that, even for the United States, proposals of this kind [international controls] involve a very real renunciation. Among other things, they involve a more or less permanent renunciation of any hope that the United States might live in relative isolation from the rest of the world."

He knew that the diplomats of many other countries were "genuinely goggle-eyed" at the sweeping nature of his proposals for international control. They involved radical sacrifices and at least a partial renunciation of sovereignty. But he now understood that the sacrifices required of the Soviet Union were of another order of magnitude. In a perceptive analysis, he noted: "That is because the proposed pattern of [international] control stands in a very gross conflict to the present patterns of state power in Russia. The ideological underpinning of that power, namely the belief in the inevitability of conflict between Russia and the capitalist world, would be repudiated by a co-operation as intense or as intimate as is required by our proposals for the control of atomic energy. Thus what we are asking of the Russians is a very far-reaching renunciation and reversal of the basis of their state power. . . ."

He knew that the Soviets were not likely to "take this great plunge." He had not given up hope that in the distant future international controls could be achieved. In the meantime, he had reluctantly decided that the United States had to arm itself. This had led him to conclude—with considerable melancholy—that the principal job of the Atomic Energy Commission would be to "provide atomic weapons and good atomic weapons and many atomic weapons." Having preached the necessity of international control and openness in 1946, Oppenheimer by 1947 was beginning to accept the idea of a defense posture supported by a multitude of nuclear weapons.

TO ALL APPEARANCES, Oppenheimer was now a member in good standing of the American Establishment. His credentials included the chairmanship of the AEC's General Advisory Committee, a coveted "Q" (atomic secrets) security clearance, the presidency of the American Physical Society and a member of Harvard University's Board of Overseers. As a Harvard Overseer, Oppenheimer rubbed shoulders with such influential men as the poet Archibald MacLeish, Judge Charles Wyzanski, Jr., and Joseph Alsop. On a warm, sunny day in early June 1947, Harvard awarded Oppenheimer an honorary degree. During the graduation ceremonies, he listened as his friend Gen. George C. Marshall unveiled the Truman Administration's plan to pour billions of dollars into a program for European economic recovery—what soon became known as the Marshall Plan.

Oppenheimer and MacLeish grew particularly close. The poet took to sending him sonnets and they corresponded frequently. He and Robert shared similar liberal values, values that they had come to believe were equally threatened by the communists on the left and the radicals on the right. In August 1949, MacLeish published an astonishingly bitter essay in the *Atlantic Monthly,* "The Conquest of America," in which he attacked the country's postwar descent into an atmosphere of dystopia, of a utopia gone awry. Although America was the most powerful nation on the globe, the American people seemed seized by a mad compulsion to define themselves by the Soviet threat. In this sense, MacLeish wryly concluded, America had been "conquered" by the Soviets, who were now dictating American behavior. "Whatever the Russians did, we did in reverse," MacLeish wrote. He harshly criticized Soviet tyranny, but lamented the fact that so many Americans were willing to sacrifice their civil liberties in the name of anticommunism.

MacLeish asked Oppenheimer what he thought of the essay. Robert's reply revealed the evolution of his own political views. He thought MacLeish's description of the "present state of affairs" was masterful. But he was troubled by MacLeish's prescription—a call for a "redeclaration of the revolution of the individual." This familiar exhortation to Jeffersonian individualism seemed somehow inadequate and not very fresh. "Man is both an end and an instrument," Oppenheimer wrote. He reminded MacLeish of the "profound part that culture and society play in the very definition of human values, human salvation and liberation." Therefore, "I think that what is needed is something far subtler than the emancipation of the individual from society; it involves, with an awareness that the past one hundred and fifty years have rendered progressively more acute, the basic dependence of man on his fellows."

Robert then told MacLeish of his midnight walk in the snow with Niels Bohr earlier that year, in which the Dane had expounded his philosophy of openness and complementarity. Bohr, he thought, provides "that new insight into the relations of the individual and society without which we can give an effective answer neither to the Communists nor to the antiquarians nor to our own confusions." MacLeish welcomed Robert's letter: "It was extraordinarily kind of you to write me at such length. The point you raise is, of course, the central point of the whole business."

Some of his friends on the left were not quite sure what to make of this transformation. But those who had all along thought of Oppenheimer as a Popular Front Democrat had no reason to think his political spots had changed. Rather, the issues had changed: With the war against fascism won (except in Franco's Spain), and the Depression over, the Communist Party was simply no longer the magnet it had once been for politically active intellectuals. To his noncommunist liberal friends like Robert Wilson, Hans Bethe and I. I. Rabi, Oppie was the same man, with the same motivations.

Significantly, Frank Oppenheimer's transformation was less abrupt. While no longer a communist, he did not think the Russians really threatened America. On this issue the two brothers had some of their most serious political arguments. Robert told his brother that he believed "the Russians were ready to march if they were given the opportunity." He favored Truman's hard line against the Soviets now, and when Frank tried to argue with him, "Robert would say that he knew things that he couldn't report, but they convinced him that the Russians could not be expected to cooperate."

In their first reunion after the war, Haakon Chevalier also noted the change in Oppie's outlook. Sometime in May 1946, Oppie and Kitty visited the Chevaliers in their new oceanfront home at Stinson Beach. Oppie made it clear that his political sympathies had moved, at least in Haakon's view, "considerably to the right." Chevalier recalled being shocked at some of the "very uncomplimentary" things he had to say about the American Communist Party and the Soviet Union. "Haakon," Oppie said, "Haakon, believe me, I am serious, I have real reason to believe, and I cannot tell you why, but I assure you I have real reason to change my mind about Russia. They are not what you believe them to be. You must not continue your trust, your blind faith, in the policies of the USSR."

Moreover, Chevalier continued to hear things about his old friend that confirmed his observation. One evening in New York, Chevalier ran into Phil Morrison on the street, and they talked about all that had happened since the outbreak of the war. Chevalier regarded Morrison as a former comrade. But he also knew Morrison as one of Oppie's closest friends before the war and as one of the key physicists who had followed him to Los Alamos.

"What about Opje?" Chevalier asked.

"I hardly see him any more," Morrison replied. "We no longer speak the same language. . . . He moves in a different circle." Morrison then related how he and Oppenheimer had been talking one day and Oppie kept referring to "George." Finally, Morrison had interrupted to ask who this George was. "You understand," Morrison said to Chevalier, "General [George C.] Marshall to me is General Marshall, or the secretary of state—not George. This is typical. . . ." Oppenheimer had changed, Morrison said: "He thinks he's God."

CHEVALIER HAD suffered numerous disappointments since he had last seen Oppenheimer in the spring of 1943. His efforts to obtain war-related work were stymied in January 1944, when the government refused him a security clearance for a job in the Office of War Information. His FBI file contained "unbelievable" allegations, said a friend working in OWI: "Someone obviously has it in for you." Mystified by this news, Chevalier stayed in New York and found occasional free-lance work as a translator and magazine writer. In the spring of 1945, he returned to his teaching post at Berkeley. But soon after the war ended, he was hired by the War Department to serve as a translator at the Nuremburg War Crimes Tribunal. He flew to Europe in October 1945 and did not return to California until May 1946. By then, Berkeley had denied him tenure. Devastated by this blow to his academic career, Chevalier decided to work full-time on a novel he had under contract with the publisher Alfred A. Knopf.

On June 26, 1946, about six weeks after his first reunion with Oppie, Chevalier was at home working on his novel when two FBI agents knocked on his door. They insisted that he accompany them to their office in downtown San Francisco. On that same summer day, at about the same hour, FBI agents also appeared at George Eltenton's home and asked him to accompany them to the FBI field office in Oakland. Chevalier and Eltenton were simultaneously questioned for about six hours. During the ensuing interrogations it became clear to both men that the agents wanted to know about the conversations that they had had regarding Oppenheimer in the early winter of 1943.

Although each was unaware of the other's interrogation, both men gave similar stories. Eltenton acknowledged that sometime late in 1942, when the Soviets were barely containing the Nazi onslaught, Peter Ivanov from the Soviet Consulate approached him and asked whether he knew Professors Ernest Lawrence and Robert Oppenheimer, and one other individual whom Eltenton could not fully recall—but he thought the name might be Alvarez. Eltenton replied that he knew only Oppenheimer, and not very well. But he volunteered that he had a friend who was close to Oppenheimer. The Russian then asked if his friend might ask Oppenheimer whether he could share information with Soviet scientists. Eltenton said he made the inquiry to Chevalier and told him that his Russian friend had assured him that such information "would be safely transmitted through his channels which involved photo reproduction. . . ." In the event, Eltenton confirmed to the FBI that a few days later, Chevalier "dropped by my house and told me that there was no chance whatsoever of obtaining any data and Dr. Oppenheimer did not approve." Further, Eltenton denied having approached any other individuals.

Chevalier confirmed to the FBI the broad outlines of Eltenton's statement. But to his surprise, the FBI agents pressed him repeatedly about approaches to three other scientists. Chevalier denied approaching anybody other than Oppenheimer. After nearly eight hours of interrogation, Chevalier reluctantly agreed to sign an affidavit: "I wish to state that to my present knowledge and recollection I approached no one except Oppenheimer to request information concerning the work of the radiation laboratory." But then he carefully qualified this categorical statement: "I may have mentioned the desirability of obtaining this information for Russia with any number of people in passing. I am certain that I never made another specific proposal in this connection." He later wrote in his memoirs that he left wondering how the FBI had heard about his conversations with Eltenton and Oppenheimer. Neither could he understand why they believed he had approached three scientists.

Some time later, perhaps in July or August 1946, Chevalier and Eltenton happened to attend the same luncheon in the Berkeley home of a mutual friend. It was the first time they had seen each other since 1943. Chevalier told him about his encounter in June with the FBI. After comparing notes, they realized that they had both been questioned on the same day. How, they wondered, had the FBI gotten wind of their conversation?

Several weeks later, Oppenheimer invited the Chevaliers to a cocktail party at Eagle Hill. They came early, as requested, so that the old friends could have a chance to visit before the other guests arrived. According to Chevalier's account in his memoir, when he broached the topic of his recent encounter with the FBI, "Opje's face at once darkened."

"Let's go outside," Robert said. Hoke took this as an indication that his friend thought his home was wiretapped. They walked into the back garden on a wooded corner of the property. As they paced, Chevalier gave a detailed account of his interrogation. "Opje was obviously greatly upset," Chevalier wrote in 1965. "He asked me endless questions." When Chevalier explained that he had been reluctant to tell the FBI about his conversation with Eltenton, Oppenheimer reassured him that it had been the right thing to do. "I had to report that conversation, you know," Oppenheimer said.

"Yes," Chevalier replied, although he wondered to himself whether it had really been necessary. "But what about those alleged approaches to three scientists, and the supposed repeated attempts to get secret information?"

In Chevalier's account, Oppenheimer gave no reply to this critical question.

As Oppenheimer stood in his Eagle Hill garden, trying to reconstruct what he had told Pash in 1943, he became more and more agitated. Chevalier thought he seemed "extremely nervous and tense."

Eventually, Kitty called, "Darling, the guests are arriving, and I think you'd better come in now." Oppie replied abruptly, saying that he would come in a minute. But he continued his pacing and had Chevalier repeat his story again. Minutes passed and Kitty came out a second time, calling out that he really must come now. When Oppie replied curtly, Kitty persisted. "Then, to my utter dismay," Chevalier wrote, "Opje let loose with a flood of foul language, calling Kitty vile names and told her to mind her goddamn business and to get the . . . hell out."

Chevalier had never seen his friend behave so intemperately. Even then, he seemed reluctant to end the conversation with Chevalier. "Something was obviously bothering him," Chevalier wrote, "but he gave no hint as to what it was."

SOON AFTER this troubling conversation with Chevalier, on September 5, 1946, agents of the FBI paid a visit to Oppenheimer's Berkeley office. Not to his surprise, they wanted to question him about his 1943 conversation with Chevalier. Gracious as always, he explained that Chevalier had informed him about Eltenton's scheme and that he had rejected it outright. He remembered telling Chevalier that "to do such a thing was treason or close to treason." He denied that Chevalier was trying to solicit information on the bomb project. On further questioning, "Oppenheimer said that due to the lapse of time since the incident, he was vague in his mind as to the exact words used by him and Chevalier in their conversation, and any present effort on his part to reconstruct their conversation would be pure guesswork, but he did definitely recollect having used either the word 'treason' or 'treasonous' to Chevalier."

When the FBI agents pressed him about three other approaches to scientists connected to the Manhattan Project, he told them that this part of the story had been "concocted" in order to protect the identity of Chevalier. "Oppenheimer stated that in previously reporting this matter to MED [Manhattan Engineer District], he tried to protect Chevalier's identity and in an effort to do so he 'concocted a completely fabricated story' which he later described as a 'complicated cock and bull story,' which was to the effect that three unidentified associates had been approached on Eltenton's behalf for information."

Why did Oppenheimer say such a thing? Why would he admit to lying about what he had said in 1943? An obvious explanation is that this version of the story was the truth; he had panicked when confronted by Pash in 1943, and had embellished his account with three fictional scientists to dramatize its importance and divert attention from himself. Another explanation is that during his garden conversation with Chevalier he learned that his friend had not approached three other scientists as he had originally thought. After all, Eltenton had mentioned Oppenheimer, Lawrence and, perhaps, Alvarez to Chevalier as potential targets, making it entirely plausible that Chevalier had related this to Oppenheimer in their kitchen conversation. Yet another possibility is that he had told some version of the truth in 1943—but now felt compelled to change his story in order to protect both Chevalier and the unnamed scientists. His enemies would insist at the 1954 security clearance hearing that this was the case, but it is the least plausible of all the explanations. He had long ago informed on Chevalier, and Lawrence and Alvarez hardly needed his protection. The only person in need of protection now was Robert Oppenheimer, and admitting to the FBI in 1946 that he had lied to military intelligence in 1943 was not the best way to protect oneself—unless it *was* the unvarnished truth. But all of these explanations—and others—would be raised again, and challenged, eight years later, during Robert's security hearing. The contradictions in these two stories would be devastating.

.   .   .

LATE IN 1946, Lewis Strauss, one of Truman's appointees to the new Atomic Energy Commission, flew out to San Francisco and was met at the airport by Ernest Lawrence and Oppenheimer. Before discussing AEC business, Strauss took Oppenheimer aside and said he had something else to talk to him about. Strauss had met Oppenheimer only once before, late in the war. Pacing about on the concrete tarmac, Strauss explained that he was a trustee of the Institute for Advanced Study in Princeton, New Jersey. At the moment, he chaired the trustees' search committee for a new director of the Institute. Oppenheimer's name was at the top of a list of five candidates, and now the trustees, Strauss said, had authorized him to offer Oppenheimer the post. Oppenheimer expressed interest in the idea, but said that he needed time to think about it.

About a month later, in late January 1947, Oppenheimer flew to Washington, and over a long breakfast he listened to Strauss pitch him the job. On the phone later that day, Oppenheimer told Kitty that he had not made up his mind but felt "rather good" about the idea. Strauss, he said, "had very nice ideas" about what Oppenheimer could do with the Institute—although they were not too realistic. Oppie remarked that there "wasn't a scientist there in any science business," but he could "soon change all that."

The Institute was most famous as the home and intellectual refuge of Albert Einstein. When Strauss had pressed Einstein to describe the ideal kind of man for the job of director, he had replied, "Ah, that I can do gladly. You should look for a very quiet man who will not disturb people who are trying to think." For his part, Oppenheimer had not always thought of it as a place for serious scholarship. After visiting the Institute for the first time in 1934, he had written derisively to his brother: "Princeton is a madhouse: its solipsistic luminaries shining in separate & helpless desolation." But now he saw it differently. "It would take some thought and some concern to do a decent job," he told Kitty, but "it was a thing he could do rather naturally." He assured her that if they moved to Princeton, they would still keep their Eagle Hill home for summers in Berkeley. Besides, he was tired of the long commutes to Washington. "It is impossible for me to live as I have been living this last winter—in airplanes." That year alone he had made fifteen transcontinental flights between Washington and California.

Still undecided, Oppenheimer consulted one of his new Washington friends, Justice Felix Frankfurter, who had himself once been a trustee of the Institute. Frankfurter discouraged Oppenheimer, saying, "You won't be free for your own creative work. Why don't you go to Harvard?" When

Oppie bristled at this suggestion, saying he knew why he shouldn't go to Harvard, Frankfurter referred him to another friend who knew Princeton well; this individual advised Oppenheimer, "Princeton was an odd sort of place, but if one had an idea of what to make of it, it was fine."

Oppenheimer was inclined to accept this new challenge. It played to his administrative talents, it promised to leave him ample time to pursue his extracurricular government responsibilities, and its location was perfect—short train rides from both Washington and New York City. Yet he took his time to mull it over, until finally, according to one report, the Oppenheimers heard a news broadcast on their car radio announcing that Robert Oppenheimer had been appointed director of the Institute for Advanced Study. "Well," Robert said to Kitty, "I guess that settles it."

The *New York Herald Tribune* applauded the appointment as "strikingly fit" in an editorial: "His name is Dr. J. Robert Oppenheimer, but his friends call him 'Oppy.' " The *Tribune*'s editorialists fairly gushed with praise, describing him as a "remarkable man," a "scientist among scientists," a "practical man" with a "streak of wit." One of the Institute's trustees, John F. Fulton, had lunch with Robert and Kitty in their home and afterwards scribbled his impressions of the new director in his diary: "In physical appearance, he is slender with rather slight features, but he has a piercing and imperturbable eye, and a quickness in repartee that gives him great force, and he would immediately command respect in any company. He is only forty-three years of age, and despite his preoccupation with atomic physics, he has kept up his Latin and Greek, is widely read in general history, and he collects pictures. He is altogether a most extraordinary combination of science and the humanities."

Lewis Strauss, however, was annoyed that Oppenheimer had taken so long to make up his mind. A self-made millionaire, Strauss had started out life as a traveling shoe salesman, with a high school education. In 1917, when he was just twenty-one years old, he landed a job as an assistant to Herbert Hoover, an engineer and up-and-coming politician with a reputation as a "progressive" Teddy Roosevelt Republican. At the time, Hoover was running President Woodrow Wilson's food relief programs for refugees in war-torn Europe. Working alongside such other Hoover protégés as Harvey Bundy, a bright young Boston Brahmin lawyer, Strauss used the food relief job as a springboard to Wall Street. After the war, Hoover helped Strauss obtain a coveted position at the New York investment banking firm of Kuhn, Loeb. Hardworking and obsequious, Strauss soon married Alice Hanauer, the daughter of a Kuhn, Loeb partner. By 1929 he himself was a full partner, making more than a million dollars a year. He survived the 1929 crash relatively unscathed. During the 1930s he became an ardent foe of the New Deal, but nine months before Pearl Harbor he persuaded the Roosevelt Administration to give him a job in the Navy Department's Bureau of Ordnance. Later he served as a special assistant to Navy Secretary James Forrestal, and he left the war with the honorary rank of rear admiral. By 1945, Strauss had used his Wall Street and Washington connections to carve out a powerful position for himself in America's post–World War II establishment. Over the next two decades, he would exercise a baleful influence over Oppenheimer's life.

Oppie's first impression of Strauss was caught on an FBI wiretap: "Regarding Strauss, I know him slightly. . . . He is not greatly cultivated but will not obstruct things." Lilienthal told Oppie he thought Strauss was "a man with an active mind, definitely conservative, apparently not too bad." Both assessments underestimated Strauss. He was pathologically ambitious, tenacious and extraordinarily prickly, a combination that made him a particularly dangerous opponent in bureaucratic warfare. One of his fellow AEC commissioners said of him, "If you disagree with Lewis about anything, he assumes you're just a fool at first. But if you go on disagreeing with him, he concludes you must be a traitor." *Fortune* magazine once described him as a man with a "rather owlish face" whose critics thought him "thin-skinned, intellectually arrogant, and rough in battle." For years, Strauss served as president of Manhattan's Temple Emanu-El—ironically, the same Reform synagogue Felix Adler abandoned in 1876 to form the Ethical Culture Society. Proud of both his Jewish and his Southern heritage, Strauss pointedly insisted on pronouncing his last name as 'Straws.' Self-righteous to a fault, he remembered every slight—and meticulously recorded them in an endless stream, each entitled "memorandum to the file." He was, as the Alsop brothers wrote, a man with a "desperate need to condescend."

KITTY WELCOMED her husband's decision to move East. The FBI's wiretap heard her telling a salesman that they "would not be gone long—only 15 or 20 years." Oppie told her that their new home in Princeton, Olden Manor, had ten bedrooms, five bathrooms and a "pleasant garden." Not surprisingly, Oppenheimer's Berkeley colleagues were disappointed. The chairman of the physics department described his departure as "the greatest blow ever suffered by the department." Ernest Lawrence was miffed to learn of Oppie's defection from a radio news report. On the other hand, Oppenheimer's East Coast friends were delighted. Isidor Rabi wrote him, "I am terribly pleased that you are coming. . . . It's a sharp break with the past for you and the perfect time of life in which to make it." His friend and former landlady, Mary Ellen Washburn, threw him a farewell party.

Oppie was leaving many old friends behind—and a lover. He had always cherished his friendship with Dr. Ruth Tolman. During the war, he had worked closely with Ruth's husband, Richard, who had served as General Groves' scientific adviser in Washington. It was Richard who had largely persuaded him to resume his teaching post at Caltech after the war. Oppenheimer counted the Tolmans among his closest friends. He had met them in Pasadena in the spring of 1928 and had always admired them both. "He was rightly very highly respected," Oppenheimer said of Richard Tolman years later. "His wisdom and broad interests, broad in physics and broad throughout, his civility, his extremely intelligent and quite lovely wife, all made a sweet island in the Southern California [locale]. . . . a friendship developed which became very close." In 1954, Oppenheimer testified that Richard Tolman had been "a very close and dear friend of mine." Frank Oppenheimer later said, "Robert loved the Tolmans—especially Ruth."

Sometime during the war—or perhaps shortly after returning from Los Alamos—Oppie and Ruth Tolman began an affair. A clinical psychologist, Ruth was nearly eleven years older than Robert. But she was an elegant and attractive woman. Another friend, the psychologist Jerome Bruner, called her "the perfect confidante, a wise woman . . . she could give a sense of personalness to anything she touched." Born in Indiana, Ruth Sherman graduated from the University of California in 1917. In 1924 she married Richard Chase Tolman, and continued her studies in psychology. Richard was by then a distinguished chemist and mathematical physicist; he was also twelve years her senior. Though the couple never had children, friends thought they were "totally suited for each other." Ruth had stimulated Richard's interest in psychology and, specifically, in the social implications of science.

Oppenheimer shared with Ruth a fascination with psychiatry. For her doctorate, Ruth had studied the psychological differences between two groups of adult criminals. In the late 1930s she had worked as a senior psychological examiner for the Los Angeles County Probation Department. And during the war, she had served as a clinical psychologist for the Office of Strategic Services. Beginning in 1946, she worked as a senior clinical psychologist with the Veterans Administration.

A career woman, Dr. Ruth Tolman possessed a formidable intellect. But by all accounts, she was also a warm, gentle and astute observer of the human condition. She seems to have known aspects of Oppie's character not visible to many others: "Remember how we have always, both of us, been miserable when we had to look more than a week ahead?"

When, in the summer of 1947, Oppenheimer was preparing to move to Princeton, he wrote Ruth a letter from his vacation at Los Pinos to complain that he was "fagged" and felt "appalled" about the future. Ruth replied, "My heart is very full of many many things I want to say. Like you, I'm grateful to be writing. Like you, I cannot yet quite accept the fact that the monthly visits will not be resumed, once the irregularities of the summer are over. From Richard I could not get very much news of you, though the impression remains that you were tired still." She urged him to visit her in Detroit while she was in that city attending a conference—and if not, then in Pasadena: "Come to us when you can, Robert. The guest house is always and completely yours."

Few of Oppenheimer's letters to Ruth Tolman survive; most were destroyed after her death. But her love letters display a deep tenderness and closeness. "I look back on your wonderful week here," she wrote in one undated letter, "with all my heart grateful, Dear. It was unforgettable. I'd give great rewards even for another day. In the meantime, you know the love and tenderness I send." On another occasion, she wrote of their plans to get together for a weekend; she promised to meet his plane and hoped "we'd go to the sea for the day." She wrote that she had recently driven by "the long stretch of beach where the sandpipers and gulls played. Oh Robert, Robert. Soon I shall see you. You and I both know how it will be." Later, after this planned outing by the sea, Oppenheimer wrote, "Ruth, dear heart . . . I write in celebration of the good day we had together which meant so very much to me. I knew that I should find you full of courage and wisdom, but it is one thing to know it, and another to be so close. . . . It was so wonderful to see you." He signed the letter, "My love, Ruth, always."

Kitty was certainly aware of Robert's long-standing friendship with the Tolmans. She knew that on his monthly travels to Pasadena, he stayed in the Tolman guest cottage while teaching his Caltech class. Frequently he would take the Tolmans, and sometimes the Bachers, out to their favorite Mexican restaurant—and often Kitty would call him from Berkeley. "I think Kitty was intensely resentful of any other person getting involved with Robert," recalled Jean Bacher. But if Kitty was naturally possessive, there is no indication that she ever learned of an affair.

Then, one Saturday night in mid-August 1948, Richard Tolman suddenly suffered a heart attack in the midst of a party he and Ruth were hosting at their home. Kitty's former husband, Dr. Stewart Harrison, was called to the scene and managed to get Richard checked into a hospital within thirty minutes. Three weeks later, Richard died. Ruth was devastated; she had dearly loved her husband of twenty-four years. But some of their friends used the tragedy to smear Robert. Ernest Lawrence, whose attitude toward Robert had by then become one of outright enmity, speculated that Richard's heart attack had been precipitated by the discovery of his wife's affair. Lawrence later told Lewis Strauss that "Dr. Oppenheimer first earned his [Lawrence's] disapproval a number of years ago when he seduced the wife of Professor Tolman at CalTech." Lawrence claimed that "it was a notorious affair which lasted for enough time for it to become apparent to Dr. Tolman who died of a broken heart."

Ruth and Robert continued to see each other after Richard's death. Four years later, Ruth wrote Robert after one such meeting, "I shall always remember the two magic chairs on the dock, with the water and the lights and the planes swooping around overhead. I suppose you realized what I did not dare to mention—that it was the anniversary—4 years—of Richard's death, and the memories of those dreadful days of August 1948, and then of many earlier sweet ones was very overwhelming to me. I felt very grateful that I could be with you that night." In another undated letter, Ruth wrote, "Robert dear—The precious times with you last week and the week before keep going through my mind, over and over, making me thankful but wistful, wishing for more. I was grateful for them, Dear, and as you knew, hungry for them, too." She went on to suggest a date for their next liaison: "How would it be if I said you had to see someone at UCLA and we'd be away for the day, [and] be back for a party at night? . . . Let's think about this." Obviously, Ruth and Robert loved each other, but neither of them intended their affair to destroy their respective marriages. Throughout these years, Ruth also managed to maintain friendly relations with Kitty and the Oppenheimer children. She was simply one of the Oppenheimer household's oldest friends—and Robert's special confidante.

BEFORE ACCEPTING the Princeton job, Oppenheimer had volunteered to Strauss that "there was derogatory information about me." At the time, Strauss had dismissed the warning. But, as mandated by the newly passed McMahon Act, the FBI was reviewing the security clearances of all Atomic Energy Commission employees and all the commissioners were obliged to read Oppenheimer's file. As an aide to J. Edgar Hoover put it, this gave the Bureau the opportunity "to conduct an open and extensive investigation of Oppenheimer since we don't have to be discreet or cautious. . . ." Agents were sent to tail Oppenheimer and interview more than a score of his associates, including Robert Sproul and Ernest Lawrence. Everyone vouched for his loyalty. Sproul told an agent that Oppenheimer had told him that he was "ashamed and embarrassed" by his left-wing past. Lawrence said Oppie "had a rash and is now immune."

Despite these testimonials to Oppenheimer's trustworthiness, Strauss and other AEC commissioners soon learned from the FBI that Oppenheimer's security clearance would be anything but a routine matter. In late February 1947, Hoover sent the White House a twelve-page summary of Oppenheimer's file, highlighting the physicist's associations with communists. On Saturday, March 8, 1947, this report was also sent to the AEC, and soon afterwards Strauss called the AEC's general counsel, Joseph Volpe, into his office. Volpe could see that Strauss was "visibly shaken" by what he had read. The two men studied the file, until finally Strauss turned to Volpe and said, "Joe, what do you think?"

"Well," Volpe replied, "if anyone were to print all the stuff in this file and say it is about the top civilian adviser to the Atomic Energy Commission, there would be terrible trouble. His background is awful. But your responsibility is to determine whether this man is a security risk *now,* and except for the Chevalier incident, I don't see anything in this file to establish that he might be."

That Monday the AEC's Commissioners met to discuss the problem. Everyone realized that withholding Oppenheimer's clearance would have serious political consequences. James Conant and Vannevar Bush told the commissioners that the FBI's allegations had been heard and dismissed years before. Still, they knew that if the AEC wished to approve Oppenheimer's security clearance, the FBI had to agree. On March 25, Lilienthal went to see the FBI chief. Hoover was still troubled by Oppenheimer's failure to report his conversation with Chevalier in a timely fashion. He nevertheless reluctantly agreed that while Oppenheimer "may at one time have bordered upon the communistic, indications [were] that for some time he [had] steadily moved away from such a position." When told that the AEC's own security officials felt the evidence was not strong enough to deny Oppenheimer a clearance, Hoover indicated he would not push the matter any further. In fact, he thought it convenient that Oppenheimer's security status was the AEC's bureaucratic responsibility, leaving the FBI free to continue its own investigation. Nevertheless, Hoover warned that Frank Oppenheimer was quite another case—the FBI, he said, would not approve a renewal of Frank's security clearance.

Afterwards, Strauss told Oppenheimer that he had examined his FBI file "rather carefully" and seen nothing in it that would bar his appointment as director of the Institute for Advanced Study. A formal clearance from the AEC Commissioners naturally took longer; it was not until August 11, 1947, that the AEC Commission formally voted Oppenheimer a top-secret "Q" clearance. The vote was unanimous; even Strauss, the most conservative commissioner, voted for the clearance.

Oppenheimer had survived his first postwar scrutiny—but he had every reason to think that he was still a marked man. Hoover persisted, despite having told Lilienthal that he would drop the case. In April 1947, a month after the AEC commissioners had decided to give Oppie his clearance, Hoover forwarded new information "specifically substantiating the fact that the Oppenheimer brothers were substantial contributors to the Communist Party in San Francisco as late as 1942." The new information came from an FBI burglary of CP offices in San Francisco that produced copies of CP financial records.

In an effort to keep the case alive, Hoover urged his agents to dig for derogatory material of any kind. In the autumn of 1947, for instance, the Bureau's San Francisco office sent Hoover and Assistant Director D. M. Ladd a confidential memo containing prurient material about the alleged sexual activities of Oppenheimer and some of his close friends. Hoover was informed that an unnamed "very reliable individual" employed at the University of California was volunteering to become a regular "confidential informant of this office." This unidentified source allegedly had known a number of Oppenheimer's Berkeley friends since 1927. The FBI's informant described one such friend, a married woman, as "an oversexed individual" inclined to bohemian tastes; the source claimed that "it was common knowledge around the campus that [this couple] were involved in a husband and wife trade with another member of the faculty and his wife. . . ." As if this weren't salacious enough, Hoover was informed that among her many affairs, this woman had attended a faculty party in 1935, become intoxicated and then disappeared with a mathematics student, Harvey Hall. Almost as a postscript, the FBI's source claimed that at the time of this seduction, Hall was living with Robert Oppenheimer. The source said it was also "common knowledge" that prior to Oppenheimer's marriage in 1940, "he had had homosexual tendencies" and that he was "having an affair with Hall."

In fact, at no time did Oppenheimer ever share quarters with Hall—and there is no evidence that Oppenheimer interrupted his socially active heterosexual life to have an affair with a man. The FBI's own source characterized these sexual escapades, probably quite accurately, as "gossip." But this did not stop Hoover from allowing the tidbit about Oppenheimer's alleged "affair" with Hall to be incorporated into some of the many summaries of Oppenheimer's FBI file. These summaries were eventually read by Strauss and many other high-ranking policy-makers in Washington. While such material no doubt titillated many officials, it also persuaded some that the information they were being passed about Oppenheimer was less than reliable. Lilienthal thought it telling, for instance, that one anonymous source was described as a twelve-year-old boy. He concluded that most of the damaging stories were little more than malicious gossip from prewar sources, many of whom clearly did not know Oppenheimer. It was an accurate assessment for much of the derogatory information in Oppenheimer's FBI dossier, but it ignored the pernicious effect of the accumulated weight that this now unevaluated information could have on readers who were not particularly sympathetic to Oppenheimer.

CHAPTER TWENTY-SEVEN

"An Intellectual Hotel"

In some sort of crude sense which no vulgarity, no humor, no overstatement can quite extinguish, the physicists have known sin; and this is a knowledge which they cannot lose.

ROBERT OPPENHEIMER

THE OPPENHEIMERS ARRIVED in Princeton in mid-July 1947, during an unusually hot and humid summer. Oppenheimer's new position, as director-elect of the Institute that had been Albert Einstein's sanctuary for nearly fifteen years, would provide him both a prestigious platform and easy access to the growing number of nuclear policy–related committees he served on in Washington. The Institute paid him a generous salary of $20,000 a year, plus rent-free use of the director's home, Olden Manor—which came with a live-in cook and a groundskeeper-handyman to tend to the house and its extensive gardens. The Institute also allowed him plenty of time to travel where and when he pleased. He would not formally assume his new responsibilities until October, and he would not preside over his first faculty meeting until December. He and Kitty—and their two young children, six-year-old Peter and Toni, age three—would have a leisurely few months to adjust to their new surroundings. Robert was just forty-three years old.

Kitty quickly fell in love with Olden Manor, a rambling, three-story white colonial home, surrounded by 265 acres of lush green woodlands and meadows. A barn and a corral stood behind the house. Robert and Kitty bought two horses which they named Topper and Step-up.

Portions of Olden Manor dated back to 1696, when the Oldens, one of Princeton's earliest pioneer families, began farming on the site. The west wing of the house had been built in 1720, and it served as a field hospital for General Washington's troops during the Battle of Princeton in early 1777. Generations of Oldens had added on to the structure, and by the late nineteenth century it had eighteen rooms. The family occupied the property until the 1930s, when it was sold to the Institute.

Painted bright white inside and out, the house had a light, roomy atmosphere. A tall central hallway cut through the entire structure, running from the front door to an arched back door that led onto a slate terrace. A formal dining room led into a large L-shaped farm kitchen. Sun poured through eight windows in the living room. Across the hallway stood a second, smaller living room, called the music room. A step down from the music room was a library dominated by a massive brick fireplace. When the Oppenheimers moved in, they found nearly every room in the house lined with bookshelves. Robert had most of them torn out, leaving only one wall in the library covered with floor-to-ceiling bookcases. Everywhere, the light-oak plank flooring creaked softly. Upstairs, the house was filled with odd nooks and crannies, hidden closets and a back staircase leading to the kitchen. A panel of numbered buzzers allowed the cook or maid to be summoned from almost any room in the house.

Soon after their arrival, Robert had an ample greenhouse built on the back of the house, near the kitchen wing. It was his birthday gift to Kitty, who filled it with dozens of varieties of orchids. The house was surrounded by acres of gardens, including a carefully manicured flower garden enclosed by four rock walls, the foundation of an ancient barn. Kitty, the trained botanist, loved gardening, and over the years she became what one friend called "an artist in the ancient magic of garden making."

"When we first moved in," Oppenheimer later told a reporter, "I thought I'd never get used to such a big house, but now we've lived in it until it has a pleasant degree of shabbiness, and I like it very much." Robert mounted one of his father's prize paintings, Vincent van Gogh's *Enclosed Field with Rising Sun* (Saint-Remy, 1889), in the living room, above the formal white fireplace. They hung a Derain in the dining room and a Vuillard in the music room. While the house was comfortably furnished, it never had a cluttered, lived-in look. Kitty kept everything neat. Oppie's austere study, with its white walls unadorned by pictures, reminded one old friend of their Los Alamos home.

From Olden Manor's back terrace, Oppenheimer could gaze south across an open field to the grounds of the Institute. Not more than a quarter mile away lay Fuld Hall, a four-story red brick building with two wings and an imposing churchlike spire. Built in 1939 at a cost of $520,000, it housed modest offices for scores of scholars, a wood-paneled library and a formal common room lined with overstuffed brown leather couches. A cafeteria and boardroom occupied the top, fourth floor. In 1947, Einstein used a corner office, Room 225, on the second floor; Niels Bohr and Paul Dirac worked in adjoining rooms on the third floor. Oppenheimer's ground-level office, Room 113, afforded him a view of the woods and meadow. His predecessor, Frank Aydelotte, a scholar of Elizabethan literature, had hung on the walls framed prints of wistful scenes of Oxford. Oppenheimer took these down and replaced them with a blackboard that ran the length of the wall. He inherited two secretaries, Mrs. Eleanor Leary, who had previously worked with Justice Felix Frankfurter, and Mrs. Katharine Russell, an efficient young woman in her twenties. Just outside his office stood a "monstrous safe," containing classified documents for his work as chairman of the AEC's General Advisory Committee (GAC). Armed guards sat twenty-four hours a day beside the locked safe.

Visitors to Fuld Hall saw a man "ablaze with power." The phone would ring and his secretary would knock on the door and announce, "Dr. Oppenheimer, General [George C.] Marshall is on the line." His colleagues could see that such phone calls would "electrify" him. He clearly relished the role history had assigned him and he tried hard to play the part well. While most of the Institute's permanent scholars walked around in sports jackets—Einstein favored a rumpled sweater—Oppenheimer often wore expensive English wool suits hand-tailored for him at Langrocks, the local tailor for Princeton's upper crust. (But he could also turn up at a party in a jacket "that looked as if it had been eaten by gerbils.") Where many scholars got around Princeton on bicycles, Oppie drove a stunning blue Cadillac convertible. Where once he'd worn his hair long and bushy, now he had it "cut like a monk's, skin-tight." At forty-three, he seemed delicate, even frail. But he was in fact quite strong and energetic. "He was very thin, nervous, jittery," Freeman Dyson recalled. "He constantly moved around; he couldn't sit still for five seconds; you had the impression of somebody who was tremendously ill at ease. He smoked all the time."

Princeton was a world away from the free-spirited, liberal, bohemian atmosphere of Berkeley and San Francisco, not to mention the lifestyle and vistas of Los Alamos. In 1947 Princeton, a suburban town of 25,000 residents, had one stoplight, at the corner of Nassau and Witherspoon streets, and no public transportation—with the exception of the "Dinky" tram that to this day ferries hundreds of daily commuters to the railroad station at Princeton Junction. From there, bankers, lawyers and stockbrokers in pinstripe suits boarded trains for the fifty-minute ride into Manhattan. Unlike most American small towns, Princeton possessed an august history and an elite sense of itself. But, as a longtime resident once observed, it was "a town with character but without soul."

ROBERT'S AMBITION was to turn the Institute into a stimulating international venue for interdisciplinary scholarship. It had been founded in 1930

by Louis Bamberger and his sister, Julie Carrie Fuld, with an initial donation of $5 million. Bamberger and his sister had sold the family business,
the Bamberger department store, to R. H. Macy & Co. in 1929, just before
the stock market crash, for the princely sum of $11 million in cash. Enam-
ored of the notion of building an institution of higher learning, Bamberger
hired Abraham Flexner, an educator and foundation executive, to be the
Institute's first director. Flexner promised that the Institute would be neither
a teaching university nor a research school: "It may be pictured as a wedge
between the two—a small university in which a limited amount of teaching
and a liberal amount of research are both to be found." Flexner told the
Bambergers that he wished to model the Institute after such European intellectual havens as Oxford's All Souls College or the Collège de France in
Paris—or Göttingen, Oppenheimer's German alma mater. It would be, he
said, "a Paradise for scholars."

In 1933, Flexner made the Institute's reputation by hiring Einstein for an
annual salary of $15,000. Other scholars were paid similarly lavish salaries.
Flexner wanted the very best people, and he wanted to ensure that none of
his scholars would ever feel compelled to supplement their income by
"writing unnecessary textbooks or engaging in other forms of hack work."
There would be "no duties, only opportunities." Throughout the 1930s,
Flexner recruited brilliant minds, mostly mathematicians like John von
Neumann, Kurt Gödel, Hermann Weyl, Deane Montgomery, Boris Podol-
sky, Oswald Veblen, James Alexander and Nathan Rosen. Flexner hailed
the "usefulness of useless knowledge." But by the 1940s, the Institute was
in danger of acquiring a reputation for coddling brilliant minds with forever
unfulfilled potential. One scientist described it as "that magnificent place
where science flourishes and never bears fruit."

Oppenheimer was determined to change all this. In his own field of theoretical physics, he hoped to do for the Institute what he had done for
Berkeley in the 1930s—turn it into a world-class center for theoretical
physics. He knew the war had suspended engagement in any truly original
work. But things were rapidly changing. "Today," he told an MIT audience
in the autumn of 1947, "barely two years after the end of hostilities, physics
is booming."

Early in April 1947, Abraham Pais, a bright young physicist with a temporary fellowship at the Institute, received a phone call from Berkeley, California. "This is Robert Oppenheimer," the caller told a startled Pais. "I have
just accepted the directorship of the Institute for Advanced Study, and I desperately hope that you will be there next year, so that we can begin building
up theoretical physics there." Flattered, Pais immediately put aside thoughts
of joining Bohr in Denmark and agreed. He would remain at the Institute

for the next sixteen years, becoming one of Oppenheimer's long-standing confidants.

Pais soon had a chance to observe Oppenheimer in action. For three days in June 1947, twenty-three of the country's leading theoretical physicists gathered at the Ram's Head Inn, an exclusive resort on Shelter Island, at the eastern tip of Long Island. Oppenheimer had taken the lead in organizing the conference. Among others, he brought Hans Bethe, I. I. Rabi, Richard Feynman, Victor Weisskopf, Edward Teller, George Uhlenbeck, Julian Schwinger, David Bohm, Robert Marshak, Willis Lamb and Hendrik Kramers to discuss "The Foundations of Quantum Mechanics." With the end of the war, theoretical physicists were finally able to shift their attention back to fundamental issues. One of Oppenheimer's doctoral students, Willis Lamb, gave the first of the conference's many remarkable presentations, outlining what would soon become known as the "Lamb shift," which in turn became a key step to a new theory of quantum electrodynamics. (Lamb would win a Nobel Prize in 1955 for his work on this topic.) Similarly, Rabi gave a groundbreaking talk on nuclear magnetic resonance.

Although Karl Darrow, secretary of the Physical Society, officially chaired the conference, Oppenheimer dominated it. "As the conference went on," Darrow noted in his diary, "the ascendancy of Oppenheimer became more evident—the analysis (often caustic) of nearly every argument, that magnificent English never marred by hesitation or groping for words (I never heard 'catharsis' used in a discourse on [physics], or the clever word 'mesoniferous' which is probably Oppenheimer's invention), the dry humor, the perpetually-recurring comment that one idea or another (including some of his own) was certainly wrong, and the respect with which he was heard." Similarly, Pais was struck by Oppenheimer's "priestly style" when speaking before an audience. "It was as if he were aiming at initiating his audience into Nature's divine mysteries."

On the third and last day, Oppenheimer led a discussion of the paradoxical behavior of mesons, a topic he had explored with Robert Serber prior to the war. Pais later remembered Oppenheimer's "masterful" performance, interrupting at all the right moments with leading questions, summarizing the discussion and stimulating others to think of solutions. "I was sitting next to Marshak," Pais later wrote, "during this discussion and can still remember how he suddenly got all red in the face. He got up and said, 'Maybe there are two kinds of mesons. One kind is copiously produced, then disintegrates into a different kind which absorbs only weakly.' " In Pais' view, Oppenheimer thus midwifed Marshak's innovative two-meson hypothesis, a breakthrough which later won the British physicist Cecil F. Powell a Nobel in 1950. The Shelter Island conference also helped Feynman and Schwinger to work out "renormalization theory," an elegant new way to calculate the interactions of an electron with its own or another electromagnetic field. Once again, if Oppenheimer himself was not the author of such discoveries, many of his peers saw him as their great facilitator.

Not everyone applauded Oppenheimer's performance. David Bohm recalled thinking that Oppie was talking too much. "He was very fluent with his words," Bohm said, "but there wasn't much behind what he was saying to back up that much talking." Bohm thought his mentor had begun to lose his insightfulness, perhaps simply because he hadn't been doing anything of any substance in physics for many years. "He [Oppenheimer] didn't sympathize with what I was doing in physics," Bohm recalled. "I wanted to question fundamentals, and he felt that one should work on using the present theory, exploiting it and trying to work out its consequences." Earlier in their relationship, Bohm had had tremendous regard for Oppenheimer. But over time he found himself agreeing with another friend who had worked with Oppenheimer, Milton Plesset, who expressed the view that Oppie was "not capable of genuine originality, but that he is very good at comprehending other people's ideas and seeing their implications."

Leaving Shelter Island, Oppenheimer hired a private seaplane to fly him to Boston, where he was scheduled to receive an honorary degree at Harvard. Victor Weisskopf and several other physicists returning to Cambridge accepted his invitation to join him on the plane. Halfway there, they ran into a storm and the pilot decided to land at a Navy base in New London, Connecticut. Civilian aircraft were forbidden to use this airdrome, and as they taxied up to the dock, the pilot could see an angry Navy captain yelling at him. Oppenheimer told the pilot, "Let me handle this." As he stepped off the plane, he announced, "My name is Oppenheimer." The Navy officer gasped and then asked, "Are you *the* Oppenheimer?" Without missing a beat, Oppie replied, "I am *an* Oppenheimer." Bowled over to be in the presence of the famous physicist, the officer went out of his way to serve Oppenheimer and his friends tea and cookies and then sent them on their way to Boston aboard a Navy bus.

THE MOST FAMOUS physicist in the United States was not doing much physics—this, despite the fact that Oppenheimer had persuaded the Institute's trustees to give him an unprecedented dual appointment as both director and "Professor of Physics." In the fall of 1946, Oppie had found the time to coauthor a paper with Hans Bethe, published in *Physical Review,* on electron scattering. That year he was nominated for the Nobel Prize in physics—but the Nobel committee evidently hesitated to give the award to someone whose name was so closely associated with Hiroshima and Nagasaki. Over the next four years, he published three more short physics papers and one paper on biophysics. But after 1950, he never published another scientific paper. "He didn't have *Sitzfleisch,*" said Murray Gell-Mann, a visiting physicist at the Institute in 1951. "Perseverance, the Germans call it *Sitzfleisch,* 'sitting flesh,' when you sit on a chair. As far as I know, he never wrote a long paper or did a long calculation, anything of that kind. He didn't have patience for that; his own work consisted of little *aperçus,* but quite brilliant ones. But he inspired other people to do things, and his influence was fantastic."

At Los Alamos, he had supervised thousands and spent millions; now he presided over an institution with just one hundred people and a budget of $825,000. Los Alamos was completely dependent on the federal government; but the Institute's trustees specifically forbade the director to solicit federal funds. The Institute was a singularly independent place. It had no official relationship with its neighbor, Princeton University. By 1948, some 180 scholars were affiliated with one of two "schools," Mathematics or Historical Studies. The Institute housed no laboratories, no cyclotrons and no more complicated apparatus than a blackboard. No courses were taught, and there were no students—only scholars. Most were mathematicians, some were physicists, and there were a few economists and humanists. The Institute was, in fact, so heavily weighted toward mathematics that some thought Oppenheimer's arrival signaled a decision by the trustees that henceforth the Institute would be devoted to mathematics/physics and nothing else.

Indeed, Oppenheimer's first appointments made it seem as if his only priority was to transform the Institute into a major center for theoretical physics. He brought with him as temporary members five research physicists from Berkeley. After coaxing Pais to stay on, he recruited another promising young English physicist, Freeman Dyson, to become a permanent member of the Institute. He persuaded Niels Bohr, Paul Dirac, Wolfgang Pauli, Hideki Yukawa, George Uhlenbeck, George Placzek, Sinitiro Tomonaga and several other young physicists to spend occasional summers or sabbaticals at the Institute. In 1949, he recruited Chen Ning Yang, a brilliant twenty-seven-year-old who would win the 1957 Nobel in physics with T. D. Lee, another Chinese-born physicist Oppenheimer brought to the Institute. "This is an unreal place," Pais wrote in his diary in February 1948. "Bohr comes into my office to talk, I look out of the window and see Einstein walking home with his assistant. Two offices away sits Dirac. Downstairs sits Oppenheimer. . . ." It was a concentration of scientific talent like no other in the world . . . except, of course, Los Alamos.

In June 1946, well before Oppenheimer's arrival at the Institute, Johnny von Neumann had begun to build a high-speed computer in the boiler room basement of Fuld Hall. Nothing so practical had ever existed at the Institute. And nothing so expensive. The trustees initially gave von Neumann $100,000 to get started. And then, in a rare departure from Institute policy, he was allowed to obtain additional funding from the Radio Corporation of America (RCA), the U.S. Army, the Office of Naval Research and the Atomic Energy Commission. In 1947, a small brick building was constructed a few hundred yards away from Fuld Hall to house the computer von Neumann envisioned.

The whole idea of building a machine was rather controversial among scholars who thought their job was to think. "There was never anything that we needed a lot of computing for," complained one mathematician, Deane Montgomery. Oppenheimer himself was of two minds about von Neumann's computer. Like many others, he thought the Institute should not be turned into a laboratory funded by defense dollars. But this was different. von Neumann was building a machine that would revolutionize research. And so he supported the project. Von Neumann agreed not to patent his machine, which soon became the model for a generation of commercial computers.

Oppenheimer and von Neumann formally unveiled the Institute computer in June 1952. At the time, it was the fastest electronic brain in the world—and its mere existence launched the computer revolution of the late twentieth century. But when the machine was surpassed by better, faster computers in the late 1950s, the permanent members of the Institute met in Oppenheimer's living room and voted to close the computer project altogether. They also passed a motion never to bring another such piece of equipment onto the grounds of the Institute.

In 1948, Oppie recruited the classicist Harold F. Cherniss, an old Berkeley friend and the country's leading scholar on Plato and Aristotle. That same year, he persuaded the trustees to establish a $120,000 "Director's Fund," which gave him personal discretion to bring in short-term scholars. Using this discretionary money, he brought his childhood friend Francis Fergusson to the Institute. Fergusson used the fellowship to write his book *The Idea of a Theatre.* At the instigation of Ruth Tolman, Oppie appointed an advisory committee on psychological scholarship. Once or twice a year, Ruth herself came to the Institute with her brother-in-law Edward Tolman, George Miller, Paul Meehl, Ernest Hilgard and Jerome Bruner. (Ed Tolman and Hilgard had both been members with Oppenheimer of Siegfried Bernfeld's monthly study group which had met in San Francisco during the years 1938–42.) Gathering in Oppenheimer's office, these eminent psychologists would brief him on the "deep questions" in their field and otherwise "keep him in the picture." Oppenheimer soon gave short-term appointments to Miller, Bruner and David Levy, a noted child psychologist. Oppenheimer loved to talk about things psychological. Bruner found him "brilliant, discursive in his interests, lavishly intolerant, ready to pursue any topic anywhere, extraordinarily lovable. . . . We talked about most anything, but psychology and the philosophy of physics were irresistible."

Soon, other such humanists were joining the Institute, including the archaeologist Homer Thompson, the poet T. S. Eliot, the historian Arnold Toynbee, the social philosopher Isaiah Berlin and, later, the diplomat and historian George F. Kennan. Oppenheimer had always admired Eliot's *The Waste Land,* and was delighted when he agreed to come to the Institute for one semester in 1948. But it didn't work out. Having a poet in residence didn't sit well with the Institute's mathematicians, some of whom snubbed Eliot, even after he was awarded the Nobel Prize for literature that year. Eliot, for his part, kept to himself and spent more time at the university than he did at the Institute. Oppenheimer was disappointed. "I invited Eliot here," he told Freeman Dyson, "in the hope that he would produce another masterpiece, and all he did here was to work on *The Cocktail Party,* the worst thing he ever wrote."

Nevertheless, Oppenheimer strongly believed it was essential that the Institute remain a home to both science and the humanities. In his speeches about the Institute, Oppenheimer continually emphasized that science needed the humanities to better understand its own character and consequences. Only a few of the senior resident mathematicians agreed with him, but their support was critical. Johnny von Neumann was almost as interested in ancient Roman history as he was in his own field. Others shared Oppenheimer's interest in poetry. He hoped that he could make the Institute a haven for scientists, social scientists and humanists interested in a multidisciplinary understanding of the whole human condition. It was an irresistible opportunity, a chance to bring together the two worlds, science and the humanities, that had engaged him equally as a young man. In this sense, Princeton would be the antithesis of Los Alamos, and perhaps a psychological antidote to it.

The Institute was as idyllic and comfortable as Los Alamos was spartan. Particularly for its lifelong members, it was a Platonic heaven. "The point of this place," Oppenheimer once said, "is to make no excuses for not doing something, for not doing good work." To outsiders, the Institute sometimes had the appearance of a pastoral asylum for the certifiably eccentric. Kurt Gödel, the renowned logician, was a painfully shy recluse. His only real friend was Einstein, and the two men were often seen walking together from town. In between bouts of severe paranoid depression—convinced that his food was being poisoned, he suffered from chronic malnutrition—Gödel spent years trying to solve the continuum problem, a mathematical conundrum involving a question of infinities. He never found an answer. Spurred on by Einstein, he also worked on general relativity, and in 1949 published a paper that described a "rotating universe" in which it was theoretically possible to "travel into any region of the past, present, and future, and back again." For most of his decades at the Institute, he was a solitary, ghostly figure, dressed in a shabby black winter coat, scribbling German shorthand into reams of notebooks.

Dirac was almost equally strange. When he was a young boy, his father had announced that he should speak to him only in French. This way, he thought, his son would quickly learn another language. "Since I found that I couldn't express myself in French," Dirac explained, "it was better for me to stay silent than to talk in English. So I became silent at that time." Wearing long rubber boots, he was often seen hacking trails through the neighboring woods with an ax. This was his form of recreational exercise, and over the years it became something of an Institute pastime. Dirac was maddeningly literal-minded. One day a reporter called to ask him about a lecture he was scheduled to give in New York. Oppenheimer had long since decided that scholars should not be distracted by having phones in their offices, so Dirac had to take the call from a hallway phone. When the reporter said he wanted a copy of the speech, Dirac put the phone down and went into Jeremy Bernstein's office to ask for advice: He feared, he said, being misquoted. So Abraham Pais, who happened to be standing there, suggested that he write "Do Not Publish in Any Form" atop a copy of the speech. Dirac absorbed this simple advice for several minutes in complete silence. Finally, he said, "Isn't 'in any form' redundant in that sentence?"

Von Neumann was unusual too. Like Oppenheimer, he was multilingual and catholic in his interests. He also loved to throw a good party, staying up well into the morning hours. And, like Edward Teller, he was rabidly anti-Soviet. One night at a party, when the conversation turned to discussion of the early Cold War, von Neumann said quite matter-of-factly that it was obvious: The United States should launch a preventive war and annihilate the Soviet Union with its atomic arsenal. "I think that the USA-USSR conflict," he wrote to Lewis Strauss in 1951, "will very probably lead to an armed 'total' collision, and that a maximum rate of armament is therefore imperative." Oppie was appalled by such sentiments, but did not allow political considerations to influence his decisions with respect to the permanent faculty.

Scholars from a wide range of disciplines were constantly amazed at the range of Oppenheimer's interests. One day a foundation executive from the Commonwealth Fund, Lansing V. Hammond, sought Oppenheimer's advice on some sixty young British applicants for scholarships to study in various American universities. The topics ranged from the liberal arts to the hard sciences. Hammond, a scholar in English literature, hoped to get Oppenheimer's advice on a few of the candidates working in math or physics. As soon as Hammond was ushered into his office, Oppenheimer surprised him by saying, "You got your doctorate at Yale in eighteenth-century English literature—Age of Johnson; was Tinker or Pottle your supervisor?" Within ten minutes, Hammond had all the information he needed to match his English physicist-applicants with suitable American universities. As he rose to leave, thinking he had taken enough of the busy director's time, Oppenheimer said, "If you have a few minutes you can spare, I'd be interested in looking at some of your applications in other fields. . . ." Over the next hour, Oppenheimer spoke at length about the strengths and weaknesses of various graduate schools around the country. "Umm . . . indigenous American music, Roy Harris is just the person for him. . . . Social psychology . . . I'd suggest looking into Vanderbilt; smaller numbers, he'd have a better opportunity of getting what he wants . . . Your field, eighteenth-century English literature; Yale is an obvious choice, but don't rule out Bate at Harvard." Hammond had never even heard of Bate. He left feeling overwhelmed. "Never before," he later wrote, "never since have I talked with such a man."

OPPENHEIMER'S RELATIONSHIP with the Institute's most famous resident was always tentative: "We were close colleagues," he later wrote of Einstein, "and something of friends." But he thought of Einstein as a living patron saint of physics, not a working scientist. (Some in the Institute suspected that Oppenheimer was the source of a statement in *Time* magazine that "Einstein is a landmark, not a beacon.") Einstein harbored a similar ambivalence about Oppenheimer. When Oppenheimer was first suggested in 1945 as a candidate for a permanent professorship at the Institute, Einstein and the mathematician Hermann Weyl wrote a memo to the faculty recommending the theoretical physicist Wolfgang Pauli over Oppenheimer. At the time, Einstein knew Pauli well, and Oppenheimer only in passing. Ironically, Weyl had tried hard in 1934 to recruit Oppenheimer to the Institute; but Oppenheimer had adamantly refused, saying, "I could be of absolutely no use at such a place." Now, however, Oppenheimer's credentials as a physicist just didn't measure up to Pauli's: "Certainly Oppenheimer has made no contributions to physics of such a fundamental nature as Pauli's exclusion principle and analysis of electronic spin. . . ." Einstein and Weyl conceded that Oppenheimer had "founded the largest school of theoretical physics in this country." But after noting that his students universally praised him as a teacher, they cautioned, "It may be that he is somewhat too dominant and [that] his students tend to be smaller editions of Oppenheimer." On the basis of this recommendation, the Institute offered the job in 1945 to Pauli—who turned it down.

Einstein eventually acquired a grudging respect for the new director, whom he described as an "unusually capable man of many-sided education." But what he admired about Oppenheimer was the man, not his physics. Still, Einstein would never count Oppenheimer as one of his close friends, "perhaps partly because our scientific opinions are fairly diametrically different." Back in the 1930s, Oppie had once called Einstein "completely cuckoo" for his stubborn refusal to accept quantum theory. All of the young physicists Oppenheimer brought to Princeton were wholly convinced of Bohr's quantum views—and uninterested in the questions that Einstein posed to challenge the quantum view of the world. They could not fathom why the great man was working indefatigably to develop a "unified field theory" to replace what he saw as the inconsistencies of quantum theory. It was lonely work, and yet he was still quite satisfied to defend "the good Lord against the suggestion that he continuously rolls the dice"—his thumbnail critique of Heisenberg's uncertainty principle, one of the foundations of quantum physics. And he didn't mind that most of his Princeton colleagues "see me as a heretic and a reactionary who has, as it were, outlived himself."

Oppenheimer deeply admired the "extraordinary originality" of the man who had formulated the general theory of relativity, "this singular union of geometry and gravitation." But he thought Einstein "brought to the work of originality deep elements of tradition." And Oppenheimer firmly believed that later in Einstein's life it was this "tradition" that misled him. To Oppenheimer's "sorrow," Einstein devoted his Princeton years to trying to prove that quantum theory was flawed by significant inconsistencies. "No one could have been more ingenious," Oppenheimer wrote, "in thinking up unexpected and clever examples; but it turned out that the inconsistencies were not there; and often their resolution could be found in earlier work of Einstein himself." What distressed Einstein about quantum theory was the notion of indeterminacy. And yet it had been his own work on relativity that had inspired some of Bohr's insights. Oppenheimer saw this as highly ironic: "He fought with Bohr in a noble and furious way, and he fought with the theory which he had fathered but which he hated. It was not the first time that this had happened in science."

These disputes did not prevent Oppenheimer from enjoying Einstein's company. One evening early in 1948, he entertained David Lilienthal and Einstein at Olden Manor. Lilienthal sat next to Einstein and "watched him as he listened (gravely and intently, and at times with a chuckle and wrinkles about his eyes) to Robert Oppenheimer describing neutrinos as 'those creatures,' and the beauties of physics." Robert still loved to be the bearer of lavish gifts. Knowing of Einstein's love of classical music, and knowing that his radio could not receive New York broadcasts of concerts from Carnegie Hall, Oppenheimer arranged to have an antenna installed on the roof of Einstein's modest home at 112 Mercer Street. This was done without Einstein's knowledge—and then on his birthday, Robert showed up on his doorstep with a new radio and suggested they listen to a scheduled concert. Einstein was delighted.

In 1949, Bohr was visiting Princeton and agreed to contribute an essay to a book celebrating Einstein's work on the occasion of his seventieth birthday. He and Einstein enjoyed each other's company, but, like Oppenheimer, Bohr could not understand why quantum theory was such a demon for Einstein. When shown the manuscript of the Festschrift, Einstein noted that the essays contained as many brickbats as words of praise. "This is not a jubilee book for me," he said, "but an impeachment." On the day of his birthday, March 14, an audience of 250 eminent scholars gathered in a Princeton auditorium to hear Oppenheimer, I. I. Rabi, Eugene Wigner and Hermann Weyl sing his praises. However strongly his colleagues may have disagreed with the old man, the air was electric with anticipation when Einstein entered the hall. After a moment of sudden silence, everyone stood to applaud the man they all knew was the greatest physicist of the twentieth century.

As PHYSICISTS, Oppenheimer and Einstein disagreed. But as humanists, they were allies. At a moment in history when the scientific profession was being bought wholesale by a Cold War national security network of weapons labs and universities increasingly dependent on military contracts, Oppenheimer had chosen another path. Though "present at the creation" of this militarization of science, Oppenheimer had walked away from Los Alamos, and Einstein respected him for attempting to use his influence to put the brakes on the arms race. At the same time, he saw that Oppenheimer used his influence cautiously. Einstein was mystified when, in the spring of 1947, Oppenheimer refused his invitation to speak at a public dinner of the newly formed Emergency Committee of Atomic Scientists. Oppenheimer explained that he felt "unprepared to make [a] public address at this time on

Atomic Energy with any confidence that the results will lead in the direction for which we all hope."

The older man clearly didn't understand why Oppenheimer seemed to care so much about maintaining his access to the Washington establishment. Einstein didn't play that game. He would never have dreamed of asking the government to give him a security clearance. Einstein instinctively disliked meeting politicians, generals or figures of authority. As Oppenheimer observed, "he did not have that convenient and natural converse with statesmen and men of power. . . ." And while Oppie seemed to relish his fame and the opportunity to mix with the powerful, Einstein was always uncomfortable with adulation. One evening in March 1950, on the occasion of Einstein's seventy-first birthday, Oppenheimer walked him back to his house on Mercer Street. "You know," Einstein remarked, "when it's once been given to a man to do something sensible, afterward life is a little strange." More than most men ever could, Oppenheimer understood exactly what he meant.

AS AT LOS ALAMOS, Oppenheimer was still uncommonly persuasive. Pais recalled meeting a senior scholar just emerging from Oppie's office. "Something odd just happened to me," said the professor. "I had gone to see Oppenheimer regarding a certain issue on which I held firm opinions. As I left, I found that I had agreed with the opposite point of view."

Oppenheimer tried to exert the same charismatic powers over the Institute's Board of Trustees—but with mixed results. In the late 1940s, the board was often stalemated between liberal and conservative factions. It was dominated by its vice chairman, Lewis Strauss. Other trustees tended to defer to his judgment, partially because he was the only board member with substantial wealth. At the same time, some of the more liberal trustees were put off by his archconservatism. One trustee grumbled that the board did not need "a Hoover Republican thinking in the last century." Although Oppenheimer had met Strauss only briefly before coming to Princeton, he was well aware of Strauss' political views and quietly made it clear that he would not welcome Strauss' elevation to the post of chairman of the board.

Oppenheimer's personal relations with Strauss were initially correct and cordial. Yet it was in these early years that the seeds of a terrible feud were sown. On his visits to Princeton, Strauss was often entertained in Olden Manor; after one such dinner, he sent Robert and Kitty a fine case of wine. But it was clear to all that both men were eager for power and willing to exercise it against each other. One day, Abraham Pais was standing outside Fuld Hall when a helicopter landed on the expansive lawn that separated the

Institute from Olden Manor. Out stepped Strauss. "I was struck by his appearance," Pais later wrote, "suave if not slick, and had the instinctive reaction: Watch out for what is behind this fellow's deportment."

Oppenheimer soon realized that Strauss had ambitions to be something of a "coadministrator." In 1948, he told Oppenheimer that he was contemplating buying a former faculty member's home on the grounds of the Institute. In a clear signal, Oppenheimer forestalled this by quickly getting the Institute to buy the house in question and renting it to another scholar. Strauss apparently got the message. As the Institute's unpublished official history notes, "The episode marks the apparent end for the time being of Mr. Strauss' hope to help govern the Institute at short range." It also established a permanent tension and mutual distrust that extended beyond the Institute. Despite this setback, Strauss exerted his influence over the Institute through his close alliance with Herbert Maas, the chairman of the Board of Trustees, and Professor of Mathematics Oswald Veblen, the only faculty trustee.

Strauss was often annoyed that Oppenheimer sometimes made politically sensitive decisions without first seeking the trustees' approval. In late 1950, Strauss temporarily blocked Oppenheimer's appointment of a mediaevalist scholar, Professor Ernst H. Kantorowicz, because he had refused to sign a California Board of Regents loyalty oath. Strauss relented only when it became clear that his was the sole dissenting vote. When Congress passed legislation requiring an FBI security clearance for scientists funded by fellowships from the AEC, Oppenheimer fired off an angry letter to the AEC. The Institute, he wrote, would no longer accept such fellowships on the grounds that the required security investigations violated its "traditions." Only a month later did Oppenheimer inform the trustees of his action. According to the minutes of the meeting, some trustees expressed the fear that the director's action might involve the Institute in a "political controversy," specifically with the FBI. Oppenheimer was told that in the future, he should consult with the Board prior to making such decisions.

In the spring of 1948, Oppenheimer gave an interview to a reporter from the *New York Times* in which he talked freely about his vision for the Institute. He said he expected to invite many more scholars—or even nonacademics with experience in business or politics—for short-term visits of a semester or a year. "Oppenheimer plans to have fewer life members," the *Times* reported. And then the reporter gave this breezy description of Oppenheimer's job: "Suppose you had funds at your disposal based on a $21,000,000 endowment. . . . Suppose you could use this fund to invite as your salaried house-guests the world's greatest scholars, scientists and creative artists—your favorite poet, the author of the book that interested you so much, the European physicist with whom you would like to mull over some speculations about the nature of the universe. That's precisely the set-up that Oppenheimer enjoys. He can indulge every interest and curiosity. . . ."

Needless to say, some of the Institute's life members winced at these words. Others took offense at the notion that their director could run the Institute according to his intellectual whims. Oppenheimer committed another indiscretion in 1948 when he quipped to *Time* magazine that, while the Institute was a place where men could "sit and think," one could only be sure of the sitting. He went on to say that the Institute had "something of the glow of a mediaeval monastery." And then he inadvertently wounded the sensibilities of the permanent faculty by suggesting that the best thing about the Institute was that it served as "an intellectual hotel." *Time* described the Institute as "a place for transient thinkers to rest, recover and refresh themselves before continuing on their way." Subsequently, the faculty told Oppenheimer that it was their "very strong opinion" that such publicity was "undesirable."

Oppenheimer's larger plans for the Institute often met with resistance— particularly from the mathematicians, who had initially thought he would favor them with appointments and an ever-larger share of the Institute's budget. The arguments could become extraordinarily petty. "The Institute is an interesting Paradise," observed his perceptive secretary, Verna Hobson. "But in an ideal society, when you remove all the everyday frictions, the frictions that are created to take their place are so much more cruel." The fights were mostly about appointments. On one occasion, Oppenheimer was presiding over a meeting when Oswald Veblen marched in and insisted on listening to the discussion. Oppenheimer told him he had to leave, and when the mathematician refused, he adjourned the meeting to another room. "It was just like little boys fighting," recalled Hobson.

Veblen frequently instigated trouble for Oppenheimer. As a trustee, he had always been something of a power broker inside the Institute. Many of the mathematicians, in fact, had expected Veblen to be named director. Instead, as one Institute professor put it, "This upstart Oppenheimer was brought in. . . ." Von Neumann had actively opposed Oppenheimer's selection as director: "Oppenheimer's brilliance is incontestable," he had written to Strauss, but he had "serious misgivings as to the wisdom of making him Director." Von Neumann and many other mathematicians had favored "replacing the directorship by a faculty committee, with a rotating one or two year chairmanship." Instead, they got precisely what they did not want: a strong-willed director with a broad, complicated agenda.

Oppie exhibited the same patience and energy at the Institute that had characterized his leadership of Los Alamos. But according to Dyson, his relations with the mathematicians were "disastrous." The Institute's mathematics school had always been first-rate, and Oppenheimer tried hard never to interfere with their business. Indeed, during his first year as director, he presided over a 60 percent increase in the number of members coming to the School of Mathematics. But instead of reciprocating, the mathematicians invariably opposed many of his appointments in the nonmathematical fields. Frustrated and angry, Oppenheimer once called Deane Montgomery, a thirty-eight-year-old mathematician, "the most arrogant, bull-headed son-of-a-bitch I ever met."

Emotions ran deep, and led to irrational outbursts. "He [Oppenheimer] was out to humiliate mathematicians," said André Weil (1906–1998), the great French mathematician who spent decades at the Institute. "Oppenheimer was a wholly frustrated personality, and his amusement was to make people quarrel with each other. I've seen him do it. He loved to have people at the Institute quarrel with each other. He was frustrated essentially because he wanted to be Niels Bohr or Albert Einstein, and he knew he wasn't." Weil was typical of the bloated egos Oppenheimer encountered at the Institute. These were not the young men he had easily led in Los Alamos by the force of his personality. Weil was arrogant, acerbic and demanding. He took an almost roguish delight in intimidating others, and he was furious that he could not intimidate Oppenheimer.

Academic politics can be notoriously petty, but Oppenheimer was confronted by several paradoxes peculiar to the Institute. By the nature of their discipline, mathematicians invariably do their best intuitive work in their twenties or early thirties—whereas historians and other social scientists often need years of studious preparation before they became capable of genuinely creative work. Thus, the Institute could easily identify and recruit brilliant, youthful mathematicians, but rarely appointed a historian who was not already well seasoned. And while the young mathematicians could read and form an opinion about a historian's work, no historian could do the same for a prospective candidate in the School of Mathematics. And herein lay the most vexing paradox: Because the mathematicians were in the nature of things quickly beyond their prime, and because they had no teaching duties, in middle age many of them tended to devote themselves to other affairs. If not distracted, the mathematicians inevitably turned every appointment into a controversy. Conversely, the nonmathematicians, being older and facing the prime productive years of their careers, had little interest or time for such academic intrigues. But, unhappily for the mathematicians, in Oppenheimer they found themselves confronted by a director who, though a physicist, was determined to balance the Institute's science culture with the humanities and social sciences. To their dismay, he recruited psychologists, literary critics and even poets.

On occasion, Oppenheimer, worn down by these territorial intrigues, vented his frustrations on those close to him. When he caught Freeman Dyson indiscreetly gossiping about an impending appointment of another physicist, Dyson quickly found himself summoned to Oppie's office. "He really flattened me," Dyson recalled. "I saw him at his most fierce. It was bad. I really felt like a worm; he convinced me that I had really betrayed all the trust that he'd ever had in me. . . . That's the way he was. He wanted to run things his own way. The Institute was his own little empire."

At Princeton, the abrasive streak in Oppenheimer that was so rarely seen at Los Alamos would sometimes appear with a ferociousness that startled even his closest friends. To be sure, most of the time Robert charmed people with his wit and gracious manners. But sometimes he seemed unable to contain his fierce arrogance. Abraham Pais recalled several occasions when Oppenheimer's unnecessarily biting comments caused young scholars to come into his office, sobbing.

Rare was the lecturer who could fend off Oppenheimer's interventions, but Res Jost did it memorably. Jost, a Swiss mathematical physicist, was giving a seminar one day when Oppenheimer interrupted to ask if he could explain a point in further detail. Jost looked up and said, "Yes," but then proceeded with his talk. Oppenheimer stopped him and said, "I meant, will you explain so and so?" This time, Jost said, "No." When Oppenheimer asked why, Jost replied, "Because you will not understand my explanation, and you will ask more questions and use up my whole hour." Robert sat quietly through the rest of Jost's lecture.

Restless, brilliant and emotionally detached, Oppenheimer always seemed an enigma to those who observed him up close. Pais, who saw him almost daily at the Institute, thought him an extraordinarily private person, "not given to showing his feelings." Rarely, a window would open up to reveal the intensity of his emotions. One evening, Pais went to Princeton's Garden Theater to view Jean Renoir's 1937 *La Grande illusion,* a classic antiwar film about comradeship, class and betrayal among World War I soldiers. After the lights went up, Pais spied Robert and Kitty sitting in the back row—and he could see that Robert had been weeping.

On another occasion in 1949, Pais invited Robert and Kitty to a party at his small apartment on Dickinson Street. During the course of the evening, Pais was inspired to pull out his guitar and urged everyone to sit on the floor and sing folk songs. Robert complied, but Pais noticed he did so with an "air of hauteur clearly indicating that he thought this was an absurd situation for him to be in." And yet, after the group had been singing for a while, Pais glanced over at Robert and was "touched to see that his attitude of superiority was gone; instead, he now looked like a man of feeling, hungry for simple comradeship."

THE PACE of life at the Institute was serene and civilized; tea was served every afternoon between three and four in the Common Room on the main floor of Fuld Hall. "Tea is where we explain to each other," Oppenheimer once said, "what we don't understand." Two or even three times a week, Oppenheimer hosted a lively seminar, often on physics but sometimes in other fields as well. "The best way to send information," he explained, "is to wrap it up in a person." Ideally, the exchange of ideas required some fireworks. "The young physicists," observed Dr. Walter W. Stewart, an economist at the Institute, "are beyond doubt the noisiest, rowdiest, most active and most intellectually alert group we have here. . . . A few days ago I asked one of them, as they came bursting out of a seminar, 'How did it go?' 'Wonderful,' he said. 'Everything we knew about physics last week isn't true!' "

On occasion, however, guest speakers found it unnerving to be subjected to what came to be called the "Oppenheimer treatment." Dyson described the experience in a letter he wrote his parents back in England: "I have been observing rather carefully his behavior during seminars. If one is saying, for the benefit of the rest of the audience, things that he knows already, he cannot resist hurrying one on to something else; then when one says things that he doesn't know or immediately agree with, he breaks in before the point is fully explained with acute and sometimes devastating criticisms. . . . he is moving around nervously all the time, never stops smoking, and I believe that his impatience is largely beyond his control." Some were unnerved by another of his tics—he'd bite the tip of his thumb, clicking his front teeth, again and again.

One day in the autumn of 1950, Oppenheimer arranged to have Harold W. Lewis present a summary of a paper he, Lewis and S. A. Wouthuysen had published in *Physical Review* on the multiple production of mesons. The paper was based on one of his last research efforts just before becoming director of the Institute, and Oppenheimer was understandably anxious to have a serious discussion of his work. Instead, the gathered physicists veered off into a discussion of *Kugelblitz* or "ball lightning," an unexplained phenomenon in which lightning has sometimes been observed in the form of a ball. As they discussed what might explain such events, Oppenheimer began to flush with fury. Finally, he rose and stalked out muttering, "Fireballs, fireballs!"

Dyson recalled that when he gave a lecture praising Dick Feynman's recent work on quantum electrodynamics, Oppenheimer "came down on me like a ton of bricks." Afterwards, he nevertheless came up to Dyson and apologized for his behavior. At the time, Oppenheimer thought Feynman's approach—done with a maximum of intuition and a minimum of mathematical calculations—was fundamentally wrong, and he simply wouldn't listen to Dyson's defense. Only after Hans Bethe came down from Cornell and gave a lecture in support of Feynman's theories did Oppenheimer allow himself to reconsider his views. When Dyson next lectured, Oppenheimer sat in uncharacteristic silence; and afterwards, Dyson found in his mailbox a very brief note: "Nolo contendere. R.O."

Dyson felt a bundle of emotions in Oppenheimer's presence. Bethe had told him that he ought to study with Oppie because he was "so much deeper." But Dyson was disappointed with Oppenheimer as a physicist— Oppie no longer seemed to have the time for doing the hard work, the calculations, that it took to be a theoretical physicist. "He may have been deeper," Dyson recalled, "but still he didn't really know what was going on!" And he was often perplexed by Oppenheimer as a man, his odd combination of philosophical detachment and driving ambition. He thought of Oppie as the kind of person whose worst temptation was to "conquer the Demon and then to save mankind."

Dyson saw Oppie as guilty of "pretentiousness." Sometimes he simply couldn't understand Oppenheimer's delphic pronouncements—and this reminded him that "incomprehensibility can be mistaken for depth." And yet, despite it all, Dyson found himself attracted to Oppenheimer.

In early 1948, *Time* magazine ran a short news item on an essay Oppenheimer had recently published in *Technology Review.* "Science's sense of guilt," *Time* reported, "was frankly admitted last week" by Dr. J. Robert Oppenheimer. The story quoted the wartime head of Los Alamos Laboratory as saying, "In some sort of crude sense which no vulgarity, no humor, no overstatement can quite extinguish, the physicists have known sin; and this is a knowledge which they cannot lose."

Oppenheimer must have understood that such words, especially coming from him, would attract controversy. Even Isidor Rabi, a close friend, thought the words ill chosen: "That sort of crap, we never talked about it that way. He felt sin, well, he didn't know who he was." The incident inspired Rabi to say of his friend that "he was full of too many humanities." Rabi knew Oppie too well to be angry with him, and he knew that one of his friend's weaknesses was "a tendency to make things sound mystical." Oppenheimer's former teacher at Harvard, Professor Percy Bridgman, told a reporter, "Scientists aren't responsible for the facts that are in nature. . . . If anyone should have a sense of sin, it's God. He put the facts there."

Oppenheimer was not, of course, the only scientist to harbor such thoughts. That year his former Cambridge tutor, Patrick M. S. Blackett (of the "poisoned apple" affair), published *Military and Political Consequences of Atomic Energy,* the first full-blown critique of the decision to use the bomb on Japan. By August 1945, Blackett argued, the Japanese were virtually defeated; the atomic bombs had actually been used to forestall a Soviet share in the occupation of postwar Japan. "One can only imagine," Blackett wrote, "the hurry with which the two bombs—the only two existing—were whisked across the Pacific to be dropped on Hiroshima and Nagasaki just in time, but only just, to insure that the Japanese Government surrendered to American forces alone." The atomic bombings were "not so much the last military act of the Second World War," he concluded, "as the first major operation of the cold diplomatic war with Russia now in progress."

Blackett suggested that many Americans were aware that atomic diplomacy had been a factor—and that this had produced an "intense inner psychological conflict in the minds of many English and American people who knew, or suspected, some of the real facts. This conflict was particularly intense in the minds of the atomic scientists themselves, who rightly felt a deep responsibility at seeing their brilliant scientific work used in this way." Blackett was describing, of course, the internal torment felt by his former pupil. He even cited the June 1, 1946, speech Oppenheimer had given at MIT in which he had bluntly said that the United States had "used atomic weapons against an enemy which was essentially defeated."

Blackett's book created a stir when it was published the following year in America. Rabi attacked it in the pages of the *Atlantic Monthly:* "The wailing over Hiroshima finds no echo in Japan." He insisted that the city was a "legitimate target." But, significantly, Oppenheimer himself never criticized Blackett's thesis—and later that year he warmly congratulated his old tutor when Blackett won the Nobel Prize in physics. Moreover, when, a few years later, Blackett published another book critical of the American decision to use the bomb, *Atomic Weapons and East-West Relations,* Oppenheimer wrote to say that while he thought some points were not "quite straight," he nevertheless agreed with the "major thesis."

THAT SPRING, a new monthly magazine, *Physics Today,* featured on its inaugural cover a black-and-white photograph of Oppie's porkpie hat slung over a metal pipe—no caption was needed to identify the owner of the famous *chapeau.* After Einstein, Oppenheimer was undoubtedly the most renowned scientist in the country—and this at a time when scientists were suddenly regarded as paragons of wisdom. His advice was eagerly sought in and out of government and his influence sometimes seemed pervasive. "He wanted to be on good terms with the Washington generals," Dyson observed, "and to be a savior of humanity at the same time."

"He Couldn't Understand
Why He Did It"

He told me that his nerve just gave way at that moment. . . .
He has this tendency when things get too much, he some-
times does irrational things.

DAVID BOHM

I N THE AUTUMN OF 1948, Robert returned to Europe, which he had
last visited nineteen years earlier. He was then a promising young
physicist from whom great work was expected. He returned as surely
the best-known physicist of his generation, the founder of the most prominent school of theoretical physics in America—and the "father of the
atomic bomb." His itinerary took him to Paris, Copenhagen, London and
Brussels, in all of which he gave talks or participated in physics conferences. As a young man, he had come of age intellectually studying in Göttingen, Zurich and Leiden, and he had eagerly anticipated the trip. But by
the end of September, he was writing his brother that he was somehow disappointed at what he had found. "The *Europa reise* is," he told Frank, "as it
was in the old days, a certain time for inventory. . . . In physics the conferences have been good, yet everywhere—Copenhagen, England, Paris, even
here [Brussels], there is the phrase 'you see, we are somewhat out of
things. . . .' " This led Robert to conclude, almost wistfully, "Above all I
have the knowledge that it is in America largely that it will be decided what
manner of world we are to live in."

Robert then turned to the primary purpose of his letter: to urge Frank to
seek "the comfort, the strength, the advice of a good lawyer." The House of
Representatives' Committee on Un-American Activities (HUAC) had been
holding hearings that summer, and Robert was worried for his brother—and
perhaps himself. "It has been hard," he wrote Frank, "since we left to follow
in detail what all is up with the [J. Parnell] Thomas Committee. . . . Even
the Hiss story seemed to me a menacing portent."

That August, a *Time* magazine editor and former communist named Whittaker Chambers had testified before HUAC that Alger Hiss, a New Deal lawyer and former high-ranking State Department official, had been a member of a secret communist cell in Washington. Chambers' accusations against Hiss quickly became the centerpiece of the Republican case that Roosevelt's New Dealers had allowed communists to worm their way into the heart of the American foreign policy establishment. Hiss sued Chambers for libel in September 1948—but by the end of the year Hiss was indicted for perjury.

Oppenheimer was right to think the Hiss case a "menacing portent." If someone of Hiss' stature could be brought down by HUAC, he feared what the Committee could do to his brother, whose Communist Party affiliation was well known. Robert knew that back in March 1947, the *Washington Times-Herald* had run a story charging that Frank had been a Party member. Frank had foolishly denied the truth of the story. Without being explicit, Robert observed that Frank had "thought about it a lot these last years. . . ." It was in this context that he gently suggested that Frank get a lawyer, and not just a good lawyer. He needed someone who knew "his way around Washington, the Congress . . . and above all the press. Why don't you consider Herb Marks, who may have all these qualifications?" Robert hoped that his brother would not be caught up in one of HUAC's witch-hunts; but clearly, Frank had to be prepared.

Now thirty-six years old, Frank Oppenheimer was standing on the brink of a rewarding career. First at the University of Rochester and now at the University of Minnesota, he was doing innovative experimental work in particle physics. By 1949, he had a reputation among his fellow physicists as one of the country's foremost experimentalists, studying high-energy particles (cosmic rays) at high altitudes. Early that year, he had shipped out to the Caribbean aboard a Navy aircraft carrier, the USS *Saipan,* from which he and his team had launched a series of helium balloons carrying a specially designed capsule containing a cloud chamber with stacks of nuclear-emulsion photo plates. Designed to climb to extremely high altitudes, the balloon-borne photo plates recorded the tracks of heavy nuclei; this data suggested that the origin of cosmic rays could be traced to exploding stars. The metal capsules had to be recovered after descending, and Frank found himself trekking through the jungles of Cuba's Sierra Maestra in search of one such capsule—which he triumphantly found perched atop a mahogany tree. But when another disappeared into the sea, Frank wrote melodramatically that his spirit was "completely broken." In fact, he loved these adventures and reveled in his work. If he had followed in Robert's footsteps through 1945, Frank was now set on an independent course as a cutting-edge experimentalist.

Worried as he was about Frank, Robert appears to have believed that his fame would neutralize his own left-wing past. In November 1948, he appeared on the cover of *Time* magazine, accompanied by a flattering profile of his life and career. *Time*'s editors told millions of Americans that Oppenheimer, a founding father of the atomic age, was an "authentic contemporary hero." When interviewed by *Time*'s reporters, he did not try to hide his radical background. He unabashedly explained that until 1936 he had been "certainly one of the most unpolitical people in the world. . . ." But then he confessed that the sight of young unemployed physicists "cracking up" and the news that his own relatives in Germany were having to flee the Nazi regime, had opened his eyes. "I woke up to a recognition that politics was a part of life. I became a real left-winger, joined the Teachers' Union, had lots of Communist friends. It was what most people do in college or late high school. The Thomas Committee [HUAC] doesn't like this, but I'm not ashamed of it; I'm more ashamed of the lateness. Most of what I believed then now seems complete nonsense, but it was an essential part of becoming a whole man. If it hadn't been for this late but indispensable education, I couldn't have done the job at Los Alamos at all."

Soon after the *Time* story was published, Oppie's good friend and sometime lawyer, Herb Marks, wrote to congratulate him on what he thought had been a "quite good" article. In what was probably a reference to Oppie's quoted remarks about his left-wing past, Marks commented, "That one 'pre-trial' touch was superb." Robert replied, "The only thing I liked was the one deliberate point you picked out, where I saw an opportunity, long solicited, but not before available." Herb's wife, Anne Wilson (Oppie's former secretary), was worried that the *Time* publicity would attract critics. Oppenheimer himself wasn't quite sure what to make of it. "I suffered from it," he wrote Herb, "in the most acute way in the first week or so, but came out of that thinking wryly that it was probably good for me."

OPPENHEIMER MAY have hoped to inoculate himself against congressional investigators, but in the spring of 1949 HUAC launched a major investigation of atomic spying at Berkeley's Rad Lab. Not only Frank but Robert himself was a potential target. Four of Oppenheimer's former students—David Bohm, Rossi Lomanitz, Max Friedman and Joseph Weinberg—were served with subpoenas requiring them to testify. HUAC's investigators knew that Weinberg had been overheard on a wiretap talking to Steve Nelson in 1943 about the atomic bomb. But while this evidence appeared to implicate Weinberg in atomic spying, HUAC's counsel knew that a warrantless wiretap would not stand up in court. On April 26, 1949, HUAC brought Weinberg face-to-face with Steve Nelson. He flatly denied having ever met

Nelson. HUAC's lawyers knew Weinberg had perjured himself—but proving it was going to be difficult. They hoped to build their case with testimony from Bohm, Friedman and Lomanitz.

Bohm was not sure whether he should testify, and if so, whether he should be willing to testify about his friends. Einstein urged him to refuse to testify, even though he might have to go to jail. "You may have to sit for a while," the scientist told him. Bohm didn't want to take the Fifth Amendment; he reasoned that being a member of the Communist Party was not illegal, and therefore there was nothing he could incriminate himself about. His instinct was to agree to testify about his own political activities but refuse to testify about others. Aware that Lomanitz had received a similar subpoena, Bohm contacted his old friend, who was teaching in Nashville at the time. Lomanitz had had a rough time since the war; each time he found a decent job, the FBI would inform his employer that Lomanitz was a communist and he would be fired. His future seemed particularly bleak, but he found the wherewithal to visit Bohm in Princeton.

Soon after his arrival, the two old friends were walking on Nassau Street when Oppenheimer emerged from a barbershop. Robert hadn't seen Lomanitz for years, but they had kept in touch. In the autumn of 1945, he had written Lomanitz. "Dear Rossi: I was glad to get your long, but very melancholy letter. When you are back in the States and free to do so please come and see me. . . . It is a hard time, and especially hard for you, but hold on—it won't last forever. With all warm good wishes, Opje." Now, after exchanging pleasantries with Oppie, Bohm and Lomanitz explained their predicament. According to Lomanitz, Oppenheimer became agitated and suddenly exclaimed, "Oh, my God, all is lost. There is an FBI man on the Un-American Activities Committee." Lomanitz thought this "paranoiac."

But Oppenheimer had every reason to worry. He, too, had been served a subpoena to testify before HUAC, and he happened to know that one member of the Committee, Illinois congressman Harold Velde, was indeed a former FBI agent, and had worked in Berkeley during the war years investigating the Rad Lab.

Oppenheimer later characterized this encounter with his former students as a brief two-minute conversation. He said he had merely advised them to "tell the truth," and they had responded, "We won't lie." In the event, Bohm testified before HUAC in May and again in June 1949. On advice of his counsel, the legendary civil liberties lawyer Clifford Durr, he refused to cooperate, citing both the First and Fifth amendments. For the time being, Princeton University, where he was then teaching, issued a statement supporting Bohm.

On June 7, 1949, it was Oppenheimer's turn to appear before a closed-door, executive session of HUAC. Six congressmen were there to interrogate him, including Representative Richard M. Nixon (R-Cal.). Oppenheimer ostensibly appeared before the Committee in his role as chairman of the General Advisory Committee of the AEC. But these hard-boiled congressmen were not there to question him on nuclear weapons policy; they wanted to know about atomic spies. Apprehensive, he nonetheless wished not to appear defensive, so he decided not to appear with a personal lawyer. Instead, he brought along Joseph Volpe, and made a point of introducing him as the AEC's general counsel. Over the next two hours, Oppenheimer was cooperative and forthcoming.

HUAC's counsel first proclaimed that the Committee was not seeking to embarrass him. But the very first question was: "You were acquainted with the fact, were you not, that a communist cell existed among certain scientists at the Radiation Laboratory?" Oppenheimer denied any such knowledge. He was then asked to talk about the political activities and views of his former students. He denied knowing before the war that Weinberg was a communist. "He was in Berkeley after the war," Oppenheimer said, "and his expressed views then were certainly not communist-line views."

HUAC's counsel then asked Oppenheimer about another of his former students, Dr. Bernard Peters. His response reflects his continuing naïveté. He appears to have assumed that because he was testifying in executive session, his comments would not become public. Was it true, asked the HUAC counsel, that Oppenheimer had told Manhattan Project security officers that Peters was "a dangerous man and quite red"? Oppenheimer admitted that he had said as much to Capt. Peer de Silva, his security officer at Los Alamos. Asked to elaborate, Oppenheimer explained that Peters had been a member of the German Communist Party and that he had fought in street battles against the Nazis. Subsequently he had been sent to a concentration camp and then miraculously escaped using "guile." He also volunteered that when Peters arrived in California, he "violently denounced" the Communist Party as "not sufficiently dedicated to the overthrow of the [U.S.] Government by force and violence." When asked how he knew Peters had been a member of the German Communist Party, Oppenheimer replied, "Among other things, he told me."

Oppenheimer seems to have been troubled about Peters. In May, just a month earlier, while he was attending a conference of the American Physical Society, his old friend Samuel Goudsmit had asked him about Peters. In his capacity as an AEC consultant, Goudsmit occasionally reviewed security cases. Peters had recently asked Goudsmit why he seemed to be in trouble—so Goudsmit had looked at his security file and read Oppenheimer's 1943 statement to De Silva in which he had said that Peters was "dangerous." When Goudsmit asked Oppie whether he still had the same opinion of

Peters, Oppenheimer surprised him by answering, "Just look at him. Can't you tell he can't be trusted?"

Oppenheimer was asked about other friends as well. When queried as to whether his old friend Haakon Chevalier was a communist, he replied that "he was the prize example of a parlor pink," but that he had no knowledge of whether or not he was a Party member. Regarding the "Chevalier affair," Oppenheimer repeated the same story he had told the FBI in 1946—that a confused and embarrassed Chevalier had told him about Eltenton's notion of "communicating information to the Soviet Government," and that in response he (Oppenheimer) had loudly and "in violent terms told him not to be confused and to have no connection with it." Chevalier had no knowledge, Oppenheimer said, of the atomic bomb until it exploded over Hiroshima. The Committee did not specifically ask about an approach to three other scientists—the version of the story he had told Pash in 1943—but he denied that any other individual had approached him for atomic information.

Regarding another of his former students, Oppenheimer briefly confirmed that Rossi Lomanitz had been dismissed from the Rad Lab and inducted into the Army owing to an "unbelievable indiscretion." He also acknowledged that Joe Weinberg was a friend of Lomanitz's and that another physics student, Dr. Irving David Fox, had been active in organizing a union inside the Rad Lab. When asked about Kenneth May, he confirmed that May was "an avowed Communist."

Oppenheimer was trying hard to please. Where he could, he was naming names. But when asked about his brother's past Party membership, Robert replied, "Mr. Chairman, I will answer the questions you put to me. I ask you not to press these questions about my brother. If they are important to you, you can ask him. I will answer, if asked, but I beg you not to ask me these questions."

In a mark of extraordinary deference, HUAC's counsel withdrew the question. Before adjourning, Congressman Nixon said that he was "tremendously impressed" with Oppenheimer and "mighty happy to have him in the position he has in our program." Joe Volpe was astonished at Oppenheimer's cool performance: "Robert seemed to have made up his mind to charm these Congressmen out of their seats." Afterwards, all six HUAC legislators came down to shake the hand of the famous scientist. Perhaps it is not surprising that Robert continued to believe that his notoriety was a protective shield.

OPPENHEIMER EMERGED unscathed from the hearings, but his former students were not so fortunate. The day after Oppenheimer's testimony, Bernard Peters spent an almost perfunctory twenty minutes before the

Committee. Peters denied that he had been a CP member in Germany or in the United States—and he denied that his wife, Dr. Hannah Peters, had ever been a Party member, or that he knew Steve Nelson.

Peters left wondering what Oppenheimer had told the Committee the previous day, so on his way back to Rochester, he stopped off in Princeton to see his mentor. Oppie quipped that "God guided their questions so that I did not say anything derogatory." One week later, however, Oppenheimer's closed-session testimony was leaked to the *Rochester Times-Union.* The headline blared: "Dr. Oppenheimer Once Termed Peters 'Quite Red.' " Peters' colleagues at the University of Rochester read that their colleague had escaped from Dachau by "guile" and had once criticized the American Communist Party as insufficiently dedicated to armed revolution.

Peters knew immediately that his job was at risk. Only the previous year, similar HUAC testimony had leaked and when the *Rochester Times-Union* published a story headlined "U of R Scientist May Face Spy Probers," Peters had sued the paper for libel. He won an out-of-court settlement of $1. With this history, Peters understood what was at stake if the allegations were resurrected. Peters promptly denied Oppenheimer's allegations, telling the *Rochester Times-Union,* "I have never told Dr. Oppenheimer or anybody that I had been a member of the Communist Party because I have not; but I did say that I greatly admired the spirited fight they put up against the Nazis . . . and also that I admired the heroes who died in the concentration camp at Dachau." Peters admitted that his political views, even today, were "not orthodox," citing his strong opposition to racial discrimination and his belief in the "desirability of socialism." But he was not a communist.

That same day, Peters wrote Oppenheimer a letter, enclosing the newspaper clip, and asked if he had indeed said these things before HUAC. "You are right that I advocated 'direct action' against fascist dictatorships. But do you know of any instances where I advocated such action in a nation where the majority of people were supporting a government of their own free choice?" He also asked, "Where did you get the dramatic story of the street battles I was in? I wish I had." Peters was outraged enough to ask his lawyer if he had sufficient cause "to sue Robert for libel."

Five days later, on June 20, Oppenheimer phoned Peters' lawyer, Sol Linowitz, and passed a message to Hannah Peters: He wanted Bernard to know that he was "very much disturbed" by the newspaper story and insisted that it misrepresented what he had said to the Committee. Robert said he was most anxious to talk with Bernard.

In short order, Oppenheimer heard from his brother Frank, Hans Bethe and Victor Weisskopf, all of them expressing pained astonishment that Oppie would attack a friend that way. Both Weisskopf and Bethe wrote that they could not understand how he could have said such things about Peters, as Weisskopf put it, and they urged him to "set this record straight and do what is in your power to prevent Peters' dismissal. . . ." Bethe wrote him that "I remember you spoke in the most friendly terms to me about the Peterses, and they certainly have considered you their friend. How could you represent his escape from Dachau as evidence for his inclination to 'direct action' rather than a measure of self-defence against mortal peril?"

Edward Condon, Oppie's friend from Göttingen days and briefly his deputy director at Los Alamos, was angry and "shocked beyond description." Now director of the U.S. Bureau of Standards, Condon was himself the occasional target of right-wing attacks on Capitol Hill. On June 23, 1949, he wrote his wife, Emilie: "I am convinced that Robert Oppenheimer is losing his mind. . . . [I]f Oppie is really becoming unbalanced, it can have very complicated consequences considering his positions, including that of originator of the Acheson-Lilienthal report on international control of atomic energy. . . . [I]f he cracks up it will certainly be a great tragedy. I only hope that he does not drag down too many others with him. Peters says the testimony of Oppie about him is full of out-and-out lies on matters where Oppie should know the truth."

Condon told his wife that he had heard from people in Princeton that "Oppie has been in a very high state of tension in the last few weeks . . . he seems to be in a great state of strain for fear he himself will be attacked. Of course he knows that he has so much of a record of leftist activities as is involved in what is brought out against the others from Berkeley. . . . It appears that he is trying to buy personal immunity from attack by turning informer. . . ."

The disheartened Condon then wrote Oppie a scathing letter: "I have lost a good deal of sleep trying to figure out how you could have talked this way about a man whom you have known so long, and of whom you know so well what a good physicist and good citizen he is. One is tempted to feel that you are so foolish as to think you can buy immunity for yourself by turning informer. I hope that this is not true. You know very well that once these people decide to go into your own dossier and make it public that it will make the 'revelations' that have been made so far look pretty tame."

Some days later, Frank Oppenheimer took Peters to see his brother, who was visiting Berkeley. Peters later described the meeting in a letter to Weisskopf: "My talk with Robert was dismal. At first he refused to tell me whether the newspaper report was true or false." When Peters insisted on the truth, Oppie confirmed the newspaper account of his testimony. "He said it was a terrible mistake," Peters wrote. Oppie tried to explain that he had not been prepared to answer these questions, and only now, seeing his words in print, did he realize that what he had said was so damaging. When Peters asked why he had misled him in their meeting in Princeton, Oppenheimer "got very red" and said he had no explanation. Peters insisted that Oppie had misunderstood him. While Peters confirmed that he had indeed attended open-air communist rallies in Germany, he swore that he had never actually joined the Party.

Oppie agreed to write a letter to the editor of the Rochester newspaper correcting his HUAC testimony. In the letter, published on July 6, 1949, Oppenheimer explained that Dr. Peters had recently given him "an eloquent denial" that he had ever been a member of the Communist Party or had advocated the violent overthrow of the U.S. government. "I believe this statement," Oppenheimer said. He went on to make a spirited defense of freedom of speech. "Political opinion, no matter how radical, or how freely expressed, does not disqualify a scientist for a high career in science. . . ."

Peters considered the letter "a not very successful piece of double-talk." Nonetheless, it managed to salvage his job at the University of Rochester. He soon realized, however, that without access to classified research and government research projects, his career in America was at a dead end. Late in 1949, the State Department refused to issue him a passport when he expressed the intention of going to India. The following year, the State Department relented and Peters accepted a teaching position at Bombay's Tata Institute of Fundamental Research. But in 1955, after the State Department refused to reissue his passport, Peters finally took German citizenship. In 1959, he and Hannah moved to Niels Bohr's institute in Copenhagen, where he spent the rest of his career.

Peters had it easy compared to Bohm and Lomanitz. More than a year later, they were both indicted for contempt of Congress; after Bohm was arrested on December 4, 1950 (and released on $1,500 bail), Princeton suspended him from all his teaching duties and even barred him from setting foot on the campus. Six months later, he was tried and acquitted. Even so, Princeton decided not to renew Bohm's teaching contract when it expired that June.

Lomanitz' fate was even worse. After his HUAC testimony, he was fired by Fisk University; he then spent two years working as a day laborer, tarring roofs, loading burlap bags and trimming trees. In June 1951 he was tried for contempt of Congress. Even after his acquittal, the only job he could find was repairing railroad tracks for $1.35 an hour. He didn't get another teaching job until 1959. Remarkably, Lomanitz never seemed to harbor resentment toward Oppenheimer. He didn't blame him for what the FBI and the political culture of the times had done to him. And yet, there was a lingering disappointment. Lomanitz had once thought of Oppenheimer as "almost a god." He didn't think Oppenheimer had been "malicious." But years later he would say that he had come to feel "sad personally about the man's weaknesses. . . ."

While there was little Oppenheimer could have done to protect his former students, he sometimes behaved as if he was truly frightened of any association with them. Their company represented a link to his political past and therefore a threat to his political future. He was clearly scared. After Bohm lost his job with Princeton, Einstein suggested that he be brought to the Institute for Advanced Study to work as his assistant. The great man was still interested in revising quantum theory, and he was heard to say that "if anyone can do it, then it will be Bohm." But Oppenheimer vetoed the idea; Bohm would be a political liability to the Institute. By one account, he also reportedly instructed Eleanor Leary to keep Bohm away. Leary was subsequently heard telling the Institute's staff, "David Bohm is not to see Dr. Oppenheimer. He is not to see him."

As a matter of expediency, Oppenheimer had every reason to distance himself from Bohm. On the other hand, when Bohm heard of a teaching opportunity in Brazil, Oppenheimer wrote him a strong letter of recommendation. Bohm spent the rest of his career abroad, first in Brazil, then in Israel and finally in England. He had once deeply admired Oppenheimer, and though over the years those feelings had turned to ambivalence, he never held Oppie responsible for his banishment from America. "I think he acted fairly to me as far as he was able to," Bohm said.

Bohm knew Oppenheimer was under a great deal of strain. Shortly after the news broke about his HUAC testimony against Peters, Bohm had a candid conversation with Oppie. He asked why he had said such things about their friend. "He told me," Bohm recalled, "that his nerve just gave way at that moment. That somehow the thing was too much for him. . . . I can't remember his words, but that's what he meant. He has this tendency when things get too much, he sometimes does irrational things. He said he couldn't understand why he did it." Of course, it had happened before—in his interview with Pash in 1943 and his meeting with Truman in 1945—and it would happen again during his security hearing in 1954. But, as Bernard Peters observed to Weisskopf, "He [Oppenheimer] was obviously scared to tears of the hearings, but this is hardly an explanation. . . . I found it a rather sad experience to see a man whom I regarded very highly in such a state of moral despair."

JUST SIX DAYS after his HUAC testimony in early June 1949, Oppenheimer returned to Capitol Hill to testify under klieg lights before an open session of the Joint Committee on Atomic Energy. The issue at hand was exports of radioisotopes for purposes of research in foreign laboratories. In a contentious four-to-one decision, the AEC commissioners had approved the exports. The lone dissenting commissioner, Lewis Strauss, was convinced that such exports were dangerous because, he believed, radioisotopes could be diverted for use in military applications of atomic energy. Shortly before, in an effort to reverse the AEC decision, Strauss had publicly testified against the exports in a hearing before the Joint Committee.

So when Oppenheimer entered the Caucus Room of the Senate Office Building, he was aware of Strauss' concerns. But he did not share them, and he now made it clear that he thought these concerns foolish. "No one can force me to say," he testified, "that you cannot use these isotopes for atomic energy. You can use a shovel for atomic energy; in fact, you do. You can use a bottle of beer for atomic energy. In fact, you do." At this, the audience murmured with laughter. A young reporter, Philip Stern, happened to be sitting in the hearing room that day. Stern had no idea who was the target of this sarcasm, but "it was clear that Oppenheimer was making a fool of someone."

Joe Volpe knew exactly who was being made a fool. Sitting next to Oppenheimer at the witness table, he glanced back at Lewis Strauss and was not surprised to see the AEC commissioner's face turning an angry beet-red. More laughter greeted Oppenheimer's next statement: "My own rating of the importance of isotopes in this broad sense is that they are far less important than electronic devices, but far more important than, let us say, vitamins, somewhere in between."

Afterwards, Oppenheimer casually asked Volpe, "Well, Joe, how did I do?" The lawyer replied uneasily, "Too well, Robert. Much too well." Oppenheimer may not have set out to humiliate Strauss over what he regarded as a minor policy disagreement. But for Oppie, condescension came easily—too easily, many friends insisted; it was part of his classroom repertoire. "Robert could make grown men feel like schoolchildren," said one friend. "He could make giants feel like cockroaches." But Strauss was not a student; he was a powerful, thin-skinned, vengeful man easily humiliated. He left the hearing room that day very angry. "I remember clearly," said Gordon Dean, another AEC commissioner, "the terrible look on Lewis' face." Years later, David Lilienthal vividly recalled, "There was a look of hatred there that you don't see very often in a man's face."

Oppenheimer's relationship with Strauss had been in steady decline since early 1948, when Oppie had made it clear that he would resist Strauss' attempts to meddle in his directorship of the Institute for Advanced Study. Prior to this hearing, they had weathered several other

AEC-related disagreements. But now Oppenheimer had made for himself a dangerous enemy who was powerful and influential in every field of Robert's professional life.

After their clashing testimonies before the Joint Committee, one of the Institute's trustees, Dr. John F. Fulton, said that he expected Strauss to resign from the Institute Board. "I don't think Robert Oppenheimer will ever feel comfortable as Director of the Institute for Advanced Study," Fulton wrote another trustee, "as long as Mr. Strauss continues on our Board of Trustees." But Strauss had allies who had recently engineered his election as president of the Institute's Board of Trustees, and he now made it clear he had no intention of resigning just because he had had the "effrontery . . . to differ with Dr. Oppenheimer on a scientific matter." Strauss was angry, and he would stay angry until he had settled the score.

THE VERY NEXT DAY, June 14, 1949, Frank Oppenheimer appeared as a witness before HUAC. Two years earlier, he had denied to a newspaper reporter that he had ever been a member of the Communist Party. He had not planned to lie about his Party membership, but a reporter for the *Washington Times-Herald* had called him late one night and explained that his paper was running a story the next morning. After reading him the article over the phone, the reporter asked for his immediate comment. "The story was full of all other kinds of allegations that were false," Frank said. "The pre-war party membership was the only true thing in it. They asked me for a statement and I simply said the whole thing was false—which was stupid of me to do. I should have just said nothing." When the story was published, authorities at the University of Minnesota pressured Frank to give them the same denial in writing. Fearing for his job, Frank had a lawyer draft a statement swearing that he had never been a member of the Communist Party.

But now, after talking it over with Jackie, Frank decided he had to tell the truth. That morning he testified under oath that he and Jackie had been members of the Communist Party for some three and a half years—from early 1937 until late 1940 or early 1941. He acknowledged that during these years his Party alias had been "Frank Folsom." On advice of his counsel, Clifford Durr, he refused to testify about the political views of others. "I cannot talk about my friends," he said. Again and again, HUAC's counsel and various congressmen pressed Frank to name names. When Congressman Velde—the ex–FBI agent—repeatedly asked him to restate his reasons for refusing to answer their questions, Frank said he would not talk about the political affiliations of his friends "because the people whom I have known throughout my life have been decent-thinking and well-meaning people. I know of no instance where they have thought, discussed or said anything which was inimical to the purposes of the Constitution or the laws of the United States." In stark contrast to his brother, Frank stood his ground; he would not name names.

He and Jackie found the whole experience surreal. Jackie had not lost her righteous anger. As she sat in the House Committee anteroom waiting to testify, she looked out the window and was startled by the contrast between Capitol Hill's marble government buildings, surrounded by manicured grounds, and the rows of tumbledown houses occupied by the city's Negro population. The children were barefoot and dressed in rags. "They all looked rachitic and most seemed undernourished. All they had to play with was junk they found in the street. As I sat there reading and listening and looking out the window, I found myself alternately worrying what the Committee was going to try to do to me and getting madder and madder at the fact that I had been called down here so that some fellow could question *me* about being Un American."

Afterwards Frank told reporters that they had joined the Party in 1937 "seeking an answer to the problems of unemployment and want in the wealthiest and most productive country in the world."

But they had left the Party in 1940, disillusioned. He had no knowledge, he said, of atomic espionage, either in Los Alamos or in Berkeley's Rad Lab: "I knew of no Communist activity, nobody ever approached me to get information and I gave none, and I worked very hard and I believe I made a valuable contribution." Barely an hour later, Frank learned from reporters that his resignation as an assistant professor of physics had been accepted by the University of Minnesota. He had lied two years earlier, and from the perspective of the university that was reason enough for his dismissal from academic life. He had literally been three months away from being awarded tenure, but in a final meeting with the president of the university, it was made clear that he was finished. Frank left the president's office in tears.

Frank was devastated. The full import of what had happened only hit him when he tried to go back to Berkeley. Naïvely, he had thought Lawrence would provide him haven, and he was shocked when Ernest turned him down.

Dear Lawrence,
What is going on? Thirty months ago you put your arms around me and wished me well. Told me to come back and work whenever I wanted to. Now you say I am no longer welcome. Who has changed, you or I? Have I betrayed my country or your lab? Of course not. I have done nothing. . . . You do not agree with my politics, but you

never have . . . so I think that you must be losing your head to the point where anybody who disagrees with you about anything is not to be tolerated. . . . I am really amazed and sore because of your action.

Sincerely,
Frank

A year earlier, Frank and Jackie had bought an 800-acre cattle ranch near Pagosa Springs, high in the Colorado mountains. They had planned to use it as a summer vacation home. In the autumn of 1949, to the surprise of many of their friends, they retreated to this spartan internal exile. "No one has offered me a job," Frank wrote Bernard Peters, "and so we are definitely planning to spend the winter here. My Christ, but it is beautiful. I think only if you have been here does staying seem to make any sense." The ranch was perched at an altitude of 8,000 feet, and the winters were unbearably cold. "Jackie would sit in the cabin," recalled Philip Morrison, "with binoculars and watch cows ready to give birth in the snow. They'd have to run out to keep the newborn calves from freezing."

For the next decade, Robert Oppenheimer's likable and brilliant younger brother eked out a living as a working rancher. They were twenty miles from the nearest town. As if to remind them of their status, FBI agents periodically showed up to question their neighbors. Occasionally they'd visit the Oppenheimer ranch and ask Frank to talk about other people in the CP. Once an agent specifically told him, "Don't you want to get a job in a university? If you do, you have to cooperate with us." Frank always turned them away. In 1950, Frank wrote: "Finally, after all these years, I have gotten wise to the fact that the FBI isn't trying to investigate me, it is trying to poison the atmosphere in which I live. It is trying to punish me for being left wing by turning my friends, my neighbors, my colleagues against me and make them suspicious of me."

Robert visited the ranch almost every summer. And while Frank had resigned himself to his situation, Robert chafed at the thought that his brother was living this kind of life. "I really felt like a rancher," Frank said, "and was a rancher. But he didn't believe I could be a rancher and was very anxious for me to get back into the academic world, although there wasn't anything he could do about it." Over the next year, Frank received tentative job offers to teach physics abroad in Brazil, Mexico, India and England—but the Department of State steadfastly refused to issue him a passport. And there were no job offers in America; he had been blacklisted. Within a few years, Frank felt compelled to sell one of his Van Goghs—*First Steps (After Millet)*—for $40,000.

In his frustration over his brother's fate, Robert talked with Supreme

Court Justice Felix Frankfurter, the Harvard overseer Grenville Clark and other legal scholars about what the Institute might be able to do by way of organizing an intellectual critique of the Truman Administration's loyalty and security programs that were supporting the sort of treatment Frank and Oppie's students were getting. He told Clark that he thought the Presidential Loyalty order, the AEC's security clearance procedures and HUAC's investigations "all lead in many individual cases to unwarranted hardship and make for an abrogation of the freedoms of inquiry, opinion and speech." Soon afterwards, Oppenheimer recruited his old friend Dr. Max Radin, dean of Berkeley's Law School, to come to the Institute for the academic year 1949–50 and write an essay on California's loyalty oath controversy.

THROUGHOUT THESE YEARS, Oppenheimer was convinced that his phones were wiretapped. One day in 1948, a Los Alamos colleague, the physicist Ralph Lapp, came to Oppie's Princeton office to discuss his (Lapp's) educational work on arms control issues. Lapp was startled when Oppenheimer suddenly rose and took him outside, muttering as they went, "Even the walls have ears." He was aware that he was under scrutiny. "He was always conscious of being followed," recalled Dr. Louis Hempelmann, his physician friend from Los Alamos and now a frequent visitor to Olden Manor. "He gave us the sense that he thought people were actually trailing him."

His phones had been monitored at Los Alamos, and his Berkeley home was wiretapped by the FBI throughout 1946–47. When he moved to Princeton, the FBI's Newark, New Jersey, field office was instructed to monitor his activities—but a decision was made that electronic surveillance was not warranted. Every effort would be made, however, "to develop confidential discreet sources close to Oppenheimer." By 1949, the bureau had recruited at least one confidential informant, a woman acquainted with Oppenheimer socially and through her university job. In the spring of 1949, the Newark office informed J. Edgar Hoover, "No additional information has been obtained or developed concerning Dr. Oppenheimer that would indicate he is disloyal." Years later, Oppenheimer claimed wryly that, "The government paid far more to tap my telephone than they ever paid me at Los Alamos."

"I Am Sure That Is Why She Threw Things at Him"

His family relationships seemed to be so terrible. And yet you never would have known it from Robert.

PRISCILLA DUFFIELD

WHILE FRANK AND JACKIE STRUGGLED to turn their Colorado spread into a working cattle ranch, Robert presided over his intellectual fiefdom in Princeton. The directorship did not absorb all his energy. He spent about a third of his time on Institute business, a third on physics or other intellectual pursuits, and a third traveling, giving speeches and attending classified meetings in Washington. One day his old friend Harold Cherniss chided him, "The time has come, Robert, for you to give up the political life and return to physics." When Robert stood silent, seeming to weigh this advice, Cherniss pressed him: "Are you like the man who has a tiger by the tail?" To this Robert finally replied, "Yes."

It was sometimes a relief to be on the road, away from Princeton—and his wife. To readers of *Life, Time* and other popular magazines, Robert's family life may have seemed idyllic. Photographs depicted a pipe-smoking father reading a book to his two young children as his pretty wife looked over his shoulder and the family's German shepherd, Buddy, lay at his feet. "He is warmly affectionate," wrote a reporter for a cover story on Oppenheimer for *Life* magazine, "with his wife and children (who are well fed and very fond of him), and attentively polite to everybody. . . ." According to *Life*, Oppenheimer walked home each evening at 6:30 p.m. to play with the children. Each Sunday, they took Peter and Toni out to hunt for four-leaf clovers. "Mrs. Oppenheimer, whose thinking is also direct, keeps her children from cluttering the house with four-leaf clovers by making them eat all they find right on the spot."

But those who knew the Oppenheimers well realized that life at Olden

Manor was difficult. "His family relationships seemed to be so terrible," said his former Los Alamos secretary Priscilla Duffield, who became a Princeton neighbor. "And yet you never would have known it from Robert."

Oppenheimer's home life was painfully complicated. Robert relied on Kitty for a great deal in his life. "She was Robert's greatest confidante and adviser," Verna Hobson said. "He told her everything. . . . He leaned on her tremendously." He took his Institute work home with him and she often became involved in his decisions. "She loved him very much and he loved her very much," Hobson insisted. But she and other close friends in Princeton knew Kitty had a relentless intensity that drained anyone near her: "What a strange person she was; all that fury and soreness and intelligence and wit. She had a constant state of the hives. She was just tensed up all the time."

Hobson got to know both Robert and Kitty as few others ever did. She and her husband, Wilder Hobson, met the Oppenheimers in 1952 at a New Year's Eve dinner hosted by their mutual friend, the novelist John O'Hara. Soon afterwards, Hobson went to work for Robert—and she stayed with him for the next thirteen years. "He was an extraordinarily demanding person to work for and Kitty demanded just as much from his secretaries, so it was like working for two demanding bosses who took you right into their lives and expected you to be at their home half the time."

Kitty, a creature of habit, presided every Monday afternoon over a gathering of women at Olden Manor; they would sit around gossiping, some drinking all afternoon. Kitty called it her "Club." The wife of a Princeton University physicist labeled these women Kitty's "crew of birds with broken wings. . . . Kitty had a ring of damaged women around her, all of them somewhat alcoholic." Kitty had drunk her fair share of martinis at Los Alamos. But now her drinking sometimes led to horrendous scenes. Hobson, who drank only in moderation, recalled, "She would get drunk sometimes to the point of falling down and not making much sense. Sometimes she passed out. But so many times I have seen her pull herself together when you didn't believe she possibly could."

Pat Sherr, Kitty's friend from Los Alamos—and the woman who had taken care of Toni as an infant for three months—was one of her regular drinking companions. The Sherrs had moved to Princeton in 1946, and soon after the Oppenheimers moved into Olden Manor, Kitty made a habit of dropping by Pat's home two or three times a week. Kitty was clearly lonely. "She would arrive at eleven in the morning," recalled Sherr, "and wouldn't leave until four in the afternoon," after having consumed a lot of Sherr's scotch. But one day Pat announced she just couldn't afford to replace the liquor. "Oh, how stupid of me," Kitty said. "I'll bring my own bottle and you'll just keep it aside for me."

Kitty's friendships were at once intense and ephemeral. She would latch onto someone and bare her soul in a torrent of intimacy. Sherr saw her do this repeatedly. She'd tell her new friend absolutely everything about herself—including her sex life. "I mean, she just had to talk about this sort of thing all the time," recalled Sherr. She could be a good friend, but she was always conscious of being a good friend. And inevitably, at some point, she would turn on her friend and publicly denigrate her. "Kitty had a certain need to hurt people," Hobson said.

Kitty had always been accident-prone, and her drinking contributed to a string of such episodes. In Princeton, she regularly had minor auto accidents. Almost every night she fell asleep in bed smoking. Her bedding was full of cigarette holes. One night she awoke startled—the room was on fire; but she put it out with a fire extinguisher that she or Robert had wisely placed in the bedroom. Oddly enough, Robert rarely intervened. He instead reacted to his wife's self-destructive behavior with stoic resignation. "He knew of Kitty's traits," observed Frank Oppenheimer, "but was unwilling to admit them—again perhaps because he couldn't admit failure."

On one occasion, Abraham Pais was talking with Oppenheimer in his office when the two men saw Kitty walking, clearly tipsy, across the lawn from Olden Manor. As she approached the door to his office, Robert turned to Pais and said, "Don't go away." It was moments like these, Pais later wrote, "when I hurt for him." In his pity for Robert, Pais nevertheless could not understand why his friend tolerated such a woman. "Quite independently from her drinking," Pais wrote, "I have found Kitty the most despicable female I have ever known, because of her cruelty."

Hobson saw past Kitty's failings and she understood why Robert loved her. He accepted her for who she was and knew she would never really change her ways. Robert once confided to Hobson that, prior to Princeton, he had consulted a psychiatrist about Kitty. In an extraordinary admission, he said he had been advised to check her into an institution, at least for a time. This he could not do. Instead, he would be Kitty's "doctor, nurse and psychiatrist." He told Hobson that he had taken this decision "with his eyes open and that he accepted the consequences of it."

Freeman Dyson had a similar observation: "Robert just liked Kitty the way she was, and he wouldn't have tried to force a different way of life on her any more than she would onto him. . . . I would say that Oppenheimer himself was certainly completely dependent on her—she was really the rock on which he stood. I think for him to have tried to treat her as a clinical case and try to reorganize her life, I think that would have been just out of character for him, and out of character for her too." Another Princeton friend, the journalist Robert Strunsky, agreed: "He was just as loyal to her as anybody could be. He really wanted to protect her as much as anything. . . . He resented any criticism of her."

Robert must have known that Kitty's drinking was a symptom of a deep pain, a pain he understood would always be there. He never tried to stop her from drinking, and neither did he sacrifice his own evening cocktail ritual. His martinis were strong and he drank them with pleasure. Unlike Kitty, he took his liquor steady and slow. Pais, who believed the cocktail hour a "barbaric custom," nevertheless thought that Robert "invariably held his liquor well." Even so, the fact that Robert continued to drink alongside his clearly alcoholic wife did not go unnoticed. "He served the most delicious and the coldest martinis," Sherr said. "Oppie made everyone drunk quite consciously." Robert himself mixed the gin martinis with just a droplet of vermouth and then poured the concoction into long-stemmed glasses he had sitting in the freezer. One faculty member renamed Olden Manor "Bourbon Manor."

Robert's passivity in the face of Kitty's drinking seemed strange to some. Whatever she did to him or to herself, he would be there for her all his life. Another old Los Alamos friend, Dr. Louis Hempelmann, admired Robert's devotion to his wife. Louis and Elinor Hempelmann visited the Oppenheimers two or three times each year and felt they knew the family well. Robert never asked him for professional advice about Kitty—but he calmly, matter-of-factly, told Hempelmann what the situation was. "He was really just a saint to her," Hempelmann recalled. "He was always sympathetic and didn't ever seem to get irritated at her. He really stuck with her very well. He was a marvelous husband."

On one occasion, however, Robert was compelled to intervene. Kitty not only drank; she often took sleeping pills to fight her insomnia. One night she accidentally took an overdose and had to be rushed to the Princeton hospital. After that, Oppenheimer asked his secretary to buy him a box with a lock on it. In the future, he said, Kitty could only get her pills by asking him for them. This arrangement lasted for a time, but over time fell by the wayside. Years later, Robert Serber insisted that Kitty "never drank excessively for a normal person." He thought Kitty's behavior could be explained by a persistent medical condition: "Kitty suffered from pancreatitis . . . and she would have to take very strong sedatives, and it gave the appearance of being drunk. I'd often seen it, staying with the Oppenheimers." Bracing herself to attend a social function, Serber said Kitty would "pull herself together at the last minute and take a Demerol to get her through the evening and then she would appear drunk. Well, it wasn't that at all."

The source of Kitty's unhappiness was no doubt rooted in her own psyche. But the pressures to play the role of the "director's wife" didn't help matters. At formal receptions, when she was required as hostess to stand and greet a long line of people, she often asked Pat Sherr to stand beside her. When Sherr asked why this was necessary, Kitty responded, "I need you at my side because when I start to fall, you're going to hold me up." Sherr realized that her friend was "very nervous and unsure of herself." Kitty could intimidate those who did not know her well. And at times she could seem perfectly animated. But it was all an act. Sherr believed that, when required to put on a performance, Kitty was "really scared out of her wits."

A free-spirited, whimsical woman, Kitty found it impossible to fit into Princeton's stiff, small-town, high-society scene. A colleague of Abraham Pais' once said of Princeton: "If you are single, you'll go crazy; or, if you are married, your wife will go crazy." Princeton drove Kitty crazy.

The Oppenheimers made no effort to accommodate Princeton society. "People left [calling] cards for them and they never returned the calls," recalled Mildred Goldberger. "They never somehow cared for that part of Princeton which in our experience was really the best part." The Goldbergers, in fact, developed a strong dislike for the Oppenheimers. Mildred literally thought Kitty a "wicked" woman, filled with "unfocused malice." Her husband, the physicist Marvin Goldberger, who later became president of Caltech, saw Robert as "an extraordinarily arrogant and difficult person to be with. He was very caustic and patronizing. . . . Kitty was just too impossible."

Kitty Oppenheimer was like a tigress caged in Princeton. If invited to the Oppenheimers' for dinner, Princetonians learned from experience not to count on anything substantial to eat; the quality of the dinner was directly related to Kitty's mood. Guests would be greeted by Robert holding a pitcher of his potent martinis. "You would sit in the kitchen," recalled Jackie Oppenheimer, "just gossiping and drinking, with not a thing to eat. Then, about ten o'clock, Kitty would throw some eggs and chili into a pan and, with all that drink, that's all you had." Neither Robert nor Kitty ever seemed hungry. One summer evening, Pais was invited over for dinner and after the usual martinis, Kitty served a bowl of vichyssoise soup. The soup was quite delicious, and Robert and Kitty "indulged in a rather extravagant exchange about its superb quality." Pais thought to himself, "Fine, now let's get on with the dinner." But no more food was forthcoming, and after a decent interval, a famished Pais politely excused himself and drove into Princeton, where he bought two hamburgers.

In her unhappiness, Kitty's marriage was everything to her. She was utterly dependent upon Robert. She tried hard to play the role of a good

housewife, "running around at his beck and call, making sure that everything was perfect for him." One evening at a party, Oppenheimer was standing in a corner of their living room, talking with a group of people, when
Kitty suddenly blurted out, "I love you." Clearly embarrassed, Oppenheimer simply nodded his head. "It was obvious," recalled Pat Sherr, "that
he wasn't terribly happy; he didn't coo over her at that point. But she would
do this kind of thing out of the blue."

Sherr had known the Oppenheimers since their years in Los Alamos,
and during their first years in Princeton she was probably Kitty's closest
friend. Kitty seems to have confided to Sherr about her marriage. "She
adored him," Sherr said. "There was no doubt about that." But in Sherr's
harsh view, Robert didn't feel the same way. "I am sure he never would have
married her had she not become pregnant. . . . I don't think that he returned
the love, and I don't think that he was capable of returning any love." By
contrast, Verna Hobson always insisted that Robert loved Kitty. "I think he
leaned on her tremendously," Hobson said. "He didn't always listen to her,
but he respected her political and intellectual capacity." Hobson tended to
observe the marriage through Robert's eyes. Both Sherr and Hobson admitted that the problem may have been one of clashing temperaments. Kitty
was extreme in her passions, whereas Robert could be surprisingly disen-
gaged. Kitty was somebody who needed to express her emotions or anger;
but Robert provided no rebound, and instead just allowed all her emotions
to be absorbed into a void. "I am sure that is why she threw things at him,"
Hobson said.

Kitty told Sherr that while she had slept with many men in her life, she
had never been unfaithful to Robert. The same, of course, was not true for
Robert. Though probably unaware of his affair with Ruth Tolman, Kitty
was nevertheless intensely jealous of Robert's affections. Another Los
Alamos friend, Jean Bacher, thought Kitty was always resentful of anyone
who got involved with Robert. Hobson reports that Robert himself confided
to her one day that part of Kitty's problem was that she "was insanely jealous of [him] and she could not stand it when he either got praise or blame
because he was in the spotlight . . . she envied him."

Kitty also confided to Sherr that "Oppie had no sense of fun and play."
According to Kitty, he was "overly fastidious." Kitty was surely right to
think him maddeningly aloof and detached. He lived his emotional life
introspectively. They were polar opposites. But that had always been the
source of their mutual attraction. If their marriage was something less than
a healthy partnership, after a decade of marriage—and two children—the
Oppenheimers had developed a bond of mutual dependency.

Soon after arriving in Princeton, Sherr was invited to Olden Manor for a picnic. After picnicking, one of the maids brought Toni, now aged three, down from her nap. Sherr hadn't seen the child—the baby that Oppie had once asked if she wanted to adopt—since she had lived with her for three months at Los Alamos. "She was a very lovely child," Sherr said. "She had Kitty's high cheekbones and very dark eyes and dark hair—but she had something of Oppie there as well." Sherr watched as Toni ran over to Oppenheimer and climbed into his lap: "She put her head on his chest and he enveloped her in his arms. And he looked at me and nodded." Teary-eyed, Sherr knew what he meant. "It was a message between us that I was right, he did love her very much."

But there seems to have been little energy left in their lives for their parental obligations. "I think to be a child of Robert and Kitty Oppenheimer," said Robert Strunsky, a Princeton neighbor, "is to have one of the greatest handicaps in the world." "On the surface," Sherr said, "he was very sweet with the children. I never saw him lose his temper." But over the years, her view of Oppenheimer changed radically. Sherr observed that Peter, aged six, was quiet and extremely shy, and to help him socialize she encouraged Kitty to take him to a child psychiatrist. But after talking to Robert about it, Kitty reported that he had no confidence in the notion of subjecting his young son to a therapist—an experience Robert himself had endured and detested. This angered Sherr, who thought Oppenheimer's attitude was that of a father who "could not have a son who needed help." She eventually concluded that she "didn't like him as a human being. . . . The more I saw of him, the more I disliked because it ended up by my feeling that he was a terrible father."

This was too harsh. Both Robert and Kitty tried to connect with their son. One day when Peter was six or seven years old, Kitty helped him build an electrical toy, a square board filled with various lights, buzzers, fuses and switches. Peter dubbed the toy his "gimmick," and two years later he still loved to play with it. One evening in 1949, David Lilienthal was visiting the Oppenheimers and observed Kitty sitting on the floor with Peter, patiently trying to fix the "gimmick." After nearly an hour, when she rose to prepare dinner in the kitchen, Robert, "looking very paternal and very loving at Peter, moved over and took his place on the floor where Kitty had previously been working with this mess of wiring." As Robert sat on the floor, a cigarette dangling from his mouth, fiddling with the wires, Peter ran to the kitchen and whispered loudly to Kitty, "Mama, is it all right to let Daddy work with the gimmick?" Everyone laughed at the notion that the man who directed the construction of the ultimate "gadget" might not be qualified to fiddle with his child's electrical toy.

Despite such moments of familial warmth, Robert was perhaps too distracted to be a very attentive father. Freeman Dyson once asked him if it wasn't a difficult thing for Peter and Toni to have such a "problematical figure for a father." Robert replied with his usual flippancy, "Oh, it's all right for them. They have no imagination." Dyson later observed of his friend that this was a man capable of "rapid and unpredictable shifts between warmth and coldness in his feelings toward those close to him." It was difficult for the children. "To an outsider like me," Pais later observed, "Oppenheimer's family life looked like hell on earth. The worst of it all was that inevitably the two children had to suffer."

Despite the "gimmick" and other indulgences, Kitty and Peter never bonded and their relationship was often quite contentious. Robert felt Kitty was the problem. "Robert thought," said Hobson, "that in their highly charged passionate falling in love that Peter had come too soon, and Kitty resented him for that." When he was about eleven years old, Peter put on some puppy fat and Kitty couldn't stop nagging him about his weight. There was never much food around the house, but now Kitty put Peter on a strict diet. Mother and son fought frequently. "She used to make Peter's life just miserable the way she went on about it," Hobson said. Sherr agreed: "Kitty was very, very impatient with him; she had absolutely no intuitive understanding of the children." Robert stood passively by, and if pressed, he invariably took Kitty's side in these arguments. "He [Robert] was very loving," recalled Dr. Hempelmann. "He didn't discipline the kids. Kitty did all that."

From all accounts, Peter was a normal rambunctious child. As a toddler, like most boys, he had been loud, active and altogether difficult to handle. But Kitty interpreted his behavior as abnormal. She once told Bob Serber that her relationship with Peter was fine until the boy turned seven years old, and then it suddenly changed and she never knew why. Peter was a great builder; like his uncle Frank, he could do marvelous things with his hands, taking things apart and putting them back together. But he never shone in school, and Kitty found this intolerable. "Peter was a terrifically sensitive child," said Harold Cherniss, "and he had a very hard time in school. . . . [But this] had nothing to do with his ability." In response to Kitty's nagging, Peter retreated into himself. Serber recalled that when Peter was five or six years old, "he seemed to be starved for affection." But as a teenager he was just very solemn. "You'd come into the Oppenheimer kitchen," Serber said, "and Peter would be a shadow . . . trying not to be noticed—that would be Peter."

Kitty treated her daughter very differently. "Her attachment to Toni," Hobson recalled, "was profound and seemed just purely loving and admiring . . . She wanted only goodness and happiness for Toni and she was just horrible to Peter." As a young girl, Toni always seemed serene and sturdy. "From when she was six or seven years old," Hobson observed, "the rest of the family relied on her to be sensible and solid and to cheer them on. . . . Toni was the one you never worried about."

Late in 1951, Toni, then seven, was diagnosed with a mild case of polio, and doctors advised the Oppenheimers to take her somewhere warm and humid. That Christmas they rented a seventy-two-foot ketch, the *Comanche,* and spent two weeks sailing around St. Croix in the U.S. Virgin Islands. The *Comanche* was owned and captained by Ted Dale, a warm and gregarious man who quickly won Robert's affection. Dale sailed the boat over to St. John, a tiny jewel of an island with pristine white beaches and turquoise waters. Anchoring in Trunk Bay, they went ashore and explored. Charmed, Robert wrote a letter to Ruth Tolman describing St. John. Ruth replied, "So the warm waters, the bright fish, the soft trade winds must all have been welcome and restoring." St. John left a deep impression upon the Oppenheimers. Toni recovered from her bout with polio; years later she would return to this lovely island paradise and make it her permanent home.

IF KITTY sometimes made family life harrowing, Robert's aloofness and detachment helped him endure. He had consciously chosen to stay in his marriage, and, to be fair to Kitty, she was perfectly capable of controlling her behavior if she wanted to. She had an iron will—with or without drink. One day when the Dysons had a sudden crisis in their household, Kitty came rushing over in her blue jeans, her hands still muddy from her garden. "She was a tower of strength to us as she was to Robert," observed Freeman Dyson. "She was in many ways the stronger of the two, and more solid in a way. You never had the feeling that she was the one who needed help. True, she got drunk from time to time, but I never thought of her as being uncontrollably alcoholic."

And if Kitty had her enemies, she also had her friends. "We always have such fun with you, and love to be in your house," wrote Elinor Hempelmann after one of her frequent visits to the Oppenheimer household. When the Oppenheimers' Los Alamos friends "Deke" and Martha Parsons visited Olden Manor, Kitty often took them on lovely picnics, serving eggs, caviar and cheeses on rye toast washed down with champagne. Parsons, a conservative Navy career man—he was by then an admiral—treasured his rambling philosophical conversations with the Oppenheimers. "Dear Oppy," he wrote after one such visit in September 1950, "As always our weekend with you and Kitty was the event of the season for us. Our little affairs and even the world problems seem more nearly soluble in such an atmosphere."

While Kitty could be outrageous, if she chose, she could also be charming and competent. She had an impish sense of humor. One evening, saying good-bye to her dinner guests, she surveyed Charley Taft's great bulk and said, "I am so glad you don't look like your brother [the very slender Senator Robert Taft]." Robert protested, raising his hands, and said, "Kitty!" Whereupon she said, to laughter all around, "I said the same thing to Allen Dulles." Like Robert, Kitty was always capable of putting on a performance. And so if there were episodes of histrionics, Kitty also set the stage for many fine performances in which she and Robert played the gracious intellectual couple.

"It was another lunch time," wrote Ursula Niebuhr, the wife of Dr. Reinhold Niebuhr, a fellow for a year at the Institute. "This one was at the Oppenheimers' house, on a beautiful spring day, and Kitty had masses of daffodils about the house." George Kennan and his wife were also guests. "Robert was at his most charming and hospitable best." After lunching, the guests adjourned for coffee to the lower level of the Oppenheimer living room. In the course of their conversation, Robert discovered that Kennan was unfamiliar with the seventeenth-century poet George Herbert. Herbert was one of Oppie's favorite poets, and so he drew out a fine old edition of Herbert from his bookcase and began reading aloud, in "that sympathetic voice of his," a Herbert poem titled "The Pulley," the theme of which was man's restlessness, a trait Oppenheimer knew he carried to a fault.

> *When God at first made man*
> *Having a glasse of blessings standing by . . .*

The poem ends with these lines:

> *Yet let him keep the rest,*
> *But keep them with repining restlessnesse;*
> *Let him be rich and wearie, that at least,*
> *If goodnesse leade him not, yet wearnesse*
> *May tosse him to My breast.*

CHAPTER THIRTY

"He Never Let On What His Opinion Was"

Our atomic monopoly is like a cake of ice melting in the sun. . . .

ROBERT OPPENHEIMER,
Time *magazine, November 8, 1948*

O N AUGUST 29, 1949, the Soviet Union secretly exploded an atomic bomb at an isolated testing site in Khazakhstan. Nine days later, an American B-29 atmospheric detection reconnaissance plane flying over the northern Pacific picked up radioactive readings on special filter paper designed expressly to detect just such an explosion. On September 9, the news was transmitted to high-ranking officials in the Truman Administration. No one wanted to believe it, and Truman himself expressed skepticism. To settle the matter, it was agreed that a panel of experts would analyze the evidence. Tellingly, the Defense Department picked Vannevar Bush to chair the panel. When called, Bush suggested it would be more reasonable if Dr. Oppenheimer chaired such a technical panel. But an Air Force general told Bush they preferred him as chairman.

Bush acquiesced, but he made sure to have Oppenheimer on the panel. Oppenheimer had just returned from Perro Caliente when Bush called to tell him the news. The panel of experts met for five hours on the morning of September 19. While Bush presided, Oppenheimer directed many of the questions, and by lunchtime everyone agreed that the evidence was overwhelming: "Joe-1" was indeed an atomic bomb test, and furthermore, it was a close copy of the Manhattan Project's plutonium bomb.

The following day, Lilienthal briefed President Truman on the detection panel's conclusions—and pleaded with him to make an immediate announcement. Lilienthal noted in his diary that he "tried every argument I knew with so little apparent headway." Truman balked, saying he was not even certain that the Soviets had a real bomb. He told Lilienthal that he would sit on the news for a few days and think about it. When Oppenheimer heard this, he was incredulous and upset; an opportunity was being missed, he told Lilienthal, to seize the initiative.

Finally, three days later, a still doubtful Truman reluctantly announced that an atomic explosion had occurred in the Soviet Union; he pointedly refused to say that it had been a bomb. A shocked Edward Teller called Oppenheimer and asked him, "What do we do now?" Oppenheimer replied laconically, "Keep your shirt on."

" 'Operation Joe' is simply the fulfillment of an expectation," Oppenheimer calmly told a reporter from *Life* magazine that autumn. He had never thought that the American monopoly would last very long. A year earlier, he had told *Time* magazine, "Our atomic monopoly is like a cake of ice melting in the sun. . . ." Now he hoped the existence of a Soviet bomb would persuade Truman to change course, and renew the efforts made in 1946 to internationalize control over all nuclear technology. But he also feared the Administration might overreact; he had heard talk of preventive war in some quarters. David Lilienthal found his friend "frantic, drawn" with nervous energy. He told Lilienthal, "We mustn't muff it this time; this could be an end of the miasma of secrecy."

Oppenheimer believed the Truman Administration's obsession with secrecy was both irrational and counterproductive. He and Lilienthal had been trying all year to nudge the president and his advisers toward more openness on nuclear issues. Now that the Soviets had the bomb, they reasoned, excessive secrecy no longer had any rationale. At a meeting of the AEC's General Advisory Committee, Oppenheimer expressed the hope that the Soviet achievement would push the United States to adopt a "more rational security policy."

Even as Oppenheimer cautioned against any drastic reaction, legislators on Capitol Hill began speaking of measures to counter the Soviet achievement. Within days, Truman endorsed a Joint Chiefs proposal for increasing the production of nuclear weapons. The U.S. stockpile of atomic weapons—which in June 1948 stood at about 50 bombs—would rise rapidly to some 300 such weapons by June 1950. This was just the beginning. AEC Commissioner Lewis Strauss circulated a memo arguing that U.S. military superiority over the Soviets would inevitably diminish; borrowing language from physics, Strauss suggested that America could only regain its absolute advantage with a "quantum jump" in technology. The nation needed a crash program to develop the Super, a thermonuclear weapon.

Truman was not even aware of the possibility of a Super until October 1949. But once apprised of it, the president was intrigued. Oppenheimer had always been skeptical. "I am not sure the miserable thing will work," he wrote Conant, "nor that it can be gotten to a target except by ox-cart," a reference to the expectation that it would be too big to be carried in an aircraft. Profoundly disturbed by the ethical implications of a weapon thousands of times more destructive than an atomic bomb, he hoped that the Super would prove technically unfeasible. More horrific than the atomic (fission) bomb, the Super (fusion) bomb would surely escalate the nuclear arms race. The physics of fusion emulated the reactions in the interior of the sun, meaning that fusion explosions had no physical limits. One could get an even larger explosion simply by adding more heavy hydrogen. Armed with Super bombs, a single airplane could kill millions of people in minutes. It was too big for any known military target; it was a weapon of mass, indiscriminate murder. The possibility of such a weapon horrified Oppenheimer as much as it excited the imaginations of various Air Force generals, their supporters in Congress and the scientists who supported Edward Teller's ambition to build a Super.

As early as September 1945, Oppenheimer had written a secret report on behalf of a special Scientific Advisory Panel composed of himself, Arthur Compton, Ernest Lawrence and Enrico Fermi. The report advised that "no such effort [on the Super or H-bomb] should be invested at the present time. . . ." To be sure, the possibility that such a weapon could be developed "should not be forgotten." But it was not an imperative. Officially, Oppenheimer raised no ethical concerns. But Compton—speaking for himself, Oppenheimer, Lawrence and Fermi—wrote Henry Wallace and explained, "We feel that this development [the H-bomb] should not be undertaken, primarily *because we should prefer defeat in war to victory obtained at the expense of the enormous human disaster that would be caused by its determined use.*" (emphasis added)

Over the next four years, much changed. Relations with the Soviet Union deteriorated, nuclear weapons emerged as the anchor of America's emerging containment policy, and the U.S. nuclear arsenal expanded to more than 100 atomic bombs, with more and larger ones on the way. The question at issue was obvious: What effect would this new, gigantic weapon, if it were built, have on American national security?

On October 9, 1949, Oppenheimer traveled to Cambridge, Massachusetts, to attend a meeting of Harvard's Board of Overseers, to which he had just been elected that spring. He stayed at Conant's home on Quincy Street, and he and Harvard's president had a "long and difficult discussion having, alas, nothing to do with Harvard." The two friends knew that they would have to grapple with a recommendation about the Super at a meeting of the General Advisory Committee later that month. So it would have been natural for them to vent their worries, and it was probably on this occasion that

Conant told Oppenheimer that the hydrogen bomb would be built "over my dead body." Conant was outraged that a civilized country would even consider using such a ghastly, murderous weapon; he thought it nothing less than a genocide machine.

Later that same month, on October 21, after being briefed on the current status of thermonuclear research, Oppie sat down and wrote "Uncle Jim" a long letter. He acknowledged that when they had last spoken, "I was inclined to think that the super might also be relevant." Technically, he still thought the Super was "not very different from what it was when we first spoke of it more than 7 years ago: a weapon of unknown design, cost, deliverability and military value." The only thing that had changed in seven years was the country's climate of opinion. He pointed out that "two experienced promoters have been at work, i.e. Ernest Lawrence and Edward Teller. The project has long been dear to Teller's heart; and Ernest has convinced himself that we must learn from Operation Joe [the Soviet atomic explosion] that the Russians will soon do the Super, and that we had better beat them to it."

Oppenheimer and all the other members of the GAC believed the technical problems associated with building an H-bomb were still formidable. But he and Conant also were deeply troubled by the political implications of the Super. "What does worry me," Oppenheimer wrote to Conant, "is that this thing appears to have caught the imagination, both of congressional and of military people, as *the answer* to the problem posed by the Russian advance [in atomic weapons]. It would be folly to oppose the exploration of this weapon. We have always known it had to be done; and it does have to be done. . . . *But that we become committed to it as the way to save the country and the peace appears to me full of dangers.*"

After noting that the Joint Chiefs were already inclined to ask the president for a crash H-bomb program, Oppie worried that "the climate of opinion among the competent physicists also shows signs of shifting." Even Hans Bethe, he wrote, was thinking of returning to Los Alamos to work on the Super on a full-time basis.

Bethe was in fact undecided and was arriving that afternoon in Princeton. He came with Edward Teller, who was already going around the country, recruiting physicists to come back to Los Alamos. According to Teller, Bethe had already said he would come. Bethe disputes this, and insists that he had come to Princeton for Oppie's advice. Instead, he found Oppenheimer "equally undecided and equally troubled in his mind about what should be done. I did not get from him the advice that I was hoping to get."

While Oppie revealed little about his own views on the Super, he did tell Bethe and Teller that Conant was opposed to a crash program. But since

Teller had arrived certain that Oppie would oppose the weapon, he left Princeton delighted that Oppenheimer seemed to be sitting on the fence. He also hoped that Bethe would now join him in Los Alamos.

But later that weekend, Bethe discussed the H-bomb with his friend Victor Weisskopf, who argued that a war fought with thermonuclear weapons would be suicidal. "We both had to agree," Bethe said, "that after such a war, even if we were to win it, the world would not be . . . like the world we want to preserve. We would lose the things we were fighting for. This was a very long conversation and a very difficult one for both of us." A few days later, Bethe phoned Teller and told him his decision. "He was disappointed," Bethe recalled. "I was relieved." Yet, despite Weisskopf's pivotal role, Teller was convinced that Oppenheimer was responsible for Bethe's volte-face.

In the meantime, Oppenheimer was having his own difficult conversations, agonizing over the issue, despite his scientific, policy and moral qualms. Taking his role as chairman of the GAC responsibly, he made a concerted effort to restrain his instincts and inclinations. He had put himself in a listening mode. But Conant felt no such restraints. Upon receiving Oppenheimer's letter of October 21, he responded sharply. He told Oppie, probably in a phone call, that if the Super ever came before the General Advisory Committee, "he would certainly oppose it as folly."

AT TWO O'CLOCK on Friday afternoon, October 28, 1949, Oppenheimer convened the eighteenth meeting (since January 1947) of the General Advisory Committee in the AEC's conference room on Constitution Avenue. Over the next three days, Isidor Rabi, Enrico Fermi, James Conant, Oliver Buckley (president of Bell Telephone Laboratories), Lee DuBridge, Hartley Rowe (a director of United Fruit Company) and Cyril Smith would listen to expert witnesses like George Kennan and Gen. Omar Bradley and carefully debate the merits of the Super. AEC commissioners Lewis Strauss, Gordon Dean and David Lilienthal also attended some of the GAC sessions. Everyone present understood that the Truman Administration had to appear to be doing something tough and concrete in response to the Soviet achievement. Lilienthal noted in his diary the day before that Ernest Lawrence and other boosters of the Super "can only be described as drooling with the prospect and 'bloodthirsty.' " These men, he wrote, believe "there's nothing to think over. . . ." Just before the GAC meeting officially convened, Oppenheimer produced a letter he had received from the chemist Glenn Seaborg, the one GAC member absent. In 1954, Oppenheimer's critics suggested that he had not shared Seaborg's views, but one of the GAC members, Cyril Smith, remembered that Oppie showed the letters to everyone before the meeting began. Seaborg was reluctantly inclined to think that the country had to develop the H-bomb. "Although I deplore the prospects of our country putting a tremendous effort into this," he wrote, "I must confess that I have been unable to come to the conclusion that we should not . . . I would have to hear some good arguments before I could take on sufficient courage to recommend not going toward such a program."

Oppenheimer made a point of not expressing his own views until everyone else had spoken. "He never let on what his opinion was at all," recalled DuBridge. "We went right around the table, and everybody gave his view of it, and they were all negative." Lilienthal heard Conant, "looking almost translucent, so gray," mutter, "We built one Frankenstein"—as if it would be madness to build another. Rabi later recalled that "Oppenheimer followed Conant's lead" throughout the weekend discussions. According to Dean, the "moral implications were discussed at great length." Lilienthal noted in his diary Saturday night that Conant argued "flatly against it [the H-bomb] on moral grounds." When Buckley suggested that there was no moral difference between an atomic bomb and a Super, Lilienthal noted, "Conant disagreed: there are grades of morality." And when Strauss pointed out that the final decision would be made in Washington and not by popular vote, Conant replied, "But whether it will stick depends on how the country views the moral issue." Conant even asked, "Can this be declassified—i.e. the fact that there is such a thing being considered . . . ?"

Rabi, presciently, observed that Washington would no doubt decide to go ahead with the project, and the only question remaining was "who will be willing to join it." During their all-day Saturday session, Fermi initially suggested that "one must explore it and do it," but that exploring the feasibility of the Super "doesn't foreclose the question: should it be made use of?" Lilienthal had made up his mind: the Super "would not further the common defense, and it might harm us, by making the prospects of the other course—toward peace—even less good than they now are."

By early Sunday, a consensus emerged among all eight GAC members present: They would oppose a crash program to develop the Super on scientific, technical and moral grounds. Rabi and Fermi qualified their opposition to the weapon—which they called "an evil thing considered in any light"—with a proposal that America "invite the nations of the world to join us in a solemn pledge" not to build the weapon. Oppenheimer toyed with signing on to this Rabi-Fermi qualification, but in the end, he and the committee's majority advised against an accelerated program to build the H-bomb on the grounds that such a weapon was neither necessary as a deterrent nor beneficial to American security.

While Oppenheimer also offered pragmatic arguments about "whether the super will be cheaper or more expensive than the fission bomb," the committee's report made it clear that nuclear weapons policies must no longer be decided in a moral vacuum. Convinced that the scientific and technical work on the Super left, at best, a fifty percent chance of such a weapon being constructed, they first made it clear why any crash program to achieve it would undermine America's security.

But to limit the issue to technical and political considerations was, in their shared view, not only a failure of responsibility, but a dereliction of duty. They were, after all, the elite veterans of the Manhattan Project, the men who had provided the scientific intelligence necessary for the creation of the atomic bomb. They had undertaken that task as enthusiastic patriots. They had followed the lead of a government determined to use the new weapon in war. Oppenheimer had worked to contain scientists like Leo Szilard and Robert Wilson who had raised moral objections to its use against Japan. But those arguments had taken place in the context of total warfare, at a time when the atomic bomb was something entirely new, and they were inexperienced in matters of state policy.

In 1949, however, circumstances were entirely different. America was not at war, the nuclear arms race had taken a new and dangerous turn with the Soviet success, and the members of the GAC were the most deeply informed and experienced atomic scientists in America. They all agreed that weapons that could annihilate life on earth could not be discussed in a military policy vacuum. Moral considerations were as relevant as technical assessments.

"The use of this weapon will bring about the destruction of innumerable human lives," Oppenheimer wrote. "It is not a weapon which can be used exclusively for the destruction of material installations of military or semi-military purposes. Its use therefore carries much further than the atomic bomb itself the policy of exterminating civilian populations."

Oppenheimer feared that the Super would simply be too big—or to put it another way, any legitimate military target for a thermonuclear device would be "too small." If the Hiroshima bomb packed an explosive yield of 15,000 tons of TNT, a thermonuclear bomb—if it proved to be feasible—might explode with the force of 100 million tons of TNT. The Super was simply too large even as a city-buster. It could easily destroy 150 to 1,000 square miles or more. As the GAC's report concluded, "a super bomb might become a weapon of genocide." Even if it was never used, the mere fact that the United States had such a genocidal weapon in its arsenal would ultimately undermine U.S. security. "The existence of such a weapon in our armory," the GAC majority report stated, "would have far-reaching effects on world opinion." Reasonable people could conclude that America was willing to contemplate an act of Armageddon. "Thus we believe that the psychological effect of the weapon in our hands would be adverse to our interest."

Like Conant, Rabi and the others, Oppenheimer hoped that the Super would "never be produced"—and that the refusal to build it would make it possible to reopen arms control negotiations with the Russians. "We believe a super bomb should never be produced," Oppenheimer wrote for the majority. "Mankind would be far better off not to have a demonstration of the feasibility of such a weapon. . . ."

As McGeorge Bundy later noted, the authors of the GAC report were essentially making the case for the kind of arms control treaties finally negotiated in the 1970s. But what if the proposal wasn't accepted? What if the Soviets were the first to obtain a Super? In that event, the Russians would have to test the weapon—H-bombs cannot be developed without being tested—and such a test was guaranteed to be detected. "To the argument that the Russians may succeed in developing this weapon, we would reply that our undertaking it will not prove a deterrent to them. Should they use the weapon against us, reprisals by our large stock of atomic bombs would be comparably effective to the use of a super."

Indeed, if the Super was not a feasible military weapon—because no target was large enough—Oppenheimer and the GAC report argued that it would be both more economical and more effective militarily to accelerate the production of fissionable materials for small, tactical atomic weapons. Together with a buildup of conventional military forces in Western Europe, such "battlefield" atomic weapons would provide the West with a deterrent that was far more effective and credible against any conceivable Soviet invasion force. It was the first serious proposal for nuclear "sufficiency," a strategic concept that proposed a nuclear arsenal designed for specific tasks rather than one amassed through an irrational race of accumulation.

Oppenheimer was pleased with the outcome of the GAC's deliberations. His personal secretary, Katherine Russell, was not so sure. After typing up the GAC's final report, she predicted, "This will cause you a lot of trouble." Oppie was nevertheless gratified to learn that on November 9, 1949, the AEC commissioners had voted three to two to endorse the GAC recommendations. Commissioners Lilienthal, Pike and Smyth had voted against a Super crash program; commissioners Strauss and Dean had voted for it.

NAÏVELY, Oppenheimer thought the battle against the Super had been won. But soon it became apparent that Teller, Strauss and other supporters of the hydrogen bomb were mounting a counteroffensive. Senator Brien McMahon told Teller that the GAC report "just makes me sick." McMahon had come to believe that war with the Soviets was "inevitable." He told a shocked Lilienthal that he thought the United States should "blow them off the face of the earth, quick, before they do the same to us. . . ." Adm. Sidney Souers warned, "It's either we make it [the H-bomb] or we wait until the Russians drop one on us without warning." Many other Washington officials had similarly apocalyptic reactions. The debate over the Super had thus crystallized the underlying hysteria of the Cold War and divided policy-makers and politicians into two permanently opposed Cold War camps—arms racers and arms controllers.

Responding to vigorous lobbying, President Truman asked AEC chairman Lilienthal, Defense Secretary Louis Johnson and Secretary of State Dean Acheson to study the issue once again and make a final recommendation. Lilienthal, of course, was staunchly opposed to the Super's development. Johnson favored it. Only Acheson was undecided. But, a man of acute political instincts, he knew what the White House wanted. After Oppenheimer briefed him on the H-bomb, the secretary of state processed Oppie's nuanced explanation of the GAC report in simplistic terms. "You know, I listened as carefully as I knew how," he told a colleague, "but I don't understand what 'Oppie' was trying to say. How can you persuade a paranoid adversary to disarm 'by example'?"

Acheson's obvious skepticism led Oppenheimer to realize how few allies he had inside the Administration. However, one firm ally was George Kennan, who that autumn was preparing to resign his post as director of the Policy Planning staff at the State Department. Although Acheson had once set great store by Kennan's advice, the two men now rarely agreed on issues of substantive policy. The architect of America's containment policy was unhappy with how militarized that policy had become. His disillusionment was complete when the Truman Administration, in reaction to Soviet intransigence, broke its agreement with the USSR and established an independent government in West Germany. So, in late September 1949, frustrated and isolated, Kennan announced his intention to leave government service altogether.

Kennan had first encountered Oppenheimer at a War College lecture in 1946. "He was dressed in the usual brown suit with trousers much too long," Kennan said. "He looked like a graduate student in physics rather than a man of distinction. He shuffled out to the edge of the platform and spoke without notes, as I recall, for 40 or 45 minutes with such startling scrupulousness and lucidity that nobody dared ask a question."

In the course of 1949–50, Kennan and Oppenheimer developed a close friendship based on mutual respect and education. Oppie had invited Kennan to Princeton for a classified seminar on nuclear weapons. Kennan had also had lengthy dealings with Oppenheimer over the question of British and Canadian access to uranium. "He kept the whole thing on a very high plane," Kennan recalled of these meetings. "He was a man who moved rapidly in the intellectual sense, and accurately and with great insight. [At these meetings] nobody wanted to engage in trivialities or to do anything but his best intellectually."

In the midst of the debate over the Super, Kennan again traveled to Princeton, arriving on November 16, 1949. He and Oppenheimer talked at length about the "present state of the atomic problem." Oppie found the visit "inspiriting." Kennan's views, he thought, were "non-doctrinaire" and "sympathetic." At the time, Kennan was suggesting that in response to the Soviet bomb, the president could propose a moratorium on building a Super. "To me," Oppenheimer wrote Kennan the next day, "the suggestions you made seemed reasonable. . . ." But he warned Kennan that in the "present climate of opinion" they would not seem so to many in Washington whose notions of safeguards "have attained a kind of rigid and absolute quality." As a measure of how politically attuned Oppenheimer had become, he warned Kennan, "We must be prepared to meet and overcome the arguments which hold that your proposals are too dangerous."

Upon receiving this warning, Kennan sat down and tried his hand at drafting a possible presidential statement announcing a decision not to build the H-bomb "at this time." In eloquent language that substantially reflected the GAC's analysis of the issue, Kennan outlined three succinct reasons not to proceed with a weapon of "almost unlimited destructive power." First, "this weapon could not conceivably have a purely military employment." Second, "there is no such thing as absolute security . . . ," and the country's current atomic arsenal was more than sufficiently power-ful to deter any kind of attack. And third, "for us to embark on such a path would certainly not deter others from doing likewise. . . ." To the contrary, to build the Super would almost certainly inspire others to do the same.

The speech was never given, but over the next six weeks Kennan fleshed out these ideas into an eighty-page formal report reexamining the entire problem of nuclear weapons. He showed an early draft of the paper to Oppenheimer, who thought it "thoroughly admirable." This prescient paper, though less well known than his famous 1947 *Foreign Affairs* essay which proposed a policy of containment, is a seminal document of the early Cold War. Kennan himself later called it "one of the most important, if not the most important, of all the documents I ever wrote in government." Knowing how controversial it would be, Kennan sent it to Acheson on January 20, 1950, as a "personal paper."

The document—"Memorandum: The International Control of Atomic

Energy"—challenged fundamental assumptions underlying the Truman Administration's view of both the bomb and the Soviet Union. Adopting Oppenheimer's perspective, Kennan argued that the atomic bomb was dangerous precisely because it was mistakenly seen as a cheap panacea for the Soviet threat. Echoing Oppie, he wrote that the "military people" had seized upon the Super as the answer to the Russian acquisition of the bomb: "I fear that the atomic bomb, with its vague and highly dangerous promise of 'decisive' results . . . of easy solutions to profound human problems, will impede understanding of the things that are important to a clean, clear policy and will carry us toward the misuse and dissipation of our national strength."

Kennan pleaded with Acheson not to support building an even more terrifying weapon of mass destruction—the Super—without first trying to negotiate a comprehensive arms control regime with the Soviets, as Oppenheimer had suggested earlier. Failing that, Kennan argued that the United States should not make the atomic weapon the centerpiece of its national defense. Instead, American officials should make it clear to the Soviets that they regarded atomic weapons "as something superfluous to our basic military posture—as something which we are compelled to hold against the possibility that they might be used by our opponents." A small number of such weapons, he wrote, would be sufficient to deter the Soviet Union from using the bomb against the West.

To this point, Kennan's memo followed the logic of the GAC's October 30, 1949, recommendations. But Kennan picked up another idea that Oppenheimer had considered recently. Instead of relying on a massive arsenal of atomic bombs, Washington should substantially augment its conventional arms, particularly in Western Europe. The Soviets, he said, must understand that the West was willing to field sufficient troops and conventional armaments in Western Europe to deter any possible invasion. Such a conventional deterrent would then permit Washington to pledge itself to a policy of "no first use" of nuclear weapons. America, he argued, should "move as rapidly as possible toward the removal of [atomic weapons] from national armaments without insisting on a deep-seated change in the Soviet system."

Kennan regarded Stalin's regime as a reprehensible tyranny—but he did not think Stalin reckless. The Soviet dictator surely was determined to defend his internal empire, but that did not mean that he intended to wage a war of aggression against the Western allies, a war that would have inevitably threatened the stability of his own regime. Stalin understood that a war with the West might well spell the ruin of the Soviet Union. "I was firmly convinced," Kennan said later, "that they had had absolutely their belly full of war. Stalin never wanted another major war."

In short, Kennan believed that it had been compelling strategic considerations, rather than the American atomic monopoly, which had deterred a
Soviet invasion of Western Europe in the years 1945–49. Now that the
Soviets had their own atomic bomb, Kennan argued that it made no sense
for the United States to get into a spiraling nuclear arms race. Like Oppenheimer, he believed that the bomb was ultimately a suicidal weapon and
therefore both militarily useless and dangerous. Besides, Kennan was confident that the Soviet Union was politically and economically the weaker of
the two adversaries, and that in the long run America could wear down the
Soviet system by means of diplomacy and the "judicious exploitation of our
strength as a deterrent to world conflict. . . ."

Kennan's eighty-page "personal document" might well have been coauthored with Oppenheimer, reflecting as it did so many of Robert's views.
Indeed, both he and Kennan took its reception as a plunging barometer,
indicating the approach of violent political storms. Circulated within the
State Department, Kennan's memo was quietly and firmly rejected by *all*
who read it. Acheson called Kennan into his office one day and said,
"George, if you persist in your view on this matter, you should resign from
the Foreign Service, assume a monk's habit, carry a tin cup and stand on the
street corner and say, 'The end of the world is nigh.' "

Acheson didn't even bother to show the document to President Truman.
By then, Oppenheimer was fully aware of which way the winds were blowing. Edward Teller was winning. But if so, Oppie still hoped that the technical obstacles to designing a thermonuclear device would prove to be
insurmountable. "Let Teller and [John] Wheeler go ahead," he was reported
to have said. "Let them fall on their faces." On January 29, 1950, he ran into
Teller at a conference of the American Physical Society in New York and
admitted that he thought Truman was going to reject his recommendation
against the Super. If so, Teller asked, would he return to Los Alamos to
work on the Super? "Certainly not," Oppie snapped.

A day later, in Washington for a meeting of the GAC, he decided to drop
in on a special meeting of the Joint Committee on Atomic Energy, called by
Senator Brien McMahon to discuss the Super. Oppenheimer knew McMahon was vigorously lobbying the president to approve a crash Super program, and he knew his views would be unwelcome. But he turned up
anyway, telling McMahon and the other legislators, "I thought it would be
cowardly for me not to come up here and let you disagree and raise questions where you thought we had missed the point." His demeanor was one
of polite resignation. Asked what would happen if the Russians got the
Super and the United States did not have it, he replied, "If the Russians have
the weapon and we don't, we will be badly off. And if the Russians have the
weapon and we do, we will still be badly off." The whole point, he

explained, was that by "going down this path ourselves, we are doing the one thing that will accelerate and insure their [Super bomb] development." When a congressman asked him if a war fought with hydrogen bombs would make the earth unfit for human habitation, Oppie interjected, "Pestiferous, you mean?" Actually, he said, he was more worried about mankind's "moral survival." He explained his position with an air of utter reasonableness, and though no one present questioned his logic, he left knowing that he had not changed anyone's mind.

The next day, January 31, 1950, Lilienthal, Acheson and Defense Secretary Louis Johnson walked across the street from the old State Department building to the White House for a meeting with the president on the Super. Lilienthal was still ardently opposed to a crash program. Acheson privately agreed with many of Lilienthal's objections, but believed that domestic political factors would compel Truman to go forward with a crash program: "The American people simply would not tolerate a policy of delaying nuclear research in so vital a matter. . . ." Johnson agreed, telling Lilienthal, "We must protect the president." It had come to that. The real issues related to national security had been rendered irrelevant by the simplifications imposed by domestic politics.

They agreed, nonetheless, that Lilienthal would be allowed to make his case. Once they were in the Oval Office, however, Lilienthal had hardly begun his presentation when Truman cut him off to ask, "Can the Russians do it?" When everyone nodded, Truman said, "In that case, we have no choice. We'll go ahead." Lilienthal noted in his diary that Truman had "clearly set on what he was going to do before we set foot inside the door." Some months earlier, Lilienthal had warned Truman that demagogues in Congress would attempt to force his hand on the Super. "I don't blitz easily," Truman had said. Walking out of the White House, Lilienthal looked at his watch. The president who couldn't be blitzed had given him exactly seven minutes. It was, Lilienthal noted, like saying " 'No' to a steamroller."

That evening, in a radio address that had no doubt been in preparation for some time, President Truman announced a program to determine the "technical feasibility of a thermonuclear weapon." At the same time, he ordered a general reexamination of the country's strategic plans. This led to a top-secret policy paper, NSC-68, largely produced by Kennan's successor as director of policy planning in the State Department, Paul Nitze. Nitze, an advocate of a large nuclear arsenal, depicted the Soviet Union as bent on world conquest. He called for "a rapid and sustained build-up of the political, economic and military strength of the free world." Circulated in April 1950, NSC-68 specifically rejected Kennan's proposal to proclaim a policy of "no first use" of nuclear weapons. To the contrary, a large arsenal of nuclear weapons was to become the foundation of U.S. defense strategy. And to that end, Truman authorized an industrial program to greatly expand the nation's capacity to build nuclear warheads of all configurations.

By the end of the decade, America's stockpile of nuclear weapons would leap from some 300 warheads to nearly 18,000 nuclear weapons. Over the next five decades, the United States would produce more than 70,000 nuclear weapons and spend a staggering $5.5 trillion on nuclear weapons programs. In retrospect—and even at the time—it was clear that the H-bomb decision was a turning point in the Cold War's spiraling arms race. Like Oppenheimer, Kennan was thoroughly "disgusted." I. I. Rabi was outraged. "I never forgave Truman," he said.

After his abbreviated meeting with Truman, David Lilienthal told Oppenheimer that the president had also demanded that all the scientists involved refrain from discussing the decision publicly: "It was like a funeral party—especially when I said we were all gagged." Sorely disheartened, Oppenheimer considered resigning his position on the GAC. Acheson, fearful that Oppenheimer and Conant would take their appeal to the American public, made a point of telling Harvard's president, "For heck's sake, don't upset the applecart."

Conant told Oppenheimer of Acheson's warning that a public debate would be "contrary to the national interest." So once again, Oppie played the role of loyal supporter. As he later testified, it did not seem responsible to resign at that time and "promote a debate on a matter which was settled." Conant wrote a friend that he and Oppenheimer "didn't [resign] (or at least I didn't) because I did not want to do anything that seemed to indicate we were not good soldiers. . . ." In retrospect, he regretted this decision—he thought they should both have immediately resigned.

How different and better Oppenheimer's life would have been had he taken that step. But he didn't, and like Conant, Oppenheimer again fell into line. Nevertheless, he could not disguise his disdain for those who had pushed through the decision. The very evening of Truman's announcement, Oppenheimer felt obligated to attend a party at the Shoreham Hotel, celebrating Strauss's fifty-fourth birthday. Finding Oppenheimer alone in a corner, a reporter walked up to him and said, "You don't look jubilant." Oppenheimer muttered in response, "This is the plague of Thebes." When Strauss tried to introduce his son and daughter-in-law to the famous physicist, Oppenheimer brusquely offered them a hand over his shoulder—and then turned away without a word. Understandably, Strauss was incensed.

THE HYDROGEN BOMB decision had been made in camera, without public debate and, Oppenheimer believed, without an honest evaluation of its consequences. Secrecy had become the handmaiden of ignorant policies, and so Oppenheimer decided to speak out against secrecy. On February 12, 1950, Strauss was angered to see Oppenheimer appear on the first telecast of Eleanor Roosevelt's Sunday morning talk show and openly challenge the manner in which the hydrogen bomb decision had been made. "These are complex technical things," Oppenheimer told the television audience, "but they touch the very basis of our morality. It is a grave danger for us that these decisions are taken on the basis of facts held secret." To Strauss, such comments signaled open defiance of the president—and he made sure the White House saw a transcript of Oppenheimer's words.

Later that summer, in the *Bulletin of the Atomic Scientists,* Oppenheimer repeated "that these decisions have been taken on the basis of facts held secret." This, he thought, was neither necessary nor wise: "The relevant facts could be of little help to an enemy; yet they are indispensable for an understanding of questions of policy." No one in the administration agreed; the trend was toward more secrecy.

FOR NEARLY five years, Oppenheimer had tried to use his prestige and status as a celebrity scientist to influence Washington's growing national security establishment from the inside. His old friends on the left, men like Phil Morrison, Bob Serber and even his own brother had warned him that this was a futile gamble. He had failed in 1946, when the Acheson-Lilienthal plan for international control over atomic bombs was sabotaged by President Truman's appointment of Bernard Baruch. And now, once again, he had failed to persuade the president and members of his Administration to turn their back on what Conant had described to Acheson as "the whole rotten business." The Administration now supported a program to build a bomb 1,000 times as lethal as the Hiroshima weapon. Still, Oppenheimer would not "upset the applecart." He would remain an insider— albeit one who was increasingly outspoken and increasingly suspect.

CHAPTER THIRTY-ONE

"Dark Words About Oppie"

How utterly nauseating—but this is like a puff of wind against the Gibraltar of your great standing in American life.

DAVID LILIENTHAL *to Robert Oppenheimer, May 10, 1950*

IN THE AFTERMATH OF WHAT HE LATER CALLED "our large and ill-managed bout with the Super," Oppenheimer retreated to Princeton, bitterly discouraged. That spring, George Kennan wrote him, "You probably do not know to what extent you have become my intellectual conscience." The debate over the Super had forged an alliance between these two formidable intellects whose instincts and sensibilities converged in opposition to a defense strategy based on the threat of nuclear war.

"What stands out in my mind when I think back on those days," Kennan recalled, "was his insistence on the desirability of openness." Oppenheimer argued that concealing information about the bomb increased the danger of misunderstandings. As Kennan recalled Oppie's argument, "You had to have the frankest possible discussions with them [the Soviets] about the problems of the future and the use of the weapon." Kennan agreed with Oppenheimer that nuclear weapons were inherently evil and genocidal: "It should have been visible to people at the time that this was a weapon from which nobody stood to gain. . . . The whole idea that you could achieve anything of a positive nature by the development of these weapons seemed to me preposterous from the start."

On a personal level, Kennan would forever feel grateful to Oppenheimer for bringing him to the Institute to begin a new career as a distinguished scholar and historian. "I, who owe to your confidence and encouragement the very opportunity to make what I could of myself as a scholar, beginning in middle age, have a special personal debt to acknowledge." Yet Kennan's appointment to the Institute was highly controversial; some questioned the credentials of this career Foreign Service officer who had published nothing that could be remotely called scholarship. Johnny von Neumann voted against the appointment, and wrote Oppenheimer that Kennan was "not, so far, an historian," and he had yet to produce any scholarly work of an "exceptional character." Most of the resident mathematicians, led as usual by Oswald Veblen, objected on the grounds that Kennan was merely a political friend of Oppie's and not an academic. "They resented Kennan," recalled Freeman Dyson, "and took the thing as an opportunity to attack Oppenheimer." But Oppenheimer, who had developed a great appreciation for Kennan's intellect, pushed the appointment through the Board of Trustees, promising to pay Kennan's $15,000 stipend out of his Director's Fund.

Kennan spent eighteen months in Princeton before leaving, reluctantly, in the spring of 1952, when Truman and Acheson pressed him to serve as the United States ambassador in Moscow. But less than six months later, he wrote Robert that he thought his tenure in Moscow might be brief, and indeed, within the next ten days his ambassadorship was aborted when he told a reporter that life in Soviet Russia reminded him of the time he had spent in Nazi Germany. Not surprisingly, the Soviets declared him persona non grata. Then, after Dwight Eisenhower won the presidential election, it became clear that the Republicans who came into office promoting "roll-back" had little use for the author of "containment." In March 1953, Kennan wrote Oppenheimer to say that he had just seen Secretary of State John Foster Dulles—who informed him that "he knew of no 'niche' for me in government at this time . . . tainted as I am with 'containment.' " Kennan therefore took early retirement and promptly moved back to Princeton, Oppie's "decompression chamber for scholars." With the exception of a slightly longer stint as ambassador to Yugoslavia in the early 1960s, Kennan would spend the rest of his life there. He was Oppenheimer's neighbor and devoted friend, and in his eyes, Oppenheimer had created a "place where the work of the mind could proceed in its highest form—gracefully, generously, and with the most exquisite scrupulousness and severity."

THE H-BOMB was not the only issue on which Oppenheimer found himself bucking the Cold War armaments buildup. By 1949, he had despaired of making progress in the foreseeable future on nuclear disarmament. He still believed Bohr's vision of global openness was mankind's only hope in the nuclear age. But developments in the early Cold War had made it clear that the negotiations in the United Nations to control nuclear weapons were at an impasse. Instead, Oppenheimer tried to use his influence to put a damper on the government's and the public's growing expectations for all things nuclear. That summer, the press quoted him as saying that "nuclear power for planes and battleships is so much hogwash." Inside the General Advisory Committee (GAC), Oppenheimer and the other scientists criticized the Air Force's Project Lexington, a program to develop nuclear-powered bomber aircraft. He also talked about the potential dangers inherent in civilian nuclear power plants. Such statements did not endear him to those in the defense establishment or the power industry who favored the development of nuclear-based technologies.

Indeed, the GAC's experiences with the military brass left all of its members increasingly uneasy about the military's nuclear weapons planning. "I know," recalled Lee DuBridge, "that there was a great deal of discussion about targets in the Soviet Union, and how many [bombs] it would take to knock out the major industrial centers. . . . At the time, we thought 50 would just about wipe out the essential things in the Soviet Union." DuBridge always thought that was a pretty good estimate. But over time, the Pentagon's representatives kept finding pretexts to push the number higher. DuBridge recalled, "We used to sometimes smile about this, that they always could seem to find targets for whatever number [of bombs] they thought they could get in the next year or two. They adjusted their target goals to the production goals."

Oppenheimer's presentations at GAC meetings were normally impeccably objective. Rarely did he reveal any emotion. One exception occurred when V. Adm. Hyman Rickover briefed the committee on the Navy's rush to develop nuclear-powered submarines. Rickover complained that the AEC was not working hard enough on reactor development. He challenged Oppenheimer by asking if he had waited until he "had all the facts" before building an atomic bomb. Oppenheimer gave him one of his ice-cold blue-eyed stares and said yes. Though the admiral was notoriously overbearing, Oppenheimer restrained himself until Rickover departed. Oppie then walked over to a table where Rickover had left a small wooden model submarine. Placing his hand around the hull, he quietly crushed it and then silently walked away.

Oppenheimer was expanding his circle of political enemies. As his old friend Harold Cherniss had observed years earlier, Oppie's remarks could be "very cruel." He was often kind and considerate to subordinates, but he could be very cutting to colleagues.

Lewis Strauss remained Oppenheimer's most dangerous political enemy. He had not forgotten how Oppenheimer had ridiculed his recommendations at a congressional hearing the previous summer. "These are not happy days for me," Strauss wrote a friend in July 1949. Having repeatedly dissented inside the AEC over various policies, Strauss felt on the defensive. With Oppenheimer and his friends in mind, he complained privately "that I have been guilty in their eyes of *lèse majesté* in having the effrontery to disagree with my colleagues." He believed that Oppenheimer's close friends Herbert Marks and Anne Wilson Marks were spreading stories "to the effect that I am an 'isolationist.' . . ." When a friend observed that some people seemed to think it "effrontery for anyone to differ with Dr. Oppenheimer on a scientific matter," Strauss wrote a memo for his files on the "theme of omniscience" in which he noted that Oppenheimer had once proposed "denaturing" uranium—a process that had since been proven impossible.

Strauss also convinced himself that Oppenheimer was consciously trying to slow work on the thermonuclear bomb. He thought of Oppenheimer as "a general who did not want to fight. Victory could hardly be expected." Early in 1951, Strauss, though no longer an AEC commissioner, went to AEC Chairman Gordon Dean and, reading from a carefully drafted memo, accused Oppenheimer of "sabotaging the project." He said "something radical" must be done, strongly implying that Oppenheimer should be fired. And, as if to underscore the political risks of taking on the scientist, Strauss ended the meeting by melodramatically throwing the memo into the fire in Dean's fireplace. Consciously or not, it was a metaphorical gesture; the security of the country demanded that Oppenheimer's influence be reduced to ashes.

Back in the autumn of 1949, just as the internal debate over the Super was heating up, Strauss was apprised of top secret information that further fueled his suspicions of Oppenheimer. In mid-October, the FBI informed him that decrypted Soviet cable traffic indicated that a Soviet spy had been operating out of Los Alamos. The crypts seemed to implicate a British physicist, Klaus Fuchs, who had arrived at Los Alamos in 1944 as a member of the British Scientific Mission. In the weeks ahead, it would become clear to Strauss and others that Fuchs had had ample access to classified information about both the atomic bomb and the Super.

While the FBI and the British investigated Fuchs, Strauss began his own investigation of Oppenheimer. He phoned General Groves and, referring to information in Oppenheimer's FBI file, asked about the Chevalier affair. In response, Groves wrote Strauss two long letters trying to explain what had happened in 1943 and why he had accepted Oppenheimer's explanation of Chevalier's activities. In his first letter, he was emphatic in his belief that Oppenheimer was a loyal American. In his second, he tried to convey the complexity of the Chevalier affair.

Groves also made it clear that he did not think that Robert's behavior in the incident was incriminating. "It is important to realize," he wrote Strauss,

Ernest Lawrence, Glenn Seaborg and Oppenheimer. "Modern Prometheans have raided Mount Olympus again," opined *Scientific Monthly*, "and have brought back for man the very thunderbolts of Zeus."

TRENDS IN AMERICAN SCIENCE by Vannevar Bush . . . . *See Page*

VOL. 1 NO 1   MAY 1948

Physics Today put Oppie's porkpie hat on its cover.

Harvard University elected Oppenheimer to its board of overseers (with James B. Conant and Vannevar Bush).

A gifted experimental physicist, Frank Oppenheimer (above) was fired in 1949 by the University of Minnesota when it was revealed that he had been a member of the Communist Party. He became a cattle rancher in Colorado.

Anne Wilson Marks was Oppie's secretary in 1945—and then she married Herbert Marks (lying on deck of the boat), his friend and lawyer.

Caltech's Richard Tolman and his wife, Ruth Tolman, a noted clinical psychologist who became one of Robert's deepest loves.

Time magazine put Oppenheimer on its cover in November 1948.

Center, Oppenheimer was chairman of the Atomic Energy Commission's General Advisory Committee. Here he is on a trip with James B. Conant, Gen. James McCormack, Harley Rowe, John Manley, I. I. Rabi and Roger S. Warner.

Bottom, Oppenheimer (far left) in 1947 receiving an honorary degree from Harvard, with Gen. George C. Marshall, Gen. Omar N. Bradley and other honorees.

Olden Manor, in Princeton, New Jersey, where the Oppenheimers lived after Robert was appointed director of the Institute for Advanced Study in 1947.

Kitty, Toni, and Peter in the greenhouse.

Robert and his children in the yard at Olden Manor.

Robert gave Kitty a greenhouse to grow her orchids. They entertained frequently. "He served the most delicious and the coldest martinis," Pat Sherr said.

Oppenheimer with mathematician John von Neumann, standing before
von Neumann's early computer.

Oppenheimer discussing physics with students at the Institute for Advanced Study
in Princeton. "The Institute was his own little empire," said Freeman Dyson.

Oppenheimer with (from left) Hans Bethe, Senator Brien McMahon, Eleanor Roosevelt, and David Lilienthal.

Oppenheimer opposed a crash program to build a hydrogen bomb. He explained to a TV audience that a "superbomb was a matter that touch[ed] the very basis of our morality. It is a grave danger for us that these decisions are taken on the basis of facts held secret."

Oppenheimer at a conference with physicist Greg Breit. "What we don't know, we explain to each other."

"Who's Being Walled Off From What?"

HERBLOCK
©1954 THE WASHINGTON POST CO.
---from Herblock's Here and Now (Simon & Schuster, 1955)

In December 1953 President Dwight Eisenhower ordered a "blank wall" between Oppenheimer and the government's nuclear secrets. Robert's ensuing security hearing was orchestrated by Atomic Energy Commission chairman Lewis Strauss (above, right), who was determined to purge Oppenheimer from government service. Oppenheimer hired lawyer Lloyd Garrison (right) to defend himself.

On April 12, 1954, Oppenheimer's security hearing opened, chaired by Gordon Gray (top, right). Only one AEC commissioner, Henry DeWolf Smyth (center, right), voted to reject the Gray Board's decision to strip Oppenheimer of his security clearance, AEC commissioner Eugene Zuckert (bottom, right) voted with the majority against Oppenheimer. Roger Robb (bottom, left) served as the Gray Board's prosecutor. Only one member of the Gray Board, Ward Evans (top, left) voted to uphold Oppenheimer's security clearance. Evans called the decision a "black mark on the escutcheon of our country."

Toni Oppenheimer on horseback. "From when she was six or seven years old," Verna Hobson observed, "the rest of the family relied on her to be sensible and solid and to cheer them on."

Oppenheimer lost his security clearance, but kept his job as director of the Institute for Advanced Study. Here, walking with Kitty in Princeton.

Robert could "just pour in the love" he felt for Peter Oppenheimer.

After the 1954 security hearing Oppenheimer "was like a wounded animal," Francis Fergusson recalled. "He retreated. And returned to a simpler way of life." He took his family to St. John in the Virgin Islands. Later he built a spartan beach cottage, and the family (below) spent many months each year on the beautiful island. He and Kitty were expert sailors.

IM-4-

Sitting with his old friend Niels Bohr, 1955.

In 1960 Oppenheimer visited Tokyo (below) where he told reporters, "I do not regret that I had something to do with the technical success of the atomic bomb. It isn't that I don't feel bad; it is that I don't feel worse tonight than I did last night."

IM-9: 91252

Oppenheimer in his office at the Institute.

In April 1962 President John F. Kennedy invited Oppenheimer to the White House. He is seen here shaking hands with Jackie Kennedy.

Frank at the Exploratorium in 1969, a science museum in San Francisco that gives visitors a "hands-on" experience with physics, chemistry, and other fields, which he founded with his wife, Jackie.

In 1963 President Lyndon B. Johnson (below) awarded Oppenheimer (left, with Kitty and Peter) the $50,000 Fermi Prize. David Lilienthal thought the whole affair "a ceremony of expiation for the sins of hatred and ugliness visited upon Oppenheimer."

Right, Edward Teller, who had testified against Oppenheimer in 1954, approached to offer his congratulations. Oppenheimer grinned and shook Teller's hand, while Kitty stood stone-faced beside her husband.

In the summer of 1966 Oppenheimer greets two beachcombers outside his waterfront cottage on St. John. He was already dying of throat cancer.

A pensive Toni inside the cottage. "Everybody loved her," June Barlas said, "but she didn't know that."

In happier days, Toni, Inga Hiilivirta, Kitty, and Doris Jadan drinking cocktails on St. John.

"that if we had eliminated promptly every man who had in the past had associations with friends who were communistically inclined, or who had been sympathetic to the Russians at one time or another, that we would have lost many of our most able scientists."

Finding Groves' defense of Oppenheimer unsatisfactory, Strauss continued his search for incriminating information. By early December he was in communication with Groves' former aide, Col. Kenneth Nichols, who loathed Oppenheimer. Over the next several years, Nichols would become one of Strauss' assistants and confidants. The two men bonded in their hostility to Oppenheimer. Now Nichols gladly provided Strauss with a copy of Arthur Compton's September 1945 letter to Henry Wallace in which Compton, allegedly speaking also for Oppenheimer, Lawrence and Fermi, had stated they would "prefer defeat in war" over a victory won by use of a genocidal weapon like the Super bomb. This view outraged Strauss, who saw in Compton's letter further evidence of Oppenheimer's dangerous influence; that Compton had written the letter, and had noted that Lawrence and Fermi supported his argument, made no difference to Strauss.

ON THE AFTERNOON of February 1, 1950, the day after Truman's endorsement of the Super, Strauss received a phone call from J. Edgar Hoover. The FBI chief informed him that Fuchs had just confessed to espionage. Although Oppenheimer had had no hand in Fuchs' transfer to Los Alamos, Strauss nevertheless held it against him that Fuchs' spying had occurred on his watch. The next day, Strauss wrote Truman that the Fuchs case "only fortifies the wisdom of your decision [on the Super]." To Strauss' way of thinking, the Fuchs case also vindicated his obsession with secrecy and his opposition to sharing nuclear technology and research isotopes with the British or anyone else. And for both Strauss and Hoover, the Fuchs revelation also demanded renewed scrutiny of Oppenheimer's left-wing past.

The day Oppenheimer learned of Fuchs' confession, he happened to be having lunch with Anne Wilson Marks in Grand Central Station's famed Oyster Bar. "Have you heard the news about Fuchs?" he asked his former Los Alamos secretary. They agreed that Fuchs had always seemed like such a quiet, lonely, even pathetic character at Los Alamos. "Robert was stunned by the news," recalled Wilson. On the other hand, he suspected that Fuchs' knowledge about the Super was probably confined to the less-than-practical "oxcart" model. That same week he told his Institute colleague Abraham Pais that he hoped Fuchs had told the Russians all he knew about the Super, because that "would set them back several years."

Just days before Fuchs' confession became public knowledge, Oppen-heimer testified in executive session before the Joint Committee on Atomic Energy. Asked for the first time specifically about his political associations in the 1930s, Oppenheimer calmly explained that he had naïvely thought the Communists possessed some answers to the problems facing the coun-try in the midst of the Depression. At home, his students had found it diffi-cult to find employment, and abroad, Hitler was a menace. While never a Party member himself, Oppenheimer volunteered that he had maintained friendships with some communists right through the war years. Gradually, however, he had discerned a "lack of honesty and integrity in the . . . Com-munist Party." By the end of the war, he said, he had become "a resolute anti-Communist, whose earlier sympathies for Communist causes would give immunity against further infection." He harshly criticized communism for its "hideous dishonesty" and "elements of secrecy and dogma."

Afterwards, a young staff member of the Joint Committee, William Lis-cum Borden, wrote Oppenheimer a letter politely thanking him for his appearance: "I . . . think it was right that you appear[ed] before the Com-mittee and I think it did lots of good."

Borden, a product of St. Albans prep school and Yale Law School, was bright, energetic—and obsessed with the Soviet menace. During the war, he was piloting a B-24 bomber on a nighttime mission when a German V-2 rocket flashed by him on its way to London. "It resembled a meteor," Borden later wrote, "streaming red sparks and whizzing past us as though the aircraft was motionless. I became convinced that it was only a matter of time until rockets would expose the United States to direct, transoceanic attack." In 1946, he wrote an alarmist book on the future risk of a "nuclear Pearl Har-bor," *There Will Be No Time: The Revolution in Strategy.* Borden predicted that in years to come, America's adversaries would possess large numbers of intercontinental rockets tipped with atomic bombs. At Yale, Borden and other conservative classmates bought a newspaper ad urging President Tru-man to issue a nuclear ultimatum to the Soviet Union: "Let Stalin Decide: atomic war or atomic peace." After spotting the incendiary ad, Senator Brien McMahon hired the twenty-eight-year-old Borden as his aide on the Joint Committee on Atomic Energy. "Borden was like a new dog on the block who barked louder and bit harder than the old dogs," wrote the Princeton physi-cist John Wheeler, who met him in 1952. "Wherever he looked, he saw con-spiracies to slow down or derail weapons development in the United States."

BORDEN HAD first met Oppenheimer in April 1949 at a GAC meeting, where he listened silently as Oppie openly disparaged Project Lexington, the Air Force's proposal to build a nuclear-powered bomber. As if that weren't controversial enough, Oppie also criticized the AEC's plan to push ahead with a program of civilian nuclear energy plants: "It is a dangerous engineering undertaking." Unconvinced, Borden left thinking that Oppenheimer was a "born leader and a manipulator."

In the wake of Fuchs' confession, however, Borden began to wonder if Oppenheimer might be something more dangerous than a mere "manipulator." Not surprisingly, his suspicions along these lines were encouraged by Lewis Strauss. By 1949, Strauss and Borden were on a first-name basis, and Strauss continued, even after leaving the AEC, to cultivate the staff director of the Senate committee responsible for oversight of AEC activities. They quickly realized that they had similar concerns about Oppenheimer's influence.

On February 6, 1950, Borden was present when FBI Director Hoover testified before the Joint Committee. Ostensibly, Hoover had come to brief the Committee about Fuchs—but he dealt at length with Oppenheimer. Sitting on the Committee that day were Senator McMahon and Congressman Henry "Scoop" Jackson (D-Wash.).

Scoop Jackson's district in Washington State was home to the Hanford nuclear facilities. He was a hard-line anticommunist and a strong proponent of nuclear weapons. He had met Oppenheimer the previous autumn, during the debate over the Super, and had invited him to dinner at the Carlton Hotel in Washington, D.C. There he had listened in disbelief as the physicist argued that building the H-bomb would only fuel an arms race and make America less secure. "I think he had a guilt complex because of his role in the Manhattan Project," Jackson would say years later.

Now, for the first time, Jackson and McMahon learned from Hoover about Haakon Chevalier's 1943 approach to Oppenheimer in which Chevalier had suggested that perhaps there was scientific information that ought to be shared with their wartime Soviet ally. Hoover reported that Oppenheimer had rejected the overture, but to Borden's suspicious mind, the incident still sounded incriminating. He began to wonder if Oppenheimer's opposition to the Super bomb was motivated by a nefarious loyalty to the communist cause.

A month later, Edward Teller told Borden that Oppenheimer had wanted to close down Los Alamos after the war. He claimed that Oppie had said, "Let's give it back to the Indians." Borden was also told that a Los Alamos security officer believed Oppenheimer had once been a "philosophical Communist." And finally, for the first time he learned that Kitty Oppenheimer had once been married to a communist who had fought and died in Spain.

Borden, McMahon and Jackson also were appalled to learn that Oppenheimer had recently begun to use his influence to make the case for battlefield, tactical nuclear weapons. To the Air Force and its congressional allies, Oppenheimer's initiative was viewed as a transparent effort to undermine the dominant role of the Strategic Air Command. Jackson and his colleagues considered SAC's ability to deliver a devastating atomic attack America's trump weapon. "Until now," Jackson said in a speech, "our atomic superiority has held the Kremlin in check. . . . Falling behind in the atomic armaments competition will mean national suicide. The latest Russian explosion means that Stalin has gone all out in atomic energy. It is high time we go all out." In the atomic era, Jackson felt America had to have absolute military superiority over any conceivable enemy. Thus, if a hydrogen bomb could be built, America should be the first to build it. His biographer, Robert Kaufman, wrote that "[h]e never forgot the experience of well-meaning but naïve scientists arguing against building the H-bomb. . . ."*

WHILE POLITICIANS like Congressman Jackson thought Oppenheimer naïve and guilty of poor judgment, Borden, as noted, was beginning to suspect him of far worse. On May 10, 1950, Borden read on the front page of the *Washington Post* that two former Communist Party members, Paul and Sylvia Crouch, had testified that Oppenheimer had once hosted a Party meeting in his Berkeley home. In testimony before the California State Senate Un-American Activities Committee, the Crouches claimed that Kenneth May had driven them to Oppenheimer's home at 10 Kenilworth Court in July 1941. Hitler had recently invaded the Soviet Union, and as chairman of the Alameda County Communist Party, Paul Crouch was supposed to explain the Party's new stand on the war. Some twenty to twenty-five people were present. Sylvia Crouch described the alleged meeting at Oppenheimer's home as a "session of a top-drawer Communist group known as a special section, a group so important that its makeup was kept secret from ordinary Communists." She said that she and her husband were not introduced to anyone in the room. She only later identified her host as Oppenheimer when she saw him in a 1949 newsreel. The Crouches further

---

*Jackson in turn influenced the neoconservatives who in 2003 shaped the Bush doctrine on preventive war. Richard Perle, who served as Jackson's top foreign policy adviser between 1969 and 1979, told Kaufman, "His [Jackson's] enthusiasm for building missile defense, his skepticism about détente and the Strategic Arms Limitation Talks (SALT), all stemmed from his previous experiences and the lessons he drew from it: that had we listened to the scientists who had opposed the Hydrogen Bomb, Stalin would have emerged with a monopoly and we would have been in deep trouble."

claimed that after being shown photographs by the FBI, they could place David Bohm, George Eltenton and Joseph Weinberg at the same meeting. Sylvia named Weinberg as "Scientist X," the individual labeled by the House Un-American Activities Committee as someone who gave atomic bomb secrets to a communist spy during the war. The California papers played these allegations as a "bombshell." Paul Crouch was described as a "West Coast Whittaker Chambers," a reference to the *Time* magazine editor and former communist whose testimony had led, on January 21, 1950, to the perjury conviction of Alger Hiss.

Oppenheimer immediately issued a written statement denying the allegation: "I have never been a member of the Communist Party. I never assembled any such group of people for any such purpose in my home or anywhere else." Oppenheimer said he didn't recognize the name "Crouch." And then he went on to say, "I have made no secret of the fact that I once knew many people in left-wing circles and belonged to several left-wing organizations. The Government has known in detail of these matters since I first started work on the atomic bomb project." His denials were widely reported in the press and seemed to put the matter to rest. His friends offered their reassurances. Having read about the "nasty thing" in the California papers, David Lilienthal wrote Oppenheimer of the Crouches' testimony, "How utterly nauseating—but this is like a puff of wind against the Gibraltar of your great standing in American life."

Lilienthal, however, was underestimating the effect of this testimony on less sympathetic minds. William Borden wrote a memo saying he found the Crouches' allegations "inherently believable." Paul and Sylvia Crouch had been extensively interviewed by the FBI weeks before their May 1950 testimony in California. By then, they were paid informants, on the payroll of the Justice Department, and testifying regularly against alleged communists in security cases around the country.

The son of a North Carolina Baptist preacher, Paul Crouch had joined the Communist Party in 1925. That same year, as an enlisted man in the U.S. Army, he wrote a letter to CP officials boasting that he had "formed an Esperanto Association as a front for revolutionary activity." The Army intercepted the letter and concluded that he had been organizing a communist cell at Schofield Barracks in Hawaii. Court-martialed on the charge of "fomenting revolution," Crouch was sentenced to an extraordinary forty years in prison. At his trial he testified, "I am in the habit of writing letters to my friends and imaginary persons, sometimes to kings and other foreign persons, in which I place myself in an imaginary position."

Curiously, Crouch was pardoned by President Calvin Coolidge after serving only three years of his forty-year sentence in Alcatraz. It is not clear whether this was the result of being turned into a double agent, as his subsequent behavior would suggest, or just incredibly good luck. But upon his release, the Communist Party hailed him as a "proletarian hero." For a short time he worked alongside Whittaker Chambers as an editorial assistant on the *Daily Worker.* And then, in 1928, the Party sent him to Moscow, where, he later claimed, he had lectured at the Lenin School and been awarded the honorary rank of colonel in the Red Army. He also asserted that he had met with Soviet Army Marshal M. N. Tukhachevsky, who had given him plans "they had formulated for the penetration of the American armed forces." Actually, his Soviet hosts thought his behavior so unhinged that they soon sent him packing. Back in America, however, the Communist Party sent him on a tour of his native South, where he sang the praises of the socialist state and Comrade Stalin. Settling in Florida, he found work as a newspaper reporter and CP organizer.

Inexplicably, one day he crossed a picket line and worked as a strike-breaker on a Miami newspaper; when his comrades discovered what he had done, Crouch fled to California, where, by 1941, he was serving as secre-tary of the Communist Party in Alameda County. He proved to be an unpop-ular comrade and an incompetent leader. "He spent a lot of his time drinking alone in bars," wrote Steve Nelson. In December 1941—or, at the latest, January 1942—local Party members demanded his dismissal when he proposed activities that many felt would invite violence at street meet-ings. Had he moved from double agent to agent provocateur? Perhaps, but in any case, at this point his Party career came to an end, and by the late 1940s he and his wife had made a remarkably smooth transition, emerging as professional witnesses against their former comrades. By 1950, Crouch was the most highly paid "consultant" on the Justice Department's payroll, and would earn $9,675 in the following two years.

Despite his bizarre career, initially Paul Crouch appeared to be a credi-ble witness against Oppenheimer. Crouch was able to describe the interior layout of Oppenheimer's Kenilworth Court home. He told the FBI that the man he later identified as Oppenheimer had asked him several questions, and that after the formal meeting ended he and Oppenheimer had spoken privately for ten minutes. As he and Kenneth May were driving home from the meeting, May told him, according to Crouch, that he "had been talking to one of the nation's leading scientists." Crouch's story had enough details to sound plausible—and highly damaging.

On the other hand, Oppenheimer had an alibi proving that he could not have hosted the CP meeting described by Crouch. Interviewed by FBI agents on April 29 and May 2, 1950, he explained that he and Kitty had been at their Perro Caliente ranch in New Mexico—1,187 miles away from Berkeley. That was the summer he and Kitty had gone to New Mexico, leaving their newborn son Peter in the care of the Chevaliers. Oppenheimer later documented that he had been kicked by a horse on July 24, 1941, and X-rayed the next day at a hospital in Santa Fe. Hans Bethe was visiting him at the time and vividly remembered the incident. Two days later, on July 26, Robert wrote a letter datelined "Cowles [N.M.]." Finally, there was also a record of the Oppenheimer car—with Kitty at the wheel—colliding with a New Mexico Fish and Game truck on the road to Pecos on July 28. All of this made it clear that Oppenheimer had been continuously in New Mexico from at least July 12 through August 11 or 13. Crouch was either mistaken, fantasizing or lying regarding his claim to have seen Robert at a Party meeting in late July in Kenilworth Court.

OVER TIME, Crouch proved to be a highly unreliable informant. In 1953, Armand Scala, an airlines worker and union leader, won a $5,000 libel judgment against the Hearst newspapers when they published one of Crouch's more outlandish allegations. He was also the source for some of Senator Joseph McCarthy's most outrageous charges—such as the claim that communists employed by the State Department had stolen blank American passports and handed them over to agents of the Soviet secret police. Later, Crouch's testimony so tainted one major Justice Department case against leading members of the Communist Party that the Supreme Court was forced in 1956 to dismiss the case.

Eventually, Crouch's lies and theatrics caught up with him. When the syndicated columnists Joseph and Stewart Alsop accused Crouch of committing perjury in a trial of Philadelphia communists, President Eisenhower's attorney general, Herbert Brownell, reluctantly announced that he would "investigate" Crouch. In response, Crouch sued the Alsop brothers for $1 million and warned Brownell that "if my reputation could be destroyed, 31 Communist leaders could get new trials. . . ." Soon, he was calling upon J. Edgar Hoover to investigate the loyalty of Brownell's aides. This prompted the *New York Times* to report that sources in Washington "could not see how the Justice Department could continue using Mr. Crouch." By the end of 1954, Crouch fled to Hawaii, where he attempted to write a memoir entitled *Red Smear Victim*. It was never published, and Crouch died before his libel suit against the Alsops could come to trial.

Yet William Liscum Borden still found Crouch believable. If Crouch was telling the truth, then Oppenheimer the enigma became Oppenheimer the Communist sympathizer. In June 1951, Borden sent one of his staff aides, J. Kenneth Mansfield, to talk with Oppenheimer. Mansfield found Oppenheimer "exceedingly ambivalent" about America's rapidly growing nuclear arsenal. Oppenheimer had explained that he believed strategic nuclear weapons—the city-busters—had only one purpose, to deter the Soviets from attacking the United States. Doubling their number, as the Truman Administration proposed, would not add to that deterrence.

Tactical nuclear warheads were a different matter, Oppenheimer had explained. In 1946, he had disparaged such weapons in a letter to President Truman. But after the Soviet detonation of an atomic bomb in 1949, Oppenheimer and his GAC colleagues had urged the Truman Administration to build more such "battlefield" weapons as an alternative to the Super. As Oppenheimer told Mansfield, the military utility of the nuclear arsenal depended more on the "wisdom of our war plan and our skill in delivery, and less on the actual number of bombs." At the time, American troops were fighting in a real war on the Korean peninsula. Oppenheimer did not advocate the use of atomic weapons in Korea, but he argued that there was an "obvious need" for small, tactical nuclear weapons that could be used on a battlefield. "Only when the atomic bomb is recognized as useful in so far as it is an integral part of military operations," he wrote in the *Bulletin of the Atomic Scientists* in February 1951, "will it really be of much help in the fighting of a war."

"I carried away the impression," Mansfield told Borden, "that Oppenheimer regards war [with the Soviet Union] as unthinkable, a game hardly worth the candle."

> I believe that he accordingly stops short of really thinking out the consequences of his policy of temperance and moderation. I also suspect that his fastidious mind finds the whole notion of strategic bombing essentially clumsy and heavy-handed. It is using the sledgehammer rather than the surgeon's scalpel; it takes no great imagination or sophistication. Couple this with his moral sensibilities of the variety especially pronounced amongst scientists, add on his deep conviction that the Russian people are essentially victims of a tyrannical . . . government, compound this with his distaste for killing noncombatants—and his frequently reiterated stress on the importance of developing tactical uses perhaps becomes more explicable.

Mansfield's June 1951 memo accurately caught the spirit and logic of Oppenheimer's thinking. But Borden appears to have set his mind against the possibility that Oppenheimer's policy recommendations could be explained by logic. He believed there were other, dark influences at work and it became clear to him that others shared that view. Later that summer,

Borden and Strauss got together to discuss their mutual suspicions about Oppenheimer. Strauss "devoted a good part of the conversation to an expression of his fear and concern over Oppenheimer," a summary of their meeting records. They talked at length about Crouch's allegation that Oppenheimer had hosted a secret Communist Party meeting.

Despite all the evidence to the contrary, both men believed Crouch's story; their minds were made up about Oppenheimer's perfidy. Yet they reluctantly concluded that the story could not be confirmed, even with the use of wiretap intelligence. Strauss told Borden, "They [Oppenheimer and his sidekicks] would now be exceedingly careful over the telephone because the 'barber' [Strauss' nickname for Joe Volpe] was in a position to know about possible telephone checks and would have passed this information on." Oppenheimer's friends in the scientific community, they thought, would always protect him, and Oppie seemed to understand that he was being watched. "I pointed out [to Strauss]," Borden noted in a memo to himself, that other officials [presumably the FBI] had the same "feeling of utter frustration about the possibility of any definite conclusions."

In their conspiratorial frame of mind, all Borden and Strauss could see was that Oppenheimer's championing of tactical nuclear weapons was a ploy to block the Super bomb. Indeed, Borden was convinced that in the years 1950–52, Oppenheimer had used all his influence against pursuing the Super's development—even after it became clear in June 1951 that Stanislaw Ulam and Teller had solved the Super's design problems. It did not seem to matter to them that Oppie had pronounced the design "technically sweet," and had formally acquiesced in its development. He and his colleagues on the GAC had repeatedly rejected Teller's proposal to build a second weapons laboratory devoted specifically to the Super, and for Borden and Strauss that was sufficient evidence of Oppenheimer's continuing resistance. But Oppie and his GAC colleagues had their reasons. They believed that dividing America's scientific talent between two weapons laboratories would impede rather than advance scientific progress.

That same year, Teller had gone to the FBI with a laundry list of accusations against Oppenheimer. The general theme of his charges was that Oppenheimer had "delayed or attempted to delay or hinder the development of the H-bomb." Interviewed at Los Alamos, Teller did his best to smear Oppenheimer by innuendo, telling the FBI that "a lot of people believe Oppenheimer opposed the development of the H-Bomb on 'direct orders from Moscow.' " To cover himself, he then said that he did not think Oppie was "disloyal." Instead, he attributed Oppenheimer's behavior to a personality defect: "Oppenheimer is a very complicated person, and an outstanding man. In his youth he was troubled with some sort of physical or mental attacks which may have permanently affected him. He has had great ambitions in science and realizes that he is not as great a physicist as he would like to be." In conclusion, Teller said that he *"would do anything possible"* to see that Oppenheimer's services to the government were terminated.

Teller was not the only H-bomb booster desperate to eliminate Oppenheimer's influence. In September 1951, David Tressel Griggs, a professor of geophysics at UCLA, was appointed chief scientist for the U.S. Air Force. As a RAND consultant in 1946, Griggs had heard rumors about Oppenheimer's security problems, and now his immediate boss, Air Force Secretary Thomas K. Finletter, told him he had "serious questions as to the loyalty of Dr. Oppenheimer." Neither Finletter nor Griggs had any new evidence, but both men believed their suspicions validated by "a pattern of activities, all of which involved Dr. Oppenheimer."

For his part, Oppenheimer questioned the sanity of the Air Force's leadership. He was appalled by their murderous schemes. In 1951, he was shown the Air Force's strategic war plan—which called for the obliteration of Soviet cities on a scale that shocked him. It was a war plan of criminal genocide. "That was the goddamnedest thing I ever saw," he later told Freeman Dyson.

Just weeks after going to work for Finletter in 1951, Griggs led an Air Force delegation to Pasadena for a conference with a group of Caltech scientists. Chaired by Caltech's president, Lee DuBridge, this group had been asked to write a highly classified report—dubbed Project Vista—on what role nuclear weapons might play in the event of a Soviet ground invasion of Western Europe. Griggs and other Air Force officials were alarmed by rumors that the Project Vista report disparaged strategic bombing. The authors of Project Vista reportedly promised to "bring the battle back to the battlefield" by giving small, tactical nuclear warheads priority over city-buster thermonuclear bombs.

Chapter Five of the report even argued that thermonuclear bombs could not be used for tactical purposes on a real battlefield—and suggested that it would serve U.S. interests if Washington publicly adopted a policy of "no first use" of nuclear weapons. The chapter also recommended that SAC receive only a third of the country's precious supply of fissionable material. The remainder would go to the Army for tactical battlefield weapons. Griggs was furious about these recommendations—and not surprised to learn that the primary author of Chapter Five was Robert Oppenheimer.

Oppenheimer had not even been a member of the Project Vista panel. But DuBridge had brought him into its deliberations to help clarify their conclusions. Characteristically, Oppie spent two days reading the panel's materials and then quickly wrote what became the controversial, but highly logical, Chapter Five. Fearing Oppenheimer's persuasive powers, Griggs and his Air Force colleagues did everything they could to bottle up the report. They were not particularly successful; just before Christmas 1951, DuBridge, Oppenheimer and the Caltech scientist Charles C. Lauritsen arrived in Paris to brief the NATO Supreme Commander, Gen. Dwight D. Eisenhower, on Project Vista's conclusions. They impressed upon Eisenhower, an Army man, what a few tactical nuclear warheads could do against a Soviet armored division. Oppie thought the briefing was a "success."

When Finletter learned of the trip, he "went straight through the roof." The Air Force did not want Eisenhower exposed to Oppenheimer's thinking, particularly since his views would support the Army's demand for a bigger share of the atomic budget. Lewis Strauss was also furious, and later wrote Senator Bourke Hickenlooper of Iowa, a conservative member of the Joint Committee on Atomic Energy, that "ever since Oppenheimer and DuBridge spent some time with Gen. Eisenhower in Paris last year I have been concerned over the probability that their visit was primarily for the purpose of indoctrinating him with their plausible but specious policy on the atomic energy situation." The Air Force Chief of Staff, Gen. Hoyt S. Vandenberg, was so alarmed at Oppenheimer's influence that he quietly removed the scientist's name from the Air Force's list of individuals cleared for access to Top Secret information.

Oppenheimer's preference for tactical nuclear weapons as an antidote to genocidal warfare had unintended consequences. By "bringing the battle back to the battlefield," he was also making it more likely that nuclear weapons would actually be used. In 1946, he had warned that atomic weapons "are not policy weapons, but . . . are themselves a supreme expression of the concept of total war." By 1951, however, he was writing in the Vista report, "It is clear that they [tactical atomic weapons] can be used only as adjuncts in a military campaign which has some other components, and whose primary purpose is a military victory. They are not primarily weapons of totality or terror, but weapons used to give combat forces help that they would otherwise lack." That they might also serve as a nuclear trip-wire which could set off an exchange of ever larger nuclear weapons was a scenario that Oppenheimer ignored in his desperation to prevent the Air Force from planning Armageddon under the guise of a rational warfighting strategy.

Griggs and Finletter were further troubled by Oppenheimer's influence over another analysis of nuclear strategy, the 1952 Lincoln Summer Study Group, a classified MIT report on how best to improve the country's air defense against a nuclear attack. The Air Force—dominated as it was by the Strategic Air Command—feared that any investment in air defense would shift resources from SAC's retaliatory forces. And that's exactly what the Lincoln Study Group proposed: to convert "the bulk of the B-47 fleet in the Strategic Air Command" to "long range interceptors, armed with relatively long-range guided missiles." Oppenheimer considered air defense a reasonable priority—but SAC's commanders—all bomber pilots—thought it sheer defeatism.

At the end of 1952, Finletter and other Air Force officials were horrified to learn that someone had slipped the summary report of the Lincoln Study Group to the Alsop brothers. Convinced that Oppenheimer was the culprit, "Finletter was filled with wrath about the collusion of Oppenheimer with the Alsop brothers."

EARLIER THAT SPRING, Griggs had told Rabi that Oppenheimer and the GAC were blocking the development of the Super. Rabi angrily defended his friend and suggested that Griggs should read the minutes of GAC's deliberations; only then, he suggested, would he understand how fairly Oppenheimer chaired these meetings. He then offered to set up a meeting in Princeton between the two antagonists. Griggs agreed.

At 3:30 p.m. on May 23, 1952, Griggs entered Oppenheimer's Princeton office and sat down for what was supposed to be an attempt at mutual understanding. Oppenheimer, however, promptly pulled out a copy of the GAC's October 1949 report with its controversial recommendation against development of the H-bomb. This was like waving a red flag. Oppie might have used his considerable charm to reassure a bureaucratic opponent, but he could not help himself. He saw in Griggs just another idiotic pretender to power, a mediocre scientist who had aligned himself with generals and an ambitious physicist, Edward Teller. He would not stoop to defend himself before such a man, and their conversation quickly became strained. When Griggs asked Oppenheimer if he had circulated a story that had Secretary Finletter boasting that with a few H-bombs the United States could rule the world, Oppenheimer lost what little patience he had maintained to that point. Staring back at Griggs, Oppie said he had heard the story and furthermore, he believed it. When Griggs insisted that he had been in the room on the occasion in question and Finletter had said no such thing, Oppie replied that he had heard it from an unimpeachable source who had also been present.

Since slander was now on the table, Oppenheimer then asked Griggs whether he thought him "pro-Russian or merely confused." Griggs replied that he wished he knew the answer to that question. Well, Oppenheimer said, have you ever assailed my loyalty? Griggs replied that he had indeed heard Oppenheimer's loyalty questioned and he had discussed Oppenheimer as a security risk with both Secretary Finletter and Air Force Chief of Staff Hoyt Vandenberg. At this, Oppenheimer pronounced Griggs "a paranoid."

Griggs left angry, and more convinced than ever that Oppenheimer was dangerous. He subsequently gave Finletter an "eyes only" account of the encounter. For his part, Oppie naïvely thought Griggs too inconsequential to do him harm. To compound his error, a few weeks later Oppenheimer repeated his Princeton performance in a luncheon with Finletter himself. The Air Force secretary's aides thought it time for the two men to meet one-on-one and talk out their differences. But Oppenheimer arrived late from testifying on the Hill and sat stony-faced throughout the lunch as Finletter—a sophisticated Wall Street lawyer—tried repeatedly to draw him out. Making no effort to disguise his contempt, Oppenheimer was "rude beyond belief." He had come to loathe these Air Force men with their commitment to building more and more bombs for the purpose of killing more and more millions of people. To his mind, they were so dangerous, so morally obtuse, that he almost welcomed them as political enemies. A few weeks later, Finletter and his people told the Joint Committee on Atomic Energy that it was an open question "whether [Oppenheimer] was a subversive."

FINLETTER'S ACCUSATIONS against Oppenheimer were a reflection of the extremes to which those engaged in the nuclear debate were driven. Oppenheimer himself was not immune to this contagion. In June 1951, he gave an off-the-record speech to the Committee on the Present Danger (of which he was a member), a private group dedicated to lobbying the government to build up its conventional defenses. Speaking without notes, he made the argument for a real defense of Western Europe, one that would "leave Europe free, not destroyed [by atomic bombs]." "In dealing with the Russians," he concluded, "we are coping with a barbarous, backward people who are hardly loyal to their rulers. Our supreme policy should ultimately be to 'get rid of this atom-stuff as a weapon.' "

As a measure of just how far his thinking had evolved, by 1952 Oppenheimer was heard speculating aloud about the possibility of preventive war, an idea he had abhorred only three years earlier. To be sure, he never actually advocated it, but on several occasions he broached its possibility. In January 1952, Oppenheimer had a discussion with the Alsop brothers, and Joe Alsop noted that "Oppie's line, to put it bluntly, was something damned close to preventive war; we can't just sit by while a potential enemy builds up the means of our certain destruction."

In February 1953, Oppenheimer gave a talk at the Council on Foreign Relations and was asked if the notion of preventive war had any meaning under present conditions. He replied, "I think it does. The general impression I have is that the United States would physically survive, damaged, but physically survive a war that not only started now, but didn't last too long. . . . That does not mean that I think it is a good idea. I believe that until you have looked the tiger in the eye you are going to be in the worst of all dangers, which is that you will back into it."

By 1952, Oppenheimer was generally fed up with Washington. President Truman had ignored his counsel so often that he now took steps to walk away from the whole business of policy-making. Early in May, he lunched at Washington's Cosmos Club with James Conant and Lee DuBridge. The three friends commiserated and gossiped about their standing in Washington. Afterwards, Conant noted in his diary: "Some of the 'boys' have their axe out for three of us on the GAC of AEC. Claim we have dragged our heels on H bomb. Dark words about Oppie!" In June, frustrated by more than a decade of dealing with "a bad business now threatening to become really bad," and aware that there was a movement afoot to remove them from the GAC, all three men submitted their resignations from that advisory committee. Oppenheimer wrote his brother that he now intended to devote himself to physics: "Physics is complicated and wondersome, and much too hard for me except as a spectator; it will have to get easy again one of these days, but perhaps not soon."

But it wasn't that easy to walk away from Washington. Even as he resigned from the GAC, the AEC's Gordon Dean persuaded him to remain available as a contract consultant. This automatically extended his top-secret Q clearance for another year. And that was not all. In April, he had agreed to Secretary of State Dean Acheson's request that he sit on a special State Department Panel of Consultants on Disarmament. Serving with him were Vannevar Bush, Dartmouth College president John Sloan Dickey, CIA Deputy Director Allen Dulles, and Joseph Johnson, president of the Carnegie Endowment for International Peace. As usual, the panel elected him chairman.

Acheson also recruited McGeorge Bundy—then a thirty-three-year-old professor of government at Harvard—to serve as the panel's recording secretary. "Mac" Bundy was the son of Henry Stimson's righthand man, Harvey Bundy, and he was eager to meet Oppenheimer. Bundy was smart, articulate and witty. As a Junior Fellow at Harvard, he had coauthored Stimson's 1948 memoir, *On Active Service in Peace and War.* And as the ghostwriter of Stimson's famous *Harper's* magazine essay of February 1947 defending the atomic bombings of Hiroshima and Nagasaki—"The

Decision to Use the Atomic Bomb"—Bundy was already familiar with some of the imponderables associated with nuclear weapons. At their first meeting, Oppenheimer took an instant liking to the precocious young Boston Brahmin. Afterwards, Bundy wrote an uncharacteristically humble note to his new friend, saying, "I find it hard to thank you enough for the patience with which you undertook my education last week; I only hope that somehow I can be useful enough to make it worth your effort." In no time at all, the two men were exchanging handwritten notes to each other addressed as "Dear Robert" and "Dear Mac," in which they discussed everything from the merits of Harvard's physics department to the health of their wives. Bundy thought Robert was "marvelous, fascinating and complicated."

Bundy would soon learn that controversy stalked his new friend. In one of their early meetings, Oppenheimer and his fellow panelists agreed that their primary question was the "problem of survival" in which the United States and Russia faced a "scorpion stalemate—which might or might not involve active war without the use of stings. . . ." Oppenheimer knew that Teller and his colleagues were hoping to test an early design for the hydrogen bomb later that autumn. So he was intrigued when Vannevar Bush suggested that before this threshold was crossed, perhaps Washington and Moscow should agree to a complete ban on the testing of any thermonuclear devices. Such a treaty would require no inspections, since any violation of the ban would immediately be detected. And without tests, the H-bomb could not be developed into a reliable military weapon. A thermonuclear arms race could be stopped before it began.

The Oppenheimer panel continued their discussions in June at a meeting hosted by Bundy in his Cambridge home, a rambling nineteenth-century house within bicycling distance of Harvard Square. James Conant joined them as an unofficial participant. Conant had soured on nuclear weapons; according to Bundy's notes, Conant complained that the "ordinary American" thought of the bomb as a weapon threatening the Soviets, "while the more significant fact was that now and in the future such blows could be delivered by others on the United States." Even without the H-bomb, Conant argued, all but the largest of U.S. cities could easily be wiped out with a single atomic weapon. No one in the room disagreed.

The public's ignorance was bad enough, but even worse, Conant said, was "the attitude of the leaders of the American military establishment." Our generals were relying almost exclusively on these weapons as "their principal hope of victory in the event of all-out war." If the country built up its conventional forces, "it would become possible for the United States to dispense with its present reliance on atomic bombs." But for this to happen,

Conant said, the generals "must be persuaded that atomic weapons in the long run are on balance a danger to the United States."

Without any prompting from Oppenheimer, Conant proposed what would become known two decades later as a "no-first-use policy." The United States, he said, should "announce officially that we would not be the first to use atomic weapons in any new war." He also agreed with Bush's proposal to announce a tacit moratorium on the testing of a thermonuclear bomb. Oppenheimer endorsed both ideas. The panel's argument on behalf of a moratorium was particularly compelling. They told Acheson:

> . . . it seems to us almost inevitable that a successful thermonuclear test will provide a heavy additional stimulus to Soviet efforts in this field. It may well be true that the Soviet level of effort in this area is already high, but if the Russians learn that a thermonuclear device is in fact possible, and that we know how to make it, their work is likely to be considerably intensified. It is also likely that Soviet scientists will be able to derive from the test [by analyzing the fallout] useful evidence as to the dimensions of the device.

Oppenheimer and his colleagues knew that the first test of a thermonuclear device—code-named "Mike"—was scheduled for the coming autumn, and that any attempt to stop it would be vigorously opposed by the Air Force. Though convinced of the soundness of their ideas, they had no means of making their views public. A veil of secrecy was tightly draped over all atomic matters, and they could not speak about their concerns without violating their security clearances. So they tried once again to convince Washington's foreign policy establishment that current nuclear weapons policies were a dead end. But on October 9, 1952, Truman's National Security Council flatly rejected the Oppenheimer panel's proposal for a moratorium on the testing of the H-bomb. Defense Secretary Robert Lovett angrily said that "any such idea should be immediately put out of mind and that any papers that might exist on the subject should be destroyed." Lovett, a powerful member of the foreign policy establishment, feared that if news of the moratorium idea leaked, Senator Joseph McCarthy would have a field day investigating the State Department and its panel of advisers.

Three weeks later, the United States exploded a 10.4-megaton thermonuclear bomb in the Pacific, vaporizing the island of Elugelab. A clearly depressed Conant told a *Newsweek* reporter, "I no longer have any connection with the atomic bomb. I have no sense of accomplishment."

A week later, Oppenheimer sat grimly with nine other members of yet another panel—the Science Advisory Committee to the Office of Defense Mobilization—debating whether or not they ought to resign in protest.

Many scientists felt the "Mike" test demonstrated that the government simply had no intention of listening to their expert advice. Oppie's old friend Lee DuBridge circulated a draft resignation letter. But in the end, the faint hope that the next administration might change course persuaded them to set the letter aside. They knew the odds were against them. At one point, James R. Killian, president of MIT, leaned over to DuBridge and whispered, "Some people in the Air Force are going to be after Oppenheimer, and we've got to know about it and be ready for it." DuBridge was shocked. He naïvely thought everyone still regarded Oppie as a hero.

In the meantime, Oppenheimer worked with Mac Bundy on drafting a final report for the State Department's disarmament panel. This document was forwarded to departing Secretary of State Acheson just as Dwight D. Eisenhower moved into the White House. At the time, of course, this paper was highly classified and circulated among only a handful of Eisenhower Administration officials. Had it been released in 1953, it surely would have created a firestorm of controversy. While Bundy was the document's wordsmith, many of the ideas were Oppenheimer's: Nuclear weapons would soon threaten all civilization. In just a few years, the Soviet Union could have 1,000 atomic bombs, and "5,000 only a few years further on." This constituted "the power to end a civilization and a very large number of people in it."

Bundy and Oppenheimer conceded that a "nuclear stalemate" between the Soviets and the United States might evolve into a "strange stability" in which both sides would refrain from using these suicidal weapons. But if so, "a world so dangerous may not be very calm, and to maintain the peace it will be necessary for statesmen to decide against rash actions not just once, but every time." They concluded that "unless the contest in atomic armaments is in some way moderated, our whole society will come increasingly into peril of the gravest kind."

In the face of such peril, the Oppenheimer panelists promoted the idea of "candor." A policy of excessive secrecy had kept Americans complacent and ignorant of the nuclear peril. To rectify matters, the new administration "should tell the story of the atomic danger. . . ." Astonishingly, the panelists even recommended that "the rate and impact of atomic production" should be revealed to the public, "and that it should direct attention to the fact that beyond a certain point we cannot ward off the Soviet threat merely by 'keeping ahead of the Russians.' "

The notion of "candor" was directly inspired by Niels Bohr, who had always insisted that security was inextricably linked to "openness." In this, Oppie was still Bohr's prophet. He no longer put any stock in the long-deadlocked UN disarmament talks. But he hoped that a new administration would see that "candor" could both alert the American people to the real perils of relying on nuclear weapons and signal to the Soviets that Americans did not intend to use these weapons in a preemptive first strike. In addition, the Disarmament Panel urged direct, continual communication with the Soviets. The Kremlin should know roughly the size and nature of the American nuclear arsenal—and that Washington strongly favored bilateral talks to reduce this arsenal.

If the recommendations of the Oppenheimer panel had been accepted by the Eisenhower Administration in 1953, the Cold War might have taken a different, less militarized trajectory. This tantalizing speculation was later advanced by Bundy in his 1982 essay in the *New York Review of Books,* "The Missed Chance to Stop the H-Bomb." And in the years since the demise of the Soviet empire, Russian archival documents have compelled historians to rethink basic assumptions about the early Cold War. The "enemy archives," as the historian Melvyn Leffler has written, demonstrate that the Soviets "did not have pre-conceived plans to make Eastern Europe communist, to support the Chinese communists, or to wage war in Korea." Stalin had no "master plan" for Germany, and wished to avoid military conflict with the United States. At the end of World War II, Stalin reduced his army from 11,356,000 in May 1945 to 2,874,000 in June 1947—suggesting that even under Stalin, the Soviet Union had neither the capability nor the intention to launch a war of aggression. George F. Kennan later wrote that he "never believed that they [the Soviets] have seen it in their interest to overrun Western Europe militarily, or that they would have launched an attack on that region generally even if the so-called nuclear deterrent had not existed."

Stalin ran a cruel police state, but economically and politically it was a totalitarian state in decay. When Stalin died in March 1953, his successors, Georgi Malenkov and Nikita Khrushchev, began a process of de-Stalinization. Both Malenkov and Khrushchev also had a sound appreciation for the inherent dangers of a nuclear arms race. Malenkov, a technocrat with an interest in quantum physics, stunned the Politburo in 1954 with a speech in which he said that the use of the hydrogen bomb in war "would mean the destruction of world civilization." Khrushchev, an erratic, mercurial leader, sometimes frightened Western audiences with his blustery rhetoric. But in practice he pursued the kind of foreign policy that would later become associated with détente, and even exhibited the first glimmers of glasnost. He renewed arms control talks with the West in 1955 and by the end of the 1950s he had sharply cut the Soviet defense budget. After receiving his first briefing on nuclear weapons in September 1953, Khrushchev later recalled, "I couldn't sleep for several days. Then I became convinced that we could never possibly use these weapons."

It would have required extraordinary efforts to persuade Khrushchev to embrace the kind of radical arms control regime Oppenheimer's panel envisioned. But the Eisenhower Administration never even tried to go down that path. Yet no less a Sovietologist than the highly regarded U.S. ambassador to Moscow Charles "Chip" Bohlen later wrote in his memoirs that Washington's failure to engage Malenkov in meaningful negotiations over nuclear weapons and other issues was a missed opportunity.

By 1953, the Cold War had frozen the policy options in Washington at least as hard as they were frozen in Moscow, and Oppenheimer's persistent efforts to somehow keep the nuclear genie attached to the bottle, if not in it, ran against the current of powerful forces at home. Now that a Republican was president, these political forces were determined to put Oppenheimer into a bottle—and throw him out to sea.

CHAPTER THIRTY-TWO

"Scientist X"

He [Oppie] had had enough of me and I had had enough of him too.

JOE WEINBERG

B Y THE SPRING OF 1950, Oppenheimer had every reason to think that the FBI, HUAC and the Justice Department were all closing in on him. Hoover was telling his agents that Oppenheimer might be indicted for perjury, and they had to continue to investigate him vigorously. Twice that spring, FBI agents had interviewed him in his Princeton office. The agents noted that while he had been "entirely cooperative," he had also "expressed great concern over the possibility of allegations concerning his past affiliation with the Communist Party being made a matter of public trial." He was deeply worried that his name would be linked to Joe Weinberg—whom the Crouches and HUAC had identified as "Scientist X," a Soviet spy. Oppenheimer had last seen Weinberg at a Physical Society conference in 1949, shortly after Weinberg's troubles with HUAC had commenced. On this occasion, Weinberg sensed a coolness in their relationship. "So there was a cloud over our relationship at that point," Weinberg recalled. "The cloud would be that Oppie wouldn't know just what I was going to do. He would have to worry that the pressure on me might eventually be turned against him in some way. . . . It was clear that he felt that there were things that were damaging to him that I could be made to say, whether or not I knew them, if I were in some way weak."

Weinberg admitted to feeling "terrified" and bewildered by what was happening to him. He knew, of course, that he was guilty of having discussed the bomb project with Steve Nelson in 1943, but he did not know that his conversation had been recorded. Nor did he believe that he had committed espionage. Recently, the *Milwaukee Journal* had published an outlandish story claiming that Weinberg had been a courier for the Soviets—and that he had even passed on a sample of uranium-235. "My God," he thought, "what connections could they have made to make up such a theory?" For a time, he felt he might crack. "I felt desperate, I felt utterly alone and broken down and beset on every side. I literally trembled. God knows what I could have been made to say if they [the FBI] had followed up."

Fortunately for Weinberg, the authorities were moving slowly. That spring a federal grand jury in San Francisco was weighing a perjury indictment against him. But the Justice Department had very little usable evidence. Weinberg had testified under oath that he had never been a Communist Party member and that he had never even met Steve Nelson. But the FBI wiretap was illegal and, therefore, inadmissible in court, and there was no other evidence that Weinberg had been a CP member. By April 1950, the Bureau had interviewed eighteen present and former Communist Party members in the San Francisco area, and none of them was able to link Weinberg to the Party. In the absence of the the wiretap evidence, the grand jury failed in 1950 to return an indictment against Weinberg.

Undeterred by this setback, the Justice Department convened a second grand jury in the spring of 1952. Their only other evidence against Weinberg was Paul Crouch's testimony that he had seen Weinberg at a Party meeting talking to Nelson. Prosecutors were well aware that Crouch's testimony might be unreliable—but they may have calculated that a trial would shake loose further evidence against Weinberg, and perhaps even Oppenheimer. By that time, Weinberg had mustered the courage to stick it out. "They were fools," Weinberg later said of his antagonists. "They waited until I got a little bit less desperate and a little bit hardened." Interviewed by the grand jury, he refused to tell them anything—and certainly nothing about Oppie. "I was not going to get Oppie into this at all," Weinberg said. "That was the thing that would happen over my dead body."

By then, Oppenheimer had been interviewed once again about the Crouches' allegation that he had hosted a Party meeting at his Kenilworth Court home in Berkeley in July 1941. On this occasion, two investigators from the Senate Judiciary Committee questioned him in the presence of his attorney, Herbert Marks. Oppenheimer again denied knowing either of the Crouches; he also denied that he had ever met Grigori Kheifets, a Soviet intelligence officer stationed in San Francisco, and he denied that Steve Nelson had ever approached him for information about the bomb project.

The interview was conducted in a less-than-friendly manner. Seeing the Senate staffers taking careful notes, Marks interrupted to say that he wanted a copy of any record they made of the conversation. When they rebuffed this request, Marks insisted that if they wished to continue to ask about Oppenheimer's affairs, "we would want a transcript." To this the Senate staffers coldly observed that last spring Oppenheimer had been under subpoena, and at that time, Oppenheimer's other lawyer, Joe Volpe, had suggested that Oppenheimer be interviewed in an "informal talk." They thought, said the Senate staffers, that they "were being nice about it." On this note, the twenty-minute interview soon ended. Such encounters convinced both Oppenheimer and Marks that the Crouches' allegations had not been put to rest.

On May 20, 1952, just three days before Weinberg's indictment, Oppenheimer arrived in Washington for yet another interrogation. The lawyers who were about to prosecute Weinberg had decided it might be useful to confront Oppenheimer with his accuser. Four years earlier, Richard Nixon and his HUAC investigators had lured an unsuspecting Alger Hiss to a room in New York City's Commodore Hotel and confronted him with his accuser, Whittaker Chambers. Hiss was now serving a prison term for perjury. Perhaps, Justice Department investigators reasoned, Nixon's tactic was worth trying on Oppenheimer.

Accompanied by his attorneys, Oppenheimer walked into the Justice Department to be interviewed by lawyers from the Criminal Division. Questioned about the alleged July 1941 meeting in his Kenilworth Court home, he once again denied the Crouches' story and insisted that he had been in New Mexico at the time. He said he did not know either Paul or Sylvia Crouch, and "that no such persons" had come to his home during that period to talk about communism or the invasion of Russia. He said that he had read Crouch's testimony before the California State Committee on Un-American Activities (the Tenney Committee) and he had no recollection of the meeting Crouch described. He volunteered that he had talked with his wife and also with Kenneth May, and "they confirmed his recollection that no such meeting occurred."

At this point, the Justice Department lawyers turned to Oppenheimer's attorneys—Herb Marks and Joe Volpe—and said that Paul Crouch was sitting in the next room. Would it be acceptable, they asked, if Crouch was brought into the room "to see if he would recognize Dr. Oppenheimer, as well as to see if Dr. Oppenheimer would recognize Crouch . . ."? With Oppenheimer's acquiescence, Marks and Volpe agreed. The door then opened and Crouch walked up to Oppenheimer, shook his hand and said, "How do you do, Dr. Oppenheimer?" He then turned melodramatically to the lawyers and said that the man with whom he had just shaken hands was the same person who had been his host at a meeting in July 1941 at 10 Kenilworth Court. Crouch reiterated that he had given a talk on the "Communist Party propaganda line to be followed after the invasion of Russia by Hitler."

If Oppenheimer was taken aback by this performance, the FBI record fails to report it. Instead, it merely notes that he quickly responded that he did not know Crouch. Prompted to describe the July 1941 meeting in greater detail, Crouch said that he remembered Oppenheimer asking him several questions at the conclusion of his hour-long presentation. To this Oppenheimer interrupted to ask what exactly he was supposed to have said in this question period. Crouch then claimed that Oppenheimer's questions involved a philosophical analysis of Russia's involvement in the war "based upon Marxian doctrine." Crouch said, "Dr. Oppenheimer stated that he could see why we should give aid to Russia but asked why we should aid Britain, who might double-cross us." Crouch claimed Oppenheimer also asked whether or not the German invasion of Russia had now created two wars: a "British-German imperialistic war" and a "Russian-German people's war." To this Oppenheimer said that such questions by him "were impossible because he had never at any time thought or advanced the suggestion of two wars."

Marks and Volpe made some attempt to trip Crouch up by asking him about Oppenheimer's appearance. Did he look substantially the same as he did in 1941? Crouch replied that he looked the same. What about his hair? asked one of the two attorneys. Crouch allowed that Oppenheimer's hair might be a little shorter than in 1941, but he hadn't really concentrated on his hair. In fact, in 1941 Oppenheimer had worn his hair in a long, bushy cut; by 1952 he kept it very short, almost a crew-cut. Still, this was a small discrepancy.

On the whole, Crouch had demonstrated that he might be a credible witness against Oppenheimer in a court trial. He had described the interior of Oppenheimer's home, and he also had seemed credible in claiming to have seen Oppenheimer in the autumn of 1941 at a housewarming party for Ken May. Oppenheimer had conceded that he remembered dancing with a Japanese girl at a party that might well have been May's housewarming party. This might be regarded as an important admission, since Crouch further claimed that he had seen Oppenheimer deep in conversation at this party with Ken May, Joseph Weinberg, Steve Nelson and Clarence Hiskey, another physics student at Berkeley.

After Crouch finally left the room, Oppenheimer turned to the Justice Department lawyers and once again stated that he had no recollection of ever having met Crouch. At that, he was excused. He left with Marks and Volpe, and afterwards the three men speculated on what the Justice Department's next step would be.

Three days later, on May 23, 1952, they learned of the indictment of Weinberg, and that the indictment made no mention of Crouch, Oppenheimer or the Kenilworth meeting. In fact, Oppenheimer's lawyers had lobbied the Justice Department, through AEC chairman Gordon Dean, to drop the Kenilworth incident from the indictment. Oppenheimer was relieved—but only momentarily.

JOE WEINBERG'S perjury trial finally commenced in the autumn of 1952, and almost immediately Oppenheimer was given notice by the government that he might be called as a witness. Herb Marks again diligently lobbied the Justice Department to keep Oppenheimer's name off the witness list. Among other things, he persuaded the AEC chairman, Gordon Dean, to write President Truman, urging him to order the Justice Department to exclude Crouch's charges from the trial proceedings. "It will be Oppenheimer's word against Crouch's word," Dean wrote the president. "Whatever the outcome of the Weinberg case, Dr. Oppenheimer's good name will be greatly impaired and much of his value to the country will be destroyed." Truman replied the very next day: "I am very much interested in the Weinberg-Oppenheimer connection. I feel as you do that Oppenheimer is an honest man. In this day of character assassination and unjustified smear tactics it seems that good men are made to suffer unnecessarily." Truman, however, gave no indication of what he would do.

Early that autumn, when the Justice Department's bill of particulars against Weinberg was filed, it made no mention of Oppenheimer. But after Dwight Eisenhower's election to the presidency in early November a harsher attitude toward security cases was instituted. A Justice Department official called Joe Volpe on November 18, 1952, and said, "Oppie will have to be brought into it." The San Francisco Chronicle, among other newspapers, picked up on wire service reports: ". . . government prosecutors said today that Dr. Joseph Weinberg attended a Communist Party meeting in Berkeley, Calif. in a 'residence believed to have been . . . occupied by J. Robert Oppenheimer.' " The very next day, Oppenheimer was served a subpoena by Weinberg's attorney to appear in court as a defense witness. Oppie let Ruth Tolman know how upset he was and she wrote him back, "Such a miserable business. Robert, I know how worrisome the prospect must be."

Marks and Volpe understood that anything could happen in a trial where it was one person's word against another's. If Weinberg was convicted of perjury, that would pave the way for an indictment of Oppenheimer himself. So, once again, Marks and Volpe scrambled to get Oppenheimer removed from the case. In a meeting with prosecutors, they argued that "it seemed a terrible thing to subject Oppenheimer to the embarrassment and grief . . . and expressed the hope that a way could be found to avoid doing this to a man who has been so important to his country. . . . [T]here would be no better way for Joe Stalin to play his game than to create suspicion about people like Oppenheimer."

In late January, soon after Eisenhower's inauguration, Volpe and Marks again approached AEC Chairman Dean and asked him if there "isn't some natural and in-channel way of getting this question considered on a higher level." But when the trial finally commenced in late February, Weinberg's lawyer announced that Oppenheimer would appear as a defense witness and that he would testify that the Kenilworth Court meeting never took place. In his opening statement, Weinberg's defense counsel dramatically announced that "this case can be cut down to whether they believe the word of a criminal [Crouch] or the word of a distinguished scientist and outstanding American. . . ."

Oppenheimer had to go to Washington to be ready to appear in court at a moment's notice. But on February 27, he was told that he probably would not have to testify; the Justice Department had suddenly agreed to drop the part of the indictment pertaining to the Kenilworth meeting. In the interest of protecting the AEC's reputation, Gordon Dean had evidently leaned on the Justice Department. Oppie took the train home on the evening of February 27 and arrived late to a party at Olden Manor hosted by Ruth Tolman, who was visiting from California. Ruth could see that he "felt so worn out and worried and frazzled." But at least he had escaped "all the miseries of subpoenas and the like."

Since the prosecution was barred from introducing the illegal FBI wiretap of Weinberg's conversation with Steve Nelson, its case was now transparently weak. The trial ended on March 5, 1953, with Weinberg's acquittal. In an extraordinary departure from legal norms, U.S. District Court Judge Alexander Holtzoff told the jury that "the court does not approve of your verdict." He went on to observe that testimony in the trial had unearthed "an amazing and shocking situation existing in the crucial years of 1939, 1940, and 1941 on the campus of a great university in which a large and active Communist underground organization was in operation."*

Nevertheless, Oppenheimer was greatly relieved. The whole business, he hoped, was finally put to rest. When David Lilienthal learned that Oppenheimer would not be called to testify in the case, he wrote his old friend, "With so many mean and unjust things happening, we're entitled to some decency even in these days." Ironically, one day when Oppenheimer happened to be up on Capitol Hill he stepped into an elevator and saw Sen-

---

*The prosecutor in the case, William Hitz, was equally outrageous. He told members of the grand jury that had indicted Weinberg, "We got enough evidence to hang the son-of-a-bitch; but it's illegal and we can't present it." In fact, the evidence of spying was ambiguous.

ator McCarthy. "We looked at each other," Robert later told a friend, "and I winked."

Joe Weinberg, now thirty-six, had a life again—though not a job. The University of Minnesota had dismissed him two years earlier when HUAC had labeled him "Scientist X." And despite his acquittal, the university's president announced that Weinberg would not be reinstated because of his refusal to cooperate with the FBI. Turning one last time to his mentor, Weinberg wrote Oppie asking for a letter of recommendation for a possible job at an optics firm. Weinberg assured him that "this will be the last time I will ever disturb you." Although Oppenheimer had every reason to believe the FBI would find out about it, as they did, he wrote a supportive letter for Weinberg, who got the job. Weinberg was grateful, but years later, when asked to reflect on his relationship with Oppie he replied, "He had had enough of me and I had had enough of him too."

The Weinberg case had been emotionally draining, and it had been an expensive ordeal. On December 30, 1952, even before the case went to trial, Oppenheimer had dropped by the offices of Lewis Strauss, telling him that he had a personal matter to discuss. His attorneys, he said, had just billed him $9,000 for their representation of him as a potential witness in the Weinberg case. The legal fees far exceeded his expectations, and he "did not know how to handle it." He then asked Strauss if he, in his capacity as chairman of the board for the Institute, would recommend that the Institute pay his legal expenses. Strauss firmly replied that this would be a "mistake." When Oppenheimer pointed out that Corning Glass Company paid the legal bills of his friend, Dr. Ed Condon, Strauss said the circumstances were not parallel. Dr. Condon's employers, he pointed out, had known of Condon's problems with HUAC before they had hired him. The Institute's trustees, Strauss said coldly, had "no indication whatever" that Oppenheimer had such problems. This, of course, was not true; in 1947, Oppenheimer had informed Strauss of his documented left-wing past. Nevertheless, Strauss suggested that his legal bills were high because his lawyers thought him "quite rich and that the traffic would bear it."

Oppenheimer replied testily that Strauss must know that was not the case since his tax returns were prepared by an Institute office manager under Strauss' supervision. Strauss said no, he had "no idea what his income situation was." To this, Oppenheimer said that he "was not wealthy—that he had a modest income aside from his Institute salary. . . ." He allowed that some people might think him wealthy because he had inherited "some quite extraordinary works of art." Clearly unsympathetic, Strauss ended the meeting by saying that he would not raise the issue with the trustees "at this time." Oppenheimer left angry and humiliated; henceforth he knew he could count on Strauss' hostility. He decided simply to go around him and send the legal bill to the Institute's trustees, hoping they would pay it. But Strauss later told the FBI that he had persuaded the "long-haired professors" on the board to reject the bill. By the spring of 1953, the enmity between the two men was palpable to everyone who knew them.

CHAPTER THIRTY-THREE

"The Beast in the Jungle"

We may be likened to two scorpions in a bottle, each capable of killing the other, but only at the risk of his own life.

J. ROBERT OPPENHEIMER, *1953*

OPPENHEIMER HAD LONG BEEN HARBORING a vague premonition that something dark and momentous lay in his future. One day in the late 1940s, he had picked up a copy of Henry James' short story "The Beast in the Jungle," a tale of obsession, tormented egotism and existential foreboding. "Utterly transfixed" by the story, Oppenheimer immediately called up Herb Marks. "He was very anxious that Herb read it," recalled Marks' widow, Anne Wilson Marks. James' central character, John Marcher, encounters a woman he met many years earlier, and she recalls his having confided in her that he was haunted by a premonition: "You said you had had from your earliest time, as the deepest thing within you, the sense of being kept for something rare and strange, possibly prodigious and terrible, that was sooner or later to happen to you, that you had in your bones the foreboding and the conviction of, and that would perhaps overwhelm you."

Marcher confesses that whatever it is, the event hasn't happened—yet: "It hasn't yet come. Only, you know, it isn't anything I'm to *do,* to achieve in the world, to be distinguished or admired for. I'm not such an ass as *that.*" When the woman asks, "It's to be something you're merely to suffer?" Marcher replies, "Well, say to wait for—to have to meet, to face, to see suddenly break out in my life; possibly destroying all further consciousness, possibly annihilating me; possibly, on the other hand, only altering everything, striking at the root of all my world and leaving me to the consequences. . . ."

Since Hiroshima, Oppenheimer had lived with just such a peculiar sense that someday his own "beast in the jungle" would emerge to alter his existence. For some years now, he had known that he was a hunted man.

And if there was a "beast in the jungle" waiting for him, it was Lewis Strauss.

ON FEBRUARY 17, 1953, some six weeks before Joe Weinberg was finally acquitted, and thus at a moment when Oppenheimer still felt vulnerable, he nevertheless gave a speech in New York that was essentially an unclassified version of the disarmament report he and Bundy had recently sent the new Eisenhower Administration urging a policy of "candor" about nuclear weapons. According to the historian Patrick J. McGrath, Oppenheimer gave the speech with the consent of Eisenhower—but he surely realized that it would anger his political enemies in Washington. His chosen audience was a closed meeting of members of the Council on Foreign Relations. Precisely because the Council was such an elite venue, his words were sure to resonate loudly throughout Washington's military and policy-making circles. Sitting in the audience that day were such luminaries of the foreign policy establishment as the young banker David Rockefeller, the *Washington Post* publisher Eugene Meyer, the *New York Times* military correspondent Hanson Baldwin and the Kuhn, Loeb investment banker Benjamin Buttenwieser. Also there that evening—Lewis L. Strauss.

Introduced by his good friend David Lilienthal, Oppie began by noting that he had entitled his talk "Atomic Weapons and American Policy." To polite laughter, he acknowledged that this was a "presumptuous title," but he begged his listeners' indulgence, explaining, "Any smaller vehicle would give an impression of clarity different than that which I wanted to communicate."

He then observed that because almost everything associated with nuclear weapons was classified, "I must reveal its nature without revealing anything." He pointed out that since the end of the war, the United States had been compelled to come to grips with the "massive evidences of Soviet hostility, and the growing evidences of Soviet power." The role of the atom in this Cold War was a simple one: American policy-makers had concluded, "Let us keep ahead. Let us be sure that we are ahead of the enemy."

Turning to the status of this race, he reported that the Soviets had produced three atomic explosions and were manufacturing substantial quantities of fissionable material. "I should like to present the evidence for this"; he said, "I cannot." But he said that he could reveal his own casual estimate of where the Soviets stood in relation to America: "I think that the USSR is about four years behind us." This might sound somewhat reassuring, but after reviewing the effects of one bomb delivered onto Hiroshima, Oppenheimer observed that both sides understood that these new weapons could become even more lethal. Vaguely alluding to missile technology, he said that technical developments would soon bring "more modern, more flexible, harder to intercept" delivery vehicles. "All of this is in train," he said. "It is my opinion that we should all know—not precisely, but quantitatively and, above all, authoritatively—where we stand in these matters."

The facts were essential to any understanding. But the facts were classified. "I cannot write of them," he said, emphasizing once again the albatross of secrecy. "What I can say is this: I have never discussed these prospects candidly with any responsible group, whether scientists or statesmen, whether citizens or officers of the government, with any group that could steadily look at the facts, that did not come away with a great sense of anxiety and somberness at what they saw." Looking a decade ahead, he said, "it is likely to be small comfort that the Soviet Union is four years behind us.... The very least we can conclude is that our twenty-thousandth bomb ... will not in any deep strategic sense offset their two-thousandth."

Without revealing the specific numbers, Oppenheimer said that the American stockpile of atomic weapons was growing rapidly. "We have from the first maintained that we should be free to use these weapons; and it is generally known we plan to use them. It is also generally known that one ingredient of this plan is a rather rigid commitment to their use in a very massive, initial, unremitting strategic assault on the enemy." This, of course, was a succinct definition of the Strategic Air Command's war plan—to obliterate scores of Russian cities in a genocidal air strike.

Atomic bombs, he continued, are "almost the only military measure that anyone has in mind to prevent, let us say, a great battle in Europe from being a continuing, agonizing, large-scale Korea." And yet, the Europeans are "in ignorance of what these weapons are, how many there will be, how they will be used and what they will do."

Secrecy in the atomic field, he charged, was leading to widespread rumor, speculation and outright ignorance. "We do not operate well when they [important facts] are known, in secrecy and in fear, only to a few men." Former president Harry Truman had recently disparaged the notion that the Soviets were developing a nuclear arsenal capable of harming continental America. Oppenheimer noted sharply: "It must be disturbing that an ex-President of the United States, who has been briefed on what we know about the Soviet atomic capability, can publicly call in doubt all conclusions from the evidence." He also ridiculed a "high officer of the Air Defense Command" for saying, only a few months ago, that "it was our policy to attempt to protect our striking force, but not really our policy to attempt to protect this country, for that is so big a job that it would interfere with our retaliatory capabilities." Oppenheimer concluded that such "follies can occur only when even the men who know the facts can find no one to talk about them, when the facts are too secret for discussion, and thus for thought."

The only remedy, Oppenheimer concluded, was "candor." Officials in Washington, D.C., had to start leveling with the American people, and tell them what the enemy already knew about the atomic armaments race.

It was an extraordinarily perceptive and brazen speech. Again and again, Oppenheimer observed that he was barred from speaking of the essential facts—and then like a Brahmin priest endowed with special knowledge, he proceeded to reveal the most fundamental secret of all—that no country could expect in any meaningful sense to win an atomic war. In the very near future, he said, "we may anticipate a state of affairs in which the two Great Powers will each be in a position to put an end to civilization and life of the other, though not without risking its own." And then, in a chilling turn of phrase that startled everyone who heard it, Oppenheimer quietly added, "We may be likened to two scorpions in a bottle, each capable of killing the other, but only at the risk of his own life."

It is hard to imagine a more provocative speech. After all, the new administration's secretary of state, John Foster Dulles, was an outspoken advocate of a defense doctrine based on massive retaliation. And yet, here was the father of the atomic era declaring that the fundamental assumptions of the country's defense policy were laced with ignorance and folly. The nation's most famous nuclear scientist was calling upon the government to release heretofore closely guarded nuclear secrets, and to discuss candidly the consequences of nuclear war. Here was a celebrated private citizen, armed with the highest security clearance, denigrating the secrecy that surrounded the nation's war plans. As word spread through Washington's national security bureaucracy of what Oppenheimer had said, many were appalled. Lewis Strauss was seething.

On the other hand, most of the lawyers and investment bankers listening to Oppenheimer's speech at the Council came away impressed. Even the new president of the United States, Dwight D. Eisenhower, when he later read the speech, was taken by the notion of candor. As a former military officer, Ike understood Oppenheimer's vivid rendering of the two major powers as "two scorpions in a bottle." Eisenhower had seen the Disarmament Panel report and he found it thoughtful and wise. Highly skeptical of nuclear weapons, he told one of his key White House aides, C. D. Jackson—who had been Henry Luce's right-hand man at Time-Life—that "atomic weapons strongly favor the side that attacks aggressively and *by surprise.* This the United States will never do; and let me point out that we never had any of this hysterical fear of *any nation* until atomic weapons appeared upon the scene." Later in his presidency, Eisenhower would feel compelled to rebuke a panel of hawkish advisers, caustically observing,

"You can't have this kind of war. There just aren't enough bulldozers to scrape the bodies off the streets."

For a time, it seemed Oppenheimer's views might influence the new president. But Lewis Strauss, who had contributed generously to Eisenhower's campaign, was appointed the president's atomic energy adviser in January 1953; and then, in July, he was elevated to the job he had bought—chairman of the Atomic Energy Commission.

Strauss, of course, violently disagreed with Oppenheimer's notion that the public should be informed about the nature of America's nuclear stockpile, or that matters of nuclear strategy should be publicly debated. Openness, he thought, would serve no purpose other than to relieve "the Soviets of trouble in their espionage activities." So Strauss now took every opportunity to sow suspicion in Eisenhower's mind about Oppenheimer. The new president later remembered someone—he thought it was Strauss—telling him that spring that "Dr. Oppenheimer was not to be trusted."

On May 25, 1953, Strauss dropped by FBI headquarters to talk with D. M. Ladd, one of Hoover's aides. Strauss was scheduled to see Eisenhower that afternoon at 3:30. He told Ladd that Oppenheimer had an appointment to brief the president and the National Security Council in a few days, and he was "very much concerned about Oppenheimer's activities." He had just learned that it had been Oppenheimer who had hired David Hawkins, a suspected communist, to work at Los Alamos in 1943. In addition, Oppenheimer, he said, had announced that he was sponsoring the appointment of Felix Browder, a brilliant young mathematician who happened to be the son of Earl Browder, former head of the Communist Party of America. Claiming that he had checked Browder's references at Boston University and found that his record there was not very favorable, Strauss told Oppenheimer that Browder's appointment would have to be put before a vote by the Board of Trustees. The trustees eventually voted six to five against Browder, but by then, Oppenheimer had already offered Browder the appointment. When Strauss challenged him on this, Oppenheimer claimed to have called Strauss' secretary and informed her that he was going to give Browder the appointment unless he heard otherwise from the board. Strauss was infuriated with Oppenheimer's high-handedness—exercised, he thought, to no other purpose than to extend a favored position to the son of America's most famous communist.*

Finally, Strauss told Ladd that he was suspicious of Oppenheimer's

---

*In the end, Oppenheimer's judgment was fully vindicated: Browder had a distinguished career, and in 1999 President Bill Clinton honored him with a National Medal of Science, the nation's highest award for science and engineering.

"contacts" with the Russians in 1942—a reference to the Chevalier affair—and the fact that he "is alleged to have delayed work on the hydrogen bomb." In view of all these facts, Strauss asked Ladd if the FBI would have any "objection" if he briefed Eisenhower on Oppenheimer's background that afternoon. Ladd quickly reassured Strauss that the Bureau had no objection to this; after all, he said, the FBI already had passed all its information on Oppenheimer to the attorney general, the AEC and "other interested Government agencies."

The initiation of Strauss' campaign to destroy Oppenheimer's reputation can thus be precisely dated; it began on the afternoon of May 25, 1953, with his appointment with the president. Ike would recall later that Strauss "came back to him time and again about the Oppenheimer matter." On this occasion, he told Eisenhower that "he could not do the job at the AEC if Oppenheimer was connected in any way with the program."

A week before Strauss' meeting with Eisenhower, Oppie had phoned the White House and explained that "he needed very badly to see the president for a short time, and that it should not be long delayed." Two days later, he was ushered into the Oval Office. After a short meeting, Eisenhower invited him to come back to brief the National Security Council on May 27. Bringing Lee DuBridge with him, Oppenheimer spent five hours lecturing and answering questions. He argued the merits of candor, and, perhaps thinking back to the 1946 Lilienthal Panel, he urged the president to create a five-member disarmament panel. According to C. D. Jackson, Oppenheimer "had everybody spellbound—except the President." Ike cordially thanked him for the briefing, but let him leave the room without tipping his hand as to what he actually thought. Perhaps Eisenhower was weighing what Strauss had told him just two days earlier—that he could not run the AEC if Oppenheimer continued to serve as a consultant. According to Jackson's account, Ike felt uncomfortable as he watched Oppenheimer exert his "almost hypnotic power over small groups." Some time later, he told Jackson that he "did not completely trust" the physicist. Strauss' first blow had found its mark.

FULLY AWARE of Oppenheimer's White House meetings, Strauss now began to orchestrate a public campaign against Robert. Over the next few months, *Time, Life* and *Fortune* magazines—all controlled by Henry Luce—published broadsides attacking Oppenheimer and the influence of scientists in defense policy. The May 1953 issue of *Fortune* magazine featured an anonymous article titled, "The Hidden Struggle for the H-Bomb: The Story of Dr. Oppenheimer's Persistent Campaign to Reverse U.S. Military Strategy." The author charged that under Oppenheimer's influence, Project Vista (the air defense study contracted out to Caltech) had been transformed into an exercise to question "the morality of a strategy of atomic retaliation." Citing Air Force secretary Finletter, the author charged that "there was a serious question of the propriety of scientists trying to settle such grave national issues alone, inasmuch as they bear no responsibility for the successful execution of war plans." After reading the *Fortune* essay, David Lilienthal spoke of it in his diary as "another nasty and obviously inspired article attacking Robert Oppenheimer. . . ."

As Lilienthal neatly summarized it, the article purported to expose how Oppenheimer, Lilienthal and Conant had tried to block development of the H-bomb, but "Strauss saved the day etc. From there on J.R.O. [Oppenheimer] is the instigator of a kind of conspiracy to defeat the idea that the strategic bombing unit of the Air Force has the answer to our defense. . . ." Lilienthal didn't know it, but the *Fortune* essay had been written by one of the magazine's editors, Charles J. V. Murphy, who was an Air Force reserve officer—and who had, moreover, an unacknowledged collaborator: Lewis Strauss.

Some time after the *Fortune* magazine attack, Oppenheimer, Rabi and DuBridge met C. D. Jackson at Washington's Cosmos Club to discuss the piece. Afterwards, Jackson reported to Luce that they were "absolutely furious" about the essay, which they described as "the unwarranted attack on Oppenheimer. . . ." He told Luce that he had tried to defend the magazine's integrity but that "I had privately felt that Murphy and [James] Shepley [*Time* magazine's Washington bureau chief] had been engaging in an unwarranted anti-Oppenheimer crusade. . . ."

OPPENHEIMER'S "candor" speech was published on June 19, 1953, in *Foreign Affairs,* having been cleared for publication by the White House. Both the *New York Times* and the *Washington Post* ran stories on the article, and Oppenheimer was quoted as saying that without "candor" the American people were going to be "talked out of reasonable defense measures." Only the president, he said, "has the authority to transcend the racket and noise, mostly consisting of lies, that have been built up about this subject of the strategic situation of the atom." *Lies!*

A smoldering Strauss hastily went to see President Eisenhower. He thought Oppenheimer's essay "dangerous and its proposals fatal." He was surprised to learn that Oppenheimer had cleared a draft of the article with the White House. The president had read Oppie's essay and found himself in general accord with its argument. In a July 8 press conference, Eisenhower indicated that he agreed with Oppenheimer's notion of the need for more "candor" about nuclear weapons. Strauss now complained to Ike that some members of the press were construing this statement as "a blanket endorsement of Dr. J. Robert Oppenheimer's recent doctrine of 'candor' and as favoring the release of information on our stockpile and production rate of weapons and our estimate of enemy capabilities."

"That's complete nonsense." Eisenhower responded. "You ought not to read what those fellows write. I am at least the one person more security-minded than you are." And then he added, "Somebody ought to do a piece to correct the Oppenheimer article." Momentarily appeased, Strauss volunteered that he might write an essay himself.

Oppenheimer's *Foreign Affairs* essay sparked a vigorous debate within the Eisenhower Administration on what the public should be told about nuclear weapons. That had been Oppie's intention. He had hoped that his blunt description of the dangers the country faced from an unfettered arms race would prompt a reconsideration of the notion of relying so heavily on nuclear weapons. Candor was necessary precisely because the public ought to be frightened at the prospect of an endless arms race. As Eisenhower and his aides wrestled with the issue, the president found himself pursuing contradictory ends. "We don't want to scare the country to death," he told Jackson after reading one of his draft "candor" speeches. And he told Strauss that he wanted to both be candid about the risks of nuclear war and yet also offer the public some "hopeful alternative."

Strauss disagreed, but astutely held his tongue. To his mounting frustration, it appeared that Ike was attracted to some of Oppenheimer's ideas—and Strauss was determined to disabuse the president of the notion of their value. Early in August 1953, Strauss had cocktails with C. D. Jackson and afterwards Jackson noted in his diary, "Very relieved to get from Strauss firm categoric denial any feuding between him and Oppenheimer and any reluctance pursue Candor, except for stockpile arithmetic." A shrewd bureaucratic infighter, Strauss had lied to Jackson. That very month he had secretly collaborated with Charles Murphy at *Fortune* on a second essay bitterly critical of Oppenheimer's call for candor on atomic secrets.

Events also conspired to help Strauss. Late that August newspaper headlines around the country blared the news, "Reds Test H-Bomb." Only nine months after the first American test of a hydrogen bomb, the Soviets had apparently been able to match that feat. At least that is what the American people were told. In fact, the Soviet test was not the technical achievement it appeared to be: It was neither truly a hydrogen bomb, nor a weapon that could be delivered in an airplane. But the impression that the Soviets were perhaps ready to surpass the American nuclear arsenal gave Strauss further political ammunition to block Oppenheimer's call for candor.

Eventually, Eisenhower found his "hopeful alternative" and presented it in a speech proposing an "Atoms for Peace" program. He suggested that the U.S. and the Soviet Union should contribute fissionable materials to an international effort to develop peaceful nuclear energy power plants. Delivered on December 8, 1953, at the United Nations, the speech was initially a public relations success—but the Soviets failed to respond. And neither had the president been candid about American nuclear weapons. Gone from the speech was any accounting of the size and nature of the nuclear arsenal, or any other information that was grist for a healthy debate. Instead of candor, Eisenhower gave America a fleeting propaganda victory.

And far from conducting any reconsideration of nuclear strategy, in the months ahead the Eisenhower Administration would begin to cut defense spending on conventional weapons while building up its nuclear arsenal. Eisenhower called this his "New Look" defense posture. The Administration had accepted the Air Force's strategy and would rely almost exclusively on air power for America's defense. A policy of "massive retaliation" appeared to be a cheap and deadly fix. It was also shortsighted, genocidal and, if initiated, suicidal. Dean Acheson called it a "fraud upon the words and upon the facts." Adlai Stevenson asked pointedly, "Are we leaving ourselves the grim choice of inaction or thermonuclear holocaust?" The "New Look" was in fact old policy, and precisely the opposite of what Oppenheimer had hoped for from the new Administration.

LEWIS STRAUSS had prevailed. The nuclear secrecy regime would remain in place and nuclear weapons would be built in dizzying numbers. Oppenheimer had once thought Strauss merely an annoyance, a man not likely to "obstruct things." Now, with a Republican administration in control of Washington, Strauss was in the driver's seat, and his right foot was pressing his political accelerator to the floor.

Oppenheimer and many of his friends were now certain that Strauss was gunning for him. In July, soon after Strauss moved in as AEC chairman, Oppenheimer's close friend and lawyer Herb Marks received a phone call from an AEC employee: "You'd better tell your friend Oppy to batten down the hatches and prepare for some stormy weather."

"I knew he was in trouble," I. I. Rabi recalled. "He had been so for a couple of years . . . he was living under this shadow . . . I knew he was being hounded." So one day Rabi told him, "Robert, you write a piece for the *Saturday Evening Post*, tell them your story, your radical connections and so on, get well paid for it—and that will kill it." Rabi thought if the story came from Robert, and it appeared in a respectable publication, the public would understand. As a matter of public relations, a frank confessional essay might well have immunized Oppenheimer from further political attacks. But as Rabi recalled, "I couldn't get him to do it."

Oppenheimer had other plans. Early that summer, Robert, Kitty and their two children all boarded the SS *Uruguay* in New York, bound for Rio de Janeiro. Traveling as a guest of the Brazilian government, Oppenheimer was scheduled to give several lectures and then return to Princeton in mid-August. While he was in Brazil, the FBI had the U.S. Embassy monitor his contacts.

While Oppenheimer enjoyed a leisurely trip to Brazil, Strauss spent the summer of 1953 feverishly preparing to finally put an end to his influence. On June 22, he visited FBI headquarters for another private meeting with Hoover. Well aware of the FBI director's extraordinary power in Washington, Strauss wanted to be sure that they maintained a "close and cordial relationship." Almost immediately, "Admiral" Strauss turned the conversation to Oppenheimer. "He stated," Hoover wrote in a memo, "that he was aware of the fact that Senator McCarthy contemplated investigating Dr. Oppenheimer and that while he, the admiral, felt that inquiry into Oppenheimer's activities might be well worth while, he hoped it would not be done prematurely."

Indeed, the Wisconsin senator and his aide Roy Cohn had visited Hoover on May 12. McCarthy said he wanted to know what Hoover's reaction would be if his Senate committee began an investigation of Robert Oppenheimer. Hoover now explained to Strauss that he had tried to divert McCarthy. Oppenheimer, he said was "quite a controversial figure," and popular among the country's scientists. He said that he had warned McCarthy that "a great deal of preliminary spade work" would have to precede any public investigation of such a formidable figure. McCarthy indicated that he had gotten the message and that he would back off from the Oppenheimer case, at least for the moment. Hoover and Strauss agreed that "this was not a case which should be prematurely gone into solely for the purpose of headlines."

Strauss now advised Hoover, "in the closest of confidence," that the syndicated columnist Joseph Alsop had recently delivered to the White House a seven-page letter urging the Eisenhower Administration to block an investigation of Oppenheimer by McCarthy. Strauss knew, of course, that Alsop was a friend of Oppenheimer's—and he wanted to be sure that Hoover understood that the scientist had influential allies. It was a good meeting between like-minded men, and Strauss left believing that he had forged an alliance with the powerful FBI chief. The task of getting rid of Oppenheimer was far too important to leave to the clownish, sensation-seeking senator from Wisconsin. It would require careful planning and skillful maneuvering.

After leaving Hoover, Strauss returned to his office and wrote to Senator Robert Taft, urging him to block McCarthy if he attempted to launch an investigation of Oppenheimer. It would be "a mistake," he wrote. "In the first place some of the evidence will not stand up. In the second place, the McCarthy committee is not the place for such an investigation and the present is not the time." Strauss would orchestrate his own investigation.

ON JULY 3, 1953, Strauss formally assumed the office of AEC chairman, taking charge, the *New Republic* reported, "as if he were flag officer on the bridge of a battleship." When he discovered that Gordon Dean, the retiring AEC chairman, had acquiesced to Oppenheimer's request that his consulting contract be renewed for another year (to enable him to lobby on behalf of greater candor), Strauss manned the battle stations. His first maneuver was to request that Hoover send him by special messenger a copy of the FBI's latest summary report on Oppenheimer. By then, Oppenheimer's FBI file ran to more than several thousand pages. The June 1953 summary alone was sixty-nine single-spaced pages, and without delay Strauss began to study it with the zeal of a prosecutor.

During the Eisenhower transition, Strauss had stayed in contact with William L. Borden, the young staff director for the Joint Committee on Atomic Energy who shared Strauss' deep suspicions of Oppenheimer. Borden was a Democrat, and had lost his job when the Republicans won control of the Senate. Yet his obsession with Oppenheimer had kept Borden working on a sixty-five-page report tracing Oppenheimer's influence in Washington. No other individual in America, he wrote, had more "detailed, precision data" about the nation's military and foreign policies than this scientist. After reviewing a résumé of Oppenheimer's postwar activities, Borden tried to convey a sense of his daily influence over Washington policy-makers.

> During a single seven-day period recently . . . Dr. Oppenheimer had talked with Dr. Charles Thomas, President of the Monsanto Chemical Corporation, concerning atomic power for industrial purposes; Dr. Oppenheimer had lunched with the Secretary of State at the latter's Maryland farm and discussed foreign policy in relation to the 1952 Fall testing operations at Eniwetok; Dr. Oppenheimer had met with the Secretary of the Air Force to discuss, among other topics, the relative merits of strategic versus tactical bombing; Dr.

Oppenheimer had met with a delegation of visiting French officials to discuss international control; Dr. Oppenheimer had talked to the President and gone to see the two 1952 presidential candidates, General Eisenhower and Governor Stevenson; and Dr. Oppenheimer, alone among Americans, may have learned from Dr. W. C. Penney, director of the British weapons laboratory equivalent to our Los Alamos, the details of Britain's bomb development. . . . It is almost universally agreed that Dr. Oppenheimer is a man of dynamic and magnetic personality, superbly articulate, and that with these qualities fortified by the prestige he enjoys among other scientists, he tends to dominate meetings in which he takes part.

In 1952, Borden hadn't come to any definite conclusions, but he couldn't get over the fact that the security file of such an influential man contained so much information that he considered derogatory. Strauss, of course, shared Borden's suspicions, and he had encouraged him to pursue them. In December 1952, just a month after Borden wrote his investigative report, Strauss sent him a four-page letter outlining his own view that the H-bomb had been delayed by three years. Not only had Oppenheimer's GAC dragged their feet on the Super, but it was now clear that the Russians had benefited from atomic espionage. "In sum," Strauss told Borden, "I think it would be extremely unwise to assume that we enjoy any lead time in the competition with Russia in the field of thermonuclear weapons." And there was no doubt in either of their minds that Oppenheimer was largely responsible for this dangerous situation.

In late April 1953, Borden visited Strauss' office to discuss their mutual concerns about Oppenheimer. Their subsequent activities suggest that during this meeting, they agreed on a plan—a conspiracy, really—to end Oppenheimer's influence. Borden would do the dirty work and Strauss would provide him access to the information he needed.

Within two weeks of their discussion, Borden had permission to check out Oppenheimer's security file from the AEC's security vault. Even though he left his government job on May 31, 1953, Borden was able to keep the file until August 18. On July 16, Strauss talked on the phone with Borden who was reading the file in the isolation of his vacation retreat in upstate New York. Within hours of its return, Strauss had Oppenheimer's dossier on his desk. He kept it for nearly three months, returning it to the AEC security vault on November 4. A few hours after Strauss returned the file, the AEC's assistant security officer, Bryan F. LaPlante, checked it out. LaPlante, a confidant of Strauss', didn't return the report until December 1.

This sequence of withdrawals and returns of Oppenheimer's file by Borden, Strauss and LaPlante was surely coordinated; it could not have been a coincidence. Clearly, Borden was working with Strauss' knowledge and encouragement to compose an indictment of Oppenheimer. When Borden completed his work and returned the dossier, Strauss retrieved it, perhaps to study the evidence himself. And when he was finished with it, he ordered LaPlante to review the report for further analysis.

Thus, during the seven months between April and December 1953, Lewis Strauss—with considerable help from William Borden—accomplished the "great deal of preliminary spade work" that he and J. Edgar Hoover had agreed was necessary before a successful assault could be launched against Oppenheimer. They had diverted Senator McCarthy from the attack, knowing that he was too unreliable to prepare the case carefully. In July 1953, according to AEC staff lawyer Harold Green, "Strauss had promised Hoover that he would purge Oppenheimer." In this instance, it appears that the AEC chairman was a man of his word.

ONE DAY in late August 1953, after Oppenheimer's return from Brazil, he phoned Strauss to say that he was going to be in Washington on Tuesday, September 1, and he wondered whether he might see him that morning. When Strauss said he was free only in the afternoon, Oppenheimer said he had an important appointment at the White House that afternoon and so couldn't make it. This news so alarmed Strauss that he immediately called the FBI and requested that the Bureau put a blanket surveillance on Oppenheimer during his visit. "The Admiral is extremely anxious," one FBI official reported, "in view of Oppenheimer's background, to find out where he will be in Washington on Tuesday afternoon, and whom he will see." Hoover authorized the surveillance, and Strauss later learned that Oppenheimer had not been to the White House; instead, he had spent the entire afternoon in a bar in the Statler Hotel with the syndicated columnist Marquis Childs. Relieved to learn that Oppenheimer was not seeing the president but only cultivating a columnist, Strauss wrote Hoover that "he was still extremely concerned about Oppenheimer's influence in the atomic energy program; and was watching the matter closely and *hoped to be able in the near future to terminate all AEC dealings with Oppenheimer*" (emphasis added).

AS STRAUSS and Borden prepared their case against Oppenheimer, Oppie spent the early autumn writing four long essays on science. Earlier in 1953, the British Broadcasting Corporation had invited him to give the prestigious Reith Lectures, a series of four talks broadcast to millions of people around the world. He and Kitty planned to stay in London for three weeks in November and then go to Paris in early December. The invitation was a considerable honor; previous Reith lecturers had included Bertrand Russell, who spoke on "Authority and the Individual," and, just the past year, Arnold Toynbee, who had lectured on the grand topic of "The World and the West."

Robert labored over his chosen theme, "to elucidate what there is new in atomic physics that is relevant, helpful and inspiriting for men to know." Most BBC listeners were probably overwhelmed by Oppenheimer's studied ambiguity. "His glittering rhetoric," wrote one critic, "held his listeners in a web of absorption that was often less attentive than trance-like." His performance was nothing if not mystical. "For all my trouble," he later admitted, "I was told I was impossibly obscure."

The Cold War was not his topic, but in an aside, he spoke briefly about the nature of communism: "It is a cruel and humorless sort of pun that so powerful a present form of modern tyranny should call itself by the very name of a belief in community, by a word, 'communism,' which in other times evoked memories of villages and village inns and of artisans concerting their skills, and of men learning [to be] content with anonymity. But perhaps only a malignant end can follow the systematic belief that all communities are one community; that all truth is one truth; that all experience is compatible with all other; that total knowledge is possible; that all that is potential can exist as actual. This is not man's fate; this is not his path; to force him on it makes him resemble not that divine image of the all-knowing and all-powerful but the helpless, iron-bound prisoner of a dying world."

Having flirted with the communist promise in the 1930s, Oppenheimer had no illusions about its reality in 1953. Like Frank, he had been attracted in those years by the vision and rhetoric of social justice promoted by the American Communist Party. Integrating public swimming pools in Pasadena, arguing for better working conditions for farm laborers, organizing a teachers' union—these were all intellectually and emotionally liberating experiences. But much had changed. Now, in pleading for a different "brave new world," he was reconstituting on an intellectual level the deepest instincts and the highest values he had been committed to as a young man. His call for an open society was, to be sure, connected to his concerns about the dangerous and stultifying effects of secrecy on American society. But it was also connected to the cause of social justice in America, a goal he had worked for before Hiroshima, before Los Alamos and before Pearl Harbor. Communism's role in America had changed; Robert's role as a responsible American citizen had changed; but his deepest values were unaltered. "The open society, the unrestricted access to knowledge, the unplanned and uninhibited association of men for its furtherance," he said in one of his Reith Lectures, "these are what may make a vast, complex, ever-growing, ever-changing, ever more specialized and expert technological world nevertheless a world of human community."

WHILE IN LONDON, Kitty and Robert had dinner one evening with Lincoln Gordon, a classmate of Frank's at the Ethical Culture School, and someone whom Robert had met in 1946 when Gordon served as a consultant to Bernard Baruch. Gordon would always remember the dinner conversation that evening. Robert was in a somber, reflective mood and when Gordon gingerly mentioned the atomic bomb, Oppenheimer spoke at some length about the decision to use the bomb. He acknowledged that he had supported the Interim Committee's decision—but he confessed that he "didn't understand to this day why Nagasaki was necessary. . . ." He said this with sadness in his voice, not anger or bitterness.

After recording the Reith Lectures in London, the Oppenheimers crossed the English Channel and went to Paris where Kitty phoned Haakon Chevalier at his Montmartre flat only to learn that Hoke was attending a conference in Rome. Informed that he might be back in a few days, Robert and Kitty took a train to Copenhagen where they visited with Bohr for three days. When they returned to Paris, Chevalier was there and he insisted that they have dinner at his apartment on their last evening in the city. It was an invitation that would have dire consequences. At Strauss' request, security officers in the U.S. Embassy in Paris followed Oppenheimer's movements about the city and obtained a list from his hotel of every phone call he made. The Paris embassy reported that, "Chevalier, who is very unfavorably known and is suspected of being a Soviet agent, is on the watch list of the French police and intelligence services."

BY DECEMBER 7, 1953, Chevalier and Oppenheimer hadn't seen each other for more than three years. Their last reunion had been at Olden Manor in the autumn of 1950, when Hoke came for solace and an extended visit following a painful divorce from Barbara. But the two old friends had maintained a warm correspondence which even included a letter of recommendation of sorts in which Robert wrote, at Hoke's request, a summary of what he had told HUAC about the Eltenton episode. The letter had not retrieved Chevalier's position at Berkeley, but he was grateful nevertheless.

In November 1950, Chevalier had moved to Paris, traveling with a French passport since the U.S. State Department had refused to issue him his American passport. In Paris, he had gradually made a life for himself, working as a translator for the United Nations and writing fiction. When he married Carol Lansburgh, a thirty-two-year-old native of California, the Oppenheimers sent them a mahogany salad bowl from the Virgin Islands as a wedding gift.

Now both men looked forward to a pleasant reunion. When Robert and Kitty arrived at Chevalier's flat at 19, Rue du Mont-Cenis, near the foot of the Sacré Coeur Cathedral, they clambered into an aged elevator cage and rose to the fourth floor. Hoke and Carol greeted them warmly, and soon the two couples were toasting each other in the small living room lined with bookcases. Chevalier cooked another of his fine dinners, and this one included a sumptuous salad tossed in the mahogany salad bowl. Over dessert, Chevalier opened a bottle of champagne and after many toasts, Oppie and Kitty autographed the champagne cork.

Oppenheimer seemed relaxed, and told wry stories about his encounters with such Washington personalities as Dean Acheson. They briefly discussed the execution earlier that year of Julius and Ethel Rosenberg, convicted of conspiracy to commit atomic espionage. And Chevalier told Oppenheimer about his current worries over his employment as a translator for UNESCO. He explained that because he had not renounced his American citizenship, it appeared he might be compelled to submit himself to a U.S. government security clearance. Oppenheimer suggested he should get some advice from Jeffries Wyman, Robert's friend from Harvard who was in Paris that year as the American Embassy's science attaché.

As the Oppenheimers rose to leave shortly after midnight, Oppie, suddenly in a laconic mood, turned to Hoke and said, "I certainly don't look forward to the next few months." Perhaps he had some intimation of trouble ahead. But if so, he made no effort to explain his remark. On their way out, Chevalier decided that his friend was not dressed warmly enough and so he quickly made him a gift of an Italian silk scarf. Neither man suspected that their friendship was about to be put on trial.

DURING OPPENHEIMER'S absence in Europe, Borden began to write a prosecutor's brief against Oppenheimer. It was based on information from Oppie's security file that Strauss had arranged for Borden to remove from the AEC's vault. Borden was both enthusiastic about his efforts and conscientious about keeping in touch with Strauss. After Borden had lost his position with the Joint Committee on Atomic Energy in late May 1953, he obtained a job in Pittsburgh with Westinghouse's nuclear submarine program. Borden had earlier thanked him profusely for his "thoughtfulness." Studying Oppenheimer's top secret AEC personnel file in the evenings, Borden had a draft of the letter by mid-October 1953—which he mailed to J. Edgar Hoover on November 7. The FBI summary reports of the same information had been lengthy and convoluted. But Borden crystallized the charges against Oppenheimer in a mere three and a half single-spaced pages with a clear focus. Its conclusion was a shocker. After marshaling the evidence of Oppenheimer's communist associations, and reviewing the history of his recommendations on nuclear weapons, Borden concluded that "more probably than not J. Robert Oppenheimer is an agent of the Soviet Union."

It is not known exactly when Strauss learned that Borden's letter was completed. He was not informed officially until Hoover forwarded it on November 27 to him, Secretary of Defense Wilson and the president. But as early as November 9 Strauss composed a note for his files that suggests he had read Borden's letter. "It is my recollection," he wrote, "that an FBI report dated 27 November 1945 on the general subject of Soviet espionage activities will record that 'as early as December 1940 surveillance showed that secret meetings of a group were held, including Steve Nelson, Haakon Chevalier, William Schneiderman, the head of the Communist organization in California, and JRO.' This information was apparently obtained by actual surveillance."

On November 30, shortly after formally receiving the letter, Strauss noted in another memo for his files that the key charge against Oppenheimer pertained to the Chevalier affair: "The important point at issue is how long after the event occurred did 'O' [Oppenheimer] report it to 'G' [Groves] and whether there was any reason to suspect that 'O' knew that 'G' had learned of it before he reported it." This was indeed an interesting question, but since there is no evidence that Groves knew anything about Oppie's conversation with Chevalier before being told about it by Oppie— and there is testimony to that effect by Groves in the FBI files—the most interesting question relates to Strauss' memo. Was he already preparing what would become the focus of the case against Oppenheimer?

BY THE AUTUMN of 1953, Washington was a city in the grip of a witch-hunt. The careers of hundreds of civil servants had come to an abrupt end on the flimsiest of charges. No one, least of all the president, seemed willing to stand up to Senator Joseph McCarthy. On November 24, 1953, the Wisconsin senator gave a blistering speech, carried on both radio and television, in

which he charged the Eisenhower Administration with "whining, whimpering appeasement." The next day, C. D. Jackson told the *New York Times'*
James Reston that he thought "McCarthy had declared war on the President." When Reston's column the next morning used the quote, attributing
it to an unnamed White House official, Jackson was roundly criticized by an
Eisenhower aide who said such talk would merely make it "more difficult to
get McCarthy and his allies to vote for [the] Presidential program." Jackson
was appalled at what he called "disastrous appeasement" in the face of
McCarthy's attacks. "All the vague feelings," he noted in his diary, "of
unhappiness I have had regarding 'lack of leadership' over the past many
months, which I have always put down, really bounced up this week, and I
am very frightened." He told the president's chief of staff, Sherman Adams,
that he hoped McCarthy's "flagrant performance will at least serve to open
the eyes of some of the President's advisers who seem to think the Senator
is really a good fellow at heart."

In this poisonous atmosphere, Defense Secretary Wilson phoned Eisenhower on December 2, 1953, and asked if he had seen J. Edgar Hoover's latest report on Dr. Oppenheimer. Ike said no. Wilson said that it was "the
worst one so far." Wilson said that Strauss had phoned him the previous
night to say that "McCarthy knows about it & might pull it on us." Eisenhower said that he wasn't going to worry about McCarthy—but the Oppenheimer case should be brought to the attention of Attorney General Herbert
Brownell. He told Wilson they "certainly will not assassinate [Oppen-
heimer's] character unless we can gain substantiating evidence." Wilson
told Ike (erroneously) that both Oppenheimer's "brother & wife *are* Communists; this fact, plus his past relations, make him a bad risk if we have
trouble with Communists."

After getting off the phone with Wilson—and before reading the document—Eisenhower noted in his diary that the new FBI report "brings forward very grave charges, some of them new in character." The attorney
general would have to judge whether an indictment was warranted, but Ike
noted, "I very much doubt that they will have this kind of evidence." But in
the meantime, he was going to cut Oppenheimer off from all contacts with
those in government. "The sad fact is that if this charge is true, we have a
man who has been right in the middle of our whole atomic development
from the very earliest days. . . . Dr. Oppenheimer was, of course, one of the
men who has strongly urged the giving of more atomic information to the
world"—a suggestion, Eisenhower failed to note in his diary, of which he
had approved.

Early the next morning, Eisenhower met with his national security
adviser, Robert Cutler, who advised him to take immediate action against

Oppenheimer. At ten o'clock that morning, Eisenhower called Strauss into the Oval Office and asked him if he had read the latest FBI report on Oppenheimer. Strauss, of course, had read the report, and the Borden letter that had prompted it. After a cursory discussion, the president directed that a "complete bar" be immediately "erected between this individual [Oppenheimer] and any information of a sensitive or classified character."

Later that day, Eisenhower noted in his diary that in the "brief time" he had to read over the "so-called 'new' charges" he had quickly realized that "they consist of nothing more than the receipt of a letter from a man named Borden. . . ." He then correctly assessed its contents: "This letter presents little new evidence. . . ." The president had been told, he confided, that the "vast bulk" of this information had been "constantly reviewed and re-examined over a number of years and that the over-all conclusion has always been that there is no evidence that implies disloyalty on the part of Dr. Oppenheimer. However, this does not mean that he might not be a security risk."

Eisenhower understood that Oppenheimer might well be the victim of scurrilous charges. But having ordered an investigation, he was not about to stop the process. Such a move would leave him vulnerable to a charge from McCarthy that the White House was shielding a potential security risk. So, the president sent a formal note to the attorney general, ordering him "to place a blank wall" between Oppenheimer and classified material.

WASHINGTON WAS a small town, and so it was no surprise that the very next day, on December 4, 1953, Oppenheimer's old Los Alamos friend and colleague, Adm. William "Deke" Parsons, learned of Eisenhower's "blank wall" directive. Parsons knew all about Oppie's left-wing associations, and thought them meaningless. Earlier that autumn, Parsons had written a "Dear Oppy" letter in which he observed, "The anti-intellectualism of recent months may have passed its peak." Now he knew otherwise. That afternoon he met his wife, Martha, at a cocktail party and she could see that he was "extremely upset." After telling her the news, he said, "I have to put a stop to it. Ike has to know what's *really* going on." At home that evening he told her, "This is the biggest mistake the United States could make!" When he said he had decided to get an appointment with the secretary of the Navy the next morning, Martha said, "Deke, you're an admiral, why can't you go to the president?"

"No," he told his wife, "the secretary of the Navy is my boss. I can't go around him."

That night, Admiral Parsons experienced chest pains. The next morning he looked so pale that Martha drove him to Bethesda Naval Hospital. He died that day of a heart attack, which Martha always believed was brought on by the news about Oppie.

Also on December 4, President Eisenhower left for a five-day trip to Bermuda, and Strauss went with him. When they returned five days later, Strauss began to choreograph the next steps in the government's case against Oppenheimer. He actually prepared several scripts of what he should say to Oppenheimer, who was scheduled to be back from Europe and in Princeton on December 13. On the following afternoon, Oppenheimer phoned and the two men exchanged mundane pleasantries. Strauss casually said that "it might be a good idea" if Oppenheimer came down to see him in two days. Oppenheimer agreed, but said he had nothing much to report: "Don't expect anything much."

As it turned out, the FBI had not completed its analysis of Borden's letter. Initially, Hoover had not taken it seriously. Borden's charges, an agent noted soon after the letter arrived, "are distorted and restated in his own words in order to make them appear more forceful than the true facts indicate." So the Bureau was now in a catch-up mode, and asked Strauss to postpone his presentation of charges to Oppenheimer. Strauss wired Oppenheimer and rescheduled their meeting for Monday, December 21.

On December 18, Strauss went to the Oval Office to discuss how he planned to handle the Oppenheimer case. Present were Vice President Richard Nixon, William Rogers, White House aides C. D. Jackson and Robert Cutler and the CIA chief, Allen Dulles. Eisenhower was out of the room, meeting with congressional leaders. Rogers briefly suggested they should simply do what Truman had done to Harry Dexter White—call Oppenheimer before an open congressional committee and grill him about the derogatory information in his security file. White, however, had dropped dead of a heart attack after the ordeal—and now Jackson and everyone else jumped all over the idea. At that, "Rogers smilingly withdrew the suggestion." Instead, they gravitated toward Strauss' notion of appointing a panel to conduct an administrative review of Oppenheimer's security clearance. It would not be a trial in the formal sense. The scientist would be offered a choice: He could quietly leave or he could appeal the suspension of his security clearance before a panel to be appointed by Strauss.

At 11:30 a.m. on the morning of December 21, 1953, as Strauss prepared to confront Oppenheimer that afternoon, he was startled to hear that Herbert Marks was outside, waiting to see him. Strauss did not believe in coincidence. Why did Oppenheimer's friend and attorney want to see him on this of all days? When Marks was ushered into his office, the lawyer announced that he urgently needed to talk with Strauss about Oppenheimer. At this, Strauss interrupted him and said that he was expecting to see Oppenheimer that afternoon and that since he was his attorney, Marks should wait until that meeting. Marks brushed this aside and said that he had just learned that the U.S. Senate's infamous Jenner Internal Security Subcommittee was proposing to investigate Oppenheimer. Pulling an old clip from the *New York Times* dated May 11, 1950, Marks read the headline—"Nixon Champions Dr. Oppenheimer"—and suggested that Vice President Nixon might be severely embarrassed if the Jenner Committee proceeded to put Oppenheimer in its spotlight. Nonplussed, Strauss calmly asked Marks if this was all that was on his mind. Marks nodded, and then Strauss asked if Oppenheimer knew of Marks' concerns. Marks said no, he had not spoken with Oppenheimer since before he had left for Europe. Marks soon departed, leaving Strauss with an overwhelming suspicion that Marks had just attempted "a polite form of blackmail."

When Oppenheimer arrived that afternoon about 3:00 p.m., Strauss and Kenneth D. Nichols, a former wartime aide to Gen. Leslie Groves and now the general manager of the AEC, were waiting for him. After briefly commenting on Admiral Parsons' sudden death, Strauss told Oppenheimer of his meeting that morning with Herb Marks. Oppenheimer expressed surprise and said he had no knowledge of the Jenner Committee's plans.

Strauss then turned to the hard business at hand. He told Oppenheimer that "we were faced with a very difficult problem pertaining to his continued clearance." President Eisenhower had issued an executive order requiring the reevaluation of all individuals whose files contained "derogatory information." When Strauss observed that Oppenheimer's file contained "a great deal of derogatory information," Oppenheimer acknowledged that he knew his security case would in due course have to be reviewed. Strauss then informed Oppenheimer that a former government official (Borden) had written a letter questioning Oppenheimer's security clearance; the president had consequently ordered an immediate investigation. Up to this point, Oppenheimer did not seem particularly surprised. But now Strauss told him that the "first step" of this review would the immediate suspension of his security clearance. And then he explained that an AEC letter had been prepared outlining the nature of the charges against him. The letter, Strauss pointedly said, had been drafted but not yet signed.

Oppenheimer was allowed to read the letter, and as he scanned through its contents, he commented that "there were many items that could be denied, some were incorrect, but that many were correct." It all seemed a familiar rehash of the mix of truths, half-truths and outright lies.

According to Nichols' notes of the meeting, it was Oppenheimer who first raised the possibility of resigning prior to any security review. However, this option seemed to be suggested by Strauss' comment that the letter of charges had not been signed—and therefore was not yet an official charge. Thinking out loud, Oppenheimer at first seemed open to this possibility, but he quickly observed that if the Jenner Committee was going to open an investigation of him anyway, a resignation now "might not be too good from a public relations point of view."

When Robert asked how long he had to decide, Strauss said he would be at home from 8:00 p.m. on to receive his answer—but that he could not in any case defer action beyond another day. When Oppie asked if he could have a copy of the letter of charges, Strauss refused, saying he could have the letter only after deciding what he was going to do. And when Oppenheimer asked if "the Hill [Congress] knew about this," Strauss said not to his knowledge, but that he doubted "such a thing could be kept from the Hill indefinitely."

Strauss finally had Oppenheimer exactly where he wanted him. Yet Oppie seems to have reacted calmly to the news, politely asking all the right questions, trying to explore his options. Thirty-five minutes after entering Strauss' office, Oppenheimer rose to leave, telling Strauss that he was going to consult with Herb Marks. Strauss offered him the use of his chauffeur-driven Cadillac and Oppenheimer—distraught (outward appearances to the contrary)—foolishly accepted.

But instead of going to Marks' office, he directed the driver to the law offices of Joe Volpe, the former counsel to the AEC who together with Marks had given him legal advice during the Weinberg trial. Soon afterwards, Marks joined them and the three men spent an hour weighing Robert's options. A hidden microphone recorded their deliberations. Anticipating that Oppenheimer would consult with Volpe, and unconcerned about violating the legal sanctity of client-lawyer privilege, Strauss had arranged in advance for Volpe's office to be bugged.*

The hidden microphones in Volpe's office allowed Strauss, through the transcripts provided to him, to monitor the discussion as to whether Oppenheimer ought to terminate his consulting contract or fight the charges in a formal hearing. Oppie was clearly undecided and anguished. Late that afternoon, Anne Wilson Marks came by and drove her husband and Robert back to their Georgetown home. On the way, Oppenheimer said, "I can't believe what is happening to me." That evening, Robert took the train back to Princeton to consult with Kitty.

Strauss had expected Oppenheimer's decision that evening, and when the next morning he still hadn't heard from him, he ordered Nichols to

---

*That same afternoon, Strauss called the FBI and repeated his December 1 request to Hoover to place telephone taps on Oppenheimer's home and office in Princeton. The phone tap was installed in Olden Manor at 10:20 a.m. on New Year's Day 1954.

phone Oppenheimer at noon that day. Oppenheimer said he needed more time to make up his mind. Nichols brusquely replied that he "could not have any more time. . . ." He gave him a three-hour ultimatum. Oppenheimer seemed to agree, but an hour later he called Nichols back and said that he wanted to come to Washington and give his answer in person. He said he would take an afternoon train and see Strauss the next morning at 9:00 a.m.

Leaving Peter and Toni in the care of his secretary, Verna Hobson, Robert and Kitty boarded a train at Trenton and arrived in Washington in the late afternoon. Heading over to the Marks home in Georgetown, they spent the evening huddled with Marks and Volpe, continuing to debate whether Robert should fight the charges.

"He was still in the same almost despairing state of mind," Anne remembered. After hours of strategizing, the lawyers finally drafted a one-page letter addressed to "Dear Lewis." Oppenheimer strongly implied that Strauss had encouraged him to resign. "You put to me as a possibly desirable alternative that I request termination of my contract as a consultant to the Commission, and thereby avoid an explicit consideration of the charges. . . ." Oppenheimer said he had earnestly considered this option. "Under the circumstances," he wrote Strauss, "this course of action would mean that I accept and concur in the view that I am not fit to serve this government, that I have now served for some twelve years. This I cannot do. If I were thus unworthy I could hardly have served our country as I have tried, or been the Director of our Institute in Princeton, or have spoken, as on more than one occasion I have found myself speaking, in the name of our science and our country."

By the end of the evening, Robert was clearly tired and despondent. After more than one drink, he rose and announced that he was retiring upstairs to the guest bedroom. A few minutes later, Anne, Herb and Kitty heard a "terrible crash" and Anne was the first to the top of the stairs. Robert was nowhere to be seen. After knocking on the bathroom door and then shouting his name, with no response, she tried to open the door. "I couldn't get the bathroom door open," she said, "and I couldn't get a response from Robert."

He had collapsed on the bathroom floor and his unconscious body was blocking the door. The three of them together gradually forced the door open, pushing Robert's limp form to one side. They then carried him to a couch and revived him. "But he sure was mumbly," Anne recalled. Robert said he had taken a sleeping pill, a prescription drug Kitty had given him. Anne called a doctor—who said, "Don't let him go to sleep." So for an hour they walked him back and forth, coaxing coffee down his throat until the doctor arrived. Oppie's "beast in the jungle" had struck; his ordeal had begun.

PART FIVE

"It Looks Pretty Bad, Doesn't It?"

*Someone must have traduced Joseph K.,
for without having done anything wrong
he was arrested one fine morning.*

FRANZ KAFKA, The Trial

AS SOON AS OPPENHEIMER INFORMED STRAUSS that he
would not resign, the AEC's general manager, Kenneth Nichols,
set in motion an extraordinary American inquisition. Nichols told
Harold Green, on the day the young AEC attorney was drafting the letter
of charges against Oppenheimer, that the physicist was "a slippery
sonuvabitch, but we're going to get him this time." In retrospect, Green
reflected that the remark was an accurate reflection of the AEC's conduct
throughout the hearing.

On Christmas Eve, two FBI agents arrived at Olden Manor and seized
control of Oppenheimer's remaining classified papers. That same day,
Oppenheimer received the AEC's letter of formal charges, dated December 23, 1953. Nichols informed Oppenheimer that the AEC now questioned
"whether your continued employment on Atomic Energy Commission
work will endanger the common defense and security and whether such
continued employment is clearly consistent with the interests of the
national security. This letter is to advise you of the steps which you may
take to assist in the resolution of this question. . . ." The charges included all
the old "derogatory" facts of Oppenheimer's associations with known and
unknown communists, his contributions to the Communist Party of California, the Chevalier affair—and "that you were instrumental in persuading
other outstanding scientists not to work on the hydrogen bomb project, and
that the opposition to the hydrogen bomb, of which you are the most experienced, most powerful, and most effective member, has definitely slowed
down its development." With the exception of this last charge—delaying
the hydrogen bomb's development—all of this information had been

reviewed previously and discounted by both General Groves and the AEC. With the full knowledge of these facts, Groves had ordered the Army to give Oppenheimer his security clearance in 1943, and the AEC had renewed it in 1947 and thereafter.

The inclusion of Oppenheimer's opposition to the Super reflected the depth of McCarthyite hysteria that had enveloped Washington. Equating dissent with disloyalty, it redefined the role of government advisers and the very purpose of advice. The AEC's charges were not the kind of narrowly crafted indictment likely to bring a conviction in a court of law. This was, rather, a political indictment, and Oppenheimer would be judged by an AEC security review panel appointed by the chairman of the AEC, Lewis L. Strauss.

A DAY OR TWO before Christmas, Oppenheimer's secretary was at her desk when Robert and Kitty walked into his office and shut the door. That was unusual: Robert almost always kept his door open. "They stayed in there a long time," Verna Hobson recalled. "It was clear that something was wrong." When they finally came out, they had a drink and offered Hobson one as well. Later, when Hobson went home, she told her husband, Wilder, "The Oppenheimers are in some kind of trouble; I do not know what it is, but I want to give them some kind of present." Wilder had just bought a record cut by a Brazilian soprano, so Verna took it into the office the next day and gave it to Robert, saying, "This is not a Christmas present, and I did not go out and buy it for you; it has been played. It's just a present I want to give you now." Robert took it and sat with his head down very still for a moment, and then he looked up and said, "How incredibly dear."

Later that afternoon, he called Hobson into his office and, closing the door, he said he wanted to tell her what had happened. For the next hour and a half, he sat there telling her not only about the charges but about the whole story of his childhood, his family and his adult life. It was all new to Hobson. And in retrospect, she thought he may have been rehearsing what he planned to say by way of answering Nichols' letter of charges. He had decided that the "items of so-called derogatory information . . . cannot be fairly understood except in the context of my life and my work."

Over the next few weeks, Robert worked feverishly to prepare a defense. The AEC had given him a thirty-day deadline to reply to the charges. First, he had to assemble a legal team. So early in January 1954, he consulted with Herb Marks and Joe Volpe. Marks strongly believed that his friend needed to be represented by a distinguished, politically connected lawyer. Volpe disagreed, and urged Oppenheimer to get a skilled trial lawyer. For a time, it was thought they might get John Lord O'Brian, a highly regarded but elderly New York attorney. O'Brian had to bow out for reasons of health. Another prominent trial lawyer, eighty-year-old John W. Davis, said he would be willing to take the case—if the AEC would agree to hold the hearing in New York City. Strauss made sure this did not happen. Eventually, Oppenheimer and Marks went to see Lloyd K. Garrison, a senior partner in the New York law firm of Paul, Weiss, Rifkind, Wharton & Garrison. Oppie had met Garrison the previous spring, when the lawyer had become a trustee at the Institute for Advanced Study, and he liked his genteel manners. Garrison's lineage was as distinguished as his own reputation. One of his great-grandfathers was the abolitionist William Lloyd Garrison, and his grandfather had served as the literary editor of *The Nation.* Garrison himself was a firm liberal and a board member of the American Civil Liberties Union. Not long after the New Year, Marks and Oppenheimer saw Garrison in his New York home and showed him General Nichols' letter of charges. After Garrison read through the document, Robert said, "It looks pretty bad, doesn't it?" Garrison replied simply, "Yes."

Garrison was sympathetic. The first thing to do, he said, was to get the AEC to extend its thirty-day deadline for Oppenheimer's response to the charges. On January 18, Garrison went to Washington and got the necessary extension. He also tried, unsuccessfully, to recruit as chief counsel a lawyer with trial experience. In the meantime, he began working with Oppenheimer on his written response to the charges. As the weeks rolled by, Garrison became by default Oppenheimer's lead counsel. Everyone realized, including Garrison, that his lack of trial experience made him a less-than-ideal choice. When, in mid-January, David Lilienthal learned from Oppenheimer that he had retained Garrison, Lilienthal noted in his diary, "I had hoped it might be an experienced trial lawyer, but the case against Robert is so weak, really, that choice of counsel isn't as important as if it were."

NEWS OF Oppenheimer's impending hearing soon began to leak all over Washington. On January 2, 1954, the FBI overheard Kitty on the phone trying, unsuccessfully, to reach Dean Acheson to see if he knew "how things stand." A few days later, Strauss reported to the FBI that he was "receiving some pressure from scientists . . . to appoint a hearing board in the Oppenheimer case which would 'whitewash' Oppenheimer." Strauss told the FBI that he "did not intend to be pressured into any action of this kind. . . ." Moreover, he said he understood that the selection of the board which would judge Oppenheimer "was most important." Vannevar Bush confronted Strauss in his office and told the AEC chairman that news of his action against Oppenheimer was "all over town." Bush bluntly informed him that this was a "great injustice," and that if he pursued the case, "it would undoubtedly result in attacks against Strauss himself." Strauss angrily replied that he "didn't give a damn" and that he wasn't going to be "blackmailed" by any such suggestions.

Strauss later portrayed himself as a man under siege, but in truth he knew he held the advantage. The FBI was feeding him daily summaries of Oppenheimer's movements and conversations with his lawyers, thereby allowing him to anticipate all of Oppenheimer's legal maneuvers. He knew Oppenheimer's FBI file contained information that Oppenheimer's lawyers would never see—because he was going to make sure that they were not given the necessary security clearance. Moreover, he was going to select the members of the hearing board. On January 16, Garrison requested a security clearance for himself and Herb Marks, and Strauss responded by denying a clearance for Marks, a former member of the AEC's legal staff. Whether or not Garrison would have received his clearance in time to help him prepare the case is an open question. But he took the position that either the entire defense team should be cleared, or none, a decision he would soon regret, and try unsuccessfully to reverse.

Late in March, however, Garrison learned that the members of the hearing board were going to spend a full week studying raw FBI investigative files on Oppenheimer. Worse, Garrison learned to his dismay that the AEC's "prosecuting" attorney would be present to help guide the board members through the derogatory items in the FBI file and answer their questions. Garrison had a "sinking feeling" that after a week's immersion in the files, the board members would become prejudiced against his client. But when he asked for the same privilege, to be present during this week-long briefing, he was flatly rebuffed. Simultaneously, Garrison tried to get an emergency security clearance for himself, so that he might at least read some of the same material. But Strauss told the Justice Department that "under no circumstances should we grant emergency clearance." In Strauss' view, neither Oppenheimer nor his lawyer had any of the "rights" afforded to a defendant in a court of law; this was an AEC Personnel Security Board Hearing, not a civil trial, and Strauss was going to be the arbiter of the rules.

Strauss was unfazed by the extraconstitutional nature of things he was doing to undermine Oppenheimer's defense. He knew, but did not care, that the FBI wiretaps were illegal, telling one agent "that the Bureau's technical coverage on Oppenheimer at Princeton had been most helpful to the AEC in that they were aware beforehand of the moves he was contemplating." Such tactics so offended Harold Green that he told Strauss "that the case was not so much an inquiry as a prosecution and that he did not want to have anything to do with it." He asked to be removed from the case.

One day, while visiting the Bachers in Washington, Robert made it clear to his hosts that he thought he was being monitored. "He'd come in the room," recalled Jean Bacher, "and before he'd do anything else, he'd lift the pictures and look under them to see where the recording device was." One night he took down a picture that was hanging on the wall and said, "There it is!" Bacher said the surveillance "terrified" Oppenheimer.

When an FBI agent in Newark suggested discontinuing the electronic surveillance on Oppenheimer's home "in view of the fact that it might disclose attorney-client relations," Hoover refused. The FBI's surveillance, moreover, was not confined to Oppenheimer alone. When Kitty's elderly parents, Franz and Kate Puening, returned by ship from a trip to Europe, the Bureau arranged to have their baggage thoroughly searched by U.S. Customs agents. They also photographed all the written material in the possession of the Puenings. Kitty's father, who was confined to a wheelchair, and Mrs. Puening were so unnerved by the treatment that they had to be hospitalized.

Strauss elevated his scheme to end Oppenheimer's influence on AEC affairs to a crusade for America's future. He told the AEC's general counsel, William Mitchell, that "if this case is lost, the atomic energy program . . . will fall into the hands of 'left-wingers.' If this occurs, it will mean another Pearl Harbor. . . . if Oppenheimer is cleared, then 'anyone' can be cleared regardless of the information against them." With the country's future at stake, Strauss reasoned, normal legal and ethical constraints could be ignored. Simply severing Oppenheimer's formal link to the AEC as a contract consultant was insufficient. Unless the physicist's reputation was smeared, Strauss feared that Oppenheimer would use his prestige to become a vocal critic of the Eisenhower Administration's nuclear weapons policies. To foreclose that possibility, he proceeded to orchestrate a "star chamber" hearing guided by rules that would assure the elimination of Oppenheimer's influence.

By the end of January, Strauss had selected Roger Robb, a forty-six-year-old native Washingtonian, to bring the case against Oppenheimer. With seven years of prosecutorial experience as an assistant U.S. attorney, Robb had a well-deserved reputation as an aggressive trial lawyer with a flair for ferocious cross-examination. He had tried twenty-three murder cases and won convictions in most of them. In 1951, as the court-appointed attorney, he successfully defended Earl Browder against charges of contempt of Congress. (Browder called him a "reactionary" but praised his legal abilities.) Robb was politically conservative in every respect; his clients included Fulton Lewis, Jr., a vitriolic right-wing columnist and radio broadcaster. Over the years, he also had "cordial contacts" with the FBI and, Hoover was informed, always had been "entirely cooperative" with

Bureau agents. On one occasion, Robb had taken the opportunity to ingratiate himself with the director by writing to congratulate him on his reply to the eminent civil libertarian Thomas Emerson, who had criticized the FBI in a *Yale Law Review* essay. It was no surprise, then, that Strauss was able to arrange a security clearance for Robb in just eight days.

As Robb prepared for the hearing in February and March, Strauss sent him information from his own notes from Oppenheimer's file that Robb might use to impeach the testimony of potential defense witnesses. "When Dr. Bradbury testifies . . . When Dr. Rabi testifies . . . When General Groves testifies . . ." And in each instance, Strauss provided Robb with a document that he thought was sure to undermine what the witness might have to say in defense of Oppenheimer. In addition, and also at Strauss' urging, the FBI provided Robb with its extensive investigative reports on Oppenheimer—including selective contents of the physicist's trash from his Los Alamos residence.

Having chosen his prosecutor, Strauss now turned his attention to selecting the judges. He needed three men to serve on the AEC security review board and he sought candidates who could be counted on to be suspicious of Oppenheimer's integrity once his left-wing past was revealed. By the end of February, he had settled on Gordon Gray to chair the board. Gray, who was then president of the University of North Carolina, had served as secretary of the Army in the Truman Administration. Strauss, an old friend, knew that Gray was a conservative Democrat who had voted for Eisenhower in the 1952 election. A Southern aristocrat whose family money came from the R. J. Reynolds Tobacco Company, Gray had no idea what he was getting into. He seemed to think the assignment would last a couple of weeks and that Oppenheimer would be cleared. Unaware of the high stakes at issue, not to mention Strauss' personal hostility to Oppenheimer, Gray naïvely suggested David Lilienthal as a prospective nominee to the security board. One can only imagine the look on Strauss' face when he heard that suggestion.

In lieu of Lilienthal, Strauss selected another reliably conservative Democrat, Thomas Morgan, chairman of the Sperry Corporation. For the third member, Strauss chose a conservative Republican, Dr. Ward Evans, whose two major qualifications were his science background—he was a professor emeritus of chemistry at Loyola and Northwestern universities—and his unblemished record of voting to deny clearances on previous AEC hearing boards. Gray, Morgan and Evans shared an ignorance of Oppenheimer's history as a fellow traveler, but they were sure to be shocked by what they would read in his security file. From Strauss' point of view, they were the perfect empty vessels.

. . .

ONE DAY in January, by coincidence, James Reston, the *New York Times'* bureau chief in Washington, boarded the flight that Oppenheimer was taking from Washington to New York City. They sat together and chatted, but afterwards Reston wrote in his notebook that Oppie seemed "unaccountably nervous in my presence and obviously under some strain." Reston began making some phone calls around Washington, asking, "What's wrong with Oppenheimer these days?" Soon the FBI wiretaps overheard Reston repeatedly trying to phone Oppie.

Oppenheimer was "highly irritated" that the suspension of his security clearance might soon become public knowledge. When he finally took one of Reston's phone calls, Reston told him of the rumors he'd heard that his security clearance was suspended and that the AEC was investigating him. Moreover, he said this information had been passed to Senator McCarthy by someone in the government. When Oppenheimer said he didn't feel that he could comment, Reston said he was on the verge of printing the story. Oppenheimer refused to comment but told him to talk with his lawyer. Reston saw Garrison in late January, and the two men came to an agreement. Knowing that the story would probably get out sooner or later, Garrison agreed to give Reston a copy of the AEC letter of charges and Oppenheimer's prepared response. In return, Reston agreed not to print the story until it appeared that the news was about to break.

OPPENHEIMER'S PREPARATION for his defense became a grueling ordeal. Most days he sat in his Fuld Hall office with Garrison, Marks and other lawyers drafting his statement and discussing fine points of the case. Each evening at five o'clock, he would leave and walk across the field to Olden Manor; often the lawyers would follow him home, where they would work late into the evening. "They were very intense days," his secretary recalled. Robert, however, seemed almost serene. "He looked as though he were holding up very well indeed," Verna Hobson said. "He had that fantastic stamina that people often have who have recovered from tuberculosis. Although he was incredibly skinny, he was incredibly tough." It was now well into February and Hobson, a loyal and highly circumspect secretary, had still not told her husband what was going on. It made her feel uncomfortable, so one day she asked Robert, "May I have your permission to tell Wilder what the trouble is?" Oppenheimer looked at her in astonishment and said, "I thought you had done so a long time ago."

Oppenheimer worked "incredibly hard" on his letter in response to the

AEC's charges. Hobson recalled that it went through "draft after draft after draft, a painful attempt to be as clear and true as possible. I can't think how many hours he put into that." Sitting in his leather swivel chair, he would think in silence for a few minutes, jot a few notes, then rise and start dictating as he paced the office. "He could dictate in rounded sentences and paragraphs for an hour straight," Hobson said. "And just when your wrist was about to give way, he would say, 'Let us take a ten-minute break.' " And then he would come back and dictate for another hour. Robert's other secretary, Kay Russell, typed Hobson's shorthand in triple-space. Robert would review it and after Kay had retyped it, Kitty would edit it. Finally, Robert would go over all the changes once more.

If Robert was working hard to defend himself, he did so almost fatalistically. Late that January, he traveled to Rochester, New York, to attend a major physics conference. All the familiar faces were there, including Teller, Fermi and Bethe. In public, Robert gave no hint of his impending ordeal, but he confided in Bethe, who clearly saw that his old friend was in "distress." Oppie confessed to Bethe his conviction that he was going to lose. Teller already had heard of Oppenheimer's suspension and so walked up to him during a break in the conference and said, "I'm sorry to hear about your trouble." Robert asked Teller if he thought there was anything "sinister" in what he (Oppenheimer) had done over the years. When Teller said no, Robert coolly suggested that he would be grateful if Teller would talk to his lawyers.

On his next visit to New York City, Teller saw Garrison and explained that, while he thought Oppenheimer had been terribly wrong about many things, in particular the H-bomb decision, he did not doubt his patriotism. Garrison sensed, however, that his feelings toward Oppenheimer were not warm: "He expressed a lack of confidence in Robert's wisdom and judgment and for that reason felt that the government would be better off without him. His feelings on this subject and his dislike of Robert were so intense that I finally concluded not to call him as a witness."

Robert had not been in touch with his brother for some time. Frank had intended to come East that winter, but work on the ranch forced a postponement. Early in February 1954, the two brothers talked on the phone and Robert revealed that he was in "considerable trouble." He hoped they could meet soon, he said, because since returning from Europe he had tried, but was unable, to compose a letter that would "adequately discuss his problem."

To his friends, Robert seemed distracted and inexplicably passive. One day, while listening to the lawyers talk about legal strategy, Verna Hobson lost her patience and began to push Robert. "I thought Robert was not fighting hard enough," she recalled. "I thought Lloyd Garrison was being too gentlemanly, I was angry. I thought we should go out and fight."

Hobson was often privy to the lawyers' discussions, and as far as she could determine, they were not helping their client. "It seemed to me that the whole story was such an obvious piece of nonsense," she said. Robert's critics in Washington "were not open to sweet reason, and whoever was doing this must be using it as a tool and the thing to do was push back, kick back, attack." Hobson was "too scared" to say what she thought before the whole group of lawyers, "but I kept muttering it at him." Finally, Oppenheimer took her aside and, as they stood on the back steps of Olden Manor, he said very gently, "Verna, I really am fighting just as hard as I know how and what seems to me to be the best way."

Hobson was not the only one who thought Garrison was not aggressive enough. Kitty, too, was unhappy with the direction the legal team was taking her husband. Kitty was a fighter. Twenty years had passed since as a young woman she had stood outside factory gates in Youngstown, Ohio, passing out communist literature. Now, perhaps for the first time since, this ordeal would require all her energy, tenacity and intelligence. Her past life, after all, was part of the indictment against her husband. She, too, would probably have to testify. It would be an ordeal for her as well as for him.

One Saturday at midday, after working all morning on his reply to the AEC charges, Oppenheimer emerged from his office, accompanied by Hobson. "I was going to drive him to his house," Hobson recalled. But as they walked out to the parking lot, Einstein suddenly appeared and Oppenheimer stopped to chat with him. Hobson sat in the car while the two men talked, and when Oppie returned to the car, he told her, "Einstein thinks that the attack on me is so outrageous that I should just resign." Perhaps recalling his own experience in Nazi Germany, Einstein argued that Oppenheimer "had no obligation to subject himself to the witch-hunt, that he had served his country well, and that if this was the reward she [America] offered he should turn his back on her." Hobson vividly remembered Oppenheimer's reaction: "Einstein doesn't understand." Einstein had fled his homeland as it was about to be overwhelmed by the Nazi contagion—and he refused ever again to set foot in Germany. But Oppenheimer could not turn his back on America. "He loved America," Hobson later insisted. "And this love was as deep as his love of science."

Einstein walked to his office in Fuld Hall, and nodding in Oppenheimer's direction, told his assistant, "There goes a *narr* [fool]." Einstein, of course, didn't think America was Nazi Germany and he didn't believe Oppenheimer needed to flee. But he was truly alarmed by McCarthyism. In early 1951 he wrote his friend Queen Elizabeth of Belgium that here in

America, "The German calamity of years ago repeats itself: People acqui-
esce without resistance and align themselves with the forces of evil." He
now feared that by cooperating with the government's security board,
Oppenheimer would not only humiliate himself but would lend legitimacy
to the whole poisonous process.

Einstein's instincts were right—and time would demonstrate that
Oppenheimer's were wrong. "Oppenheimer is not a gypsy like me," Einstein confided to his close friend Johanna Fantova. "I was born with the skin
of an elephant; there is no one who can hurt me." Oppenheimer, he thought,
clearly was a man who was easily hurt—and intimidated.

IN LATE FEBRUARY — just as Oppenheimer was putting the final touches
on his letter responding to the AEC charges—his old friend Isidor Rabi
attempted to broker a deal whereby Robert could avoid a hearing altogether.
Earlier in the year, having heard that Rabi was trying to see President Eisenhower about the case, Strauss had successfully blocked this attempt. Now
Rabi proposed directly to Strauss that if he and Nichols would withdraw the
formal letter of charges and restore Oppenheimer's suspended security
clearance, Oppenheimer would quickly resign his AEC consultancy. It
wasn't as if the AEC were using Oppenheimer's time very much—during
the last two years he had racked up a grand total of only six days on his consulting contract.

Soon after this meeting, on March 2, 1954, Garrison and Marks themselves appeared in Strauss' office and confirmed that Oppenheimer was
willing to accept such a compromise. But Strauss, confident of victory, dismissed this solution as "out of the question." AEC regulations, he insisted,
called for the case to be heard by a hearing board. He countered that if
Oppenheimer would indicate his desire to resign in writing, "the AEC
would give it further consideration." This was a very thin reed, and later that
day Garrison and Marks revisited Strauss to say they had talked to their
client on the phone and had decided to "fight his case before the hearing
board."

Consequently, on March 5, 1954, Oppenheimer's response to the
charges, written in the form of an autobiography, was delivered to the AEC.
It ran to forty-two typed pages.

AS A WIDER circle of Oppenheimer's friends in the scientific community
became aware of what was happening, many called to express their concern. On March 12, 1954, Lee DuBridge phoned from Washington and

asked if there was anything he could do. Oppenheimer bitterly observed, "I think there are things that the White House might do if they wanted to, but I don't think they are ready to. . . . I don't need to tell you that I think the whole thing is damn nonsense."

"It's more troublesome than that," DuBridge replied. "If it were only nonsense, we might fight it, but it is deeper than that." Robert seemed to agree and said he had resigned himself to just having to go through the "rigamarole." Another friend, Jerrold Zacharias, reassured him that "You have nothing personal to fear—really not—and your stand is so important for the nation. I guess all I mean is, give them hell."

On April 3, Robert phoned his old love, Ruth Tolman, and told her what was about to happen. It was the first time they had talked in months. "It was incredibly good to hear your voice this morning," Tolman wrote in a letter to him. "I suppose you have felt too harassed and confused to write. . . . You have been constantly in my thoughts, Dear, and with, of course, much concern. . . . Oh Robert, Robert, how often it has been this way for us: that we have felt powerless to help when we wanted to so deeply."

A few days later, the Oppenheimers sent Peter and Toni by train to their old Los Alamos friends the Hempelmanns. The children would remain in Rochester, New York, for the duration of the hearings. Just before Robert and Kitty themselves departed for Washington, Robert received a letter from his old friend Victor Weisskopf who, having learned of his predicament wrote to express support and encouragement: "I would like you to know that I and everybody who feels as I do are fully aware that you are fighting here our own fight. Somehow Fate has chosen you as the one who has to bear the heaviest load in this struggle. . . . Who else in this country could represent better than you the spirit and the philosophy of all that for which we are living. Please think of us when you are low. . . . I beg you to remain what you always have been, and things will end well."

It was a nice thought.

"I Fear That This Whole Thing Is a Piece of Idiocy"

The proceeding was skewed from the outset.

ALLAN ECKER
Oppenheimer defense team

L EWIS STRAUSS WAS ANXIOUS to have the security board proceedings commence. For one thing, he actually feared that his
quarry might flee the country. Hoping that Oppenheimer's passport
could be confiscated, Strauss warned the Justice Department that "if he
decided to defect while the AEC charges were pending against him, it
would be most unfortunate." He also worried that Senator McCarthy might
interfere with his plans. On April 6, McCarthy—replying to an attack on
him by CBS television commentator Edward R. Murrow—charged that
America's hydrogen bomb project had been deliberately sabotaged.
Clearly, there was a real danger that the unpredictable senator could go public with what he knew about the Oppenheimer case.

So Strauss was relieved when the hearing board finally convened on
Monday, April 12, 1954, in Building T-3, a dilapidated two-story temporary
structure built during the war on the Mall near the Washington Monument
at 16th Street and Constitution. It housed the office of the AEC's director of
research, but for this occasion, Room 2022 had been turned into a bare-
bones courtroom. At one end of the long, dark, rectangular room, the three
board members—Chairman Gordon Gray and his two colleagues, Ward
Evans and Thomas A. Morgan—sat behind a large mahogany table stacked
with black binders containing classified FBI documents. One of Garrison's
assistants, Allan Ecker, recalled how stunned Robert's attorneys were to see
that each member of the security review board had those bound books in
front of them. "This was the shock of the day," Ecker recalled, "and the
shock of the case, because the classical notion of the legal system is the tab-

ula rasa. There is nothing in front of the judge except that which is put in front of the judge openly and with an opportunity of the person accused or charged to respond. . . . They had examined [those books] in advance; they knew what was in there. We did not know what was in there. We did not have a copy; we had no opportunity to challenge whatever documents were not brought forward. . . . So I thought that the proceeding was skewed from the outset."

The opposing teams of lawyers sat across from each other at two long tables positioned to form a "T." On one side sat the AEC's lawyers, Roger Robb and Carl Arthur Rolander, Jr., the AEC's deputy director of security. Facing them were Oppenheimer's defense team, Lloyd Garrison, Herbert Marks, Samuel J. Silverman and Allan B. Ecker. At the bottom of the "T" was placed a single wooden chair, where the defendant or other witnesses sat facing the judges. When Oppenheimer was not testifying, he sat on a leather couch against the wall, behind the witness chair. Over the next month, Oppenheimer would spend some twenty-seven hours in the witness chair—and many more hours languishing on the couch, alternately chain-smoking cigarettes or filling the room with the aroma of his walnut pipe tobacco.

That very first morning, Oppenheimer and his lawyers had arrived nearly an hour late. A few days earlier, Kitty had had another one of her accidents. This time, she had fallen down the stairs and her leg was in a cast. Hobbling on crutches, she slowly made her way to the leather couch, where she sat down with her husband and waited for the proceedings to begin. Robert appeared subdued and almost resigned to his fate. "We made a pretty bedraggled kind of spectacle," Garrison recalled. "Her appearance didn't add much to the smoothness of things." The board seemed "pretty irritated" by the delay. Garrison apologized for their tardiness. Vaguely alluding to the fact that the press might be on to the story, he said they were delayed because they had been keeping their "fingers in the dike."

Gray spent the morning reading aloud the AEC's letter of "indictment" and Oppenheimer's reply. Over the next three and a half weeks, Gray repeatedly insisted that the proceedings were an "inquiry," not a trial. But no one could listen to the AEC's letter of charges without thinking that Robert Oppenheimer was on trial. His alleged crimes included joining numerous Communist Party front organizations; being "intimately associated" with a known communist, Dr. Jean Tatlock; associating with such other "known" communists as Dr. Thomas Addis, Kenneth May, Steve Nelson, and Isaac Folkoff; being responsible for the employment in the atom bomb project of such known communists as Joseph W. Weinberg, David Bohm, Rossi Lomanitz (all former students of Oppenheimer's) and David

Hawkins; contributing $150 per month to the Communist Party in San Francisco; and, perhaps most ominously, failing to report promptly his conversation with Haakon Chevalier in early 1943 about George Eltenton's proposal to funnel information about the Radiation Laboratory to the Soviet Consulate in San Francisco.

Oppenheimer's letter of response acknowledged the truth of his friendships with Tatlock, Addis and other left-wingers—but he denied there was anything nefarious about these relationships. "I liked the new sense of companionship," he said of those associations. He freely admitted being a fellow traveler in the 1930s and acknowledged that he had made financial contributions to a variety of causes through the Communist Party. He could not remember saying, as claimed by the AEC indictment, that he had "probably belonged to every Communist-front organization on the west coast." The quotation, he now said, was not true, but if he had ever said something like it, "it was a half-jocular overstatement." (In point of fact, these were Col. John Lansdale's words, posed to Oppenheimer as a question in 1943— "You've probably belonged to every front organization on the coast"—and at the time he had merely replied, "Just about.") He denied that he had been responsible for the employment of his former students by Ernest Lawrence in the Radiation Laboratory. And as to the Chevalier affair, Oppenheimer acknowledged that Chevalier had spoken to him about Eltenton's suggestion: "I made some strong remark to the effect that this sounded terribly wrong to me. The discussion ended there. Nothing in our long-standing friendship would have led me to believe that Chevalier was actually seeking information; and I was certain that he had no idea of the work on which I was engaged." As to the delay in reporting this conversation, Oppenheimer acknowledged that he should have reported it at once. But he pointed out that he had eventually volunteered the information about Eltenton to a security officer—and he doubted that this story would ever have become known "without my report."

On the whole, Oppenheimer's replies seemed credible. If judged by his whole life, the charges lodged against him involved behavior not at all unusual for a New Deal liberal in the 1930s committed to supporting and working for racial equality, consumer protection, labor union rights and free speech. But there was one more allegation in the AEC indictment that would prove to be almost as difficult to deal with as the Chevalier affair. The indictment claimed that "during the period 1942–45 various officials of the Communist Party, including Dr. Hannah Peters, organizer of the professional section of the Communist Party, Alameda County, Calif., Bernadette Doyle, secretary of the Alameda County Communist Party, Steve Nelson, David Adelson, Paul Pinsky, Jack Manley and Katrina Sandow are reported to have made statements indicating that you were then a member of the Communist Party; that you could not be active in the party at that time; that your name should be removed from the party mailing list and not mentioned in any way; that you have talked the atomic-bomb question over with party members during this period; and that several years prior to 1945 you had told Steve Nelson that the Army was working on an atomic bomb."

What was the source of these specific allegations? These individuals had not talked to the authorities. When summoned before HUAC, Nelson and others always had refused to name names. Obviously, these charges were based on illegal FBI wiretaps that were transcribed in those black binders stacked on the table before the hearing panel judges. Not admissible in a court of law, these unevaluated transcripts would be used with impunity in the Gray Board's "inquiry." All three Board members had read the FBI's summary of these ten-year-old conversations—yet Oppenheimer's lawyers were barred from seeing them and therefore were unable to challenge their contents.

Garrison and Marks should have realized that, presented as it was, this charge of secret membership in the Communist Party in the indictment made it impossible to mount a defense. Oppenheimer denied the allegations. "Your letter," he wrote, "sets forth statements made in 1942–45 by persons said to be Communist Party officials to the effect that I was a concealed member of the Communist Party. I have no knowledge as to what these people might have said. What I do know is that I was never a member of the party, concealed or open. Even the names of some of the people mentioned are strange to me, such as Jack Manley and Katrina Sandow. I doubt that I met Bernadette Doyle, although I recognize her name. Pinsky and Adelson I met at most casually. . . ." In a court of law, such evidence would be unacceptable and dismissed as double hearsay—third parties recounting what they heard from others about a defendant. But in this "inquiry," Oppenheimer's judges would always believe that the FBI had recorded the voices of well-informed communists whose claims that Oppenheimer was one of their own were valid.

Some of the information in those binders was even manipulated to appear more damaging to Oppenheimer. The source of one key allegation was two FBI informants, Dickson and Sylvia Hill, who had infiltrated the Montclair branch of the Communist Party in California. In November 1945, this husband-and-wife team walked into the FBI office in San Francisco and reported on a CP meeting they had attended shortly after the bombing of Hiroshima. Sylvia Hill said she heard a Communist Party official, Jack Manley, refer to Oppenheimer as "one of our own men." Mrs. Hill, however, went on to say that "Manley's statement concerning the subject

[Oppenheimer] did not necessarily mean to her that the subject was a card-carrying member of the CP. She believed her impression at the time was that the subject was probably not an actual member but went along with Communist ideas." Put in this context, Sylvia Hill's information did not buttress the AEC charge that known communists had been overheard calling Oppenheimer a Party member. But this level of nuance was lost when the FBI highlighted Hill's information in its summaries of Oppenheimer's file. What amounted to hearsay thus rose to the level of "derogatory" information.

HAVING READ THE INDICTMENT and Oppenheimer's reply, Chairman Gray asked Oppenheimer if he wished to "testify under oath in this proceeding?" He did, and Gray administered the standard oath to tell the truth and nothing but the truth required by any court of law. The inquiry had begun. Oppenheimer took the witness chair and spent the rest of the afternoon being questioned gently by his defense counsel.

ON THE NEXT MORNING, Tuesday, April 13, 1954, the *New York Times* broke the story in a front-page exclusive written by James Reston. The headline read:

DR. OPPENHEIMER SUSPENDED BY A.E.C. IN SECURITY REVIEW; SCIENTIST DEFENDS RECORD; HEARINGS STARTED; ACCESS TO SECRET DATA DENIED NUCLEAR EXPERT—RED TIES ALLEGED

The newspaper published the full text of both General Nichols' letter of charges and Oppenheimer's response. Reston's story was picked up by newspapers around the country and abroad. Millions of readers were exposed for the first time to intimate details of Oppenheimer's political and private life.

The news had an instant polarizing effect; liberals were aghast that such an eminent man could be attacked in such a manner. Drew Pearson, the liberal syndicated columnist, noted in his diary: "Strauss and the Eisenhower people are certainly getting petty. I can conceive of no move more calculated to bolster McCarthy and to encourage witch-hunting than this throwback to the prewar years and this attempt to search under the bed of Oppenheimer's past to see whom he was talking to or meeting with in 1939 or 1940. . . ." On the other hand, conservative commentators like Walter

Winchell had a field day with the story. Just two days earlier, Winchell had announced on his Sunday telecast that Senator McCarthy would soon reveal that a "key atomic figure had urged that the H-bomb not be built at all." This famous atomic scientist, Winchell claimed, has been "an active Communist Party member" and the "leader of a Red cell including other noted atomic scientists."

Chairman Gray was furious over Reston's report. Addressing Garrison, he said, "You said you were late yesterday because you had your 'fingers in the dike.'" Garrison explained that Reston had known of Oppenheimer's security suspension since mid-January. But Gray brushed this aside and grilled Garrison on when he had given the reporter copies of the AEC letter of charges. Oppenheimer interrupted to say, "These documents were given to Mr. Reston by my counsel Friday night, I believe. . . ." This only heightened Gray's anger: "So that you knew when you made the statement here yesterday morning that you were keeping the finger in the dike that these documents . . . were already in the possession of the *New York Times*?"

"Indeed we did," Oppenheimer replied.

Clearly annoyed with both Oppenheimer and his lawyers, Gray blamed them for the leaks. He never knew that his ire should have been directed at Lewis Strauss. The chairman of the AEC had known all along about Reston's phone calls to Oppenheimer, and it was Strauss, not Garrison, who had given the *New York Times* the green light to publish. Fearing that McCarthy would release the news first, Strauss calculated that it was time for the story to come out—particularly if he could blame the leak on Oppenheimer's lawyers. Eisenhower's press secretary, James C. Hagerty, agreed. So on April 9, Strauss called the publisher of the *New York Times,* Arthur Hays Sulzberger, and released him from their previously arranged agreement to keep a lid on the story.

Strauss also feared that there was a danger now of the whole case "being tried in the press," and that a lengthy hearing would work to Oppenheimer's advantage. The longer it dragged on, he calculated, the more time Oppenheimer's allies would have to "propagandize" the scientific community. A quick decision was essential. So later that week, he sent a note to Robb urging him to expedite the hearing.

A FEW DAYS earlier in Princeton, Abraham Pais had learned that the *New York Times* was about to break the story. Knowing that reporters would pester Einstein for a comment, he drove over to the physicist's house on Mercer Street. When Pais explained his mission, Einstein chuckled loudly, and then said, "The trouble with Oppenheimer is that he loves a woman who doesn't love him—the United States government. . . . [T]he problem was simple: All Oppenheimer needed to do was go to Washington, tell the officials that they were fools, and then go home." Privately, Pais may have agreed, but he felt this would not serve as a statement to the press. So he persuaded Einstein to draft a simple statement in support of Oppenheimer—"I admire him not only as a scientist but also as a great human being"—and got him to read it to a United Press reporter over the phone.

On Wednesday, April 14, day three of the hearing, Oppenheimer began the morning in the witness stand, answering questions posed to him by Garrison about his brother, Frank. Oppenheimer was very concerned that the AEC letter of charges included language stating that "Haakon Chevalier thereupon approached you either directly or through your brother, Frank Friedman Oppenheimer, in connection with this matter." So when Garrison asked him whether Frank was involved with the Chevalier approach, he replied, "I am very clear on this. I have a vivid and I think certainly not fallible memory. He had nothing whatever to do with it. It would not have made any sense, I may say, since Chevalier was my friend. I don't mean that my brother did not know him, but this would have been a peculiarly roundabout and unnatural thing." This made perfect sense, but Strauss, Robb and Nichols believed it was a lie and, without any proof, they would insist that Oppenheimer had lied to the hearing board.

GARRISON'S DIRECT EXAMINATION of Oppenheimer thus concluded as it had begun: as a reinforcement of his responses to the AEC's letter of charges. It had gone well, Oppenheimer and his lawyers believed. But as Robb began his cross-examination, it became clear that he had a carefully worked out strategy to reverse that good impression. Having spent nearly two months immersed in the FBI files, he was well prepared. "I had been told that you can't get anywhere cross-examining Oppenheimer," Robb later said. "He's too fast and he's too slippery. So I said, 'Maybe so, but then he's not been cross-examined by me before.' Anyway, I sat down and planned my cross-examination most carefully, the sequences to it and the references to the FBI reports and so on, and my theory was that if I could shake Oppenheimer at the beginning, he would be apt to be more communicative thereafter."

Wednesday, April 14, was perhaps the most humiliating day in Oppenheimer's life. Robb's interrogation was relentless and exacting. It was the sort of grilling that Oppenheimer had never experienced and was totally unprepared for. Robb began by leading Oppenheimer to admit that close association with the Communist Party was "inconsistent with work on a secret war project." Robb then asked him about former members of the

Communist Party. Would it be appropriate, Robb asked, for such a person to work on a secret war project?

Oppenheimer: "Are we talking about now or then?"

Robb: "Let us ask you now, and then we will go back to then."

Oppenheimer: "I think that depends on the character and the totality of the disengagement and what kind of a man he is, whether he is an honest man."

Robb: "Was that your view in 1941, 1942, and 1943?"

Oppenheimer: "Essentially."

Robb: "What test do you apply and did you apply in 1941, 1942, and 1943 to satisfy yourself that a former member of the party is no longer dangerous?"

Oppenheimer: "As I said, I knew very little about who was a former member of the party. In my wife's case, it was completely clear that she was no longer dangerous. In my brother's case, I had confidence in his decency and straightforwardness and in his loyalty to me."

Robb: "Let us take your brother as an example. Tell us the test that you applied to acquire the confidence that you have spoken of?"

Oppenheimer: "In the case of a brother you don't make tests, at least I didn't."

ROBB'S INTENTIONS were twofold: first, to catch Oppenheimer in contradictions with the written record to which Robert and his lawyers had been denied access; second, to place those things that Oppenheimer admitted into a context which implied that Robert had directed Los Alamos irresponsibly at best—or, worse, that he had hired communists consciously and purposefully. Robb's aim at every turn was to humiliate the witness, often merely by making him repeat what he had already admitted. "Doctor, I notice in your answer on page 5 you use the expression 'fellow travelers.' What is your definition of a fellow traveler, sir?"

Oppenheimer: "It is a repugnant word which I used about myself once in an interview with the FBI. I understood it to mean someone who accepted part of the public program of the Communist Party, who was willing to work with and associate with Communists, but who was not a member of the party."

Robb: "Do you think that a fellow traveler should be employed on a secret war project?"

Oppenheimer: "Today?"

Robb: "Yes, sir."

Oppenheimer: "No."

Robb: "Did you feel that way in 1942 and 1943?"

Oppenheimer: "My feeling then and my feeling about most of these things is that the judgment is an integral judgment of what kind of man you are dealing with. Today I think association with the Communist Party or fellow-traveling with the Communist Party manifestly means sympathy for the enemy. In the period of the war, I would have thought that it was a question of what the man was like, what he would and wouldn't do. Certainly fellow-traveling and party membership raised a question and a serious question."

Robb: "Were you ever a fellow traveler?"

Oppenheimer: "I was a fellow traveler."

Robb: "When?"

Oppenheimer: "From late 1936 or early 1937, and then it tapered off, and I would say I traveled much less fellow after 1939 and very much less after 1942."

While preparing for the hearing, Robb had seen numerous references in the FBI files to Oppenheimer's 1943 interview with Lt. Col. Boris Pash. The files indicated this interview had been recorded. "Where are those recordings?" Robb asked. The FBI soon retrieved the ten-year-old Presto disks and Robb listened to Oppenheimer's first description of the Chevalier incident. It differed markedly from what he had told the FBI in 1946. Obviously Oppenheimer had lied in one of these interviews, and so Robb came prepared to exploit the contradictory stories. Oppenheimer, of course, had no idea that his conversation with Pash had been recorded. So when Robb turned to the Chevalier incident, he knew the details far better than Oppenheimer could now recall them.

Robb began by reminding Oppenheimer of his brief interview with Lieutenant Johnson in Berkeley on August 25, 1943.

Oppenheimer: "That is right. I think I said little more than that Eltenton was somebody to worry about."

Robb: "Yes."

Oppenheimer: "Then I was asked why did I say this. Then I invented a cock-and-bull story."

Unfazed by this startling admission, Robb focused on what Oppenheimer had told Lt. Col. Boris Pash on the following day, August 26.

Robb: "Did you tell Pash the truth about this thing?"

Oppenheimer: "No."

Robb: "You lied to him?"

Oppenheimer: "Yes."

Robb: "What did you tell Pash that was not true?"

Oppenheimer: "That Eltenton had attempted to approach members of the project—three members of the project—through intermediaries."

A few moments later, Robb asked, "Did you tell Pash that X [Chevalier] had approached three persons on the project?"

Oppenheimer: "I am not clear whether I said there were 3 X's or that X approached 3 people."

Robb: "Didn't you say that X had approached 3 people?"

Oppenheimer: "Probably."

Robb: "Why did you do that, Doctor?"

Oppenheimer: "Because I was an idiot."

"An idiot"? Why did Oppenheimer say such a thing? According to Robb, Oppenheimer was in a state of anguish, cornered, as it were, by the clever prosecutor. After the hearing, Robb dramatized the moment to a reporter, saying that as Oppenheimer said these words he was "hunched over, wringing his hands, white as a sheet. I felt sick. That night when I came home I told my wife, 'I've just seen a man destroy himself.' "

This description was nonsense, self-serving publicity designed to promote Robb's courtroom image, and his humanity ("I felt sick . . ."). It is a measure of how cleverly Robb and Strauss manipulated the aftermath of the Oppenheimer hearings that journalists and historians have heretofore accepted Robb's interpretation of this moment. But contrary to what Robb claimed, Oppenheimer's "I was an idiot" comment was simply meant to eliminate the ambiguities surrounding the Chevalier incident. He was making it clear that he had no rational explanation as to why he had said that X (Chevalier) had approached three people. Robert knew that everyone knew he was not an idiot. He was using a colloquial phrase in a self-deprecating attempt to disarm his interrogator. Within minutes, however, it would become clear to him that he had not succeeded in disarming anyone—he was facing an adversary bent on destroying him.

Robb had only begun. Oppenheimer had admitted lying. Now Robb was going to confront him with the evidence and in painful detail dramatize the lie. Pulling out a transcript of Colonel Pash's encounter with Oppenheimer on August 26, 1943, Robb said, "Doctor . . . I will read to you certain extracts from the transcript of that interview." He then read a portion from the eleven-year-old transcript in which Oppenheimer asserted that someone in the Soviet Consulate was ready to transmit information "without any danger of a leak or scandal. . . ."

When Robb asked if he recalled saying this to Pash, Oppenheimer said he certainly didn't recall saying such a thing. "Would you deny you said it?" Robb asked. Realizing, of course, that Robb had in his hand a transcript, Oppenheimer replied, "No."

Robb melodramatically announced, "Doctor, for your information, I might say we have a record of your voice."

"Sure," Oppenheimer replied. But he went on to say that he was fairly certain that Chevalier had not mentioned someone from the Soviet Consulate when he told him about Eltenton's idea. But he had given this detail to Colonel Pash and had also told Pash that there had been "several"—not one—approaches to scientists.

Robb: "So you told him specifically and circumstantially that there were several people that were contacted?"

Oppenheimer: "Right."

Robb: "And your testimony now is, that was a lie?"

Oppenheimer: "Right."

Robb continued reading from the 1943 transcript: "Of course," Oppenheimer had told Pash, "the actual fact is that since it is not a communication that ought to be taking place, it is treasonable."

"Did you say that?" Robb asked.

Oppenheimer: "Sure. I mean I am not remembering the conversation, but I am accepting it."

Robb: "You did think it was treasonable anyway, didn't you?"

Oppenheimer: "Sure."

Robb, quoting the transcript again: "But it was not presented in that method. It is a method of carrying out a policy which was more or less a policy of the Government. The form in which it came out was that couldn't an interview be arranged with this man Eltenton who had a very good contact with a man from the Embassy attached to the Consulate who is a very reliable guy and who had a lot of experience in microfilm or whatever."

"Did you tell Colonel Pash," Robb asked, "that microfilm had been mentioned to you?"

Oppenheimer: "Evidently."

Robb: "Was that true?"

Oppenheimer: "No."

Robb: "Then Pash said to you: 'Well, now, I may be getting back to a little systematic picture. These people whom you mention, two are down with you now [in Los Alamos]. Were they contacted by Eltenton direct?' You answered 'No.' "

Pash then said, "Through another party?"

Oppenheimer: "Yes."

"In other words," Robb summed up, "you told Pash that X [Chevalier] had made these other contacts, didn't you?"

Oppenheimer: "It seems so."

Robb: "That wasn't true?"

Oppenheimer: "That is right. This whole thing was a pure fabrication except for the one name Eltenton."

With his client now genuinely squirming, Garrison finally interrupted this painful interrogation to ask Gray, "Mr. Chairman, could I just make a short request at this point?"

Gray: "Yes."

Garrison politely wondered "if it would not be within the proprieties of this kind of proceeding when counsel reads from a transcript for us to be furnished with a copy of the transcript as he reads from it. This, of course, is orthodox in a court of law. . . ."

After some discussion, Gray and Robb agreed that perhaps at the end of the day a classification officer could make a determination about the release of the document—which of course, Robb was already selectively reading into the record.

Garrison's intervention was long overdue and overly solicitous—and it did nothing to help release his client from the trap that Robb had set.

Soon Robb was back to quoting the Pash-Oppenheimer transcript with evident relish. "Dr. Oppenheimer . . . don't you think you told a story in great detail that was fabricated?"

Oppenheimer: "I certainly did."

Robb: "Why did you go into such great circumstantial detail about this thing if you were telling a cock-and-bull story?"

Oppenheimer: "I fear that this whole thing is a piece of idiocy. I am afraid I can't explain why there was a consul, why there was microfilm, why there were three people on the project, why two of them were at Los Alamos. All of them seem wholly false to me."

Robb: "You will agree, would you not, sir, that if the story you told to Colonel Pash was true, it made things look very bad for Mr. Chevalier?"

Oppenheimer: "For anyone involved in it, yes, sir."

Robb: "Including you?"

Oppenheimer: "Right."

Robb: "Isn't it a fair statement today, Dr. Oppenheimer, that according to your testimony now, you told not one lie to Colonel Pash, but a whole fabrication and tissue of lies?"

Feeling cornered, and perhaps panicky, Oppenheimer carelessly replied, "Right."

Robb's relentless questioning had backed Robert into a corner. He didn't recall his conversation with Pash at the level required to respond adequately to Robb's interrogation. And so he accepted his tormenter's selective presentation of the transcript. Had Garrison been an experienced trial-room counsel he would have insisted earlier that his client answer no further questions about his interview with Pash until he had had an opportunity to review the transcript, and he also would have objected to Robb's strategic use of the transcript to ambush Oppenheimer. But Garrison left the door to the interview wide open, and Oppenheimer stoically walked through it.

But Oppenheimer need not have capitulated so easily. There was an explanation for the convoluted story he had told Pash that was far less damaging than the interpretation that Robb maneuvered him into accepting. Recall that Eltenton told the FBI in 1946 that the Russian consular official, Peter Ivanov, had initially suggested that he contact three scientists associated with the Berkeley Rad Lab: Oppenheimer, Ernest Lawrence and Luis Alvarez. Eltenton knew only Oppenheimer, and not well enough to ask him about sharing information with the Russians. But it seems entirely reasonable to suppose that Eltenton would have mentioned the three names to Chevalier—and that Chevalier might very well have specifically mentioned them to Oppenheimer, or at least noted that Eltenton had mentioned two (unspecified) others.

So in recounting to Pash what he knew about Eltenton's activities, Oppenheimer referred to three scientists. Of all the interpretations of Oppenheimer's "cock-and-bull story," this notion appears to make the most sense, supported as it is by evidence from the FBI's own files. Tellingly, the official historians of the AEC, Richard G. Hewlett and Jack M. Holl, reached a similar conclusion: "Oppenheimer's story, although misleading, was accurate as far as it went; unfortunately, thereafter, it became confused and twisted."

Why?

The clearest and most convincing explanation of why Oppenheimer presented Pash with such an elaborately confused representation of his kitchen conversation with Chevalier was offered by Oppenheimer himself the day before his security hearing was concluded. His explanation not only conforms with the most compelling known facts, but it also conforms with Oppenheimer's character—especially, as he had confessed to David Bohm five years earlier, "his tendency when things get too much" to say "irrational things." Responding to Chairman Gray's query whether he might have been telling the truth in 1943 to Pash and Lansdale, and was, in fact, fabricating today about the Chevalier incident, Oppenheimer replied:

> The story I told Pash was not a true story. There were not three or more people involved on the project. There was one person involved. That was me. I was at Los Alamos. There was no one else at Los Alamos involved. There was no one at Berkeley involved. . . . I testified that the Soviet consulate had not been mentioned by Chevalier. That is the very best of my recollection. It is conceivable that I knew of Eltenton's connection with the consulate, but I believe I can do no

more than say the story told in circumstantial detail, and which was elicited from me in greater and greater detail during this was a false story. It is not easy to say that. Now when you ask me for a more persuasive argument as to why I did this than that I was an idiot, I am going to have more trouble being understandable. I think I was impelled by two or three concerns at that time. One was the feeling that I must get across the fact that if there was, as Lansdale indicated, trouble at the Radiation Laboratory, Eltenton was the guy that might very well be involved and it was serious. Whether I embroidered the story in order to underline the seriousness or whether I embroidered it to make it more tolerable, that I would not tell the simple facts, namely Chevalier had talked to me about it, I don't know. There were no other people involved, the conversation with Chevalier was brief, it was in the nature of things not utterly casual, but I think the tone of it and his own sense of not wishing to have anything to do with it, I have correctly communicated.

Oppie went on to elaborate,

I should have told it [the story] at once and I should have told it completely accurately, but that it was a matter of conflict for me and I found myself, I believe, trying to give a tip to the intelligence people without realizing that when you give a tip you must tell the whole story. When I was asked to elaborate, I started off on a false pattern. . . . The notion that he [Chevalier] would go to a number of project people to talk to them instead of coming to me and talking it over as we did would have made no sense whatever. He was an unlikely and absurd intermediary for such a task . . . there was no such conspiracy. . . . When I did identify Chevalier, which was to General Groves, I told him of course that there were no three people, that this had occurred in our house, that this was me. So that when I made this damaging story, it was clearly with the intention of not revealing who was the intermediary.

THE NEXT TOPIC Robb turned to was certain to humiliate Robert—his love affair with Jean Tatlock.

"Between 1939 and 1944, as I understand it," Robb asked, "your acquaintance with Miss Tatlock was fairly casual; is that right?

Oppenheimer: "Our meetings were rare. I do not think it would be right to say that our acquaintance was casual. We had been very much involved with one another and there was still very deep feeling when we saw each other."

Robb: "How many times would you say you saw her between 1939 and 1944?"

Oppenheimer: "That is 5 years. Would 10 times be a good guess?"

Robb: "What were the occasions for your seeing her?"

Oppenheimer: "Of course, sometimes we saw each other socially with other people. I remember visiting her around New Year's of 1941."

Robb: "Where?"

Oppenheimer: "I went to her home or to the hospital. I don't know which, and we went out for a drink at the Top of the Mark. I remember that she came more than once to visit our home in Berkeley."

Robb: "You and Mrs. Oppenheimer."

Oppenheimer: "Right. Her father lived around the corner not far from us in Berkeley. I visited her there once. I visited her, as I think I said earlier. In June or July of 1943."

Robb: "I believe you said in connection with that you had to see her."

Oppenheimer: "Yes."

Robb: "Why did you have to see her?"

Oppenheimer: "She had indicated a great desire to see me before we left. At that time I couldn't go. For one thing, I wasn't supposed to say where we were going or anything. I felt she had to see me. She was undergoing psychiatric treatment. She was extremely unhappy."

Robb: "Did you find out why she had to see you?"

Oppenheimer: "Because she was still in love with me."

Robb: "Where did you see her?"

Oppenheimer: "At her home."

Robb: "Where was that?"

Oppenheimer: "On Telegraph Hill."

Robb: "When did you see her after that?"

Oppenheimer: "She took me to the airport, and I never saw her again."

Robb: "That was in 1943?"

Oppenheimer: "Yes."

Robb: "Was she a Communist at that time?"

Oppenheimer: "We didn't even talk about it. I doubt it."

Robb: "You have said in your answer that you knew she had been a Communist?"

Oppenheimer: "Yes. I knew that in the fall of 1937."

Robb: "Was there any reason for you to believe that she wasn't still a Communist in 1943?"

Oppenheimer: "No."

Robb: "Pardon?"

Oppenheimer: "There wasn't, except that I have stated in general terms what I thought and think of her relation with the Communist Party. I do not know what she was doing in 1943."

Robb: "You have no reason to believe she wasn't a Communist, do you?"

Oppenheimer: "No."

Robb: "You spent the night with her, didn't you?"

Oppenheimer: "Yes."

Robb: "Did you think that consistent with good security?"

Oppenheimer: "It was, as a matter of fact. Not a word—it was not good practice."

Robb: "Didn't you think that put you in a rather difficult position had she been the kind of Communist that you have described her[e] or talk[ed] about this morning?"

Oppenheimer: "Oh, but she wasn't."

Robb: "How did you know?"

Oppenheimer: "I knew her."

Having suffered the indignity of testifying to an affair with Tatlock three years into his marriage to Kitty, Oppenheimer was then asked by Robb to name the names of his lover's friends, and to state which of them were Communists and which were merely fellow travelers. It was a pointless question with respect to the purpose of the hearing, but it was not a question without a point. This was 1954, the apogee of the McCarthy years, and forcing former communists, fellow travelers and left-wing activists called before congressional committees to name names was precisely the McCarthyites' political game. It was a humiliating experience in a culture that despised a "snitch," a Judas, and that was the point: to destroy a witness' sense of personal integrity.

Oppenheimer gave Robb the names: Dr. Thomas Addis he thought was close to the Party, but he didn't know if he had ever been a member; Chevalier was a fellow traveler; Kenneth May, John Pitman, Aubrey Grossman and Edith Arnstein were communists. Well aware of the degrading nature of the exercise he was being subjected to, Oppenheimer sarcastically asked Robb, "Is the list long enough?" As was often the case, the names were known. Robb's relentless hammering was taking its toll. He was beginning to respond unthinkingly, "the way a soldier does in combat, I suppose," he later recalled to a reporter. "So much is happening or may be about to happen that there is no time to be aware of anything except the next move. Like something in a fight—and this was a fight. I had very little sense of self."

Years later, Garrison would recall Oppenheimer's mood during these torturous days: "From the beginning, he had a quality of desperation about him. . . . I think we all felt oppressed by the atmosphere of the time but Oppenheimer particularly so. . . ."

ROBB GAVE STRAUSS daily reports on what was happening inside the privileged hearing room, and the AEC chairman was pleased with the way things were going. He wrote President Eisenhower: "On Wednesday, Oppenheimer broke and admitted, under oath, that he had lied. . . ." Gleefully anticipating victory, he informed Ike that "an extremely bad impression toward Oppenheimer has already developed in the minds of the Board." Ike cabled him in reply from his Augusta, Georgia, retreat, thanking him for his "interim report." He also informed Strauss that he had burned his interim report, apparently not wanting to leave any evidence that he or Strauss was inappropriately monitoring the security hearing.

ON THE MORNING of Thursday, April 15—four days into the hearing—Gen. Leslie Groves was sworn as a witness. Questioned by Garrison, Groves praised Oppenheimer's wartime performance at Los Alamos, and when asked whether he was capable of consciously committing a disloyal act, he said emphatically, "I would be amazed if he did. . . ." When asked specifically about the Chevalier incident, Groves testified: "I have seen so many versions of it, I don't think I was confused before, but I am certainly starting to become confused today. . . . My conclusion was that there was an approach made, that Dr. Oppenheimer knew of this approach. . . ."

Groves went on to explain that when he first learned of the story, he thought Robert's reticence could be explained by "the typical American schoolboy attitude that there is something wicked about telling on a friend. I was never certain as to just what he was telling me. I did know this: that he was doing what he thought was essential, which was to disclose to me the dangers of this particular attempt to enter the project, namely, it was concerned with the situation out there near Berkeley—I think it was the Shell Laboratory at which Eltenton was supposedly one of the key members—and that was a source of danger to the project and that was the worry. I always had the impression that Dr. Oppenheimer wanted to protect his friends of long standing, *possibly* his brother. It was always my impression that he wanted to protect his brother, and that his brother *might* be involved in having been in this chain. . . ."

Groves' testimony "possibly" expanded the cast of characters associated with the Chevalier affair. Frank "might be involved," Groves speculated, surely without malice and probably without a full realization of the potential consequences of his hypothesis. For if Frank *had* been involved, then not only had Robert lied to Pash in 1943, but he had lied to the FBI in 1946 and was lying now to the hearing board in 1954. Regardless of the extenuating circumstances—Robert's desire to protect his younger brother, whom he knew to be innocent of any wrongdoing—Groves' conjecture further undermined Robert's veracity and, in the end, despite the lack of any evidence pointing to Frank's participation, it deepened the mystery surrounding—and therefore the hearing board's interest in—the Chevalier affair.

Any effort to explain the source and tentative nature of Groves' testimony connecting Frank to Chevalier leads back to what was recorded in Oppenheimer's FBI dossier during the war. From there our attention will fast-forward ten years to a series of FBI interviews conducted in December 1953 in preparation for Oppenheimer's appearance before the AEC's Personnel Security Board. The interviewees were John Lansdale and William Consodine, wartime assistants to General Groves, Groves himself, and Corbin Allardice, who had succeeded William Borden as staff director of Congress' Joint Committee on Atomic Energy (JCAE).

These interviews played a critical role in shaping Groves' testimony, for both Consodine and Lansdale reported to him what they had told the Bureau's agents. Their recollections were disconcerting for Groves, who in several important ways had a different memory of what Oppie had told him. Furthermore, their communications with the FBI put him in a compromising position which forced him to acknowledge to the hearing board that in 1954 he could not support renewing Robert's security clearance.

As noted earlier, the first documented reference to Frank's association with Chevalier to appear in the FBI's files was in a memo of March 5, 1944, by agent William Harvey. Harvey had no independent information about the Chevalier affair but in composing a summary of it he identified Frank as the "one person" approached by Chevalier. However, Harvey failed to cite any evidence for this conclusion, an oversight that would baffle senior agents a decade later, when they reported to Hoover: "File review failed to reflect any info that Frank Oppenheimer was approached for data concerning MED [Manhattan Engineer District] project or that such info was ever reported by J. Robert Oppenheimer to MED or Bureau."

But on December 3, 1953—several weeks after Borden's letter had been mailed—Frank's name was again brought to the attention of the FBI by another purveyor of hearsay. Corbin Allardice, who was an AEC employee prior to replacing Borden at the JCAE, was apparently encouraged by someone unfriendly to Oppenheimer to rekindle the suspicion that Frank was Chevalier's contact. Allardice reported having been "informed by a source whom he believed to be extremely reliable that J. Robert Oppenheimer had stated that his contact in the Eltenton–Haakon Chevalier espionage apparatus had been his own brother, Frank Oppenheimer." Allardice further stated—which suggests that his informant had some familiarity with Oppenheimer's FBI dossier—that he didn't think this information was in the FBI's record on the case. He suggested that if the FBI wished to check out his tip, they should interview John Lansdale, who was then practicing law in Cleveland.

Lansdale was interviewed on December 16. But the day before, another of Groves' wartime assistants, William Consodine (Allardice's friend and therefore most likely his "reliable" informant), spoke to an FBI agent.

The FBI summary, written on December 18, has Consodine telling the following story:

The day after General Groves returned from Los Alamos, "where he had induced [Oppenheimer] to identify [Eltenton's] intermediary," he held a conference in his office with Lansdale and Consodine. After announcing to them "that the intermediary had been identified by Oppenheimer, General Groves pushed a yellow pad toward both Consodine and Lansdale and asked them to write down three guesses as to the identity of the intermediary. Lansdale wrote down three names which Consodine cannot now recall. Consodine stated he wrote down one name only, that of Frank Oppenheimer. General Groves expressed surprise at this guess and said it was correct. General Groves asked Consodine how he selected the name Frank Oppenheimer. Consodine said he explained to the general that he thought it was Frank Oppenheimer because J. Robert Oppenheimer would probably be more likely to be reluctant to involve his brother.

"According to Consodine, General Groves then informed [them] that he had obtained the admission after J. Robert Oppenheimer exacted a promise that the general would not identify Frank Oppenheimer as the intermediary to the FBI. In concluding Consodine stated . . . that he had not been in communication with Lansdale concerning this matter but that he had discussed the matter on the telephone with General Groves during the past few days."

On December 16 Lansdale told a modified version of Consodine's story to his FBI interviewer. He clearly had no recollection of Consodine's "yellow pad" story (and neither did Groves). What Lansdale did recall was an impression he received from the general that after Groves asked Oppenheimer to fully disclose Eltenton's contacts, "Oppenheimer told Groves that an approach had been made to Frank Oppenheimer by Haakon Chevalier." In conclusion, however, "Lansdale stated that *General Groves was of the opinion that an approach had been made directly to J. Robert Oppenheimer* but Lansdale felt that the approach was made to Frank Oppenheimer. Lansdale advised that to his knowledge, only he and General Groves knew about the incident." When Lansdale was asked point-blank by Garrison whether it was possible that Groves, "told you that he *thought* it was Frank—rather than it *was* Frank," Lansdale conceded, "Yes, it is possible."

On December 21, 1953—the day on which Oppenheimer was informed that his security clearance was suspended—another FBI agent interviewed Groves in his Darien, Connecticut home.

Until then, Groves had refused to talk with the FBI about Oppenheimer and the Chevalier affair. He had not even bothered to reply to the FBI's initial queries on this topic in 1944. And then, in June 1946, as the Bureau was about to interview both Chevalier and Eltenton, FBI agents asked Groves what he knew about the affair. Groves brushed them off, saying he really couldn't talk about it because Oppenheimer had talked to him in "strict confidence." Groves said that "he could not break faith with 'Oppie' and tell us the name of the man that the Shell Development representative approached." The FBI agents replied that they knew the Shell man was Eltenton and that they were about to interview him. In an extraordinary demonstration of his continuing loyalty to Oppenheimer, Groves said he "did not want us to confront Eltenton with this matter as it might get back to Oppenheimer and Oppenheimer would know Groves had broken confidence with him." Groves bluntly told the FBI agents that he was "hesitant to furnish any further information."

Hoover must have been astonished to learn that an American army general was refusing to cooperate with an FBI investigation. On June 13, 1946, Hoover personally wrote Groves, asking him to reveal what Oppenheimer had told him about George Eltenton. Groves replied on June 21, politely declining to furnish this information, "as it would endanger" his relationship with Oppenheimer. Not many men in Washington defied a direct request from the director of the FBI, but in 1946 Groves had a lot of prestige and self-confidence.

But now, in 1953, forewarned by Consodine and Lansdale of their having informed the FBI that Frank was the contact in the Eltenton-Chevalier incident, Groves felt compelled to incorporate their recollections into his own account. The problem was that he himself couldn't remember what exactly Oppenheimer had told him in 1943–44. But, prompted by his former assistants, Groves now told his interviewer that late in 1943 he had finally ordered Oppenheimer to "make a full disclosure" about who had approached him for information about the project. To encourage Robert to be forthcoming, Groves had assured him that he would not make a formal report on the incident, or "to put it very bluntly, it would not get to the FBI." With that promise, Groves reported that Robert told him that "Chevalier had made the approach to Frank Oppenheimer," and that Frank had asked Robert what he should do. According to Groves, Robert had told his brother "to have nothing to do" with Eltenton, and he had also spoken directly to Chevalier and given him his "comeuppance." Groves further explained that "it was Eltenton who wanted the information and that the intermediaries [Chevalier and Frank] were innocent of the intent to commit espionage."*

Groves said further that he thought "it was natural and proper for Frank Oppenheimer to do what he did despite the fact that he should have notified the local security officers." The Oppenheimer brothers were very close, and it only made sense that the younger brother—"much perturbed about the visit" from Chevalier—would immediately contact the older brother and tell him about the incident. "He [Groves] said it was a technical violation of security to have handled it the way he [Frank] did, but that he had in fact done all that could be reasonably expected. . . . The General said it was obvious that the subject [Oppenheimer] wanted to protect his brother, Chevalier and the subject [Robert himself]."

But then Groves went on to "speculate" whether Robert had "invented Frank as a party in order to justify his delay in reporting the original approach or whether Frank had, in fact, been involved." In other words, while Groves clearly had said something in 1943 about Frank which led Lansdale and Consodine to believe that Chevalier had contacted Frank, Groves himself had serious doubts on this point. Groves' confusion about Frank's role never abated. As late as 1968, he confessed to a historian, "Of course, I wasn't sure just who the man was he [Oppenheimer] was protecting. Today I would *guess* it was probably his brother. He didn't want his brother involved."

Groves appears to have been convinced of two things: first, that Chevalier had approached Robert on Eltenton's behalf; second, that Robert had said something in 1943 designed to make it clear to him, Groves, that Frank had promptly reported to Robert some sort of inappropriate inquiry from Chevalier. Anything more specific is lost to history. After all, Groves himself said, "I was never certain as to just what he [Robert] was telling me." And, in an earlier letter, "It was very difficult to tell how much Frank was involved and how much Robert was involved." The most probable explanation of why Lansdale and Consodine believed that Frank was Chevalier's contact is that Groves had told them about his conversation with Robert without making clear his doubts about Frank's involvement.

No other explanation seems possible when all of the interviews and documents are read together. Frank simply could not have been either

*When the FBI asked Frank Oppenheimer about this, he categorically denied that Chevalier had ever approached him, or that he had ever talked with his brother about a query from Eltenton.

Eltenton's or Chevalier's contact in the "Chevalier affair." By all accounts—
Chevalier's and Eltenton's simultaneous FBI interviews in 1946, Barbara
Chevalier's unpublished memoirs, Kitty's recollection to Verna Hobson,
Frank's statement to the FBI in early January 1954 and finally Robert's
statements to the FBI in 1946 and in his concluding testimony—it was
Haakon who approached Robert.

Nevertheless, for having trusted in Oppenheimer's "story"—and for
having promised to keep it from the FBI—Groves now found himself personally compromised. The historian Gregg Herken makes a credible case
that both Strauss and J. Edgar Hoover thought they could use the fact that
Groves had implicated himself in a "cover-up" to exert pressure on the general to testify against Oppenheimer in the upcoming security hearing. One
of Hoover's key aides, Alan Belmont, implicitly suggested this when he
wrote his boss that it was "readily apparent that Groves has attempted to
withhold and conceal important information concerning espionage conspiracy violation from the FBI. Even now Groves is behaving with a certain
amount of coyness in his dealings and admissions to the Bureau."

While embarrassed by the FBI's discovery, Groves was unapologetic
about having promised Oppenheimer that he would not reveal Frank's name
to the Bureau. Moreover, it was a promise he still defended: "The General
said he did not feel that he was violating the spirit of the promise to Oppenheimer in having the present interview with the Agent because the matter
was already known to the authorities. He said he wanted this noted in the
record, because it was possible that a friend of Oppenheimer might some
day see this file and consider that 'I had broken my promise after all.' " If
Groves had at any time thought for a minute that Oppenheimer was actually
protecting a spy, he would certainly have gone to the FBI. He obviously was
confident of Oppenheimer's loyalty.

This, of course, was not how Strauss saw things. What could have been
interpreted as exculpatory evidence was ignored. Instead, Strauss pursued
Groves, and asked him in February to come to Washington for another
interview. By then, Groves understood that he would be asked to testify
against Oppenheimer and, if he refused, he could be accused of participating in a coverup.

ASTONISHINGLY, Robb failed to follow-up on Groves' speculations
about Frank, no doubt because to do so would portray Robert as someone
who was taking the fall for his brother. Neither did Robb reveal to the Gray
Board, or to Oppenheimer's lawyers, that Groves had promised not to
reveal Frank's name to the FBI. This too would have diverted the spotlight
from Robert. This part of the story would remain classified in the FBI doc-

uments for twenty-five years. Under Robb's cross-examination, Groves made it clear that while he still thought his decision to give Oppenheimer a clearance in 1943 was the right judgment then, today things might be different. When Robb asked him point-blank: ". . . would you clear Dr. Oppenheimer today?" Groves waffled. "I think before answering that I would like to give my interpretation of what the Atomic Energy Act requires." Read literally, he said, the act specified that the AEC must determine that people given access to restricted data "will not endanger the common defense or security . . ." In Groves's view, there was no wiggle room. "It is not a case of proving that a man is a danger," he said. "It is a case of thinking, well, he might be a danger . . ." On this basis, and given Oppenheimer's past associations, "I would not clear Dr. Oppenheimer today if I were a member of the Commission on the basis of this interpretation." That's all Robb wanted or needed the general to say. And why had Groves turned against the man he had hitherto defended so resolutely? Strauss knew. He had made it clear to the general, in a not so subtle fashion, that he, Strauss, would make certain that there would be grave consequences for Groves if he did not cooperate.

THE NEXT DAY, Friday, April 16, Robb resumed his cross-examination of Oppenheimer. He grilled him about his relationships with the Serbers, David Bohm and Joe Weinberg, and late in the day he got around to asking the physicist about his opposition to the development of the hydrogen bomb. After nearly five full days of intense interrogation, Oppenheimer must have been physically and mentally exhausted. But on this day—his last in the witness chair—he nevertheless mustered his razor-sharp wit. Wary from experience at being ambushed, and crystal clear about the issue, he was more adept at parrying Robb's questions.

Robb: "Did you subsequent to the President's decision in January 1950 ever express any opposition to the production of the hydrogen bomb on moral grounds?"

Oppenheimer: "I would think that I could very well have said this is a dreadful weapon, or something like that. I have no specific recollection and would prefer it, if you would ask me or remind me of the context or conversation that you have in mind."

Robb: "Why do you think you could very well have said that?"

Oppenheimer: "Because I have always thought it was a dreadful weapon. Even [though] from a technical point of view it was a sweet and lovely and beautiful job, I have still thought it was a dreadful weapon."

Robb: "And have said so?"

Oppenheimer: "I would assume that I have said so, yes."

Robb: "You mean you had a moral revulsion against the production of such a dreadful weapon?"

Oppenheimer: "This is too strong."

Robb: "Beg pardon?"

Oppenheimer: "That is too strong."

Robb: "Which is too strong, the weapon or my expression?"

Oppenheimer: "Your expression. I had a grave concern and anxiety."

Robb: "You had moral qualms about it, is that accurate?"

Oppenheimer: "Let us leave the word 'moral' out of it."

Robb: "You had qualms about it."

Oppenheimer: "How could one not have qualms about it? I know no one who doesn't have qualms about it."

Later in the day, Robb produced a letter written by Oppenheimer to James Conant dated October 21, 1949. The document came from Oppenheimer's own files—papers confiscated by the FBI the previous December. Addressed to "Dear Uncle Jim," the letter complained that "two experienced promoters have been at work, i.e. Ernest Lawrence and Edward Teller," lobbying on behalf of the hydrogen bomb. In a testy exchange, Robb asked Oppenheimer, "Would you agree, Doctor, that your references to Dr. Lawrence and Dr. Teller . . . are a little bit belittling?"

Oppenheimer: "Dr. Lawrence came to Washington. He did not talk to the Commission. He went and talked to the joint congressional committee and to members of the Military Establishment. I think that deserves some belittling."

Robb: "So you would agree that your references to those men in this letter were belittling?"

Oppenheimer: "No. I pay my great respects to them as promoters. I don't think I did them justice."

Robb: "You used the word 'promoters' in an invidious sense, didn't you?"

Oppenheimer: "I have no idea."

Robb: "When you use the word now with reference to Lawrence and Teller, don't you intend it to be invidious?"

Oppenheimer: "No."

Robb: "You think that their work of promotion was admirable, is that right?"

Oppenheimer: "I think they did an admirable job of promotion."

BY FRIDAY, it was clear to everyone that Robb and Oppenheimer despised each other. "My feeling was," Robb recalled, "that he was just a brain and as cold as a fish, and he had the iciest pair of blue eyes I ever saw."

Oppenheimer felt only revulsion in Robb's presence. During a brief recess one day, the two men happened to be standing near each other when Oppenheimer suddenly had one of his coughing spells. As Robb indicated his concern, Oppenheimer cut him off angrily and said something that caused Robb to turn on his heel and walk away.

At the end of each day, Robb closeted himself with Strauss and took stock of the day's events. They had little doubt about the outcome. Strauss told an FBI agent that he was "convinced that in view of the testimony to date the board could take no other action but to recommend the revoking of Oppenheimer's clearance."

Oppenheimer's lawyers felt much the same way. To escape the scrutiny of the press corps, the Oppenheimers were now spending each night in the Georgetown home of Randolph Paul, a law partner of Garrison's. The press did not discover their location for a week, but FBI agents staked out the house and reported that Oppenheimer was staying up late and pacing the room.

Garrison and Marks spent several hours most evenings in Paul's home, planning the next day's strategy. "All we had the energy for was preparation," Garrison said, "we were too weary to do much post-morteming. Of course, Robert was in the most overwrought state imaginable—so was Kitty—but Robert even more so."

Paul listened with growing unease as the Oppenheimers described each day's events to him. Their recounting sounded a lot more like a trial than an administrative hearing. So on the evening of Easter Sunday, April 18, Paul invited Garrison and Marks to his home for a consultation with Joe Volpe. After drinks were served, Oppenheimer turned to the AEC's former general counsel and said, "Joe, I would like to have these fellows describe to you what's going on in the hearing." Over the next hour, Volpe listened with rising outrage as Marks and Garrison summarized Robb's adversarial tactics and the general tone of Oppie's daily ordeal. Finally, he turned to Oppenheimer and said, "Robert, tell them to shove it, leave it, don't go on with it because I don't think you can win."

Oppenheimer had heard this advice before, from Einstein among others. But this time it came from an experienced attorney who had helped write the rules for AEC hearings, and in whose opinion both the spirit and the letter of those rules were being outrageously violated. Even so, Oppenheimer decided he had no choice now but to see the process through to a conclusion. It was a stoical and rather passive reaction, not unlike his quiet acceptance all those years before when as a young boy he had been locked in the camp icehouse.

CHAPTER THIRTY-SIX

"A Manifestation of Hysteria"

> *I am very distressed, as I assume you are, over the Oppen-*
> *heimer matter. I feel that it is somewhat like inquiring into*
> *the security risk of a Newton or a Galileo.*
>
> JOHN J. MCCLOY *to President Dwight D. Eisenhower*

AFTER OPPENHEIMER WAS EXCUSED from the witness chair on
Friday, Garrison was allowed to call a parade of more than two
dozen defense witnesses to vouch for Oppenheimer's character and
loyalty. They included Hans Bethe, George Kennan, John J. McCloy, Gordon Dean, Vannevar Bush and James Conant, among other eminent figures
from the worlds of science, politics and business. By far one of the most
interesting of these was John Lansdale, the Manhattan Project's former
chief of security, and now a partner in a Cleveland law firm. That the
Army's key security officer during the Los Alamos years was testifying for
the defense should have carried great weight with the hearing panel. Moreover, unlike Oppenheimer, Lansdale immediately knew how to fend off
Robb's aggressive tactics. Under cross-examination, Lansdale said he
"strongly" felt Oppenheimer to be a loyal citizen. And then he added, "I am
extremely disturbed by the current hysteria of the times, [of] which this
seems to be a manifestation."

Robb could not possibly let this pass, and asked him, "You think this
inquiry is a manifestation of hysteria?"

Lansdale: "I think—"

Robb: "Yes or no?"

Lansdale: "I won't answer that question 'Yes' or 'No.' If you are tending
to be that way—if you will let me continue, I will be glad to answer your
question."

Robb: "All right."

Lansdale: "I think the hysteria of the times over communism is
extremely dangerous." He then explained that at the same time in 1943

when he was handling Oppenheimer's security clearance, he had also been grappling with the sensitive question of whether to commission as Army officers known communists who had volunteered to fight the Spanish fascists in Republican Spain. Because he had "dared to stop the commissioning" of a group of fifteen or twenty such communists, Lansdale said he had been "vilified" by his superiors. His decision was overruled by the White House—and Lansdale said he blamed Mrs. Roosevelt "and those around her in the White House" for creating an atmosphere in which communists were given officer commissions.

Having thus established his anticommunist credentials, Lansdale went on to say that, "We are going through today the other extreme of the pendulum, which is in my judgment equally dangerous. . . . Now, do I think this inquiry is a manifestation of hysteria? No. I think the fact that so much doubt and so much—let me put it this way. I think the fact that associations in 1940 are regarded with the same seriousness that similar associations would be regarded today is a manifestation of hysteria."

JOHN J. MCCLOY, now chairman of Chase National Bank, agreed with Lansdale. A member of Eisenhower's private "kitchen cabinet," McCloy was also chairman of the Council on Foreign Relations, and he sat on the boards of the Ford Foundation and a half dozen of the richest corporations in the country. On the morning of April 13, 1954, when McCloy read Reston's story about the Oppenheimer case, he found the news profoundly "disturbing." "I didn't give a damn if he was sleeping with a mistress who was a communist," he recalled later.

McCloy had been seeing Oppie regularly at the Council on Foreign Relations and had no real doubts about his loyalty—an opinion he did not hesitate to share immediately with Eisenhower: "I am very distressed, as I assume you are, over the Oppenheimer matter," he wrote the president. "I feel that it is somewhat like inquiring into the security risk of a Newton or a Galileo. Such people are themselves always 'top secret.' " Ike lamely replied that he hoped the "distinguished" Gray Board would exonerate the scientist.

McCloy felt strongly enough about the whole matter that at the end of April he was easily persuaded by Garrison—who had known McCloy since their Harvard Law School years—to attend the hearing as a last-minute defense witness. McCloy's testimony produced some memorable exchanges as he tried to raise issues that bore directly on the very legitimacy of the hearing. He began his defense of Oppenheimer by questioning the Gray Board's definition of security:

"I don't know just exactly what you mean by a security risk. I know that I am a security risk and I think every individual is a security risk. . . . I think there is a security risk in reverse. . . . We are only secure if we have the best brains and the best reach of mind. If the impression is prevalent that scientists as a whole have to work under such great restrictions and perhaps great suspicion in the United States, we may lose the next step in this [nuclear] field, which I think would be very dangerous for us."

When Garrison asked him about the Chevalier incident, McCloy responded that the Gray Board ought to weigh Oppenheimer's willingness to lie in order to protect a friend against his value to the country as a theoretical physicist. This line of argument, of course, greatly unsettled the Gray Board, for it suggested that there could be no absolutes in matters of security, that a value judgment had to be made on the merits of each individual—which AEC security regulations did, in fact, recommend. During his cross-examination of McCloy, Robb countered with a clever analogy: Did the chairman of Chase National Bank employ anyone who for some time had associated with bank robbers? "No," said McCloy, "I don't know of anyone." And if a Chase branch manager had a friend who volunteered that he knew some people who planned to rob the bank, wouldn't McCloy expect his branch manager to report the conversation? McCloy, of course, had to answer yes.

McCloy understood that this exchange had damaged Oppie's case, and the more so when Gray returned to the analogy a short time later: "Would you leave someone in charge of the vaults about whom you have any doubt in your mind?"

No, said McCloy, but then he quickly interjected that if an employee of doubtful background nevertheless "knew more about . . . the intricacies of time locks than anybody else in the world, I might think twice before I let him go, because I would balance the risks in this connection." When it came to the mind of Dr. Oppenheimer, he said, "I would accept a considerable amount of political immaturity in return for this rather esoteric, this rather indefinite, theoretical thinking that I believe we are going to be dependent on for the next generation."

SUCH DRAMATIC EXCHANGES were not unusual. The drab hearing room at 16th and Constitution had rapidly become a stage upon which an extraordinary cast of actors addressed Shakespearean themes. How should a man be judged, by his associations or by his actions? Can criticism of a government's policies be equated with disloyalty to country? Can democracy survive in an atmosphere that demands the sacrifice of personal relationships to state policy? Is national security well served by applying narrow tests of political conformity to government employees?

Oppenheimer's character witnesses offered eloquent and sometimes poignant testaments. George Kennan was unequivocal: In Oppenheimer, he said, we were faced with "one of the great minds of this generation of Americans." Such a man, he suggested, could not "speak dishonestly about a subject which had really engaged the responsible attention of his intellect. . . . I would suppose that you might just as well have asked Leonardo da Vinci to distort an anatomical drawing as that you should ask Robert Oppenheimer to speak . . . dishonestly."

This provoked Robb to ask Kennan under cross-examination if he meant to suggest that different standards should be used when judging "gifted individuals."

Kennan: "I think the church has known that. Had the church applied to St. Francis the criteria relating solely to his youth, it would not have been able for him to be what he was later. . . . it is only the great sinners who become the great saints and in the life of the Government, there can be applied the analogy."

One member of the Gray Board, Dr. Ward Evans, interpreted this to mean that "all gifted individuals were more or less screwballs."

Kennan politely demurred: "No, sir; I would not say that they are screwballs, but I would say that when gifted individuals come to a maturity of judgment which makes them valuable public servants, you are apt to find that the road by which they have approached that has not been as regular as the road by which other people have approached it. It may have zigzags in it of various sorts."

Seeming to agree, Dr. Evans responded, "I think it would be borne out in the literature. I believe it was Addison, and someone correct me if I am wrong, that said, 'Great wits are near to madness, closely allied and thin partitions do their bounds divide.' "

At this, Dr. Evans took note that "Dr. Oppenheimer is smiling. He knows whether I am right or wrong on that. That's all."

LATER that same day, Tuesday April 20, David Lilienthal followed Kennan into the witness chair. Kennan had emerged unscathed. But Robb had prepared a trap for the new witness. The previous day, Lilienthal had received permission to review his own AEC papers in order to refresh his memory. But as Robb began his cross-examination, it soon became clear that Robb had some documents in hand that had been kept from Lilienthal. After leading him on to recount his memory of Oppenheimer's 1947 security review, Robb suddenly produced memorandums that made it clear that Lilienthal had himself recommended "the establishment of an evaluation board of distinguished jurists to make a thorough review" of Oppenheimer's case.

Robb: "In other words, you recommended in 1947 that the exact step which is now being taken, be taken then?"

Flustered and angry, Lilienthal foolishly admitted that had been the case, when, in fact, he had suggested something quite different from the star chamber proceeding that was now under way. As Robb pressed him relentlessly, Lilienthal at one point protested that ". . . a simple way to secure the truth and accuracy would have been to have given me these files yesterday, when I asked for them, so that when I came here, I could be the best possible witness and disclose as accurately as possible what went on at that time."

Garrison interrupted at this point to complain once again that "the surprise production of documents is not the shortest way to arrive at the truth. It seems to me more like a criminal trial than it does like an inquiry and I just regret that it has to be done here." And once again, Chairman Gray brushed aside Garrison's protest. And once again, Garrison fell silent.

At the end of this very long day, Lilienthal went home and noted in his diary that he had difficulty sleeping, "so steamed up was I over the 'entrapment' tactics . . . and sadness and nausea at the whole spectacle."

WHERE LILIENTHAL EMERGED chastened and angered by the experience, the inimitable and unflappable Isidor Rabi walked out of the hearing room defiant and unscathed. In one of the more memorable statements of the entire hearing, Rabi said, "I never hid my opinion from Mr. Strauss that I thought this whole proceeding was a most unfortunate one. . . . That the suspension of the clearance of Dr. Oppenheimer was a very unfortunate thing and should not have been done. In other words, there he was; he is a consultant, and if you don't want to consult the guy, you don't consult him, period. Why you have to then proceed to suspend clearance and go through all of this sort of thing, he is only there when called and that is all there was to it. So it didn't seem to me the sort of thing that called for this kind of proceeding at all against a man who had accomplished what Dr. Oppenheimer has accomplished. There is a real positive record, the way I expressed it to a friend of mine. We have an A-bomb and a whole series of *it* . . . [deleted classified material] and what more do you want, mermaids? This is just a tremendous achievement. If the end of that road is this kind of hearing, which can't help but be humiliating, I thought it was a pretty bad show. I still think so."

Upon cross-examination, Robb attempted to shake Rabi's self-confidence by posing yet another hypothetical question about the Chevalier incident. If Rabi had been put in such circumstances, Robb asked, he would have told the "whole truth about it, wouldn't you?"

Rabi: "I am naturally a truthful person."

Robb: "You would not have lied about it?"

Rabi: "I am telling you what I think now. The Lord alone knows what I would have done at that time. This is what I think now."

A few moments later, Robb asked, "Of course, Doctor, you don't know what Dr. Oppenheimer's testimony before this board about the incident may have been, do you?"

Rabi: "No."

Robb: "So perhaps in respect of passing judgment on that incident, the board may be in a better position to judge than you?"

Never at a loss for words, Rabi parried, "It may be. On the other hand, I am in possession of a long experience with this man, going back to 1929, which is 25 years, and there is a kind of seat-of-the-pants feeling [on] which I myself lay great weight. In other words, I might even venture to differ from the judgment of the board without impugning their integrity at all. . . .

"You have to take the whole story," Rabi insisted. "That is what novels are about. There is a dramatic moment and the history of the man, what made him act, what he did, and what sort of person he was. That is what you are really doing here. You are writing a man's life."

In the midst of Rabi's testimony, Oppenheimer excused himself from the hearing room and upon his return a few minutes later, Chairman Gray noted his presence: "You are back now, Dr. Oppenheimer."

Oppenheimer replied laconically, "This is one of the few things I am really sure of."

Rabi was both stunned by the hostile atmosphere in the hearing room and struck by Oppenheimer's metamorphosis. Robert had walked into Room 2022 an eminent, proud and self-assured scientist-statesman—but he was now playing the role of political martyr. "He was a very adaptable fellow," Rabi later observed. "When he was riding high, he could be very arrogant. When things went against him, he could play the victim. He was a most remarkable fellow."

IF THE PROCEEDING SEEMED SURREAL, it was nevertheless high theater, bristling at times with profound emotion. On Friday, April 23, Dr. Vannevar Bush was called to testify and was asked about Oppenheimer's opposition in the summer and autumn of 1952 to the testing of the early hydrogen bomb. Bush explained, "I felt strongly that that test ended the possibility of the only type of agreement that I thought possible with Russia at that time, namely, an agreement to make no more tests. For that kind of an agreement would have been self-policing in the sense that if it was violated, the violation would be immediately known. I still think that we made a grave error in conducting that test at that time." His conclusion was uncompromising: "I think history will show that was a turning point, that when we entered into the grim world that we are entering right now, that those who pushed that thing through to a conclusion without making that attempt have a great deal to answer for."

Regarding the whole controversy over Oppenheimer's opposition to the crash development of the hydrogen bomb, Bush bluntly said that it appeared to most scientists around the country that Oppenheimer was "now being pilloried and put through an ordeal because he had the temerity to express his honest opinions." As to the written charges against Oppenheimer, Bush bluntly said it was a "poorly written letter," and one which the Gray Board should have rejected from the outset.

Chairman Gray interjected at this point that, setting aside the allegations about the hydrogen bomb, there were "items of so-called derogatory information," items which did not relate to the mere expression of opinion.

"Quite right," Bush said, "and the case should have [been] tried on those."

Chairman Gray: "This is not a trial."

Bush: "If it were a trial, I would not be saying these things to the judge, you can well imagine that. . . ."

Dr. Evans: "Dr. Bush, I wish you would make clear just what mistake you think the Board made. I did not want this job when I was asked to take it. I thought I was performing a service to my country."

Bush: "I think the moment you were confronted with that letter, you should have returned the letter, and asked that it be redrafted so that you would have before you a clear-cut issue. . . . I think this board or no board should ever sit on a question in this country of whether a man should serve his country or not because he expressed strong opinions. If you want to try that case, you can try me. I have expressed strong opinions many times, and I intend to do so. They have been unpopular opinions at times. When a man is pilloried for doing that, this country is in a severe state. . . . Excuse me, gentlemen, if I become stirred, but I am."

ON MONDAY, April 26, Kitty Oppenheimer took the witness chair and testified about her communist past. She acquitted herself easily, coolly and precisely answering each question. Although she confided to her friend Pat Sherr that she was nervous, to the Gray Board she appeared forthright and unflustered. As a young girl, Kitty had been trained by her German-born parents to sit still without fidgeting, and now she drew upon this training to put on a performance of tremendous self-control. When Chairman Gray asked her if a distinction could be drawn between Soviet communism and the Communist Party of America, Kitty answered, "There are two answers to that as far as I am concerned. In the days that I was a member of the Communist Party, I thought they were definitely two things. The Soviet Union had its Communist Party and our country had its Communist Party. I thought that the Communist Party of the United States was concerned with problems internal. I now no longer believe this. I believe the whole thing is linked together and spread all over the world."

When Dr. Evans asked her if there were two kinds of communists, "an intellectual Communist and just a plain ordinary Commie," Kitty had the good sense to say, "I couldn't answer that one."

"I couldn't either," replied Dr. Evans.

MOST OF THE WITNESSES called in Oppenheimer's defense were close friends and professional allies. Johnny von Neumann was different. Though they had always maintained friendly relations personally, von Neumann and Oppenheimer had strong disagreements politically. For this reason, von Neumann was potentially a particularly persuasive defense witness. A fervid supporter of the hydrogen bomb program, von Neumann explained that while Oppenheimer had tried to persuade him to his views—and von Neumann had done the same with Oppenheimer—he could not say that Oppenheimer had ever interfered with his work on the Super. When asked about the Chevalier incident, von Neumann cheerfully explained, "This would affect me the same way as if I would suddenly hear about somebody that he has had some extraordinary escapade in his adolescence." And when Robb pressed him with the usual hypothetical about lying to the security officers in 1943, von Neumann replied, "Sir. I don't know how to answer this question. Of course, I hope I wouldn't [lie]. But—you are telling me now to hypothesize that somebody else acted badly, and you ask me would I have acted the same way. Isn't this a question of when did you stop beating your wife?"

At this point, the Gray Board members jumped in and tried to get von Neumann to answer the same hypothetical.

Dr. Evans: "If someone had approached you and told you he had a way to transport secret information to Russia, would you have been very much surprised if that man approached you?"

Dr. von Neumann: "It depends who the man is."

Dr. Evans: "Suppose he is a friend of yours. . . . Would you have reported it immediately?"

Dr. von Neumann: "This depends on the period. I mean, before I got conditioned to security, possibly not. After I got conditioned to security, certainly yes. . . . What I am trying to say is this, that before 1941, I didn't even know what the word 'classified' meant. So God only knows how intelligently I would have behaved in situations involving this. I am quite sure that I learned it reasonably fast. But there was a period of learning during which I may have made mistakes or might have made mistakes. . . ."

Perhaps sensing that von Neumann was scoring points, Robb resorted to one of the oldest ploys in a prosecutor's bag of tactics: asking only one question on cross-examination. "Doctor," he asked, "you have never had any training as a psychiatrist, have you?" Von Neumann was one of the most brilliant mathematicians of his time. He knew Oppenheimer both professionally and socially. But no, he was not a psychiatrist—and therefore, in Robb's not-so-subtle view, von Neumann was not qualified to judge Oppenheimer's behavior in the Chevalier affair.

MIDWAY THROUGH THE HEARING, Robb had announced that, "unless ordered to do so by the board, we shall not disclose to Mr. Garrison in advance the names of the witnesses we contemplate calling." Garrison had revealed his list of witnesses at the very beginning of the hearing, thus allowing Robb to prepare detailed questions, often based on classified documents. But Robb now explained that he could not extend the same courtesy to his adversary because, "I will be frank about it, that in the event that any witnesses from the scientific world should be called, they would be subject to pressure." Perhaps, but it was a transparent rationalization that should have been vigorously challenged by Garrison. In the first instance, it was obvious to everyone that Edward Teller would be called, and so whatever pressure his colleagues intended to apply would be applied. Ernest Lawrence and Luis Alvarez were also likely candidates—and the list goes on. The irony of this professed concern on the part of the prosecutor lies in the fact that the producer of this show trial, Lewis Strauss, was indefatigable in his pursuit of hostile witnesses.

A week after testifying, Rabi ran into Ernest Lawrence at Oak Ridge and asked him what he was going to say about Oppenheimer. Lawrence had agreed to testify against him. He was truly fed up with his old friend. Oppie had opposed him on the hydrogen bomb and opposed the building of a second weapons lab at Livermore. And more recently, Ernest had come home from a cocktail party outraged upon being told that Oppie had years before had an affair with Ruth Tolman, the wife of his good friend Richard. He was angry enough to accede to Strauss' request to testify against Oppenheimer in Washington. But the night before his scheduled appearance, Lawrence fell ill with an attack of colitis. The next morning, he called Strauss to tell him he could not make it. Sure that Lawrence was making excuses, Strauss argued with the scientist and called him a coward.

Lawrence did not appear to testify against Oppenheimer. But Robb had interviewed him earlier and now made sure that the Gray Board—though not Garrison—saw the transcript of this interview. Lawrence's conclusion, therefore, that Oppenheimer was guilty of so much bad judgment that "he should never again have anything to do with the forming of policy" went unseen and unchallenged by Oppenheimer's lawyers. Surely this was the sort of violation of the rules of due process that would have constituted grounds for halting the proceedings.

UNLIKE LAWRENCE, Edward Teller had no hesitations about testifying. On April 22, six days before his testimony, Teller had an hour-long conversation with an AEC public information officer, Charter Heslep. In the course of the conversation, Teller expressed his deep animosity to Oppenheimer and the "Oppie machine." A way had to be found, Teller believed, to destroy Oppenheimer's influence. Heslep's report to Strauss includes the following paragraph: "Since the case is being heard on a security basis, Teller wonders if some way can be found to 'deepen the charges' to include a documentation of the 'consistently bad advice' that Oppenheimer has given, going all the way back to the end of the war in 1945." Heslep added that "Teller feels deeply that this 'unfrocking' must be done or else— regardless of the outcome of the current hearing—scientists may lose their enthusiasm for the [atomic weapons] program."

Heslep's memo to Strauss lays out the full political motivations behind the Oppenheimer case:

> Teller regrets the case is on a security basis because he feels it is untenable. He has difficulty phrasing his assessment of Oppie's philosophy except a conviction that Oppie is not disloyal but rather— and Teller put this somewhat vaguely—more of a "pacifist."
>
> Teller says what is needed . . . and the job is most difficult, was to show his fellow scientists that Oppie is not a menace to the program but simply no longer valuable to it.
>
> Teller said "only about one per cent or less" of the scientists know of the real situation and that Oppie is so powerful "politically" in scientific circles that it will be hard to "unfrock him in his own church." (This last phrase is mine and he agrees it is apt.)

Teller talked at length about the "Oppie machine," running through many names, some of which he listed as "Oppie men" and others as not being "on his team" but under his influence. . . .

On April 27, Teller met with Roger Robb, who wanted to be sure that the mercurial physicist was still ready to testify against his old friend. Teller later claimed that this meeting occurred the next day, only minutes before he was sworn in, but his memory is contradicted by a handwritten note he later sent to Strauss in which he stated he had met with Robb the evening before his testimony. According to Teller's account, Robb bluntly asked, "Should Oppenheimer be cleared?" "Yes, Oppenheimer should be cleared," he replied. Whereupon Robb pulled out a transcript and had Teller read that part of Oppenheimer's testimony in which he had admitted to inventing a "cock-and-bull story." Claiming to have been astounded that Oppenheimer had so brazenly confessed to lying, Teller later said that he left Robb uncertain about whether he would testify that Oppenheimer deserved to be cleared.

Teller's recounting of this incident is disingenuous. For more than a decade, he had deeply resented Oppenheimer's influence and popularity among his fellow scientists. By 1954, he desperately wanted to "defrock him in his own church." What Robb had shown him from the still secret hearing transcript simply made it easier for him to testify against Oppie.*

THE NEXT AFTERNOON, with Oppenheimer sitting on his couch a few steps away, Teller took the witness chair. Robb let him testify at considerable length about Oppenheimer's attitude toward the development of the H-bomb, and other issues. Finally, aware that Teller wished to appear ambivalent, Robb gently guided him to say only what was necessary.

Robb: "To simplify the issues here, perhaps, let me ask you this question: Is it your intention in anything that you are about to testify to, to suggest that Dr. Oppenheimer is disloyal to the United States?"

Teller: "I do not want to suggest anything of the kind. I know Oppenheimer as an intellectually most alert and a very complicated person, and I think it would be presumptuous and wrong on my part if I would in any way

---

*Nor was Teller the only prosecution witness to be so prepped by Robb. One night Garrison's assistant, Allan Ecker, was working late in the hearing room when he was distracted by loud voices across the hallway. "I could hear a tape being played," Ecker said. And then he saw Robb and a number of people who were later to be witnesses leaving the room. "Mr. Robb had brought in people who were afterwards to be witnesses, and they had listened to a tape of an interrogation [Colonel Pash's August 1943 interrogation of Oppenheimer]."

analyze his motives. But I have always assumed, and now assume, that he is loyal to the United States. I believe this, and I shall believe it until I see very conclusive proof to the opposite."

Robb: "Now, a question which is a corollary of that. Do you or do you not believe that Dr. Oppenheimer is a security risk?"

Teller: "In a great number of cases I have seen Dr. Oppenheimer act—I understood that Dr. Oppenheimer acted in a way which for me was exceedingly hard to understand. I thoroughly disagreed with him in numerous issues, and his actions, frankly, appeared to me confused and complicated. To this extent, I feel that I would like to see the vital interests of this country in hands which I understand better and therefore trust more."

Under cross-examination by Chairman Gray, Teller amplified his statement by saying, "If it is a question of wisdom and judgment, as demonstrated by actions since 1945, then I would say one would be wiser not to grant clearance. I must say that I am myself a little bit confused on this issue, particularly as it refers to a person of Oppenheimer's prestige and influence. May I limit myself to these comments?"

Robb needed nothing more said. Excused from the witness chair, Teller turned around, and walking past Oppenheimer, who was sitting on the leather couch, he offered him a hand and said, "I'm sorry."

Oppie shook his hand and replied laconically, "After what you've just said, I don't know what you mean."

Teller would pay dearly for what he had said. Later that summer, on a visit to Los Alamos, Teller spotted an old friend, Bob Christy, in the dining hall. Walking over to greet him with outstretched hand, Teller was stunned when Christy refused to shake hands and abruptly turned his back. Standing close by was a furious Rabi, who said, "I won't shake your hand, either, Edward." Stunned, Teller went back to his hotel room and packed his bags.

AFTER TELLER'S testimony the hearing dragged on anticlimactically for another week. On May 4—some three weeks into the hearing—Kitty was called back to the witness chair. Chairman Gray and Dr. Evans pressed her again about when she had broken with the Communist Party. Kitty again said that after 1936, "I stopped having anything to do with the Communist Party." Their exchange then turned fairly testy.

Chairman Gray: "Would it be fair to say that Dr. Oppenheimer's contributions in the years as late as possibly 1942 meant that he had not stopped having anything to do with the Communist Party? I don't insist that you answer that yes or no. You can answer that any way you wish."

Kitty Oppenheimer: "I know that. Thank you. I don't think that the question is properly phrased."

Chairman Gray: "Do you understand what I am trying to get at?"

Kitty: "Yes; I do."

Chairman Gray: "Why don't you answer it that way?"

Kitty: "The reason I don't like the phrase 'stopped having anything to do with the Communist Party.' . . . It is because I don't think Robert ever had anything to do with the Communist Party as such. I know he gave money for Spanish refugees; I know he gave it through the Communist Party."

Chairman Gray: "When he gave money to Isaac Folkoff, for example, this was not necessarily for Spanish refugees, was it?"

Kitty Oppenheimer: "I think so."

Chairman Gray: "As late as 1942?"

Kitty Oppenheimer: "I don't think it was that late. . . ."

When Gray reminded her that her husband had used that date, she responded, "Mr. Gray, Robert and I don't agree about everything. He sometimes remembers something different than the way I remember it."

One of Oppenheimer's lawyers tried to enter the conversation at this point, but Gray insisted on pursuing his line of questioning. What he was trying to get at, he said, was, when did her husband's associations with communists cease?

Kitty Oppenheimer: "I do not know, Mr. Gray. I know that we still have a friend of whom it has been said that he is a Communist." (She meant, of course, Chevalier.) Startled by this casual admission, Robb interjected, "I beg your pardon?" But Gray forged ahead, and asked again about the "mechanics" by which one becomes "clearly disassociated" from the Communist Party. Kitty answered quite sensibly, "I think that varies from person to person, Mr. Gray. Some people do the bump, like that, and even write an article about it. Other people do it quite slowly. I left the Communist Party. I did not leave my past, the friendships, just like that. Some continued for a while. I saw Communists after I left the Communist Party."

The questions kept coming. Dr. Evans asked her to define the difference between a communist and a fellow traveler. Kitty replied simply, "To me, a Communist is a member of the Communist Party who does more or less precisely what he is told."

When Robb asked her about their subscription to the *People's World* Kitty quite plausibly explained that she doubted they had ever subscribed to the newspaper. "I did not subscribe to it," Kitty said. "Robert says he did. I sort of doubt it. The reason I have for that is that I know we [in Ohio] often sent the *Daily Worker* to people that we tried to get interested in the Communist Party without their having subscribed to it."

Kitty did not give an inch. Not even Robb could touch her. Calm and yet alert to every nuance, she was undoubtedly a better witness than the husband she was defending.

ON MAY 5, the final day of the hearing, as Oppenheimer was about to be excused from the witness chair for the last time, he asked to make one further comment. After enduring almost four weeks of excruciating humiliation, Oppenheimer played the last act of Garrison's strategy of conciliation and thanked his tormentors: "I am grateful to, and I hope properly appreciative of, the patience and consideration that the board has shown me during this part of the proceedings." It was a demonstration of deference designed to prove to the Gray Board that Robert Oppenheimer was a reasonable, cooperative person, a member of the establishment who could be worked with and trusted. Chairman Gray was unimpressed. "Thank you very much, Dr. Oppenheimer," he responded.

THE NEXT MORNING, Garrison spent three hours on his summation of the case. He again protested, less gently this time, the way in which the "hearing" had been turned into a "trial." He reminded the Gray Board that they had spent a full week before the hearing ever began reading FBI materials on Oppenheimer. "I remember a kind of sinking feeling," Garrison said, "that I had at that point—the thought of a week's immersion in FBI files which we would never have the privilege of seeing. . . ." But sensing that he shouldn't protest too harshly, Garrison immediately backed off. While it was true, he said, they had found themselves "unexpectedly in a proceeding which seemed to us to be adversary in nature. . . . I do want to say in all sincerity that I recognize and appreciate very much the fairness which the members of the board have displayed. . . ."

If Garrison was embarrassingly submissive, he was also eloquent in his summation. He warned the Gray Board against the "illusion of a foreshortening of time here which to me is a grisly matter, and very, very misleading." What happened in the 1943 Chevalier incident must be judged by the atmosphere of that time: "Russia was our so-called gallant ally. The whole attitude toward Russia, toward persons who were sympathetic with Russia, everything was different from what obtains today." As to Oppenheimer's personal character and integrity, Garrison reminded the Board, "You had three and a half weeks now with the gentleman on the sofa. You have learned a lot about him. There is a lot about him, too, that you haven't learned, that you don't know. You have not lived any life with him."

Garrison continued: "There is more than Dr. Oppenheimer on trial in this room. . . . The Government of the United States is here on trial also." In a veiled reference to McCarthyism, Garrison spoke of the "anxiety abroad in the country." Anticommunist hysteria had so infected the Truman and Eisenhower administrations that the security apparatus was now behaving "like some monolithic kind of machine that will result in the destruction of men of great gifts. . . . America must not devour her own children." On this note, having pleaded once again that the Gray Board should "judge the whole man," Garrison ended his summation.

THE TRIAL WAS OVER, and on the evening of May 6, 1954, the defendant returned to Princeton to await the board's judgment.

As Garrison had tried to show, belatedly, the Gray Board hearings were patently unfair and outrageously extrajudicial. The primary responsibility for the proceedings lay with Lewis Strauss. But as chairman of the board, Gordon Gray could have ensured that the hearing was conducted properly and fairly. He did not do his job. Instead of taking control of the hearing to maintain fairness, which would have required him to rein in Robb's illicit tactics, he allowed Robb to control the proceedings. Prior to the hearing, Gray permitted Robb to meet exclusively with the board to review the FBI files, a direct violation of the AEC's 1950 "Security Clearance Procedures." He accepted Robb's recommendation that Garrison be denied a similar meeting; he acquiesced to Robb's refusal to reveal his witness list to Garrison; he did not share Lawrence's damaging written testimony with the defense; he did nothing to expedite a security clearance for Garrison. The Gray Board was, in sum, a veritable kangaroo court in which the head judge accepted the prosecutor's lead. As AEC commissioner Henry D. Smyth would insist, any objective legal review of how the hearing was conducted surely would result in its nullification.

CHAPTER THIRTY-SEVEN

"A Black Mark on the Escutcheon of Our Country"

It is sad beyond words. They are so wrong, so terribly wrong, not only about Robert, but in their concept of what is required of wise public servants. . . .

DAVID LILIENTHAL

OPPENHEIMER RETURNED TO OLDEN MANOR tired and irritable. He knew things had gone badly, and there was not much he could do but wait for the Gray Board's judgment. He thought it would be weeks before it reached a decision. The FBI wiretap overheard him telling a friend that even then, "he believes he will never be through with the situation. He does not believe the case will come to a quiet end as *all the evil of the times is wrapped in this situation.*" A few days later, the FBI reported that Oppenheimer was "very depressed at the present time and has been ill-tempered with his wife."

As they awaited the panel's judgment, he and Kitty spent hours in front of their black-and-white television set, watching the Army-McCarthy Senate hearings. This extraordinary drama had begun on April 21, 1954, in the middle of Oppenheimer's own ordeal, and as the hearings dragged on through May, an estimated 20 million Americans tuned in each day to watch as Senator McCarthy and the Army's counsel, Boston lawyer Joseph Nye Welch, traded barbs. Like many Americans, Oppenheimer was transfixed by this live television drama; but for him it must have been a painfully personal reminder of the star chamber nature of the hearings he had just endured. Could he have helped but think that things might have gone better for him if he had been represented by Welch, or someone like him?

GORDON GRAY thought things had gone splendidly. The day after the proceedings ended, he dictated a private memo for his files summing up his initial reactions: "It is my present conviction that up to this point the proceedings have been as fair as circumstances permit. My reason for the qualification is that, of course, Dr. Oppenheimer and his counsel are not privileged to see certain documents such as FBI reports and other classified material. . . ." Gray also confessed that "I was mildly uncomfortable about Mr. Robb's cross-examination and his piecemeal and surprise references to and quotations from documents." But in the end, he rationalized to himself, "that there was no damage to Dr. Oppenheimer's interests if the proceedings are viewed as a whole."

From Gray's informal discussions with his fellow panel members, there seemed little doubt of the outcome. Oppenheimer, in his view, was certainly guilty of putting "loyalty to an individual above loyalty or obligation to Government." Or, as Gray had told Morgan and Evans one morning earlier that week, Dr. Oppenheimer had a "repeated tendency to put his own judgment about a situation ahead of the considered and official judgment in many cases of people whose responsibility and duty it was to have such judgments." Gray cited the Chevalier affair, Oppenheimer's defense of Bernard Peters, the hydrogen bomb debate and several of Oppenheimer's other atomic policy positions. Morgan and Evans indicated their agreement—and Dr. Evans specifically commented that "Oppenheimer certainly was guilty of very bad judgment."

Upon his return from a ten-day recess, therefore, Gray was shocked to learn that Dr. Evans had penciled a draft dissent supporting Oppenheimer. Gray had thought Evans disposed "from the beginning" to rule that Oppenheimer's clearance should not be reinstated. Evans had told him privately that in his experience "almost without exception those who turned up with subversive backgrounds and interests were Jewish." Bluntly put, Gray thought Evans' anti-Semitism would prejudice his judgment. Throughout the month-long proceedings, Gray noted, "my impression grew that both of my colleagues were pretty well committed to a view." But now, upon his return from Chicago, "Dr. Evans clearly had undergone a complete reversal of view." Evans said he had simply reviewed the record and decided that there was nothing new in the charges. The FBI thought "someone had 'gotten to' him."

Strauss became frantic when he learned of this development. He and Robb had wiretapped Oppenheimer's lawyers, they had blocked Garrison's attempt to get a security clearance, they had ambushed witnesses with classified documents, they had prejudiced the Gray panel with hearsay evidence from the FBI files—and despite all their efforts to assure a guilty verdict, now it seemed possible that Oppenheimer would be exonerated.

Fearing that Evans might influence one of the two other panel members, Strauss called Robb. The two men agreed that something had to be done and Robb, with Strauss' approval, called the FBI and asked for Hoover's intercession. Robb told Bureau agent C. E. Hennrich that he thought "it extremely important that the Director discuss this matter with the Board. . . . Robb said that he feels it will be a tragedy if the decision of the Board goes the wrong way and that he considers this a matter of extreme urgency." Almost at the same moment, Strauss was on the phone to A. H. Belmont, one of Hoover's personal assistants, begging him to get the director to intervene. He said things were "touch and go" and that "a slight tip of the balance would cause the Board to commit a serious error."

Agent Hennrich observed: "This all boils down, it seems to me, to a situation where Strauss and Robb, who want the Board to make a finding that Oppenheimer is a security risk, are doubtful that the Board will find so at this point. . . . It is my feeling that the Director should not see the Board."

Any such intervention on Hoover's part would have been considered highly prejudicial if it ever became public—and Hoover knew it. He told his aides, "I think it would be highly improper for me to discuss [the] Oppenheimer case. . . ." He would not see the Gray Board.

Years later, when Robb was confronted with an FBI memo documenting his attempt to get Hoover's intercession, he denied that he had tried to get the FBI director to influence the board's judgment. He told the filmmaker and historian Peter Goodchild, "I specifically and categorically deny that I ever encouraged a meeting between the Board and the Director for the purpose of having the Director influence the Board. . . . I also deny that I ever told Hennrich that I considered this 'to be a matter of extreme urgency' because unless the Board talked with Mr. Hoover it might decide in favor of Oppenheimer." But the documentary record is clear—he was lying.

Ironically, Gray thought Evans' brief so badly written that he asked Robb to rewrite it. "I didn't want 'Doc' Evans' opinion to be too vulnerable," Robb explained. "If it was, it would look as though he was just a plant on the Board, do you follow me, it would look as though we put a nincompoop on the Board."

ON MAY 23, the Gray Board returned its formal verdict. By a vote of two to one, the board deemed Oppenheimer a loyal citizen who was nevertheless a security risk. Accordingly, Chairman Gray and board member Morgan recommended that Oppenheimer's security clearance *not* be restored. "The following considerations," Gray and Morgan wrote, "have been controlling in leading us to our conclusion:

1. We find that Dr. Oppenheimer's continuing conduct and association have reflected a serious disregard for the requirements of the security system.

2. We have found a susceptibility to influence which could have serious implications for the security interests of the country.

3. We find his conduct in the hydrogen bomb program sufficiently disturbing as to raise a doubt as to whether his future participation, if characterized by the same attitudes in a Government program relating to the national defense, would be clearly consistent with the best interests of security.

4. We have regretfully concluded that Dr. Oppenheimer has been less than candid in several instances in his testimony before this Board.

Their reasoning was tortured. They did not accuse Oppenheimer of violating any laws or even security regulations. But his associations gave evidence of a certain indefinable ill-judgment. His studied lack of deference to the security apparatus was particularly damning in their eyes. "Loyalty to one's friends is one of the noblest of qualities," Gray and Morgan wrote in their majority opinion. "Being loyal to one's friends above reasonable obligations to the country and to the security system, however, is not clearly consistent with the interests of security." Among other deviations, Oppenheimer was guilty of excessive friendship.

Evans' dissent on the other hand, was a clear, unambiguous critique of his fellow board members' verdict. "Most of the derogatory information," Evans observed in his dissent, "was in the hands of the Committee when Dr. Oppenheimer was cleared in 1947."

They apparently were aware of his associations and his left-wing policies: yet they cleared him. They took a chance because of his special talents and he continued to do a good job. Now when the job is done, we are asked to investigate him for practically the same derogatory information. He did his job in a thorough and painstaking manner. There is not the slightest vestige of information before this Board that would indicate that Dr. Oppenheimer is not a loyal citizen of his country. He hates Russia. He had communistic friends, it is true. He still has some. However, the evidence indicates that he has fewer of them than he had in 1947. He is not as naïve as he was then. He has more judgment; no one on the Board doubts his loyalty— even the witnesses adverse to him admit that—and he is certainly less a security risk than he was in 1947, when he was cleared. To deny him clearance now for what he was cleared for in 1947, when we must know he is less of a security risk now than he was then, seems hardly the procedure to be adopted in a free country. . . .

I personally think that our failure to clear Dr. Oppenheimer will be a black mark on the escutcheon of our country. His witnesses are a considerable segment of the scientific backbone of our Nation and they endorse him.

Whether Evans' dissent was written entirely by his own hand or edited by Robb, it is a remarkable document. In the two short paragraphs quoted, it demolishes points 1, 2 and 4 of the "considerations" above that Gray and Morgan presented as the basis for their verdict. Nonetheless, it fails to confront point 3, the issue that precipitated this "train wreck," as Oppenheimer later referred to his ordeal. "We find his conduct on the hydrogen bomb program sufficiently disturbing . . . ," Gray and Morgan wrote.

Why was his conduct with respect to the hydrogen bomb program disturbing? Oppenheimer had opposed a crash program to develop a hydrogen bomb, but so had seven other members of the GAC; and they all had explained their reasons clearly. What Gray and Morgan were actually saying was that they opposed Oppenheimer's judgments and they did not want his views represented in the counsels of government. Oppenheimer wanted to corral and perhaps even reverse the nuclear arms race. He wanted to encourage an open democratic debate on whether the United States should adopt genocide as its primary defense strategy. Apparently, Gray and Morgan considered these sentiments unacceptable in 1954. More, they were asserting in effect that it was not legitimate, not permissible, for a scientist to express strong disagreement on matters of military policy.

Strauss was relieved that the panel had narrowly handed down the equivalent of a guilty verdict—but now he feared the possibility that Evans' dissent could persuade the AEC commissioners to reverse it. The verdict, after all, was only a recommendation, which the AEC commissioners had the option of confirming or rejecting. Oppenheimer's lawyers assumed that standard procedures would be followed and the AEC's general manager, Kenneth Nichols, would merely pass to the commissioners the Gray Board's report. But Nichols—who viewed Oppenheimer as a "slippery sonuvabitch"—sent the commissioners a letter that was actually a fullfledged brief. Nichols' letter written under the guidance of Strauss, Charles Murphy (the *Fortune* magazine editor), and Robb, put an entirely new spin on the panel's report.

The Nichols letter presented an entirely new argument for why Oppenheimer's security clearance should not be reinstated. His speculations went far beyond the Gray Board's verdicts. Drawing on Strauss' research in Oppenheimer's FBI dossier during the three months he had kept it in his office, Nichols argued, first, that Oppenheimer was not merely a "parlor pink" fellow traveler. "His relations with these hardened Communists were such that they considered him to be one of their number." Citing the cash contributions that Oppenheimer had passed through the Communist Party, Nichols concluded, "The record indicates that Dr. Oppenheimer was a Communist in every respect except for the fact that he did not carry a party card."

Although the Gray Board's verdict had emphasized Oppenheimer's opposition to a crash program to develop the H-bomb, Nichols dismissed this politically awkward part of the indictment and astutely added that it was not the intention of the AEC to question the right of a scientist like Dr. Oppenheimer to express his "honest opinions."

Instead, Nichols shifted the emphasis to the Chevalier affair. But he embraced an interpretation of this murky business quite different from the one presented by the Gray Board. The panel had accepted Oppenheimer's admission that he had lied to Colonel Pash in 1943 when he first spoke of the Chevalier-Eltenton incident. Nichols rejected this conclusion and, in an astonishing and perhaps even extralegal maneuver, completely reinterpreted the incident. In effect, Nichols retried Oppenheimer, dismissed the Gray Board's majority opinion, and presented the AEC commissioners with an entirely new basis for removing Oppenheimer's security clearance.

After reviewing the sixteen-page transcript of that fateful encounter between Oppenheimer and Colonel Pash on August 26, 1943, Nichols argued, "it is difficult to conclude that the detailed and circumstantial account given by Dr. Oppenheimer to Colonel Pash was false and that the story now told by Dr. Oppenheimer is an honest one." Why, asked Nichols, would Oppenheimer "tell such a complicated false story to Colonel Pash?" Rejecting Oppenheimer's quite plausible explanation, that he had sought to divert attention from both Chevalier and himself, Nichols pointed out that Oppenheimer "did not give his present version of the story until 1946, shortly after he had learned from Chevalier what Chevalier himself had told the FBI about the incident. . . ." Withholding from the commissioners the critical fact that Eltenton's interview with the FBI—conducted simultaneously with the FBI's interview with Chevalier—had irrefutably confirmed the 1946 Chevalier–Oppenheimer version of the Chevalier affair, Nichols concluded that Oppenheimer had lied in 1946 to the FBI and again in the 1954 hearings.

Nichols had unearthed no additional facts; indeed, he had suppressed facts. He merely asserted that Oppenheimer lied to protect his brother, a theory that, as we have seen, has scant evidence to support it. Curiously, the

Gray Board made no effort to obtain testimony from Frank Oppenheimer—nor, for that matter, from the two principals, Haakon Chevalier and George Eltenton. (Chevalier was then living in Paris and Eltenton had long since returned to England, but both men could have been interviewed abroad.)

Nichols' letter contained only a supposition, a personal interpretation, and one that had not been raised by the Gray Board. Why at this late date was he introducing another theory? The answer is obvious: Arguing that Oppenheimer had lied in 1954, to the hearing board, was far more damning than the claim that he had lied eleven years earlier to a lieutenant colonel.

Since it is impossible to imagine that Nichols presented this radical interpretation without Strauss' approval, it is clear that Strauss feared that the ambiguities in the majority's decision, combined with the clarity of Evans' dissent, might lead the AEC commissioners to overrule the Gray Board.

Oppenheimer's lawyers knew nothing of Nichols' letter. Garrison might have learned of it if he had been given the opportunity to present an oral argument before the AEC commissioners. The one commissioner sympathetic to Garrison's request, Dr. Henry D. Smyth, warned, "If we give Dr. Oppenheimer's attorneys no opportunity to comment on Nichols' letter, we will be open to grave criticism when the letter is published." But once again, Strauss prevailed, and Garrison's request was flatly turned down without explanation.

OPPENHEIMER'S LAWYERS briefly hoped that the five AEC commissioners would reverse the Gray Board's recommendation. There were, after all, three Democrats (Henry De Wolf Smyth, Thomas Murray and Eugene Zuckert) and only two Republicans (Lewis Strauss and Joseph Campbell) on the Commission. Initially, Strauss himself feared a three-to-two vote in Oppenheimer's favor. But as chairman, Strauss was in a position to influence his fellow commissioners. He understood how power worked in Washington, and he had no qualms about offering his colleagues tangible rewards for seeing things his way. He treated them to lavish lunches and talked to Smyth about lucrative employment opportunities in private industry. At one point, Smyth wondered whether Strauss was trying to buy his vote. Harold P. Green, the AEC lawyer who had been called upon to write the original letter of charges against Oppenheimer, thought Strauss was playing hardball. Green knew that Zuckert was initially inclined to find Oppenheimer innocent. In fact, on May 19, Strauss was informed that, "Gene Zuckert would welcome the opportunity not to stand up and be counted on the vote making final disposition of the security case." But at some point, Zuckert flipped. He was scheduled to resign his post as a commissioner of the AEC on June 30—the day after signing on with the majority decision against Oppenheimer—to start a private law practice in Washington. Green firmly believed that something untoward was happening, especially after he learned that Strauss subsequently transferred a lot of his legal business to Zuckert. Green didn't know it, but Zuckert also signed a contract with Strauss to serve as the latter's "personal adviser and consultant."

By the end of June, Strauss had the votes of all but one commissioner. The only scientist on the Commission, Professor Smyth had made it clear that he thought Oppenheimer's security clearance should be restored. As the author of the 1945 "Smyth Report," an unclassified scientific history of the Manhattan Project, Smyth was familiar with both Oppenheimer and the security issues at stake. On a personal level, he didn't particularly care for Oppenheimer; they had been Princeton neighbors for ten years, and Oppenheimer had always struck him as a vain and pretentious man. What mattered was that Smyth didn't find the evidence convincing. In early May, he and Strauss had lunch and proceeded to argue about the verdict. At the end of their lunch, Smyth said, "Lewis, the difference between you and me is that you see everything as either black or white and to me everything looks gray."

"Harry," Strauss snapped back, "let me recommend you to a good oculist."

A few weeks later, Smyth told Strauss he was determined to write a dissenting report. Working late each night until midnight, Smyth waded through the Gray Report and the hearing transcript, a stack of papers four feet high. To help him in this task, he requested the assistance of two AEC staff aides. Nichols warned one of these aides, Philip Farley, that the job would harm his career, but Farley courageously went to work for Smyth anyway. By June 27, Smyth had produced a draft of his dissenting opinion—only to learn that the final majority opinion had been so completely rewritten as to require him to redraft his own.

Beginning at 7:00 p.m. on Monday, June 28, Smyth and his assistants began writing a completely new dissent. He had merely twelve hours to meet the AEC's self-imposed deadline for submission of the final opinion. As they worked through the night, Smyth could see through the window a car parked outside his house; two men were sitting inside the car, watching the house. Smyth thought someone from the AEC or the FBI had sent them to intimidate him. "You know it's funny I should be going to all this trouble for Oppenheimer," he told one of his assistants late that night. "I don't even like the guy much."

At ten that morning, Farley took Smyth's dissenting opinion downtown to the AEC office and stood by to make sure that it was reproduced in full.

That afternoon, Smyth's dissent and the majority opinions were made available to the press. The commissioners voted four to one that Oppenheimer was loyal and four to one that he was a security risk. Gone from the majority opinion was any reference to the hydrogen bomb issue—even though that had been a central theme of the Gray Board's decision. Drafted by Strauss, the majority decision focused on Oppenheimer's "fundamental defects" of character. Specifically, the Chevalier affair and his past associations with various students in the 1930s who had been communists took center stage. "The record shows that Dr. Oppenheimer has consistently placed himself outside the rules which govern others. He has falsified in matters wherein he was charged with grave responsibilities in the national interest. In his associations he has repeatedly exhibited a willful disregard of the normal and proper obligations of security."

OPPENHEIMER'S SECURITY CLEARANCE was thus rescinded just one day before it was due to expire. After reading the AEC commissioners' verdicts, David Lilienthal noted in his diary: "It is sad beyond words. They are so wrong, so terribly wrong, not only about Robert, but in their concept of what is required of wise public servants. . . ." Einstein, disgusted, quipped that henceforth the AEC should be known as the "Atomic Extermination Conspiracy."

Earlier in June, using as an excuse that a copy of the transcript had been stolen from a train (it was soon located in New York's Pennsylvania Station's lost-and-found office), Strauss persuaded his fellow commissioners to have all 3,000 typewritten pages of the hearing transcript published by the Government Printing Office. This violated the Gray Board's promise to all the witnesses that their testimony would remain confidential. But Strauss felt that he was not winning the public relations battle and so he brushed aside this concern.

Comprising some 750,000 words in 993 densely printed pages, *In the Matter of J. Robert Oppenheimer* soon became a seminal document of the early Cold War. To be certain that the initial news stories embarrassed Oppenheimer, Strauss had the AEC staff highlight the most damaging testimony for reporters. Walter Winchell—the right-wing mudslinging syndicated columnist—obligingly wrote: ". . . Oppenheimer's testimony (which most people skip over) included the name of his mistress (the late Jean Tatlock), a fanatical 'Redski' with whom he admitted associations after his marriage 'of the most intimate kind.' . . . This when he was working on the Big Bomb and knew his Doll was an active member of a Commy apparatus. . . ."

Radically conservative organs such as the *American Mercury* hailed the downfall of this "longtime glamour-boy of the atomic scientists" and criticized Oppenheimer's supporters as men who would "coddle potential traitors." When the Commission's ruling was announced on the floor of the House of Representatives, some congressmen stood and applauded.

IN THE LONG RUN, however, Strauss' strategy backfired; the transcript revealed the inquisitorial character of the hearing, and the corruption of justice during the McCarthy period. Within four years, the transcript would destroy the reputation and government career of Lewis Strauss.

Ironically, publicity surrounding the trial and its verdict enhanced Oppenheimer's fame both in America and abroad. Where once he was known only as the "father of the atomic bomb," now he had become something even more alluring—a scientist martyred, like Galileo. Outraged and shocked by the decision, 282 Los Alamos scientists signed a letter to Strauss defending Oppenheimer. Around the country, more than 1,100 scientists and academics signed another petition protesting the decision. In response, Strauss replied that the AEC's decision was "a hard one, but the proper one." The broadcaster Eric Sevareid noted, "He [Oppenheimer] will no longer have access to secrets in government files, and government, presumably, will no longer have access to secrets that may be born in Oppenheimer's brain."

Oppenheimer's friend the syndicated columnist Joe Alsop was outraged by the decision. "By a single foolish and ignoble act," he wrote Gordon Gray, "you have cancelled the entire debt that this country owes you." Joe and his brother Stewart soon published a 15,000-word essay in *Harper's* lambasting Lewis Strauss for a "shocking miscarriage of justice." Borrowing from Emile Zola's essay on the Dreyfus affair, "J'Accuse," the Alsops titled their essay "We Accuse!" In florid language they argued that the AEC had disgraced, not Robert Oppenheimer, but the "high name of American freedom." There were obvious similarities: Both Oppenheimer and Capt. Alfred Dreyfus came from wealthy Jewish backgrounds and both men were forced to stand trial, accused of disloyalty. The Alsops predicted that the long-term ramifications of the Oppenheimer case would echo those of the Dreyfus case: "As the ugliest forces in France engineered the Dreyfus case in swollen pride and overweening confidence, and then broke their teeth and their power on their own sordid handiwork, so the similar forces in America, which have created the climate in which Oppenheimer was judged, may also break their teeth and power in the Oppenheimer case."

After news of the verdict was published, John McCloy wrote Supreme Court Justice Felix Frankfurter: "What a tragedy that one who contributed so much—more than half the bemedaled generals I know—to the security of the country should now after all these years be designated a security risk. I understand the Admiral [Lewis Strauss] is annoyed at my testimony but great God what does he expect? I was there when Oppie's massive contribution was rendered and know there is so much more to say, but what's the use?"

Frankfurter tried to reassure his old friend, writing to him that "you opened a good many minds to a realization of the profound importance of your 'concept of an affirmative security.' " Both Frankfurter and McCloy agreed that the chief culprit in the whole sad case was Strauss.

AT THE APEX of the McCarthyite hysteria, Oppenheimer had become its most prominent victim. "The case was ultimately the triumph of McCarthyism, without McCarthy himself," the historian Barton J. Bernstein has written. President Eisenhower appeared satisfied with the outcome—but unaware of the tactics Strauss had used to obtain it. In mid-June, seemingly oblivious to the nature and import of the hearing, Ike wrote Strauss a short note suggesting that Oppenheimer be put to work solving the problem of the desalinization of seawater. "I can think of no scientific success of all time that would equal this in its boon to mankind. . . ." Strauss quietly ignored his suggestion.

Lewis Strauss, with the help of his like-minded friends, had succeeded in "defrocking" Oppenheimer. The implications for American society were enormous. One scientist had been excommunicated. But all scientists were now on notice that there could be serious consequences for those who challenged state policies. Shortly before the hearing, Oppenheimer's MIT colleague Dr. Vannevar Bush had written a friend that "the problem of how far a technical man working with the military is entitled to speak out publicly is quite a question. . . . I kept in channels rather religiously, perhaps too much so." From experience, Bush believed he would only destroy his usefulness if he talked publicly about internal government deliberations. On the other hand, "when an individual citizen sees his country going down a path which he thinks is likely to be disastrous he has some obligation to speak out." Bush shared many of Oppenheimer's critical instincts about Washington's growing reliance on nuclear weapons. But unlike Oppenheimer, he had never really spoken out. Oppenheimer had—and now all his colleagues could see him punished for his courage and patriotism.

The scientific community remained traumatized for years. Teller became a pariah to many of his former friends. Three years after the case, Rabi still couldn't control his anger at those who had judged his friend. Bumping into Gene Zuckert at New York City's Place Vendôme, an upscale French restaurant, Rabi launched into a tirade of abuse, his voice rising to a fervent pitch. He loudly denounced Zuckert for the decision he had rendered as an AEC commissioner in the case. Mortified, Zuckert beat a hasty retreat and later complained to Strauss about Rabi's behavior.

Lee DuBridge wrote Ed Condon that "it is probably quite impossible for anything to be done about the Oppenheimer case itself. The term 'security risk' is such a broad one that you can start out accusing a fellow of treason and end up by convicting him of fibbing, but still impose the same punishment. I guess there is no doubt that Robert did do some fibbing, and in the public mind now anybody who fibbed and also once was a 'Communist' is clearly an unforgivable character."

FOR A FEW YEARS after World War II, scientists had been regarded as a new class of intellectuals, members of a public-policy priesthood who might legitimately offer expertise not only as scientists but as public philosophers. With Oppenheimer's defrocking, scientists knew that in the future they could serve the state only as experts on narrow scientific issues. As the sociologist Daniel Bell later observed, Oppenheimer's ordeal signified that the postwar "messianic role of the scientists" was now at an end. Scientists working within the system could not dissent from government policy, as Oppenheimer had done by writing his 1953 *Foreign Affairs* essay, and still expect to serve on government advisory boards. The trial thus represented a watershed in the relations of the scientist to the government. The narrowest vision of how American scientists should serve their country had triumphed.

For several decades, American scientists had been leaving the academy in droves for corporate jobs in industrial research laboratories. In 1890, America had only four such labs; by 1930 there were over a thousand. And World War II had only accelerated this trend. At Los Alamos, of course, Oppenheimer had been central to the process. But afterwards, he had taken an alternative course. In Princeton, he was not part of any weapons laboratory. Increasingly alarmed by the development of what President Eisenhower would someday call the "military-industrial complex," Oppenheimer had tried to use his celebrity status to question the scientific community's increasing dependency on the military. In 1954, he lost. As the science historian Patrick McGrath later observed, "Scientists and administrators such as Edward Teller, Lewis Strauss, and Ernest Lawrence, with their full-throated militarism and anti-communism, pushed American scientists and their institutions toward a nearly complete and subservient devotion to American military interests."

Oppenheimer's defeat was also a defeat for American liberalism. Liberals were not on trial during the Rosenberg atom spy case. Alger Hiss was accused of perjury, but the underlying accusation was espionage. The Oppenheimer case was different. Despite Strauss' private suspicions, no evidence emerged to suggest that Oppenheimer had passed any secrets. Indeed, the Gray Board had exonerated him of any such accusations. But like many Roosevelt New Dealers, Oppenheimer had once been a man of the broad Left, active in Popular Front causes, close to many communists and to the Party itself. Having evolved into a liberal disillusioned with the Soviet Union, he had used his iconic status to join the ranks of the liberal foreign policy establishment, counting as personal friends men like Gen. George C. Marshall, Dean Acheson and McGeorge Bundy. Liberals had then embraced Oppenheimer as one of their own. His humiliation thus implicated liberalism, and liberal politicians understood that the rules of the game had changed. Now, even if the issue was not espionage, even if one's loyalty was unquestioned, challenging the wisdom of America's reliance on a nuclear arsenal was dangerous. The Oppenheimer hearing thus represented a significant step in the narrowing of the public forum during the early Cold War.

CHAPTER THIRTY-EIGHT

"I Can Still Feel the Warm Blood
on My Hands"

*It achieved just what his opponents wanted to achieve; it
destroyed him.*

I. I. RABI

THE OPPENHEIMERS WERE DELUGED WITH LETTERS — supportive letters from admirers, abusive letters from cranks, and anguished letters from close friends. Jane Wilson, the wife of the Cornell physicist Robert Wilson, wrote Kitty, "Robert and I have been shocked from the onset, & each new development fills us with nausea and disgust. Uglier little comedies have probably been played in the course of history, but I can't recall them." Robert tried to make light of the whole affair, telling his cousin Babette Oppenheimer Langsdorf, "Aren't you tired of reading about me? I am!" But then the bitterness would seep out in wry comments like "They paid more to tap my phone than they paid me to run the Los Alamos Project."

In a phone conversation with his brother, Robert said he had known "all the time the way the affair would turn out. . . ." Though certainly disheartened, he was already trying to think of his ordeal as history. He told Frank in early July that he had spent $2,000 for extra copies of the hearing transcripts "so that historians and scholars might study them."

Some of his closest friends thought he had aged noticeably in the previous six months. "One day he would indeed look drawn and haggard," said Harold Cherniss. "Another day he was as robust and as beautiful as ever." Robert's childhood friend Francis Fergusson was startled by his appearance. His short-cropped, speckled-gray hair had turned silver white. He had just turned fifty, but now, for the first time in his life, he looked older than his age. Robert confessed to Fergusson that he had been a "damn fool" and that he probably deserved what had happened to him. Not that he had been guilty of anything, but he had made real mistakes, "like claiming to know things that he didn't know." Fergusson thought his friend knew by now that "some of his most depressing mistakes were due to his vanity." "He was like a wounded animal," Fergusson recalled. "He retreated. And returned to a simpler way of life."

Reacting with the same stoicism he had displayed at the age of fourteen, Oppenheimer refused to protest the verdict. "I think of this as a major accident," he told a reporter, "much like a train wreck or the collapse of a building. It has no relation or connection to my life. I just happened to be there." But six months after the trial, when the writer John Mason Brown compared his ordeal to a "dry crucifixion," Oppenheimer answered with a thin smile, "You know, it wasn't so very dry. I can still feel the warm blood on my hands." Indeed, the more he tried to trivialize the ordeal—as a "major accident" with "no connection to my life"—the more heavily it weighed on his spirit.

Robert did not plunge into a deep depression or suffer any visible blows to his psyche. But some of his friends noticed a change in tenor. "Much of his previous spirit and liveliness had left him," Hans Bethe said. Rabi later said of the security hearing, "I think to a certain extent it actually almost killed him, spiritually, yes. It achieved what his opponents wanted to achieve; it destroyed him." Robert Serber always thought that in the aftermath of the hearings, Oppie was "a sad man, and his spirit was broken." But later that year, when David Lilienthal encountered the Oppenheimers at a party in New York, hosted by the socialite Marietta Tree, he noted in his diary that Kitty looked "radiant" and that Robert was "looking actually happy, something I can't remember ever thinking about him." A close friend like Harold Cherniss "thought that both Robert and Kitty had come through the hearings amazingly well." Indeed, if Robert had changed at all, Cherniss thought it was a change for the better. After his ordeal, Cherniss said, Robert listened more and displayed "a greater understanding of others."

Oppenheimer was devastated and yet simultaneously capable of remarkable equanimity. He could pass off what had happened as an absurd accident, but such diffidence left him without the energy and anger that a different kind of man might have used to fight back. Perhaps the diffidence was a deep-rooted survival strategy, but if so it came at considerable cost.

For a time, Oppenheimer wasn't even sure whether the Institute's trustees would permit him to keep his job. He knew Strauss would like to see him ousted as director. In July, Strauss told the FBI that he believed eight of the Institute's thirteen trustees were ready to dismiss Oppenheimer—but he had decided to postpone a vote on the matter until the autumn so it would not appear that Strauss as chairman was acting out of personal vindictiveness. This proved to be a miscalculation, because the delay gave members of the faculty time to organize an open letter in support of Oppenheimer. Every member of the Institute's permanent faculty signed the letter, an impressive show of solidarity for a director who had bruised more than a few egos over the years. Strauss was forced to back off, and later that autumn the trustees voted to keep Oppie as director. Angry and frustrated, Strauss continued to clash with Oppenheimer at Institute board meetings. Strauss never relinquished his obsession with Oppenheimer, filling his files with memoranda that obsessively detailed Robert's alleged infractions. "He cannot tell the truth," he wrote in January 1955 about a minor dispute over a faculty sabbatical payment. Over the years, he filed away vindictive notes on Oppie's friends and defenders: He called Justice Frankfurter "an unconscionable liar" and took delight in passing around rumors that Joe Alsop's sexual preferences made him "vulnerable to Soviet blackmail."*

IF OPPENHEIMER was showing the strain of the last few months, so, too, was his immediate family. Although Kitty had given a stellar performance before the security panel, her friends could see that she was visibly distressed. One night at 2:00 a.m., she phoned her old friend Pat Sherr. "We were sound asleep," Sherr recalled, "and she was obviously quite drunk; her speech was slurred and she was saying things that were sort of disconnected." In early July, just after the AEC's decision to uphold the ruling, an illegal FBI phone tap picked up the information that Kitty had just suffered a severe attack of an unidentified illness and had to be attended to by a physician at Olden Manor.

Nine-year-old Toni seemed to take it all in stride. But according to Harold Cherniss, Peter, thirteen, had "a very difficult time in school during Robert's ordeal." One day he came home from school and told Kitty that a classmate had said, "Your father is a communist." Always a sensitive child, Peter now became more reticent. One day early that summer, after watching some of the televised Army-McCarthy hearings, Peter went upstairs and wrote on the blackboard mounted in his bedroom: "The American Government is unfair to Accuse Certain People that I know of being unfair to them. Since this true, I think that Certain People, and may I say, only

---

*In 1957, Alsop was confronted by Soviet secret police with photographic evidence of a homosexual tryst. Strauss made sure that letters documenting the incident were preserved in CIA director Allen Dulles' personal safe.

Certain People in the U.S. government, should go to HELL. Yours truly
Certain People"

Understandably, Robert thought a long vacation might be good for
everyone. He and Kitty decided to return to the Virgin Islands, but while
they were making their plans, Robert told Kitty she shouldn't send a wire to
St. Croix because he thought his communications were still being monitored. Fearing that the authorities might interfere, he said "if that corner
isn't loused up already, it will be by doing that." Kitty disregarded this
advice and sent the cable anyway, reserving a seventy-two-foot sailing
ketch, the *Comanche,* owned by their friend Edward "Ted" Dale.

FBI technical surveillance had been withdrawn in early June. But a
month later, after the AEC commissioners released the final verdict against
Oppenheimer, Strauss had again pressed the FBI to keep Robert under surveillance. Illegal, warrantless phone taps were reinstalled in early July, and
at the same time the Bureau assigned six agents to keep Oppenheimer under
tight physical surveillance from 7:00 a.m. to midnight every day. Both
Strauss and Hoover feared that he might make a run for it. Strauss had
visions of a Soviet submarine surfacing in the warm Caribbean waters and
spiriting Oppenheimer off behind the Iron Curtain.

Oppenheimer himself was amused to read a report in *Newsweek* that
"key security officials have been alerted against a Communist effort to get
Dr. J. Robert Oppeneheimer to visit Europe and then coax him into doing a
Ponti Corvo [*sic*]," a reference to Bruno Pontecorvo, an Italian physicist
who had defected to the Soviets in 1950. The FBI's wiretaps picked up
Herb Marks advising Oppenheimer that under the circumstances, he probably ought to write a letter to J. Edgar Hoover, informing him of his vacation
plans. "The letter," the FBI summary of their conversation noted, "will be
predicated on the foolish rumors being circulated to the effect that Dr.
Oppenheimer may leave the country, may be kidnapped, may be met by a
Russian submarine, is planning a European vacation, etc." Oppenheimer
obligingly sent Hoover a letter, informing him of his plan to spend a three-
or four-week sailing vacation in the Virgin Islands.

Robert and his family boarded a flight for St. Croix on July 19, 1954,
and from there they made their way to St. John, a pristine Caribbean island
about the size of Manhattan (21 square miles), with no more than 800 residents—ten percent of whom were "continentals." In 1954, there might have
been a couple of sailing sloops anchored in the bay. The island's one village
and only commercial port, Cruz Bay, had several hundred people, mostly
descendants of St. John's slave population. The only bar in the village,
Mooie's, wouldn't be built for two years. The largest building, Meade's Inn,
was a one-story West Indian gingerbread cottage. Peacocks and donkeys
roamed the unpaved streets.

Stepping off the ferry, the Oppenheimers found a jeep taxi to take them over dirt roads along the island's northern coast. Seeking anonymity, they passed by Caneel Plantation, the island's only upscale resort, developed by Laurance S. Rockefeller, and drove to Trunk Bay's Guest House, a primitive bed-and-breakfast lodge run by a longtime resident, Irva Boulan Thorpe. There were no phones, no electricity, and rooms for no more than a dozen guests. Seeking a solitary refuge, they had come to the right place. "They were sort of in a state of shock," recalled Irva Claire Denham, the daughter of the proprietor. "It was isolated enough so that people couldn't get at them. They were being careful about who they even talked to. . . . Kitty was very protective. She was like a tigress when anybody approached him, because he was willing to talk." When Kitty was in a foul mood, she often threw things—and the next morning Robert would go to see the Boulans and pay handsomely for the damage. Using Cruz Bay as their home port, the Oppenheimers spent the next five weeks sailing the *Comanche* in the waters around St. John and the neighboring British Virgin Islands.

As late as August 25, 1954, the Bureau was still worried about a communist plot, dubbed "Operation Oppenheimer," to whisk the Oppenheimers behind the Iron Curtain. "According to the plan," an FBI report reads, "Oppenheimer will first travel to England, from England he will travel to France, and while in France he will vanish into Soviet hands."

The FBI found it impossible to keep Oppenheimer under surveillance while he was on St. John. So when he finally flew back to New York on August 29, 1954, FBI agents accosted him and requested that he accompany them to a private room in the airport terminal. Oppenheimer agreed, but insisted that his wife be present. When they got inside the room, the agents bluntly asked if he had been approached by Soviet agents in the Virgin Islands and asked to defect. The Russians, he said, "were damn fools," but he didn't think they "were foolish enough to approach him with such an offer." He volunteered that if this ever happened, he would promptly notify the FBI. After this short interrogation, the Oppenheimers left the airport. Agents followed their car to Princeton, and the next day the FBI once again placed a wiretap on their home phone.

Incredibly, the FBI sent another team of agents back down to St. John in March 1955—six months after Oppenheimer had left. The agents went around asking residents whom Oppenheimer had talked to while he was on the island.

ABROAD, foreign opinion reacted to the trial with incredulity. European intellectuals saw it as further evidence that America was gripped by irrational fears. "How can the independent experimental mind survive in such an atmosphere?" asked R.H.S. Crossman in *The New Statesman and Nation*, Britain's leading liberal weekly. In Paris, when Chevalier received his copy of the hearing transcript—shipped to him by Oppenheimer himself—he read portions of the document out loud to André Malraux. Both men were struck by Oppenheimer's strange passivity in the face of his interrogators. Malraux was particularly troubled that Oppenheimer had freely answered questions about the political views of his friends and associates. The hearing had turned him into an informer. "The trouble was," Malraux told Chevalier, "he accepted his accusers' terms from the beginning. . . . He should have told them, at the very outset, 'Je suis la bombe atomique!' He should have stood on the ground that he was the builder of the atom bomb—that he was a scientist, and not an informer."

Initially, Oppenheimer seemed destined to become a pariah, at least in mainstream circles. For nearly a decade, he had been more than just a famous scientist. Once a ubiquitous and influential public figure, now he was suddenly gone—still alive, but disappeared. As Robert Coughlan later wrote in *Life* magazine, "After the security hearings of 1954, the public character ceased to exist. . . . He had been one of the most famous men in the world, one of the most admired, quoted, photographed, consulted, glorified, well-nigh deified as the fabulous and fascinating archetype of a brand new kind of hero, the hero of science and intellect, originator and living symbol of the new atomic age. Then, suddenly, all the glory was gone and he was gone, too. . . ." In the media, Teller replaced Oppenheimer as the face of the archetypical scientific statesman. "The glorification of Teller in the 1950s was accompanied," Jeremy Gundel wrote, "perhaps inevitably, by the defamation of the man who had been his chief rival, J. Robert Oppenheimer."

While Oppenheimer was excommunicated from government circles, he nevertheless quickly became a symbol to liberals of everything that was wrong with the Republican Party. That summer, the *Washington Post* ran a series of articles by the newspaper's assistant managing editor, Alfred Friendly, which the FBI observed "slanted favorably toward Oppenheimer. . . ." In one article, headlined DRAMA PACKS AMAZING OPPENHEIMER TRANSCRIPT, Friendly called the hearing an "Aristotelian drama," "Shakespearean in richness and variety," with "Eric Ambler allusions to espionage," "a plot more intricate than *Gone With the Wind*," and "with half again as many characters as *War and Peace*."

Many Americans began to regard Oppenheimer as a scientist-martyr, a victim of the era's McCarthyite excesses. At the end of 1954, Columbia University invited him to give an address on the occasion of its bicentennial; the lecture was broadcast to a national audience. His message was bleak and pessimistic. Earlier, in his Reith Lectures, he had extolled the virtues of science in communitarian endeavors, but now he dwelled on the solitary condition of intellectuals, embattled by the fierce winds of popular emotions. "This is a world," he said, "in which each of us, knowing his limitations, knowing the evils of superficiality, will have to cling to what is close to him, to what he knows, to what he can do, to his friends and his tradition and his love, lest he be dissolved in a universal confusion and know nothing and love nothing. . . . If a man tells us that he sees differently than we, or that he finds beautiful what we find ugly, we may have to leave the room, from fatigue or trouble. . . ."

A few days later, millions of Americans watched as Edward R. Murrow interviewed Oppenheimer on his national television show *See It Now.* Robert had not wanted to do the show; at the last minute, he tried to back out. Murrow's own network had serious misgivings, but the famous broadcaster nevertheless prevailed on Oppenheimer to sit for a taping in his Institute office.

Murrow edited his two-and-a-half-hour conversation with Oppenheimer down to a twenty-five-minute segment that aired on January 4, 1955. Oppenheimer used the occasion to talk about the debilitating effects of secrecy. "The trouble with secrecy," he said, "is that it denies to the government itself the wisdom and resources of the whole community. . . ." Murrow never directly brought up the security hearing—no doubt, because Robert had insisted that it not be raised. Instead, he gently asked Oppenheimer if scientists had become alienated from the government. "They like to be called in and asked for their counsel," Oppenheimer replied obliquely. "Everybody likes to be treated as though he knew something. I suppose that when the government behaves badly in the field you're working close to, and when decisions that look cowardly or vindictive, or short-sighted, or mean are made . . . then you get discouraged and you may—may—you may recite George Herbert's poem *I Will Abroad.* But that's human rather than scientific." Asked whether humanity now had the capability to destroy itself, Oppenheimer replied, "Not quite. Not quite. You can certainly destroy enough of humanity so that only the greatest act of faith can persuade you that what's left will be human."

Just a few weeks after his appearance on *See It Now,* Oppenheimer's name again surfaced in the national press, this time in a controversy over academic freedom. In 1953, the University of Washington had offered Oppenheimer a short-term visiting professorship. Because of the security hearing, Oppenheimer had postponed the appointment. But in late 1954, the physics department renewed the invitation—only to have it canceled by the university's president, Henry Schmitz. When the *Seattle Times* got wind of Schmitz' decision, the news stirred a national debate on academic freedom. Some scientists announced that they were going to boycott the University of Washington. The *Seattle Post-Intelligencer* editorialized in support of President Schmitz: "The notion that 'academic freedom' is involved . . . is emotional and juvenile balderdash." Those supporting Oppenheimer's presence on campus, the newspaper insisted, were "apologists for totalitarianism."

Oppenheimer tried to stay above the fray. When asked by a reporter if the cancellation of his visit was an impingement of academic freedom, he said, "That's not my problem." But when the reporter followed up by asking whether the scientists' boycott would bring some embarrassment to the university, he replied sharply, "It seems to me that the university has already embarrassed itself."

Such incidents reinforced Oppenheimer's new image. His public transformation from Washington insider to exiled intellectual was complete. And yet, this did not mean that the private Oppenheimer thought of himself as a dissident. Nor was he inclined to play the role of an activist public intellectual. Gone were the days when he might organize a fund-raiser for some good cause—or even sign a petition. Indeed, some of his friends thought him oddly passive now, even deferential, in the face of authority. His friend and admirer David Lilienthal was struck by a conversation he had with Oppenheimer in March 1955, less than a year after the security hearing. The occasion was a board meeting of the Twentieth Century Fund, a liberal foundation whose trustees included Lilienthal, Oppenheimer and Adolph Berle, as well as Jim Rowe and Ben Cohen—both former assistants to Franklin Roosevelt—and Francis Biddle, FDR's former attorney general. After their foundation business was concluded, Berle turned the conversation to a discussion of the current crisis between Communist China and Chiang Kai-shek's Taiwan over the Formosa Straits. Berle thought war was imminent, and that it might well begin with "little A-bombs, and where does it go from there?" He added that he knew that some generals believed "we should destroy the Chinese with atomic weapons now, before they get any stronger. . . ." This touched off a vigorous discussion about what should be done, and in due course a consensus emerged that they should all sign a public statement warning the country against any precipitate military action.

But then, to Lilienthal's surprise, Oppenheimer spoke up and "explained that he didn't think he should sign the statement, though agreeing with it, because of the to-do this would cause." He went on to throw cold water over the whole notion of protesting the Eisenhower Administration's drift toward war. After all, he said, a war over Formosa (Taiwan) was not necessarily worse than a peace under any circumstances, and if it came to war, the limited use of tactical A-bombs might not lead inexorably to the wholesale bombings of cities. He even argued that any statement—which he agreed with but would not sign—should not imply that "thoughtful and careful and intelligent attention to the relevant issues was not already being given, in Washington." Robert had always been persuasive with any audience—and by the end of the meeting, they all agreed that perhaps a public statement was not in order. Lilienthal came away wondering "whether those of us—such as myself—who have been under terrific attack don't go out of our way to be conservative in discussing the position of our country and our Government, lest we be thought less than pro-American."

It seems obvious that Robert was determined to prove that he was a reliable patriot, that his critics had been wrong to question his devotion to the country. He was steering clear of all public policy confrontations, especially those that had any relationship to nuclear weapons. He disapproved of self-appointed pundits—like the young Henry Kissinger, who had turned himself into a nuclear strategist. "A lot of nonsense," he privately told Lilienthal, waving his unlit pipe around in the air. "To think that these are troubles that can be solved by the theory of games or behavioral research!" But he would not publicly condemn Kissinger or any other nuclear strategist.

That same spring, Oppenheimer turned down an invitation from Bertrand Russell to attend the inaugural session of the Pugwash Conference, a gathering of international scientists organized by the industrialist Cyrus Eaton, Russell, Leo Szilard and Joseph Rotblat, the Polish-born physicist who had left Los Alamos in the autumn of 1944. Oppenheimer wrote Russell that he was "somewhat troubled when I look at the proposed agenda. . . . Above all, I think that the terms of reference 'the hazards arising from the continuous development of nuclear weapons' prejudges where the greatest hazards lie. . . ." Nonplussed, Russell replied, "I can't think that you would deny that there are hazards associated with the continued development of nuclear weapons."

Citing this and other exchanges, the science sociologist Charles Robert Thorpe has argued that while Oppenheimer may have been "excommunicated from the inner circle of the nuclear state," he nevertheless "remained in spirit a supporter of the fundamental direction of its policies." In Thorpe's eyes, Oppenheimer was slipping back into his "earlier role as scientific-military strategist of the winnable nuclear war and apologist for the powers-that-be." It seemed that way to some. Oppenheimer was certainly not willing to throw in his lot with political activists like Lord Russell, Rotblat, Szilard, Einstein and others who frequently signed petitions protesting the American-led arms race. Indeed, his name was conspicuously absent from one such open letter, dated July 9, 1955, and signed by not only Russell, Rotblat and Einstein, but also such former teachers and friends as Max Born, Linus Pauling and Percy Bridgman.

But Oppenheimer was still capable of being a critic; he just wanted to stand alone and with far more ambiguity than his fellow scientists. He was consumed with the deep ethical and philosophical dilemmas posed by nuclear weapons, but at times it seemed that, as Thorpe puts it, "Oppenheimer offered to weep for the world, but not help to change it."

In truth, Oppenheimer very much wanted to change the world—but he knew he was barred from pulling on the levers of power in Washington, and he no longer had the spirit for public activism that had motivated him in the 1930s. His excommunication had not freed him to enter the great debates of the day; it had inclined him, rather, to censor himself. Frank Oppenheimer thought his brother felt enormously frustrated that he could not find a way back into official circles. "He wanted to get back into that, I think," Frank said. "I don't know why, but I think it's one of these things where there's a—when you get the taste of it, it's hard to not want it."

On occasion, however, he spoke publicly about Hiroshima and did so with a vague sense of regret. In June 1956, he told the graduating class of the George School—attended by his son, Peter—that the Hiroshima bombing may have been "a tragic mistake." America's leaders, he said, "lost a certain sense of restraint" when they used the atomic bomb on the Japanese city. A few years later, he gave a hint of his feelings to Max Born, his former professor in Göttingen, who had made it clear that he rather disapproved of Oppenheimer's decision to work on the atomic bomb. "It is satisfying to have had such clever and efficient pupils," Born wrote in his memoirs, "but I wish they had shown less cleverness and more wisdom." Oppenheimer wrote Born, "Over the years, I have felt a certain disapproval on your part for much that I have done. This has always seemed to me quite natural, for it is a sentiment that I share."

IF OPPENHEIMER was unwilling to enter publicly the roiling debates of the mid-1950s over the Eisenhower Administration's nuclear policies, he had no hesitation about speaking on cultural and scientific issues. Only a year after the security hearings, he published a collection of essays under the title *The Open Mind*. It included eight lectures he had given since 1946, all speaking to the issue of the relationship between atomic weapons, science and postwar culture. Published by Simon & Schuster, and widely reviewed, the book served to present him as a modern seer, a thoughtful, enigmatic philosopher of the role of science in the modern world. In these essays, he pleaded for an "open mind" as a necessary component for an open society. He made the case for "the minimization of secrecy," and he observed, "We seem to know, and seem to come back again and again to this knowledge, that the purposes of this country in the field of foreign policy cannot in any real or enduring way be achieved by coercion." In an implicit rebuke to those who thought that a powerful, nuclear-armed America could act unilaterally, Oppenheimer intoned, "The problem of doing justice to the implicit, the imponderable, and the unknown is of course not unique in politics. It is always with us in science, it is with us in the most trivial of personal affairs, and it is one of the great problems of writing and of all forms of art. The means by which it is solved is sometimes called style. It is style which complements affirmation with limitation and with humility; it is style which makes it possible to act effectively, but not absolutely; it is style which, in the domain of foreign policy, enables us to find a harmony between the pursuit of ends essential to us, and the regard for the views, the sensibilities, the aspirations of those to whom the problem may appear in another light; it is style which is the deference that action pays to uncertainty; it is above all style through which power defers to reason."

In the spring of 1957, Oppenheimer was invited by the philosophy and psychology departments of Harvard University to give the prestigious William James Lectures. His friend McGeorge Bundy, then dean of Harvard, extended the invitation which, predictably, sparked considerable controversy. A group of Harvard alumni led by Archibald B. Roosevelt threatened to withhold donations if Oppenheimer was allowed to speak. "We don't believe people who tell lies," said Roosevelt, "should lecture at a place whose motto is 'Veritas.' " Dean Bundy listened to the protests and then made a point of attending the April 8 lecture.

Oppenheimer titled his series of six public lectures "The Hope of Order." At the inaugural talk, 1,200 people packed Harvard's largest lecture hall, Sanders Theater. Another 800 people listened to the lecture piped into a nearby hall. Anticipating protests, armed police stood at the doors. A large American flag hung on the wall behind the lectern, giving the scene an oddly cinematic aura. By coincidence, Senator Joe McCarthy had died four days earlier and his remains were lying in state that very afternoon in the Capitol. As Oppenheimer rose to speak, he hesitated, and then walked over to a blackboard and wrote, "R.I.P." As some in the audience murmured with comprehension at the audacity of this silent rebuke to the dead senator, Oppenheimer walked back to the lectern stony-faced and began his talk. Edmund Wilson attended one of the lectures and afterwards described his impressions in his diary. As Harvard's president, Nathan Pusey, was introducing him, Oppenheimer sat alone on the platform, "nervously shifting his arms and feet in an ungainly Jewish way; but when he began to speak, he had the whole audience riveted; there was scarcely a sound throughout. He spoke very quietly but with piercing point. Extraordinary how terse and precise he was, speaking merely from notes—as in his description of William James, in which he touched on his relation to Henry. The opening was quite thrilling—he did nothing to make it dramatic, but he was raising terrific questions that were painfully in everyone's mind and one felt, as Elena said, his feeling of intense responsibility. We were both of us moved and stimulated."

But afterwards, Wilson began to wonder whether Oppenheimer was "a brilliant man who had been beaten by the age, who knew no more what to do about it than anybody, who was as incapable of leading it as anybody; his humility now seemed to me hangdog." Like many who heard Oppenheimer speak, Wilson came away from the experience with a troubled sense of the man's fragile ambiguities.

From his perch at the Institute, and in numerous other speeches around the country, Oppenheimer was carving out a new role for himself. Once he had been the scientific insider; now he was becoming a distant but charismatic intellectual outsider. David Lilienthal, who saw him frequently, thought he had mellowed. Certainly, he had aged; by 1958, Robert's lanky, fifty-four-year-old frame had the forward stoop of an old man. But Lilienthal thought the lines of care in his face had "given way to a kind of 'success' calm. He has weathered one of the most violent, bitter storms that any human being ever went through."

OPPENHEIMER CONTINUED to preside over the Institute with deftness and sensitivity. He could take pride in his creation. Like Berkeley in the 1930s, the Institute had become one of the world's foremost centers for theoretical physics—and much more. It was a haven for brilliant scholars, young and old, in numerous disciplines. John Nash was one such young scholar, a brilliant mathematician who held a fellowship at the Institute in 1957.* Having read Werner Heisenberg's 1925 paper on the "uncertainty principle," Nash began questioning veteran physicists about some of the unresolved contradictions of quantum theory. Like Einstein, Nash was troubled by the neatness of the theory. In the summer of 1957, when he raised such heresies with Oppenheimer, the director impatiently dismissed his questions. But Nash persisted and Oppenheimer soon found himself drawn into a serious argument. Afterwards, Nash wrote him an apology but insisted that most physicists were "quite too dogmatic in their attitudes."

*Nash was portrayed in *A Beautiful Mind,* by Sylvia Nasar, and later in a film by the same title.

Nash left that summer, and for many years afterwards he struggled with a debilitating mental illness that for a time required him to be institutionalized. Oppenheimer was sympathetic with Nash's psychiatric ordeals, and invited him back to the Institute when he had recovered from one of his severest bouts with schizoid symptoms. Robert had a forgiving instinct for the frailty of the human psyche, an awareness of the thin line between insanity and brilliance. So when Nash's doctor called Oppenheimer in the summer of 1961 to ask whether Nash was still sane, he replied, "That's something no one on earth can tell you, doctor."

Oppenheimer could be embarrassingly opaque about his own complicated personal life. When twenty-seven-year-old Jeremy Bernstein arrived at the Institute in 1957, he was informed that Dr. Oppenheimer wanted to see him right away. As Bernstein walked into the director's office, Oppenheimer greeted him jauntily, "What is new and firm in physics?" Before Bernstein could muster a reply, the phone rang and Oppenheimer motioned for him to stay as he took the call. When he hung up, he turned to Bernstein, someone he had barely met, and said casually, "It's Kitty. She has been drinking again." With that, he invited the young physicist to come by Olden Manor to see some of his "pictures."

Bernstein spent two years at the Institute and found Oppenheimer "endlessly fascinating." The man could be by turns sharply intimidating and charmingly disarming. When called to Oppenheimer's office one day for one of his periodic "confessionals" with the director, Bernstein happened to remark that he was reading Proust. "He looked at me kindly," Bernstein later wrote, "and said that when he was about my age he had taken a walking trip on Corsica and had read Proust at night by flashlight. He was not bragging. He was sharing something."

IN 1959, Oppenheimer attended a conference in Rheinfelden, West Germany, sponsored by the Congress on Cultural Freedom. He and twenty other world-renowned intellectuals gathered in the luxurious Saliner Hotel on the banks of the Rhine near Basel to discuss the fate of the Western industrialized world. Safe in this cloistered environment, Oppenheimer broke his silence on nuclear weapons and spoke with uncharacteristic clarity about how they were seen and valued in American society. "What are we to make of a civilization which has always regarded ethics as an essential part of human life," he asked, but "which has not been able to talk about the prospect of killing almost everybody except in prudential and game-theoretical terms?"

Oppenheimer deeply empathized with the Congress' liberal anticommunist message. As someone who had once surrounded himself with communists, Oppenheimer was now in the company of intellectuals dedicated to dispelling the illusions of "frivolous fellow-travelers." He enjoyed the company of the men he met at its annual sessions. These included such writers as Stephen Spender, Raymond Aron and the historian Arthur Schlesinger, Jr. He and the Congress' executive director, Nicolas Nabokov, became good friends. Nabokov, a cousin of the novelist, was a well-regarded composer who divided his time between Paris and Princeton. He certainly knew that the Congress was receiving funding from the Central Intelligence Agency. And so, too, did Oppenheimer. "Who didn't know, I'd like to know? It was a pretty open secret," recalled Lawrence de Neufville, a CIA officer stationed in Germany. When the *New York Times* broke this news in the spring of 1966, Oppenheimer joined Kennan, John Kenneth Galbraith, and Arthur Schlesinger, Jr., in a joint letter to the editor defending the Congress' independence and the "integrity of its officials." They didn't bother to deny the CIA link. Later that year, Oppenheimer wrote Nabokov, assuring him that he regarded the Congress as one of the "great and benign influences" of the postwar era.

As time went by, Oppenheimer became more visible as an international celebrity. He began to travel abroad more often. In 1958, he visited Paris, Brussels, Athens and Tel Aviv. In Brussels, he and Kitty were greeted by the Belgian royal family—Kitty's distant relations. In Israel, his host was Prime Minister David Ben-Gurion. In 1960, he visited Tokyo, where reporters greeted him at the airport with a barrage of questions. "I do not regret," he said softly, "that I had something to do with the technical success of the atomic bomb. It isn't that I don't feel bad; it is that I don't feel worse tonight than I did last night." The translation of that ambiguously loaded sentiment into Japanese could not have been easy. The following year, he toured Latin America, sponsored by the Organization of American States, garnering headlines in local newspapers as "El Padre de la Bomba Atomica."

LILIENTHAL, who so admired Oppenheimer's intellect, was saddened by what he observed of Robert's family life. There was, he later said, a "contradiction between Oppenheimer's brilliant mind and his awkward personality. . . . He did not know how to deal with people, his children especially." Lilienthal later harshly concluded that Oppenheimer "ruined" his children's lives. "He kept them on a tight leash." Peter grew up to become a shy but highly sensitive and intelligent young man. But he lived estranged from his mother. Francis Fergusson knew that Robert loved his son, but he saw that

Robert seemed incapable of protecting Peter from his mother's volatile moods. In 1955, Robert and Kitty sent Peter, fourteen, to the George School, an elite Quaker boarding school in Newtown, Pennsylvania, hoping that a little distance would ease tensions between his son and his wife.

A crisis occurred in 1958 when Robert was offered a visiting professorship in Paris for a semester. He and Kitty decided to pull Toni, twelve, out of her private school in Princeton and bring her with them. But they decided that Peter, seventeen, should remain behind at the George School. Robert wrote his brother that Peter had expressed the desire to visit Frank on his ranch and maybe try to get a summer job on one of the dude ranches in New Mexico. "He is still in a very volatile mood," Robert wrote, "and I am afraid I cannot predict what will happen in June with any kind of certitude."

Robert's personal secretary, Verna Hobson, disapproved: "What a slap to leave him behind. He [Peter] was enormously sensitive. I felt tremendously on his side." Hobson told Robert what she thought, but it was clear that Kitty had made up her mind. Hobson saw it as a real turning point in Peter's relationship with his father. "There came a time," Hobson said, "when Robert had to choose between Peter—of whom he was very fond—and Kitty. She made it so it had to be one or the other, and because of the compact he had made with God or with himself, he chose Kitty."

"It Was Really Like a Never-Never-Land"

Robert was a very humble man. I adored him.

INGA HIILIVIRTA

BEGINNING IN 1954, the Oppenheimers spent several months each year living on the tiny island of St. John in the Virgin Islands. Surrounded by the stunning, primordial beauty of the island, Robert relished this self-imposed exile, living as if he were a social outcast. In the words of a poem he had written as a young man at Harvard, he was fashioning in St. John "his separate prison," and the experience seemed to rejuvenate him now as his summers in New Mexico had reinvigorated him decades earlier. During their first few visits, the Oppenheimers returned to the small guest house at Trunk Bay on the north shore of the island, owned by Irva Boulan. But in 1957, Robert bought two acres of land on Hawksnest Bay, a beautiful cove on the northwest tip of the island. The site lay just below a towering hump-shaped outcropping of rock known ironically, at least for Robert, as "Peace Hill." Palm trees dotted the cove's gently sloping white beach and the turquoise waters were filled with parrotfish, blue tang, grouper and the occasional school of barracuda.

In 1958, Robert hired the eminent architect Wallace Harrison—who had helped design such landmarks as Rockefeller Center, the United Nations building and Lincoln Center—to design a spartan beach cottage, something of a Caribbean version of Perro Caliente. However the contractor Robert hired for the project poured the foundation in the wrong spot—perilously close to the water's edge. (He claimed a donkey had eaten the surveyor's plans.) When finally built, the cottage consisted of one large rectangular room, some sixty or seventy feet long, sitting atop a slab of concrete. The room was divided only by a four-foot-high wall, setting off the sleeping area from the rest of the cottage. The floor was covered with pretty terracotta tiles. A well-equipped kitchen and a small bathroom occupied the back of the structure. Shuttered windows let sunlight pour into the cottage from three sides. But the front of the cottage, facing the cove, was completely open—to the cove and to the island's warm trade winds. The house thus had only three walls, with a tin roof designed to roll down to cover the front of the structure during the hurricane season. They called it "Easter Rock," after the large, egg-shaped rock that sat perched atop Peace Hill.

A hundred yards up the beach lived their only neighbors, Robert and Nancy Gibney, who had reluctantly sold them the beach property, after much gentle cajoling by Robert. The Gibneys had been living on the island since 1946, when they had bought for a paltry sum seventy acres around Hawksnest Bay. A former editor at *The New Republic,* Bob Gibney had literary ambitions, but the longer he lived on the island, the less he wrote.

Gibney's wife, Nancy, came from a wealthy Boston family. An elegant woman, she had once worked as an editor at *Vogue.* With three young children, and little regular income, the Gibneys were land-rich and cash-poor. Nancy Gibney had first met the Oppenheimers in 1956, during a lunch at Trunk Bay's guest house. "They were got up in routine tourist garb," she later wrote, "cotton shirts and shorts and sandals, but they looked like nothing human, too thin and frail and pale for earthly life. . . . Kitty was the more humanoid of the two, although she seemed to have no features except for her dark eyes. Her voice was too deep and hoarse to emanate from her tiny chest. . . ."

Upon being introduced, Kitty said to Nancy, "Aren't you hot with all that hair?" It was a remark Nancy considered "staggeringly rude." But initially she liked Robert. He looked "astoundingly like Pinocchio, and he moved as jerkily as a marionette on strings. But there was nothing wooden about his manner: he exuded warmth and sympathy and courtesy along with the fumes of his famous pipe." When Robert politely asked what her husband did, Nancy explained that on occasion he worked for Laurance Rockefeller at his Caneel Bay hotel.

"He worked for Rockefeller?" Oppenheimer said, puffing on his pipe. And then lowering his voice, he quipped, "I, too, have taken money for doing harm."

Nancy was awed. She had never met such exotic people. The next year, Oppenheimer persuaded the Gibneys to sell him the land for a cottage—and then in the spring of 1959, while a construction crew was still putting up the new house, Kitty wrote to Nancy Gibney, telling her that they wanted to come down to St. John in June but had no place to stay. Against her better judgment, Gibney offered them a room in their large rustic beach house.

A few weeks later, the Oppenheimers showed up, together with fourteen-year-old Toni and a schoolmate, Isabelle. Kitty said the two girls would sleep in a tent they had brought. And then she announced they couldn't possibly stay the whole summer, but might manage a month. Nancy Gibney was stunned; she had thought they would be staying for a few days. Thus began what Nancy later called "seven hideous, hilarious weeks," marked by disagreements, misunderstandings and worse.

To say the least, the Oppenheimers were not easy houseguests. Kitty was invariably up half the night, often groaning with pain from what she called her "pancreas attacks." These only got worse with her drinking. Both Kitty and Robert "were great believers in drinking and smoking in bed." Each night the Gibneys heard Kitty rummaging around in the kitchen, getting more precious ice for her drink. Nancy Gibney was sometimes awakened by Robert's "frequent nightmares." Insomniacs, the Oppenheimers often would not rise until noon.

One night in August, Nancy was awakened for the third time by Kitty banging about in the kitchen, looking for ice with a flashlight. Rising to investigate, Nancy finally exploded with anger: "Kitty, no one who drinks all night needs ice. You get back in that room and you close the doors and stay in there if it kills you."

Kitty looked at her for a moment and then hit Nancy as hard as she could with the flashlight. The blow just grazed Nancy's cheek. "I got a good grip on her shoulder," Gibney later wrote, "and gave her the bum's rush into 'their room' and slammed and barricaded all the doors." The next morning, Gibney left to visit her mother in Boston, telling her children that she would return only "when those lunatics go." The Oppenheimers finally left in mid-August.

The next year, they returned to their now finished beach cottage—but, not surprisingly, their relations with the Gibneys never recovered. Never again on speaking terms with the Oppenheimers, Nancy Gibney routinely provoked Kitty by sticking "Private Property" signs on her side of the beach. The Gibney children remember Kitty marching up and down the beach, ripping out the signs.

Nancy Gibney fought with Kitty—but she reserved her real dislike for Robert. "I came to have a sneaking fondness and respect for Kitty, although I took care not to show it. At her worst, she was absolutely without guile, brave as a little lion, and fiercely loyal to her own team." Robert, she thought—despite her originally favorable impression of him—was the devious one. Nancy's perception of Oppenheimer was uniquely hostile. In her essay on that summer sojourn, she relates that August 6—the fourteenth anniversary of the atomic bombing of Hiroshima—"was a day of fond nostalgia for our guests, a day of smirks and excitable recall. No one observing

Robert Oppenheimer *en famille* that day could question what had been his finest hour . . . he transparently loved the Bomb and his lordly role in its creation."

Robert never raised his voice. Indeed, no one ever saw him angry—with one memorable exception. Several years after moving into their new beach cottage, Robert and Kitty were hosting a raucous New Year's Eve party when one of their guests, Ivan Jadan, burst into gusty, operatic song. The singing was too much for Bob Gibney, who came storming down to the Oppenheimer beach in a rage. He had brought a gun with him, and, apparently in an effort to get everyone's attention, he fired several shots in the air. Robert turned on him ferociously and shouted, "Gibney, never come to my house again!" Thereafter, the Gibneys and the Oppenheimers had nothing to do with each other. They hired lawyers and squabbled over beach rights. The feud became a legend on the island.

THE GIBNEYS' view of the Oppenheimers was not shared by other natives of St. John. Ivan and Doris Jadan, a colorful couple who had lived on the island since 1955, adored Robert. "You never felt uncomfortable around him," Doris recalled, "which was a tribute to the kind of poise he had." Born in 1900 in Russia, Ivan Jadan was the Bolshoi's premier lyric tenor in the late 1920s and '30s. Despite his status, Jadan had refused to join the Communist Party, and in 1941, when the Germans invaded, he and a dozen Bolshoi friends walked toward the German lines and surrendered themselves. They were soon packed into cattle cars and sent to Germany. In 1949, he managed to emigrate from West Germany to the United States. He married Doris in 1951, and when the couple visited St. John in June 1955, Ivan announced, "I stay here."

Introduced to the Oppenheimers, the Jadans were delighted to learn that these newcomers spoke German. Ivan's English was always rudimentary and he and Doris usually spoke Russian to each other. Boisterous and outspoken, Ivan could break into song at the slightest pretext. He could also be rather prickly; he'd get up and leave the table if he found himself in disagreement with someone. Ivan was as profoundly anti-Soviet as anyone could be—but while he knew all about Robert's trial, he detected nothing in Oppenheimer's moral sensibilities that was not profoundly right. Ivan rarely talked politics, but with Robert he was drawn to the subject. They made an odd pair—but he and Robert obviously enjoyed each other's company.

"Kitty, of course, was something else," recalled Doris Jadan. "She was disturbed. But they [she and Robert] were both very protective of each other, even when she was not herself. . . . She could be quite mischievous. The devil had struck through part of her, and she knew it." Doris nevertheless liked her. One day Kitty told Doris, "You know, Doris, you and I have something in common. We are both married to totally unique people, and it is for us a responsibility different than other people's."

Everyone drank on the island, and while Kitty drank a great deal, she could also be cold sober for days on end. "I don't remember Kitty, or only a few times, being what you'd call drunk," recalled Sabra Ericson, a neighbor of the Oppenheimers. "She was the great trouble in his life," Doris Jadan said, "and she knew it. But she knew that he would not have gone through what he had done, I think, except for her. . . . She loved Robert. There's no doubt about that. But she was a tangled person. . . . I think in fairness to her, she may have been as good a wife as he could have had." As for Robert, "He treated her with total devotion," said another St. John resident, Sis Frank. "She could do no wrong in his eye."

Kitty occupied herself for hours on end with her gardening. St. John was a paradise for her orchids. "There might be a dead spot in the garden," Frank observed, "and in a week's time it was growing beautifully. She was wonderful with the orchids." But she dreaded the thought of stopping by the cottage if Kitty was there alone. Inevitably, Kitty would make some caustic, "malicious" remark about something unpleasant. "I learned to overlook those things because a lot of the time she wasn't herself. . . . I knew her moves. I knew what to anticipate. What a ghastly life, to be that unhappy."

"ROBERT WAS a very humble man," recalled Inga Hiilivirta, a beautiful young Finnish woman who had been visiting the island since 1958. "I adored him. I thought he was kind of saintly. His blue eyes were just marvelous. It looked like he could read what you were thinking." She and her husband, Immu, met the Oppenheimers at a Christmas party on December 22, 1961. Walking into the beach house on Hawksnest Bay, the twenty-five-year old Inga was impressed that such a famous man was living in such rustic circumstances. But then she noted that they had all the good things in life too. When Robert asked her, "Would you care for a little wine?" he brought out a bottle of expensive champagne. The Oppenheimers bought their champagne by the case.

A few days later, Robert and Kitty hosted a New Year's Eve party; they hired "Limejuice" Richards, an elderly black native of the island, to transport guests over the winding dirt road from Cruz Bay in their light-green Land Rover. That night the Oppenheimers served lobster salad and champagne. Limejuice and his "scratchy band" played calypso music. Robert danced the calypso with Inga and afterwards everyone went swimming. "It was really like a never-never-land," said Inga, "like a dream." Later that evening, they walked on the beach and Robert pointed out various constellations.

Limejuice became the Oppenheimers' caretaker and gardener. When they weren't on the island, he had the use of their Land Rover, which he employed as a taxi to drive tourists around the island. Robert clearly liked the old man and wanted to help him, even to the point of turning a blind eye to his use of the Land Rover to bootleg Tortola rum.

One evening in early 1961, Ivan Jadan caught a small hawksbill turtle while swimming in Maho Bay—and later, over dinner he displayed the squirming turtle and announced his intention to cook it. Wincing, Robert pleaded for the turtle's life, telling everyone that it "brought back to him the horrible memories of what happened to all the little creatures after the [Trinity] test in New Mexico." So Ivan carved his initials on the turtle's shell and then released it. Inga was touched: "It made me feel even more fond of Robert."

On another occasion, the Oppenheimers were visiting the Jadans at their house perched above Cruz Bay, watching a brilliant sunset. Turning to Sis Frank, Robert rose from his chair and said, "Sis, come with me to the edge of the hill. Tonight you're going to see the green flash." And sure enough, just as the sun sank behind the horizon, Sis saw a flash of green light. Robert quietly explained the physics behind what Sis had seen: As viewed from St. John, layers in the earth's atmosphere functioned like a prism, creating for just a second a flash of green. Sis was thrilled by the sight, and charmed by Robert's patient explanation.

"He was an unassuming man," recalled Sabra Ericson. Each September, the Oppenheimers mailed out three dozen invitations to their island friends for a New Year's Eve party. All sorts of people came—blacks and whites, educated and uneducated. Robert made no distinction. "They were real human beings that way," said Ericson.

The Gibneys aside, the gentlest part of Robert's nature was unfurled daily on St. John. Gone were his cutting comments about others. "He was the gentlest, kindest man I think I have ever met," said John Green. "I have never known anyone who felt or expressed less ill will to any other person." Rarely did he refer even obliquely to his ordeal. But one day, when the conversation turned to President Kennedy's promise to send a man to the moon, someone asked him, "Do you think you'd like to go to the moon?" Robert replied, "Well, I sure know some people I'd like to send there."

Robert and Kitty spent more and more time on the island, often flying in for Easter week, Christmas and a good part of every summer. One Easter week they invited Robert's childhood friend Francis Fergusson to accompany them. Robert unfortunately caught a bad cold and spent most of the week curled up in bed. Kitty, however, acted the perfect hostess and took Fergusson on long walks on the beach, using her training as a botanist to point out the island's spectacular flora. Kitty always made a point of liking Robert's childhood friends, but on this occasion, Fergusson thought her behavior a tad bizarre. "She was trying to flirt with me," he recalled.

Kitty made a pretense of being a good cook, but this meant their meals had style but little substance. Robert had a fish pot in the bay and they ate a lot of seafood salad, octopus and barbecued shrimp. Like the natives, they chewed on raw whelks, a West Indian snail they could harvest from the beach. One Christmas dinner they served their guests champagne and Japanese seaweed. Robert ate practically nothing. "My God," Doris Jadan recalled, "if the man ate a thousand calories a day it was a miracle."

PETER SELDOM came to St. John; as a young man he preferred the rugged mountains of New Mexico. But Toni made the island her spiritual home. "She was very sweet," said one longtime resident. She took on native ways, and in time acquired a near-perfect command of West Indian Calypso, the Creole English common in the islands. She loved the island's steel band music. As a young adolescent, she was "a dead-serious child, with beautiful smooth features, tragic dark eyes, long lustrous dark hair, and the condescending politeness of a princess." Extremely shy, she hated to have her photograph taken. She told friends on St. John that she had always hated the popping flashes of cameras pointed at her whenever she traveled in public with her famous father. St. John was a perfect place for someone who so treasured her privacy.

"Toni was very pliable and very demure," recalled Inga Hiilivirta, who became a good friend. "Toni would do anything she was told to do. She rebelled later." Kitty depended upon her heavily, often treating her like a handmaid, asking her to fetch her cigarettes. Toni was always picking up after her mother, and inevitably, as a teenager she began to fight with her. "Toni and her mother were at each other's throats all the time," recalled Sis Frank.

One neighbor on St. John recalled that "Robert didn't pay too much attention to Toni. He was nice to her, but he just didn't pay too much attention to her. She could have been anybody's child." On the other hand, another neighbor, Steve Edwards, thought Robert had "a deep regard for his daughter . . . you could just tell he was proud of Toni." At seventeen, Toni struck most people as very bright, but also reserved, sensitive and gentle: a very old-fashioned family girl. For a time, Alexander Jadan, Ivan's

Russian-born son, pursued her. "Alex was crazy about Toni," recalled Sis Frank. But when Toni began to show a serious interest in Alex, Robert intervened, insisting that Alex was too old for her.

As a result of her friendship with the Jadans, Toni decided to study Russian in a serious way. An excellent linguist like her father, she majored in French, but by the time she finished Oberlin College, she could speak Italian, French, Spanish, German and Russian, which she used for her diary entries.

Robert, Kitty and Toni were all expert sailors—or "rag people," as the islanders called those who preferred sailboats to motorboats. They'd go off on sailing expeditions for three or four days at a time. One day Robert was sailing alone into Cruz Bay's tiny marina at sunset; with the brim of his old straw hat pulled low over his forehead, he failed to see the bow's breadth of another boat anchored in the harbor, and he crashed into it, demasting his own boat. Fortunately, no one was hurt, but thereafter it became a family joke to "keep your hat brim up when sailing into port."

Robert lived a casual life, sailing by day and entertaining a diverse group of island friends at night. Life on Hawksnest Bay could be dangerously primitive. Robert was alone one day when a wasp stung his hand just as he was pouring kerosene into a lantern. Startled, he dropped the jug; it shattered on the tile floor, driving a piece of the broken pottery into his right foot like a dagger. Robert extracted the shard, but by the time he limped down to the ocean to wash the blood away, he realized he could no longer move his big toe. His small sailboat was already rigged and anchored at the beach, so he decided to sail it around to Cruz Bay's clinic. When the doctor examined him, he discovered the pottery shard had cut cleanly through the tendon in his foot; no longer properly attached, the tendon had receded up into his leg. Robert suffered without complaint as the doctor retrieved the tendon, pulled it taut and stitched it back into place. "Out of your mind," the doctor admonished him. "Sailing across the bay . . . lucky you won't lose the whole foot."

After a morning of sailing or walking the beach, Robert would invite anyone he met to come over for drinks. He still served martinis, but they didn't seem to affect him. "I never saw Robert drunk," recalled Doris Jadan. The drinks would turn into dinner and Robert would often begin reciting poetry. In a low whisper of a voice, he would recite Keats, Shelley, Byron and sometimes Shakespeare. He loved *The Odyssey,* and had memorized long passages of it in translation. He had become the simple philosopher king, adored by his ragtag followers of expatriates, retirees, beatniks and natives. Despite his cultivated aura of otherworldliness, he fit comfortably into their island world. On St. John, the father of the atomic bomb had somehow found just the right refuge from his inner demons.

CHAPTER FORTY

"It Should Have Been Done the Day After Trinity"

I think it is just possible, Mr. President, that it has taken some charity and some courage for you to make this award today.

ROBERT OPPENHEIMER *to President Lyndon Johnson, December 2, 1963*

BY THE EARLY 1960S, with the return of Democrats to the White House, Oppenheimer was no longer a political pariah. The Kennedy Administration was not going to bring him back into government, but liberal Democrats nevertheless thought of him as an honorable man martyred by Republican extremists. In April 1962, McGeorge Bundy—the former Harvard dean and now national security adviser to President Kennedy—had Oppenheimer invited to a White House dinner honoring forty-nine Nobel laureates. At this gala affair, Oppie rubbed elbows with such other luminaries as the poet Robert Frost, the astronaut John Glenn and the writer Norman Cousins. Everyone laughed when Kennedy quipped, "I think this is the most extraordinary collection of talent, of human knowledge, that has ever been gathered together at the White House, with the possible exception of when Thomas Jefferson dined alone." Afterwards, Oppenheimer's old friend from his GAC days, Glenn Seaborg—now chairman of the AEC—asked if he would be willing to endure another hearing to get his security clearance reinstated. "Not on your life," Robert snapped.

Oppenheimer continued to give public lectures, most often in university settings, and usually he dwelled on broad themes related to culture and science. Since he had been deprived of any status associated with the government, the power of his persona now was entirely that of the public intellectual. He presented himself as a diffident humanist, pondering man's survival in an age of weapons of mass destruction. When the editors of

Christian Century asked him in 1963 to list some of the books that had shaped his philosophical outlook, Oppenheimer named ten. At the top of the list was Baudelaire's *Les fleurs du mal,* and then came the Bhagavad-Gita . . . and last was Shakespeare's *Hamlet.*

IN THE SPRING of 1963, Oppenheimer learned that President Kennedy had announced his intention to give him the prestigious Enrico Fermi Prize, a $50,000 tax-free award and medal for public service. Everyone understood that this was a highly symbolic act of political rehabilitation. "Disgusting!" cried one Republican senator when he heard the news. Republican staffers on the House Un-American Activities Committee circulated a fifteen-page summary of the 1954 security charges against Oppenheimer. On the other hand, the veteran CBS broadcaster Eric Severeid described Oppenheimer as "the scientist who writes like a poet and speaks like a prophet"—and approvingly suggested that the award signaled Oppenheimer's rehabilitation as a national figure. When reporters pressed Oppenheimer for his reaction, he demurred, saying, "Look, this isn't a day for me to go shooting my mouth off. I don't want to hurt the guys who worked on this." He knew his friends inside the Administration, McGeorge Bundy and Arthur Schlesinger, Jr., were no doubt responsible.

Edward Teller, who had received the same prize the year before, immediately wrote Oppenheimer his congratulations: "I have been tempted often to say something to you. This is the one time I can do so with full conviction and knowing that I am doing the right thing." Actually, many physicists had quietly campaigned to have the Kennedy Administration restore Oppenheimer's security clearance. They wanted a real vindication for their old friend, not merely a symbolic rehabilitation. But Bundy thought the political price too high. Indeed, even after the Administration announced that Oppenheimer would be given the Fermi Award, Bundy waited to gauge the response from Republicans before deciding that the president would personally award the prize in a White House ceremony.

On November 22, 1963, Oppenheimer was sitting in his office, working on a draft acceptance speech for the December 2 White House ceremony, when he heard knocking on his outer office door. It was Peter, who said that he had just heard on his car radio that President Kennedy had been shot in Dallas. Robert looked away. At that moment, Verna Hobson dashed in, exclaiming, "My God, did you hear?" Robert looked at her and said, "Peter just told me." When others arrived, Robert turned to Peter and asked his twenty-two-year-old son if he'd like a drink. Peter nodded, and Robert walked over to Verna's large walk-in closet, where he knew some liquor was kept. But then Peter observed that his father just stood there, "his arm hanging down by his side, fourth finger repetitively rubbing his thumb, gazing downward toward the little collection of liquor bottles." Finally, Peter mumbled, "Well, never mind then." As they walked out together, past his secretary's desk, Verna Hobson heard Robert say, "Now things are going to come apart very fast." Later, he told Peter that "nothing since Roosevelt's death had felt to him like that afternoon." For the next week, Oppenheimer, like much of the nation, sat in front of a television and watched the tragedy further unfold.

On December 2, President Lyndon Johnson went ahead with the Fermi Award ceremony, as scheduled. Standing next to Johnson's hulking figure in the Cabinet Room of the White House, Oppie seemed almost diminutive. He stood like a "figure of stone, gray, rigid, almost lifeless, tragic in his intensity." By contrast, Kitty was positively exultant, "a study in joy." David Lilienthal thought the whole affair "a ceremony of expiation for the sins of hatred and ugliness visited upon Oppenheimer. . . ." With Peter and Toni looking on, Johnson said a few words and then handed Robert a medal, a plaque and a check for $50,000.

In his acceptance speech, Oppenheimer mentioned that an earlier president, Thomas Jefferson, "often wrote of the 'brotherly spirit of science.' . . . We have not, I know, always given evidence of that brotherly spirit of science. This is not because we lack vital common or intersecting scientific interests. It is in part because, with countless other men and women, we are engaged in this great enterprise of our time, testing whether men can both preserve and enlarge life, liberty and the pursuit of happiness, and live without war as the great arbiter of history." And then he turned to Johnson and said, "I think it is just possible, Mr. President, that it has taken some charity and some courage for you to make this award today. That would seem to me a good augury for all our futures."

Johnson then responded with a gracious reference to Kitty as the "lady who shares honors with you today—Mrs. Oppenheimer." And then, to laughter, he quipped, "You may observe she got hold of the check!"

Teller was in the audience that day, and everyone watched with mounting tension as the two men came face to face. With Kitty standing stone-faced beside him, Oppenheimer grinned and shook Teller's hand. A *Time* magazine photographer caught the moment with his camera.

Afterwards, John F. Kennedy's grieving widow sent word that she wanted to see Robert in her private quarters. Robert and Kitty went upstairs and were greeted by Jackie Kennedy. She said she wanted him to know just how much her late husband had wanted to give him this award. Robert, in describing the moment later, confided that he had been deeply touched.

Oppenheimer, however, was still a polarizing figure in Washington. At least one Republican politician, Senator Bourke B. Hickenlooper, had publicly announced that he would boycott the White House ceremony, and in response to Republican criticism, the Johnson Administration agreed the following year to reduce the Fermi prize money to $25,000. Lewis Strauss, of course, was mortified by Robert's semi-rehabilitation, and wrote an angry letter to *Life* magazine, suggesting that the award to Oppenheimer had "dealt a severe blow to the security system which protects our country. . . ."

Strauss' enmity toward Oppenheimer had only deepened since the 1954 trial. And then all the old wounds had been reopened in 1959, when President Eisenhower nominated Strauss as his commerce secretary. In the bitter confirmation battle, in which the Oppenheimer hearing was a central factor, Strauss narrowly lost, by a vote of 49–46. Strauss correctly blamed Senator Clinton Anderson, and then Senator John F. Kennedy—who had been lobbied by Oppenheimer defenders like McGeorge Bundy and Arthur Schlesinger, Jr. When Kennedy protested, "It would require an extreme case to vote against the president," Mac Bundy responded, "Well, this is an extreme case." Bundy laid out for Kennedy Strauss' reprehensible conduct in the Oppenheimer case. Convinced, Kennedy switched his vote and Strauss lost the confirmation. "It's a lovely show—never thought I'd live to see my revenge," cabled Bernice Brode to Oppie. "In unchristianly spirit, enjoy every squirm and anguish of victim. Having wonderful time—wish you were here!" Even seven years later, Strauss thought he saw Oppenheimer's influence at work, complaining that "Oppenheimer's partisans are continuing their reprisals against individuals who did their duty." The case would follow both Strauss and Oppenheimer to their graves.

EVEN AFTER Robert won the Fermi award, Kitty's resentments against Teller and others remained unshakable. One late afternoon in the spring of 1964, she and Robert had drinks with David Lilienthal. Robert had just recovered from a terrible bout of pneumonia; he had finally given up cigarettes but still smoked a pipe. He and Kitty had aged. Robert still wore his signature flat porkpie hat and he drove around Princeton in a Cadillac convertible that had seen better days. When Lilienthal remarked that the last time he had seen them had been at the White House Fermi award ceremony, Kitty's dark eyes smoldered. "That was awful," she snapped, "there were some awful things about it." Robert sat there with his head bowed and murmured softly, "There were some very sweet things said." But a moment later Robert lost his "kindly, almost rabbinical posture" when Teller's name was mentioned, and his eyes flashed with real anger. The wounds, Lilienthal noted, were "still sore." Lilienthal completed his diary entry with the observation that "She [Kitty] burns with an intensity of feeling one rarely sees,

mostly with a deep resentment against all those who had any part in the torture Robert had to undergo."

For a man who had been so politically engaged in the 1930s and '40s,
Oppenheimer was oddly disconnected from the turmoil of the 1960s. At the
beginning of the decade, as many Americans dug atomic bomb shelters in
their backyards, Oppenheimer never spoke out against such hysteria. When
pressed by Lilienthal, he explained, "There is nothing I can do about what is
going on; I would be the worst person to speak out about them in any case."
Similarly, as the Vietnam War escalated in 1965–66, he had nothing to say
in public—though privately, when he discussed it with Peter, it was evident
that he was skeptical of the Administration's escalating commitment.

IN 1964, Oppenheimer received an advance copy of a book with a startling new interpretation of the decision to use the bomb on Hiroshima.
Using such newly opened archival sources as former secretary of war Henry
L. Stimson's diaries and State Department materials related to former secretary of state James F. Byrnes, Gar Alperovitz argued that atomic diplomacy
against the Soviet Union was a factor in President Truman's decision to use
the bomb against a Japanese enemy that appeared to be defeated militarily.
*Atomic Diplomacy: Hiroshima and Potsdam: The Use of the Atomic Bomb
and the American Confrontation with Soviet Power* created a storm of controversy. When Alperovitz asked for his comments, Oppenheimer wrote
him that much of what he had written had "been largely unknown to
me. . . ." He pointedly added, however, "[B]ut I do recognize your Byrnes,
and I do recognize your Stimson." He would not be drawn into the controversy over the book—but clearly, as with P. M. S. Blackett's 1948 book
Fear, War and the Bomb, he still thought the Truman Administration had
used atomic weapons on an enemy already essentially defeated.

That same year, a German playwright and psychiatrist, Heinar Kipphardt, wrote a play, *In the Matter of J. Robert Oppenheimer.* Drawing heavily from the transcripts of the 1954 security board hearing, Kipphardt's
drama was first shown on German television and then produced for live theater audiences in West Berlin, Munich, Paris, Milan, and Basel. These
European audiences were mesmerized by Kipphardt's portrayal of Oppenheimer standing frail and lean before his accusers, like a modern Galileo, a
scientist-hero martyred by the authorities in America's anticommunist
witch-hunt. Acclaimed by reviewers, the drama won five major awards.

But when Oppenheimer finally read the script, he so disliked it that he
wrote Kipphardt an angry letter threatening legal action. (Strauss and Robb,
who followed the reviews of the play closely, also briefly considered suing

the Royal Shakespeare Company in London for defamation—but their lawyers persuaded them they didn't have a case.) Oppenheimer particularly disliked the play's concluding monologue, where the playwright had him expressing guilt for having built the atomic bomb: "I begin to wonder whether we were not perhaps traitors to the spirit of science. . . . We have been doing the work of the Devil. . . ." Such melodrama somehow cheapened the character of his ordeal. In short, he thought the script poor drama precisely because it lacked ambiguity.

Audiences disagreed. In October 1966, a British production opened in London, with the actor Robert Harris playing the role of Oppenheimer, and became wildly popular. A British reviewer wrote that the drama "causes one furiously to think." Harris wrote Oppenheimer to report that "audiences have been attentive and enthusiastic—especially the young ones—which both surprised and pleased us."

Oppenheimer later grudgingly agreed that the playwright was guilty of nothing more than dramatic license. He liked a French production of Kipphardt's drama better because it drew almost exclusively from the security hearing transcripts—but even then, he complained that both productions "turned the whole damn farce into a tragedy." Whatever its merits, Kipphardt's play reintroduced Oppenheimer to a new generation of European and American audiences. The play eventually premiered in New York and inspired a BBC TV docudrama and other film renderings of Oppenheimer's life.

There were other media projects that attempted to delve into Oppenheimer's life. In 1965, on the twentieth anniversary of the Hiroshima bombing, NBC television aired a documentary, *The Decision to Use the Atomic Bomb,* narrated by Chet Huntley, which featured Robert's recollection of the July 16 Trinity test and his recitation from the Bhagavad-Gita: "Now I am become Death, the Destroyer of Worlds." On another occasion, when an interviewer asked him on camera what he thought of Senator Robert Kennedy's recent proposal that President Johnson initiate talks with the Soviet Union to halt the proliferation of nuclear weapons, Oppenheimer puffed hard on his pipe and said, "It's twenty years too late. . . . It should have been done the day after Trinity."

Around this time, Oppenheimer learned that a well-connected and sympathetic journalist, Philip M. Stern, was working on a book about his 1954 security hearing. But even though mutual friends vouched for Stern, Oppenheimer decided not to be interviewed. "The subject of the book," he explained, "is one on which I do not manage to have a total sense of detachment, and on which I have very large and central areas of ignorance. I cannot think of a more poisonous brew." Stern would write a better book, he thought, "without my collaboration, suggestions, or implied approval." Stern's book, *The Oppenheimer Case: Security on Trial,* was published in 1969 to critical acclaim.*

IN THE SPRING of 1965, Oppenheimer was gratified to see the completion of a new library for the Institute. It was built adjacent to a large artificial pond and surrounded by acres of green lawn, and Robert regarded it as one of his legacies. Designed by Wallace Harrison—the same architect who had designed his St. John beach cottage—the library had an innovative roof that used glass louvers set at an angle. In daytime, this provided ample sunlight. But at night, the library's electric lighting shone upward. From a distance the whole sky seemed to be lit up by a great fire. When David Lilienthal praised the beauty of the new library's setting and the spectacle it created at night, Robert gave him a "little-boy grin," and said, "The library is beautiful, and the setting. It is also an illustration of how we don't anticipate the most obvious consequences. This happened to us in a major way with the bomb in Los Alamos. As for the ceiling for the library, we wanted the best light, the light in just the right way. . . . In the daylight it turned out to be wonderful. But no one, not one of us, foresaw that not only would light come in, but it would go out—into the sky."

His pleasure with the new library only partly compensated for his ongoing clashes with various members of the mathematics faculty. The Institute's petty politics sometimes provoked him to angry outbursts. "The trouble is that Robert loves controversy," reported one trustee to Lewis Strauss, "and essentially hates people. He ought to be asked to leave." Strauss relished such reports, but he still lacked the votes to oust Oppenheimer.

But then, in the spring of 1965, Oppenheimer told the Institute's trustees that he had resolved that the time had come for him to resign, and he suggested that he should leave in June 1966, at the end of that academic year. Strauss was present to hear the news. Oppenheimer gave three reasons for his decision. First, he was just two years away from the statutory retirement age of sixty-five, and there was no point in "simply waiting for the bell to toll." Second, he explained that Kitty had been "suffering from an illness which the doctors have pronounced incurable. . . ." (In his memo for his files, Strauss wickedly labeled Kitty's disease "dipsomania"—the uncon-

---

*Stern's book remains the most complete account of the Oppenheimer security hearing. Other good treatments include John Major, *The Oppenheimer Hearing* (New York: Stein & Day, 1971); Barton J. Berstein, "The Oppenheimer Loyalty-Security Case Reconsidered," *Stanford Law Review* 42 (July 1990): 1383–1484; and Charles P. Curtis, *The Oppenheimer Case: Trial of a Security System* (New York: Chilton, 1964).

trollable craving for alcohol.) Robert said that this was now making it impossible for him to entertain visitors or members of the faculty. Third, he said his relations with some members of the faculty, particularly in the mathematics faculty, were "intolerable and worsening."

Robert had wanted to make this decision public later that year, perhaps in the autumn, but that very night he had some faculty members over for dinner and Kitty spilled the beans. Since the news was now bound to leak, the trustees quickly drafted a press release and the story appeared in newspapers around the country on Sunday morning, April 25, 1965.

Oppenheimer had few regrets about leaving. But one was the fact that he would have to move out of Olden Manor, his and Kitty's home for nearly two decades. Robert consoled himself that the trustees had voted to build a new house for him on the grounds of the Institute—or otherwise provide them housing. The Oppenheimers had hired an architect, Henry A. Jandel, and created a model of the new home, a modern glass-and-steel one-story structure to be built on a lot two hundred yards down the road from Olden Manor. But in what can only be described as a characteristic act of personal vengeance, Strauss used his still considerable influence as a trustee to block the project. On December 8, 1965, Strauss told his fellow trustees that he took a "dim view" of these plans. It was a "mistake," he argued, to have Oppenheimer living on campus, let alone next door to Olden Manor. Another trustee, Harold K. Hochschild, interrupted to say that "even Princeton was too close." In short order, Strauss persuaded the trustees to rescind their promise. When Oppenheimer was informed the next day, he was "enraged." If that was the board's firm decision, he said, he would leave Princeton altogether. If Robert was understandably angered, a furious Kitty vented her outrage on another trustee and his wife, who reported to Strauss that "a very unpleasant conversation had ensued." Strauss kept his hand invisible in all this, leaving the Oppenheimers only with their suspicions. That was how things stood in December. But by February 1966, Oppenheimer somehow persuaded the trustees to reverse themselves yet again. To Strauss' disgust, Oppenheimer was allowed to build the house on the site he wanted. Construction began in September 1966 and the house was completed the following spring. But he would never live in it.

IN THE autumn of 1965, Oppie visited his doctor for a physical checkup. It was not something he did very often, but he came home that day and announced that he had been given a clean bill of health. "I am going to outlive every one of you," he said jocularly. But two months later, his smoker's cough became noticeably worse. In St. John that Christmas, he complained to Sis Frank of a "terrible sore throat," and mused, "Maybe I'm smoking too much." Kitty thought he just had a bad cold. Finally, in February 1966, she took him to a doctor in New York. The diagnosis was clear and devastating. Kitty phoned Verna Hobson with the news: "Robert has cancer," she whispered.

Four decades of heavy tobacco smoke had taken its toll on his throat. When Arthur Schlesinger, Jr., heard the "dreadful news," he immediately wrote him, "I can only dimly imagine how hard these next months will be for you. You have faced more terrible things than most men in this terrible age, and you have provided all of us with an example of moral courage, purpose and discipline."

Though no longer a chain-smoker, Oppenheimer was still seen puffing on his pipe. In March, he underwent a painful and inconclusive operation on his larynx—and then he began receiving cobalt radiation therapy at the Sloan-Kettering Institute in New York. He talked quite candidly about his cancer with friends. He told Francis Fergusson that he had a "faint hope that it could be stopped where it was." By late May, however, all could see that he was "wasting away."

On a beautiful spring day in 1966, Lilienthal went by Olden Manor and found Anne Marks, Robert's Los Alamos secretary, visiting the Oppenheimers. Lilienthal was shocked by Robert's appearance. "For the first time Robert himself is 'uncertain about the future,' as he says, so white and— scared." Walking alone with Kitty around the garden, Lilienthal asked her how he was getting along. Kitty froze, biting her lip; uncharacteristically, she seemed at a loss for words. When Lilienthal bent down and gently kissed her cheek, she uttered a deep moan and began to cry. A moment later, she straightened up, wiped her tears away and suggested they should go back inside and join Anne and Robert. "I have never admired the strength of a woman more," Lilienthal noted in his diary that evening. "Robert is not only her husband, he is her past, the happy past and the tortured one, and he is her hero and now her great 'problem.'"

In June 1966, Robert accepted an honorary degree at Princeton's commencement, where he was hailed as a "physicist and sailor, philosopher and horseman, linguist and cook, lover of fine wine and better poetry." But he appeared exhausted and spent; suffering from a pinched nerve, he couldn't walk without a cane and a leg brace.

Frail and clearly battered by his illness, Robert nevertheless somehow seemed to grow in stature. Freeman Dyson observed that "his spirit grew stronger as his bodily powers declined. . . . He accepted his fate gracefully; he carried on with his job; he never complained; he became quite suddenly simple and no longer trying to impress anybody." He had been a man with a talent for self-dramatization, but now, Dyson noticed, "he was simple, straightforward, and indomitably courageous." At times, Lilienthal noted, Robert seemed "vigorous and almost gay."

In mid-July, his doctor found no traces of the malignancy in his throat. The radiation treatment had tired him out, but it appeared to have done the job. So on July 20, he and Kitty returned to St. John. Friends on the island who had not seen him in a year thought he looked like a "ghost, an absolute ghost." He quietly complained that while he wanted to go swimming, the always warm waters around St. John now made him feel cold. Instead he managed some walks along the beach and was courteous and patient with everyone whom he met—even strangers. Learning that Sis Frank's husband, Carl, was recuperating from a serious heart operation, Robert went to visit him. "Robert was so kind to him," recalled Sis, "trying to get him over this terrible trauma."

Robert was on a liquid diet at that point, supplemented with protein powder. He told Sis Frank, "You don't know what I would give you if I could have that chicken salad sandwich." Invited to dinner at Immu and Inga Hiilivirta's new home, Robert couldn't eat the lamb chops and managed to get down only a glass of milk. "I felt very sorry for him," Inga said.

After nearly five weeks, he and Kitty returned to Princeton in late August. Robert felt better. He still had a sore throat, but he thought himself stronger. His doctors again examined his throat and found no trace of cancer. "They were, in fact, convinced that I was cured," Oppenheimer wrote one friend. After only five days back in Princeton, he flew out to Berkeley and spent a week seeing old friends. Upon his return in September, he complained to his doctors of continued soreness, "but they were not very thorough and attributed my discomfort to radiation. . . ."

Early that autumn, the Oppenheimers had to move out of their beloved Olden Manor to make way for the Institute's new director, Carl Kaysen. Temporarily, Robert and Kitty decided to move into a house at 284 Mercer Road formerly occupied by the physicist C. N. Yang. Unoccupied for some years, it was a rather dreary place. Their neighbors were Freeman and Imme Dyson. The Dysons' young son, George, recalled growing up on the grounds of the Institute during the years of Oppenheimer's directorship: "He [Oppenheimer] was a very, very strong presence—a benevolent but mysterious ruler of the world in which we lived." But when Oppenheimer moved next door, "To us children he seemed like a ghost, deprived of his kingdom, pacing around the yard next door, very pale and thin."

Robert didn't see his doctor again until October 3. "By then," Oppenheimer wrote "Nico" Nabokov, his friend from the Congress for Cultural Freedom, "the cancer was very manifest and had spread into the palate, the base of the tongue, and the left Eustachian tube." It was not operable, and so his doctors prescribed thrice-weekly radiation treatments, this time with a betatron: "Everybody knows that reradiation with a still ulcerated throat is no great joy. It is not bad yet, but I cannot be very sure of the future."

He faced the prospect of an early death with resignation. In mid-October, Lilienthal dropped by and learned the news. Robert's once brilliant blue eyes now seemed bleary with pain. "The last mile for Robert Oppenheimer," Lilienthal wrote in his diary afterwards, "and it may be a very short one. . . . Kitty had all she could do to suppress the tears." In November, Robert wrote a friend, "I am much less able to speak and eat now." He had hoped to visit Paris in December, but his doctors insisted they wanted to continue with regular radiation treatments until Christmas. Instead, he stayed at home, seeing such old friends as Francis Fergusson and Lilienthal. Early in December, Frank visited from Colorado.

In early December 1966, Oppenheimer heard from his former student, David Bohm, who had spent most of his career in Brazil and later, England. Bohm wrote to say that he had seen the Kipphardt play and a television program on Los Alamos in which Oppenheimer had been interviewed. "I was rather disturbed," Bohm wrote, "especially by a statement you made, indicating a feeling of guilt on your part. I feel it to be a waste of the life that is left to you for you to be caught up in such guilt feelings." He then reminded Oppenheimer of a play by Jean-Paul Sartre "in which the hero is finally freed of guilt by recognizing responsibility. As I understand it, one feels guilty for past actions, because they grew out of what one *was* and still *is*." Bohm believed that mere guilt feelings are meaningless. "I can understand that your dilemma was a peculiarly difficult one. Only you can assess the way in which you were responsible for what happened. . . ."

Oppenheimer replied promptly: "The play and such things have been rattling around for a long time. What I have never done is to express regret for doing what I did and could at Los Alamos; in fact, on varied and recurrent occasions, I have reaffirmed my sense that, with all the black and white, that was something I did not regret." And then, in words he edited out before mailing the letter, he wrote, "My principal remaining disgust with Kipphardt's text is the long and totally improvised final speech I am supposed to have made, which indeed affirms such regret. My own feelings about responsibility and guilt have always had to do with the present, and so far in this life that has been more than enough to occupy me."

Oppenheimer may well have had this exchange with Bohm in mind when Thomas B. Morgan—a *Look* magazine journalist—dropped by to interview him at his Institute office in early December. Morgan found him gazing at the autumn woods and the pond outside his window. On the wall of his office there now hung an old photograph of Kitty jumping her horse gracefully over a fence. Morgan could see that he was dying. "He was very frail and no longer the lean, lank man who impressed you as a cowboy genius. There were deep lines in his face. His hair was hardly more than a white mist. And yet, he prevailed with that grace." As their conversation turned philosophical, Oppenheimer stressed the word "responsibility"— and when Morgan suggested he was using the word in an almost religious sense, Oppenheimer agreed it was a "secular device for using a religious notion without attaching it to a transcendent being. I like to use the word 'ethical' here. I am more explicit about ethical questions now than ever before—although these were very strong with me when I was working on the bomb. Now, I don't know how to describe my life without using some word like 'responsibility' to characterize it, a word that has to do with choice and action and the tension in which choices can be resolved. I am not talking about knowledge, but about being limited by what you can do. . . . There is no meaningful responsibility without power. It may be only power over what you do yourself—but increased knowledge, increased wealth, leisure are all increasing the domain in which responsibility is conceivable."

After this soliloquy, Morgan wrote, "Oppenheimer then turned his palms up, the long, slender fingers including his listener in his conclusion. 'You and I,' he said, 'neither of us is rich. But as far as responsibility goes, we are both in a position right now to alleviate the most awful agony in people at the starvation level.'"

This was only a different way of saying what he had learned from reading Proust forty years earlier in Corsica: that "indifference to the sufferings one causes . . . is the terrible and permanent form of cruelty." Far from being indifferent, Robert was acutely aware of the suffering he had caused others in his life—and yet he would not allow himself to succumb to guilt. He would accept responsibility; he had never tried to deny his responsibility. But since the security hearing, he nevertheless no longer seemed to have the capacity or motivation to fight against the "cruelty" of indifference. In that sense, Rabi had been right: "They achieved their goal. They killed him."

On January 6, 1967, Robert's doctor told him that the radiation therapy was proving ineffective against his cancer. The next day, he and Kitty had some friends over for lunch, including Lilienthal. They served a very expensive goose liver, and Kitty acted like a perfect hostess. But as Lilienthal was leaving, Robert helped him with his coat and confided, "I don't feel very gay; the doctor gave us bad news yesterday." Kitty then walked Lilienthal outside the house and suddenly broke down into sobs. "Impending death is no new story," Lilienthal recorded that evening, "but this is one that seems so wasteful and cruel. But Robert, in my presence at least, looks at it with those eyes of the doomed, that seem to look inward, rigid, caught up in the final reality."

On January 10 he wrote Sir James Chadwick, a friend from the Los Alamos years, to acknowledge that he was "battling a cancerous throat . . . with only indifferent success." He added, "It reminds me of the virulent strictures of Ehrenfest on the evils of smoking. We did live in a lucky time, didn't we, to have even our critics so full of love and light?"

One day late that January, Robert called in his secretary of fourteen years, Verna Hobson, and gently encouraged her to leave Princeton. Hobson had intended to retire when he stepped down as director. But she had delayed, knowing that he was sick and that Kitty was still very dependent on her. "I knew what he was saying was that he was dying soon," said Hobson, "and that if I didn't go then, it would be so difficult for me to leave Kitty that I'd never make it."

By mid-February 1967, Robert knew the end was near. "I am in some pain . . . my hearing and my speech are very poor," he wrote a friend. His doctors had decided he couldn't take any more radiation, so they ordered a strong regimen of chemotherapy. But he remained at home, and sent word to a few friends that he would welcome a visit. Nico Nabokov came by the house repeatedly and urged other friends to visit Robert.

On Wednesday, February 15, Robert made a supreme effort to attend a committee meeting at the Institute to select the candidates for the following year's visiting fellows. It was the last time Freeman Dyson saw him. But like everyone else, Oppenheimer had done his homework, reading through scores of applications. "He could speak only with great difficulty," Dyson later wrote, but he nevertheless "remembered accurately the weak or strong points of the various candidates. The last words I heard him say were, 'We should say yes to Weinstein. He is good.' "

The next day, Louis Fischer dropped by. In recent years, Fischer and Oppenheimer had become casual, respectful friends. An acclaimed, globe-trotting journalist, Fischer was the author of more than two dozen books—including such popular volumes as *The Life of Mahatma Gandhi* (1950) and *The Life and Death of Stalin* (1953). Robert particularly liked his 1964 biography of Lenin. Kitty had encouraged Fischer to bring along some chapters from his current book project to divert Robert.

But when Fischer rang the doorbell, he waited in silence for several minutes—and, giving up, started to walk away when he heard knocking on an upstairs window. Looking up, he saw Robert motioning for him to return. A moment later, Robert opened the front door. He had lost much of his hearing and so hadn't heard the ringing doorbell. Robert tried awkwardly to help Fischer out of his coat, and then the two friends sat down on opposite sides of a bare table. Fischer remarked that he had recently talked with Toni, who was using her Russian-language skills to do some research for George Kennan. When Robert tried to talk, "he mumbled so badly that I suppose I understood about one word out of five." But he managed to convey that Kitty was napping—she had been sleeping badly at night—and no one else was in the house.

When Fischer handed Robert two chapters of his manuscript, he began reading a few pages and asked a question about Fischer's source material. "From Berlin?" he said. Fischer pointed to a footnote on the page. "He gave me a very sweet smile at this point," Fischer later wrote. "He looked extremely thin, his hair was sparse and white, and his lips were dry and cracked. As he read, and at other times too, he kept moving his lips as if to speak but did not speak, and, probably realizing that this made a bad impression, he held his bony hand in front of his mouth; his fingernails were blue."

After some twenty minutes, Fischer thought it time for him to leave. On his way out, he spotted a packet of cigarettes lying on the second step of the stairs leading to the second floor. Three cigarettes had fallen out of the pack and were lying on the carpet nearby, so Fischer reached down to put them back in the pack. When he stood up, Robert was by his side; reaching into his pocket, he brought out a lighter and lit it. He knew Fischer didn't smoke and was on his way out of the house, but the gesture was instinctive. He had always been the first to light a guest's cigarette. "I have a strong impression," Fischer wrote a few days later, "that he knew his mind was failing and that he probably wanted to die." After insisting on helping Fischer with his coat, Robert opened the door and said with a thick tongue, "Come again."

Francis Fergusson dropped by the house on Friday, February 17. He could see that Robert was pretty far gone. He could still walk, but he now weighed under a hundred pounds. They sat together in the dining room, but after a short time, Fergusson thought Robert looked so feeble that he ought to take his leave. "I walked him into his bedroom, and there I left him. And the next day I heard that he had died."

Robert died in his sleep at 10:40 p.m. on Saturday, February 18, 1967. He was only sixty-two years old. Kitty later confided to a friend, "His death was pitiful. He turned into a child first, then an infant. He made noises. I couldn't go into the room; I had to go into the room, but I couldn't. I couldn't bear it." Two days later his remains were cremated.

LEWIS STRAUSS sent Kitty a cable, claiming that he was "grieved at the news of Robert's death. . . ." Newspapers at home and abroad published long, admiring obituaries. The *Times* of London described him as the quintessential "Renaissance man." David Lilienthal told the *New York Times*: "The world has lost a noble spirit—a genius who brought together poetry and science." Edward Teller had less fulsome remarks: "I like to remember that he did a magnificent job and a very necessary job . . . in organizing [the Los Alamos Laboratory]." In Moscow, the Soviet news agency Tass reported the death of an "outstanding American physicist." *The New Yorker* remembered him as "a man of exceptional physical elegance and grace, an aristocrat with an enduring touch of the intellectual bohemian about him." Senator Fulbright gave a speech on the floor of the Senate, and said of the late physicist, "Let us remember not only what his special genius did for us; let us also remember what we did to him."

After the memorial service in Princeton on February 25, 1967, Oppenheimer was memorialized once again in the spring at a special session of the American Physical Society in Washington. Isidor Rabi, Bob Serber, Victor Weisskopf and several others spoke. Rabi later wrote an introduction for the speeches, which were subsequently collected and published in book form. "In Oppenheimer," he wrote, "the element of earthiness was feeble. Yet it was essentially this spiritual quality, this refinement as expressed in speech and manner, that was the basis of his charisma. He never expressed himself completely. He always left a feeling that there were depths of sensibility and insight not yet revealed."

KITTY TOOK her husband's ashes in an urn to Hawksnest Bay, and then, on a stormy, rainy afternoon, she, Toni and two St. John friends, John Green and his mother-in-law, Irva Clair Denham, motored out toward Carval Rock, a tiny island in sight of the beach house. When they got to a point in between Carval Rock, Congo Cay and Lovango Cay, John Green cut the motor. They were in seventy feet of water. No one spoke, and instead of scattering Robert's ashes into the sea, Kitty simply dropped the urn overboard. It didn't sink instantly, so they circled the boat around the bobbing urn and watched silently until it finally disappeared below the choppy sea. Kitty explained that she and Robert had discussed it, and "That's where he wanted to be."

Epilogue:

"There's Only One Robert"

Within a year or two of Oppie's death, Kitty began living with Bob Serber, Robert's close friend and former student. When a friend mistakenly called Serber "Robert," Kitty reprimanded her sharply: "Don't you call him Robert—there's only one Robert." In 1972, Kitty bought a magnificent fifty-two-foot teak ketch, christened the Moonraker. The name refers to the top-most sail on a large sailing vessel—or to someone touched with madness. Kitty persuaded Serber to sail with her around the world in May 1972. But they didn't make it very far. Off the coast of Colombia, Kitty became so ill that Serber turned the boat around and made for port at Panama. Kitty died of an embolism on October 27, 1972, in Panama City's Gorgas Hospital. Her ashes were scattered near Carval Rock, in the same spot off the coast of St. John where Robert's urn had been sent to the sea's bottom in 1967.

In 1959, ten years after his banishment, Frank Oppenheimer finally made it back into academia when the University of Colorado gave him an appointment in the physics department. In 1965, he won a prestigious Guggenheim Fellowship to do bubble chamber research at University College in London. While in Europe that year, he and Jackie visited a number of science museums; they were particularly impressed by the Palais de la Découverte, which used models to demonstrate basic scientific concepts. Upon their return to America, he and Jackie began to develop plans for a science museum that would give children and adults a "hands-on" experience with physics, chemistry and other scientific fields. The idea took hold, and in August 1969, with grants from various foundations, Frank and Jackie Oppenheimer's Exploratorium opened its doors on the grounds of San Francisco's renovated Palace of Fine Arts, a monumental exhibition hall built in 1915. The Exploratorium quickly became a showcase in the "participatory museum movement," and Frank became its charismatic director. Jackie and their son Michael worked closely with Frank, and the museum became a family endeavor—and possibly the world's most interesting pedagogical museum of science.

Robert would have been proud of Frank. Everything the two brothers had learned in two lives devoted to science, art and politics was brought together in the Exploratorium. "The whole point of the Exploratorium," Frank said, "is to make it possible for people to believe they can understand the world around them. I think a lot of people have given up trying to comprehend things, and when they give up with the physical world, they give up with the social and political world as well. If we give up trying to understand things, I think we'll all be sunk." If Frank ran his Exploratorium as a "benevolent despot" until his death in 1985, it was always with the egalitarian notion that "human understanding will cease to be an instrument of power . . . for the benefit of a few, and will instead become a source of empowerment and pleasure to all."

Peter Oppenheimer moved to New Mexico, living in his father's Perro Caliente cabin overlooking the Sangre de Cristo Mountains. Over the years, he raised three children. Twice divorced, he eventually settled in Santa Fe, and made a living as a contractor and carpenter. Peter never advertised his familial connections to the father of the atomic bomb—even when he occasionally went canvassing door-to-door as an environmental activist, lobbying against nuclear waste hazards in the region.

After her father's death, Toni floundered. "Toni always felt inferior to Kitty," recalled Serber. "Kitty managed her life so much that Toni never became independent." Her strong-willed mother had pressured her into going to graduate school, but after a while she dropped out. She lived alone in a small apartment in New York City for a time, but she had few close friends. Eventually she moved out of her apartment and lived in a back room of Serber's large Riverside Drive apartment. Using her facility for languages, she got a temporary job in 1969 as a trilingual translator for the United Nations. "She could shift from one language to another without any problem whatsoever," recalled Sabra Ericson. "But somehow or other, she was always getting slapped in the face." The position required a security clearance. The FBI opened a full field investigation—and dredged up all the old charges about her father. In what must have been a painful and ironic blow to a tender ego, the security clearance never came through.

Toni eventually returned to St. John, resigned to making the island her home. "She made the mistake of staying on St. John," Serber said. "I mean, it's so limited. There was nobody there she could talk to, really . . . nobody her own age." Twice married and twice divorced, Toni enjoyed only fleeting happiness. Denied her chosen career by the FBI, she never seemed to recover her footing.

After her second divorce, she became good friends with another recent arrival on the island, June Katherine Barlas, a woman eight years older. With Barlas and others, Toni rarely talked about her parents. "But when she did mention her father," recalled Barlas, "it was always lovingly." She often wore a ponytail holder that had been given to her by Robert—and she'd become very upset if she ever misplaced it. She avoided discussing the 1954 hearing, other than to say on occasion "that those men had destroyed her father."

But clearly, she still had issues with her parents. For a time, she saw a psychiatrist in St. Thomas, and she told her friend Inga Hiilivirta that this experience had helped her to understand "her resentment toward her parents from the way she had been treated as a young child." She suffered from fits of depression. One day, determined to drown herself, she started swimming out from Hawksnest Bay toward Carval Rock, where Robert's ashes rested on the sea bottom in an urn. She swam for a long time straight out across the ocean—and then, as she later confided to a friend—she suddenly felt better and turned back to shore.

On a Sunday afternoon in January 1977, she hanged herself in the beach cottage Robert had built on Hawksnest Bay. Her suicide was clearly premeditated. On her bed Toni had left a $10,000 bond and a will deeding the house to "the people of St. John." She was beloved throughout the island. "Everybody loved her," Barlas said, "but she didn't know that." Hundreds came to the funeral—so many, in fact, that scores had to stand outside the small church in Cruz Bay.

The cottage on Hawksnest Bay is now gone, swept away by a hurricane, but in its place is a community house standing on what is now called Oppenheimer Beach.

Author's Note and Acknowledgments

"My Long Ride with Oppie"

BY MARTIN J. SHERWIN

ROBERT OPPENHEIMER was an accomplished horseman, and so it was not entirely bizarre that in the summer of 1979 I sought to give new meaning to the scholarly concept of *Sitzfleisch* (sitting flesh) by starting my research for his biography on horseback. My adventure began at the Los Pinos Ranch, located ten miles above Cowles, New Mexico, from which in the summer of 1922, Oppie had first explored the beautiful Sangre de Cristo Mountains. I had not ridden for decades and, to say the least, the prospect of the long ride ahead—actually and metaphorically—was daunting. My destination, several hours by horseback from Los Pinos, over the 10,000 foot summit of Grass Mountain, was the "Oppenheimer ranch," Perro Caliente, the spare cabin on 154 acres of spectacular mountainside that Oppie had leased in the 1930s and purchased in 1947.

Bill McSweeney, the owner of Los Pinos, was our trail guide and local historian. Among other things, he told us (my wife and children were with me) about the tragic death—during a burglary of her Santa Fe home in 1961—of Oppie's good friend, Katherine Chaves Page, the ranch's previous owner. Oppie had met Katherine during his first visit to New Mexico and his youthful infatuation with her was one of the strong inducements repeatedly pulling him back to this beautiful country. After purchasing his own ranch, Oppie rented several of Katherine's horses each summer, for himself, his younger brother, Frank (and, after 1940, his wife, Kitty), and their stream of guests, mostly physicists who had never mounted anything more independent-minded than a bicycle.

My trip had two purposes. The first was to share in a small way the experience that Oppie had so often shared with his friends, the liberating joy of riding on horseback through this awesome wilderness. The second purpose was to talk with his son, Peter, who was living in the family cabin.

As I helped him build a corral, we talked for over an hour about his family and his life. It was a memorable beginning.

A few months earlier, I had signed a contract with the publisher Alfred A. Knopf for a biography of Robert Oppenheimer—physicist, founder in the 1930s of America's leading school of theoretical physics, erstwhile political activist, "father of the atomic bomb," prominent government adviser, director of the Institute for Advanced Study, public intellectual and the most prominent victim of the McCarthy era. The manuscript would be completed in four or five years, I assured my then editor, Angus Cameron, who is one of the dedicatees of this book.

During the next half-dozen years I traveled across the country and abroad, propelled from introduction to introduction, conducting many more interviews with those who had known Oppenheimer than I had imagined possible. I visited scores of archives and libraries, gathered tens of thousands of letters, memoranda and government documents—10,000 pages from the FBI alone—and eventually came to understand that any study of Robert Oppenheimer must necessarily encompass far more than his own life. His personal story, with all its public aspects and ramifications was more complicated, and shed vastly more light on the America of his day, than either Angus or I had anticipated. It is an indication of this complexity, this depth and wider resonance—of Oppenheimer's iconic standing—that since his death, his story has taken on a new life, as books, movies, plays, articles and now an opera (Dr. Atomic), have etched his shadow ever more sharply on the pages of American and world history.

Twenty-five years after I started out on that ride to Perro Caliente, the writing of Oppenheimer's life has given me a new understanding of the complexities of biography. It has been sometimes an arduous journey but always an exhilarating one. Five years ago, soon after my good friend Kai Bird completed *The Color of Truth,* a joint biography of McGeorge and William Bundy, I invited him to join me. Oppenheimer was big enough for both of us and I knew my pace would be quicker with Kai as my partner. Together we have finished what turned out to be a very long ride.

We both have many people who shared our journey and nurtured the dream of this book. Another worthy dedicatee of *American Prometheus* is the late Jean Mayer, president of Tufts University, a man whom I deeply admired. In 1986, Mayer appointed me the founding director of the Nuclear Age History and Humanities Center (NAHHC), an organization devoted to the study of the dangers associated with the nuclear arms race that Oppenheimer had confronted. Oppenheimer's life story also inspired the Global Classroom project, an American-Soviet program that from 1988 to 1992 connected students at universities in Moscow and Tufts University to discuss the nuclear arms race and other pressing issues. Several times a year our discussions were linked by TV satellite, and broadcast throughout the Soviet Union and on selected PBS stations in the United States. Oppenheimer's ideas shaped many of these remarkable moments in the evolution of glasnost.

We'd also like to thank two talented and accomplished women, our long suffering wives, Susan Sherwin and Susan Goldmark; they also have shared our long ride—and kept us in our respective saddles. We love them, respect them, and thank them for their special blends of patience and exasperation with our obsession for this book.

We also thank Ann Close, a seasoned Knopf editor whose Southern patience and attention to the smallest of details has enriched this book. She expertly shepherded a long manuscript to publication under an incredibly tight schedule. Our copy editor, the legendary Mel Rosenthal, sharpened our focus, improved our prose, and taught us how not to dangle our modifiers. We also thank Millicent Bennett for making sure that nothing got lost. Stephanie Kloss executed an elegant design for the book's jacket. We thank the Washington, D.C. artist Steve Frietch for initially proposing the Alfred Eisenstadt portrait of Oppenheimer for the cover.

We are also deeply grateful to another wonderful editor, Bobbie Bristol, who nurtured and protected this book for decades before she retired and passed it on to Ann. But even under Bobbie's protective care it could not have been sustained for a quarter of a century were it not for the serious intellectual culture and respect for authors that characterizes the publishing house of Alfred A. Knopf.

Gail Ross is both a lawyer and a book agent—and we are grateful to her for renegotiating the terms of a twenty-year-old contract with Knopf—and for many future lunches at La Tomate!

The "wily" Victor Navasky has been a friend and mentor to us both— and he deserves credit for having introduced us more than two decades ago. We are grateful for his wisdom and his friendship, and for his wonderful wife, Annie.

We are indebted to several eminent scholars who took the time to carefully read early versions of our manuscript. Jeremy Bernstein, also an Oppenheimer biographer, is an accomplished physicist and writer who did his patient best to correct our wrong-headed apprehensions of quantum physics.

Richard Polenberg, the Goldwin Smith Professor of American History at Cornell University, ruined his summer on our behalf by meticulously reading the entire manuscript and sharing with us both his knowledge of the Oppenheimer security case and his artful sensibility as a writer of history.

James Hershberg, William Lanouette, Howard Morland, Zygmunt Nagorski, Robert S. Norris, Marcus Raskin, Alex Sherwin and Andrea Sherwin Ripp also read all or parts of the manuscript and we are grateful for their insights and comments.

Over the years, we have benefited from the willingness of such formidable scholars as Gregg Herken, S. S. Schweber, Priscilla McMillan, Robert Crease, and the late Philip Stern to challenge us with their own ideas and scholarship about the controversial issues surrounding Oppenheimer's life. Both of these fine historians have graciously shared documents and interview sources. Max Born's biographer, Nancy Greenspan, generously shared the fruits of her research. We are indebted to Jim Hijiya for his scholarly interpretation of Oppenheimer's fascination with the Bhagavad-Gita. More recently, we have encountered the work of the British historian of science, Charles Thorpe, and we thank him for permission to quote from his doctoral dissertation—a version of which will soon be published.

We wish to thank Drs. Curtis Bristol and Floyd Galler and the psychoanalyst Sharon Alperovitz for their psychological insights about Oppenheimer's early life. Dr. Jeffrey Kelman graciously helped us to interpret the autopsy report and other medical records pertaining to the death of Dr. Jean Tatlock. Dr. Daniel Benveniste shared with us his insights on Oppenheimer's study of psychoanalysis with Dr. Siegfried Bernfeld. We are indebted to the late Alice Kimball Smith and to Charles Weiner whose superbly annotated collection of Oppenheimer's correspondence inspired many of our interpretations. We similarly owe a debt to Richard G. Hewlett and Jack Holl for their assistance during the earliest stages of this book, and for their excellent official histories of the Atomic Energy Commission.

Many dedicated archivists went out of their way to guide us through many thousands of pages of official documents and private papers. We wish to thank in particular Linda Sandoval and Roger A. Meade at the Los Alamos National Laboratory Archives; Ben Primer at Princeton University; Dr. Peter Goddard, Georgia Whidden and Christine Ferrara, Rosanna Jaffin at the Institute for Advanced Study; John Stewart and Sheldon Stern at the John F. Kennedy Presidential Library; Spencer Weart at the American Institute of Physics; John Earl Haynes at the Library of Congress; and the many others who assisted us at the libraries and archives listed on pages 601 and 602.

These and many other archivists at the Library of Congress, the National Archives, and archives at Harvard, Princeton, and the University of California's Bancroft Library are working hard to preserve our history.

As both American citizens and historians we salute all who have supported and sustained the Freedom of Information/Privacy Act. It has not only made access to FBI, CIA and other previously closed government investigative files available to historians and journalists, but more importantly, it has contributed to sustaining our democracy.

No book of this scope can be researched without the assistance of young and energetic students of history. A select group of them associated with the Nuclear Age History and Humanities Center (NAHHC) at Tufts University prepared chronologies, analyzed and organized documents, researched articles and transcribed hundreds of hours of interviews. Susanne LaFeber Kahl and Meredith Mosier Pasciuto, both Tufts graduates, and both brilliantly efficient administrators, organized this work and contributed research of their own.

A remarkable group of research assistants and graduate students at NAHHC contributed in numerous ways. Miri Navasky, now a talented documentary filmmaker, spent many long hours searching out documents and creating a chronology of Kitty Oppenheimer's life. Jim Hershberg constantly asked probing questions and enthusiastically shared documents that he had gathered for his magisterial biography of James Conant. Debbie Herron Hand efficiently transcribed interviews. Tanya Gassel, Hans Fenstermacher, Gerry Gendlin, Yaacov Tygiel, Dan Lieberfeld, Philip Nash, and Dan Hornig all provided intellectual and moral support.

Peter Schwartz did some of the early spadework in San Francisco Bay Area archives. Erin Dwyer and Cara Thomas typed corrections into the final chapters, Patrick J. Tweed, Pascal van der Pijl and Euijin Jung also assisted us in the research of this book.

Many other friends and colleagues have sustained us over the years it has taken to write this biography.

Kai wishes particularly to thank his parents, Eugene and Jerine Bird for nurturing his passion for history, and his son, Joshua Kodai Bird, for patiently allowing him to read aloud large portions of the manuscript at bedtime. He also thanks Joseph Albright and Marcia Kunstel; Gar Alperovitz; Eric Alterman; Scott Armstrong; Wayne Biddle; Shelly Bird; Nancy Bird and Karl Becker; Norman Birnbaum; Jim Boyce and Betsy Hartmann; Frank Browning; Avner Cohen and Karen Gold; David Corn; Michael Day; Dan Ellsberg; Phil and Jan Fenty; Thomas Ferguson; Helma Bliss Goldmark; Richard Gonzalez and Tara Siler; Neil Gordon; Mimi Harrison; Paul Hewson; Congressman Rush Holt; Brennon Jones; Michael Kazin and Beth Horowitz; Jim and Elsie Klumpner; Lawrence Lifschultz and Rabia Ali; Richard Lingeman; Ed Long; Priscilla Johnson McMillan; Alice McSweeney; Christina and Rodrigo Macaya; Paul Magnuson and Cathy Trost; Emily Medine and Michael Schwartz (and their mountain sanctuary); Andrew Meier; Branco Milanovic and Michelle de Nevers; Uday Mohan;

Dan Moldea; John and Rosemary Monagan (and all our friends at his writers' group); Jacques and Val Morgan of Idle Time Books; Anna Nelson; Paula Newberg; Nancy Nickerson; Tim Noah and the late Marjorie Williams; Jeffery Paine; Jeff Parker; David Polazzo; Lance Potter (who found the epigraph on Prometheus); William Prochnau and Laura Parker; Tim Rieser; Caleb Rossister and Maya Latynski; Arthur Samuelson; Nina Shapiro; Alix Shulman; Steve Solomon; John Tirman; Nilgun Tolek; Abigail Wiebenson; Don Wilson; Adam Zagorin, and Eleanor Zelliot.

Kai is particularly indebted to Lee Hamilton, Rosemary Lyon, Lindsay Collins, Dagne Gizaw, Janet Spikes and all his other friends at the Woodrow Wilson Center for listening to his long-winded stories about Oppie.

Martin adds his thanks to his many mutual friends above and wishes particularly to acknowledge his children, Alex Sherwin and Andrea Sherwin Ripp for their love and their bemused willingness to share so many years of their lives and their living space with the enormous collection of boxes, file cabinets and bookshelves that were dedicated to "Oppie's cocoon." His sister Marjorie Sherwin and her partner Rose Walton did not have to live with the cocoon, but they frequently visited it and never lost hope that a butterfly would emerge. That it finally did is in no small way due to the encouragement and support of three wonderful mentors who taught and sustained him through graduate school at UCLA—and beyond: Keith Berwick, Richard Rosecrance and "RD."

Martin also thanks and acknowledges the support and intellectual encouragement—and in many cases the hospitality during research trips—of many old friends and colleagues: Hiroshima's Mayor, Tadoshi Akiba; Sam Ballen; Joel and Sandy Barkan; Ira and Martha Berlin (and *The Wisconsin Magazine of History*); Richard Challener; Lawrence Cunningham; Tom and Joan Dine; Carolyn Eisenberg; Howard Ende; Hal Feiveson; Owen and Irene Fiss; Lawrence Friedman; Gary Goldstein; Ron and Mary Jean Green; Sol and Robyn Gittleman; Frank von Hippel; David and Joan Hollinger; Michele Hochman; Al and Phyllis Janklow; Mikio Kato; Nikki Keddie; Mary Kelley; Robert Kelley; Dan and Bettyann Kevles; David Kleinman; Martin and Margaret Kleinman; Barbara Kreiger; Normand and Marjorie Kurtz; Rodney Lake; Mel Leffler; Alan Lelchuk; Tom and Carol Leonard; Sandy and Cynthia Levinson; Dan Lieberfeld; Leon and Rhoda Litwack; Marlaine Lockheed; Janet Lowenthal and Jim Pines; David Lundberg; Gene Lyons; Lary and Elaine May; David Mizner; Bob and Betty Murphy; Arnie and Sue Nachmanoff; Bruce and Donna Nelson; Arnold and Ellen Offner; Gary and Judy Ostrower; Donald Pease; Dale Pescaia; Constantine Pleshakov; Phil Pochoda; Ethan Pollock; the late Leonard Rieser; Del and Joanna Ritchhardt; John Rosenberg; Michael and Leslie Rosenthal;

Richard and Joan Rudders; Lars Ryden; Pavel Sarkisov; Ellen Schrecker; Sharan Schwartzberg; Edward Segel; Ken and Judy Seslowe; Saul and Sue Singer; Rob Sokolow; Christopher Stone; Cushing and Jean Strout; Natasha Tarasova; Stephen and Francine Trachtenberg; Evgeny Velikhov; Charlie and Joanne Weiner; Dorothy White; Peter Winn and Sue Gronwald; Herbert York; Vladislav Zubok.

Over the many years that this book has been in preparation many scholar-friends have sent us unsolicited Oppenheimer documents discovered while doing their own research. For these acts of generosity and fellowship we wish to thank Herbert Bix, Peter Kuznick, Lawrence Wittner, and Poland's eminent historian and Ambassador to the United States, Przemyslaw Grudzinski. We also acknowledge the many kindnesses that Peter, Charles and Ella Oppenheimer and Brett and Dorothy Vanderford extended to us in the course of our research. We are grateful to Barbara Sonnenberg for permission to reprint some of her Oppenheimer family photographs. The current owners of One Eagle Hill in Berkeley, Dr. David and Kristin Myles, graciously gave us a tour of Oppenheimer's lovely home overlooking San Franciso Bay.

There is also a long list of interviewees on pages 697–699 to whom we are deeply indebted. Thank you for your time, your stories and your patience with us; this book could not have been written without your help.

Scholars cannot live on documents alone and this book could not have been written without the financial support of numerous foundations. Martin is grateful for the support extended to him by Arthur Singer and the Alfred P. Sloan Foundation, the John Simon Guggenheim foundation, Ruth Adams and the John D. and Catherine T. MacArthur Foundation, the National Endowment for the Humanities, Tufts University and the George Washington University President's James Madison Fund. Kai wishes to thank the Woodrow Wilson International Center for Scholars; Cindy Kelly of the Atomic Heritage Foundation; and Ellen Bradbury-Reid, executive director of Recursos in Santa Fe, New Mexico.

We both wish to acknowledge the percipiency of Susan Goldmark and Ronald Steel, who independently and simultaneously suggested to us that "American Prometheus" would be an excellent title for our book. Finally, our thanks to Lynn Nesbit who, a quarter century ago, negotiated the original contract for this book.

NOTES

Our research files—including those designated in the notes as the "Bird Collection" and "Sherwin Collection"—will be distributed to appropriate archives and libraries. The details of this distribution will be posted on our websites, *www.HistoryHappens.net* and *www .AmericanPrometheus.org*.

ABBREVIATIONS

AEC	Atomic Energy Commission
AIP	American Institute of Physics (Niels Bohr Library)
APS	American Philosophical Society
Caltech	California Institute of Technology
CU	Clemson University Archives
CUL	Cornell University Library
DCL	Dartmouth College Library
DDEL	Dwight D. Eisenhower Presidential Library
ECS	Ethical Culture Society archives
FBI	Federal Bureau of Investigation Reading Room
FDRL	Franklin D. Roosevelt Presidential Library
FRUS	Foreign Relations of the United States, U.S. State Department
HBSL	Harvard Business School Library
HHL	Herbert Hoover Presidential Library
HSTL	Harry S. Truman Presidential Library
HU	Harvard University archives
HUAC	U.S. House Un-American Activities Committee
IAS	Institute for Advanced Study (Princeton)
JFKL	John F. Kennedy Presidential Library
JRO	J. Robert Oppenheimer
JRO FBI File	J. Robert Oppenheimer FBI file number 100-17828
JRO Hearing	United States Atomic Energy Commission, In the Matter of J. Robert Oppenheimer: Transcript of Hearing before Personnel Security Board and Texts of Principal Documents and Letters. Forward by Philip M. Stern. Cambridge, MA: MIT Press, 1971.
JRO Papers	J. Robert Oppenheimer Papers, Library of Congress
LANL	Los Alamos National Laboratory Archives
LBJL	Lyndon B. Johnson Presidential Library
LOC	Library of Congress (Manuscript Reading Room)
MED	Manhattan Engineer District
MIT	Massachusetts Institute of Technology archives
NA	National Archives
NBA	Niels Bohr Archive, Copenhagen
NBL	Niels Bohr Library, American Institute of Physics
NYT	*New York Times*
PUL	Princeton University Library (Mudd Manuscript Library)
SU	Stanford University Libraries

UC	University of Chicago Archives
UCB	University of California at Berkeley (Bancroft Library)
UCSDL	University of California at San Diego Library
UM	University of Michigan Library
WP	*Washington Post*
WU	Washington University archives
YUL	Yale University, Sterling Library

Preface

xii **"We have had the bomb":** E. L. Doctorow, "The State of Mind of the Union," *The Nation,* 3/22/86, p. 330.

Prologue

3 **The Nobelists included:** Murray Schumach, "600 at a Service for Oppenheimer," NYT, 2/26/67.

4 **"He did more than":** Ibid.

4 **"Such a wrong":** *Bulletin of the Atomic Scientists,* October 1967.

5 **"In the dark days":** Schumach, NYT, 2/26/67; Abraham Pais, *A Tale of Two Continents,* p. 400.

5 **Oppenheimer was an enigma:** Jeremy Bernstein, *Oppenheimer: Portrait of an Enigma,* pp. vii–xi.

5 **"a symbol of the tragedy":** NYT, 2/20/67.

5 **"very wise":** I. I. Rabi, interview by Sherwin, 3/12/82, p. 11.

5 **"a Faustian bargain":** Freeman Dyson, interview by Jon Else, 12/10/79, pp. 5, 9–10.

Chapter One: "He Received Every New Idea as Perfectly Beautiful"

10 **"an almost medieval":** Oppenheimer family tree, folder 4–24, box 4, Frank Oppenheimer Papers, UCB; JRO interview by Kuhn, 11/18/63, APS, p. 3. The third brother also immigrated to New York but returned permanently to Germany after a brief stay. One of the three sisters came to the United States at some point but returned to Germany, where she died. Hedwig Oppenheimer Stern, the youngest of the three sisters, immigrated to the United States in 1937 and settled in California. (Babette Oppenheimer Langsdorf, interview by Alice Smith, 12/1/76, Sherwin Collection.) Babette, Emil Oppenheimer's daughter, was a couple of years younger than Robert Oppenheimer. The U.S. Census of 1900 records, perhaps incorrectly, that Julius Oppenheimer was born in August 1870, and emigrated from Germany in 1888; Julius listed his occupation as traveling salesman. (1900 Census, New York, N.Y., roll 1102, vol. 149, enumeration 455, sheet 8, line 27, NA.)

10 **"You have a way":** Ella Friedman to Julius Oppenheimer, undated, circa March 1903, folder 4–10, box 4, Frank Oppenheimer Papers, UCB.

11 **"an exquisitely beautiful":** Dorothy McKibbin, interview by Jon Else, 12/10/79, p. 21. McKibbin is quoting Katherine Chaves Page. See also Miss Frieda Altschul to JRO, 12/9/63, describing Ella's eyes.

11 **The glove covering:** Alice Kimball Smith and Charles Weiner, *Robert Oppenheimer: Letters and Recollections,* p. 2; Frank Oppenheimer, interview by Alice Smith, 3/17/75, p. 58.

11 **"a gentle, exquisite":** Lincoln Barnett, "J. Robert Oppenheimer," *Life,* 10/10/49.

11   **Upon her return:** Frank Oppenheimer oral history, 2/9/73, AIP, p. 2.

11   **"I do so":** Ella Friedman to Julius Oppenheimer, 3/10/03, folder 4–10, box 4, Frank Oppenheimer Papers, UCB.

11   **"Julius Robert Oppenheimer":** FBI File 100–9066, 10/10/41, and File 100–17828–3, citing Oppenheimer's birth certificate, no. 19763.

11   **Sometime after Robert's:** Frank Oppenheimer, interview by Alice Smith, 3/17/75, p. 34; 1920 U.S. Census.

12   **Over the years:** Frank Oppenheimer, interview by Alice Smith, 3/17/75, p. 54; Else Uhlenbeck, interview by Alice Smith, 4/20/76, p. 2. Oppenheimer's cousin Babette Oppenheimer Langsdorf later described Ella as a "talented painter" and a "connoisseur" (Mrs. Walter Langsdorf to Philip M. Stern, 7/10/67, Stern Papers, JFKL; George Boas to Alice Smith, 11/28/76, Smith correspondence, Sherwin Collection; Smith and Weiner, *Letters,* p. 138). Julius acquired Van Gogh's *First Steps (After Millet)* in 1926, and Frank Oppenheimer inherited it in 1935. For the provenance of the Oppenheimer family's Van Gogh collection, see "Vincent van Gogh: The Complete Works," a CD-ROM database, copyright David Brooks (Sharon, MA: Barewalls Publications, 2002). Julius bought Picasso's *Mother and Child* in 1928, and Frank Oppenheimer sold it in 1980 for $1,050,000 (see Dr. Joseph Baird, Jr., to Frank Oppenheimer, 4/12/80, folder 4–46, box 4; Jack Tanzer to Frank Oppenheimer, 5/13/80, folder 4–46, box 4, Frank Oppenheimer Papers, UCB).

12   **"My mother didn't":** JRO, interview by T. S. Kuhn, 11/18/63, p. 10. The 1920 U.S. Census listed three live-in maids in the Oppenheimer household: Nellie Connolly, age 87, of Ireland; Henrietta Rosemund, age 21, of Germany; and Signe McSorley, age 29, of Sweden (1920 Census, vol. 244, enumeration 702, sheet 13, line 37, roll 1202, NA).

12   **"It was lovely":** Smith and Weiner, *Letters,* p. 34; Frank Oppenheimer, interview by Alice Smith, 3/17/75, p. 26.

12   **"Robert was doted":** Harold F. Cherniss, interview by Sherwin, 5/23/79, p. 3.

12   **"He [Julius] was jolly":** Francis Fergusson, interview by Sherwin, 6/8/79, p. 7.

13   **A family friend:** Julius Oppenheimer to Frank Oppenheimer, 3/11/30, folder 4–11, box 4, Frank Oppenheimer Papers, UCB; Boas to Alice Smith, 11/28/76, Smith correspondence, Sherwin Collection.

13   **Ella, by contrast:** Fergusson, interview by Alice Smith, 4/23/75, p. 10.

13   **"She [Ella] was a very":** Peter Goodchild, *J. Robert Oppenheimer,* p. 11.

13   **Four years after:** Jeremy Bernstein, *Oppenheimer,* p. 6; Frank Oppenheimer, oral history, 2/9/73, p. 4, AIP.

13   **Ella encouraged Robert:** Frank Oppenheimer to Denise Royal, 2/25/67, Frank Oppenheimer Papers, carton 4, UCB.

13   **"Julius's articulate and":** Ruth Meyer Cherniss, interview by Alice Smith, 11/10/76; Herbert Smith, interview by Charles Weiner, 8/1/74, pp. 12, 16–17.

14   **"Just as I do":** Oppenheimer may indeed have had a brief bout of polio. See Alice Smith to Frank Oppenheimer, 8/6/79, carton 4, Frank Oppenheimer Papers, UCB; Peter Michelmore, *The Swift Years,* p. 4.

14   **"It was clear":** JRO, interview by Kuhn, 11/18/63, APS, pp. 1–4; *Time,* 11/8/48, p. 70.

14   **Robert recounted that:** JRO, interview by Kuhn, 11/18/63, p. 1.

14   **From the ages of:** Denise Royal, *The Story of J. Robert Oppenheimer,* p. 13.

15   **"They adored him":** Quotes in this paragraph are taken from Smith and Weiner, *Letters,* p. 5; JRO, interview with Kuhn, p. 3; Babette Oppenheimer Langsdorf to Philip M. Stern, 7/10/67, Stern Papers, JFKL.

15   **At some point:** Frank Oppenheimer, oral history, 2/9/73, AIP, p. 1.

15   **"If we had":** Frank Oppenheimer, oral history, 2/9/73, AIP, p. 4.

15   **"I repaid my parents' ":** Denise Royal, *The Story of J. Robert Oppenheimer,* p. 16.

16 **He and Ella:** Board of Trustees, 1912, Ethical Culture Archives, New York Society for Ethical Culture.

16 **"Deed, not Creed,":** *Time,* 11/8/48, p. 70.

16 **"Man must assume":** Richard Rhodes, "I Am Become Death . . ." *American Heritage,* vol. 28, no. 6 (1987).

16 **The son of Rabbi:** Horace L. Friess, *Felix Adler and Ethical Culture,* p. 194.

17 **"Ethical Culture" was:** Stephen Birmingham, *The Rest of Us,* pp. 29–30.

17 **"emancipated Jews":** Friess, *Felix Adler and Ethical Culture,* p. 198.

17 **"Zionism itself is":** Benny Kraut, *From Reform Judaism to Ethical Culture,* pp. 190, 194, 205. Perhaps this explains why Oppenheimer himself never displayed any particular interest in Zionism.

18 **"artistic gifts to":** Friess, *Felix Adler and Ethical Culture,* pp. 136, 122.

18 **"I must square":** Friess, *Felix Adler and Ethical Culture,* pp. 35, 100, 153, 141.

19 **"ethical imagination":** Felix Adler, "Ethics Teaching and the Philosophy of Life," *School and Home,* a publication of the Ethical Culture School P.T.A., November 1921, p. 3.

19 **"and after he came":** Smith and Weiner, *Letters,* p. 3; Frank Oppenheimer, oral history, 4/14/76, AIP, p. 56.

19 **"undivided allegiance":** Friess, *Felix Adler and Ethical Culture,* pp. 131, 201–2.

20 **"a witty saint":** Robin Kadison Berson, *Marching to a Different Drummer,* pp. 101–5.

20 **"I did not know":** John Lovejoy Elliott to Julius Oppenheimer, 10/23/31, archives of the New York Society of Ethical Culture.

20 **"Negro problem"; "sex relations":** Friess, *Felix Adler and Ethical Culture,* p. 126; Yvonne Blumenthal Pappenheim, interview by Alice Smith, 2/16/76.

20 **"the ethics of loyalty":** *The Course of Study in Moral Education,* (New York: Ethical Culture School, 1912, 1916 [pamphlet], p. 22); Kevin Borg, "Debunking a Myth: J. Robert Oppenheimer's Political Philosophy," unpublished paper, University of California, Riverside, 1992.

21 **"I was an unctuous":** *Time,* 11/8/48; Denise Royal, *The Story of J. Robert Oppenheimer,* pp. 15–16.

21 **"tortured":** Herbert Smith, interview by Alice Smith, 7/9/75, p. 1; Denise Royal, *The Story of J. Robert Oppenheimer,* p. 23; Smith and Weiner, *Letters,* p. 6; Rhodes, "I Am Become Death . . ." *American Heritage,* p. 73.

22 **"He received every":** Smith and Weiner, *Letters,* p. 4; "Remembering J. Robert Oppenheimer," *The Reporter,* Ethical Culture Society, 4/28/67, p. 2.

22 **"Ask me a question":** Stern, *The Oppenheimer Case,* pp. 11–12; Ruth Meyer Cherniss, interview by Alice Smith, 11/10/76; Cassidy, *J. Robert Oppenheimer and the American Century,* pp. 33–46.

22 **"We were thrown":** Stern, *The Oppenheimer Case,* pp. 11–12.

22 **"rather gauche":** Harold F. Cherniss, interview by Sherwin, 5/23/79, p. 3.

22 **"It's no fun":** Barnett, "J. Robert Oppenheimer," *Life,* 10/10/49.

22 **"special friend":** Jeanette Mirsky, interview by Alice Smith, 11/10/76.

23 **"magnificent prose style":** Herbert Smith, interview by Weiner, 8/1/74, p. 3; JRO, interview by Kuhn, 11/18/63, p. 3.

23 **"very, very kind":** Smith and Weiner, *Letters,* p. 5.

23 **"He was marvelous"** *and subsequent quotes:* JRO, interview by Kuhn, 11/18/63, p. 2.

23 **"He blushed":** Jane Kayser, interview by Weiner, 6/4/75, p. 34; Smith and Weiner, *Letters,* pp. 6–7.

23 **"He was just":** Francis Fergusson, interview by Sherwin, 6/8/79, p. 4.

23 **"He still took":** Peter Michelmore, *The Swift Years,* p. 9; Gregg Herken, *Brotherhood of the Bomb,* p. 338, note 55.

24   **"Roberty, Roberty":** Michelmore, *The Swift Years,* pp. 8–9.
24   **"It was a blowy":** Francis Fergusson, interview by Sherwin, 6/8/79, p. 6.
24   **Robert graduated:** As a child, Oppenheimer had his fair share of illnesses. At the age of six he had a tonsillectomy and an adenoidectomy; in 1916 he had an appendectomy; and in 1918 he had scarlet fever. J. Robert Oppenheimer, medical physical, Presidio of San Francisco, 1/16/43; box 100, series 8, MED, NA.
25   **"his Jewishness":** Smith and Weiner, *Letters,* p. 9.
25   **"We all did":** Jeanette Mirsky, interview by Alice Smith, 11/10/76; Smith and Weiner, *Letters,* p. 61.
25   **Frank Oppenheimer said:** Smith and Weiner, *Letters,* p. 40.
25   **"this great troika":** Smith and Weiner, *Letters,* p. 9.
26   **Robert was intensely:** Frank Oppenheimer, interview by Alice Smith, 4/14/76, p. 12. In 1961, Katherine Chaves Page (Cavanaugh) was stabbed to death in her bed during an apparent robbery by a young Mexican-American neighbor (Dorothy McKibbin, interview by Alice Smith, 1/1/76).
26   **"reigning princess":** Herbert Smith, interview by Weiner, 8/1/74, p. 6.
26   **"his very good friend":** Francis Fergusson, interview by Sherwin, 6/8/79, p. 3, and 6/18/79, p. 8.
26   **"Thank God I won":** Herbert Smith, interview by Weiner, 8/1/74, pp. 15–16.
27   **"I never heard":** Herbert Smith, interview by Weiner, 8/1/74, pp. 6–10.
27   **Robert told Smith:** Herbert Smith, interview by Weiner, 8/1/74, p. 1.
27   **"He looked at me sharply":** Smith and Weiner, *Letters,* p. 9.
27   **"For the first time":** Smith and Weiner, *Letters,* p. 10.
28   **"beautiful and savage country":** Emilio Segrè, *Enrico Fermi: Physicist,* p. 135.
28   **The ranch school stood:** *Los Alamos: Beginning of an Era 1943–45,* Los Alamos National Laboratory, 1986, p. 9.
28   **"Of course I am insanely jealous":** Smith and Weiner, *Letters,* p. 22 (JRO to Herbert Smith, 2/18/23).

Chapter Two: *"His Separate Prison"*

29   **"because I could":** JRO, interview by Kuhn, 11/18/63, p. 14; William Boyd, interview by Alice Smith, 12/21/75, p. 5.
29   **His eyes were:** Robert Oppenheimer, U.S. Army physical, 1/16/43, box 100, series 8, MED, NA.
30   **"He [Robert] found":** Smith and Weiner, *Letters,* p. 61.
30   **"Robert had bouts":** Smith and Weiner, *Letters,* p. 9.
30   **"What intolerable heat":** Michelmore, *The Swift Years,* p. 15, and Jeffries Wyman, interview by Charles Weiner, 5/28/75, p. 14; JRO, interview by Kuhn, 11/18/63, p. 6.
31   **"He went to bed":** Frederick Bernheim, interview by Weiner, 10/27/75, pp. 7, 16.
31   **"We had lots of":** Smith and Weiner, *Letters,* p. 33.
31   **"he was pretty careful":** Smith and Weiner, *Letters,* p. 45; William Boyd, interview by Alice Smith, 12/21/75, p. 4.
31   **"I was very fond":** Smith and Weiner, *Letters,* p. 34.
32   **"The notion that":** Barnett, "J. Robert Oppenheimer," *Life,* 10/10/49.
32   **"When I am inspired":** Smith and Weiner, *Letters,* p. 59.
32   **"The dawn invests":** Robert Oppenheimer, "Le jour sort de la nuit ainsi qu'une victoire," Oppenheimer poems received from Francis Fergusson, Alice Smith Collection (now in Sherwin Collection).
33   **"will look like Jews":** Richard Norton Smith, *The Harvard Century,* p. 87; *Harvard Crimson,* 12/13/24 and 1/17/23.

33 **In March 1923:** "Liberals Take Stand Against Restriction," *Harvard Crimson,* 3/14/23.

33 **"asinine pomposity of":** John Trumpbour, ed., *How Harvard Rules,* p. 384; *The Gadfly,* December 1922, published by the Student Liberal Club, Harvard University; JRO, interview by Kuhn, 11/18/63, p. 9; Smith and Weiner, *Letters,* p. 15; Michelmore, *The Swift Years,* p. 15. John Edsall, interview by Weiner, 7/16/75, p. 6.

33 **"I can't recall":** JRO, interview by Kuhn, 11/18/63, pp. 7, 9.

33 **"Obviously, if he [Oppenheimer]":** JRO, interview by Kuhn, 11/18/63, p. 8; Smith and Weiner, pp. 28–29.

33 **"I found Bridgman":** *Time,* 11/8/48, p. 71.

34 **"I judge from":** Gerald Holton, "Young Man Oppenheimer," *Partisan Review,* 1981, vol. XLVIII, p. 383; *Time,* 11/8/48, p. 71. The temple of Segesta was probably built in the years 430–420 B.C.

34 **When the famous:** William Boyd, interview by Alice Smith, 12/21/75, p. 7.

34 **"it would be hard to exaggerate":** Pais, *Niels Bohr's Times,* pp. 541, 253; *Time,* 11/8/48, p. 71.

34 **"To this day":** JRO, interview by Kuhn, 11/18/63, pp. 5, 9.

34 **"I had a very exciting":** JRO, interview by Kuhn, 11/18/63, pp. 5, 9.

35 **"[Oppenheimer] has grown":** Smith and Weiner, *Letters,* p. 48.

35 **"Leonardos and Oppenheimers":** Paul Horgan, *A Certain Climate,* p. 5.

35 **"Generously, you ask what I do":** Smith and Weiner, *Letters,* p. 54.

35 **Dark wit aside:** William Boyd, interview by Alice Smith, 12/21/75, p. 9.

36 **"My feeling about myself":** Smith and Weiner, *Letters,* pp. 60–61, 19; *Time,* 11/8/48, p. 71.

36 **"too much in love":** Smith and Weiner, *Letters,* p. 60.

36 *"Tonight she wears":* JRO, "Neophyte in London," Oppenheimer poems received from Francis Fergusson, Alice Smith Collection.

36 *"No, I know":* JRO, "Viscount Haldome in Robbins," Oppenheimer poems received from Francis Fergusson, Alice Smith Collection. In the margins of this typed poem, Oppenheimer has scrawled, "My first love poem."

37 **"What has soothed me most":** Smith and Weiner, *Letters,* p. 62.

37 **"The job and people":** Smith and Weiner, *Letters,* pp. 32–33.

38 **He made the dean's list:** *Harvard Crimson,* 11/18/24, 3/9/25.

38 **"Even in the last stages":** Smith and Weiner, *Letters,* p. 60.

38 **"Although I liked to work":** Robert Oppenheimer's Harvard transcript, 1922–25, Alice Smith Collection; Smith and Weiner, *Letters,* p. 68; JRO, interview by Kuhn, 11/18/63, p. 10.

38 **"Boyd and I got plastered":** Smith and Weiner, *Letters,* p. 74; Michelmore, *The Swift Years,* p. 15.

38 **"more near the center":** JRO, interview by Kuhn, 11/18/63, p. 14.

39 **"As appears from his name":** Smith and Weiner, *Letters,* p. 77.

39 **"We hit the divide":** Smith and Weiner, *Letters,* pp. 80–81.

40 **Pipe tobacco and cigarettes:** Michelmore, *The Swift Years,* p. 14.

40 **"Rutherford wouldn't have me":** JRO, interview by Kuhn, 11/18/63, p. 14.

Chapter Three: "I Am Having a Pretty Bad Time"

41 **In mid-September 1925:** Smith and Weiner, *Letters,* p. 86.

41 **"When I met him":** Francis Fergusson, "Account of the Adventures of Robert Oppenheimer in Europe," memo, February 26 (no full year, but quite likely February 1926), attached to Fergusson, interview by Alice Smith, 4/21/76, Sherwin Collection.

41 **"was completely at a loss"**: Fergusson, interview with Sherwin, 6/18/79, p. 1.

41 **"I was cruel"**: Fergusson, "Account of the Adventures of Robert Oppenheimer in Europe."

42 **Quantum theory**: John Gribbin, *Q Is for Quantum*, pp. 284, 321–22.

42 **"I was still"**: JRO, interview by Kuhn, 11/18/63, p. 11.

42 **"I met [Patrick M. S.] Blackett"**: Smith and Weiner, *Letters*, p. 89; JRO, interview by Kuhn, 11/18/63, p. 16.

43 **"the place is very rich"**: Smith and Weiner, *Letters*, pp. 87–88.

43 **"The point is"**: Goodchild, *J. Robert Oppenheimer*, p. 17.

43 **"that he felt so miserable"**: Michelmore, *The Swift Years*, p. 17; Wyman, interview by Weiner, 5/28/75, p. 22.

43 **On another occasion**: Pais, *Inward Bound*, p. 367. Rutherford told this story to Paul Dirac, who conveyed it to Pais.

43 **Neither was it any comfort**: Fergusson, interview by Alice Smith, 4/21/76, p. 36.

43 **"There are some terrible"**: Smith and Weiner, *Letters*, p. 88.

43 **"In a way"**: Frederick Bernheim, interview by Weiner, 10/27/75, p. 20.

43 **Robert had always been fond**: Smith and Weiner, *Letters*, p. 19; Herbert Smith, interview by Weiner, 8/1/74, p. 19.

44 **"first class case of depression"**: Fergusson, "Account of the adventures of Robert Oppenheimer in Europe.

44 **"He found himself in"**: Ibid.

45 **Ella "saw to it"**: Fergusson, interview by Sherwin, 6/18/79.

45 **He "did a very good"**: Alice Smith, notes on Fergusson, 4/23/75, p. 4.

45 **"There they lay"**: Fergusson, interview by Sherwin, 6/18/79, p. 1; Fergusson, "Account of the adventures of Robert Oppenheimer in Europe," p. 3.

45 **"Really I have been engaged"**: Smith and Weiner, *Letters*, p. 90.

45 **Among other complaints**: Edsall, interview by Weiner, 7/16/75, p. 27.

46 **"Whether or not"**: Wyman, interview by Weiner, 5/28/75, p. 23.

46 **"He had kind of poisoned"**: Fergusson, interview by Sherwin, 6/18/79, pp. 4–6.

46 **"He was retained"**: Herbert Smith, interview by Weiner, 8/1/74, p. 16.

46 **"further analysis would"**: Edsall, interview by Weiner, 7/16/75, p. 19. Edsall later said that in June 1926 Oppenheimer told him of the analyst's diagnosis, but in Edsall's memory, the psychiatrist in question was in Cambridge. Edsall was astonished that a doctor would say such a cruel thing to a patient. Prominent disciples of Freud such as Dr. Ernest Jones dominated the psychiatric profession in London during the mid-1920s; indeed, it is entirely plausible that Jones was the psychiatrist who treated Oppenheimer. Julius Oppenheimer always sought out the best for his son. Dr. Jones was not only the most famous Freudian practicing in England, but he was also one of only four analysts who maintained an office on Harley Street. Furthermore, though he was undoubtedly a devoted disciple of Freud's—and later became his biographer—Jones was notorious in the profession for misdiagnoses. Jones could easily have mis-diagnosed Oppenheimer with dementia praecox. [See *International Journal of Psychoanalysis*, vol. 8, part 1, courtesy of Dr. Daniel Benveniste, e-mail 4/19/01 to Bird re: Harley Street analysts. Dr. Curtis Bristol is our source for Dr. Jones' predilection for misdiagnosis.]

46 **"He looked crazy"**: Fergusson, interview by Sherwin, 6/18/79, p. 2; Smith and Weiner, *Letters*, p. 94.

47 **"I was on the point"**: *Time*, 11/8/48, p. 71.

47 **"My reaction was dismay"**: Fergusson interview by Sherwin, 6/18/79, p. 5.

47 **"began to get very queer"**: Fergusson claimed that the Paris psychiatrist referred Robert to a high-class prostitute, a woman experienced in dealing with young men about

their sexual needs. According to Fergusson, Robert didn't like the idea much, but he went to see the woman. "Robert couldn't get to first base with her at all," Fergusson said. "She was an older woman, an experienced, intelligent woman. But nothing would click." Fergusson, interview by Alice Smith, 4/21/76, p. 39; see also Fergusson, interview by Sherwin, 6/18/79, pp. 1–4, 7.

47 **"I leaned over":** Fergusson, interview by Sherwin, 6/18/79, pp. 7–9; Fergusson, "Account of the Adventures of Robert Oppenheimer in Europe." Fergusson's engagement to Keeley was later broken off.

48 **"deep glares that Robert":** Fergusson, "Account of the Adventures of Robert Oppenheimer in Europe."

48 **"I've a notion":** Smith and Weiner, *Letters,* p. 86.

48 **"He knew that I knew"; "You should have":** Smith and Weiner, *Letters,* pp. 91–98.

48 *"awful fact of excellence":* Smith and Weiner, *Letters,* pp. 91–98.

49 **"He put me in a room":** Fergusson, interview by Sherwin, 6/18/79, pp. 7–9.

49 **"[Dr.] M has decided":** Edsall, interview by Weiner, 7/16/75, pp. 18–20.

49 **"gave the psychiatrist":** Herbert Smith, interview by Weiner, 8/1/74, p. 16.

49 **For ten days:** "Talk of the Town," *The New Yorker,* 3/4/67.

49 **"The scenery was magnificent":** Bernheim to Alice Smith, 8/3/76, Alice Smith correspondence A–Z, Sherwin Collection.

50 **"The kind of person":** Edsall, interview by Weiner, 7/16/75, pp. 26, 31.

50 **"It's a great place":** Smith and Weiner, *Letters,* p. 95.

50 **"One day"; "I never knew":** Wyman, interview by Weiner, 5/28/75, pp. 21–23.

50 **"he [Robert] spoke of it":** Edsall, interview by Weiner, 7/16/75, pp. 20, 27.

50 **"All this happened":** Alice Kimball Smith and Charles Weiner, speculated, "Perhaps the apple symbolized a scientific paper containing a suddenly recognized error." Smith and Weiner, *Letters,* p. 93; Denise Royal, *The Story of J. Robert Oppenheimer,* p. 36; Fergusson, interview by Sherwin, 6/18/79, pp. 4–6; Fergusson, interview by Alice Smith, 4/23/75, pp. 36–37.

51 **"The psychiatrist was a prelude":** He went on to explain to Davis why he wished the event to remain inscrutable: "My reason for telling you? Those loyalty hearings that the government held on me in 1954. The records printed in so many hundreds of pages of fine print in 1954. My big year, I've heard people say, and my life story complete in those records. But it isn't so. Almost nothing that was important to me came out there, almost nothing that meant anything to me is in those records. You see, don't you, that I'm proving this point to you now. With something important to me not in those records." (Nuel Pharr Davis, *Lawrence and Oppenheimer,* pp. 21–22.)

51 **So what actually happened:** Some historians, including S. S. Schweber and Abraham Pais, have speculated that Oppenheimer may have been wrestling with latent homosexuality. We think this speculation is groundless. Pais, who knew Oppenheimer as a friend and colleague, wrote in his 1997 memoirs that in the early 1950s he "was convinced that a strong, latent homosexuality was an important ingredient in Robert's emotional makeup." And yet, the friend who knew him best in those years, Francis Fergusson, insisted, "I never found in him any homosexual tendencies. I don't think it bothered him at all. He was just frustrated with his inability to make it with women at the time and his frustrations with his work." Similarly, Robert's Harvard roommate, Frederick Bernheim, explained, "He felt he was very inadequate with girls, and he would resent very much if I went out with a girl. . . . There was no homosexuality at all. . . . I had no sexual feelings for him or he for me, as far as I know, but he had—I don't know why—he had sort of [a] feeling that we should make a unit." See Pais, *A Tale of Two Continents,* p. 241. See also Schweber, *In the Shadow of the Bomb,* pp. 56, 203. For some hearsay of Oppenheimer's latent homosexuality, see JRO FBI security

file, V. P. Keay to Mr. Ladd, 11/10/47, where he is rumored to have had "an affair with Harvey Hall, . . . a mathematics student at the University, who was an individual of homosexual tendencies and at the time was living with Robert Oppenheimer" (FBI security file, microfilm, reel 1; see also Schweber, p. 203). However, Harvey Hall never lived with Oppenheimer. Hall and Oppenheimer did collaborate on at least one paper, published in *Physical Review* (Haakon Chevalier, *Oppenheimer,* p. 12). Fergusson, interview by Sherwin, 6/18/79, pp. 3–4, 7; Bernheim, interview by Weiner, 10/27/75, p. 16.

51 **The book was:** Haakon Chevalier, interview by Sherwin, 6/29/82, p. 6.

52 **"I felt much kinder":** Royal, *The Story of J. Robert Oppenheimer,* p. 36.

52 **"Another cackle":** JRO, interview by Kuhn, 11/18/63, p. 16.

53 **"In a rudimentary way":** Smith and Weiner, *Letters,* p. 96; JRO, interview by Kuhn, 11/18/63, p. 17.

53 **Dirac's work; "he didn't think":** Smith and Weiner, *Letters,* p. 96; Wyman, interview by Weiner, 5/28/75, p. 18.

53 **"reading books interfered":** Pais, et al., *Paul Dirac,* p. 29.

53 **"Not often in life"; "his God":** Rhodes, *The Making of the Atomic Bomb,* pp. 53–54; Wyman, interview by Weiner, 5/28/75, p. 30.

53 **"At that point I forgot":** JRO, interview by Kuhn, 11/18/63, p. 17.

53 **"How is it going?":** JRO, interview by Kuhn, 11/18/63, p. 21.

54 **Bohr vividly remembered:** Pais, *Niels Bohr's Times,* p. 495.

54 **"are the problems mathematical"** *and subsequent quotes:* JRO, interview by Kuhn, 11/20/63, pp. 1–2.

54 **"It was wonderful":** Smith and Weiner, *Letters,* p. 97.

54 **"Oppenheimer seemed to me":** Royal, *The Story of J. Robert Oppenheimer,* p. 36.

55 **"very great misgivings":** JRO, interview by Kuhn, 11/18/63, p. 21.

Chapter Four: "I Find the Work Hard, Thank God, & Almost Pleasant"

56 **German physicists at the time:** "Talk of the Town," *The New Yorker,* 3/4/67.

56 **It was Oppenheimer's good fortune:** Pais, *The Genius of Science,* pp. 32–33.

57 **He was the ideal mentor:** Gribbin, *Q Is for Quantum,* pp. 55–57; "Obituary: Prof. Max Born," *The Times* of London, 1/7/70.

57 **"We got along":** Smith and Weiner, *Letters,* p. 97.

57 **"had the typical bitterness":** Smith and Weiner, p. 100.

58 **"Although this [university]":** JRO, interview by Kuhn, 11/20/63, p. 5.

58 **"He was," said Uhlenbeck:** Pais, *The Genius of Science,* pp. 307–8.

58 **"I was part of a little community":** JRO, interview by Kuhn, 11/20/63, p. 4.

59 **"professors at Princeton":** Smith and Weiner, *Letters,* p. 100.

59 **"You would like":** Smith and Weiner, *Letters,* pp. 100–101.

59 **One day, Paul Dirac:** Pais, *Inward Bound,* p. 367. Pais cites a private communication from Dirac.

59 **"I still was not entirely":** JRO, interview by Kuhn, 11/20/63, p. 6.

59 **"highly neurotic":** Helen C. Allison, interview by Alice Smith, 12/7/76. The Hogness couple followed Oppenheimer to Berkeley in 1929.

59 **On occasion:** Max Debruck, "In Memory of Max Born," Debruck Papers, 37.8, Caltech Archives, courtesy of Nancy Greenspan.

60 **"He was a man"; "To make this more certain":** Max Born, *My Life,* p. 229; Goodchild, *Oppenheimer,* p. 20.

60 **"I couldn't find":** Born, *My Life,* p. 234; Royal, *The Story of J. Robert Oppenheimer,* p. 38.

60   **"Almost all of the theorists":** Smith and Weiner, *Letters,* p. 102.

61   **"On the classical":** Smith and Weiner, *Letters,* pp. 104–5.

61   **"The trouble is":** Michelmore, *The Swift Years,* p. 20.

61   **"All right, we'll leave":** Michelmore, *The Swift Years,* p. 21.

62   **"A little late":** Smith and Weiner, *Letters,* p. 104; Margaret Compton, interview by Alice Smith, 4/3/76.

62   **"The most exciting time":** JRO, interview with Kuhn, 11/20/63, p. 6.

62   **"How can you":** Michelmore, *The Swift Years,* p. 21; Pais, *The Genius of Science,* p. 54.

62   **by contrast:** Pais, *The Genius of Science,* p. 67; Luis Alvarez, *Adventures of a Physicist,* p. 87; Leo Nedelsky, interview by Alice Smith, 12/7/76.

62   **He still loved:** Smith and Weiner, *Letters,* p. 101; Davis, *Lawrence and Oppenheimer,* p. 22.

63   **"When your ancestors":** Thomas Powers, *Heisenberg's War,* pp. 84–85; James W. Kunetka, *Oppenheimer,* p. 12.

63   **As fate would have it:** Houtermans' political loyalties were to the Left; he would spend two and a half years in Stalin's prisons before he was repatriated to Germany in April 1940. For more on Houtermans' fascinating story, see Powers, *Heisenberg's War,* pp. 84, 93, 103, 106–7, and David Cassidy, *The Uncertainty Principle.*

63   **"Great ideas were":** Helge Kragh, *Quantum Generations,* p. 168.

64   **"Heisenberg has laid":** Gribbin, *Q Is for Quantum,* pp. 174, 417–18.

64   **"An inner voice":** Daniel J. Kevles, *The Physicists,* p. 167; Albrecht Fölsing, *Albert Einstein,* pp. 730–31. In 1929, Einstein qualified his critique by explaining that he believed "in the profound truth contained in this theory, except that I think that its restriction to statistical laws will be a temporary one." But shortly later he hardened his views, insisting that it was "not possible to get to the bottom of things by this semiempirical means." (Fölsing, *Albert Einstein,* pp. 566, 590.)

64   **"Einstein is completely":** Smith and Weiner, *Letters,* p. 190 (JRO to Frank Oppenheimer, 1/11/35). Oppenheimer first met Einstein at Caltech in 1930 (JRO to Carl Seelig, 9/7/55, JRO Papers).

64   **"harmonious, consistent and intelligible":** JRO, interview by Kuhn, 11/20/63, p. 7.

64   **"We have here":** Smith and Weiner, *Letters,* p. 103.

65   **"There are three":** Kevles, *The Physicists,* p. 217.

65   **Robert got into the habit:** Schweber, *In the Shadow of the Bomb,* p. 64.

65   **He walked around:** Royal, *The Story of J. Robert Oppenheimer,* p. 42.

65   **"life at the centers":** Hans Bethe, review of Robert Jungk's *Brighter Than a Thousand Suns,* in *Bulletin of the Atomic Scientists,* vol. 12, pp. 426–29; Schweber, *In the Shadow of the Bomb,* p. 100.

66   **"[i]n 1926 Oppenheimer had":** Hans Bethe, Ibid.

Chapter Five: "I Am Oppenheimer"

68   **"He's too much":** Michelmore, *The Swift Years,* p. 23.

69   **"My brother and I":** Smith and Weiner, *Letters,* p. 108.

69   **"Never having heard":** Frank Oppenheimer, oral history, 2/9/73, AIP, p. 5.

69   **"We all got":** Goodchild, *Oppenheimer,* p. 22.

69   **"Is the Ritz":** Michelmore, *The Swift Years,* p. 24.

69   **But though Charlotte:** Else Uhlenbeck, interview by Alice Smith, 4/20/76, p. 2; Michelmore, *The Swift Years,* pp. 24–25.

70   **"It was evening":** Smith and Weiner, *Letters,* p. 110; *Hound and Horn: A Harvard Miscellany,* vol. 1, no. 4 (June 1928), p. 335.

70 **"separate prison"**: JRO, "Le jour sort de la nuit ainsi qu'une victoire," Oppenheimer poems received from Francis Fergusson, Alice Smith Collection.

70 **"I have had trouble"**: Smith and Weiner, *Letters,* p. 113.

71 **"Don't worry about girls"**: Smith and Weiner, *Letters,* p. 113.

72 **As they rode:** *Time,* 11/8/48, p. 72.

72 **"For the former group"**: Frank Oppenheimer to Denise Royal, 2/25/67, folder 4–23, box 4, Frank Oppenheimer Papers, UCB.

72 **The kitchen had a sink:** Robert Serber, *Peace and War,* p. 38.

73 **"Hot Dog!"**: Royal, *The Story of J. Robert Oppenheimer,* p. 44; Michelmore, *The Swift Years,* pp. 26–27; Smith and Weiner, *Letters,* pp. 118, 126, 163–65.

73 **"We had a variety of mishaps"**: Frank Oppenheimer, oral history, 2/9/73, AIP, p. 18.

73 **Robert fractured his right arm:** JRO medical physical, Presidio of San Francisco, 1/16/43, box 100, series 8, MED, NA.

73 **"sipping from a bottle"**: Frank Oppenheimer to Denise Royal, 2/25/67, folder 4–23, box 4, Frank Oppenheimer Papers, UCB.

73 **"I am Oppenheimer"**: Smith and Weiner, *Letters,* p. 119 (citing an interview of Frank Oppenheimer by Smith, 4/14/76); Royal, *The Story of J. Robert Oppenheimer,* p. 50; Davis, *Lawrence and Oppenheimer,* p. 24.

73 **he "thought it"**: JRO, interview by Kuhn, 11/20/63, p. 18.

74 **In the event, Ehrenfest:** In 1933 Ehrenfest shot and killed his mentally retarded son and then turned the gun on himself. John Archibald Wheeler with Kenneth Ford, *Geons, Black Holes, and Quantum Foam,* p. 260.

74 **"Oppenheimer is now"**: Max Born to Paul Ehrenfest, 7/26/27, and 8/7 or 17/27, Ehrenfest letters, Archives of the History of Quantum Physics, NBL, AIP, courtesy of Nancy Greenspan, Born's biographer.

74 **Only six weeks:** Barnett, "J. Robert Oppenheimer," *Life,* 10/10/49.

74 **His Dutch friends:** Serber, *Peace and War,* p. 25; Rabi, et al., *Oppenheimer,* p. 17. According to Peter Michelmore, it was Paul Ehrenfest who nicknamed Robert "Opje" (Michelmore, *The Swift Years,* p. 37).

74 **"For the development"**: Victor Weisskopf, *The Joy of Insight,* p. 85.

74 **"that Bohr with his largeness"**: JRO, interview by Kuhn, 11/20/63, pp. 20–21. *Herausprugeln* means an inner thrashing or disciplining (courtesy of Helma Bliss Goldmark). Ehrenfest once teased Oppenheimer about his philosophical bent, cheerily telling him, "Robert, the reason you know so much about ethics is that you have no character" (Herken, *Brotherhood of the Bomb,* p. 15).

75 **He had ignored:** JRO letter to James Chadwick, 1/10/67, JRO papers, box 26, LOC.

75 **"prefers to live"**: Smith and Weiner, *Letters,* p. 127.

75 **"At first [we] thought"**: JRO, interview by Kuhn, 11/20/63, pp. 22–23.

75 **"He was such a good"**: Royal, *The Story of J. Robert Oppenheimer,* p. 45.

75 **"so young and already"**: Ed Regis, *Who Got Einstein's Office?,* p. 195.

76 **"His ideas are always"**: Michelmore, *The Swift Years,* p. 28.

76 **"nim-nim-nim man"**: Regis, *Who Got Einstein's Office?,* p. 133.

76 **"His strength," Pauli soon:** Wolfgang Pauli, *Scientific Correspondence,* vol. I, p. 486.

76 **"We got along"**: Jeremy Bernstein, "Profiles: Physicist," *The New Yorker,* 10/13/75 and 10/20/75.

76 **"Even in casual conversation"**: Rigden, *Rabi,* p. 19; Bernstein, *Oppenheimer,* p. 5.

76 **"never got to be an integrated"**: Rigden, *Rabi,* pp. 228–29.

77 **"Oppenheimer? A rich spoiled Jewish brat"**: Pais, *The Genius of Science,* p. 276.

77 **"I didn't think"**: Rabi, interview by Sherwin, 3/12/82, pp. 7, 12–13.

77 **"I was never in the same class"**: Rigden, *Rabi,* p. 214.

77 **"Rabi was a great"**: Rigden, *Rabi,* p. 215.

78 **"We felt a certain":** Rigden, *Rabi,* pp. 218–19.

78 **"air of easy nonchalance"** *and subsequent quotes:* Royal, *The Story of J. Robert Oppenheimer,* pp. 45–46; Rabi, et al., *Oppenheimer,* p. 5 (Introduction).

78 **"The time with Pauli":** JRO, interview by Kuhn, 11/20/63, p. 22.

78 **By the time Robert left:** Rabi, et al., *Oppenheimer,* pp. 12, 72.

79 **"[Quantum mechanics] describes":** Brian Greene, *The Elegant Universe,* p. 111.

Chapter Six: "Oppie"

80 **"open up the place":** Smith and Weiner, *Letters,* pp. 126–27.

80 **Frank needed no:** Twenty-five years later, Robert would testify that Dr. Roger Lewis was one of those friends from whom he felt estranged since the war because "there has been a sense of hostility which I identified with their remaining close to the [Communist] party." Smith and Weiner, *Letters,* p. 132; JRO hearing, p. 190.

80 **"It was a great spree":** Frank Oppenheimer to Alice Smith, July 16 (no year), folder 4–24, box 4, Frank Oppenheimer Papers, UCB.

81 **"My two great loves":** Royal, *The Story of J. Robert Oppenheimer,* p. 49.

81 **On one trip:** Frank Oppenheimer, interview by Weiner, 2/9/73, p. 51.

81 **"We'd get sort of drunk":** *The Day After Trinity,* dir. Jon Else, transcript, pp. 5–6; Uhlenbeck, interview by Alice Smith, 4/20/76, p. 9; Frank Oppenheimer, interview by Weiner, 2/9/73, p. 52.

81 **"I think we probably":** Frank Oppenheimer, interview by Weiner, 2/9/73, p. 51.

81 **"So we set the horses":** Frank Oppenheimer to Alice Smith, July 16 (no year), folder 4–24, box 4, Frank Oppenheimer Papers, UCB.

81 **"The reason why":** JRO to Frank Oppenheimer, 3/12/30, folder 4–12, box 1, Frank Oppenheimer Papers, UCB.

81 **"Your tales of a burro":** Smith and Weiner, *Letters,* p. 132.

82 **"The undergraduate college":** Smith and Weiner, *Letters,* p. 133.

82 **"I had for":** JRO, interview by Kuhn, 11/20/63, p. 29.

82 **"I'm going so slowly":** Royal, *The Story of J. Robert Oppenheimer,* p. 54.

82 **"I was a very difficult":** JRO, interview by Kuhn, 11/20/63, p. 30.

83 **"We were always"; "Well, Robert":** Goodchild, *Oppenheimer,* p. 25; Royal, *The Story of J. Robert Oppenheimer,* p. 55.

83 **"Robert's blackboard manners":** Smith and Weiner, *Letters,* p. 149; Leo Nedelsky, interview by Alice Smith, 12/7/76.

83 **"tendency to answer your question":** Rabi, et al., *Oppenheimer,* p. 18; Royal, *The Story of J. Robert Oppenheimer,* p. 56.

83 **"He could . . . be":** Harold Cherniss, interview by Sherwin, 5/23/79, pp. 2–3.

83 **"somewhat obscurely":** Smith and Weiner, *Letters,* p. 149.

84 **"She went on a hunger":** Smith and Weiner, *Letters,* p. 149; Nedelsky, interview by Alice Smith, 12/7/76.

84 **"It is easy":** Barnett, "J. Robert Oppenheimer," *Life,* 10/10/49, p. 126.

84 **"Oppenheimer was interested":** Lillian Hoddeson, et al., eds., *The Rise of the Standard Model,* p. 311; Rabi, et al., *Oppenheimer,* p. 18.

85 **"I didn't start":** JRO, interview by Kuhn, 11/20/63.

85 **In 1934:** Serber, *Peace and War,* p. 28.

85 **"unbelievable vitality":** Herbert Childs, *An American Genius,* p. 143.

85 **Even after Lawrence:** Herken, *Brotherhood of the Bomb,* p. 51. Lawrence also had in mind his other good friend, Robert Cooksey.

86 **By early 1931:** Rhodes, *The Making of the Atomic Bomb,* p. 148; Davis, *Lawrence and Oppenheimer,* pp. 17, 30–31.

86 **"an activity that":** Patrick J. McGrath, *Scientists, Business, and the State,* pp. 36, 64.

86 **Building cyclotrons with:** Gray Brechin, *Imperial San Francisco,* pp. 312, 354.

87 **"like marriage and poetry":** Nedelsky, interview by Alice Smith, 12/7/76.

87 **"Pauli thought it was nonsense":** JRO, interview by Kuhn, 11/20/63, p. 25.

87 **Anderson's discovery came:** Schweber, *In the Shadow of the Bomb,* p. 66; Gribbin, *Q Is for Quantum,* pp. 266, 107.

88 **"It was amazing":** Serber, interview by Sherwin, 1/9/82, p. 14.

88 **"his work is apt":** Nedelsky, interview by Alice Smith, 12/7/76; Schweber, *In the Shadow of the Bomb,* p. 68.

88 **"His physics was good":** Regis, *Who Got Einstein's Office?,* p. 147.

88 **Robert did not have:** Serber, interview by Sherwin, 1/9/82, p. 15. Willis Lamb earned his Ph.D. in physics in 1938 under Oppenheimer. See Gribbin, *Q Is for Quantum,* pp. 203–4.

89 **"He was an idea man":** Melba Phillips, interview by Sherwin, 6/15/79, p. 5.

89 **His interest in astrophysics:** Rabi, et al., *Oppenheimer,* p. 16.

89 **"white dwarfs":** *Physics Review,* 10/1/38.

89 **"one of the great":** *Physics Review,* 9/1/39; Bernstein, *Oppenheimer,* p. 48.

90 **"Oppenheimer's work with Snyder":** Marcia Bartusiak, *Einstein's Unfinished Symphony,* pp. 60–61; Bernstein, *Oppenheimer,* pp. 48–50.

90 **Characteristically, however, Oppenheimer:** Gribbin, *Q Is for Quantum,* pp. 45, 266.

90 **"Oppie was always":** Serber, interview by Sherwin, 1/9/82, p. 15.

90 **Having made the initial:** Rabi, et al., *Oppenheimer,* pp. 13–17.

90 **"Robert's own knowledge":** Nedelsky, interview by Alice Smith, 12/7/76.

90 **"Oppenheimer was a very":** Edwin Uehling, interview by Sherwin, 1/11/79, pp. 5–6.

91 **"The work is fine":** Smith and Weiner, *Letters,* p. 159 (JRO to Frank Oppenheimer, fall 1932).

91 **Unlike many European theorists:** Rigden, *Rabi: Scientist and Citizen,* p. 7.

91 **"Had Oppenheimer gone":** Decades later, Oppenheimer himself thought all copies of these syllabus/lecture notes had disappeared. JRO, interview by Kuhn, 11/20/63, p. 28; Royal, *The Story of J. Robert Oppenheimer,* pp. 64–65. Actually, Sherwin obtained a copy from Herve Voge. It will be donated to an appropriate archive.

91 **"I need physics":** Smith and Weiner, *Letters,* p. 135 (letter of 10/14/29).

91 **When Julius found out:** Smith and Weiner, *Letters,* p. 138.

91 **"Natalie was a dare-devil":** Smith and Weiner, *Letters,* pp. 172, 191; Helen Campbell Allison, correspondence with Alice Smith, undated (circa 1976), Alice Smith interview notes. Natalie Raymond died in 1975.

92 **"young wives falling for Robert":** Helen C. Allison, interview by Alice Smith, 12/7/76.

92 **"Everyone wants rather":** JRO to Frank Oppenheimer, 10/14/29; Smith and Weiner, *Letters,* p. 135.

92 **"the jeunes filles Newyorkaises":** JRO to Frank Oppenheimer, 10/14/29; Smith and Weiner, *Letters,* p. 135.

93 **"His mere physical appearance":** Cherniss, interview by Sherwin, 5/23/79, pp. 1–2.

Chapter Seven: "The Nim Nim Boys"

94 **The stock market crash:** Cassidy, *J. Robert Oppenheimer and the American Century,* p. 123.

94 **"reconstructed Chrysler emitted":** Julius Oppenheimer to Frank Oppenheimer,

3/11/30, folder 4–11, box 4, Frank Oppenheimer Papers, UCB; Michelmore, *The Swift Years,* p. 33.

94 **"How far is it wise":** Smith and Weiner, *Letters,* p. 139 (3/12/30).

94 **"We did everything":** Uehling, interview by Sherwin, 1/11/79, pp. 2, 9.

94 **"I'll be back presently** *and subsequent quotes: San Francisco Chronicle,* 2/14/34, p. 1; Serber, *Peace and War,* p. 27; Serber, interview by Jon Else, 12/15/79, p. 26.

95 **In 1934:** Royal, *The Story of J. Robert Oppenheimer,* p. 63; Serber, *Peace and War,* p. 25; Smith and Weiner, *Letters,* pp. 149, 186; Herken, *Brotherhood of the Bomb,* p. 13; Robert Serber, interview by Jon Else, 12/15/79, p. 23.

95 **"the most beautiful harbor":** Smith and Weiner, *Letters,* p. 143 (JRO to Frank Oppenheimer, 8/10/31). For the description of the Shasta house, see Edith A. Jenkins, *Against a Field Sinister,* p. 28, and Robert Serber, interview by Jon Else, 12/15/79, p. 23.

96 **His long bony fingers:** Chevalier, *Oppenheimer,* pp. 20–21.

96 **"They copied his gestures":** Rabi, et al., *Oppenheimer,* p. 20; Rigden, *Rabi,* p. 213.

96 **"He [Oppie] was like a spider":** Jeremy Bernstein, *Oppenheimer,* p. 62.

96 **"We weren't supposed":** Uehling, interview by Sherwin, 1/11/79, p. 15.

96 **"He read a good deal":** Harold Cherniss, interview by Sherwin, 5/23/79, p. 10.

97 **"so that Robert":** Herbert Smith, interview by Weiner, 8/1/74, p. 14.

97 **"knew all the best restaurants":** Harold Cherniss, interview by Sherwin, 5/23/79, p. 8.

97 **He always picked up:** Serber, *Peace and War,* pp. 29–31.

97 **"The world of good food":** Royal, *The Story of J. Robert Oppenheimer,* p. 63, quoting Serber.

97 **Most people left:** Uehling, interview by Sherwin, 1/11/79, p. 15.

97 **"Frank has done this work":** Phillips, interview by Sherwin, pp. 9–11. Carlson later taught physics at Princeton and several other universities; in 1955 he committed suicide.

97 **Each spring:** Rabi, et al., *Oppenheimer,* p. 19.

98 **"Shut up, Pauli":** Smith and Weiner, *Letters,* p. 141.

98 **During the summer:** Frank Oppenheimer to Royal, 2/25/67, folder 4–23, box 4, Frank Oppenheimer Papers, Bancroft Library.

98 **"Mother critically ill":** Smith and Weiner, *Letters,* pp. 144–45 (JRO to Ernest Lawrence, 10/12/31, 10/16/31).

98 **"I'm the loneliest man":** Herbert Smith, interview by Weiner, 8/1/74, p. 12; Michelmore, *The Swift Years,* p. 33; Royal, *The Story of J. Robert Oppenheimer,* pp. 61–62.

99 **"They are very good fun":** Smith and Weiner, *Letters,* pp. 152–53 (Julius Oppenheimer to Frank Oppenheimer, 1/18/32).

99 **"Nobody could make":** Uehling, interview by Sherwin, 1/11/79, p. 31.

99 **"He is an astounding person":** Cherniss, interview by Sherwin, 5/23/79, p. 5; Smith and Weiner, *Letters,* pp. 143, 165; *Time,* 11/8/48, p. 75.

99 **"He liked things that":** Cherniss, interview by Sherwin, 5/23/79, p. 11.

99 **"It is very easy":** Smith and Weiner, *Letters,* pp. 143, 165; Royal, *The Story of J. Robert Oppenheimer,* p. 64.

99 **Robert was so enraptured:** Smith and Weiner, *Letters,* p. 164; Michelmore, *The Swift Years,* p. 39.

100 **Like many Western:** For an exploration of the influence of the Bhagavad-Gita on Western intellectuals, see Jeffery Paine, *Father India.*

100 **"Therefore," he concluded:** Smith and Weiner, *Letters,* pp. 155–56 (JRO to Frank Oppenheimer, 3/12/32).

100 **"Why not the Talmud?":** Rabi, interview by Sherwin, 3/12/82.

101 **"The Meghaduta I read":** James A. Hijiya, "The Gita of J. Robert Oppenheimer"; Smith and Weiner, *Letters,* p. 180.

101 *"Vanquish enemies at arms":* Hijiya, "The Gita of J. Robert Oppenheimer," p. 146; Barbara Stoler Miller, trans., *Bhartrihari: Poems,* p. 39.

101 **"From conversations with him":** Friess, *Felix Adler and Ethical Culture,* p. 124; Rabi, et al., *Oppenheimer,* p. 4.

101 **"I may, as we all have to":** We are indebted to James Hijiya for suggesting this interpretation of Oppenheimer's fascination with the Gita (Hijiya, "The Gita of J. Robert Oppenheimer," *Proceedings of the American Philosophical Society* vol. 144, no. 2 (2000), pp. 161–64; JRO, *Flying Trapeze,* p. 54).

102 **In June 1934:** Serber, *Peace and War,* pp. 25–29.

102 **Charlotte took her politics:** JRO FBI file, doc. 241, p. 12, 1/31/51, declassified 2001.

102 **"no definite evidence":** Ibid.; Barton J. Bernstein, "Interpreting the Elusive Robert Serber," p. 12.

103 **"one of the few really first-rate":** Bernstein, "Interpreting the Elusive Robert Serber," p. 11; Bernstein cites JRO to Ernest Lawrence, 7/20/38, box 16, Lawrence Papers, UCB.

103 **"For the first few days":** Serber, *Peace and War,* pp. 38–39.

104 **The next morning:** Else Uhlenbeck, interview by Alice Smith, 4/20/76, pp. 11–12.

104 **"To many of my friends":** JRO hearing, p. 8.

104 **"an unworldly, withdrawn un-esthetic person":** Robert Serber, 1972 J. Robert Oppenheimer Memorial Prize acceptance speech, biographical file, Oppenheimer Memorial Prize, AIP Archives.

104 **"active member of the Communist Party":** JRO FBI file, doc. 241, p. 13, 1/31/51, declassified 2001.

105 **A young professor:** Chevalier, *Oppenheimer,* p. 29.

105 **"Never since the Greek tragedies":** Jenkins, *Against a Field Sinister,* pp. 23, 27. Serber, *Peace and War,* p. 43.

105 **"We were not political":** Phillips, interview by Sherwin, 6/15/79, p. 1. In 1947, the FBI's J. Edgar Hoover claimed that Phillips had "reportedly distributed Communist pamphlets" at Brooklyn College (Hoover to Commerce Secretary Averell Harriman, 9/6/47, folder: *Arms Control, 1947,* Harriman Papers, Kai Bird Collection). In the early 1950s, Phillips was subpoenaed for questioning by the McCarran Committee. She refused to cooperate with the committee and was dismissed from Brooklyn College and the Columbia Radiation Laboratory. In 1987, Brooklyn College publicly apologized.

105 **"I know three people":** Nedelsky, interview by Alice Smith, 12/7/76; Smith and Weiner, *Letters,* p. 195.

105 **He immediately agreed:** Smith and Weiner, *Letters,* p. 173.

105 **Similarly, Max Born:** "Obituary: Prof. Max Born," *The Times* of London, 1/7/70.

106 **Although Sinclair lost:** Stephen Schwartz, *From West to East,* pp. 226–46.

106 **"We were sitting up high":** Serber, *Peace and War,* p. 31.

106 **"It was very nice":** Frank Oppenheimer oral history, interview by Weiner, 2/9/73.

107 **friendship as "very close":** Smith and Weiner, *Letters,* pp. 194–95.

107 **"made a sweet island":** JRO, interview by Kuhn, 11/18/63, p. 19.

107 **"one Jew in the department":** Serber, *Peace and War,* pp. 42, 50.

107 **"I could be":** JRO, interview by Kuhn, 11/20/63, p. 31; Smith and Weiner, *Letters,* pp. 181, 190. The mathematician Hermann Weyl made the offer to Oppenheimer about joining the Institute for Advanced Study.

Chapter Eight: "In 1936 My Interests Began to Change"

111 **"he began to court her":** Jenkins, *Against a Field Sinister,* p. 23; JRO hearing, p. 8.

111 **"like an old Irish princess":** Priscilla Robertson, undated ltr. entitled "Promise," circa January 1944 addressed to the deceased Jean Tatlock, Sherwin Collection. Edith Jenk-

ins reports that Tatlock had blue eyes (p. 28), but the coroner's death certificate for Tatlock described them as hazel. Michelmore reports them as a "luminous green" (*The Swift Years*, p. 47).

111   **Five feet, seven inches:** City and County of San Francisco Coroner's Office, Coroner's report for Jean Tatlock, 1/6/44; secret FBI memo, "Subject: Jean Tatlock," 6/29/43, file A, RG 326, entry 62, box 1, NA.

111   **She had but one:** Jenkins, *Against a Field Sinister*, p. 28.

111   **"Jean was very private":** Jenkins, *Against a Field Sinister*, p. 21; Michelmore, *The Swift Years*, p. 52.

111   **Over lunch at the Faculty Club:** Chevalier, *Oppenheimer*, p. 13; Nuel Pharr Davis, a not always reliable source, claimed that Professor Tatlock "did not care for Jews." He also quotes Mrs. Tatlock saying, "I must go to pick up my fascist husband and radical daughter" (Davis, *Lawrence and Oppenheimer*, p. 82). On the other hand, in 1938 Prof. Tatlock joined Oppenheimer, Chevalier, and other Berkeley professors in raising $1,500 in support of the East Bay chapter of the Medical Bureau to Aid Spanish Democracy, an act highly unlikely for a fascist or a conservative (*People's Daily World*, 1/29/38, p. 3).

111   **"Batter my heart":** Jenkins, *Against a Field Sinister*, p. 24.

111   **Jean owned a roadster:** Ibid., p. 26.

112   **"the most promising girl":** Priscilla Robertson, "Promise," seven-page letter, circa January 1944.

112   **"having gotten by nature":** Ibid.

112   **"I just wouldn't want":** Ibid.

112   **"It was this social conscience":** Her poor grades that year perhaps reflect the time she gave to the Party. She received an A in psychology—but mostly C's in her premed courses (University of California, Berkeley, Graduate School transcript, 1935–36; Jean Tatlock to Priscilla Robertson, undated, circa 7/15/35.)

113   **"I find it impossible":** The Berkeley chapter of the Communist Party routinely harassed any of its members who went into analysis. When Frances Behrend Burch, a friend of the Chevaliers', joined the Party in 1942, she simultaneously began seeing Donald MacFarlane, a Freudian analyst and a good friend of the Oppenheimers'. When Party officials learned of her analysis, they attempted to persuade her to end the sessions. (Kent Mastores and Constance Rowell Mastores, e-mail to Kai Bird, 5/6/04. Constance is Burch's daughter.)

113   **"a feeling for the sanctity":** Tatlock to Robertson, circa 7/15/35.

113   **"worthy of Robert":** Royal, *The Story of J. Robert Oppenheimer*, p. 69.

113   **"All of us were":** Jenkins, *Against a Field Sinister*, p. 22.

114   **"You must remember":** Ibid.

114   **"There were a half dozen":** Serber, interview by Sherwin, 1/9/82, pp. 9–10. See also Serber, *Peace and War*, p. 46.

114   **"Jean was Robert's":** Haakon Chevalier, interview by Sherwin, 5/9/80.

114   **"Beginning in late 1936":** JRO hearing, p. 8.

114   **"He manifested deep interest":** Avram Yedidia to Sherwin, 2/14/80.

115   **"If ever a revolution was due"** *and subsequent quotes:* Harvey Klehr, *The Heyday of American Communism*, pp. 270, 413; Ellen Schrecker, *Many Are the Crimes*, p. 15; Edward L. Barrett, Jr., *The Tenney Committee*, p. 1; *The Nation*, 9/12/34, cited by Dorothy Healey, *Dorothy Healey Remembers*, pp. 40, 59; Steve Nelson, et al., *American Radical*, p. 262.

115   **"I liked the new sense":** JRO hearing, p. 8.

115   **"opened the door":** The phrase "opened the door" comes from Oppenheimer's first draft of his autobiographical statement for the 1954 hearing. He cut the phrase in the final version. See Goodchild, *Oppenheimer*, p. 233.

116 **After three terrifying months:** "Dr. Peters Replies to Oppenheimer," *Rochester Times Union,* 6/15/49; hearings before the HUAC, 7/8/49, p. 9, Bernard Peters Papers, NBA. Peters testified, "I was transferred to a prison in Munich and then I was released." Peters also testified at this time that neither he nor his wife, Hannah, had ever been a member of the Communist Party.

116 **"died in my hands":** Bernard Peters, "Report of a Prisoner at the Concentration Camp at Dachau, Near Munich," written by Peters in 1934 in New York; Peters, "War Crimes," 5/11/45, Peters Papers, NBA.

116 **"a little different from most of us":** Schweber, *In the Shadow of the Bomb,* p. 120.

117 **When Peters displayed:** Ibid., pp. 120, 220.

117 **"strengthened a conviction":** Dr. Hannah Peters to Mrs. Ruth B. Shipley, chief, Passport Division, Department of State, 8/28/51, Peters Papers, NBA. Appealing Shipley's refusal to issue her a passport, Peters flatly denied she had ever been a member of the Communist Party. She said she had been a member of the Anti-Fascist Refugee Committee.

117 **Hannah also insisted:** JRO to the editors of the *Rochester Democrat and Chronicle,* 6/30/49, Peters Papers, NBA. In September 1943, Oppenheimer told Col. Lansdale and Gen. Groves that he thought Hannah Peters was a member of the CP; Herken, *Brotherhood of the Bomb,* p. 111; JRO FBI file, memo 4/28/54, document 1320; See also AEC report on JRO (*Rochester Times Union,* 7/7/54, folder 11, Bernard Peters Papers, NBA).

117 **He was favorably:** Stern, *The Oppenheimer Case,* p. 19.

117 **"I suppose somewhere":** Cherniss, interview by Sherwin, 5/23/79, p. 5.

117 **"better read":** Chevalier's diary notation is dated 7/20/37—but his friend "E." reported that Oppenheimer had read *Das Kapital* the previous summer. See Chevalier, *Oppenheimer,* p. 16; Steve Nelson was told the same story: Steve Nelson, et al., *American Radical,* p. 269.

118 **Born in 1901:** Haakon Chevalier FBI file (100-18564), part 1 of 2, background report, pp. 2, 16.

118 **"He was a terribly charismatic":** Larken Bradley, "Stinson Grand Dame Barbara Chevalier Dies," *Point Reyes Light,* 7/24/03.

118 **Frequently partying late:** Haakon Chevalier, *Oppenheimer,* p. 30; Barbara Chevalier "diary," 8/8/81, courtesy of Gregg Herken, *www.brotherhoodofthebomb.com.*

118 **"gave shelter and moral support":** Jenkins, *Against a Field Sinister,* p. 25.

119 **"to witness the transition":** Chevalier, *Oppenheimer,* pp. 8–9.

119 **"the new vision":** Chevalier, *Oppenheimer,* p. 8; Axel Madsen, *Malraux,* p. 195.

119 **Over these years:** Robert A. Rosenstone, *Crusade of the Left,* p. vii; Schrecker, *Many Are the Crimes,* p. 15.

120 **"anxious to do something":** Chevalier, *Oppenheimer,* p. 16.

120 **"A group of people":** JRO hearing, p. 156; memo to FBI director, 1/17/58, regarding a term paper written by Mrs. Fred Airy, formerly Helen A. Lichens, entitled, "Term Report: Teachers' Union of Berkeley and Oakland, Spring 1936." Mrs. Airy explained to the FBI that she had written this term paper while a student at Berkeley in 1936. In the course of researching her paper, she attended many of the union meetings and interviewed its officers.

120 **"hallucinatory feeling"** *and subsequent quotes:* Chevalier, *Oppenheimer,* pp. 16–19, 21–22.

121 **"Oh for God's sake":** Michelmore, *The Swift Years,* p. 49.

121 **"a good friend":** JRO hearing, pp. 155, 191. When in 1950 the FBI questioned Oppenheimer about Dr. Addis, Oppenheimer refused to discuss the doctor, saying he was "dead and couldn't defend himself" about "being close to the Communist Party." By then, Addis' widow told Linus Pauling that she did not want her late husband's politi-

cal views discussed in a memorial essay for the National Academy of Sciences, because she and her two children "feared for their own safety" Kevin V. Lemley and Linus Pauling, "Thomas Addis," *Biographical Memoirs,* p. 3.

121 **Even as a young doctor:** Richard M. Lippman, M.D., to Linus Pauling, 2/1/55, Addis Memorial Committee, box 60, Linus Pauling Papers, Oregon State University.

121 **In 1944 he was elected:** Lemley and Pauling, "Thomas Addis," p. 6.

121 **Even as he was building:** Ibid., p. 5; see also Dr. Frank Boulton e-mail to Kai Bird, 4/27/04, and Herken, website, *www.brotherhoodofthebomb.com* (endnotes for chapter 2, note 33).

122 **He was a friend:** Frank Boulton, "Thomas Addis (1881–1949)," *Journal of the Royal College of Physicians of Edinburgh,* vol. 33, pp. 135–42; Lemley and Pauling, "Thomas Addis," p. 28.

122 **In 1935, Addis:** Herbert Romerstein and Eric Breindel, *The Venona Secrets,* pp. 265–66. Romerstein and Breindel cite "Comintern Archives, Moscow, Fond 515, Opis 1, Delo 3875." They also cite a 1944 FBI report that described Addis as "active in 27 Communist Front organizations in the San Francisco Bay Area during the last ten years." Addis: San Francisco field report, 5/17/44, sect. 4, Federation of Architects, Engineers, Chemists and Technicians (FAECT) file, no. 61-723, FBI.

122 **"an act of faith":** Lippman to Pauling, 2/1/55, with attached draft memoir essay on Addis, Addis Memorial Committee, box 60, Pauling Papers, Oregon State University. Lemley and Pauling, "Thomas Addis," p. 29.

122 **"a great man":** Pauling to Donald Tresidder (president, Stanford University), box 77, Pauling Papers; Dr. Horace Gray to Pauling, 4/5/57, Addis Memorial Committee, box 60, Linus Pauling Papers, Oregon State University.

122 **"close to one":** JRO hearing, p. 1004.

122 **"Injustice or oppression":** Dr. Frank Weymouth (chair of physiology dept., Stanford University) to Addis Memorial Committee, box 60, Linus Pauling Papers, Oregon State University.

122 **"instrumental":** Thomas Addis, ltr. addressed to "Dear Friend," September 1940, Addis correspondence with Pauling, 1040–42, box 59, Pauling Papers, Oregon State University. Other sponsors included Helen Keller, Dorothy Parker, George Seldes, and Donald Ogden Stewart.

123 **"reached into his work":** Ibid.; Boulton, "Thomas Addis (1881–1949)," p. 24.

123 **"You are giving":** JRO hearing, pp. 183, 185, 9.

123 **His annual donations:** According to the Bureau of Labor Statistics' Consumer Price Index Adjuster, a dollar in 1938 had the purchasing power of $12.42 in 2001.

123 **Robert's last such:** JRO hearing, pp. 5, 9, 157; Stern, *The Oppenheimer Case,* p. 22.

123 **"He was a respected":** Nelson, interview by Sherwin, 6/17/81, p. 14; Nelson, et al., *American Radical,* p. 258; Haakon Chevalier FBI file (100-18564), part 1 of 2, SF 61-439, p. 37.

123 **"I doubt that":** JRO hearing, p. 9.

124 **"formulation of issues":** Ibid., p. 157; Stern, *The Oppenheimer Case,* p. 22.

124 **Late in January 1938:** Oppenheimer's donation was given to the American Medical Bureau to Aid Spanish Democracy (see *Daily People's World,* 1/29/38, p. 3, cited in FBI background report on Oppenheimer, 2/17/47). The U.C. Berkeley fund-raising committee included Oppenheimer, Chevalier, Rudolph Schevill, Robert Brady, G. C. Cook, Frank Oppenheimer, John S. P. Tatlock, A. G. Brodeur, R. D. Calkins, H. G. Eddy, E. Gudde, W. M. Hart, S. C. Morley, G. R. Hoyes, A. Perstein, M. I. Rose, F. M. Russell, L. B. Simpson, P. S. Taylor, A. Torres-Rioseco, R. Tryon, and T. K. Whipple.

124 **That spring, Robert:** *Daily People's World,* 4/26/38; *ACLU News,* vol. IV, no. 1, San Francisco, January 1939, p. 4; JRO hearing, p. 3.

124 **"It was a time":** Chevalier, interview by Sherwin, 6/29/82, p. 3.

124 **He nevertheless stood up:** Chevalier, *Oppenheimer*, pp. 32–33; Chevalier, interview by Sherwin, 6/29/82, p. 4. In the spring of 1939 Oppenheimer served as chairman of Local 349's Educational Policy Committee. Arthur Brodeur was president, and other committee chairmen included Chevalier and Philip Morrison (Joseph E. Fontrose, Secretary of Local 349, to Irvin R. Kuenzli, 4/27/39, reproduced from the collections of archives of Labor and Urban Affairs, Wayne State University, courtesy of John Cortesi).

124 **"Somehow one always knew":** Jenkins, *Against a Field Sinister,* p. 22.

125 **"As long as she":** Smith and Weiner, *Letters,* p. 202.

125 **An eloquent teacher:** Petteri Pietikainen, "Dynamic Psychology, Utopia, and Escape from History: The Case of C. G. Jung," *Utopian Studies,* vol. 12, no. 1 (1/1/01), p. 41.

125 **"fear of castration":** Siegfried Bernfeld Papers, "Psychoanalytic Committee—San Francisco," box 9, LOC, contains invitation lists and various topics discussed by the committee.

126 **"Some psychological damage":** Gerald Holton, "Young Man Oppenheimer," *Partisan Review,* 1981, vol. XLVIII, p. 385.

126 **"Bernfeld was one":** Siegfried Bernfeld Papers, "Psychoanalytic Committee—San Francisco," box 9, LOC; Dr. Robert S. Wallerstein, phone interview, 3/19/01; see also Daniel Benveniste, "Siegfried Bernfeld in San Francisco," unpublished essay, 5/20/93, and Benveniste's interview with Dr. Nathan Adler, courtesy of Dr. Benveniste. Bernfeld was analyzing Wolff and possibly other members of the group, which raises the question of whether Oppenheimer himself was undergoing analysis with Dr. Bernfeld. While Oppenheimer's name does not appear on a partial list of Dr. Bernfeld's patients, Bernfeld later told Adler that one of his patients was a physicist at Berkeley who had played a central role in designing the cyclotron.

126 **"seemed to treat physics":** Rabi, et al., *Oppenheimer,* p. 5.

126 **Things metaphysical:** Siegfried Bernfeld Papers, "Psychoanalytic Committee—San Francisco," box 9, LOC; Dr. Wallerstein phone interview, 3/19/01. Dr. Wallerstein said that he knew Oppenheimer had been "intensely interested" in psychoanalysis and for this reason had regularly attended Dr. Bernfeld's seminars; Dr. Stanley Goodman, a student of Dr. Bernfeld's, e-mail, 3/20/01; Ernest Jones, *The Life and Work of Sigmund Freud,* vol. 3, p. 344; Reuben Fine, *A History of Psychoanalysis,* p. 108.

127 **"You're too good a physicist":** Herbert Childs, *An American Genius,* pp. 266–67.

Chapter Nine: "[Frank] Clipped It Out and Sent It In"

128 **Julius' fortune:** J. Edgar Hoover to the president, FBI memo, 2/28/47, JRO FBI file.

128 **But as if:** JRO hearing, p. 8.

129 **"youthful cockiness":** Frank Oppenheimer, interview by Alice Smith, 3/17/75, p. 37.

129 **"Frank himself is a sweet":** Leona Marshall Libby, *The Uranium People,* p. 106.

129 **"He is a much finer person":** Herken, *Brotherhood of the Bomb,* p. 54; Herken's source is a letter from Clifford Durr to Frank Oppenheimer, 12/10/69, Durr folder, box 1, Frank Oppenheimer Papers, UCB.

129 **"I don't think you":** Smith and Weiner, *Letters,* p. 95.

129 **At Hopkins, he:** William L. Marbury to Allen Weinstein, 3/11/75, James Conant Papers, HU, courtesy of James Hershberg.

129 **"we had a fine holiday":** Smith and Weiner, *Letters,* p. 147. Frank's friend Roger Lewis persuaded him to go to Johns Hopkins rather than Harvard. See Frank Oppenheimer, interview by Alice Smith, 3/17/75, p. 10.

130 **"I know very well surely":** Smith and Weiner, *Letters,* p. 155.

130   **"You know how happy"**: Smith and Weiner, *Letters,* p. 163.

130   **"There has seldom been a time"**: Smith and Weiner, *Letters,* pp. 169–70.

130   **He loved tinkering**: Frank Oppenheimer, interview by Alice Smith, 3/17/75, p. 15.

130   **"reducing a specific"**: Paul Preuss, "On the Blacklist," *Science,* June 1983, p. 35.

130   **Robert "did something"**: Frank Oppenheimer oral history, told to Judith R. Goodstein, 11/16/84, p. 12, Caltech Archives.

130   **In the laboratory**: Frank Oppenheimer oral history, 2/9/73, AIP, pp. 38, 40.

130   **Whereas Robert took**: FBI background file on Frank Friedman Oppenheimer, 7/23/47, from D. M. Ladd to the director.

131   **"Jackie prided herself"**: Robert Serber, interview by Sherwin, 3/11/82, p. 11.

131   **They arrived in a brand-new**: Frank Oppenheimer to Alice Smith, July 16 (no year), folder 4–24, box 4, Frank Oppenheimer Papers, UCB.

131   **"It was an act"**: Michelmore, *The Swift Years,* p. 47; Goodchild, *J. Robert Oppenheimer,* p. 34.

131   **"The three of us saw"**: Frank Oppenheimer to Alice Smith, July 16 (no year), folder 4–24, box 4, Frank Oppenheimer Papers, UCB.

131   **"She could drive you crazy"**: Hans "Lefty" Stern, interview by Kai Bird, 3/4/04.

131   **As an undergraduate**: Frank Oppenheimer oral history, told to Goodstein, 11/16/84, p. 32, Caltech Archives.

132   **"I used to tell people"**: Frank Oppenheimer oral history, told to Goodstein, 11/16/84, pp. 9–11, Caltech Archives; William L. Marbury, *In the Catbird Seat,* p. 107.

132   **Upon his return**: Frank Oppenheimer oral history, told to Weiner, 2/9/73, p. 46, AIP.

132   **"I clipped it out"**: Frank Oppenheimer testimony, 6/14/49, "Hearings Regarding Communist Infiltration of Radiation Laboratory and Atomic Bomb Project at the University of California, Berkeley, Calif.," HUAC, p. 365; FBI report, 8/20/47, citing a *Minneapolis Star* article of 7/12/47. In 1938 his book number was 60439 and in 1939 it was 1001.

133   **"The intellectuals who were drawn"**: Frank Oppenheimer to Denise Royal, 2/25/67, folder 4–23, box 4, Frank Oppenheimer Papers, UCB.

133   **"We tried to integrate"**: Frank Oppenheimer, interview by Sherwin, 12/3/78; Frank Oppenheimer oral history, interviewed by Goodstein, 11/16/84, Caltech Archives, pp. 14–15. Jackie Oppenheimer testimony, 6/14/49, "Hearings Regarding Communist Infiltration of Radiation Laboratory and Atomic Bomb Project at the University of California, Berkeley, Calif.," HUAC, p. 377.

133   **"was essentially a secret group"**: Jackie Oppenheimer testimony, 6/14/49; Frank Oppenheimer oral history, interviewed by Goodstein, 11/16/84, p. 15.

133   **"I remember a friend"**: Frank Oppenheimer oral history, interviewed by Weiner, 2/9/73, AIP, p. 46.

133   **The Stanford physicist**: Frank Oppenheimer, interview by Sherwin, 12/3/78.

133   **One day Ernest Lawrence**: Michelmore, *The Swift Years,* p. 115.

134   **"made a rather pathetic impression"**: FBI summary memo on Frank Oppenheimer, 7/23/47, p. 2; JRO hearing, pp. 101–2.

134   **"We spent a lot of time"**: Frank Oppenheimer, interview by Sherwin, 12/3/78.

134   **"He frequently spoke"**: FBI summary memo on Frank Oppenheimer, 7/23/47, p. 3.

134   **"He was passionately fond"**: JRO hearing, p. 102.

134   **"I was quite upset"**: JRO hearing, pp. 186–87.

135   **"in his opinion Frank"**: FBI summary memo on Frank Oppenheimer, 7/23/47, pp. 3–4.

135   **"very brief and very intense"**: JRO, interview by John Lansdale, 9/12/43; JRO hearing, pp. 871–86.

136   **"In those days . . . the Party"**: Jessica Mitford, *A Fine Old Conflict,* p. 67.

137 **"How shall we dissipate"**: Klehr, *The Heyday of American Communism,* p. 413.
137 **"We/he initiated it"**: Haakon Chevalier, interview by Sherwin, 6/29/82, pp. 3, 4, 6, 7; see also Chevalier, *Oppenheimer,* p. 19. Many years after her divorce, Barbara Chevalier noted in her unpublished memoir that Opje and Haakon had "joined a secret unit of the Communist Party. There must have been only six or eight members—a doctor, a wealthy businessman (maybe)." Barbara noted that she had deliberately not wanted to remember the names of those involved (Barbara Chevalier manuscript, 8/8/81, courtesy of Gregg Herken).
137 **For almost a year the FBI:** Born in Russia in 1905, Schneiderman came to the United States when he was three years old. In 1939, government prosecutors attempted to revoke his citizenship and deport him. The case was still under appeal at the time of his meeting with Oppenheimer; in 1943 the Supreme Court upheld Schneiderman's citizenship (Klehr, *The Heyday of American Communism,* p. 484).
137 **"the big boys"**: FBI report, 5/19/41, document 2, and FBI teletype, 10/16/53, San Francisco bureau to FBI director, Haakon Chevalier, FBI file, part 1 of 2. The cable reports that when Schneiderman and Folkoff arrived, "there were observed parked in Chevalier's driveway cars registered to [blank] and J. Robert Oppenheimer."
138 **"persons to be considered"**: N.J.L. Piper to FBI director, 3/28/41, JRO FBI file, sect. 1, doc. 1.
138 **Another FBI document:** FBI report, 6/18/54, by Joe R. Craig, with attachment, "Excerpts from 97-1 (C-14)." The attachment is undated, but judging from the context of the excerpts, it must have been written sometime after August 1941, when Oppenheimer moved into his home at One Eagle Hill, Berkeley. Oppenheimer had met Helen Pell through their joint activities on behalf of the Committee to Aid Democratic Spain. (Pell was also a good friend of Steve Nelson; see Nelson, interview by Sherwin, p. 13.) Dr. Addis, of course, was Jean Tatlock's friend and the man who initially funneled Oppenheimer's donations on behalf of the Spanish Republic to the Communist Party. Alexander Kaun was a Berkeley professor who rented Oppenheimer his house for a time. In 1943 Oppenheimer told Lt. Col. Lansdale that he knew Kaun was a member of the American Soviet Council—but that he did not know whether he was a Party member (JRO hearing, p. 877). George Andersen was identified as the "official Communist Party Attorney" in San Francisco. Aubrey Grossman and Richard Gladstein were attorneys for the union leader Harry Bridges.
138 **Morrison, of course:** See Philip Morrison testimony, 5/7–8/53, "Subversive Influence in the Educational Process," 83rd U.S. Congress, Senate Committee on the Judiciary, part 9, pp. 899–919.
138 **When asked about Chevalier's:** Morrison, interview by Sherwin, 6/21/02.
138 **"What made you a member?"** *and subsequent quotes:* Haakon Chevalier, interviews by Sherwin, 6/29/82, pp. 6–7, and 7/15/82, p. 5.
138 **"I don't know that I could"**: Nelson, interview by Sherwin, 6/17/81, p. 14.
139 **"My own estimate"**: Nelson, interview by Sherwin, 6/17/81, p. 22.
140 **"liaison with the Faculty group"**: Griffiths, "Venturing Outside the Ivory Tower: The Political Autobiography of a College Professor," unpublished manuscript, LOC. Griffiths produced two versions of this typed manuscript; the shorter, untitled manuscript names Oppenheimer as a member of the closed unit. Oppenheimer's name is not used in the longer manuscript; apparently, when Griffiths began to circulate the manuscript for possible publication, a friend persuaded him that he should not disclose Oppenheimer's name. We are quoting here from the shorter manuscript, p. 26.
141 **he "did not consider"**: Gordon Griffiths, "Venturing Outside the Ivory Tower," unpublished manuscript, shorter version, LOC, p. 26; FBI report of interview with Kenneth O. May, 3/5/54, JRO FBI file.

141 **Once a graduate student:** Kenneth May, confidential letter to Dr. Lawrence M. Gould, president of Carleton College, 9/25/50, Carleton College Archives, courtesy of college archivist Eric Hilleman. May wrote an article in the *New Masses* entitled "Why My Father Disinherited Me." David Hawkins, interview by Sherwin, 6/5/82, p. 15.

142 **"agree with CP aims":** FBI report of interview with Kenneth May, 3/5/54. May left the Party sometime during World War II. In 1946 he finally obtained his Ph.D. in mathematics, and later that year he joined the mathematics department at Carleton College in Northfield, MN. Interviews with John Dyer-Bennett, May's roommate at Berkeley, and Miriam May, May's third wife, by Bird, 5/15/01.

Chapter Ten: "More and More Surely"

143 **"I know Charlie":** Smith and Weiner, *Letters*, p. 211.

143 **No issue of the day was:** Maurice Isserman, *Which Side Were You On?*, pp. 32–54.

143 **"fantastic falsehood that":** *The Nation* reprinted this open letter (Schwartz, *From West to East*, p. 290).

143 **"that Opje proved himself":** Chevalier, *Oppenheimer*, pp. 31–32. In his 1959 novel, *The Man Who Would Be God,* Chevalier has his Oppenheimer character defend the Stalin-Hitler Pact with these words: " 'Even in the worst situation,' he said in a low voice, 'there is a right move, and there are many wrong ones. Since the Western powers violated their pledge to Czechoslovakia in Munich, Russia's situation has been dangerously exposed. This is surely the right move. Because it's the one move that foils the plot of a united attack on the Soviet Union by Germany and a coalition of Western nations—France and England, with American support. . . . The pact is not an alliance with Germany. It is a quarantining of Germany against any combination with the West. . . . This is going to be beastly to explain' " (Chevalier, *The Man Who Would Be God*, pp. 21–22).

143 **at a time when:** Numerous historians have lent credence to this argument (see Alexander Werth, *Russia At War*, pp. 3–39, and Peter Calvovoressi and Guy Wint, *Total War*, p. 82.

144 **"seasoned liberals into reactionaries"** *and subsequent quotes:* Chevalier, *Oppenheimer*, p. 33.

144 **Robert was not himself:** Maurice Isserman, *Which Side Were You On?*, pp. 38, 42. In 1941 the newly created "Fact-Finding Committee on Un-American Activities"—chaired by California state senator Jack B. Tenney—held hearings to investigate allegations that the League of American Writers was in fact a Communist front (see Edward L. Barrett, Jr., *The Tenney Committee*, p. 125).

144 **Not surprisingly, their discovery:** Herken, *Brotherhood of the Bomb*, p. 31; Chevalier, interview by Sherwin, 6/29/82, pp. 6–7; Chevalier, *Oppenheimer*, pp. 35–36.

144 **"They were printed":** Gordon Griffiths, "Venturing Outside the Ivory Tower," unpublished manuscript, shorter version, LOC, pp. 27–28.

144 **"The outbreak of war":** The pamphlets came to the attention of the university's president, Robert G. Sproul, who placed them in his presidential papers in a folder marked "Communists, 1940." During the course of an interview, Chevalier brought out copies of the pamphlets, and Sherwin read excerpts into a tape recorder (Chevalier, interview by Sherwin, 7/15/82).

144 **"you can recognize his style":** Chevalier, interview by Sherwin, 7/15/82.

145 **"The elementary test":** "Report to Our Colleagues: II," 4/6/40, "Communism," Office of the President (Robert Sproul), 1940, UCB.

146 **"something of a progressive":** Ibid.

146 **If Oppenheimer had:** JRO to Edwin and Ruth Uehling, 5/17/41; Smith and Weiner, *Letters*, p. 217.

147 **"I may be out of a job":** Smith and Weiner, *Letters,* p. 216. We see no record that JRO was questioned by any investigative committee at this time, so perhaps he was not called.

147 **"The University of California":** Martin D. Kamen, interview by Sherwin, 1/18/79, p. 27.

147 **While his friend:** Chevalier, interview by Sherwin, 7/15/82. *Daily Worker,* 4/28/38. Chevalier was joined in this statement by nearly 150 prominent intellectuals, including Nelson Algren, Dashiell Hammett, Lillian Hellman, Dorothy Parker, and Malcolm Cowley.

147 **"It was an absolutely":** During World War II, Weissberg was eventually shipped to an extermination camp in Poland. He jumped from a truck, however, and managed to escape into the woods, where he became active in the Polish underground (Victor Weisskopf, interview by Sherwin, 3/23/79, p. 5).

147 **"It's worse than you":** Michelmore, *The Swift Years,* pp. 57–58.

147 **"What they reported":** JRO hearing, p. 10.

148 **Oppie "still believed":** Weisskopf, *The Joy of Insight,* p. 115.

148 **"He really had":** Weisskopf, interview by Sherwin, 3/23/79, pp. 3–7.

148 **"I know that these conversations":** Weisskopf, interview by Sherwin, 3/23/79, p. 10.

148 **"Opje said he came":** Edith Arnstein Jenkins, *Against a Field Sinister,* p. 27. Edith chose as her Party alias the name of Mary Shelley's mother, Mary Wollstonecraft. She said that no one was a "card-carrying communist" in their own name: "It was too dangerous." From 1936 to 1938, Arnstein was the official secretary and dues collector for a closed unit of the CP at Berkeley—but she left this position in 1938 when she quit law school. The professional section of the Communist Party at Berkeley, she said, was composed of several units; with about eight individuals in each unit. She later said that Oppenheimer was certainly not a member of her closed unit, though she could not speak to this point for the years after 1938. Jenkins also remembered that Oppenheimer had once given her a small sum of money as a contribution to the Young Communist League (YCL) (Edith Arnstein Jenkins, interview by Herken, 5/9/02; Jenkins, interview by Bird, 7/25/02).

148 **"Opje is fine":** Schweber, *In the Shadow of the Bomb,* p. 108; Bloch to Rabi, 11/2/38, box 1 (general correspondence), Bloch Papers, SU.

148 **That evening he presented:** Childs, *An American Genius,* p. 307.

149 **"beautifully eloquent speech":** Schweber, *In the Shadow of the Bomb,* p. 108.

149 **"He had sympathies":** Bernstein, *Hans Bethe,* p. 65.

149 **"Our little group":** Chevalier, interview by Sherwin, 6/29/82, p. 10; Chevalier, *Oppenheimer,* p. 46.

149 **"[W]e shared the ideal":** Chevalier, *Oppenheimer,* p. 187.

150 **"Sebastian would meet":** Chevalier, *The Man Who Would Be God,* pp. 14–15.

150 **"It was his baby":** Ibid., pp. 88–89.

150 **the "novel's underlying tone":** *Time,* 11/2/59, p. 94.

151 **"Your letter asks":** Chevalier to JRO, 7/23/64, and JRO to Chevalier, 8/7/64, folder "Chevalier, Haakon—Reference to Case," box 200, JRO Papers, LOC.

151 **"discussion group":** Chevalier, *Oppenheimer,* pp. 19, 46.

151 **"to be a Communist":** John Earl Haynes and Harvey Klehr, *In Denial,* p. 39. John Haynes later wrote, "Oppenheimer, of course, would have been regarded by any party officer with any sense as a highly valuable ally. Further, he had no dependence on the party for organizational or other assistance. He was highly valuable to the party, but the party was not valuable to Oppenheimer except to the extent of his belief in its goals and objectives and whatever personal/fraternal ties he had developed with others in the movement. No skilled party leader would impose 'discipline' on someone like Oppenheimer; instead of giving orders, he would persuade, convince, cajole, ask politely and

even plead if necessary" (John Haynes, e-mail to Gregg Herken, 4/26/04, courtesy of Herken).

152 **In short, Oppenheimer:** As one of the FBI's informants put it, "although Oppenheimer may not have been actually brought into the Communist Party, the effort to bring him to acceptance of Communist Party philosophy and to secure his support for Communist aims was regarded by the Communists as successful." This FBI informant was Louis Gibarti, a Hungarian-born communist who spent the years 1923 to 1938 as a Comintern agent. Gibarti, whose real name was Laszlo Dobos, left the Party in 1938 and then worked as a journalist. There is no evidence that Gibarti ever knew Oppenheimer, or for that matter had any evidence to support his supposition quoted above. In 1950 he became an informant for the FBI (J. Edgar Hoover to Lewis Strauss, 6/25/54, JRO FBI file, sect. 44, doc. 1800).

Chapter Eleven: "I'm Going to Marry a Friend of Yours, Steve"

153 **"We were at least twice":** JRO to Maj. Gen. K. D. Nichols, 3/4/54.

153 **"No more flowers, please":** Michelmore, *The Swift Years,* p. 49.

153 **"she disappeared for weeks":** Goodchild, *J. Robert Oppenheimer,* p. 35.

154 **"mostly very attractive":** Chevalier, interview by Sherwin, 6/29/82, p. 9; Chevalier, *Oppenheimer,* p. 30; Herken, *Brotherhood of the Bomb,* p. 345.

154 **Bob Serber recalled:** Serber, interview by Sherwin, 1/9/82, p. 10. Interestingly, Sandra Dyer-Bennett must have been a decade or more older than Robert. She was the mother of the folk musician Richard Dyer-Bennett, born in 1913.

154 **"I fell in love with Robert"** *and subsequent quotes:* Serber, interview by Sherwin, 1/9/82; Goodchild, *J. Robert Oppenheimer,* p. 39; Chevalier, interview by Sherwin, 6/29/82, p. 9; Chevalier, *Oppenheimer,* p. 31; Michelmore, *The Swift Years,* p. 63; JRO to Niels Bohr, 11/2/49, box 21, JRO Papers.

154 **"Kitty was related":** Robert Serber, interview by Sherwin, 3/11/82.

155 **Kitty had been born:** Katherine Oppenheimer FBI file (100-309633-2), FBI memo, 8/7/51.

155 **"a prince of a small principality":** Serber, interview by Jon Else, 12/15/79, p. 9.

155 **The Blonays served:** *www.swisscastles.ch/Vaud/chateau/blonay.html.*

155 **Kaethe Vissering was beautiful:** Wilhelm Keitel, *Mein Leben,* pp. 19–20. Keitel's German language memoirs describe the noble ancestry of his grandparents, Bodewin Vissering and Johanna Blonay. (Portions of this memoir were published in English, translated by David Irving, *The Memoirs of Field-Marshal Keitel* [New York, Stein and Day, 1966]. But this version excludes material about Keitel's family background.) For Keitel's temporary engagement to Kaethe Vissering, see JRO hearing, p. 277.

155 **"Her Highness, Katherine":** Serber, interview by Sherwin, 3/11/82, p. 13.

155 **"She was wild as hell":** Pat Sherr, interview by Sherwin, 2/20/79, p. 10; Serber, interview by Sherwin, 3/11/82, p. 14.

155 **"I spent little time":** Goodchild, *J. Robert Oppenheimer,* p. 37.

156 **Several months into the marriage:** Sherr, interview by Sherwin, 2/20/79, p. 10.

156 **"The consensus was":** JRO hearing, p. 571; Goodchild, *J. Robert Oppenheimer,* p. 38.

156 **"He was a handsome":** Steve Nelson, interview by Sherwin, 6/17/81, p. 39.

156 **"an utter misfit"; "It is difficult to tell":** Robert A. Karl, "Green Anti-Fascists: Dartmouth Men and the Spanish Civil War," unpublished Dartmouth College research paper, 9/21/00, p. 42, DCL.

156 **Determined to "throttle":** Karl, "Green Anti-Fascists," pp. 43–44; Hugh Thomas, *The Spanish Civil War,* p. 473; Marion Merriman and Warren Lerude, *American Commander in Spain,* p. 124. For Dallet's Jewish background, see Margaret Nelson, inter-

view by Sherwin, 6/17/81, p. 34, and *Dartmouth Alumni,* December 1937, Dallet's alumni file, DCL.

157 **By 1932, Dallet:** Peer de Silva, unpublished manuscript, p. 2, courtesy of Gregg Herken; *Daily Worker,* 10/27/37; Fifth Report of the Senate Fact-Finding Committee on Un-American Activities in California, 1949, p. 553.

157 **"The house had a kitchen":** Michelmore, *The Swift Years,* p. 61; Goodchild, *J. Robert Oppenheimer,* p. 38.

157 **Joe "was a bit dogmatic":** Steve Nelson, interview by Sherwin, 6/17/81, p. 4.

157 **"The poverty became":** Sherr, interview by Sherwin, 2/20/79, p. 25; JRO hearing, p. 572; Goodchild, *J. Robert Oppenheimer,* p. 38.

158 **"I was like a third wheel":** Steve Nelson, interview by Sherwin, 6/17/81, pp. 3, 6.

158 **"I adore you":** Joe Dallet, *Letters from Spain,* pp. 56–57; Dallet to Kitty Dallet, 4/9/37, 4/22/37, and 7/25/37, reprinted in Cary Nelson and Jefferson Hendricks, eds., *Madrid 1937: Letters of the Abraham Lincoln Brigade from the Spanish Civil War,* pp. 71–74, 77–78.

158 **"We had a nice few days":** Margaret Nelson, interview by Sherwin, 6/17/81, p. 28. Nelson read this letter into Sherwin's tape recorder.

159 **"Man, what a feeling":** Dallet, *Letters from Spain,* p. 45.

159 **"near hatred":** Sandor Voros, *American Commissar,* pp. 338–40.

159 **"A percentage of the men":** Merriman and Lerude, *American Commander in Spain,* pp. 124–25. FBI doc. 263; FBI doc. 49, 10/9/37, contained in Harvey Klehr, John Earl Haynes, and Fridrikh Igorevich Firsov, *The Secret World of American Communism,* pp. 184–86; Schwartz, *From West to East,* p. 360; Peter Carroll, *The Odyssey of the Abraham Lincoln Brigade,* pp. 164–65.

160 **"The attack started":** Voros, *American Commissar,* p. 342. Vincent Brome, *The International Brigades,* 1966, p. 225. "We lost some good men in the attack," wrote Bob Merriman to his wife on 10/16/37, "including Joe Dallet"; Merriman and Lerude, *American Commander in Spain,* p. 175; FBI doc. 158, p. 3; Rosenstone, *Crusade of the Left: The Lincoln Battalion in the Spanish Civil War,* pp. 234–36.

160 **"She was crushed":** Steve Nelson, interview by Sherwin, 6/17/81, pp. 8–9; Nelson, et al., *American Radical,* pp. 232–33; JRO hearing, p. 574. FBI doc. 284, p. 5.

160 **given "themselves completely":** Allen Guttmann, *The Wound in the Heart,* p. 142; *Daily Worker,* 10/27/37.

160 **Kitty spent a couple of months:** FBI memo 5/6/52, Katherine Oppenheimer FBI file (100-309633). Kitty met Browder only once, when he came to Youngstown, Ohio, to see Joe Dallet; they had dinner together (FBI memo on Katherine Oppenheimer, 4/23/52, JRO file, sect. 12).

160 **"She seemed to be":** Margaret Nelson, interview by Sherwin, 6/17/81, p. 32; Sherr, interview by Sherwin, 2/20/79, p. 10.

161 **"an impossible marriage":** Jean Bacher, interview by Sherwin, 3/29/83, p. 4; Goodchild, *J. Robert Oppenheimer,* p. 39; JRO FBI file, doc. 108, p. 4.

161 **At twenty-nine, Kitty:** JRO hearing, p. 574. Kitty was enrolled at UCLA from September 1939 through June 1940 and lived at 553½ Coronado Street, Los Angeles.

161 **"He would ride up":** Dr. Louis Hempelmann, interview by Sherwin, 8/10/79, p. 26.

162 **Just a day or two:** Serber, *Peace and War,* pp. 59–60. Frank and Jackie Oppenheimer also spent some time that summer on the ranch, bringing with them eleven-year-old Hans "Lefty" Stern, the son of their cousins, Dr. Alfred and Lotte Stern.

162 **"he and the Oppenheimers":** JRO FBI file, doc. 154, p. 7.

162 **Even though Bob Serber:** Serber, *Peace and War,* p. 60.

162 **"Kitty Dallet!":** Steve Nelson, interview by Sherwin, 6/17/81, p. 12; Nelson, et al., *American Radical,* p. 268.

163   **By the time the newlyweds:** Herken, *Brotherhood of the Bomb*, p. 52.
163   **At the end of November:** D. M. Ladd to FBI director, 8/11/47, JRO FBI file, doc. 159, p. 7. Ladd is quoting Nelson, apparently from an 8/7/45 wiretap.
163   **Kitty immediately invited:** Kitty Oppenheimer to Margaret Nelson, undated, circa 11/29/40, in Margaret Nelson, interview by Sherwin, 6/17/81, p. 30.
163   **An FBI wiretap:** Herken, *Brotherhood of the Bomb*, p. 56.
163   **"With all of his brilliance":** Margaret Nelson, interview by Sherwin, 6/17/81, p. 31; Steve Nelson, et al., *American Radical*, p. 268.
163   **"He was gentle, mild":** Sabra Ericson, interview by Sherwin, 1/13/82.
163   **"She could not stand"** *and subsequent quotes:* Frank and Jackie Oppenheimer, interview by Sherwin, 12/3/78; Goodchild, *J. Robert Oppenheimer*, pp. 39–40; Serber, interview by Sherwin, 3/11/82, p. 15; Chevalier, interview by Sherwin, 6/29/82, p. 2.
164   **"Bombsight":** Michelmore, *The Swift Years*, p. 65.
164   **"A certain stuffiness":** *Time*, 11/8/48, p. 76.
164   **"their bill for liquor":** Margaret Nelson, interview by Sherwin, 6/17/81, p. 33.
164   **"I felt that he obviously":** Smith and Weiner, *Letters*, p. 215; Edsall, interview by Weiner, 7/16/75, p. 40.
164   **Kitty told some of her friends:** Sherr, interview by Sherwin, 2/20/79, p. 11.
165   **"Deeply flattered":** Chevalier, *Oppenheimer*, p. 42.
165   **"Kitty seemed quite":** Ruth Meyer Cherniss, interview by Alice Smith, 11/10/76; Harold Cherniss, interview by Smith, 4/21/76, p. 20.
165   **Robert felt reinvigorated:** Stern, *The Oppenheimer Case*, pp. 33–34. Dorothy McKibbin found a hospital record for the X rays, dated July 25 (FBI memo, 11/18/52, p. 46, JRO FBI file, series 14; FBI doc. 327, pp. 17–18); Michelmore, *The Swift Years*, p. 65; Goodchild, *J. Robert Oppenheimer*, p. 40, JRO hearing, p. 336.
165   **Upon their return:** See correspondence of July 1941 in box 232, "Real Estate" folder, JRO Papers.
165   **A Spanish-style, one-story villa:** Bird and Sherwin toured the house on 4/23/04; Chevalier, *Oppenheimer*, p. 43.

Chapter Twelve: "We Were Pulling the New Deal to the Left"

166   **Alvarez "stopped the barber"** *and subsequent quotes:* Luis W. Alvarez, *Alvarez*, pp. 75–76.
166   **"The U business":** Smith and Weiner, *Letters*, pp. 207–8. Richard Rhodes credibly suggests that this letter was written on 2/4/39—and not 1/28/39 as Smith and Weiner conjectured (Rhodes, *The Making of the Atomic Bomb*, p. 812, note 274).
167   **"So I think it really":** Smith and Weiner, *Letters*, p. 209. Oppenheimer also wrote a letter to Serber about the fission discovery: "The news had just gotten to Berkeley and he wrote me. I gave a seminar on it that same day. . . . And I think even in the first letter he mentioned the possibility of making a bomb" (*The Day After Trinity*, dir. Jon Else, transcript, p. 12). Serber later destroyed all his letters from Oppenheimer (Serber, interview by Sherwin, 3/11/82, p. 21).
167   **"It was first names":** Joseph Weinberg, interview by Sherwin, 8/23/79, pp. 4–5.
168   **"a drawing—a very bad":** Rhodes, *The Making of the Atomic Bomb*, p. 275.
168   **"That was the only":** Weinberg, interview by Sherwin, 8/23/79, p. 10.
169   **"Bohr was God":** Ibid., pp. 6, 15–16.
169   **"He gave us the usual":** Ibid., p. 13.
170   **"He was very keenly":** Ibid., p. 8.
170 · **Oppenheimer gave no final exams:** Ed Geurjoy, "Oppenheimer as a Teacher of

Physics and Ph.D. Advisor," speech delivered at Atomic Heritage Foundation conference, Los Alamos, 6/26/04.

170   **"[The student] was a very genial"** *and subsequent quotes:* Joseph Weinberg, interview by Sherwin, 8/23/79, p. 15.

171   **"self-conscious and daring":** Schrecker, *No Ivory Tower,* p. 133.

171   **Joe Weinberg was probably:** Hawkins, interview by Sherwin, 6/5/82, p. 14. Hawkins says Weinberg was in his Berkeley Party group: "I think maybe at some time, yes."

171   **born in 1915:** Schrecker, *No Ivory Tower,* pp. 149, 41; Hawkins, interview by Sherwin, 6/5/82, p. 16.

171   **"We were all close to communism":** Bohm, interview by Sherwin, 6/15/79, p. 5.

172   **"No one can feel":** Weinberg, quoted in F. David Peat, *Infinite Potential,* p. 60.

172   **"I had the feeling":** Bohm, interview by Sherwin, 6/15/79, p. 17.

172   **"many people who were not":** Schrecker, *No Ivory Tower,* pp. 38, 47, 49, 56.

172   **"He was very persuasive":** Hawkins, interview by Sherwin, 6/5/82, p. 6.

173   **"We were pretty secretive":** Ibid., p. 14.

173   **"The centralization of":** Ibid., p. 12.

173   **"Not that I know":** Ibid., p. 15.

173   **Martin D. Kamen was:** Kamen and Ruben made their carbon-14 discovery in 1940. Yet another chemist, Willard Libby, won the 1960 Nobel Prize in chemistry for developing the technique of carbon dating (Kamen, *Radiant Science, Dark Politics,* pp. 131–32).

173   **"It was like Mecca":** Kamen, interview by Sherwin, 1/18/79, p. 20.

174   **"Everyone sort of regarded":** Ibid., pp. 2, 6.

174   **"So we drove up and down":** Ibid., pp. 6–7.

174   **"When he spoke":** Herve Voge, interview by Sherwin, 3/23/83, p. 19.

175   **Fifteen people were present:** JRO hearing, pp. 131, 135.

175   **"leftwandering activities":** Childs, *An American Genius,* p. 319. Oppenheimer later testified that they debated at this meeting whether it would be a good idea to set up a branch of the Association of Scientific Workers. "We concluded negatively, and I know my own views were negative." (JRO hearing, pp. 131, 135.)

175   **"If he would just":** Kamen, interview by Sherwin, 1/18/79, pp. 24–28; Kamen, *Radiant Science, Dark Politics,* pp. 184–86. Kamen eventually lost his job at the Rad Lab—largely due to a series of misunderstandings that led authorities to think he had acted as a spy for the Soviets. The false allegations haunted him for years; in 1951 Senator Bourke B. Hickenlooper accused Kamen of being an "atom spy." Depressed and beleaguered, Kamen attempted suicide, recovered, and decided to sue the *Chicago Tribune* for libel; eventually Kamen won the suit and was awarded $7,500 in compensatory damages. (Kamen, *Radiant Science, Dark Politics,* pp. 248, 288.)

176   **"It seemed like":** Rossi Lomanitz, interview by Sherwin, 7/11/79, part 2, p. 2.

176   **"It was a title":** Max Friedman, interview by Sherwin, 1/14/82. Friedman later changed his name to Ken Max Manfred.

176   **"an organization known to be":** Peat, *Infinite Potential,* pp. 62–63. A 1947 report of the California Joint Fact-Finding Committee on Un-American Activities in California contained a long report by R. E. Combs "charging that the International Federation of Architects, Engineers, Chemists and Technicians had been used as a front for communist espionage in connection with atomic research in the University of California Radiation Laboratory" (Barrett, *The Tenney Committee,* pp. 54–55).

177   **"Oppenheimer has important":** Smith and Weiner, *Letters,* pp. 222–23.

177   **"but it was not until":** JRO hearing, p. 11.

177   **"there will be no further":** JRO to Ernest Lawrence, 11/12/41, Smith and Weiner, *Letters,* p. 220.

177 **But though Oppenheimer ceased:** Smith and Weiner, *Letters,* pp. 217–18; Schrecker, *No Ivory Tower,* pp. 76–83.

178 **"teachers who were communists":** Smith and Weiner, *Letters,* pp. 218–19.

178 **"Everything that happened":** Kamen, interview by Sherwin, 1/18/79, p. 21.

178 **"that I had had about enough":** JRO hearing, p. 9.

Chapter Thirteen: "The Coordinator of Rapid Rupture"

179 **"that extremely powerful bombs":** Martin J. Sherwin, *A World Destroyed,* p. 27.

180 **"Uranium Committee":** Ibid., pp. 36–37.

180 **Oppenheimer "would be a tremendous":** Herken, *Brotherhood of the Bomb,* p. 51.

181 **"essential point is to enlist":** Smith and Weiner, *Letters,* pp. 226–27.

181 **"We were together":** Serber, interview by Sherwin, 1/9/82, p. 20.

181 **"Oh geez, look":** Weinberg, interview by Sherwin, 8/23/79, part 3, p. 17.

181 **"As Chairman," Edward Teller later wrote:** Bernstein, *Hans Bethe,* pp. 65, 78.

182 **"We were forever":** Rhodes, *The Making of the Atomic Bomb,* p. 420.

182 **While Oppenheimer soon concluded:** Richard G. Hewlett and Oscar E. Anderson, Jr., *The New World,* vol. 1, p. 104.

182 **"we would do better":** JRO to John Manley, 7/14/42, box 50, JRO Papers.

183 **"I didn't believe it":** Rhodes, *The Making of the Atomic Bomb,* p. 418.

183 **"I'll never forget":** Arthur H. Compton, *Atomic Quest,* p. 127.

183 **In the event:** Edward Teller had a different memory of this incident: "The question of igniting the atmosphere, if it was mentioned at all, was not discussed in any detail at the summer conference. It was not an issue" (Teller, with Judith Shooley, *Memoirs,* p. 160).

183 **According to Oppenheimer:** Rhodes, *The Making of the Atomic Bomb,* pp. 418–21.

183 **"only an atomic bomb":** Teller, *Memoirs,* p. 161.

184 **"I'm cutting off":** Compton, *Atomic Quest,* p. 126.

184 **"turned thumbs down":** Herken, *Brotherhood of the Bomb,* p. 349, note 26 (memorandum of conversation, 8/18/42, box 1, JRO, AEC, record group 326, NA).

184 **In this beautiful setting:** Vincent C. Jones, *Manhattan: The Army and the Atomic Bomb,* pp. 70–71.

184 **"decidedly left-wing":** James Hershberg, *James B. Conant,* pp. 165–66; Goodchild, *J. Robert Oppenheimer,* p. 49.

184 **"Oh, that thing":** Leslie M. Groves, *Now It Can Be Told,* p. 4.

185 **"Groves is a bastard":** Herbert Smith, interview by Weiner, 8/1/74, p. 7.

185 **"General Groves is the biggest S.O.B.":** Nichols, *The Road to Trinity,* p. 108; Goodchild, *J. Robert Oppenheimer,* pp. 56–57.

185 **"Take this and find":** Robert S. Norris, *Racing for the Bomb,* pp. 179–83; Serber, *The Los Alamos Primer,* p. xxxii.

185 **"overweening ambition"** *and subsequent quote:* Norris, *Racing for the Bomb,* pp. 240–42; Rhodes, *The Making of the Atomic Bomb,* p. 449.

186 **"we could begin":** JRO hearing, p. 12; Lillian Hoddeson, et al., *Critical Assembly,* p. 56.

186 **A week after:** Norris, *Racing for the Bomb,* p. 241.

186 **"[his political] background included":** Groves, *Now It Can Be Told,* p. 63.

186 **"It was not obvious":** Hans Bethe later claimed that Ernest Lawrence had wanted his Rad Lab colleague Edwin McMillan appointed director of Los Alamos. "Groves very wisely decided that the director had to be Oppenheimer," Bethe told Jeremy Bernstein. (Bernstein, *Hans Bethe,* p. 79.)

186 **"I had no support":** Groves to Victor Weisskopf, March 1967, Weisskopf folder, box 6, RG 200, NA, Papers of Leslie Groves, courtesy of Robert S. Norris.

186 **As much as he admired:** Herken, *Brotherhood of the Bomb*, p. 71.

186 **"He was a very impractical"; "He couldn't run":** Charles Thorpe and Steven Shapin, "Who Was J. Robert Oppenheimer?" *Social Studies of Science,* August 2000, p. 564; Bernstein, *Experiencing Science*, p. 97.

187 **"was a real stroke of genius":** Jon Else, *The Day After Trinity*, transcript, p. 11.

187 **"It is about time":** JRO to Hans Bethe, 10/19/42, Bethe folder, box 20, JRO Papers.

187 **"He talked very fast":** John McTernan, phone interview by Bird, 6/19/02.

187 **"many people all around":** Bohm, interview by Sherwin, 6/15/79, p. 15.

187 **"various radical study groups":** Betty Friedan, *Life So Far,* pp. 57–60.

187 **"They were all working":** Ibid., p. 60; Friedan, interview by Bird, 1/24/01.

188 **"Many of us thought":** Lomanitz, interview by Sherwin, 7/11/79, part 1, p. 17.

188 **"I've heard some of these":** Lomanitz, interview by Sherwin, 7/11/79, part 2, p. 5. For an argument about why a "second front" was not opened up in 1943, see John Grigg, *1943: The Victory That Never Was.*

188 **"respected him a great deal":** Lomanitz, interview by Sherwin, 7/11/79.

188 **"I was responsible":** Steve Nelson, *American Radical,* pp. 268–69.

188 **By the early spring of 1943:** Steve Nelson–Joseph Weinberg transcript, 3/29/43, entry 8, box 100, RG 77, MED, NA, College Park, MD.

191 **"cunning and shrewd":** Anonymous review of *The Alsos Mission,* by Boris T. Pash (1969), in *Intelligence in Recent Public Literature,* Winter 1971. The author of this book review reports that he is a close friend of Pash's.

191 **Pash quickly leaped:** Herken, *Brotherhood of the Bomb*, pp. 96–98. Shortly after Nelson's wiretapped conversation with "Joe," the FBI observed Nelson meeting Peter Ivanov, the Soviet vice counsel in San Francisco. They were seen talking on the grounds of the St. Francis Hospital—and then a few days later a Soviet diplomat stationed in Washington visited Nelson in his home and paid him ten bills of unknown denomination. As a result, J. Edgar Hoover himself wrote a letter to Harry Hopkins in the White House to report that Nelson was trying to infiltrate Communist Party members into "industries engaged in secret war production" (Report on Atomic Espionage [Nelson-Weinberg and Hiskey-Adams Cases], 9/29/49, HUAC, pp. 4–5; J. Edgar Hoover to Harry Hopkins, 5/7/43, reprinted in Benson and Warner, *Venona,* p. 49. Hoover claimed this transaction occurred on 4/10/43. Haynes and Klehr, *Venona,* pp. 325–26).

192 **"we had an unidentified man":** JRO hearing, pp. 811–12.

192 **"Pressure was brought to bear":** Herken, *Brotherhood of the Bomb*, p. 106.

192 **"Lehmann advised Nelson":** FBI doc. 100-17828-51, 3/18/46, JRO background. According to the FBI, in May 1943, John V. Murra, a veteran of the Abraham Lincoln Brigade, arrived in San Francisco and contacted Bernadette Doyle. Murra reportedly told Doyle that he wanted to get in touch with Mrs. Oppenheimer. Presumably, Murra had known Joe Dallet in Spain. In response, Doyle directed Murra to call the Joint Anti-Fascist Committee or the University of California at Berkeley. According to the FBI document, Doyle stated that Robert Oppenheimer was a Party member but that his name should be removed from any mailing lists in Murra's possession and he should not be mentioned in any way. There is no indication that Murra ever saw Kitty, who was by then in Los Alamos. We see this story as evidence that some members of the CP thought of Oppenheimer as a comrade—not that he was in fact a Party member.

193 **He spent the war years:** Peat, *Infinite Potential,* p. 64.

193 **Max Friedman was called:** Friedman, interview by Sherwin, 1/14/82.

193 **Army intelligence:** In 1949, Irving David Fox—by then a teaching assistant in physics at Berkeley—was called to testify before HUAC. He refused to name names and subsequently was called before the university regents to explain his political

beliefs. Fox frankly explained that while he had attended some Communist-sponsored meetings, he had never joined the Party. Fox was nevertheless fired, an action that precipitated a furious controversy over loyalty oaths at Berkeley for several years. (Griffiths, "Venturing Outside the Ivory Tower," unpublished manuscript, shorter version, LOC, pp. 18–19.)

193 **As for Weinberg:** Joseph Albright and Marcia Kunstel, *Bombshell*, p. 106.

194 **"He appeared excited":** Steve Nelson, interview by Sherwin, 6/17/81, p. 17; Steve Nelson, et al., *American Radical*, p. 269.

Chapter 14: "The Chevalier Affair"

196 **"He was visibly disturbed":** Chevalier, *Oppenheimer*, p. 55; Chevalier said Kitty never entered the kitchen while he and Oppenheimer discussed Eltenton's proposition (Chevalier, interview by Sherwin, 6/29/82, p. 2).

196 **"I saw George Eltenton recently":** JRO hearing, p. 130.

197 **"But that would be treason!":** Verna Hobson, interview by Sherwin, 7/31/79, p. 22. Hobson, Oppenheimer's secretary at the Institute for Advanced Study and a friend of Kitty's, observed that the "treason" comment "sounds like Kitty and doesn't sound like Robert."

197 **"I was not, of course":** Barbara Chevalier "diary," 8/8/81, 2/19/83 and 7/14/84, courtesy of Gregg Herken, *www.brotherhoodofthebomb.com*.

197 **Oppenheimer knew Eltenton:** JRO hearing, p. 135.

197 **a thin, Nordic-featured:** Oppenheimer told Col. Pash on 8/27/43 that Eltenton was "certainly very far 'left,' whatever his affiliations" (JRO hearing, p. 846). There is no hard evidence that Eltenton was a member of the Communist Party although Priscilla McMillan in *The Ruin of J. Robert Oppenheimer* asserts that he was; see chap. 18. Herve Voge thought Eltenton's wife, Dolly, "was probably more radical than he was" (Voge, interview by Sherwin, 3/23/83, p. 9). In 1998, Dolly privately published a memoir, *Laughter in Leningrad*, of their five years in Leningrad. While working at the Leningrad Institute of Chemical Physics, Eltenton became friends with many Russian scientists, including Yuli Borisovich Khariton, a nuclear physicist who later helped to develop the Soviet Union's first atomic and hydrogen bombs.

197 **Chevalier had first met:** Haakon Chevalier FBI file, part 1 of 2, SF 61-439, p. 33; Haynes and Klehr, *Venona*, p. 233.

198 **"I told him [Ivanov]"** *and subsequent quotes:* FBI (Newark) synopsis of facts, 2/12/54, pp. 19–22 (Eltenton and Chevalier signed statements, 6/26/46), contained in JRO FBI file, doc. 786.

199 **In 1947, when the details:** Interestingly, he kept up his friendship with Chevalier, and even attended Chevalier's eightieth birthday party in Berkeley, as did Frank Oppenheimer (Herken, *Brotherhood of the Bomb*, p. 333). Sherwin contacted Eltenton in London in the early 1980s, but he declined to be interviewed.

200 **"I would like Russia to win":** Voge, interview by Sherwin, 3/23/83, p. 3.

200 **"a dupe of the Russian consulate"** . . . **"We were never able to convince":** Voge, interview by Sherwin, 3/23/83, p. 18. Voge read portions of this FBI document into Sherwin's tape recorder.

201 **"If he'd really been":** Voge, interview by Sherwin, 3/23/83, pp. 4, 8. The historians John Earl Haynes and Harvey Klehr flatly state that Eltenton was a "concealed Communist," but they offer no evidence of this beyond an FBI report that he met on several occasions with the GRU officer Peter Ivanov (Haynes and Klehr, *Venona*, p. 329). Voge said he doubted that Eltenton was a Communist but that "it's conceivable" (Voge, interview by Sherwin, 3/23/83, p. 10). Eltenton's son, Mike Eltenton, later wrote, "So far as

I know, neither of my parents ever became members of the Communist party—though their views on several issues came close to the party line" (Dorothea Eltenton, *Laughter in Leningrad*, p. xii).

Chapter 15: "He'd Become Very Patriotic"

205 **a "lovely spot":** Smith and Weiner, *Letters*, p. 236.

206 **"We were arguing about this":** Gen. John H. Dudley, "Ranch School to Secret City," public lecture, 3/13/75, in Lawrence Badash, et al., eds., *Reminiscences of Los Alamos, 1943–45;* Norris, *Racing for the Bomb*, pp. 243–44; Lawren, *The General and the Bomb*, p. 99; Marjorie Bell Chambers and Linda K. Aldrich, *Los Alamos, New Mexico*, p. 27; John D. Wirth and Linda Harvey Aldrich, *Los Alamos*, p. 155.

206 **It was already late:** Founded in 1917, the Los Alamos Ranch School recruited no more than forty-four boys from wealthy families in the East and subjected them to a strenuous life. Its alumni include a Colgate (Colgate products), Burroughs (Burroughs adding machines), Hilton (Hilton hotels), and Douglas (Douglas aircraft). Each boy had his own horse and was responsible for its maintenance. Gore Vidal, who attended in the 1939–40 school year, later wrote that "reading was discouraged at Los Alamos in the interest of strenuousness" (Gore Vidal, *Palimpsest*, pp. 80–81).

206 **"This is the place":** John H. Manley, "A New Laboratory Is Born," unpublished manuscript, p. 13, Sherwin Collection; Edwin McMillan, *Early Days of Los Alamos*, unpublished manuscript, p. 7, Sherwin Collection; Dudley, "Ranch School to Secret City," in Badash, et al., eds., *Reminiscences of Los Alamos*. See also Leslie Groves to Victor Weisskopf, March 1967, Weisskopf folder, box 6, RG 200, Papers of Leslie Groves, courtesy of Robert S. Norris.

206 **Within two days:** The Los Alamos Ranch School probably would have closed down even if Oppenheimer had not chosen it as a site for the new laboratory. See Fred Kaplan's description of the school in his biography *Gore Vidal*, pp. 99–112.

206 **"Suddenly we knew the war":** Sterling Colgate, interview by Jon Else, 11/12/79, pp. 2–3; Peggy Pond Church, *The House at Otowi Bridge*, p. 84.

206 **Soon afterwards, an armada:** Edwin McMillan, *Early Days of Los Alamos*, p. 8.

207 **"I am responsible for":** Wirth and Aldrich, *Los Alamos*, p. viii. JRO said this to Wirth's grandfather in 1955.

207 **"What we were trying to do":** Manley, "A New Laboratory Is Born," unpublished manuscript, p. 18.

207 **Robert assured Hans Bethe:** Smith and Weiner, *Letters*, pp. 244–45; JRO to Hans and Rose Bethe, 12/28/42.

208 **"He was something of an eccentric":** Raymond T. Birge, "History of the Physics Department," vol. 4, unpublished manuscript, UCB, p. xiv; Robert R. Wilson, interview by Owen Gingrich, 4/23/82, p. 3.

208 **"wondering whether we":** Hershberg, *James B. Conant*, p. 167.

208 **"I was somewhat frightened":** Manley, interview by Sherwin, 1/9/85, p. 23; Manley, "A New Laboratory Is Born," unpublished manuscript, p. 21.

209 **Stunned, Wilson and Manley:** Robert R. Wilson, "A Recruit for Los Alamos," *Bulletin of the Atomic Scientists*, March 1975, p. 45; Goodchild, *Oppenheimer*, p. 72.

209 **"So it was quite a change":** Mary Palevsky, *Atomic Fragments*, pp. 128–29.

209 **"He had style":** Robert R. Wilson, interview by Gingrich, 4/23/82, p. 4.

209 **"when I was with him":** Palevsky, *Atomic Fragments*, pp. 134–35; Wilson, interview by Gingrich, 4/23/82, p. 4, Sherwin Collection.

209 **through these early planning:** Dudley, "Ranch School to Secret City," in Badash, et al., eds., *Reminiscences of Los Alamos*, Sherwin Collection.

210 **When Los Alamos opened:** For security reasons, the total population at Los Alamos was regarded as highly classified information; a census was not taken until April 1946. Different sources use different figures: See Thorpe and Shapin, "Who Was J. Robert Oppenheimer?" *Social Studies of Science,* August 2000, p. 585; Kunetka, *City of Fire,* pp. 89, 130; Kunetka uses a figure of 4,000 for Los Alamos' "scientific population" (p. 65). According to Edith C. Truslow's *Manhattan District History* (1991), by the end of 1944 Los Alamos had a population of 5,675. She reports a sharp increase in 1945 to a total of 8,200. Norris, *Racing for the Bomb,* p. 246, uses similar figures.

210 **"the above physical defects":** JRO medical physical, Presidio of San Francisco, 1/16/43, box 100, series 8, MED, NA; Herken, *Brotherhood of the Bomb,* p. 75. This medical record reported that Oppenheimer was five feet ten inches tall, that he weighed 128 pounds, and that he had a 28-inch waist. He registered a regular blood-pressure of 128 over 78. He had 20/20 vision and perfectly normal hearing—but he was missing five of his original teeth. Oppenheimer told the army doctors that he had no history of mental illness.

210 **"Oppie would get":** Jane Wilson, ed., *All in Our Time,* 1974, p. 147; Libby, *The Uranium People,* p. 197; Wilson, "A Recruit for Los Alamos," *Bulletin of the Atomic Scientists,* March 1975, pp. 42–43.

211 **"he was very foolish":** Rabi, interview by Sherwin, 3/12/82, p. 11.

211 **By the end of that month:** Smith and Weiner, *Letters,* pp. 247–49.

211 **"I think that you have":** Hans Bethe to JRO, 3/3/43, Bethe folder, box 20, JRO Papers, LOC.

211 **"Without Rabi":** Rigden, *Rabi,* p. 149.

211 **"I was strongly opposed":** Ibid., p. 152.

212 **"I thought it over":** Rhodes, *The Making of the Atomic Bomb,* p. 452.

212 **"I think if I believed":** Smith and Weiner, *Letters,* p. 250.

212 **"I never went on the payroll":** Rigden, *Rabi,* p. 146.

212 **"the nerve center":** JRO to Rabi, 2/26/43; Rabi to JRO, 3/8/43, and Rabi to JRO, "Suggestions for Interim Organization and Procedure," 2/10/43, Rabi folder, box 59, JRO Papers.

213 **Feynman was touched:** James Gleick, *Genius,* p. 159.

213 **"We should start now":** JRO to John H. Manley, 10/12/42, box 50, Manley folder, JRO Papers.

213 **"very nearly unique":** JRO to Robert Bacher, memo, 4/28/43, box 18, Bacher folder, JRO Papers.

213 **"I saw a man walking":** McKibbin was also an old friend of Luvie Pearson's, the wife of the influential syndicated columnist, Drew Pearson (Nancy C. Steeper, *Gatekeeper to Los Alamos,* p. 73 of draft manuscript).

214 **"gatekeeper to Los Alamos":** Dorothy McKibbin, interview by Jon Else, 12/10/79, p. 2, Sherwin Collection; Peggy Corbett, "Oppie's Vitality Swayed Santa Fe," McKibbin folder, JRO Papers; Steeper, *Gatekeeper to Los Alamos,* p. 3.

214 **"He had the bluest eyes":** McKibbin, interview by Jon Else, 12/10/79, pp. 21–23.

215 **That first spring in 1943:** Bernice Brode, *Tales of Los Alamos,* p. 8.

215 **"I was rather shocked":** Bethe, interview by Jon Else, 7/13/79, p. 7.

215 **"We could gaze beyond":** Brode, *Tales of Los Alamos,* p. 15.

215 **"Nobody could think straight":** Davis, *Lawrence and Oppenheimer,* p. 163.

215 **Everyone had to change:** Brode, *Tales of Los Alamos,* p. 37.

216 **The astonished MP:** Elsie McMillan, "Outside the Inner Fence," in Badash, et al., eds., *Reminiscences of Los Alamos,* p. 41.

216 **"refrain from flying":** Leslie Groves to JRO, 7/29/43, Groves folder, box 36, JRO Papers.

216 **"I don't recall":** Brode, *Tales of Los Alamos,* p. 33.

216 **"His porkpie hat"**: Eleanor Stone Roensch, *Life Within Limits,* p. 32. (Oppie's phone number was 146.)

216 **"several times Dr. Oppenheimer"**: Ed Doty to his parents, 8/7/45 (Los Alamos Historical Museum), cited by Thorpe and Shapin, "Who Was J. Robert Oppenheimer?," p. 575.

216 **who "demanded attention"**: Roensch, *Life Within Limits,* p. 32.

216 **When the young physicist**: Kunetka, *City of Fire,* p. 59; Brode, *Tales of Los Alamos,* p. 37.

217 **Oppenheimer himself had been**: McKibbin, interview by Jon Else, 12/10/79, p. 19.

217 **"The work was terribly"**: Bethe, interview by Jon Else, 7/13/70, p. 7.

217 **Scientists accustomed**: Thorpe and Shapin, "Who Was J. Robert Oppenheimer?," p. 546; see also Charles Thorpe, "J. Robert Oppenheimer and the Transformation of the Scientific Vocation," dissertation, pp. 302–3.

217 **"Feynman sort of materialized"**: Bernstein, *Hans Bethe,* p. 60.

217 **"No, no, you're crazy"**: Badash, et al., eds., *Reminiscences of Los Alamos,* p. 109; James Gleick, *Genius,* p. 165.

217 **"Oppenheimer at Los Alamos"**: Bethe, interview by Jon Else, 7/13/79, p. 9.

218 **"very easily and naturally"**: Eugene Wigner, *The Recollections of Eugene P. Wigner,* p. 245.

218 **"never dictated what"**: Bethe, "Oppenheimer: Where He Was There Was Always Life and Excitement," *Science,* vol. 155, p. 1082.

218 **"In his presence"**: Wilson, "A Recruit for Los Alamos," *Bulletin of the Atomic Scientists,* March 1975, p. 45.

218 **"The power of his personality"**: John Mason Brown, *Through These Men,* p. 286.

218 **"He could read a paper"**: Lee DuBridge, interview with Sherwin, 3/30/83, p. 11.

218 **"One would listen"**: Thorpe and Shapin, "Who Was J. Robert Oppenheimer?," p. 574.

218 **"He made you do"**: McKibbin, interview by Jon Else, 12/10/79, pp. 21–23.

219 **"I think he had"**: Manley, interview by Sherwin, 1/9/85, p. 24; Smith and Weiner, *Letters,* p. 263; Manley, interview by Alice Smith, 12/30/75, pp. 10–11.

219 **"The background of our work"**: JRO to Enrico Fermi, 3/11/43, box 33, Fermi, JRO Papers.

219 **"Security was terrible"**: Serber, *Peace and War,* p. 80.

219 **"The object of the project"**: Serber, *The Los Alamos Primer,* p. 1.

220 **Some of the physics**: Rhodes, *The Making of the Atomic Bomb,* p. 460.

220 **"That once plutonium"**: Bethe, interview by Jon Else, 7/13/79, p. 1.

220 **"the pieces might be mounted"**: Serber, *The Los Alamos Primer,* pp. xxxii, 59; Rhodes, *The Making of the Atomic Bomb,* p. 466.

221 **"I believe your people"**: Davis, *Lawrence and Oppenheimer,* p. 182.

221 **"In this connection"**: Barton J. Bernstein, "Oppenheimer and the Radioactive-Poison Plan," *Technology Review,* May–June 1985, pp. 14–17; Rhodes, *The Making of the Atomic Bomb,* p. 511; JRO to Fermi, 5/25/43, box 33, JRO Papers.

222 **"I doubt that you will"**: JRO to Weisskopf, 10/29/42, box 77, Weisskopf folder, JRO Papers; Sherwin, *A World Destroyed,* p. 50.

222 **Much later, Groves seriously**: Norris, *Racing for the Bomb,* p. 292. See also Nicholas Dawidoff, *The Catcher Was a Spy,* pp. 192–94.

Chapter Sixteen: "Too Much Secrecy"

223 **He thought of himself**: Edward Condon to Raymond Birge, 1/9/67, box 27, Condon folder, JRO Papers; Jessica Wang, "Edward Condon and the Cold War Politics of Loyalty," *Physics Today,* December 2001.

223 **"I join every organization":** Wheeler, *Geons, Black Holes, and Quantum Foam,* p. 113.

223 **An idealist with strong:** In just a few years, the House Un-American Activities Committee would label Condon "one of the weakest links" in atomic security (*New York Sun,* 3/5/48, "Law to Dig Out Condon's Files May Be Asked," box 27, Condon folder, JRO Papers).

224 **"Compartmentalization of knowledge":** Thorpe and Shapin, "Who Was J. Robert Oppenheimer?," *Social Studies of Science,* August 2000, p. 562.

224 **"The thing which upsets me":** Edward Condon to JRO, April 1943, reprinted in Groves, *Now It Can Be Told,* pp. 429–32.

225 **"Basically his way":** Thorpe, "J. Robert Oppenheimer and the Transformation of the Scientific Vocation," p. 251.

225 **"matter of policy":** Serber, *Peace and War,* p. 73; Norris, *Racing for the Bomb,* p. 243. Norris writes that Groves "treated Oppenheimer delicately, like a fine instrument that needed to be played just right. . . . Some men if pushed too hard will break."

225 **"He had his hair":** Hempelmann, interview by Sherwin, 8/10/79, pp. 26, 27.

225 **"my old anxiety":** Teller to JRO, 3/6/43, box 71, Teller, JRO Papers.

226 **"I know General Groves":** JRO hearing, p. 166.

226 **"While I may have":** JRO hearing, p. 166.

226 **In May 1943:** Thorpe, "J. Robert Oppenheimer and the Transformation of the Scientific Vocation," dissertation, p. 229.

226 **He told Groves:** Ibid., pp. 233–34.

226 **"My view about the whole":** JRO FBI file, doc. 159, D. M. Ladd to FBI director, 8/11/47. Ladd is quoting a statement made by Oppenheimer to Col. Boris Pash on 8/26/43. See JRO hearing, p. 849.

227 **"the drive which accompanied":** Morrison to JRO, 7/29/43, with attached letter to Roosevelt, 7/29/43, box 51, JRO Papers; Sherwin, *A World Destroyed,* p. 52 and ch. 2.

227 **"Recent reports both through":** Bethe and Teller to JRO, memo, 8/21/43, box 20, Bethe, JRO Papers.

228 **With the title of scientific director:** Norris, *Racing for the Bomb,* pp. 245–46.

228 **Security was always:** Brode, *Tales of Los Alamos,* p. 16.

228 **"Emilio, you left":** Serber, *Peace and War,* p. 80.

229 **"If they had their way":** Serber, interview by Sherwin, 1/9/82, p. 19.

229 **"Oppenheimer volunteered information":** Peer de Silva, FBI interview, 2/24/54, RG 326, entry 62, box 2, file C (FBI report), NA.

229 **"center for all gossip":** Jane S. Wilson and Charlotte Serber, eds., *Standing By and Making Do,* pp. 65, 70.

229 **"Therefore," said Oppie:** JRO to Groves, 4/30/43, Groves, box 36, JRO Papers; Jane S. Wilson and Charlotte Serber, eds., *Standing By and Making Do,* p. 62; Robert Serber, *Peace and War,* p. 79; *The Day After Trinity,* Jon Else.

230 **His antics became:** Richard P. Feynman, "Los Alamos from Below," Badash, et al., eds., *Reminiscences of Los Alamos,* pp. 105–32, 79; Gleick, *Genius,* pp. 187–89.

230 **"Try to satisfy":** Kunetka, *City of Fire,* p. 71; Thorpe "J. Robert Oppenheimer and the Transformation of the Scientific Vocation," dissertation, pp. 201, 249.

230 **"I have a complaint":** Hawkins, interview by Sherwin, 6/5/82, p. 19.

230 **"He was profoundly":** Hawkins, interview by Sherwin, 6/5/82, p. 18.

230 **"He complained constantly":** Robert R. Wilson, "A Recruit for Los Alamos," *Bulletin of the Atomic Scientists,* March 1975, p. 43.

231 **Walker later confirmed:** G. C. Burton to Ladd, FBI memo, 3/18/43; J. Edgar Hoover to SAC San Francisco, 3/22/43, re: report from Gen. Strong that the Army now has full-time technical and physical surveillance on Oppenheimer; see also Goodchild, *Oppenheimer,* p. 87, for report by Andrew Walker.

231 **"There isn't anybody"**: Powers, *Heisenberg's War*, p. 216; Smith and Weiner, *Letters*, p. 261.

231 **"we had been very much involved"**: JRO hearing, pp. 153–54; Bob Serber was driving home one night when he spotted Oppie and Jean walking in the neighborhood, deep in conversation. "It surprised me that he was still seeing her," Serber said. "And then later on, Kitty told me that she knew all about it, that Robert would tell her that Jean was in trouble and he was going to see what he could do." Later Serber learned that Jean had phoned Oppie "not frequently, but at least several times . . . in desperation." (Robert Serber, interview by Sherwin, 1/9/82, p. 11.)

232 **"had she not been so mixed up"**: Fervent Communist Party activists, the Jenkinses had named their baby daughter Margaret Ludmilla Jenkins after Ludmilla Pavlichenko, the woman sniper who is alleged to have killed 180 Nazis during the siege of Stalingrad (see Jenkins, *Against a Field Sinister*, pp. 30–31).

232 **She was a pediatric:** *Directory of Physicians and Surgeons, Naturopaths, Drugless Practitioners, Chiropodists, Midwives*, 3/3/42 and 3/3/43, published by the Board of Medical Examiners of the State of California. The directory lists Dr. Jean Tatlock as having graduated in 1941 from Stanford University School of Medicine.

232 **She implored him:** Michelmore, *The Swift Years*, p. 89. Michelmore does not provide citations or quotes, and no such letters have been found.

232 **"she was extremely unhappy"**: JRO hearing, p. 154.

232 **"On June 14, 1943, Oppenheimer"**: Secret FBI memo, "Subject: Jean Tatlock," 6/29/43, file A, RG 326, entry 62, box 1, and also found in AEC PSB record of JRO hearing, box 1, NA. See also Rhodes, *The Making of the Atomic Bomb*, p. 571; JRO hearing, p. 154. The military intelligence agents watched the darkened apartment building at least until 1:00 a.m. But by another account they may also have been able to eavesdrop on the couple electronically. The alleged transcript reportedly has Oppenheimer and Tatlock talking for a long time in the living room before they adjourned to the bedroom. See Goodchild, *J. Robert Oppenheimer*, p. 90. Goodchild cites two unnamed sources who claim to have seen a transcript showing that the FBI had managed to bug Tatlock's apartment. No such transcript has been declassified.

233 **"Did you find out"**: JRO hearing, p. 154.

233 **"might either use her"**: FBI memo for Mr. E. A. Tamm (Hoover's assistant), 8/27/43, 101-6005-8, Jean Tatlock FBI file, 100-190625-308.

233 **"it has been determined"**: An FBI document obtained under the Freedom of Information Act reveals that a wiretap was placed on Tatlock's home phone on 9/10/43 (FBI radiogram NR 070305, 9/10/43). But Tatlock's FBI file contains phone index cards dated August 1943, suggesting that perhaps Army Counter-Intelligence had already begun the phone surveillance (Jean Tatlock FBI file, FOIA no. 0960747-000/190-HQ-1279913, San Francisco (SF) 100-18382). There are index cards for at least some of her phone calls (dozens of pages are still classified). We learn little of substance. For instance, on 8/25/43, an unidentified woman who is apparently in the Marines calls Jean from New York City. Jean tells her that she is flying to Washington, D.C., on September 11 for a vacation. (Jean Tatlock FBI file, FOIA no. 0960747-000/190-HQ-1279913, SF 100-18382; Hoover, "Memorandum for the Attorney General," 9/1/43, FBI doc. 100-203581574, found in Jean Tatlock FBI file; Hawkins, interview by Sherwin, 6/5/82.)

233 **Oppenheimer "may still be connected"**: JRO FBI file, doc. 51, 3/18/46, JRO background; JRO FBI file, doc. 1320, 4/28/54.

233 **"he may be making"**: Col. Boris Pash to Lt. Col. Lansdale, memo on JRO, 6/29/43, reprinted in JRO hearing, pp. 821–22.

233 **Pash naturally wondered:** FBI confidential memo SF 101-126, p. 4. The FBI knew, for instance, that as late as 10/29/42, Tatlock was still a subscriber to *People's World*.

The FBI also thought it suspicious that two other people living in Tatlock's small apartment building had close associations with the Communist Party. Emil Geist was a subscriber to *People's World*. Another neighbor, David Thompson, was identified as literature director of the North Beach section of the Communist Party. (Secret FBI memo, "Subject: Jean Tatlock," 6/29/43, File A, RG 326, entry 62, box 1, NA.)

234 **"It is the opinion":** Pash to Lansdale memo on JRO, 6/29/43, reprinted in JRO hearing, pp. 821–22.

234 **"you should tell":** Lansdale to Gen. Groves, 7/6/43, RG 77, entry 8, box 100, NA.

235 **"specially trained Counter Intelligence":** Pash to Lansdale, memo on JRO, 6/29/43, reprinted in JRO hearing, pp. 821–22.

Chapter Seventeen: "Oppenheimer Is Telling the Truth . . ."

236 **"irrespective of the information":** Stern, *The Oppenheimer Case,* p. 49.

236 **"In the future":** Nichols, *The Road to Trinity,* p. 154; Richard G. Hewlett and Jack M. Holl, *Atoms for Peace and War,* p. 102.

237 **"indiscretions which could":** JRO hearing, p. 276.

237 **"trying to indicate":** JRO hearing, p. 276 (Lansdale to Groves, memo, 8/12/43).

238 **General Groves later told:** FBI from Newark SAC, to FBI director, 12/22/53, doc. 565, p. 2, JRO FBI file.

238 **"This is a pleasure"** *and subsequent quotes from Pash and Oppenheimer:* JRO hearing, pp. 845–48 (Pash-Oppenheimer interview, 8/25/43).

239 **Perhaps Chevalier had mentioned:** Hewlett and Holl, *Atoms for Peace and War,* p. 97.

241 **"this man Eltenton":** JRO hearing, pp. 845–48.

242 **"association with the Communist":** Ibid., p. 847.

242 **"I think that a lot":** Ibid.

243 **"a boy called [David] Fox":** Ibid., p. 852.

243 **"My motto is":** Ibid., p. 853.

244 **"I want to say this":** Ibid., pp. 871–86.

246 **"O.K. sir":** Ibid.

246 **"believed that Oppenheimer":** Ibid., p. 167; A. H. Belmont to D. M. Ladd, FBI synopsis memo, p. 5, 12/29/53, JRO FBI file.

247 **"In other words":** Memo to file, 9/10/43, conversation between James Murray, investigative officer, DSM project, Berkeley, and General Groves, Groves file, Lewis L. Strauss Papers, HHL. This memo was passed to Teeple by Murray in September 1954. Teeple passed it to Strauss.

248 **"gave him hell":** Belmont to Ladd, FBI synopsis memo, p. 5, 12/29/53, JRO FBI file; Haakon Chevalier FBI file part 1 of 2, doc. 110, memo to director, 3/2/54, p. 3.

248 **"the original source of the story:"** Belmont to Ladd, FBI synopsis memo, p. 7, 12/29/53, JRO FBI file.

Chapter Eighteen: "Suicide, Motive Unknown"

250 **"lying on a pile of pillows":** City and County of San Francisco, Coroner's Office, Necropsy Department, CO-44-63, 1/6/44, 9:30 a.m.

250 **"I am disgusted with everything":** *San Francisco Chronicle,* 1/7/44, p. 9; *San Francisco Examiner,* 1/6/44, front page; *San Francisco Examiner,* 1/7/44, p. 3. Michelmore, *The Swift Years,* p. 50. The suicide note was not preserved in the San Francisco medical examiner's file on the Tatlock death. No handwriting analysis was made of the note.

250 **"Tatlock sympathized with"**: Peter Goodchild reports that John Tatlock was well-known in Berkeley for his right-wing views (Goodchild, *J. Robert Oppenheimer*, p. 31). According to Phil Morrison, this is incorrect. See also FBI documents re: Richard Combs: "Extract from Memorandum on Communist Activities, Los Angeles, Calif., 15 Oct. 38."

251 **"acute edema of the lungs"**: Necropsy Department report, 1/6/44, Coroner's Office, City and County of San Francisco, CO-44-63; Pathological Department report, CO-44-63; Toxicological Department report, case no. 63; 1/13/44, Certificate of Death, 1/8/44; Coroner's Register, record of death of Jean Tatlock.

251 **"Suicide, motive unknown"**: *San Francisco Chronicle*, 1/7/44, p. 9. Dr. Siegfried Bernfeld was listed as a witness on the coroner's "record of death" for Jean Tatlock. Beside his name are scrawled the words "15 calls in Nov," perhaps indicating that he had seen her for fifteen sessions of analysis in November.

251 **"For you were never starved"**: Priscilla Robertson, undated letter, circa 1944, "Promise," p. 28, Sherwin Collection.

251 **Jackie Oppenheimer later reported:** Goodchild, *J. Robert Oppenheimer*, p. 35.

251 **"had slept with every 'bull' "**: Edith Jenkins, interview by Herken, 5/9/02. Tatlock's bisexuality was attested to by Mildred Stewart and Dorothy Baker, two literary figures in California who wrote about the lesbian community (Mildred Stewart oral history, p. 34, Special Collections, SU).

252 **"Jean seemed to need"**: Jenkins, *Against a Field Sinister*, p. 28. Hilda Stern Hein—the granddaughter of Oppenheimer's aunt Hedwig Stern—later said she knew that Washburn and Tatlock were "more than friends" (Hans "Lefty" Stern, phone interview by Bird, 3/4/04).

252 **Unable to come:** Edith Jenkins, interview by Herken, 5/9/02; Barbara Chevalier, interview by Herken, 5/29/02. Chevalier said Washburn had told her this story.

252 **The day after:** Cap. Peer de Silva, the Los Alamos security officer whose job it was to know everything about Oppenheimer's personal life, later claimed to have been the one who first gave him the news. Robert wept openly, De Silva wrote. (Peer de Silva, unpublished manuscript, p. 5.) De Silva's manuscript contains many incorrect assertions, such as the notion that Tatlock became Steve Nelson's mistress or that she served in an ambulance corps in the Spanish Civil War. He also wrongly asserts that Tatlock cut her own throat in the bathtub. De Silva described Oppenheimer's reaction to Tatlock's death in an interview with the FBI in February 1954. He wrote in his unpublished manuscript, "He [Robert] then went on at considerable length about the depth of his emotion for Jean, saying there was really no one else to whom he could speak." In what De Silva felt was a "sincere display of emotion," Oppenheimer confessed he had been "deeply devoted" to Tatlock and "had resumed a very close intimate association with her after his marriage and until the time of her death." De Silva is not a reliable observer, and it is not credible that Oppenheimer would confide in him. (FBI interview of Peer de Silva, 2/24/54, RG 326, entry 62, box 2, file C [FBI report], NA.)

252 **"He was deeply grieved"**: Robert Serber, interview by Sherwin, 1/9/82, p. 11. Michelmore, *The Swift Years*, p. 50; Serber, *Peace and War*, p. 86.

253 **"No direct action"**: Confidential FBI teletype from San Francisco to director, date censored, 100203581-1421, Jean Tatlock FBI files 100-18382-1 and 100-190625-20.

253 **In the years since:** Schwartz, *From West to East*, p. 380; see also private detective report from Keith Patterson of Josiah Thompson Investigations to Stephen Rivele, 7/12/91, regarding an investigation into Tatlock's death.

253 **"If you were clever"**: Dr. Jerome Motto, interview by Bird, 3/14/01. Dr. Jeffrey Kelman, interview by Bird, 2/3/01. Dr. Kelman suggests that one might be able to know if Tatlock was murdered if the coroner had reported the exact levels of chloral hydrate in

her blood. If these levels were too low—in other words, if she had been given just enough to knock her out in the form of a "Mickey Finn"—then someone would have had to hold her head under the water. The death certificate merely reports "a faint trace of chloral hydrate." Arguably, "a faint trace" may militate against suicide. But if so, does this mean that the suicide note was forged? Unfortunately, records from the brief formal inquest into the Tatlock death seem not to have been preserved.

253 **Some investigators:** Dr. Hugh Tatlock filed a Freedom of Information Act request to the FBI for information about his sister (Dr. Hugh Tatlock, interview by Sherwin, February 2001). The FBI released about eighty pages of highly censored material—but several documents suggest that "technical surveillance" was initiated on Jean Tatlock's phone line on 9/10/43.

254 **"in the Russian manner":** Herken, *Brotherhood of the Bomb,* p. 106.

254 **delegated responsibility for assassinations:** Church Committee Final Report, Book IV, pp. 128–29; William R. Corson, *The Armies of Ignorance,* pp. 362–64; Warren Hinckle and William W. Turner, *The Fish Is Red,* p. 29.

254 **Clearly, Col. Boris Pash:** After the war, Col. Pash was decorated for his leadership of the top secret Alsos Mission, raiding teams that seized dozens of top German scientists and 70,000 tons of Axis uranium ore in 1944 and 1945 (Christopher Simpson, *Blowback,* pp. 152–53).

Chapter Nineteen: "Would You Like to Adopt Her?"

255 **"We had no invalids":** Brode, *Tales of Los Alamos,* p. 13.

255 **"as if we shut":** Thorpe, "J. Robert Oppenheimer and the Transformation of the Scientific Vocation," dissertation, p. 188.

255 **"Only the oldest":** Church, *The House at Otowi Bridge,* p. 126.

255 **"We found that on the mesas":** Ibid., p. 98.

256 **Just before arriving:** Wilson, "A Recruit for Los Alamos," *Bulletin of the Atomic Scientists,* March 1975, p. 41.

256 **"Here at Los Alamos":** Thorpe and Shapin, "Who Was J. Robert Oppenheimer?," *Social Studies of Science,* August 2000, p. 547.

256 **"island in the sky":** Thorpe, "J. Robert Oppenheimer and the Transformation of the Scientific Vocation," dissertation, p. 182; Wilson and Serber, eds., *Standing By and Making Do,* p. 5.

256 **"Everyone in your house":** Brode, *Tales of Los Alamos,* p. 39.

256 **When the local theater:** Smith and Weiner, p. 265; Brode, *Tales of Los Alamos,* pp. 72, 23.

256 **"The alcohol hits you":** Dr. Louis Hempelmann, interview by Sherwin, 8/10/79, p. 29.

257 **"Kitty was strictly":** Anne Wilson Marks, interview by Bird, 3/5/02.

257 **"She was awful bossy":** Hempelmann, interview by Sherwin, 8/10/79, pp. 8, 24.

257 **"If you were":** Brode, *Tales of Los Alamos,* pp. 72, 23; Hempelmann, interview by Sherwin, 8/10/79, p. 30; Dorothy McKibbin, interview by Jon Else, 12/10/79, p. 22.

257 **On Sundays, many residents:** Hempelmann, interview by Sherwin, 8/10/79, p. 10; Brode, *Tales of Los Alamos,* pp. 56, 88–93; McKibbin, interview by Jon Else, 12/10/79, p. 20; Wirth and Aldrich, *Los Alamos,* p. 261.

258 **"He was always":** Hempelmann, interview by Sherwin, 8/10/79, p. 22.

258 **"I think he only picked":** Marks, interview by Bird, 3/5/02.

258 **The habit had so callused:** Peer de Silva, unpublished manuscript, p. 1, courtesy of Gregg Herken.

258 **Gradually, life on the mesa:** Brode, *Tales of Los Alamos,* pp. 28, 33, 51–52.

259 **"It seems that the girls"**: Wilson, "A Recruit for Los Alamos," *Bulletin of the Atomic Scientists*, March 1975, p. 47.

259 **"the best dry martinis"**: Nancy Cook Steeper, *Gatekeeper to Los Alamos*, p. 83.

259 **"Dorothy loved Robert"**: Steeper, *Gatekeeper to Los Alamos*, pp. 60, 83. Steeper is citing her 1999 interview of David Hawkins. Steeper wrote of "the many quiet evenings Robert spent at Dorothy's house, an oasis from the unsightly settlement of Los Alamos and a respite from the urgency and relentless stress of building the bomb. What a comfort Dorothy must have been to him and how she delighted in his friendship." (Steeper, *Gatekeeper to Los Alamos*, p. 125.)

260 **"She wanted to make big talk"**: Pat Sherr, interview by Sherwin, 2/20/79.

260 **"She seemed to be"**: Joseph Rotblat, interview by Sherwin, 10/16/89, p. 8.

260 **"She was a very intense"**: Goodchild, *J. Robert Oppenheimer*, p. 127.

260 **"She would give me"**: Sherr, interview by Sherwin, 2/20/79.

260 **"He would give her"**: Goodchild, *J. Robert Oppenheimer*, p. 127.

260 **"It never seemed"**: Hempelmann, interview by Sherwin, 8/10/79, p. 18.

260 **"He [Oppenheimer] stopped"**: Marks, interview by Bird, 3/5/02.

261 **"Our thin, ascetic Director"**: Wilson and Serber, eds., *Standing By and Making Do*, p. 50.

261 **"I have to tell you"**: Marks, interview by Bird, 3/14/02.

262 **"He really was"**: Marks, interview by Bird, 3/5/02.

262 **"They were terribly close"**: Hempelmann, interview by Sherwin, 8/10/79, p. 25.

262 **"Her background was not good"**: JRO hearing, p. 266; Goodchild, *J. Robert Oppenheimer*, p. 88.

262 **"I was sure she'd been"**: Davis, *Lawrence and Oppenheimer*, p. 156.

263 **"Dr. Oppenheimer was"**: Goodchild, *J. Robert Oppenheimer*, p. 90.

263 **So many young couples**: By June 1944, one fifth of all the married women in Los Alamos were pregnant. Thorpe, "J. Robert Oppenheimer and the Transformation of the Scientific Vocation," dissertation, p. 276; Wilson and Serber, eds., *Standing By and Making Do*, p. 92; Robert Serber, *Peace and War*, p. 83.

263 **On December 7, 1944**: Brode, *Tales of Los Alamos*, p. 22.

263 **"Kitty had begun"**: Sherr, interview by Sherwin, 2/20/79.

263 **"She must have felt"**: Frank and Jackie Oppenheimer, interview by Sherwin 12/3/78; Goodchild, *J. Robert Oppenheimer*, p. 128.

264 **"It seemed to me"**: Pat Sherr, interview by Sherwin, 2/20/79. Pat's husband, Rubby Sherr, confirmed that his wife took care of Toni Oppenheimer (Rubby Sherr, e-mail to Bird, 7/11/04).

265 **"It was known"**: Jackie Oppenheimer, interview by Sherwin 12/3/78; Goodchild, *J. Robert Oppenheimer*, p. 128.

265 **"She drank somewhat"**: Pat Sherr, interview with Sherwin, 2/20/79, p. 4.

265 **"She certainly didn't drink"**: Hempelmann, interview by Sherwin, 8/10/79, pp. 11, 20.

265 **the "place where the water"**: Steeper, *Gatekeeper to Los Alamos*, p. 34.

265 **"We had tea"**: JRO to Mrs. Fermor S. Church, 11/21/58, box 76, JRO Papers.

265 **"slim and wiry hero"**: Church, *The House at Otowi Bridge*, p. 86.

265 **Her life was simple**: Pettitt, *Los Alamos Before the Dawn;* Church, *The House at Otowi Bridge*, pp. 12, 86; Church, *Bones Incandescent*, p. 30.

266 **"Don't do it"**: Dorothy McKibbin to Alice Smith, 10/17/75, Smith correspondence, Sherwin Collection; Smith and Weiner, *Letters*, p. 280; McKibbin interview, 1/1/76.

266 **"Along about April"**: Church, *The House at Otowi Bridge*, pp. 123–24.

266 **"Kitty and I understood"**: Peter Miller to JRO, 4/27/51, box 76, JRO Papers.

266 **"Mr. Nicholas Baker"**: JRO to Groves, 11/2/43, Groves folder, box 36, JRO Papers.

267 **"in gratitude for":** Church, *The House at Otowi Bridge,* pp. 95–98; Peter Miller to JRO, 4/27/51, box 76, JRO Papers. Miller was quoting Warner's words about Bohr and Oppenheimer on her deathbed.

267 **But it was only Oppie's:** Church, *The House at Otowi Bridge,* p. 130; Brode, *Tales of Los Alamos,* pp. 120–27.

267 **"Not the smallest part":** Church, *The House at Otowi Bridge,* pp. 98–99, 130. In her 1945 Christmas letter, Miss Warner wrote, "I had not known what was being done up there, though in the beginning I had suspected atomic research."

Chapter Twenty: "Bohr Was God, and Oppie Was His Prophet"

268 **The "race" for the atomic:** Rhodes, *The Making of the Atomic Bomb,* pp. 523–24; Sherwin, *A World Destroyed,* p. 106.

269 **"To Bohr":** JRO, "Niels Bohr and Atomic Weapons," *New York Review of Books,* 12/17/64; Powers, *Heisenberg's War,* pp. 237–38.

269 **Groves' displeasure had:** Powers, *Heisenberg's War,* pp. 239–40.

269 **To prevent this:** Sherwin, *A World Destroyed,* pp. 90–114.

270 **"General, I can't stand it":** Powers, *Heisenberg's War,* p. 247.

270 **"whispering mumble":** Norris, *Racing for the Bomb,* p. 252.

270 **"within five minutes":** JRO hearing, p. 166.

270 **"Is it really big enough?":** JRO, "Niels Bohr and Atomic Weapons," *New York Review of Books,* 12/17/64.

270 **"That is why":** Sherwin, *A World Destroyed,* p. 91.

271 **"require a terrific technical effort":** Robert Jungk, *Brighter Than a Thousand Suns,* p. 103; Powers, *Heisenberg's War,* p. 253.

271 **"On the other hand":** See February 2002 release of Bohr letters by the Niels Bohr Institute, doc. 10. See the website of the Niels Bohr Archive: *www.nba.nbi.dk;* see also Michael Frayn's play *Copenhagen,* and Powers, "What Bohr Remembered," *New York Review of Books,* 3/28/2002.

271 **"Bohr had the impression":** JRO, "Niels Bohr and Atomic Weapons," *New York Review of Books,* 12/17/64. See also Powers, *Heisenberg's War,* pp. 120–28; Cassidy, *Uncertainty;* Jungk, *Brighter Than a Thousand Suns,* pp. 102–4.

272 **One glance, however, persuaded:** Robert Serber, *Peace and War,* p. 86. The sketch was probably Bohr's and depicted what Heisenberg had shown him. It has since disappeared.

272 **"My God," Bethe said:** Powers, *Heisenberg's War,* p. 253.

272 **"be a quite useless":** Powers, *Heisenberg's War,* p. 254; JRO to Groves, 1/1/44, MED RG 77E 5, box 64, 337.

272 **"it is easy":** JRO, "Niels Bohr and Atomic Weapons," *New York Review of Books,* 12/17/64.

272 **"Bohr at Los Alamos"** *and subsequent quotes:* JRO to Groves, 1/17/44, Groves folder, box 36, JRO Papers; JRO, "Three Lectures on Niels Bohr and His Times: Part III, The Atomic Nucleus," Pegram Lecture, August 1963, box 247, JRO Papers; JRO, "Niels Bohr and Atomic Weapons," *New York Review of Books,* 12/17/64.

272 **"this bomb may":** Victor Weisskopf, interview by Sherwin, 4/21/82.

272 **"it is already evident":** Bohr, "Confidential comments on the project of exploiting the latest discoveries in atomic physics for industry and warfare," 4/2/44, box 34, Frankfurter-Bohr folder, JRO Papers.

273 **Finally, Bohr concluded:** Sherwin, *A World Destroyed,* pp. 93–96; Goodchild, *J. Robert Oppenheimer,* p. 92. See also Margaret Gowing, *Britain and Atomic Energy, 1939–1945.*

273 **"[He] knew Bohr"**: Weisskopf, interview by Sherwin, 4/21/82.

274 **"My God, suppose"**: Powers, *Heisenberg's War,* p. 255.

274 **"the implication was"**: Palevsky, *Atomic Fragments,* p. 117. Years later, Oppenheimer told friends that he wanted someday to write a play to explore the notion of what would have happened if Roosevelt had lived into the postwar period.

274 **"Complementarity"**: Gribbin, *Q Is for Quantum,* pp. 85, 88.

274 **"Bohr was not satisfied"**: Bernstein, *Cranks, Quarks, and the Cosmos,* p. 44.

274 **join them in their "scientific work"**: Peter Kapitza to Bohr, 10/28/43, box 34, Frankfurter-Bohr folder, JRO Papers.

275 **"to propose to the rulers"**: David Lilienthal, *The Journals of David E. Lilienthal,* vol. 2, p. 456 (diary entry of 2/3/49).

275 **To Bohr's thinking**: Sherwin, *A World Destroyed,* p. 106.

275 **"It seemed to me"**: Palevsky, *Atomic Fragments,* p. 134; Robert Wilson, interview by Owen Gingrich, 4/23/82, p. 5 (Sherwin Collection); Wilson, "Niels Bohr and the Young Scientists," *Bulletin of the Atomic Scientists,* August 1985, p. 25.

275 **"there was never"**: JRO hearing, p. 173.

275 **"How did he [Bohr]"**: Sherwin, *A World Destroyed,* pp. 107–10. Bohr met with Churchill in mid-May 1944 and with Roosevelt on 8/26/44. The meeting with Churchill was brief and disappointing: "We did not even speak the same language," Bohr later said. By contrast, Bohr came away from his meeting with Roosevelt with the impression that the president was strongly sympathetic to his views.

275 **"was at times a thorn"**: Groves to JRO, 12/7/64, Groves folder, box 36, JRO Papers.

276 **"from leakage regarding"**: Bohr, "Confidential comments on the project of exploiting the latest discoveries in atomic physics for industry and warfare," 4/2/44, box 34, Frankfurter-Bohr folder, JRO Papers.

276 **"were committed to building"**: Powers, *Heisenberg's War,* p. 257.

Chapter Twenty-one: "The Impact of the Gadget on Civilization"

277 **"He was present"**: Thorpe and Shapin, "Who Was J. Robert Oppenheimer?," *Social Studies of Science,* August 2000, p. 573.

277 **Hans Bethe recalled**: Bethe, "Oppenheimer: Where He Was There Was Always Life and Excitement," *Science,* 3/3/67, p. 1082.

277 **"It [the rubber tube]"**: McAllister Hull, interview by Charles Thorpe, 1/16/98, in Thorpe, "J. Robert Oppenheimer and the Transformation of the Scientific Vocation," dissertation, p. 250.

278 **hope "that the production"**: Jones, *Manhattan: The Army and the Atomic Bomb,* pp. 176, 182; Richard G. Hewlett and Oscar E. Anderson, Jr., *The New World, 1939–1946,* p. 168.

278 **Indeed, any such attempt**: Jones, *Manhattan: The Army and the Atomic Bomb,* p. 509.

279 **"One could have separated"**: Hoddeson, et al., *Critical Assembly,* p. 242.

279 **"We have investigated"**: Ibid., pp. 241–43.

280 **"became terribly impatient"**: Davis, *Lawrence and Oppenheimer,* p. 219.

280 **"Oppenheimer lit into me"**: Goodchild, *J. Robert Oppenheimer,* p. 116.

280 **"Parsons was furious"**: Thorpe, "J. Robert Oppenheimer and the Transformation of the Scientific Vocation," dissertation, p. 326; Goodchild, *J. Robert Oppenheimer,* p. 118.

280 **"The kind of authority"**: Thorpe, "J. Robert Oppenheimer and the Transformation of the Scientific Vocation," dissertation, pp. 263–64.

281 **"Who were the German"**: Rigden, *Rabi,* pp. 154–55.

281 **"The only way":** Studs Terkel, *The Good War,* p. 510.

281 **In a decision critical:** George B. Kistiakowsky, "Reminiscences of Wartime Los Alamos," Badash, et al., eds., *Reminiscences of Los Alamos,* p. 54; Jones, *Manhattan: The Army and the Atomic Bomb,* p. 510.

282 **"He was a leader":** Smith and Weiner, *Letters,* p. 264.

282 **"On the Construction":** Sherwin, *A World Destroyed,* p. 34.

282 **"He was the first person":** Sir Rudolf Peierls, interviews by Sherwin, 6/6/79, p. 12, and 3/5/79.

282 **"he could stand up":** Peierls, interview by Sherwin, 6/6/79, pp. 6, 10.

282 **"I was not happy":** Teller, *Memoirs,* pp. 85, 176–77.

282 **Every morning Teller:** Serber, *The Los Alamos Primer,* p. xxxi.

283 **"God protect us":** Teller, *Memoirs,* p. 222.

283 **"Edward essentially went":** JRO to Groves, 5/1/44, MED, record group 77, box 201, Rudolf Peierls folder; see also Herken, *Brotherhood of the Bomb,* p. 86, and Goodchild, *J. Robert Oppenheimer,* p. 105. In his memoir, Teller has a slightly different account of why he walked out of this meeting, claiming that Oppenheimer had rudely ordered him to talk about a problem related to the Super that Teller felt he wasn't ready to talk about (see Teller, *Memoirs,* p. 193). See also Thorpe, "J. Robert Oppenheimer and the Transformation of the Scientific Vocation," dissertation, p. 255.

283 **"a disaster to any organization"; "somewhat wild":** Serber, *The Los Alamos Primer,* p. xxx. Peierls, interview by Sherwin, 6/6/79, p. 14.

283 **"It could have":** Peierls, interview by Sherwin, 3/5/79, p. 1.

283 **"Dear Rab":** JRO to Rabi, 12/19/44, box 59, Rabi, JRO Papers; Rigden, *Rabi,* p. 168.

284 **"I hope it has not":** Smith and Weiner, *Letters,* pp. 273–74.

284 **"Only one man paused":** Palevsky, *Atomic Fragments,* p. 173; Dyson, *From Eros to Gaia,* p. 256.

284 **"You realize of course":** Rotblat, interview by Sherwin, 10/16/89. Stunned, Rotblat related the dinner-table conversation to one person, a fellow physicist, Martin Deutsch.

284 **"Until then I had thought":** Rotblat, interview by Sherwin, 10/16/89, p. 16; Albright and Kunstel, *Bombshell,* p. 101.

285 **"anti-government agitation":** Ted Morgan, *Reds,* p. 278.

285 **"among the two or three":** Robert Chadwell Williams, *Klaus Fuchs,* p. 32.

286 **"If he was a spy":** Ibid., p. 76.

286 **Oppenheimer heard that Hall:** Albright and Kunstel, *Bombshell,* pp. 62, 119.

286 **"it seemed to me":** Ibid., p. 90.

286 **His sole purpose:** Ted Hall, interview by Sherwin; Joan Hall, "A Memoir of Ted Hall," posted at *www.historyhappens.net.*

286 **"It used to be":** Albright and Kunstel, *Bombshell,* pp. 86–87. Rotblat later turned against Oppenheimer. "Gradually things came to my knowledge," Rotblat said. "I felt, this is not the way a hero of mine should behave. Gradually he became an antihero. For example, the fact that he agreed that the bomb should be used on the cities. He could have said no. And at the time, he was powerful enough that his voice might have prevailed." Palevsky, *Atomic Fragments,* p. 171.

287 **"quite long discussions":** Palevsky, *Atomic Fragments,* pp. 135–36; Wilson told the same story to Owen Gingrich (Robert Wilson, interview by Gingrich, 4/23/82, p. 6, Sherwin Collection).

287 **"All right. So what?":** Robert Wilson, interview by Gingrich, 4/23/82, p. 6; see also Robert Wilson, "Niels Bohr and the Young Scientists," *Bulletin of the Atomic Scientists,* August 1985, p. 25, and Robert Wilson, "The Conscience of a Physicist," in Richard Lewis and Jane Wilson, eds., *Alamogordo Plus Twenty-five Years,* pp. 67–76.

287   **"I can remember":** Robert Wilson, interview by Gingrich, 4/23/82, p. 6. Wilson told Jon Else that he thought thirty to fifty people attended the meeting (*The Day After Trinity,* Jon Else, transcript, p. 37).

287   **"whether the country":** Louis Rosen, interview by Sherwin, 1/9/85, p. 1.

288   **Oppenheimer argued:** Badash, et al., eds., *Reminiscences of Los Alamos,* p. 70.

288   **"caused us to think":** Weisskopf, interview by Sherwin, 4/21/82, p. 5.

288   **"the thought of quitting":** Weisskopf, *The Joy of Insight,* pp. 145–47. Robert Wilson also describes this meeting in similar terms in a 1958 review of Robert Jungk's book, *Brighter Than a Thousand Suns.* But on this, the first occasion in which he told this story, Wilson wrote that the meeting occurred in 1944, not 1945. (Robert Wilson, "Robert Jungk's Lively but Debatable History of the Scientists Who Made the Atomic Bomb," *Scientific American,* December 1958, p. 146.) See also Alice Smith, *A Peril and a Hope,* p. 61. Another Harvard-trained physicist, Roy Glauber, remembered the meeting Wilson organized to discuss the impact of the gadget (see Albright and Kunstel, *Bombshell,* p. 87).

288   **"You know, you're the director":** Palevsky, *Atomic Fragments,* pp. 135–36.

288   **When Oppenheimer took:** Robert Wilson, interview by Gingrich, 4/23/82, p. 7.

289   **They had to forge:** *The Day After Trinity,* Jon Else, transcript, p. 37.

289   **"I thought that":** Palevsky, *Atomic Fragments,* pp. 136–37.

289   **"My feeling about":** Ibid., p. 138.

Chapter Twenty-two: "Now We're All Sons-of-Bitches"

290   **"Sunday morning found":** Smith and Weiner, *Letters,* p. 287.

290   **"We have been living":** Ibid., p. 288.

290   **"Roosevelt was a great":** Palevsky, *Atomic Fragments,* p. 116.

291   **The resulting firestorm:** Mark Selden, "The Logic of Mass Destruction," in Kai Bird and Lawrence Lifschultz, eds., *Hiroshima's Shadow,* pp. 55–57.

291   **"I remember":** Len Giovannitti and Fred Freed, *The Decision to Drop the Bomb,* p. 36. The authors interviewed Oppenheimer. Some Americans did criticize the fire-bombings. See *Commonweal,* 6/22/45 and 8/24/45.

291   **"We have been too late":** Emilio Segrè, *A Mind Always in Motion,* p. 200.

291   **"our 'demonstration' "** *and subsequent quotes:* William Lanouette, *Genius in the Shadows,* pp. 261–62; Leo Szilard to JRO, 5/16/45, Szilard folder, box 70, JRO Papers.

292   **"General Groves tells me"** *and subsequent quotes:* Lanouette, *Genius in the Shadows,* pp. 266–67.

293   **"a Frankenstein which":** Minutes of the Interim Committee mtg., 5/31/45, in Sherwin, *A World Destroyed,* pp. 299–301 (appendices); also pp. 202–210.

295   **"We might say that a great":** Ibid.

296   **"feel free to tell their people":** Sherwin, *A World Destroyed,* pp. 295–304 (Appendix L, Notes of the Interim Committee Meeting, 5/31/45); Giovannitti and Freed, *The Decision to Drop the Bomb,* pp. 102–5.

297   **"It may be very difficult":** Alice K. Smith, *A Peril and a Hope,* p. 25; Sherwin, *A World Destroyed,* p. 211. "The Political Implications of Atomic Weapons," (Frank report), pp. 323–32 (Appendix S).

297   **"feeling that we can trust":** Giovannitti and Freed, *The Decision to Drop the Bomb,* p. 115.

298   **"He should have":** Palevsky, *Atomic Fragments,* p. 142; *The Day After Trinity,* Jon Else, transcript, p. 20.

298   **"in a large urban area":** Sherwin, *A World Destroyed,* pp. 229–30; Thorpe, "J. Robert Oppenheimer and the Transformation of the Scientific Vocation," dissertation, p. 344.

Thorpe cites Major J. A. Derry and Dr. N. F. Ramsey, memo for General L. R. Groves, "Summary of Target Committee Meetings on 10 and 11 May 1945," also cited in Jones, *Manhattan: The Army and the Atomic Bomb,* pp. 529–30.

298 **"I thought even leaflet"** *and subsequent quotes:* Palevsky, *Atomic Fragments,* pp. 84, 252; Norris, *Racing for the Bomb,* pp. 382–83.

299 **"I set forth":** Alice Smith, *A Peril and a Hope,* p. 50; Goodchild, *J. Robert Oppenheimer,* p. 143.

299 **"There was not sufficient:"** Gar Alperovitz, *The Decision to Use the Atomic Bomb,* p. 189.

300 **"We didn't know beans":** JRO hearing, p. 34.

300 **"unconditional surrender":** After meeting with President Truman, Grew recorded in his diary on 5/28/45: "The greatest obstacle to unconditional surrender by the Japanese is their belief that this would entail the destruction or permanent removal of the Emperor." Joseph C. Grew, *Turbulent Era,* vol. 2, 1952, pp. 1428–34; Sherwin, *A World Destroyed,* p. 225; Alperovitz, *The Decision to Use the Atomic Bomb,* pp. 48, 66, 479, 537, 712, 753.

300 **"They wanted to keep":** Allen Dulles, foreword to Per Jacobsson's pamphlet "The Per Jacobsson Mediation," Balse Centre for Economic and Financial Research, ser. C, no. 4, circa 1967, on file in Allen Dulles Papers, box 22, John J. McCloy 1945 folder, Princeton University.

300 **"It is my opinion":** William D. Leahy diary, 6/18/45, William D. Leahy Papers, LOC, reprinted in Bird and Lifschultz, eds., *Hiroshima's Shadow,* p. 515.

300 **"question of whether":** Walter Mills, ed., *The Forrestal Diaries,* p. 70; "Extracts from Minutes of Meeting Held at the White House 18 June 1945," in Sherwin, *A World Destroyed,* pp. 355–63 (Appendix W).

301 **According to McCloy:** James V. Forrestal diary, 3/8/47, President's Secretary's files, HSTL, reprinted in Bird and Lifschultz, eds., *Hiroshima's Shadow,* p. 537.

301 **"The delivery of a warning":** John J. McCloy diary, 7/16–17/45, DY box 1, folder 18, John J. McCloy Papers, Amherst College.

301 **"the Japanese were ready":** "Ike on Ike," *Newsweek,* 11/11/63, p. 107. Some historians question Eisenhower's account. See Robert S. Norris, *Racing for the Bomb,* pp. 531–32; Barton J. Bernstein, "Understanding the Atomic Bomb and the Japanese Surrender: Missed Opportunities, Little-Known Near Disasters, and Modern Memory," *Diplomatic History* 19, no. 2 (1995).

301 **"telegram from Jap Emperor":** Harry S. Truman, *Off the Record,* ed. Robert H. Ferrell, p. 53; Sherwin, *A World Destroyed,* p. 235.

301 **"it was ever present":** James F. Byrnes, interview by Fred Freed for NBC television, circa 1964, transcript found in Herbert Feis Papers, box 79, LOC. At Potsdam on 7/29/45, Ambassador Joseph E. Davies noted in his diary, "Byrnes was disgusted with Molotov's stubbornness, and said 'The New Mexico situation' (Atomic Bomb) had given us great power, and that in the last analysis, it would control" (Joseph E. Davies diary, 7/29/45, Chron file, box 19, Davies Papers, LOC).

301 **"Believe Japs will fold":** Truman, *Off the Record,* ed. Ferrell, pp. 53–54.

301 **"President, Leahy, JFB [Byrnes] agreed Japs":** Walter Brown diary, 8/3/45, Special Collections, Robert Muldrow Cooper Library, CU, reprinted in Bird and Lifschultz, eds., *Hiroshima's Shadow,* p. 546.

301 **Isolated in Los Alamos:** For further evidence on the debate over the bomb in Washington in the summer of 1945, see the documents reprinted in Bird and Lifschultz, eds., *Hiroshima's Shadow,* pp. 501–50. For a different perspective on the question of whether the Japanese were attempting to surrender see Richard Frank, *Downfall: The End of the Imperial Japanese Empire* (Random House, 1999); Herbert Bix, *Hirohito*

and the Making of Modern Japan (Harper Collins, 2000); and Barton J. Bernstein, "The Alarming Japanese Buildup on Southern Kyushu," *Pacific Historical Review,* November 1999.

302 **"the United States shall":** Bird and Lifschultz, eds., *Hiroshima's Shadow,* pp. 553–54, 558.

302 **Teller claims in his memoirs:** Teller to Szilard, 7/2/45, Teller folder, box 71, JRO Papers; Teller, *Memoirs,* pp. 205–7.

303 **He was convinced:** Alice Smith, *A Peril and a Hope,* pp. 53, 63.

303 **"since an opportunity"** *and subsequent quotes:* Szilard to JRO, 5/16/45 and 7/10/45; Edward Creutz to Szilard, 7/13/45, Szilard folder, box 70, JRO Papers.

303 **"The enclosed note":** Szilard Papers 21/235; NND-730039, NA 201 E Creutz; Groves diary, 7/17/45, NA, courtesy of William Lanouette. Both Szilard and Lapp confirmed in interviews that Oppenheimer decided that the petition "could not be circulated" (Alice Smith, *A Peril and a Hope,* p. 55).

303 **"There was tension":** Church, *The House at Otowi Bridge,* p. 129.

303 **"incompleteness of our knowledge":** Norris, *Racing for the Bomb,* p. 395.

303 **"the planning of the use":** Jones, *Manhattan: The Army and the Atomic Bomb,* p. 511.

304 **Here the army staked:** Peer de Silva, unpublished manuscript, p. 12; Rhodes, *The Making of the Atomic Bomb,* p. 652.

304 **"Batter my heart":** JRO to Groves, 10/20/62, box 36, JRO Papers; Hijiya, "The Gita of J. Robert Oppenheimer," *Proceedings of the American Philosophical Society,* vol. 144, no. 2, June 2000, pp. 161–64; Szasz, *The Day the Sun Rose Twice,* p. 41; Norris, *Racing for the Bomb,* p. 397.

304 **"I believe we were under":** JRO hearing, p. 31.

304 **By the end of June:** Norris, *Racing for the Bomb,* pp. 399–400; Morrison, "Blackett's Analysis of the Issues," *Bulletin of the Atomic Scientists,* February 1949, p. 40.

305 **To Robert's delight:** *The Day After Trinity,* Jon Else, transcript, p. 7.

305 **The FBI and army:** In June 1944, while Frank was stationed at the Oak Ridge, Tennessee, uranium separation plant, Jackie had written him soon after the Allied landings in France, "Well, well, the D-Day has come. I think that is wonderful. . . . But as you have predicted, and as I more or less [illegible], the battle against Russia (propaganda) has already started. . . . it's insidious." To Jackie this was "pure, unadulterated American Fascism." (Jackie Oppenheimer to Frank Oppenheimer, undated, circa June 1944, folder 4–13, box 4, Frank Oppenheimer Papers, UCB.)

305 **"We spent several days":** Frank Oppenheimer, interview by Weiner, 2/9/73, p. 56.

305 **"You can change the sheets":** Goodchild, *J. Robert Oppenheimer,* p. 151.

305 **"Oppenheimer became so emotional":** George Kistiakowsky, "Trinity: A Reminiscence," *Bulletin of the Atomic Scientists,* June 1980, p. 21.

305 **"In battle, in forest":** Vannevar Bush, *Pieces of the Action,* p. 148.

306 **"The weather is whimsical":** Lansing Lamont, *Day of Trinity,* p. 184.

306 **"Funny how the mountains":** Ibid., p. 193.

306 **To relieve the tension:** *The Day After Trinity,* Jon Else, transcript, p. 12.

306 **"All the frogs":** Frank Oppenheimer, interview by Weiner, 2/9/73, p. 57.

306 **"There could be":** Lamont, *Day of Trinity,* p. 210; *The Day After Trinity,* Jon Else, transcript, p. 12.

307 **"obviously confused":** Szasz, *The Day the Sun Rose Twice,* p. 73.

307 **Worried that some:** Norris, *Racing for the Bomb,* pp. 403–4; Lamont, *Day of Trinity,* p. 210.

307 **"If we postpone":** Lamont, *Day of Trinity,* pp. 212, 220.

307 **"a big ball of orange":** Feynman, "*Surely You're Joking, Mr. Feynman!,*" p. 134.

307 **"something had gone wrong":** Hershberg, *James B. Conant,* p. 232.

308 **"I could feel the heat"**: Serber, *Peace and War,* pp. 91–93.

308 **"All of a sudden"**: Badash, et al., *Reminiscences of Los Alamos,* pp. 76–77.

308 **Frank Oppenheimer was:** *The Day After Trinity,* Jon Else, transcript, p. 47.

308 **"the light of the first"**: Frank Oppenheimer, interview by Weiner, 2/9/73, AIP, p. 56; *The Day After Trinity,* Jon Else, transcript, p. 14.

308 **"Lord, these affairs"**: Lamont, *Day of Trinity,* p. 226.

308 **"Dr. Oppenheimer . . . grew"**: General Thomas Farrell, "Memorandum for the Secretary of War," 7/18/45, reprinted in Groves, *Now It Can Be Told,* pp. 436–37; NYT, 8/7/45, p. 5; Hijiya, "The Gita of J. Robert Oppenheimer," *Proceedings of the American Philosophical Society,* vol. 144, no. 2 (June 2000), p. 165.

308 **"I think we just said"**: *The Day After Trinity,* Jon Else, transcript, pp. 15–16.

308 **"I'll never forget"**: Davis, *Lawrence and Oppenheimer,* p. 242; *The Day After Trinity,* Jon Else, transcript, p. 50; Frank Oppenheimer, interview by Jon Else, 1980; Szasz, *The Day the Sun Rose Twice,* p. 89.

309 **"Lots of boys"**: William L. Laurence, NYT, 9/27/45, p. 7.

309 **"We knew the world"**: *The Day After Trinity,* Jon Else, transcript, pp. 79–80. Some Sanskrit scholars suggest that a better translation of this line would be "I am become Time, destroyer of worlds."

309 **"priestly exaggerations"**: Pais, *The Genius of Science,* p. 273.

309 **"The big boom"**: Alice Smith, *A Peril and a Hope,* p. 76; NYT, 9/26/45, pp. 1, 16.

309 **Oppie pulled out:** Lamont, *Day of Trinity,* p. 237; Kistiakowsky, "Trinity: A Reminiscence," *Bulletin of the Atomic Scientists,* June 1980, p. 21.

309 **"Now we're all sons"**: Years later, Oppenheimer remembered Bainbridge's remark and told David Lilienthal that he agreed with it: "I guess that is just about right" (Lilienthal, *The Journals of David E. Lilienthal,* vol. 6, p. 89, diary entry for 2/13/65).

309 **"Tell her she can"**: Lamont, *Day of Trinity,* pp. 242–43; Anne Wilson, his secretary, said she had no such recollection (Anne Wilson Marks, phone interview by Bird, 5/22/02). While Richard Feynman got out his bongo drums and beat them with elation, he later said of the moment, "You stop thinking, you know; you just stop." Robert Wilson, who was not elated, had said to Feynman, "It's a terrible thing that we made." Feynman, *"Surely You're Joking, Mr. Feynman!,"* pp. 135–36.

309 **But curiously, he didn't:** Hijiya, "The Gita of J. Robert Oppenheimer," *Proceedings of the American Philosophical Society,* vol. 144, no. 2 (June 2000), pp. 123–24.

Chapter Twenty-three: "Those Poor Little People"

313 **"They were picking"; "Those poor little"**: Anne Wilson Marks, interview by Bird, 3/5/02.

314 **"Don't let them bomb"**: Lt. Col. John F. Moynahan, *Atomic Diary,* p. 15. The bombardiers followed Oppenheimer's instructions, dropping the bomb visually on the center of Hiroshima. But Nagasaki was bombed "largely by radar," because of cloud cover and because the bomber was running low on fuel. (See Norman Ramsey to JRO, dated "after August 20, 1945," box 60, JRO Papers.)

315 **"what actually occurred"**: Alice Smith, *A Peril and a Hope,* p. 53; see also Hershberg, *James B. Conant,* p. 230.

315 **"the visible effects"; "Why the hell"**: Manley, "A New Laboratory Is Born," Badash, et al., eds., *Reminiscences of Los Alamos,* p. 37.

315 **"I'm proud of you"**: Groves and JRO, transcript of phone conversation, 8/6/45, RG 77, entry 5, MED files, 201 Groves, box 86, gen. correspondence 1942–45, telephone conversation file.

315 **"Attention please, attention"**: *The Day After Trinity,* Jon Else, transcript, p. 58.

316 **"This last 24 hours"**: Ed Doty to parents, 8/7/45, Los Alamos Historical Museum.

316 **"too early to determine"**: Sam Cohen, *The Truth About the Neutron Bomb*, p. 22; Hijiya, "The Gita of J. Robert Oppenheimer," *Proceedings of the American Philosophical Society*, vol. 144, no. 2 (June 2000), p. 155. Hijiya cites Cohen for the claim that Oppenheimer clasped his hands together like a prizefighter, but this detail is not in Cohen's book. It is found, however, in Lawren, *The General and the Bomb*, p. 250.

316 **"That night we"**: Phil Morrison's radio talk, ALAS series for station KOB (Albuquerque), no. 3, Federation of American Scientists Records, XXII, p. 2. "The Atom Bomb Scientists Report Number Three: Death of Hiroshima," p. 1, Special Collections, UC.

316 **"certainly no one [at Los Alamos] celebrated"** *and subsequent quotes:* Ed Doty to parents, 8/7/45, Los Alamos Historical Museum; Smith, *A Peril and a Hope*, p. 77. Smith wrote only that Oppenheimer saw a "young group leader being sick in the bushes." Thomas Powers identifies the young group leader as Robert Wilson (Powers, *Heisenberg's War*, p. 462). See also *The Day After Trinity*, Jon Else.

317 **"I felt betrayed"**: Robert Wilson, "Robert Jungk's Lively but Debatable History," *Scientific American*, December 1958, p. 146; Palevsky, *Atomic Fragments*, pp. 140–41.

317 **"People were going"**: *The Day After Trinity*, Jon Else, transcript, pp. 59–60; Palevsky, *Atomic Fragments*, p. 141.

317 **"As the days passed"; "Oppie says"**: Smith, *A Peril and a Hope* (1971 edition), p. 77; Robert Serber, *Peace and War*, p. 142.

317 **"nervous wreck"**: Herken, *Brotherhood of the Bomb*, p. 139; FBI memo, 4/18/52, sect. 12, JRO FBI file.

318 **"unconditional surrender"**: Hershberg, *James B. Conant*, pp. 279–304; Alperovitz, *The Decision to Use the Atomic Bomb*, pp. 417–20; see also Barton J. Bernstein, "Seizing the Contested Terrain of Early Nuclear History;" Uday Mohan and Sanho Tree, "The Construction of Conventional Wisdom," and the essays by Norman Cousins, Reinhold Niebuhr, Felix Morley, David Lawrence, Lewis Mumford, Mary McCarthy and other early critics of the bombings, reprinted in Bird and Lifschultz, *Hiroshima's Shadow*, pp. 141–97, 237–316.

318 **Lawrence tried to reassure**: Childs, *An American Genius*, p. 366; Herken, *Brotherhood of the Bomb*, p. 140.

318 **"it is our firm opinion"**: Smith and Weiner, *Letters*, pp. 293–94 (JRO to Stimson, 8/17/45).

319 **"no alternative to"**: Ibid., pp. 300–1; JRO to Ernest Lawrence, 8/30/45.

319 **"You will believe"**: Ibid., pp. 297–98; JRO to Herbert Smith, 8/26/45; JRO to Frederick Bernheim, 8/27/45.

319 **"Dear Opje"**: Chevalier, *Oppenheimer*, p. xi.

319 **"Circumstances are heavy"**: *The Day After Trinity*, Jon Else, transcript, p. 65, JRO to Haakon Chevalier, 8/27/45, *The Day After Trinity*, supplemental files; Herken, *Brotherhood of the Bomb*, p. 142.

320 **"I have a sense"**: JRO to Conant, 9/29/45, JRO Papers.

320 **Incredibly, a formal offer**: Smith and Weiner, *Letters*, p. 300.

320 **"I have very mixed"**: Ibid., pp. 301–2.

320 **"Kitty didn't often"**: Jean Bacher, interview by Sherwin, 11/5/87, pp. 3–4. Didisheim quote contained in a letter from Herbert Smith to Frank Oppenheimer, 9/19/73, folder 4–23, box 4, Frank Oppenheimer Papers, UCB.

321 **"But Phil was"**: Bacher, interview by Sherwin, 11/5/87, p. 2.

321 **"Virtually everyone in the street"**: A transcript of Phil Morrison's radio talk can be found in the ALAS series for station KOB (Albuquerque), no. 3, Federation of American Scientists (FAS) XXII, p. 2. "The Atom Bomb Scientists Report Number Three: Death of Hiroshima," p. 5, Special Collections, UC.

321 **"At one point"**: Serber, *Peace and War*, p. 129.

321 **"We circled finally"**: Smith, *A Peril and a Hope,* p. 115; a transcript of Morrison's radio talk can be found in the ALAS series for station KOB (Albuquerque), no. 3, FAS XXII, p. 2.

321 **"Much was now explained"**: Church, *The House at Otowi Bridge,* pp. 130–31; Church, *Bones Incandescent,* p. 38.

322 **"We took this tree"**: Michael A. Day, "Oppenheimer on the Nature of Science," *Centaurus,* vol. 43 (2001), p. 79; *Time,* 11/8/48.

322 **"The war had made it"**: Weisskopf, note on physics in the postwar years, December 1962, box 21, "JRO and Niels Bohr," JRO Papers.

322 **"very belatedly"**: JRO, "Three Lectures on Niels Bohr and His Times," Pegram Lectures, Brookhaven National Laboratory, August 1963, p. 16, filed in Louis Fischer Papers, box 9, folder 3, PUL. Henry Stimson diary, 9/21/45, p. 3, YUL.

322 **"Now it is in your"**: Ibid.

Chapter Twenty-four: "I Feel I Have Blood on My Hands"

323 **"Hats off to the men"** *and subsequent quotes:* Paul Boyer, *By Bomb's Early Light,* pp. 266–67; Pais, *The Genius of Science,* p. 274.

323 **"We have made"**: JRO, "Atomic Weapons," *Proceedings of the American Philosophical Society,* January 1946. He gave this speech on 11/16/45 in Philadelphia, where it was entitled "Atomic Weapons and the Crisis in Science," filed in folder 168.1, Lee DuBridge Papers, courtesy of James Hershberg.

324 **"I thought that he"**: Cherniss, interview by Sherwin, 5/23/79, p. 11.

324 **On September 9, Oppenheimer:** Smith and Weiner, *Letters,* p. 304; JRO to Harrison, 9/9/45.

325 **"the situation looked"**: Smith, *A Peril and a Hope,* pp. 116–17.

325 **"the suppression of the document"**: Ibid., p. 120.

325 **"not merely a super"**: Herken, *Brotherhood of the Bomb,* p. 150.

325 **"armaments manufacturer"**: Barnett, "J. Robert Oppenheimer," *Life,* 10/10/49.

325 **"I neither can"**: Teller and Brown, *The Legacy of Hiroshima,* p. 23.

326 **Unbeknownst even to Washington:** Henry Wallace diary, 10/19/45, reprinted in John Morton Blum, *The Price of Vision,* p. 497.

326 **"The hope of civilization"**: Truman, *Memoirs,* vol. 1, p. 532.

326 **Leo Szilard was outraged:** Lanouette, *Genius in the Shadows,* p. 286.

326 **"I believe that"**: Smith, *A Peril and a Hope,* p. 167; Hewlett and Anderson, *The New World,* vol. 1, p. 432.

327 **"The Johnson bill"**: Smith, *A Peril and a Hope,* p. 153; Thorpe, "J. Robert Oppenheimer and the Transformation of the Scientific Vocation," dissertation, pp. 401–2.

327 **"a masterpiece"**: Lanouette, *Genius in the Shadows,* p. 293.

327 **"oblique attack"**: Smith, *A Peril and a Hope,* p. 154.

327 **"He said there wasn't"**: *The Day After Trinity,* Jon Else, transcript, p. 68; Goodchild, *J. Robert Oppenheimer,* p. 178.

327 **"Mailing it was"**: Thorpe, "J. Robert Oppenheimer and the Transformation of the Scientific Vocation," dissertation, pp. 395–96; Wilson, "Hiroshima: The Scientists' Social and Political Reaction," *Proceedings of the American Philosophical Society,* September 1996, p. 351.

328 **"Oppie's suggestions [should]"**: Thorpe, "J. Robert Oppenheimer and the Transformation of the Scientific Vocation," dissertation, p. 409.

328 **Within the month:** Smith, *A Peril and a Hope,* pp. 197–200.

328 **"his eyes were glazed"**: Steeper, *Gatekeeper to Los Alamos,* p. 111.

329 **"*Today that pride*"**: Smith and Weiner, *Letters,* pp. 310–11.

329 **"That day he was us":** Eleanor Jette, *Inside Box 1663*, p. 123.

329 **"[i]t would seem wrong":** Smith and Weiner, *Letters*, p. 306.

329 **"how much we pay":** Herken, *Brotherhood of the Bomb*, p. 149.

331 **"The guilt consciousness":** Henry Wallace diary, 10/19/45, reprinted in Blum, ed., *The Price of Vision*, pp. 493–97. For more on Byrnes' atomic diplomacy, see Alperovitz, *The Decision to Use the Atomic Bomb*, p. 429.

331 **"The first thing is":** Murray Kempton, "The Ambivalence of J. Robert Oppenheimer," *Esquire*, December 1983, reprinted in Kempton, *Rebellions, Perversities, and Main Events*, p. 121. Kempton mistakenly places this conversation in 1946. Another version of this story appears in Davis, *Lawrence and Oppenheimer*, p. 260. Davis provides no date or citation—but according to President Truman's appointment calendar, the president met with Oppenheimer on only four occasions: 10/25/45, 4/29/48, 4/6/49, and 6/27/52.

331 **"incomprehension it showed":** Davis, *Lawrence and Oppenheimer*, p. 261.

332 **"I feel I have blood"** *and subsequent quotes:* Truman to Dean Acheson, memo, 5/7/46, box 201 PSF, HSTL. See also Merle Miller, *Plain Speaking*, p. 228, and Boyer, *By Bomb's Early Light*, p. 193. Boyer places Dean Acheson in the room, but the Truman Presidential Appointment Calendar notes the presence only of Robert Patterson, Oppenheimer, and Truman (Matthew J. Connelly files, Presidential Appointment Calendar, 10/25/45, HSTL). Herken, *Brotherhood of the Bomb*, p. 150. Herken is citing Davis, *Lawrence and Oppenheimer*, p. 258; Michelmore, *The Swift Years*, pp. 121–22, and Lilienthal, *The Journals of David E. Lilienthal*, vol. 2, p. 118.

333 **"He was not a man":** Rabi, interview by Sherwin, 3/12/82, p. 9.

333 **"a simple man, prone":** John J. McCloy diary, 7/20/45, DY box 1, folder 18, McCloy Papers, Amherst College.

333 **"the fix we are in":** Smith and Weiner, *Letters*, pp. 315–25.

333 **"I remember Oppie's":** Ibid., p. 315.

334 **"I know that many":** Ibid., pp. 315–25.

334 **"sacred trust":** Truman, *Memoirs*, vol. 1, p. 537.

335 **"Dear Mr. Opp.":** Smith and Weiner, *Letters*, pp. 325–26.

Chapter Twenty-five: "People Could Destroy New York"

336 **"Since the use of":** JRO FBI file, sect. 1, doc. 20, Hoover to Byrnes, memo, 11/15/45, and Hoover to Brig. Gen. Harry H. Vaughan, military aide to the president, memo, 11/15/45.

336 **"Isn't it nice?":** JRO FBI file, sect. 4, doc. 108, p. 9.

337 **"appears to leave some doubt":** Herken, *Brotherhood of the Bomb*, p. 160; see Herken's website *www.brotherhoodofthebomb.com* for chapter 9's extended endnote 7: Menke, FBI memo to file, 3/14/47, box 2, JRO/AEC.

337 **"close to Oppenheimer":** JRO FBI file, doc. 51 (3/18/46, p. 6) and doc. 159 (Ladd to FBI director, 8/11/47, p. 7).

337 **"done nothing towards":** JRO FBI file, doc. 134, "Julius Robert Oppenheimer: Background," 1/28/47, p. 7.

338 **A wiretap on:** Memo to FBI director, 5/23/47, JRO FBI file, serial 6. Hoover also authorized "microphone surveillance."

338 **"OK if Father will keep quiet":** Upon receiving this news, Hoover ordered no further contact with Wilson (JRO FBI file, sect. 1, doc. 25, 3/26/46); Anne Wilson Marks, phone interview by Bird, 10/21/02.

338 **"What the hell":** Joseph Weinberg, interview by Sherwin, 8/23/79, p. 17.

338 **"reliable informants":** Hoover to George E. Allen, 5/29/46, PSF Box 167, folder: FBI Atomic Bomb, HSTL; Bird, *The Chairman*, p. 281.

339 **"So it came to me":** Rabi, interview by Sherwin, 3/12/82, pp. 2–5; Rigden, *Rabi,* pp. 196–97.

339 **Four weeks later:** Hewlett and Anderson, *The New World,* vol. 1, p. 532.

340 **"He walked back and forth":** Lilienthal, *The Journals of David E. Lilienthal,* vol. 2, p. 13; Lilienthal to Herb Marks, 1/14/48, Lilienthal letters to JRO, box 46, JRO Papers.

340 **"Everybody genuflected":** Goodchild, *J. Robert Oppenheimer,* p. 178.

340 **"almost musically delicate mind":** Bird, *The Chairman,* p. 277.

340 **"All the participants":** Dean Acheson, *Present at the Creation,* p. 153.

341 **"Our bewildered questions":** Ibid., see also JRO hearing, pp. 37–40.

341 **"This is a brilliant":** Joseph I. Lieberman, *The Scorpion and the Tarantula,* p. 255.

341 **"without world government":** JRO, "Atomic Explosives." Folder: United Nations, AEC, box 52, Bernard Baruch Papers, PUL.

342 **"Oppenheimer screwed it up":** Rabi, interview by Sherwin, 3/12/82, p. 6; Herken, *Brotherhood of the Bomb,* p. 164.

342 **"Only something as drastic":** Lieberman, *The Scorpion and the Tarantula,* p. 246.

342 **Soon afterwards:** "A Report on the International Control of Atomic Energy—Prepared for the Secretary of State's Committee on Atomic Energy by a Board of Consultants: Chester I. Barnard, Dr. J. R. Oppenheimer, Dr. Charles A. Thomas, Harry A. Winne, David E. Lilienthal, Chairman," Washington, D.C., 3/16/46.

342 **"favorably impressed":** James F. Byrnes, *Speaking Frankly,* p. 269. For Byrnes' business ties to Baruch, see Burch, *Elites in American History,* vol. 3, pp. 60, 62; see also David Robertson, *Sly and Able,* p. 118, for a description of Byrnes' close friendship with Baruch.

343 **"When I read the news":** Lilienthal, *The Journals of David E. Lilienthal,* vol. 2, p. 30; Bird, *The Chairman,* p. 279.

343 **"We're lost":** Herken, *Brotherhood of the Bomb,* p. 165. Oppenheimer later said of Baruch's appointment, "That was the day I gave up hope, but that was not the day for me to say so publicly" (Davis, *Lawrence and Oppenheimer,* p. 260).

343 **"winning weapon":** Herken, *The Winning Weapon,* p. 366. Herken also cites a letter from Fred Searls to Byrnes, 1/17/48 (Searls folder, Byrnes manuscripts), to show that Searls wanted Byrnes to help protect Newmont Corporation's tax status. Newmont Mining Corporation was founded in 1921 by "Colonel" William Boyce Thompson, a friend and business associate of Baruch. (Baruch, *My Own Story,* p. 238.) See also Allen, *Atomic Imperialism,* p. 108. The fact that Fred Searls was head of Newmont Mining Corp. is cited in Baruch, *The Public Years,* p. 363. Searls had also served as Byrnes' assistant during the war.

343 **"Don't let these associates":** Lieberman, *The Scorpion and the Tarantula,* p. 273.

344 **"despised Baruch":** Rabi, interview by Sherwin, 3/12/82, p. 6.

344 **"It is too bad":** Lilienthal, *The Journals of David E. Lilienthal,* vol. 2, p. 70 (diary entry for 7/24/46).

344 **Baruch was right:** Hershberg, *James B. Conant,* p. 270.

344 **"follow Oppenheimer's activities":** Hoover to SAC Los Angeles, JRO FBI file, sect. 1, doc. 23, 3/13/46.

344 **"an attempt to box":** SAC San Francisco, FBI memo to Hoover, 5/14/46, regarding surveillance of Oppenheimer telephone conversation with Kitty on 5/10/46 (JRO FBI file, docs. 45, 46). Almost a year later, the FBI wiretap was still active, and Kitty knew it. On 3/25/47, she told a friend, "Be careful what you say on the phone." When asked why, she replied, "The FBI, you know." (JRO FBI file, doc. 148, 3/25/47.)

345 **"as of possible interest":** FBI teletype to director, 5/8/46, JRO FBI file, doc. 33.

345 **"if a major power":** Hewlett and Anderson, *The New World,* vol. 1, pp. 562–66.

345 **Baruch nevertheless insisted:** Bird, *The Chairman,* p. 281.

346 **"I am told that"**: Ibid., p. 282.

346 **"still very heavy of heart"**: JRO to Lilienthal, 5/24/46, Lilienthal Papers, cited in Lieberman, *The Scorpion and the Tarantula*, pp. 284–85.

346 **"The American disposition"**: Lilienthal, *The Journals of David E. Lilienthal*, vol. 2, p. 70 (diary entry for 7/24/46).

346 **"The American proposal"**: Ibid., pp. 69–70 (Lilienthal diary entry for 7/24/46).

346 **"They worry me like hell"**: FBI wiretap excerpt, 6/11/46, Lewis Strauss Papers, HHL.

347 **"It proposes that"**: JRO, "The Atom Bomb as a Great Force for Peace," *New York Times Magazine*, 6/9/46.

347 **"What do we do"**: Weinberg, interview by Sherwin, 8/23/79, p. 25.

347 **"the quick and the dead"**: Hewlett and Anderson, *The New World*, p. 590.

347 **"knowing it was a damn fool"**: FBI wiretap of Kitty and Robert Oppenheimer phone conversation, 6/20/46, JRO FBI file, doc. 68.

348 **"It was his [Baruch's] ball"**: Dean Acheson oral history, n.d., PPF, HSTL; Bird, *The Chairman*, p. 282; Goodchild, *J. Robert Oppenheimer*, p. 181.

348 **"I cannot tell"**: JRO, "Three Lectures on Niels Bohr and His Times," Pegram Lectures, Brookhaven National Laboratory, August 1963, p. 15, Louis Fischer Papers, box 9, folder 3, PUL.

348 **"He is really a tragic"**: Lilienthal, *The Journals of David E. Lilienthal*, vol. 2, p. 69 (diary entry for 7/24/46).

348 **"Every American knows"**: JRO, "The International Control of Atomic Energy," *Bulletin of the Atomic Scientists*, 6/1/46.

349 **"Of course it could be done"**: Bird and Sherwin, "The First Line Against Terrorism," WP, 12/12/01; see also John von Neumann to Lewis Strauss, 10/18/47, Strauss Papers, HHL; Herken, *Counsels of War*, p. 179. See also Herken, *Brotherhood of the Bomb*, chapter 18, footnote 92 (web version only), where Herken reports that the project to investigate the dangers of nuclear terrorism was code-named "Cyclops." He cites Matteson to Stassen, 9/8/55, box 16, USSD; Panofsky interview by Herken (1993). A few years later, Oppenheimer persuaded the Atomic Energy Commission to have two physicists, Robert Hofstadter and Wolfgang Panofsky, write a report on the problem. The resulting top secret report recommended the installation of radiation detectors at all airports and ports. For a time, this was actually done in a few major airports. The Hofstadter-Panofsky report—known in the intelligence community as the "Screwdriver Report"—is still classified.

349 **"the only way in which this country"**: JRO speech, "Atomic Energy as a Contemporary Problem," 9/17/47, reprinted in JRO, *The Open Mind*, p. 25.

349 **He was conspicuously absent**: General Groves issued instructions to the effect that while Oppenheimer would be invited to witness the Bikini tests, he would not be permitted to evaluate the results (Herken, *The Winning Weapon*, p. 224). See also *Radio Bikini,* (documentary film).

350 **"cry-baby scientist"**: Truman, memo to Acheson, 5/7/46, "Atomic Tests" folder, PSF Box 201, HSTL (courtesy of archivist Dennis E. Bilger).

Chapter Twenty-six: "Oppie Had a Rash and Is Now Immune"

351 **"I did actually"**: JRO hearing, p. 35; JRO, interview by Kuhn, 11/18/63, p. 32.

351 **"I owe you"**: JRO FBI file, doc. 102, phone transcript, 10/23/46.

352 **"it is not a major issue"**: Hershberg, *James B. Conant*, p. 308; phone conversation between Kitty and Robert Oppenheimer, FBI memo, 12/14/46, doc. 120, JRO FBI file; Hewlett and Duncan, *Atomic Shield*, vol. 2, pp. 15–16.

352 **"given up all hope":** JRO hearing, p. 327.

352 **"He wanted me":** Ibid., p. 41. Acheson's statement to JRO makes it clear that the Truman Doctrine was the American Government's opening move in the emerging cold war.

352 **To Osborn's surprise:** Hewlett and Duncan, *Atomic Shield,* vol. 2, p. 268. See also James G. Hershberg, "The Jig Was Up: J. Robert Oppenheimer and the International Control of Atomic Energy, 1947–49," paper presented at Oppenheimer Centennial Conference, Berkeley, April 22–24, 2004.

353 **"encirclement and their need":** JRO hearing, p. 40.

353 **"terribly depressed":** Keith G. Teeter, FBI memo to file, 3/3/54, SF 100-3132.

353 **he "commented on the fact":** JRO FBI file doc. 159, Ladd to director, 8/11/47, p. 13.

353 **"It is clear":** JRO, *The Open Mind,* pp. 26–27. See also Thorpe, "J. Robert Oppenheimer and the Transformation of the Scientific Vocation," dissertation, pp. 446–47.

354 **"take this great plunge":** JRO hearing, p. 69.

354 **To all appearances, Oppenheimer:** Joseph Alsop to JRO, 7/29/48, Alsop folder, box 15, JRO Papers.

354 **"Whatever the Russians did":** Scott Donaldson, *Archibald MacLeish: An American Life,* p. 400.

355 **"that new insight":** JRO to MacLeish, 9/27/49; MacLeish to JRO, 10/6/49; JRO to MacLeish, 2/14/49. All in MacLeish folder, box 49, JRO Papers.

355 **Significantly, Frank Oppenheimer's:** In February 1947, two CP functionaries visited Frank in his home and spent two hours coaxing him to renew his prewar contributions to the Party. They left empty-handed, and the FBI later heard from an informant that one of the CP officials complained, "I think we lost about ten G's." JRO FBI file, doc. 149, 4/23/47.

355 **"the Russians were ready":** Frank Oppenheimer, interview by Sherwin, 12/3/78.

356 **"Haakon, believe me":** Chevalier, *Oppenheimer,* pp. 69, 74; Barbara Chevalier diary, 7/14/84, notes taken by Gregg Herken. See Herken's website, *www.brotherhoodofthebomb.com.* An FBI wiretap reports that Chevalier phoned Kitty Oppenheimer on 6/3/46 to confirm that he would visit the Oppenheimers the next evening (JRO FBI file, sect. 2, doc. 56, 6/3/46). This suggests that Chevalier met with Oppenheimer not twice but three times in the spring and summer of 1946: May 1946 at Stinson Beach; June 4, 1946, at Eagle Hill; and sometime between 6/26/46 (the day of Chevalier's FBI interrogation) and 9/5/46, the day of Oppenheimer's FBI interview. In addition, Kitty agreed to spend the weekend of June 22–23 at the Chevaliers' home. But she later postponed this visit to the following weekend. (6/21/46 memo.)

356 **"What about Opje?"** *and subsequent quotes:* Chevalier claims that a day later he outlined the plot for his 1959 novel, *The Man Who Would Be God* (Chevalier, *Oppenheimer,* pp. 79–80).

356 **"Someone obviously has it":** Chevalier, *Oppenheimer,* p. 58.

356 **On June 26, 1946:** FBI background report on JRO, 2/17/47, p. 10; Goodchild, *J. Robert Oppenheimer,* p. 70.

357 **"would be safely transmitted":** FBI (Newark) synopsis of facts, 19–22. Eltenton and Chevalier signed statements 6/26/46, document 786, JRO FBI files.

357 **"I wish to state":** Chevalier, affidavit for the FBI, 6/26/46, Chevalier FBI file part 1, also read into a tape recorder by Sherwin during an interview with Chevalier, 7/15/82, pp. 10–11.

357 **Some time later:** Chevalier, *Oppenheimer,* p. 68.

358 **"Opje's face at once":** Ibid., pp. 69–70; JRO hearing, p. 209.

358 **"Then, to my utter dismay":** Chevalier, *Oppenheimer,* pp. 69–70.

359 **"to do such a thing":** JRO FBI file, sect. 12, doc. 287, 4/18/52, "Allegation of Espionage Activity on the Part of George Charles Eltenton," p. 20 (declassified 1996).

360 **Oppenheimer expressed interest:** Strauss, *Men and Decisions,* p. 271.

360 **there "wasn't a scientist":** JRO FBI file, sect. 1, 1/29/47 and 2/2/47, summaries of wiretap conversations between Kitty and Robert Oppenheimer.

360 **"Ah, that I can do":** Strauss, *Men and Decisions,* p. 271.

360 **"Princeton is a madhouse":** Smith and Weiner, *Letters,* p. 190.

360 **"It is impossible for me":** Barnett, "J. Robert Oppenheimer," *Life,* 10/10/49.

360 **"You won't be free":** JRO FBI file, sect. 1, 1/29/47 and 2/2/47, summaries of wire-tapped conversations between Kitty and Robert Oppenheimer.

361 **"I guess that settles it":** Michelmore, *The Swift Years,* p. 142.

361 **"His name is":** *New York Herald Tribune,* 4/19/47.

361 **"In physical appearance":** Beatrice M. Stern, "A History of the Institute for Advanced Study, 1930–1950," p. 613, unpublished manuscript, IAS Archives.

361 **Lewis Strauss, however, was:** Richard Pfau, *No Sacrifice Too Great,* p. 93; Strauss, *Men and Decisions,* pp. 7, 84.

362 **"Regarding Strauss":** JRO FBI file sect. 3, doc. 103, FBI wiretap of JRO phone conversation with David Lilienthal and Robert Bacher, 10/23–24/46.

362 **"If you disagree":** Joseph and Stewart Alsop, *We Accuse,* p. 19; Duncan Norton-Taylor, "The Controversial Mr. Strauss," *Fortune,* January 1955; Brown, *Through These Men,* p. 275.

362 **"would not be gone long":** Herken, *Brotherhood of the Bomb,* p. 174; JRO FBI file, 5/9/47.

362 **"pleasant garden":** JRO FBI file, sect. 6, 5/7/47, contained in wiretap summary, 5/27/47.

362 **"the greatest blow":** JRO FBI file, sect. 6, newspaper clipping, 4/28/47.

362 **"I am terribly pleased":** Rabi to JRO, undated, Sunday afternoon, circa April 1947, Rabi correspondence, box 59, JRO Papers.

363 **His friend and former:** JRO FBI file, sect. 6, phone transcript, 2/27/47.

363 **"His wisdom and broad interests":** JRO, interview by Kuhn, 11/20/63, p. 19.

363 **"a very close":** JRO hearing, p. 957.

363 **"Robert loved the Tolmans":** Frank Oppenheimer, interview by Sherwin, 12/3/78.

363 **"totally suited for":** Jerome Seymour Bruner, *In Search of Mind,* pp. 236–38; John R. Kirkwood, Oliver R. Wolff and P. S. Epstein, "Richard Chase Tolman, 1881–1948," National Academy of Sciences of the United States of America, Biographical Memoirs, vol. 27, Washington, D.C., National Academy of Sciences, 1952, pp. 143–44.

363 **And during the war:** *Who Was Who in America,* vol. 3, 1951–1960 (Chicago: A. N. Marquis Co., 1966), p. 857.

363 **"Remember how we":** Ruth Tolman to JRO, 4/16/49, Ruth Tolman folder, box 72, JRO Papers.

364 **"My heart is very full":** Ruth Tolman to JRO, 8/24/47, Ruth Tolman folder, box 72, JRO Papers.

364 **"I look back":** Ruth Tolman to JRO, August 1 (1947?), Ruth Tolman folder, box 72, JRO Papers.

364 **"we'd go to the sea":** Ruth Tolman to JRO, undated (November 1948?) Thursday night, Pasadena, Ruth Tolman folder, box 72, JRO Papers.

364 **"Ruth, dear heart":** JRO to Ruth Tolman, 11/18/48, Ruth Tolman folder, box 72, JRO Papers.

364 **"I think Kitty":** Jean Bacher, interview by Sherwin, 3/29/83. When asked by Sherwin about rumors of an affair between Tolman and Oppenheimer, Bacher became flustered, and insisted, "There certainly was never any sexual interest in the relationship; it was very supportive." She then made it clear that further questions about an affair would conclude the interview.

365   **"Dr. Oppenheimer first earned":** "Memorandum for the Files of Lewis L. Strauss,"
       12/9/57, box 67, Strauss Papers, HHL. Strauss' secretary, Virginia Walker, told the his-
       torian Barton J. Bernstein that her boss was very upset when he learned of Oppen-
       heimer's affair with Tolman (Walker, interview by Barton Bernstein, 11/7/02). Bernstein
       also reports an interview with James Douglas, an aircraft company executive who
       claimed that he had visited the Tolman house one morning during the war and saw
       Oppenheimer and Ruth Tolman alone, wearing only dressing gowns. See also Herken,
       Brotherhood of the Bomb, pp. 290, 404; Herken cites a 1997 interview with Lawrence's
       wife, Molly, who remembered her husband coming home in a rage from a cocktail
       party hosted by Gloria Gartz, a neighbor and psychologist who knew Ruth Tolman.
       Gartz apparently told Lawrence of the affair at this party, which took place sometime
       prior to the 1954 Oppenheimer hearings. When Herken asked Molly if Richard Tolman
       was still alive at the time of the affair, Molly answered, "I know he was."

365   **"I shall always remember":** Ruth Tolman to JRO, undated, Tuesday (spring 1949?),
       Ruth Tolman folder, box 72, JRO Papers. Ruth Tolman's papers were destroyed at her
       instructions upon her death (Alice Smith to Beatrice Stern, 12/14/76, Smith correspon-
       dence, Sherwin Collection). A friend of Ruth's later said that Ruth herself destroyed
       her letters from Robert. Dr. Milton Pleoset, interview by Sherwin, 3/28/83, p. 11.
       Pleoset recalled, "She was very close to Oppenheimer."

365   **"there was derogatory":** JRO hearing, p. 27.

365   **"to conduct an open"** *and subsequent quotes:* Barton J. Bernstein, "The Oppenheimer
       Loyalty-Security Case Reconsidered," *Stanford Law Review,* July 1990, p. 1399.

366   **"Joe, what do you think?":** Stern, *The Oppenheimer Case,* p. 104.

366   **Oppenheimer "may at one time":** Stern, *The Oppenheimer Case,* pp. 104–5; Bern-
       stein, "The Oppenheimer Loyalty-Security Case Reconsidered," *Stanford Law Review,*
       July 1990, p. 1399; Herken, *Brotherhood of the Bomb,* p. 179.

366   **"rather carefully":** Stern, *The Oppenheimer Case,* p. 104.

367   **"specifically substantiating the fact":** FBI to Lilienthal, JRO FBI file, doc. 149,
       4/23/47; see also Herken, *Brotherhood of the Bomb,* p. 179.

367   **"he had had homosexual tendencies":** JRO FBI file, doc. 165, 10/30/47, SAC San
       Francisco to FBI director, declassified 6/28/96. The "extremely derogatory" story
       about Hall and Oppenheimer was regurgitated in another FBI memo to Mr. Ladd on
       11/10/47. S. S. Schweber cites this FBI document in his book *In the Shadow of the
       Bomb,* p. 203.

367   **Lilienthal thought it telling:** Herken, *Brotherhood of the Bomb,* pp. 179, 377.

Chapter Twenty-seven: "An Intellectual Hotel"

369   **The Oppenheimers arrived:** Regis, *Who Got Einstein's Office?,* p. 138; Michelmore,
       The Swift Years, p. 141.

370   **Robert had most of them torn out:** Anne Wilson Marks to Kai Bird, 5/11/02.

370   **Soon after their arrival:** *Time,* 11/8/48, p. 76.

370   **"an artist in the ancient":** Lilienthal, *The Journals of David E. Lilienthal,* vol. 6, p.
       130.

370   **"When we first moved":** Morgan, "A Visit with J. Robert Oppenheimer," *Look,*
       4/1/58, p. 35.

370   **Robert mounted one:** Oppenheimer sold this painting in 1965 for $350,000; twenty
       years later it was sold to a private collector at Sotheby's for $9 million.

370   **They hung a Derain:** Brown, *Through These Men,* p. 286.

370   **Oppie's austere study:** Hempelmann, interview by Sherwin, 8/10/79, pp. 16–17.

370   **Oppenheimer's ground-level office:** Pais, *A Tale of Two Continents,* p. 198.

371 **Oppenheimer took these:** Regis, *Who Got Einstein's Office?*, p. 139.

371 **"monstrous safe":** Freeman Dyson, interview by Sherwin, 2/16/84, p. 8; Pais, *A Tale of Two Continents*, p. 240. By 1953, the classified documents had been moved to a vault in the basement. But the AEC was still spending $18,755 a year on five guards to maintain twenty-four-hour security. (F. J. McCarthy, Jr., to Strauss, memo, 7/7/53, Strauss Papers, HHL.)

371 **"ablaze with power":** Pais, *A Tale of Two Continents*, p. 241.

371 **"that looked as if":** Jeremy Bernstein, e-mail to Sherwin, April 2004.

371 **Oppie drove a stunning blue:** Bernstein, *The Merely Personal,* p. 164; Bernstein, *The Life It Brings,* p. 100; Pais, *A Tale of Two Continents*, p. 255.

371 **"cut like a monk's":** Lilienthal, *The Journals of David E. Lilienthal,* vol. 3, p. 173 (diary entry of 6/6/51).

371 **"He was very thin":** Freeman Dyson, interview by Jon Else, 12/10/79, p. 9.

371 **"a town with character":** Pais, *A Tale of Two Continents*, p. 322.

372 **In 1933, Flexner:** Ibid., p. 196.

372 **"writing unnecessary textbooks"** *and subsequent quotes:* Regis, *Who Got Einstein's Office?*, pp. 26–27; Abraham Flexner, *Harper's,* October 1939; Pais, *A Tale of Two Continents,* pp. 194–96, 223.

372 **"Today," he told:** JRO, "Physics in the Contemporary World," Second Annual Arthur Dehon Little Memorial Lecture at MIT, 11/25/47, p. 7.

372 **"This is Robert"** *and subsequent quotes:* Pais, *A Tale of Two Continents,* pp. 224, 230, 221. Pais is citing K. K. Darrow's diary for 6/3/47, on file at NBL.

373 **"I was sitting next to":** Pais, *A Tale of Two Continents,* pp. 232, 234.

374 **"renormalization theory":** Weisskopf, *The Joy of Insight,* p. 171.

374 **"Let me handle this":** Ibid., p. 167.

374 **"Professor of Physics":** Regis, *Who Got Einstein's Office?,* p. 140.

375 **"He didn't have *Sitzfleisch*":** Ibid., p. 147.

375 **The Institute was a singularly:** Stern, "A History of the Institute for Advanced Study, 1930–1950," p. 642. Stern's unpublished manuscript was commissioned by Oppenheimer in 1964, but never published (IAS Archives).

375 **"This is an unreal place":** Pais, *A Tale of Two Continents,* pp. 248–49.

376 **"There was never":** Regis, *Who Got Einstein's Office?,* p. 113.

376 **At the time:** Von Neumann's machine is on display in the Smithsonian Museum.

377 **"brilliant, discursive in his interests":** Bruner, *In Search of Mind,* pp. 44, 111, 238; JRO, "Report of the Director, 1948–53," IAS, 1953, p. 25. Much later, Oppenheimer used the Director's Fund to bring the linguist Noam Chomsky to the institute in 1958–59.

377 **Soon, other such:** JRO, "Report of the Director, 1948–53," IAS, 1953; Pais, *A Tale of Two Continents,* pp. 235–38.

377 **"I invited Eliot":** Dyson, *Disturbing the Universe,* p. 72; Stern, "A History of the Institute for Advanced Study, 1930–1950," p. 662, unpublished manuscript, IAS Archives.

377 **Nevertheless, Oppenheimer:** Harold Cherniss, interview by Sherwin, 5/23/79, p. 20.

377 **"The point of this":** Regis, *Who Got Einstein's Office?,* p. 280.

378 **"rotating universe":** Ibid., pp. 62–63.

378 **"Since I found":** Ibid., p. 193.

378 **"Isn't 'in any form' ":** Bernstein, *The Merely Personal,* p. 155.

378 **Von Neumann was unusual:** Pais, *A Tale of Two Continents,* p. 207.

378 **"I think that":** Fred Kaplan, *The Wizards of Armageddon,* p. 63.

379 **"You got your doctorate"** *and subsequent quotes:* Lansing V. Hammond, "A Meeting with Robert Oppenheimer," written October 1979, courtesy of Freeman Dyson.

379 **"We were close":** JRO, "On Albert Einstein," *New York Review of Books,* 3/17/66.

379 **"Einstein is a landmark":** *Time,* 11/8/48, p. 70.

379 **When Oppenheimer's name:** Regis, *Who Got Einstein's Office?,* p. 135.

379 **"I could be":** Smith and Weiner, *Letters,* p. 190.

379 **"Certainly Oppenheimer has made":** Regis, *Who Got Einstein's Office?,* p. 136.

380 **"unusually capable man":** Fölsing, *Albert Einstein,* p. 734.

380 **"completely cuckoo":** Smith and Weiner, *Letters,* p. 190.

380 **"the good Lord":** Fölsing, *Albert Einstein,* p. 730.

380 **"see me as a heretic":** Ibid., p. 735.

380 **"extraordinary originality"** *and subsequent quotes:* JRO, "On Albert Einstein," *New York Review of Books,* 3/17/66.

381 **"watched him as he":** Lilienthal, *The Journals of David E. Lilienthal,* vol. 2, p. 298.

381 **Oppenheimer arranged to have:** Georgia Whidden, interview by Bird, 4/25/03.

381 **"This is not a jubilee":** Denis Brian, *Einstein: A Life,* p. 376.

381 **"unprepared to make":** JRO to Einstein, undated (reply to Einstein's letter of 4/15/47), JRO Papers.

382 **"He did not have":** Ronald W. Clark, *Einstein: The Life and Times,* p. 719.

382 **"You know," Einstein told him:** JRO, "On Albert Einstein," *New York Review of Books,* 3/17/66.

382 **"Something odd just":** Pais, *A Tale of Two Continents,* p. 240.

382 **"a Hoover Republican":** Stern, "A History of the Institute for Advanced Study, 1930–1950," pp. 613–14, unpublished manuscript, IAS Archives.

383 **"I was struck":** Pais, *A Tale of Two Continents,* p. 327.

383 **"The episode marks":** Stern, "A History of the Institute for Advanced Study, 1930–1950," pp. 672–73, 688, unpublished manuscript, IAS Archives.

383 **"political controversy":** Ibid., pp. 679–80, 691.

383 **"Oppenheimer plans to have":** Harry M. Davis, "The Man Who Built the A-Bomb," *New York Times Magazine,* 4/18/48, p. 20.

384 **"an intellectual hotel":** "The Eternal Apprentice," *Time,* 11/8/48, p. 70.

384 **"very strong opinion":** Stern, "A History of the Institute for Advanced Study, 1930–1950," p. 651, unpublished manuscript, IAS Archives.

384 **"The institute is":** Verna Hobson, interview by Sherwin, 7/31/79, p. 14.

384 **"This upstart Oppenheimer":** John von Neumann to Lewis Strauss, 5/4/46, Strauss Papers, HHL. The founding director of the institute, Dr. Abraham Flexner, also strongly opposed Strauss's selection of Oppenheimer (Strauss, *Men and Decisions,* p. 271).

385 **"disastrous":** Freeman Dyson, interview by Sherwin, 2/16/84, p. 18.

385 **Indeed, during his first:** Stern, "A History of the Institute for Advanced Study, 1930–1950," p. 654, unpublished manuscript, IAS archives.

385 **"the most arrogant":** Regis, *Who Got Einstein's Office?,* pp. 151.

385 **"He [Oppenheimer] was out to humiliate":** Ibid., p. 152.

385 **Academic politics can:** Stern, "A History of the Institute for Advanced Study, 1930–1950," pp. 667–69, unpublished manuscript, IAS Archives.

386 **"He really flattened me":** Dyson, interview by Sherwin, 2/16/84, p. 17.

386 **Abraham Pais recalled:** Pais, *A Tale of Two Continents,* p. 240.

386 **"I meant, will you explain":** Bernstein, *Oppenheimer,* pp. 184–85.

386 **"air of hauteur":** Pais, *A Tale of Two Continents,* p. 241.

387 **"Tea is where":** Wheeler, *Geons, Black Holes, and Quantum Foam,* p. 25.

387 **"The best way to send":** *Time,* 11/8/48, p. 81.

387 **"The young physicists":** Barnett, "J. Robert Oppenheimer," *Life,* 10/10/49.

387 **"I have been observing":** Dyson, *Disturbing the Universe,* p. 73; John Manley, interview by Sherwin, 1/9/85, p. 27.

387   **"Fireballs, fireballs!":** Murray Gell-Mann, *The Quark and the Jaguar,* p. 287.

388   **"came down on me":** Dyson, *Disturbing the Universe,* pp. 55, 73–74.

388   **"so much deeper":** Dyson, interview by Sherwin, 2/16/84, p. 3.

388   **"conquer the Demon":** Dyson, *Disturbing the Universe,* p. 80.

388   **"incomprehensibility can be mistaken":** Dyson, interview by Sherwin, 2/16/84, p. 5.

388   **"Science's sense of guilt":** *Time,* 2/23/48, p. 94.

388   **"That sort of crap":** Rabi, interview by Sherwin, 3/12/82, p. 11.

388   **"Scientists aren't responsible":** Barnett, "J. Robert Oppenheimer," *Life,* 10/10/49.

389   **"One can only imagine"** *and subsequent Blackett quotes:* P. M. S. Blackett, *Fear, War, and the Bomb,* pp. 135, 139–40. This is the American edition of the original British publication.

389   **"The wailing over Hiroshima":** Thorpe, "J. Robert Oppenheimer and the Transformation of the Scientific Vocation," dissertation, pp. 433–35. Philip Morrison wrote a highly favorable review of Blackett's book in the February 1949 issue of the *Bulletin of the Atomic Scientists.* JRO to Blackett, cable, 11/6/48; JRO to Blackett, 12/14/56, JRO Papers.

389   **That spring:** *Physics Today,* vol. 1, no. 1 (May 1948).

390   **"He wanted to be":** Dyson, *Disturbing the Universe,* p. 87.

Chapter Twenty-eight: "He Couldn't Understand Why He Did It"

391   **"The *Europa reise* is":** JRO to Frank Oppenheimer, 9/28/48, Alice Smith Collection, Sherwin Collection.

392   **"completely broken":** Preuss, "On the Blacklist," *Science,* June 1983, p. 33.

393   **"authentic contemporary hero":** *Time,* 11/8/48, p. 70; *Time*'s cover photo showed Oppenheimer standing before a blackboard filled with mathematical formulae; Dyson, *Disturbing the Universe,* p. 74.

393   **"I woke up to a recognition":** *Time,* 11/8/48, p. 76.

393   **"quite good":** Herbert Marks to JRO, 11/12/48; JRO to Marks, 11/18/48, box 49, JRO Papers.

394   **"You may have to":** Peat, *Infinite Potential,* p. 92.

394   **"Dear Rossi: I was glad":** JRO to Lomanitz, 10/30/45, Sherwin Collection.

394   **"Oh, my God":** Lomanitz, interview by Sherwin, 7/11/79. Lomanitz wrote Peter Michelmore that Oppenheimer had "seemed inordinately worried" (Lomanitz to Michelmore, 5/21/68, Sherwin Collection).

394   **But Oppenheimer had:** Walter Goodman, *The Committee,* pp. 239, 273. HUAC's chief investigator, Louis Russell, was another former FBI agent.

394   **"We won't lie":** JRO hearing, p. 151.

395   **"a dangerous man":** Hearings before the HUAC, 6/7/49, Records of the U.S. House of Representatives, RG 233 HUAC Executive Session Transcripts, box 9, JRO folder, pp. 8–9, 21.

396   **"Just look at him":** Stern, *The Oppenheimer Case,* pp. 124–25.

396   **"tremendously impressed":** Hearings before the HUAC, 6/7/49, Records of the U.S. House of Representatives, RG 233 HUAC Executive Session transcripts, box 9, JRO folder, Robert Oppenheimer, p. 42.

396   **"Robert seemed to have":** Stern, *The Oppenheimer Case,* p. 120.

397   **Peters denied that he had been:** Hearings before the HUAC, 6/8/49, pp. 1–9, Bernard Peters Papers, NBA.

397   **"God guided their questions":** FBI file 100-205953, report made in Buffalo, New York, 3/5/54, by Charles F. Ahern, Sherwin Collection. The FBI obtained this quote from a 6/23/49 letter intercepted between Ed Condon and his wife, Emilie (*New York*

Herald Tribune, 4/20/54). By one account, Peters replied, "What do you mean? What if God had not guided their questions, would you have said something derogatory about me?" (Stern notes and questions for Harold Green, Philip Stern Papers, JFKL.)

397 **"Dr. Oppenheimer Once Termed":** Stern, *The Oppenheimer Case,* p. 125; *Rochester Times Union,* 6/15/49.

397 **Peters knew immediately:** Sol Linowitz, a lawyer—later a high-ranking official in the Carter Administration—represented Peters. See Linowitz to Peters, 11/29/48, and attached legal document, Peters Papers, NBAC.

397 **"I have never told":** *Rochester Times-Union,* 6/15/49; Peters was apparently arrested on a warrant of the State Secret Police of Munich, issued on 5/13/33 on suspicion of illegal communist activities. Another police order, dated 10/14/33, charged him with communist activities and barred him from further academic studies. (*Rochester Times-Union,* 7/8/54, contained in folder 11, Peters Papers, NBAC.) Peters was Jewish and the Nazis were in power, suggesting that these charges should be taken with a grain of salt.

397 **"You are right that I":** Bernard Peters to JRO, 6/15/49, Peters Papers, NBAC.

397 **"to sue Robert":** Bernard Peters to Hannah Peters, 6/26/49, Bernard Peters Papers, NBAC.

397 **"very much disturbed":** JRO FBI file, sect. 7, doc. 175, 7/5/49, p. 18. The FBI is citing an Oppenheimer phone conversation dated 6/20/49. See also Hannah Peters to Bernard Peters, 6/20/49, Bernard Peters Papers, NBAC.

398 **"set this record straight":** JRO hearing, p. 212; Schweber, *In the Shadow of the Bomb,* pp. 123–27.

398 **"I remember you":** Hans Bethe to JRO, 6/26/49, Peters Papers, NBAC.

398 **"shocked beyond description":** Condon's letter to his wife was intercepted by the FBI, and in 1954 it was leaked to the press. See *New York Herald Tribune,* 4/20/54.

398 **"Oppie has been":** Paul Martin, "Oppenheimer Testimony on Dr. Peters Draws Charges of 'Immunity Buying,' " *Rochester Times-Union,* 7/9/54, folder 11, Peters Papers, NBAC.

398 **"I have lost":** Stern, *The Oppenheimer Case,* p. 126. "The thing that horrified me most," Condon later said, "was he [Oppenheimer], a Jewish boy, so soon after the six million had been cremated—and this was his personal protégé, also a Jewish boy—he said to this scoundrelly committee, 'I'm not sure how far I would trust Peters, because he resorted to guile in escaping from Dachau' " (see Thorpe, "J. Robert Oppenheimer and the Transformation of the Scientific Vocation," dissertation, p. 486).

398 **"My talk with Robert"** *and subsequent quotes:* Schweber, *In the Shadow of the Bomb,* p. 127; Schweber cites Peters, ltr. to Victor Weisskopf, 7/21/49, folder 42, box 3, Weisskopf Papers, MIT.

399 **"I believe this statement":** JRO hearing, p. 214.

399 **"a not very successful":** Schweber, *In the Shadow of the Bomb,* p. 127.

399 **Nonetheless, it managed to salvage:** The University of Rochester remained remarkably steadfast in its support for Dr. Peters. The university sponsored his trip to India in 1950 and the following year promoted him to associate professor. (Donald W. Gilbert, provost, to Bernard Peters, 5/29/51, folder 13, Peters Papers, NBAC.)

399 **Lomanitz' fate:** Lomanitz, interview by Sherwin, 7/11/79.

400 **"sad personally about":** Lomanitz to Peter Michelmore, 5/21/68, Sherwin Collection.

400 **"if anyone can do it":** Peat, *Infinite Potential,* pp. 104, 337; Peat cites a newspaper article, "After 40 Years, Professor Bohm Re-emerges," by H. K. Fleming, *Baltimore Sun,* April 1990.

400 **"I think he acted fairly":** Bohm, interview by Sherwin, 6/15/79.

400 **"He told me":** Ibid.

400 **"He [Oppenheimer] was obviously":** Schweber, *In the Shadow of the Bomb,* p. 127. Schweber quotes from Peters, ltr. to Victor Weisskopf, 7/21/49, folder 42, box 3, Weisskopf Papers, MIT.

401 **A young reporter:** In 1969, Philip Stern would write a brilliant book on the 1954 Oppenheimer security trial (see Stern, *The Oppenheimer Case,* p. 131).

401 **"Well, Joe, how did I do?"** *and subsequent quotes:* Stern, *The Oppenheimer Case,* pp. 129–31; Herken, *Brotherhood of the Bomb,* pp. 196–97.

402 **"I don't think Robert":** Dr. John F. Fulton to Herbert H. Maas, 8/1/49, quoted in Beatrice M. Stern, "A History of the Institute for Advanced Study, 1930–1950," p. 676, unpublished manuscript, IAS Archives.

402 **"effrontery . . . to differ":** Strauss, memo to file, 9/30/49, LLS Papers, HHL. In September 1953, Strauss learned that the request for the isotopes in question had been made by Norway's military on behalf of a Dr. Ivan Th. Rosenquist, who had later been dismissed by the Norwegians as a communist. Feeling vindicated, Strauss noted this fact in a memo to file, undated, Strauss Papers, HHL.

402 **"The story was full":** Frank Oppenheimer, interview by Weiner, 2/9/73, p. 72.

402 **"I cannot talk":** Frank Oppenheimer testimony, 6/14/49, "Hearings Regarding Communist Infiltration of Radiation Laboratory and Atomic Bomb Project at the University of California, Berkeley," HUAC, pp. 355–73.

403 **"They all looked rachitic":** Frank Oppenheimer, undated memo, folder 3–37, box 4, Frank Oppenheimer Papers, UCB.

403 **"I knew of no Communist":** Frank Oppenheimer, interview by Weiner, 5/21/73, p. 2.

403 **"What is going on?":** Frank Oppenheimer to Ernest Lawrence, undated, circa 1949, folder 4–34, box 4, Frank Oppenheimer Papers, UCB. Frank Oppenheimer may not have mailed this letter.

404 **"No one has offered":** Frank Oppenheimer to Bernard Peters, undated, autumn 1949, Peters Papers, NBAC. Oppenheimer was tentatively offered a job by the Tata Institute in Bombay, India—but the State Department denied him a passport (Ed Condon to Bernard Peters, 12/27/49, folder 12, Peters Papers, NBAC).

404 **"Jackie would sit":** Preuss, "On the Blacklist," *Science,* June 1983, p. 37.

404 **"Don't you want":** Frank Oppenheimer, interview by Weiner, 2/9/73, p. 73.

404 **"Finally, after all these":** Frank Oppenheimer, "The Tail That Wags the Dog," unpublished manuscript, folder 4–39, box 4, Frank Oppenheimer Papers, UCB; Preuss, "On the Blacklist," *Science,* June 1983, p. 34.

404 **"I really felt like":** Frank Oppenheimer, interview by Weiner, 5/21/73, pp. 11–12.

404 **Over the next year:** JRO to Dr. Harold C. Urey, box 74, JRO Papers.

404 *First Steps (After Millet):* Dalzell Hatfield to Frank Oppenheimer, 2/2/54, folder 4–45, box 4, Frank Oppenheimer Papers, UCB.

405 **"all lead in many":** JRO to Grenville Clark, 5/17/49, Grenville Clark Papers, sect. 13, box 17, DCL.

405 **"Even the walls have ears":** Stern, *The Oppenheimer Case,* p. 113.

405 **"He was always conscious":** Hempelmann, interview by Sherwin, 8/10/79, p. 20.

405 **By 1949, the bureau:** JRO FBI file 100-17828, doc. 162, 10/24/47; FBI SAC to Hoover, 4/13/49, JRO FBI file, 100-17828, doc. 173.

405 **"No additional information":** JRO FBI file 100-17828, sect. 6, doc. 156, 6/27/47, and doc. 176, 4/13/49.

Chapter Twenty-nine: "I Am Sure That Is Why She Threw Things at Him"

406 **He spent about:** Verna Hobson, interview by Sherwin, 7/31/79, p. 15.

406 **"The time has come":** Michelmore, *The Swift Years,* p. 143.

406 **"He is warmly affectionate":** Barnett, "J. Robert Oppenheimer," *Life,* 10/10/49.

406 **"Mrs. Oppenheimer, whose thinking":** Rhodes, *Dark Sun,* p. 309; *Life,* vol. 29, no. xii (1947), p. 58.

407 **"His family relationships":** Priscilla Duffield, interview by Alice Smith, 1/2/76, p. 11 (MIT Oral History Laboratory).

407 **"He was an extraordinarily":** Verna Hobson, interview by Sherwin, 7/31/79, pp. 3–4, 8, 18.

407 **"crew of birds":** Mildred Goldberger, interview by Sherwin, 3/3/83, pp. 5, 13.

407 **"She would get drunk":** Verna Hobson, interview by Sherwin, 7/31/79, p. 3.

407 **"She would arrive":** Pat Sherr, interview by Sherwin, 2/20/79, p. 15.

408 **"I mean, she just":** Ibid., p. 25.

408 **"He knew of Kitty's":** Goodchild, *J. Robert Oppenheimer,* p. 272.

408 **"Don't go away":** Pais, *A Tale of Two Continents,* pp. 242–43.

408 **"doctor, nurse and psychiatrist":** Verna Hobson, interview by Sherwin, 7/31/79, p. 19.

408 **"Robert just liked":** Dyson, interview by Sherwin, 2/16/84, p. 16.

409 **"He was just as loyal":** Robert Strunsky, interview by Sherwin, 4/26/79, p. 11.

409 **"barbaric custom":** Sherr, interview by Sherwin, 2/20/79, p. 18; Pais, *A Tale of Two Continents,* p. 242.

409 **"He was really just":** Hempelmann, interview by Sherwin, 8/10/79, pp. 12–13.

409 **On one occasion:** Verna Hobson, interview by Sherwin, 7/31/79, p. 20.

409 **"never drank excessively":** Robert Serber, interview by Sherwin, 3/11/82, p. 16. Serber's explanation is somewhat misleading. Typically, alcoholism is a primary cause for pancreatitic attacks. According to Dr. Hempelmann, Kitty developed pancreatitis in the late 1950s. Her doctors prescribed very strong painkillers that didn't mix with alcohol.

410 **"I need you":** Sherr, interview by Sherwin, 2/20/79, p. 14.

410 **"If you are single":** Pais, *A Tale of Two Continents,* p. 322.

410 **"People left [calling] cards":** Mildred Goldberger, interview by Sherwin, 3/3/83, pp. 9–10.

410 **"wicked" woman:** Ibid., pp. 5, 16; Marvin Goldberger, interview by Sherwin, 3/28/83, p. 3.

410 **"You would sit in":** Goodchild, *J. Robert Oppenheimer,* p. 272.

410 **"indulged in a rather":** Pais, *A Tale of Two Continents,* p. 242.

411 **"running around at his":** Sherr, interview by Sherwin, 2/20/79, pp. 25–26.

411 **"I think he leaned on her":** Verna Hobson, interview by Sherwin, 7/31/79, p. 19. Hobson never actually saw Kitty throw anything at Robert, but she saw him come into the office with abrasions, and more so as the years went by.

411 **Kitty told Sherr:** Sherr, interview by Sherwin, 2/20/79, p. 25.

411 **Another Los Alamos friend:** Jean Bacher, interview by Sherwin, 3/29/83, p. 1.

411 **"was insanely jealous":** Verna Hobson, interview by Sherwin, 7/31/79, p. 6.

412 **"She was a very lovely":** Sherr, interview by Sherwin, 2/20/79, p. 12.

412 **"I think to be a child":** Strunsky, interview by Sherwin, 4/26/79, p. 11.

412 **"On the surface":** Sherr, interview by Sherwin, 2/20/79, p. 17.

412 **"could not have a son":** Ibid., pp. 16–17.

412 **"looking very paternal":** Lilienthal, *The Journals of David E. Lilienthal,* vol. 2, p. 456 (diary entry 2/3/49).

413 **"problematical figure for a father":** Dyson, *Disturbing the Universe,* p. 79.

413 **"To an outsider":** Pais, *A Tale of Two Continents,* p. 243.

413 **"Robert thought," said Hobson:** Verna Hobson, interview by Sherwin, 7/31/79, p. 18.

413 **"Kitty was very, very":** Sherr, interview by Sherwin, 2/20/79.

413 **"He [Robert] was very loving":** Hempelmann, interview by Sherwin, 8/10/79, p. 19.

413 **From all accounts:** Ibid., p. 14.

413 **"he seemed to be starved":** Robert Serber, interview by Sherwin, 3/11/82, p. 20.

413 **"Her attachment to Toni":** Verna Hobson, interview by Sherwin, 7/31/79, p. 18.

414 **"So the warm waters":** Ruth Tolman to JRO, 1/15/52, box 72, JRO Papers.

414 **"She was a tower":** Freeman Dyson to Alice Smith, 6/1/82, Alice Smith correspondence, Sherwin Collection; Dyson, interview by Sherwin, 2/16/84, p. 15.

414 **"We always have such":** Elinor Hempelmann to Kitty Oppenheimer, undated, circa 1949–50, JRO Papers.

414 **"Dear Oppy":** Al Christman, *Target Hiroshima*, p. 242.

415 **"I am so glad":** Lilienthal, *The Journals of David E. Lilienthal,* vol. 3, pp. 381–82 (diary entry of 3/28/53).

415 **"When God at first made man":** Dyson, *From Eros to Gaia,* p. 256. Dyson quotes Mrs. Ursula Niebuhr in a draft, book-review copy sent to Sherwin. George Herbert wrote with almost morbid sensitivity about his inner moods—which may explain Oppenheimer's attraction.

Chapter Thirty: "He Never Let On What His Opinion Was"

416 **When called, Bush:** JRO hearing, p. 910.

416 **"tried every argument":** Lilienthal to JRO, 9/23/49, box 46, JRO Papers; Lilienthal, *The Journals of David E. Lilienthal,* vol. 2, pp. 571–72. Hewlett and Duncan, *Atomic Shield,* vol. 2, p. 367.

417 **"Keep your shirt on":** Teller, *Memoirs,* p. 279.

417 **" 'Operation Joe' is simply":** Lincoln Barnett, "J. Robert Oppenheimer," *Life,* 10/10/49, p. 121.

417 **"Our atomic monopoly":** *Time,* 11/8/48, p. 80.

417 **But he also feared:** Around this time, Einstein wrote the Harvard astronomer Harlow Shapley, "I now feel sure that the people in power in Washington are pushing systematically toward preventive war" (William L. Shirer, *Twentieth Century Journey,* p. 131).

417 **"We mustn't muff it":** Lilienthal to JRO, 9/23/49, box 46, JRO Papers (Lilienthal quotes Oppenheimer in this letter). See also Lilienthal, *The Journals of David E. Lilienthal,* vol. 2, pp. 570, 572.

417 **a "more rational":** Hewlett and Duncan, *Atomic Shield,* p. 368.

417 **The U.S. stockpile:** Melvyn P. Leffler, *A Preponderance of Power,* p. 324.

417 **"quantum jump":** Strauss to AEC commissioners Lilienthal, Pike, Smyth, and Dean, memo 10/5/49, memorandum for the record, 1949–1950, box 39, Strauss Papers, HHL; McGeorge Bundy, *Danger and Survival,* p. 204; Hewlett and Duncan, *Atomic Shield,* p. 373; Herbert York, *The Advisors,* pp. 41–56.

417 **Truman was not:** McGeorge Bundy, *Danger and Survival,* p. 201; Herken, *Brotherhood of the Bomb,* p. 204.

417 **"I am not sure":** JRO to James Conant, 10/21/49, reprinted in JRO hearing, p. 242.

418 **The physics of fusion:** Hewlett and Duncan, *Atomic Shield,* vol. 2, p. 383.

418 **"no such effort":** Bernstein, "Four Physicists and the Bomb," *Historical Studies in the Physical Sciences,* vol. 18, no. 2 (1988), pp. 243–44 (italics are ours). See also Bernstein and Galison, "In Any Light: Scientists and the Decision to Build the Superbomb, 1952–1954," HSPS, vol. 19, no. 2 (1989), pp. 267–347.

418 **"long and difficult discussion"; "over my dead body":** Hershberg, *James B. Conant,* pp. 470–71.

419 **"What does worry me":** JRO hearing, pp. 242–43; Herken, *Brotherhood of the Bomb,* p. 204.

419 **"the climate of opinion":** JRO hearing, p. 242 (JRO to James Conant, 10/21/49).

419   **"equally undecided":** JRO hearing, p. 328.

420   **"We both had to agree":** Rhodes, *Dark Sun,* p. 393.

420   **"he would certainly":** JRO hearing, p. 76.

420   **At two o'clock:** Hershberg, *James B. Conant,* p. 473.

420   **" 'bloodthirsty' ":** Lilienthal, *The Journals of David E. Lilienthal,* vol. 2, p. 582 (diary entry of 10/30/49); see also Hewlett and Duncan, *Atomic Shield,* vol. 2, pp. 381–85.

421   **"Although I deplore":** Rhodes, *Dark Sun,* p. 395. Rabi believes Seaborg would have changed his mind had he been present. "If he had been there," Rabi said, "and stood out against it, I would have been very astonished." (Rabi, interview by Sherwin, 3/12/82, p. 8.) See also Herken, *Brotherhood of the Bomb,* p. 384.

421   **"He never let on":** Lee DuBridge, interview by Sherwin, 3/30/83, p. 21; see also DuBridge testimony in JRO hearing, p. 518.

421   **"looking almost translucent":** Lilienthal, *The Journals of David E. Lilienthal,* vol. 2, p. 581.

421   **"Oppenheimer followed Conant's lead":** Hershberg, *James B. Conant,* p. 478.

421   **"flatly against it":** Lilienthal, *The Journals of David E. Lilienthal,* vol. 2, pp. 580–83; Schweber, *In the Shadow of the Bomb,* p. 158; Hershberg, *James B. Conant,* p. 474.

421   **"who will be willing":** Schweber, *In the Shadow of the Bomb,* p. 158.

421   **"one must explore it":** Lilienthal, *The Journals of David E. Lilienthal,* vol. 2, p. 582.

422   **"The use of this weapon":** "The GAC Report of October 30, 1949," reprinted in York, *The Advisors,* pp. 155–62; Bernstein, "Four Physicists and the Bomb: The Early Years, 1945–1950," p. 258.

422   **"too small":** JRO hearing, p. 236; Hershberg, *James B. Conant,* pp. 467–68.

423   **"To the argument":** "The GAC Report of October 30, 1949," reprinted in York, *The Advisors,* pp. 155–62.

423   **Indeed, if the Super:** York, *The Advisors,* p. 160; Bundy, *Danger and Survival,* pp. 214–19.

423   **"This will cause you":** Michelmore, *The Swift Years,* p. 173.

424   **"blow them off the face":** Lilienthal, *The Journals of David E. Lilienthal,* vol. 2, pp. 584–85; York, *The Advisors,* p. 60.

424   **"You know, I listened":** Gordon R. Arneson, "The Decision to Drop the Bomb," interview transcript by NBC News film, 3/1/86, courtesy of Nancy Arneson, part 1, p. 13; Rhodes, *Dark Sun,* p. 405; Hershberg, *James B. Conant,* p. 481.

424   **His disillusionment was complete:** See Carolyn Eisenberg, *Drawing the Line;* Bird, "Stalin Didn't Do It," *The Nation,* 12/16/96.

424   **Kennan had first encountered:** David Mayers, *George Kennan and the Dilemmas of US Foreign Policy,* p. 241.

424   **"He was dressed":** George Kennan, interview by Sherwin, 5/3/79.

425   **"He kept the whole thing":** Ibid., p. 3.

425   **"present state of the atomic":** JRO to Kennan, 11/17/49, box 43, JRO Papers.

425   **"this weapon could not":** Untitled draft speech, initialed "GFKennan," 11/18/49, box 43, JRO Papers.

425   **"thoroughly admirable":** JRO to Kennan, 1/3/50, box 43, JRO Papers.

426   **"I fear that the atomic bomb":** Mayers, *George Kennan and the Dilemmas of US Foreign Policy,* pp. 307–8; FRUS 1950, vol. 1, pp. 22–44, George Kennan, *Memoirs, 1925–1950,* p. 355; George Kennan, "Memorandum: International Control of Atomic Energy," 1/20/50.

426   **"as something superfluous":** Walter L. Hixson, *George F. Kennan,* p. 92.

426   **"move as rapidly as possible":** Ibid.

426   **"I was firmly convinced":** Kennan, interview by Sherwin, 5/3/79, p. 13.

427   **"judicious exploitation":** Mayers, *George Kennan and the Dilemmas of US Foreign Policy,* p. 308. In retrospect, Kennan argued, "our stance toward the Russians should

have been: Look here, so long as there are no arrangements for international controls, we are going to hold enough of these weapons—a small amount—to make it no temptation for anybody else to use them against us; but we deplore their very existence; we are anxious to get on with agreements to rule them out entirely, and we are not going to base our defense posture on them, nor our diplomacy" (Kennan, interview by Sherwin, 5/3/79, p. 10).

427 **"George, if you persist"**: Gordon R. Arneson, "The Decision to Drop the Bomb," interview transcript by NBC News film, 3/1/86, courtesy of Nancy Arneson, part 2, p. 2.

427 **"Let them fall"**: Wheeler, *Geons, Black Holes, and Quantum Foam,* p. 200.

427 **"Certainly not"**: Teller, *Memoirs,* p. 289.

427 **"I thought it would be"**: Transcript of executive meeting, JCAE 1/30/50, doc. 1447, RG 128, courtesy of Gregg Herken. See also Herken, *Brotherhood of the Bomb,* p. 216.

428 **"The American people"**: Acheson, *Present at the Creation,* p. 349.

428 **"We must protect"**: Patrick J. McGrath, *Scientists, Business, and the State, 1890–1960,* p. 124.

428 **"Can the Russians do it?"**: Lilienthal, *The Journals of David E. Lilienthal,* vol. 2, pp. 594, 601 (diary entry of 11/7/49).

428 **" 'No' to a steamroller"**: Ibid., pp. 630–33 (diary entry of 1/31/50).

429 **By the end of the decade**: David Alan Rosenberg, "The Origins of Overkill: Nuclear Weapons and American Strategy, 1945–60," *International Security,* no. 7 (Spring 1983), p. 23; Stephen Schwartz, ed., Introduction, *Atomic Audit,* pp. 3, 33;

429 **"I never forgave"**: Rhodes, *Dark Sun,* p. 408. The "secret" of the H-bomb could not be kept secret. As Hans Bethe later wrote, "Of course in the long run this secret will be discovered by any nation which tries hard" (Bethe to Philip M. Stern, 7/3/69, Stern Papers, JFKL).

429 **"It was like a funeral"**: Lilienthal, *The Journals of David E. Lilienthal,* vol. 2, p. 633.

429 **"For heck's sake"**: Hershberg, *James B. Conant,* p. 481.

429 **"promote a debate"**: JRO hearing, p. 898.

429 **"didn't [resign]"**: Hershberg, *James B. Conant,* p. 482 (Conant to William L. Marbury, 6/30/54).

429 **"You don't look jubilant"**: Goodchild, *J. Robert Oppenheimer,* p. 204; Pfau, *No Sacrifice Too Great,* p. 123. Pfau cites an interview with Strauss for this incident.

430 **"These are complex"**: Lewis Strauss to R. Adm. Sidney Souers in the White House, 2/16/50, folder "H-bomb," AEC series, box 39, Strauss Papers, HHL.

430 **"that these decisions"**: *Bulletin of the Atomic Scientists,* July 1950, p. 75.

430 **"the whole rotten business"**: Acheson, *Present at the Creation,* p. 346.

Chapter Thirty-one: *"Dark Words About Oppie"*

431 **"our large and ill-managed"**: Davis, *Lawrence and Oppenheimer,* p. 316.

431 **"You probably do not"**: Kennan to JRO, 6/5/50, box 43, JRO Papers.

431 **"What stands out"**: Kennan, interview by Sherwin, 5/3/79, pp. 4, 6.

431 **"I, who owe to your"**: Kennan to JRO, 6/26/66, box 43, JRO Papers.

432 **"not, so far, an historian"**: John von Neumann to JRO, 11/1/55, Strauss Papers, HHL.

432 **"They resented Kennan"**: Freeman Dyson, interview by Sherwin, 2/16/84, p. 19; Harold Cherniss, interview by Sherwin, 5/23/79, p. 14. Stern, "A History of the Institute for Advanced Study, 1930–1950," p. 683, unpublished manuscript, IAS archives.

432 **But less than six months**: Kennan to Barklie Henry, 9/9/52, box 43, JRO Papers (Kennan asked Henry to forward a copy of this letter to Oppenheimer); Kennan to JRO, 10/14/52, box 43, JRO Papers.

432 **"he knew of no 'niche' "**: Hixson, *George F. Kennan,* p. 117.

433 **"nuclear power for planes":** Stern, *The Oppenheimer Case,* p. 133.

433 **"I know," recalled Lee DuBridge:** DuBridge, interview by Sherwin, 3/30/83, p. 16.

433 **"had all the facts":** Norman Polmar and Thomas B. Allen, *Rickover,* p. 138.

433 **Placing his hand:** John Manley, interview by Alice Smith, 12/30/75, p. 12; Herken, *Brotherhood of the Bomb,* p. 195.

433 **"very cruel":** Cherniss, interview by Sherwin, 5/23/79, p. 3.

433 **"These are not happy":** Strauss to William T. Golden (Strauss' assistant in the AEC), 7/21/49, Strauss Papers, HHL.

434 **"to the effect":** Strauss to Golden, 9/15/49, Strauss Papers, HHL.

434 **"effrontery for anyone":** Strauss, memos for the record, 1949–1950, box 39, Strauss Papers, HHL.

434 **"a general who did not":** Pfau, *No Sacrifice Too Great,* p. 132; Bernstein, "The Oppenheimer Loyalty-Security Case Reconsidered," *Stanford Law Review,* p. 1414; McGrath, *Scientists, Business, and the State, 1890–1960,* p. 146.

434 **"It is important to realize":** Leslie Groves to Strauss, 10/20/49 and 11/4/49, Strauss Papers, HHL.

435 **"prefer defeat in war":** Strauss to Kenneth Nichols, 12/3/49, Strauss Papers, HHL.

435 **On the afternoon:** Strauss, memo to file, 2/1/50, box 39, Strauss papers, HHL.

435 **"only fortifies the wisdom":** Robert Chadwell Williams, *Klaus Fuchs,* pp. 116, 137.

435 **"Have you heard":** Anne Wilson Marks, interview by Bird, 3/5/02.

435 **"would set them back":** Pais, *A Tale of Two Continents,* p. 258.

436 **"lack of honesty":** Bernstein, "The Oppenheimer Loyalty-Security Case Reconsidered," *Stanford Law Review,* July 1990, p. 1408.

436 **"I . . . think it was right":** Ibid.

436 **"It resembled a meteor":** Herken, *Counsels of War,* pp. 10–14; Herken, *Brotherhood of the Bomb,* p. 194.

436 **"Borden was like a new dog":** Wheeler, *Geons, Black Holes, and Quantum Foam,* p. 284.

437 **"It is a dangerous":** Herken, *Brotherhood of the Bomb,* p. 195.

437 **By 1949, Strauss and Borden:** See Lewis Strauss correspondence with William L. Borden, 2/4/49, 2/24/49, 12/10/52, 10/11/54, and 2/3/58, and other letters, William L. Borden, box 10, AEC series.

437 **"I think he had":** William W. Prochnau and Richard W. Larsen, *A Certain Democrat,* p. 114.

437 **Now, for the first time** *and subsequent notes:* Bernstein, "The Oppenheimer Loyalty-Security Case Reconsidered," *Stanford Law Review,* July 1990, pp. 1409–10.

438 **"Until now," Jackson said:** Robert G. Kaufman, *Henry M. Jackson,* p. 55.

438 **"[h]e never forgot":** Ibid., p. 56.

438 **"session of a top-drawer":** Stern, *The Oppenheimer Case,* p. 164; FBI memo, 8/18/50, pp. 18–20, sect. 10, JRO FBI file.

439 **"West Coast Whittaker Chambers":** Newspaper clippings from *San Francisco News, San Francisco Call-Bulletin,* and *Oakland Tribune,* 5/9/50, contained in JRO FBI file, sect. 8. For more on the Hiss case, see Sam Tanenhaus, *Whittaker Chambers;* Allen Weinstein, *Perjury;* Alger Hiss, *Recollections of a Life;* Victor Navasky, "The Case Not Proved Against Alger Hiss," *The Nation,* 4/8/78; John Lowenthal, "Venona and Alger Hiss," *Intelligence and National Security* 15, no. 3 (2000); and Tony Hiss, *The View from Alger's Window: A Son's Memoir.*

439 **"I have never been":** Statement by JRO, 9:45 p.m., 5/9/50, JRO FBI file sect. 8.

439 **"How utterly nauseating":** Lilienthal to JRO, 5/10/50, box 46, JRO Papers.

439 **"inherently believable":** Borden, memo to file, 8/13/51, JCAE records, doc. 3464,

cited in Barton J. Bernstein, "The Oppenheimer Loyalty-Security Case Reconsidered," *Stanford Law Review,* July 1990, pp. 1409–11.

439   **"I am in the habit":** Victor Navasky, *Naming Names,* p. 14.

439   **Curiously, Crouch was pardoned:** Memo re: Herbert Marks, 12/1/50, sect. 44, doc. 1817, JRO FBI file.

440   **"they had formulated":** *Oakland Tribune,* 5/9/50; Navasky, *Naming Names,* p. 14. Marshall Tukhachevsky was executed on 6/12/37, during one of Stalin's early purges.

440   **"He spent a lot of his time":** Cedric Belfrage, *The American Inquisition,* pp. 16, 168; Nelson, et al., *American Radical,* p. 332. Fred J. Cook, *The FBI Nobody Knows,* 388; Joseph and Stewart Alsop, WP, 7/4/54. Crouch testified against Harry Bridges, the famous union leader who had been indicted on perjury charges. In the course of the 1949–50 trial, Bridges' lawyer presented evidence that Crouch had perjured himself. (Charles P. Larrowe, *Harry Bridges,* pp. 311, 322.)

440   **"had been talking":** FBI memo, 4/18/50 (Paul Crouch interview), JRO FBI file, sect. 8; see also Paul Crouch, unpublished memoir, chapter 29, Crouch Papers, Hoover War Institute Archives, Stanford, CA, courtesy of Andrew Meier.

441   **Oppenheimer later documented:** Dorothy McKibbin found a hospital record for the X-ray dated July 25 (FBI memo, 11/18/52, p. 46, JRO FBI file, sect. 14).

441   **Crouch was either mistaken:** Herken, *Brotherhood of the Bomb,* p. 231. Herken speculates that Oppenheimer may have had a reason to drive the 2,200-mile round-trip journey between his ranch and Berkeley during the three-day window between Friday, July 25, and Monday afternoon, July 28—when Kitty crashed the car. Even today, the trip would take more than eighteen hours of straight driving in each direction. In 1941 such a drive would have taken considerably longer. Dorothy McKibbin found bills from a Santa Fe grocery store charged to the Oppenheimers for July 12, 14, 25, 28 and 29, 1941—indicating that the Oppenheimers had not left New Mexico in late July (FBI memo, 11/18/52, JRO FBI file, sect. 14, p. 45). Furthermore, Oppenheimer was at that time negotiating to buy a home at One Eagle Hill in Berkeley. On 7/26/41, Oppenheimer signed a letter sent from Cowles, New Mexico, to the real estate agent, Robinson, saying, "As for the furniture, we would, I think, be just as content to have everything taken from the house." So this indicates that they did not comply with the home owner's cabled request to meet them July 26 or 27 to dispose of the furniture. Oppie also says, "There is a chance that we shall be back in Berkeley before we had planned, perhaps within a week. . . . If you do not hear from us by Wednesday, you may assume that we shall be back about the 13th of August." Finally, on 8/11/41, the Title Insurance Co. received a check for $22,163.87 in payment for the Eagle Hill house. Kitty is identified as the "deliverer of the check" (sect. 44, doc. 1805, 6/25/54, JRO FBI file).

441   **Over time, Crouch:** Fred J. Cook, *The Nightmare Decade,* p. 388; Cedric Belfrage, *The American Inquisition,* pp. 208, 221–22.

441   **Later, Crouch's testimony:** Robert Justin Goldstein, *Political Repression in Modern America,* p. 348; Navasky, *Naming Names,* p. 14.

441   **Eventually, Crouch's lies and theatrics:** When Crouch named as communists the well-known lawyer and former FCC commissioner Clifford Durr and his wife, Virginia (Justice Hugo Black's sister-in-law), Virginia responded that Crouch was a "grinning, lying dog." Years later, she described Crouch as "a dirty piece of Kleenex about to disintegrate—such a wreck of a man that even while he was destroying you anyone would feel sorry for him." The usually mild-mannered Clifford Durr was so incensed by what Crouch said about his wife that he once tried to punch Crouch in the nose. Navasky, *Naming Names,* p. 14.

441   **"if my reputation":** Belfrage, *American Inquisition, 1945–1960,* pp. 227–28; Edwin M. Yoder, Jr., *Joe Alsop's Cold War,* p. 129.

441 **"exceedingly ambivalent":** Bernstein, "The Oppenheimer Loyalty-Security Case Reconsidered," *Stanford Law Review,* July 1990, p. 1415.

442 **"wisdom of our war plan":** Ibid.

442 **"I carried away":** Ibid.

443 **Strauss "devoted a good part":** Extract from JCAE staff memo written by Borden, concerning conversation with Commissioner Strauss, 8/13/51, Philip M. Stern Papers, JFKL. See also Bernstein, "The Oppenheimer Loyalty-Security Case Reconsidered," *Stanford Law Review,* July 1990, pp. 1413–14.

443 **"technically sweet":** Wheeler, *Geons, Black Holes, and Quantum Foam,* p. 222.

443 **"delayed or attempted to delay":** FBI memo, Albuquerque, 5/15/52, declassified 9/9/85 and 10/23/96, JRO FBI file.

444 ***"would do anything possible":*** Edward Teller, interview by FBI, report made at Albuquerque, 5/15/52, nine pages, declassified 10/23/96, JRO FBI file.

444 **"serious questions as to":** JRO hearing, p. 749.

444 **"That was the goddamnedest":** Dyson, *Weapons and Hope,* p. 137.

444 **Chapter Five of the report:** Stern, *The Oppenheimer Case,* pp. 182–85.

445 **briefing was a "success":** Ruth Tolman to JRO, 1/15/52, box 72, JRO Papers. In an early draft of chapter 5, Oppenheimer advanced the ethical argument that tactical weapons should replace strategic weapons—but this passage was eventually deleted (Herken, *Counsels of War,* p. 67).

445 **"went straight through":** Stern, *The Oppenheimer Case,* p. 185.

445 **"ever since Oppenheimer":** Lewis Strauss to Senator Bourke Hickenlooper, 9/19/52, "H-bomb," AEC series, box 39, Strauss Papers, HHL.

445 **The Air Force's:** William L. Borden, memo to JCAE chairman, 11/3/52, p. 2, box 41, JCAE, no. DCXXXV, RG 128, NA.

445 **"bringing the battle back":** Oppenheimer was right to regard the ten- and twenty-megaton hydrogen bombs carried by SAC aircraft as both genocidal and militarily useless weapons. But he did not realize that, in just a few years, technical developments would make it possible to design low-yield hydrogen weapons small enough to mount on intercontinental ballistic missiles—or in an artillery shell (Herbert York, e-mail to Howard Morland, 3/5/03).

445 **"are not policy weapons":** Thorpe, "J. Robert Oppenheimer and the Transformation of the Scientific Vocation," dissertation, pp. 450–51.

446 **"the bulk of the B-47 fleet":** Steven Leonard Newman, "The Oppenheimer Case: A Reconsideration of the Role of the Defense Department and National Security," dissertation, New York University, February 1977, p. 48.

446 **"Finletter was filled with wrath":** Ibid., p. 53. Newman's source is a letter to him from Col. Charles J. V. Murphy, 9/17/74. Murphy was the author of the *Fortune* magazine attack on JRO.

446 **"pro-Russian or merely confused":** Stern, *The Oppenheimer Case,* pp. 190–91.

447 **"rude beyond belief":** Ibid., pp. 191–92.

447 **"whether [Oppenheimer] was a subversive":** Herken, *Brotherhood of the Bomb,* p. 253.

447 **"In dealing with the Russians":** William L. Clayton Papers, 6/7/51, p. x, HSTL; see also "A Statement on the Mutual Security Program," April 1952, Committee on the Present Danger, Averell Harriman Papers, Kai Bird Collection.

447 **"Oppie's line":** Stewart Alsop to Martin Sommers, 2/1/52, "Sat. Evening Post Jan.–Nov. 1952" folder, box 27, Alsop Papers, LOC. Yoder, *Joe Alsop's Cold War,* p. 121; JRO hearing, p. 470.

448 **"I think it does":** "Meeting for Dr. J. Robert Oppenheimer," 2/17/53, p. 28, Council on Foreign Relations Archives.

448 **"Some of the 'boys' ":** Hershberg, *James B. Conant,* p. 600.

448 **"Physics is complicated"**: Herken, *Brotherhood of the Bomb,* p. 251; JRO to Frank Oppenheimer, 7/12/52, "Weinberg Perjury Trial, 1953" folder, box 237, JRO Papers.

449 **"I find it hard to thank you"**: Bird, *The Color of Truth,* p. 113; Bundy correspondence, box 122, JRO Papers.

449 **"problem of survival"**: Minutes, mtg. of 5/16–18/52, Panel of Consultants on Arms and Policy, Princeton, box 191, JRO Papers; Bird, *The Color of Truth,* p. 113.

449 **"while the more significant fact"**: Hershberg, *James B. Conant,* pp. 602–4, 902; Bird, *The Color of Truth,* p. 114.

450 **"it seems to us"**: David Holloway, *Stalin and the Bomb,* p. 311.

450 **"any such idea"**: Hershberg, *James B. Conant,* p. 605; minutes of mtg., NSC, 10/9/52, FRUS 1952–54, vol. 2, pp. 1034–35.

450 **"I no longer have"**: Hershberg, *James B. Conant,* p. 605.

450 **Oppenheimer sat grimly:** Herken, *Brotherhood of the Bomb,* p. 257.

451 **"Some people in the Air Force"**: Lee DuBridge, interview by Sherwin, 3/30/83, p. 23.

451 **This document was forwarded:** Mac Bundy published the declassified version of this report in the journal *International Security* (Fall 1982) under the title "Early Thoughts on Controlling the Nuclear Arms Race." See also Bundy's essay "The Missed Chance to Stop the H-Bomb," *New York Review of Books,* 5/13/82, p. 16.

451 **"should tell the story"**: Bird, *The Color of Truth,* p. 115.

452 **"The Missed Chance"**: McGeorge Bundy, "The Missed Chance to Stop the H-Bomb," *New York Review of Books,* 5/13/82, p. 16.

452 **"enemy archives"**: Leffler, "Inside Enemy Archives: The Cold War Re-Opened," *Foreign Affairs,* Summer 1996.

452 **"never believed that"**: Bird, "Stalin Didn't Do It," *The Nation,* 12/16/96, p. 26; Alperovitz and Bird, "The Centrality of the Bomb," *Foreign Policy,* Spring 1994, p. 17. See also Arnold A. Offner, *Another Such Victory,* and Carolyn Eisenberg, *Drawing the Line.*

452 **"would mean the destruction"**: Vladislav Zubok and Constantine Pleshakov, *Inside the Kremlin's Cold War,* pp. 166–68.

452 **But in practice:** David S. Painter, *The Cold War,* p. 41.

452 **"I couldn't sleep"**: Holloway, *Stalin and the Bomb,* pp. 340–45, 370; William Taubman, *Khrushchev,* p. xix.

453 **Yet no less a Sovietologist:** Charles E. Bohlen, *Witness to History,* pp. 371–72.

Chapter Thirty-two: "Scientist X"

454 **"entirely cooperative"**: JRO, interview by FBI, 5/3/50, sect. 8, JRO FBI files.

454 **"So there was a cloud"**: Joseph Weinberg, interview by Sherwin, 8/23/79, pp. 20–21.

454 **"My God," he thought:** Ibid., p. 22.

455 **Fortunately for Weinberg:** J. Edgar Hoover, FBI memo, 5/8/50, JRO FBI files, sect. 8.

455 **By April 1950:** A. H. Belmont to D. M. Ladd, FBI memo, 4/14/50, Crouch affair, JRO FBI file.

455 **"They were fools"**: Weinberg, interview by Sherwin, 8/23/79, pp. 22, 30.

455 **"we would want a transcript"**: Transcript of conference between Oppenheimer, Marks, Arens, and Connors, 12/13/51, box 237, JRO Papers.

456 **"that no such persons"**: Keith G. Teeter, FBI memo, 11/18/52, re: 5/20/52 interview of JRO and Crouch, JRO FBI file, sect. 14, p. 3. Oppenheimer did volunteer that he vaguely recalled that someone, perhaps Ken May, had asked permission to use his home for "a meeting of young people." But he could not recall whether he had given his permission or even where he had been living at the time of this request.

456 **"to see if he would recognize"**: Ibid. The FBI memo claims that Crouch was not fore-

warned of Oppenheimer's presence. According to Crouch, he had not seen Oppenheimer since their one encounter in July 1941. Even so, anyone who read the newspapers would have seen photographs of Oppenheimer.

457   **"Dr. Oppenheimer stated":** Ibid.

457   **This might be regarded:** The FBI later learned that Hiskey was employed until 8/28/41 by the TVA in Knoxville, TN; TVA records showed that Hiskey had not left Knoxville until the end of August (A. H. Belmont to D. M. Ladd, FBI memo, 7/10/52, declassified 7/22/96, JRO FBI file).

457   **In fact, Oppenheimer's lawyers:** Excerpts from Gordon Dean's diary, 5/16/52 to 2/25/53, History Division, Department of Energy.

458   **"It will be Oppenheimer's word":** Dean to Truman, 8/25/52, and Truman to Dean, 8/26/52, D folder, PSF general file, box 117, HSTL.

458   **"Oppie will have to":** Gordon Dean diary, 11/18/52, History Division, Department of Energy.

458   **"government prosecutors said":** Bernstein, "The Oppenheimer Loyalty-Security Case Reconsidered," *Stanford Law Review,* July 1990, p. 1426; *San Francisco Chronicle,* 12/2/52.

458   **"Such a miserable":** Ruth Tolman to JRO, 1/2/53, box 72, JRO Papers.

458   **"it seemed a terrible thing":** Bernstein, "The Oppenheimer Loyalty-Security Case Reconsidered," *Stanford Law Review,* July 1990, p. 1426.

459   **"isn't some natural":** Ibid., pp. 1426–27.

459   **"this case can be cut":** Gordon Dean diary, 2/25/53.

459   **Oppenheimer had to go:** Criminal docket, U.S. District Court for the District of Columbia, criminal no. 829-52, chronology of *United States v. Joseph W. Weinberg.*

459   **"felt so worn out":** Ruth Tolman to JRO, Sunday, 3/1/53, box 72, JRO Papers.

459   **Since the prosecution was:** Affidavit of Joseph A. Fanelli, *United States v. Joseph W. Weinberg,* criminal no. 829-52, U.S. District Court for the District of Columbia, filed 11/4/52.

459   **"the court does not":** NYT, 3/6/53, p. 14.

459   **"With so many mean":** Lilienthal to JRO, 3/1/53, box 46, JRO Papers, LOC, cited in Barton J. Bernstein, "The Oppenheimer Loyalty-Security Case Reconsidered," *Stanford Law Review,* July 1990, p. 1427.

460   **"We looked at each other":** Sis Frank, interview by Sherwin, 1/18/82, p. 5.

460   **And despite his acquittal:** NYT, 3/6/53.

460   **"this will be the last time":** JRO to Bernard Spero, 4/27/53, box 237, JRO Papers; Weinberg, interview by Sherwin, 8/23/79, p. 25. Weinberg said his prospective employer told him he'd need some cover to hire him and that he'd accept a letter from Robert Oppenheimer.

460   **he "did not know"** *and subsequent quotes:* Lewis Strauss, memo to file, 1/6/53, box 66, Strauss Papers, HHL. Oppenheimer's final legal bill in connection with the Weinberg case was $14,780 (Katherine Russell to Strauss, 4/28/53, HHL). That the board finally rejected Oppenheimer's legal bill can be found in A. H. Belmont, to D. M. Ladd, FBI memo, 6/19/53, sect. 14, JRO FBI file.

Chapter Thirty-three: "The Beast in the Jungle"

461   **"Utterly transfixed":** Anne Wilson Marks to Bird, 5/11/02.

461   **"It hasn't yet come":** Henry James, *The Beast in the Jungle and Other Stories,* pp. 39, 70.

462   **According to the historian:** Hewlett and Holl, *Atoms for Peace and War,* p. 44; McGrath, *Scientists, Business, and the State, 1890–1960,* p. 155.

463 **His chosen audience:** "Meeting for Dr. J. Robert Oppenheimer," 2/17/53, Council on Foreign Relations Archives.

463 **"Atomic Weapons and American Policy"** *and subsequent quotes:* "Armaments and American Policy: A Report of a Panel of Consultants on Disarmament of the Department of State," January 1953, top secret, declassified 3/10/82, White House Office of Special Assistant for National Security Affairs, NSC series, Policy Papers subseries, Disarmament folder, box 2, DDEL.

464 **Atomic bombs, he continued** *and subsequent quotes:* "We may be likened to": JRO, "Atomic Weapons and American Policy," Council on Foreign Relations speech, 2/17/53, reprinted in JRO's *The Open Mind,* pp. 61–77. Oppenheimer may have borrowed the "two scorpions in a bottle" phrase from a speech Vannevar Bush gave at Princeton. See McGrath, *Scientists, Business, and the State, 1890–1960,* p. 151.

465 **On the other hand:** Later that evening Oppenheimer dined alone with Lilienthal, who thought the speech was quite eloquent (Lilienthal, *The Journals of David E. Lilienthal,* vol. 3, p. 370).

465 **two major powers as "two scorpions":** A draft copy of Oppenheimer's speech, dated March 1953, was sent to C. D. Jackson. It was published in *Foreign Affairs* in July 1953 (JRO, "A Note on Atomic Weapons and American Policy," Atomic Energy folder, box 1, C. D. Jackson Papers, DDEL).

465 **"atomic weapons strongly favor":** Eisenhower to C. D. Jackson, 12/31/53, DDE diary, Ann Whitman file, December 1953 folder (1), box 4, DDEL.

466 **"You can't have":** Herken, *Counsels of War,* p. 116.

466 **For a moment:** Stephen E. Ambrose, *Eisenhower,* p. 132. See also "Chronology: Candor-Wheaties," 9/30/54, Ann Whitman Admin. Series, Atoms for Peace folder, box 5, DDEL.

466 **"the Soviets of trouble":** Strauss, *Men and Decisions,* p. 356. Eisenhower appointed Strauss on 3/9/53 to be his "special assistant" on atomic energy issues. In July 1953, Strauss became chairman of the AEC.

466 **"Dr. Oppenheimer was not":** Eisenhower diary, 12/2/53, Ann Whitman file, box 4, folder Oct.–Dec. 1953, DDEL. Eisenhower noted, "When I first came to this office some one individual (I cannot now recall who it was) stated that in his opinion Dr. Oppenheimer was not to be trusted. Whoever it was—and I think it was probably Admiral Strauss—later told me that he had reason to revise his opinion."

466 **"Finally Strauss told":** JRO to Strauss, 5/18/53, re: Felix Browder; Strauss to JRO, 5/12/53, JRO correspondence, IAS Archives. Browder taught at Princeton, Yale, the University of Chicago, and Rutgers University. He later won both a prestigious Guggenheim Fellowship and a Sloan Fellowship and was elected president of the American Mathematical Society.

467 **"is alleged to have delayed":** D. M. Ladd to Hoover, 5/25/53, sect. 14, JRO FBI file.

467 **Strauss "came back to him":** Newman, "The Oppenheimer Case," dissertation, chapter 4, footnote 127. Newman is citing an Eisenhower quote in a letter from Philip Stern to Gen. Robert L. Schulz, 7/21/67, box 1, Stern Papers, JFKL.

467 **"he could not do the job":** Ladd to Hoover, 5/25/53, sect. 14, JRO FBI file 100-17828.

467 **"he needed very badly":** Newman, "The Oppenheimer Case," chapter 2, footnotes 18, 21, 24.

467 **"almost hypnotic power":** Ibid., chapter 4, footnote 165. Newman is citing Jackson, memo to Henry Luce, 10/12/54, box 66, Jackson Papers, DDEL.

467 **"The Hidden Struggle":** Herken, *Counsels of War,* p. 69.

468 **"another nasty":** Lilienthal, *The Journals of David E. Lilienthal,* vol. 3, pp. 390–91; Stern, *The Oppenheimer Case,* p. 203; Herken, *Brotherhood of the Bomb,* p. 263.

468 **"absolutely furious":** Newman, "The Oppenheimer Case," chapter 4, footnote 69.

468 **"talked out of reasonable":** Ibid., chapter 2, footnote 30 (Newman is citing Gertrude Samuels, "A Plea for Candor About the Atom," *New York Times Magazine*, 6/21/53, pp. 8, 21); Hewlett and Holl, *Atoms for Peace and War*, p. 53.

468 **"dangerous and its proposals":** Pfau, *No Sacrifice Too Great*, p. 145.

469 **"That's complete nonsense":** Lewis Strauss, "Memorandum of Conversation with the President," 7/22/53, Strauss Papers, AEC memos to AEC Commissioners, box 66, HHL.

469 **"We don't want":** Ambrose, *Eisenhower*, p. 133.

469 **"Very relieved to":** Jackson diary, 8/4/53, box 56, log 1953 (2), Jackson Papers, DDEL; Hewlett and Holl, *Atoms for Peace and War*, p. 57.

469 **Events also conspired:** Hewlett and Holl, *Atoms for Peace and War*, pp. 58–59.

470 **"New Look":** Ambrose, *Eisenhower*, p. 171; Strauss, *Men and Decisions*, pp. 356–62.

470 **"fraud upon the words":** Newman, "The Oppenheimer Case," chapter 2, footnote 102.

470 **"obstruct things":** JRO FBI file, sect. 3, doc. 103, FBI wiretap of JRO phone conversations with David Lilienthal and Robert Bacher, 10/23–24/46.

470 **"You'd better tell":** Stern, *The Oppenheimer Case*, p. 208.

470 **"I knew he was in trouble":** Rabi, interview by Sherwin, 3/12/82, p. 13.

471 **While he was in Brazil, the FBI:** Hoover to legal attaché, Rio de Janeiro, 6/18/53, sect. 14, JRO FBI file, doc. 348.

471 **"close and cordial":** Hoover to Tolson and Ladd, memo, 6/24/53, sect. 14, JRO FBI file.

471 **"in the closest of confidence":** Ibid.; Hoover to Tolson, Ladd, Belmont, and Nichols, memo, 5/19/53, sect. 14, JRO FBI file.

472 **"In the first place":** Strauss crossed out the word *much* and replaced it with *some*. Lewis Strauss to Senator Robert Taft, draft ltr., 6/22/53, Taft folder, Strauss Papers, HHL.

472 **"as if he were flag":** Roland Sawyer, "The Power of Admiral Strauss," *New Republic*, 5/31/54, p. 14.

472 **His first maneuver:** Belmont to Ladd, memo, 6/5/53, sect. 14, JRO FBI file 100-17828; FBI summary of Oppenheimer file, 6/25/53, sect. 14, JRO FBI file. Strauss, memo for Gen. Robert Cutler and C. D. Jackson, 12/17/53, Strauss Papers, HHL.

472 **During the Eisenhower:** Hewlett and Holl, *Atoms for War and Peace*, p. 45.

472 **"During a single seven-day":** William L. Borden, memo to JCAE chairman, 11/3/52, pp. 8–9, box 41, JCAE, no. DCXXXV, RG 128, NA.

473 **"I think it would be":** Strauss to Borden, 12/10/52, William Borden, box 10, AEC series, NA. For a discussion of other influences on Borden's pursuit of Oppenheimer, see Priscilla McMillan, *The Ruin of J. Robert Oppenheimer*, Chap. 15.

473 **This sequence of withdrawals:** There is a cover page to the Oppenheimer dossier which records the names and dates of previous users of the file. See John A. Waters memo to file, 5/14/53 and Gordon Dean letter to the Attorney General, 5/20/53, AEC files. As Jack Holl has written, "Publicly Borden always claimed that he acted alone and without consultation . . . Privately, he later told a Commission official that he had discussed the case with 'one individual who is intimately familiar with the atomic program,' whose name he preferred not to give, and whose name was not revealed." That individual was certainly Lewis Strauss. Jack A. Holl, "In the Matter of J. Robert Oppenheimer: Origins of the Government's Security Case," a December 1975 paper presented to the American Historical Association, pp. 7–8. See also Hewlett and Holl, *Atoms for Peace and War*, pp. 45–47, 63. For more on Strauss' meeting with Borden, see also McMillan, *The Ruin of J. Robert Oppenheimer*, Chap. 15.

474 **"Strauss had promised":** Harold P. Green, "The Oppenheimer Case: A Study in the Abuse of Law," *Bulletin of the Atomic Scientists*, September 1977, p. 57.

474 **"The Admiral is extremely anxious":** Belmont to Ladd, memo, 9/10/53, JRO FBI file, sect. 14.

475 **"to elucidate what":** Goodchild, *J. Robert Oppenheimer,* pp. 219–20.

475 **"For all my trouble":** Michelmore, *The Swift Years,* pp. 199–200.

475 **"It is a cruel":** Reith Lectures, 1953, boxes 276–278, JRO Papers, LOC.

476 **"The open society":** Michelmore, *The Swift Years,* pp. 202–3.

476 **he "didn't understand":** Lincoln Gordon, phone interview by Bird, 5/18/04. At the time, Gordon was stationed in the U.S. Embassy in London. Later he served as U.S. ambassador to Brazil.

476 **"Chevalier, who is very":** Secret cable from U.S. Legation, Paris, to FBI director, 2/15/54, JRO FBI file, doc. 797, declassified 7/11/01.

476 **By December 7, 1953:** According to Chevalier, he had seen Oppenheimer two or three times in the autumn of 1946, five or six times in 1947, four or five times in 1949, twice in September and October 1950—and once in December 1953 (Chevalier to Philip Stern, 6/15/68, Stern Papers, JFKL).

477 **Oppenheimer suggested he:** Stern, *The Oppenheimer Case,* pp. 213–14. Chevalier later met with Wyman, who informally tried to get him good advice on what he should do about his American citizenship. But Chevalier never reapplied for a U.S. passport. By early 1954, he was "denied any type of employment by UNESCO because of his refusal to comply with U.S. executive order 10422." Issued on 1/9/53, this executive order required American employees of the U.N. to pass a security investigation. (Chevalier FBI file, 100-18564, part 2, doc. dated 3/17/54.)

477 **"I certainly don't":** Chevalier, *Oppenheimer,* pp. 86–87. The next morning, Chevalier took Oppie and Kitty to visit the French novelist André Malraux.

478 **"thoughtfulness":** Borden to Strauss, 11/19/52, Lewis Strauss folder, box 52, AEC, JCAE Papers, NA.

478 **"more probably than not":** JRO hearing, pp. 837–38.

478 **"It is my recollection":** Strauss, "Memorandum for Oppenheimer File," 11/9/53, Strauss Papers, HHL.

478 **"The important point":** Lewis L. Strauss memo, 11/30/53; Barton J. Bernstein, "The Oppenheimer Loyalty-Security Case Reconsidered," *Stanford Law Review,* July 1990, p. 1442.

479 **"whining, whimpering":** Thomas C. Reeves, *The Life and Times of Joe McCarthy,* p. 530.

479 **"All the vague feelings":** C. D. Jackson diary, 11/27/53, log 1953 (2), box 56, DDEL. Jackson later told a White House staff meeting that "this Three Little Monkeys act was not working and would not work, and that appeasing McCarthy in order to save his 7 votes for this year's legislative program was poor tactics, poor strategy and . . . unless the President stepped up to bat on this one soon, the Republicans would have neither a program, nor 1954, nor 1956."

479 **"flagrant performance":** C. D. Jackson to Sherman Adams, 11/25/53, Sherman Adams folder, box 23, C. D. Jackson Papers, DDEL.

479 **"the worst one so far":** Eisenhower, telephone calls, 12/2/53, Phone calls folder, July–Dec. 1953 (1), box 5, DDE Diary Series, Ann Whitman file, DDEL.

479 **Early the next morning:** Pfau, *No Sacrifice Too Great,* p. 151; Strauss, *Men and Decisions,* p. 267.

480 **"they consist of nothing":** Eisenhower diary, 12/2/53 and 12/3/53, "Oct.–Dec. 1953," folder box 4, Ann Whitman file, DDEL.

480 **Eisenhower's "blank wall":** Eisenhower, "Memorandum for the Attorney General," 12/3/53, Strauss Papers, HHL.

480 **"The anti-intellectualism":** Christman, *Target Hiroshima,* pp. 249–50; Royal, *The Story of J. Robert Oppenheimer,* p. 155.

481 **"it might be a good idea":** Record of phone conversation (JRO calling Strauss), 3:05 p.m., 12/14/53, Strauss Papers, HHL.

481 **"are distorted and restated":** Belmont to Ladd, FBI memo, 11/19/53, doc. 549, JRO FBI file, cited in Bernstein, "The Oppenheimer Loyalty-Security Case Reconsidered," *Stanford Law Review,* July 1990, p. 1440.

481 **"Rogers smilingly withdrew":** C. D. Jackson diary, 12/18/53, log 1953 (2), box 56, DDEL.

482 **"a polite form of":** Strauss, memo to file, 12/21/53, 12/22/53, box 66, Strauss Papers, HHL.

482 **According to Nichols' notes:** Kenneth D. Nichols, confidential memo, 12/21/53, Strauss Papers, HHL; FBI memo to Belmont, 12/21/53, JRO FBI file, sect. 16, doc. 512.

483 **A hidden microphone recorded:** Stern, *The Oppenheimer Case,* p. 234; Stuart H. Loory, "Oppenheimer Wiretapping Is Disclosed," WP, 12/28/75.

483 **"I can't believe":** Stern, *The Oppenheimer Case,* p. 235.

483 **Strauss had expected:** JRO FBI file, sect. 16, doc. 574–575, Belmont, memo to Ladd, 12/22/53.

483 **The phone tap was finally:** Ladd to Hoover, memo, 12/21/53, JRO FBI file, sect. 16, doc. 514. This memo indicates that Strauss requested the wiretaps and surveillance on 12/17/53. Curiously, an internal FBI memo warned their agents that "according to the AEC, Oppenheimer keeps a .22 caliber pistol on a chair near the front door." See Belmont to Ladd, memo, 12/22/53, JRO FBI file, doc. 513.

484 **"Dear Lewis":** JRO to Strauss, 12/22/53, Strauss Papers, HHL.

484 **"terrible crash":** Anne Marks, interview by Bird, 3/14/02.

Chapter Thirty-four: "It Looks Pretty Bad, Doesn't It?"

487 **"a slippery sonuvabitch":** Bernstein, "The Oppenheimer Loyalty-Security Case Reconsidered," *Stanford Law Review,* July 1990, p. 1449.

487 **"whether your continued":** JRO hearing, pp. 3, 6.

488 **"They stayed in there":** Verna Hobson, interview by Sherwin, 7/31/79, p. 4.

488 **"items of so-called":** JRO hearing, p. 7.

489 **"It looks pretty bad":** Stern, *The Oppenheimer Case,* p. 520.

489 **"I had hoped":** Lilienthal, *The Journals of David E. Lilienthal,* vol. 3, p. 462.

489 **"how things stand":** Belmont to Ladd, FBI memo, 1/7/54, sect. 17, doc. 605, JRO FBI file.

490 **"all over town":** Belmont to Ladd, FBI memo, 1/15/54, sect. 18, JRO FBI file.

490 **On January 16, Garrison:** Strauss to Hoover, 1/18/54, Strauss Papers, HHL.

490 **"under no circumstances":** Stern, *The Oppenheimer Case,* p. 257; Strauss, memo to file, 1/29/54, Strauss Papers, HHL.

490 **"that the Bureau's":** Goodchild, *J. Robert Oppenheimer,* p. 227.

490 **"that the case was":** Strauss, memo to file, 2/15/62, Harold Green folder, 1957–1976, box 36, Strauss Papers, HHL. Strauss learned this from Green, who said Herbert Marks had told him of the wiretaps at the time.

491 **"He'd come in the room":** Bacher, interview by Sherwin, 3/29/83.

491 **"in view of the fact":** FBI cable, 3/17/54, sect. 24, doc. 1024, JRO FBI file.

491 **"if this case is lost":** Belmont to Ladd, FBI memo, 1/26/54, sect. 19, doc. 704, JRO FBI file. Not all historians agree that Strauss was uncompromising in his pursuit of Oppenheimer. For a slightly different view, see Bernstein, "The Oppenheimer Loyalty-Security Case Reconsidered," *Stanford Law Review,* July 1990, p. 1385.

491 **To foreclose that possibility:** Thorpe, "J. Robert Oppenheimer and the Transformation of the Scientific Vocation," dissertation, p. 562.

491 **Browder called him:** Stern, *The Oppenheimer Case,* p. 242; Goodchild, *J. Robert Oppenheimer,* p. 230.

491 **"cordial contacts":** Belmont to Ladd, FBI memo, 1/29/54, JRO FBI file, sect. 19, doc. 716,

492 **"When Dr. Bradbury testifies":** Strauss to Robb, 2/23/54, Strauss Papers, HHL; Belmont to Ladd, FBI memo, 2/25/54, sect. 21, doc. 824, JRO FBI file.

492 **In addition and also at Strauss':** Hewlett and Holl, *Atoms for Peace and War,* p. 86.

493 **"unaccountably nervous":** James Reston, *Deadline: A Memoir,* p. 221–26; Richard Polenberg, *In the Matter of J. Robert Oppenheimer,* p. xxvii.

493 **"highly irritated":** FBI to Lewis Strauss, 2/2/54, sect. 19, doc. 741, JRO FBI file (declassified 1997).

493 **When he finally took:** FBI summary for 1/29/54, sect. 19, doc. 720, JRO FBI file.

493 **In return, Reston:** Stern, *The Oppenheimer Case,* p. 531.

493 **"They were very intense":** Verna Hobson, interview by Sherwin, 7/31/79, p. 8.

493 **"May I have your":** Ibid., p. 5.

494 **Oppie confessed to Bethe:** Jeremy Bernstein, *Oppenheimer,* p. 96; Bernstein cites a phone interview with Bethe.

494 **"I'm sorry to hear":** Robert Coughlan, "The Tangled Drama and Private Hells of Two Famous Scientists," *Life,* 12/13/63; Teller, *Memoirs,* p. 373.

494 **"He expressed a lack":** Stern, *The Oppenheimer Case,* p. 516.

494 **"considerable trouble":** FBI summary for 2/6/54 (wiretap), sect. 19, doc. 760, JRO FBI file.

495 **"It seemed to me":** Verna Hobson, interview by Sherwin, 7/31/79, p. 5.

495 **"I was going to drive":** Ibid., p. 10; Hobson, review of *In the Matter of J. Robert Oppenheimer,* a play by Heinar Kipphardt, *Princeton History,* no. 1, 1971, pp. 95–97.

495 **"There goes a *narr*":** Seymour Melman told this story to Marcus Raskin. Melman heard it from Einstein's assistant, Bruria Kaufmann.

496 **"The German calamity":** Alice Calaprice, ed., *The Expanded Quotable Einstein,* p. 55.

496 **"Oppenheimer is not a gypsy":** NYT, 4/24/04; Holton, *Einstein, History, and Other Passions,* pp. 218–20.

496 **In late February:** Belmont to Ladd, FBI memo, 1/15/54, sect. 18, JRO FBI file.

496 **Now Rabi proposed:** Thorpe, "J. Robert Oppenheimer and the Transformation of the Scientific Vocation" dissertation, p. 496.

496 **"out of the question":** Belmont to Boardman, FBI memo, 3/4/54, sect. 21, doc. 844, JRO FBI file. Herken, *Brotherhood of the Bomb,* p. 281.

496 **It ran to forty-two:** Stern, *The Oppenheimer Case,* p. 253.

497 **"I think there are things":** FBI wiretap, 3/12/54, sect. 24, doc. 1037, JRO FBI file.

497 **"You have nothing personal":** Jerrold Zacharias to JRO, 4/6/54, Philip M. Stern Papers, JFKL.

497 **"It was incredibly good":** Ruth Tolman to JRO, 4/3/54, Ruth Tolman folder, box 72, JRO Papers.

497 **The children would remain:** Louis Hempelmann, interview by Sherwin, 8/10/79, p. 11.

497 **"I would like you to know":** Stern, *The Oppenheimer Case,* p. 258.

Chapter Thirty-five: "I Fear That This Whole Thing Is a Piece of Idiocy"

498 **"if he decided":** Belmont to Boardman, FBI memo, 3/2/54 and 3/1/54, Strauss-Rogers phone conversation, sect. 21, doc. 834, JRO FBI file.

498 **"This was the shock of the day":** Ecker interview by Sherwin, 7/16/91, p. 7.

499 **The opposing teams of lawyers:** Rhodes, *Dark Sun,* p. 543; Herken, *Brotherhood of the Bomb,* p. 286; Goodchild, *J. Robert Oppenheimer,* p. 236; Stern, *The Oppenheimer Case,* pp. 260, 268; Polenberg, ed., *In the Matter of J. Robert Oppenheimer,* p. xxix.

499 **"We made a pretty"**: Goodchild, *J. Robert Oppenheimer,* p. 237.

499 **"fingers in the dike"**: JRO hearing, p. 53.

499 **"inquiry," not a trial**: Polenberg, ed., *In the Matter of J. Robert Oppenheimer,* p. 29. Polenberg's edited and abbreviated version of the Oppenheimer hearing transcript is superb, but generally we have cited the full transcript published by MIT Press.

500 **"I liked the new"**: JRO hearing, pp. 8 and 876.

500 **"without my report"**: Ibid., p. 14.

500 **"during the period"**: Ibid., p. 5.

501 **"Your letter"**: Ibid., pp. 10–11.

501 **"one of our own men"**: Keith Teeter, FBI memo, 3/24/54, sect. 24, doc. 980, JRO FBI file.

502 **"Strauss and the Eisenhower people"**: Drew Pearson, *Diaries 1949–1959,* p. 303.

503 **"key atomic figure"**: Excerpt of Walter Winchell telecast, 4/11/54, Strauss Papers, HHL.

503 **"You said you were late"**: JRO hearing, pp. 53–55.

503 **So on April 9 Strauss**: memo to file, 4/9/54, Strauss Papers, HHL; Hewlett and Holl, *Atoms for Peace and War,* pp. 89, 91.

503 **"being tried in the press"**: Bernstein, "The Oppenheimer Loyalty-Security Case Reconsidered," *Stanford Law Review,* July 1990, p. 1463; Strauss to Roger Robb, memo 4/16/54, Strauss Papers, HHL.

503 **"The trouble with Oppenheimer"**: Pais, *A Tale of Two Continents,* p. 326; Robert Serber, *Peace and War,* pp. 183–84.

504 **"I am very clear on this"**: JRO hearing, p. 103.

504 **"I had been told"**: Goodchild, *J. Robert Oppenheimer,* p. 231.

505 **"In the case of a brother"**: JRO hearing, p. 111.

505 **"Doctor, I notice"** *and subsequent quotes:* Ibid., pp. 113–14.

506 **While preparing for the hearing**: Goodchild, *J. Robert Oppenheimer,* p. 231; Herken, *Brotherhood of the Bomb,* p. 287.

507 **"Because I was an idiot"**: JRO hearing, p. 137.

507 **" 'I've just seen a man' "**: Stern, *The Oppenheimer Case,* p. 283; Robert Coughlan, "The Tangled Drama and Private Hells of Two Famous Scientists," *Life,* 12/13/63, p. 102.

507 **"Doctor . . . I will read to you"**: JRO hearing, p. 144.

509 **Feeling cornered**: Ibid., pp. 146–49.

510 **"Oppenheimer's story, although misleading"**: Hewlett and Holl, *Atoms for Peace and War, 1953–1961,* p. 96.

510 **"The story I told Pash"**: JRO hearing p. 888.

511 **" 'I should have told it' "**: JRO hearing, pp. 888–89.

513 **"I knew her"**: JRO hearing, pp. 153–54.

513 **It was a humiliating experience**: Navasky, *Naming Names,* p. 322.

513 **"Is the list long enough?"**: JRO hearing, p. 155.

513 **"the way a soldier does"**: Coughlan, "The Tangled Drama and Private Hells of Two Famous Scientists," *Life,* 12/13/63.

514 **"From the beginning"**: Goodchild, *J. Robert Oppenheimer,* p. 228.

514 **"On Wednesday, Oppenheimer broke"**: Strauss to President Eisenhower, 4/16/54; Eisenhower to Strauss, cable, 4/19/54, Strauss Papers, Eisenhower folders, box 26D, AEC series, HHL.

514 **"I would be amazed"**: JRO hearing, p. 167; Polenberg, ed., *In the Matter of J. Robert Oppenheimer,* pp. 77–78.

515 **"File review failed"**: FBI memo to Hoover, 12/23/53, sect. 16, doc. 563, JRO FBI file (for Harvey memo, see p. 248).

516 **Lansdale was interviewed**: Herken, *Brotherhood of the Bomb,* p. 400, note 47.

517 **"Yes, it is possible":** JRO hearing, p. 265.

517 **"as it would endanger":** Hoover to Groves, 6/13/46, and Groves to Hoover, 6/21/46, RG 77 (MED files) entry 8, box 100, NA.

518 **When the FBI asked Frank:** Frank Oppenheimer was interviewed by the FBI on 12/29/53 at his Colorado ranch. He refused to sign an affidavit. Strauss was given a copy of the FBI interview on 1/7/54. (Herken, *Brotherhood of the Bomb*, pp. 272, 400.)

518 **But then Groves went on:** FBI memo to Hoover, 12/22/53, sect. 16, doc. 557, 565, JRO FBI file.

518 **As late as 1968:** Leslie Groves oral history interview by Raymond Henle, 8/9/68, p. 17, HHL.

518 **"It was very difficult":** Groves to Strauss, 10/20/49 and 11/4/49, box 75, Strauss Papers, HHL.

519 **The historian Gregg Herken:** Gregg Herken, *Brotherhood of the Bomb*, p. 280. The historian Barton J. Bernstein disagrees with Herken's view. See Barton J. Bernstein, "Reconsidering the Atomic General: Leslie R. Groves," *The Journal of Military History*, July 2003: 899.

519 **"The General said":** FBI memo to Hoover, 12/22/53, sect. 16, doc. 565, JRO FBI file.

519 **By then, Groves:** Gregg Herken, *Brotherhood of the Bomb*, p. 281.

519 **This part of the story:** Hewlett and Holl, *Atoms for Peace and War*, p. 98.

520 **"would you clear Dr. Oppenheimer":** Polenberg, ed., *In the Matter of J. Robert Oppenheimer*, pp. 80–81.

521 **"How could one not have qualms?":** JRO hearing, p. 229.

521 **"I think they did an admirable":** Polenberg, ed., *In the Matter of J. Robert Oppenheimer*, pp. 107–8.

521 **"My feeling was":** Goodchild, *J. Robert Oppenheimer*, pp. 248–49.

522 **"convinced that in view of the testimony":** Polenberg, ed., *In the Matter of J. Robert Oppenheimer*, p. xxv. Belmont to Boardman, 4/17/54, JRO FBI file.

522 **The press did not discover:** Stern, *The Oppenheimer Case*, p. 303; Herken, *Brotherhood of the Bomb*, p. 288.

522 **"All we had the energy for":** Goodchild, *J. Robert Oppenheimer*, p. 249.

522 **"Robert, tell them to shove it":** Stern, *The Oppenheimer Case*, pp. 303–4; Goodchild, *J. Robert Oppenheimer*, p. 244.

Chapter Thirty-six: "A Manifestation of Hysteria"

523 **After Oppenheimer was excused:** At the time, Conant was serving in the Eisenhower Administration as its high commissioner to West Germany, and Secretary of State John Foster Dulles tried to persuade Conant not to testify. Conant refused and noted in his diary, "Told him I had no choice but to testify at Oppenheimer's hearings. He said I should know that this might destroy my usefulness in government." (James Conant diary, 4/19/54, cited in Bernstein, "The Oppenheimer Loyalty-Security Case Reconsidered," *Stanford Law Review*, July 1990, p. 1459.)

524 **"I didn't give a damn":** John J. McCloy, interview by Bird, 7/10/86.

524 **"I am very distressed:** Bird, *The Chairman*, p. 423; McCloy, to Eisenhower, 4/16/54 and 4/23/54, DDEL.

525 **"I don't just know exactly"** *and subsequent quotes:* Bird, *The Chairman*, pp. 424–25.

526 **"one of the great minds":** JRO hearing, p. 357; Polenberg, ed., *In the Matter of J. Robert Oppenheimer*, pp. 140–41.

526 **"Dr. Oppenheimer is smiling":** JRO hearing, p. 372; Polenberg, ed., *In the Matter of J. Robert Oppenheimer*, pp. 147–48.

527 **"In other words":** Polenberg, ed., *In the Matter of J. Robert Oppenheimer*, pp. 162–63.

527 **"the surprise production"**: JRO hearing, pp. 419–20; Polenberg, ed., *In the Matter of J. Robert Oppenheimer,* p. 165.

527 **"so steamed up"**: Polenberg, ed., *In the Matter of J. Robert Oppenheimer,* p. 156.

527 **"I never hid my opinion"**: JRO hearing, p. 468.

528 **"I am naturally a truthful"** *and subsequent quotes:* Ibid., pp. 469–70; Polenberg, ed., *In the Matter of J. Robert Oppenheimer,* pp. 178–79.

528 **"You are back now"**: Polenberg, ed., *In the Matter of J. Robert Oppenheimer,* p. 173.

528 **"He was a very adaptable"**: Bernstein, *Oppenheimer,* p. 62.

528 **"I felt strongly"** *and subsequent quotes:* JRO hearing, pp. 560–67.

529 **As a young girl:** Verna Hobson, interview by Sherwin, 7/31/79, p. 18.

530 **"There are two answers"**: JRO hearing, p. 576.

530 **"This would affect me"** *and subsequent quotes:* JRO hearing, pp. 643–56; Polenberg, ed., *In the Matter of J. Robert Oppenheimer,* pp. 231–37.

531 **"unless ordered to do so"**: Polenberg, ed., *In the Matter of J. Robert Oppenheimer,* p. 196.

532 **Sure that Lawrence was making:** Herken, *Brotherhood of the Bomb,* p. 291 (Herken is citing Childs' interview with Luis Alvarez, box 1, Childs Papers).

532 **"he should never again"**: Hewlett and Holl, *Atoms for Peace and War,* p. 87.

532 **"Teller regrets the case"**: Charter Heslep to Lewis Strauss, memo 5/3/54, Teller folder, AEC Series, box 111, Strauss papers, HHL.

533 **"defrock him in his own"**: Teller, *Memoirs,* pp. 374–81; Hewlett and Holl, *Atoms for Peace and War,* p. 93; Herken, *Brotherhood of the Bomb,* pp. 292–93.

533 **"To simplify the issues"** *and subsequent quotes:* JRO hearing, pp. 710, 726.

533 **"I could hear a tape"**: Ecker interview by Sherwin, 7/16/91, p. 13.

534 **"After what you've just said"**: Goodchild, *J. Robert Oppenheimer,* pp. 254–55.

534 **"I won't shake"**: Ibid., p. 286; Herken, *Brotherhood of the Bomb,* p. 298.

534 **"I stopped having"** *and subsequent quotes:* JRO hearing, pp. 915–18.

535 **"I did not subscribe"**: Ibid., p. 919.

536 **"I am grateful to"**: Ibid., p. 961.

536 **"I remember a kind of sinking"**: Ibid., pp. 971–72; Polenberg, ed., *In the Matter of J. Robert Oppenheimer,* p. 347.

536 **"Russia was our so-called"**: JRO hearing, pp. 971–92; Polenberg, ed., *In the Matter of J. Robert Oppenheimer,* p. 351.

537 **"There is more than"**: Polenberg, ed., *In the Matter of J. Robert Oppenheimer,* pp. 351–52.

537 **"Security Clearance Procedures"**: U.S. AEC, Security Clearance Procedures, Code of Federal Regulations, title 10, chap. 1, pa. 4, adopted 9/12/50, *Federal Register,* 9/19/50, p. 6243, cited in Newman, "The Oppenheimer Case," dissertation, chapter 5, note 60; McMillan, *The Ruin of J. Robert Oppenheimer,* Chap. 21.

Chapter Thirty-seven: "A Black Mark on the Escutcheon of Our Country"

538 **"he believes he will never"**: Polenberg, ed., *In the Matter of J. Robert Oppenheimer,* p. xv; FBI summary of wiretap for 5/7/54 and 5/12/54, doc. 1548, JRO FBI file.

539 **"It is my present conviction"** *and subsequent quotes:* "Memorandum for Mr. Gordon Gray's files re: Oppenheimer Case," 5/7/54, Oppenheimer Correspondence Dictation folder, box 4, Gordon Gray Papers, DDEL.

539 **"from the beginning"**: Ibid.

540 **thought "it extremely important"**: C. E. Hennrich to Belmont, FBI memo, 5/20/54, doc. 1690, JRO FBI file; Goodchild, *J. Robert Oppenheimer,* pp. 259–61.

540 **"I didn't want"**: Goodchild, *J. Robert Oppenheimer,* p. 261.

540 **"The following considerations":** JRO hearing, p. 1019.

541 **"Loyalty to one's friends":** Polenberg, ed., *In the Matter of J. Robert Oppenheimer,* p. 361.

541 **"Most of the derogatory":** Ibid., p. 1020; Polenberg, ed., *In the Matter of J. Robert Oppenheimer,* p. 365.

543 **"His relations with these":** Polenberg, ed., *In the Matter of J. Robert Oppenheimer,* p. 372.

544 **"If we give":** Hewlett and Holl, *Atoms for Peace and War,* p. 103.

544 **At one point, Smyth wondered:** Goodchild, *J. Robert Oppenheimer,* p. 265.

544 **"Gene Zuckert would welcome":** Handwritten note from McKay Dunkin, 5/19/54, Zuckert folder, Strauss Papers, HHL; Harold P. Green, interview by Barton J. Bernstein, 1984 (Bernstein, phone interview by Bird, 2/13/04). See also Bernstein, "The Oppenheimer Loyalty-Security Case Reconsidered," *Stanford Law Review,* p. 1477. Zuckert later said, "I had a difficult time under Lewis Strauss." He called the Oppenheimer hearing a "dog fight. . . . It was not a pleasant year. I still consider myself a friend of Lewis' but it was no fun." (Eugene Zuckert oral history interview, 9/27/71, HSTL.) See also Burch, *Elites in American History,* vol. 2, p. 178.

545 **"personal adviser and consultant":** In May 1959, Strauss confirmed to Smyth that "Mr. Zuckert signed a contract with me as my personal adviser and consultant, after his term of office expired" (LLS Confirmation folder, series 3, box 2, Smyth Papers, American Philosophical Society, Philadelphia, cited by Herken, notes for chap. 18, note 16, posted at *www.brotherhoodofthebomb.com*). See also, McMillan, *The Ruin of J. Robert Oppenheimer,* postlude.

545 **"Lewis, the difference":** Strauss, memo to file, 5/4/54, "Memos for the Record, 1954," box 66, Strauss Papers, HHL.

545 **"You know, it's funny":** Goodchild, *J. Robert Oppenheimer,* pp. 264–65.

546 **"The record shows":** JRO hearing, p. 1050.

546 **"It is sad beyond words":** Lilenthal, *The Journals of David E. Lilienthal,* vol. 3, p. 528.

546 **"Atomic Extermination Conspiracy":** NYT, 4/24/04.

546 **"Oppenheimer's testimony":** Walter Winchell, 6/7/54, *New York Mirror;* FBI memo, 6/8/54, sect. 40, doc. 1691, JRO FBI file.

547 **"longtime glamour-boy":** Thorpe, "J. Robert Oppenheimer and the Transformation of the Scientific Vocation," dissertation, p. 587.

547 **When the Commission's ruling:** Eric Sevareid, *Small Sounds in the Night,* p. 224.

547 **Ironically, publicity surrounding the trial:** For example, see "Le Risque de Securité," *Le Monde,* 6/8/54, p. 1.

547 **"a hard one, but":** "We the undersigned . . . ," 6/7/54, petition to AEC, doc. 1804, sect. 44, JRO FBI file; *New York Post,* 7/10/54. Hewlett and Holl, *Atoms for Peace and War,* p. 111. The decision generated such controversy that Attorney General Herbert Brownell quietly asked Assistant Attorney General Warren Burger to review the record. The future Supreme Court chief justice did so and reported back that he had come to "the personal conclusion that if we were at war, Oppenheimer should have been hung." (Strauss, memo to file, 3/27/69; Warren Burger to Strauss, 5/14/69, Strauss Papers, HHL.)

547 **"He [Oppenheimer] will no longer":** Sevareid, *Small Sounds in the Night,* p. 223.

547 **"By a single foolish":** Joe Alsop to Gordon Gray, 6/2/54, Miscellaneous Correspondence, 1951–57 folder, box 1, Gordon Gray Papers, DDEL.

547 **"We accuse!":** Joseph and Stewart Alsop, *We Accuse,* p. 59; Robert W. Merry, *Taking on the World,* pp. 262–63.

548 **"you opened a good many"**: Bird, *The Chairman*, p. 425.

548 **"The case was ultimately"**: Bernstein, "The Oppenheimer Loyalty-Security Case Reconsidered," *Stanford Law Review*, July 1990, p. 1388.

548 **"I can think of no"**: Eisenhower to Strauss, 6/16/54, Ann Whitman DDE Diaries, June 1954 folder (1), box 7, DDEL.

548 **"the problem of how far"**: McGrath, *Scientists, Business, and the State, 1890–1960*, p. 167.

549 **Mortified, Zuckert beat**: Strauss, memo to file, 12/5/57, box 67, Strauss Papers, HHL.

549 **"it is probably quite impossible"**: Thorpe, "J. Robert Oppenheimer and the Transformation of the Scientific Vocation," dissertation, p. 588.

549 **"messianic role of the scientists"**: Daniel Bell, *The Coming of Post-Industrial Society*, p. 400; Thorpe, "J. Robert Oppenheimer and the Transformation of the Scientific Vocation," dissertation, p. 551.

549 **"Scientists and administrators"**: Ambrose, *Eisenhower*, p. 612; McGrath, *Scientists, Business, and the State, 1890–1960*, p. 4.

Chapter Thirty-eight: "I Can Still Feel the Warm Blood on My Hands"

551 **"Robert and I"**: Jane Wilson to Kitty Oppenheimer, 6/20/54, Robert Wilson folder, box 78, JRO Papers.

551 **"Aren't you tired"**: Babette Oppenheimer Langsdorf to Philip Stern, 7/10/67, Stern Papers, JFKL.

551 **known "all the time"**: FBI "Summary for July 8, 1954," sect. 45, doc. 1858, JRO FBI file.

551 **"One day he would"**: Harold Cherniss, interview by Alice Smith, 4/21/76, p. 24.

551 **"damn fool"**: Francis Fergusson, interview by Sherwin, 6/23/79, pp. 6–8.

552 **"He was like"**: Ibid.

552 **"dry crucifixion"**: Brown, *Through These Men*, p. 288.

552 **"Much of his previous"**: *The Day After Trinity*, Jon Else, transcript, p. 76, Sherwin Collection.

552 **"a sad man"**: Serber, *Peace and War*, p. 183.

552 **"looking actually happy"**: Lilienthal, *The Journals of David E. Lilienthal*, vol. 3, p. 594 (diary entry of 12/24/54).

552 **"a greater understanding of"**: Harold Cherniss, interview by Alice Smith, 4/21/76, p. 23.

552 **In July, Strauss told the FBI**: Roach to Belmont, FBI memo, 7/14/54, sect. 46, doc. 1866, JRO FBI file.

553 **This, however, proved to be**: Oppenheimer's old friend Harold Cherniss took a lead in organizing the petition effort. After talking with a couple of trustees, Cherniss had realized that Oppie's job was in doubt. (Cherniss, interview by Sherwin, 5/23/79, p. 16.)

553 **"He cannot tell the truth"**: Strauss, memo to file, 1/5/55, Strauss Papers, HHL.

553 **"an unconscionable liar"**: Strauss, memos to file, 5/7/68 and 5/12/67, Strauss Papers, HHL; Merry, *Taking on the World*, pp. 360–63; Yoder, *Joe Alsop's Cold War*, pp. 153–55.

553 **"We were sound asleep"**: Sherr, interview by Sherwin, p. 24.

553 **In early July**: Hoover, ltr., 7/15/54, sect. 46, doc. 1869, JRO FBI file.

553 **"a very difficult time"**: Harold Cherniss, interview by Alice Smith, 4/21/76, p. 19; Stern, *The Oppenheimer Case*, p. 393.

553 **"The American Government is unfair"**: Peter wrote these words (spelling corrected) on 6/9/54; Brown, *Through These Men*, p. 228.

554 **"if that corner isn't"**: FBI memo, 7/14/54, sect. 46, doc. 1888, JRO FBI file.

554 **FBI technical surveillance:** Newark FBI bureau, memo to Hoover, 7/13/54, sect. 46, doc. 1880, JRO FBI file.

554 **"key security officials":** FBI summary of surveillance, 7/15/54, sect. 46, doc. 1893, JRO FBI file.

554 **"The letter," the FBI summary:** JRO to Hoover, 7/15/54, doc. 1891; FBI summary of surveillance, 7/17/54, 1899, sect. 46, JRO FBI file.

554 **The island's one village:** Susan Barry, "Sis Frank," *St. John People,* pp. 89–90.

555 **"They were sort of":** Irva Clair Denham, interview by Sherwin, 2/20/82, p. 4.

555 **When Kitty was in a foul mood:** Inga Hiilivirta, interview by Sherwin, 1/16/82, p. 19.

555 **"According to the plan":** FBI, JRO files, sect. 49, 8/23/54 and 8/25/54.

555 **"were damn fools":** FBI, JRO files, 8/30/54, sect. 49, docs. 1981, 2002.

555 **Incredibly, the FBI:** Lilienthal, *The Journals of David E. Lilienthal,* vol. 3, p. 615. Lilienthal had visited St. John that spring and learned of the FBI's visit from Ralph Boulon, coproprietor of the hotel on Trunk Bay.

556 **"How can the independent":** Ferenc M. Szasz, "Great Britain and the Saga of J. Robert Oppenheimer," *War in History,* vol. 2, no. 3 (1995), p. 327; *News Statesman and Nation,* 10/23/54, p. 525. The French press reacted equally critically. On June 8, 1954, *Le Monde* editorialized, "The obsession with security is in the process of leading the United States toward a mental and moral crisis of the first order. It is pushing them to forge the chains of that totalitarianism that they wish precisely to combat. No one wants to risk being accused of being soft on communism. And the views of Senator McCarthy unconsciously have wound up being imposed upon the majority."

556 **"The trouble was":** Chevalier, *Oppenheimer,* p. 116.

556 **"After the security hearings":** Coughlan, "The Equivocal Hero of Science: Robert Oppenheimer," *Life,* February 1967, p. 34A; see also Thorpe, "J. Robert Oppenheimer and the Transformation of the Scientific Vocation," dissertation, p. 572.

556 **"The glorification of Teller":** Jeremy Gundel, "Heroes and Villains: Cold War Images of Oppenheimer and Teller in Mainstream American Magazines" (July 1992), Occasional Paper 92-1, Nuclear Age History and Humanities Center, Tufts University, p. 56.

556 **"slanted favorably toward":** W. A. Branigan to Belmont, FBI memo, 7/27/54, sect. 47, doc. 1912, JRO FBI file; WP, 7/25/54.

557 **"This is a world":** Thorpe, "J. Robert Oppenheimer and the Transformation of the Scientific Vocation," dissertation, p. 608; JRO, *The Open Mind,* pp. 144–45.

557 **"The trouble with secrecy":** *See It Now,* transcript, 1/4/55, CBS News Documentary Library, New York City.

558 **"apologists for totalitarianism"** *and subsequent quotes:* Thorpe, "J. Robert Oppenheimer and the Transformation of the Scientific Vocation," dissertation, pp. 581–84; Jane A. Sanders, "The University of Washington and the Controversy Over J. Robert Oppenheimer," *Pacific Northwest Quarterly,* January 1979, pp. 8–19.

558 **But then to Lilienthal's surprise:** Lilienthal, *The Journals of David E. Lilienthal,* vol. 3, pp. 618–19.

559 **"A lot of nonsense":** Ibid., vol. 5, p. 156.

559 **"somewhat troubled":** Bertrand Russell to JRO, 2/8/57; JRO to Russell, 2/18/57; Russell to JRO, 3/11/57, box 62, JRO Papers; Lanouette, *Genius in the Shadows,* p. 369.

559 **"excommunicated from the inner":** Thorpe, "J. Robert Oppenheimer and the Transformation of the Scientific Vocation," dissertation, pp. 619–20.

559 **Indeed, his name was conspicuously:** Max Born, et al., "The Peril of Universal Death," 7/9/55, reprinted in Bird and Lifschultz, eds., *Hiroshima's Shadow,* pp. 485–87.

560 **"Oppenheimer offered to weep":** Thorpe, "J. Robert Oppenheimer and the Transformation of the Scientific Vocation," dissertation, pp. 617–18.

560 **"He wanted to get back":** *The Day After Trinity,* Jon Else, transcript, p. 76.

560 **"a tragic mistake":** "A-Bomb Use Questioned," 6/9/56, United Press International.

560 **"It is satisfying":** Max Born, *My Life and My Views,* p. 110; JRO to Born, 4/16/64, courtesy of Nancy Greenspan.

560 **Only a year after:** JRO, *The Open Mind,* pp. 50–51.

560 **"the minimization of secrecy":** Ibid., p. 54.

561 **"We don't believe":** *New York Herald Tribune,* 3/26/56; Bird, *The Color of Truth,* p. 147. Professor Morton White of the philosophy department initiated the invitation. M. White interview by Sherwin, 10/27/04.

561 **As Oppenheimer rose:** "Requiescat," *Harvard Magazine,* May–June 2004.

561 **"nervously shifting his arms":** Edmund Wilson, *The Fifties,* pp. 411–12. Bernstein, *Oppenheimer,* p. 174.

562 **"given way to a kind":** Lilienthal, *The Journals of David E. Lilienthal,* vol. 4, p. 259.

562 **"quite too dogmatic":** Nasar, *A Beautiful Mind,* pp. 220–21.

563 **"That's something no one":** Ibid., pp. 221, 294. Oppenheimer had Nash back to the institute in 1961–62 and 1963–64.

563 **"What is new and firm":** Bernstein, *Oppenheimer,* pp. 187–88.

563 **"endlessly fascinating":** Ibid., p. 189; Jeremy Bernstein to Sherwin, memo, April 2004.

563 **"What are we to make":** Peter Coleman, *The Liberal Conspiracy,* pp. 120–21.

564 **"Who didn't know":** Frances Stonor Saunders, *The Cultural Cold War,* pp. 378–79, 394–95; NYT, 5/9/66; Coleman, *The Liberal Conspiracy,* pp. 177, 297.

564 **"I do not regret":** Michelmore, *The Swift Years,* pp. 241–42. For Japanese newspaper articles on Oppenheimer's visit, we thank Mikio Kato of International House, Tokyo, Japan.

564 **"contradiction between Oppenheimer's":** Lilienthal, interview by Sherwin, 10/17/78.

564 **"He kept them on a tight":** Ibid.

565 **Francis Fergusson:** Francis Fergusson, interview by Sherwin, 7/7/79, p. 10.

565 **"He is still in a very":** JRO to Frank Oppenheimer, 4/2/58, Alice Smith Collection.

565 **"What a slap":** Verna Hobson, interview by Sherwin, 7/31/79; Francis Fergusson, interview by Sherwin, 7/7/79, p. 8.

Chapter Thirty-nine: *"It Was Really Like a Never-Never-Land"*

566 **In 1958, Robert hired:** Nancy Gibney, "Finding Out Different," in *St. John People,* p. 151.

566 **When finally built:** Sabra Ericson, interview by Sherwin, 1/13/82, p. 6; Francis Fergusson, interview by Sherwin, 7/7/79, p. 1.

567 **"Easter Rock":** Sis Frank, interview by Sherwin, 1/18/82, p. 1.

567 **A hundred yards up:** Nancy Gibney initially sold one acre to a couple from St. Louis—who then sold their acre to Oppenheimer. A year later, Oppenheimer persuaded the Gibneys to sell him a second acre. (Eleanor Gibney, interview by Bird, 3/27/01.)

567 **The Gibneys had been living:** Ericson, interview by Sherwin, 1/13/82, p. 6.

567 **A former editor:** Ibid., p. 7; Irva Claire Denham, interview by Sherwin, 2/20/82, p. 20.

568 **"seven hideous, hilarious weeks"** *prior and subsequent quotes:* Gibney, "Finding Out Different," in *St. John People,* pp. 153–55.

568 **"Private Property" signs:** Ed Gibney, interview by Bird, 3/26/01.

568 **"I came to have":** Gibney, "Finding Out Different," in *St. John People,* pp. 150–67.

569 **"Gibney, never come to my":** Doris and Ivan Jadan, interview by Sherwin, 1/18/82, p.

14; Inga Hiilivirta, interview by Sherwin, 1/16/82, p. 8; Ericson, interview by Sherwin, 1/13/82, p. 8. The feud ended only after both Robert and Kitty were dead. Toni thought the whole thing was ridiculous, so one day she got Sabra Ericson to take her next door to see Nancy Gibney and got the whole thing settled.

569 **"You never felt uncomfortable"**: Doris Jadan, interview by Sherwin, 1/18/82, pp. 1–4. Ivan Jadan never left the island; he died in 1995.

569 **"Kitty, of course"**: Doris Jadan, interview by Sherwin, 1/18/82, p. 3.

570 **"I don't remember Kitty"**: Ericson, interview by Sherwin, 1/13/82, pp. 14, 19.

570 **"She was the great trouble"**: Doris Jadan, interview by Sherwin, 1/18/82, p. 6.

570 **"He treated her"**: Sis Frank, interview by Sherwin, 1/18/82, p. 7.

570 **"There might be a dead spot"**: Sis Frank, interview by Sherwin, 1/18/82, pp. 2, 8.

570 **"Robert was a very humble"** *and subsequent quotes:* Hiilivirta, interview by Sherwin, 1/16/82, pp. 3–5; Hiilivirta, interview by Bird, 3/26/01.

571 **Limejuice became:** Hiilivirta, interview by Sherwin, 1/16/82, p. 4.

571 **"brought back to him"**: Ibid., p. 5.

571 **"Sis, come with me"**: Sis Frank, interview by Sherwin, 1/18/82, p. 2.

571 **"He was an unassuming"**: Ericson, interview by Sherwin, 1/13/82, pp. 14–15.

571 **"He was the gentlest"**: John Green, interview by Sherwin, 2/20/82, p. 15.

572 **"She was trying"**: Francis Fergusson, interview by Sherwin, 7/7/79, p. 2.

572 **"My God"**: Fiona and William St. Clair, interview by Sherwin, 2/17/82, p. 9; Hiilivirta, interview by Sherwin, 1/16/82, p. 4; Doris Jadan, interview by Sherwin, 1/18/82, p. 4.

572 **Peter seldom came down:** John Green, interview by Sherwin, 2/20/82, p. 21.

572 **"She was very sweet"**: Hiilivirta, interview by Bird, 3/26/01.

572 **"a dead-serious child"**: Gibney, "Finding Out Different," in *St. John People,* p. 157.

572 **Extremely shy:** Hiilivirta, interview by Sherwin, 1/16/82, p. 17.

572 **"Toni was very pliable"**: Ibid., p. 2. Sis Frank, interview by Sherwin, 1/18/82, p. 5; Ericson, interview by Sherwin, 1/13/82, p. 9.

572 **"Robert didn't pay"**: Ericson, interview by Sherwin, 1/13/82, p. 11.

572 **"a deep regard"**: Steve Edwards, interview by Sherwin, 1/18/82, p. 4.

573 **"Alex was crazy about Toni"**: Sis Frank, interview by Sherwin, 1/18/82, p. 7.

573 **But when Toni:** Hiilivirta, interview by Sherwin, 1/16/82, pp. 1–2.

573 **"rag people"**: John Green, interview by Sherwin, 2/20/82, p. 12.

573 **"keep your hat brim"**: Betty Dale, interview by Sherwin, 1/21/82, pp. 2–3.

573 **"Out of your mind"**: Michelmore, *The Swift Years,* p. 240.

573 **"I never saw Robert drunk"**: Doris Jadan, interview by Sherwin, 1/18/82, p. 8.

573 **He loved *The Odyssey:*** Ericson, interview by Sherwin, 1/13/82, p. 14.

Chapter Forty: "It Should Have Been Done the Day After Trinity"

574 **"Not on your life"**: Glenn T. Seaborg, *A Chemist in the White House,* p. 106; Goodchild, *J. Robert Oppenheimer,* p. 275.

574 **When the editors:** Thorpe, "J. Robert Oppenheimer and the Transformation of the Scientific Vocation," dissertation, p. 593.

575 **"Disgusting!" cried one:** "Dr. J. Robert Oppenheimer," 6/26/63, folder 2 of Oppenheimer file, HUAC name file, RG 233, NA.

575 **"the scientist who writes:** Szasz, "Great Britain and the Saga of J. Robert Oppenheimer," *War in History,* vol. 2, no. 3 (1995), p. 329.

575 **"Look, this isn't a day"**: Michelmore, *The Swift Years,* p. 247–48.

575 **"I have been tempted"**: Ibid., p. 248; Teller claimed in his memoirs that he submitted Oppenheimer's name for the 1963 Fermi Prize (Teller, *Memoirs,* p. 465).

575 **Actually, many physicists:** NYT, 11/22/63; Herken, *Cardinal Choices,* pp. 307–8.

575 **"My God, did you hear?":** Peter Oppenheimer, e-mail to Bird, 9/7/04; Michelmore, *The Swift Years,* p. 249.

576 **"figure of stone":** Lilienthal, *The Journals of David E. Lilienthal,* vol. 5, p. 529.

576 **"I think it is just possible":** White House press release, "Remarks of President Johnson, Seaborg, and Oppenheimer," 12/2/63, Philip M. Stern Papers, JFKL; Seaborg, *A Chemist in the White House,* p. 186; Lilienthal, *The Journals of David E. Lilienthal,* vol. 5, p. 530.

576 **Teller was in the audience:** Goodchild, *J. Robert Oppenheimer,* pp. 276–77.

576 **Afterwards, John F. Kennedy's grieving:** David Pines, interview by Bird, 6/26/04.

577 **"dealt a severe blow":** Herken, *Brotherhood of the Bomb,* p. 331.

577 **"It would require":** Bird, *The Color of Truth,* p. 151.

577 **"It's a lovely show":** Herken, *Brotherhood of the Bomb,* p. 330.

577 **"Oppenheimer's partisans":** Strauss, memo to file, 1/21/66, Strauss Papers, HHL.

577 **"That was awful":** Lilienthal, *The Journals of David E. Lilienthal,* vol. 6, p. 22.

578 **"There is nothing":** Ibid., vol. 5, p. 275.

578 **though privately, when he discussed:** Peter Oppenheimer, e-mail to Bird, 9/10/04.

578 **"[B]ut I do recognize your Byrnes":** JRO to Gar Alperovitz, 11/4/64, courtesy of Alperovitz; Alperovitz, *The Decision to Use the Atomic Bomb,* p. 574.

579 **"I begin to wonder":** Heinar Kipphardt, *In the Matter of J. Robert Oppenheimer,* pp. 126–27.

579 **"causes one furiously to think":** Szasz, "Great Britain and the Saga of J. Robert Oppenheimer," *War in History,* vol. 2, no. 3 (1995), p. 330.

579 **"turned the whole damn farce":** Ibid., p. 329.

579 **"It's twenty years too late":** *The Day After Trinity,* Jon Else, transcript, p. 77, Sherwin Collection.

579 **"The subject of the book":** JRO to Dr. Jerome Wiesner, 6/6/66, Stern Papers, JFKL.

580 **"The library is beautiful":** Lilienthal, *The Journals of David E. Lilienthal,* vol. 6, p. 173.

580 **"The trouble is that Robert":** Strauss, memo to file, 4/22/63, Strauss Papers, HHL.

580 **"simply waiting for the bell":** Ibid., 4/29/65, Strauss Papers, HHL.

581 **"even Princeton was too close":** Ibid., 12/14/65, Strauss Papers, HHL.

581 **Construction began in September:** Georgia Whidden (IAS), e-mail to Bird, 2/24/04.

581 **"I am going to outlive"** *and subsequent quotes:* Sis Frank, interview by Sherwin, 1/18/82, p. 3; Verna Hobson, interview by Sherwin, 7/31/79, p. 26.

582 **"dreadful news":** Arthur Schlesinger, Jr., to JRO, 2/21/66, box 65, JRO Papers.

582 **"faint hope":** Francis Fergusson, interview by Sherwin, 6/23/79, p. 10.

582 **"For the first time Robert":** Lilienthal, *The Journals of David E. Lilienthal,* vol. 6, p. 255.

582 **"physicist and sailor":** Pais, *A Tale of Two Continents,* p. 399; Goodchild, *J. Robert Oppenheimer,* p. 279; Michelmore, *The Swift Years,* p. 253.

582 **"his spirit grew":** Dyson, interview by Jon Else, 12/10/79, p. 4; Dyson, *Disturbing the Universe,* p. 81.

583 **"vigorous and almost gay":** Lilienthal, *The Journals of David E. Lilienthal,* vol. 6, p. 234.

583 **In mid-July his doctor:** JRO to Nicolas Nabokov, cable, 7/11/66, Nabokov folder, box 52, JRO Papers.

583 **"ghost, an absolute ghost":** Sabra Ericson, interview by Sherwin, 1/13/82, pp. 16, 21; Sis Frank, interview by Sherwin, 1/18/82, p. 4.

583 **"You don't know what I'd":** Hiilivirta, interview by Sherwin, 1/16/82, pp. 9, 12.

583 **"They were, in fact":** JRO to Nicolas Nabokov, 10/28/66, Nabokov folder, box 52, JRO Papers.

583 **"He [Oppenheimer] was a very":** George Dyson, e-mail to Bird, 5/23/03.
583 **"the cancer was very manifest":** JRO to Nicolas Nabokov, 10/28/66, Nabokov folder, box 52, JRO Papers.
584 **"The last mile":** Lilienthal, *The Journals of David E. Lilienthal,* vol. 6, pp. 299–300.
584 **"I am much less able":** Michelmore, *The Swift Years,* p. 254.
584 **Early in December:** 1966 desk book, box 13, JRO Papers.
584 **"I was rather disturbed"** *and subsequent quotes:* David Bohm to JRO, 11/29/66; JRO to Bohm, draft letter, 12/2/66; and JRO to Bohm, 12/5/66, Bohm file, box 20, JRO Papers.
585 **"Oppenheimer then turned":** Thorpe, "J. Robert Oppenheimer and the Transformation of the Scientific Vocation," dissertation, pp. 629–30; Thomas B. Morgan, "With Oppenheimer, on an Autumn Day," *Look,* 12/27/66, pp. 61–63.
585 **"indifference to the sufferings":** Chevalier, *Oppenheimer,* pp. 34–35.
585 **"They achieved their goal":** *The Day After Trinity,* Jon Else.
585 **"I don't feel very gay":** Lilienthal, *The Journals of David E. Lilienthal,* vol. 6, p. 348.
586 **"battling a cancerous throat":** JRO letter to James Chadwick, 1/10/67, box 26, JRO Papers.
586 **"I knew what he":** Verna Hobson, interview by Sherwin, 7/31/79, p. 10.
586 **"I am in some pain":** Michelmore, *The Swift Years,* p. 254.
586 **"He could speak only":** Dyson, *Disturbing the Universe,* p. 81. Marvin Weinstein was a Columbia University–trained physicist who spent the years 1967 to 1969 as a fellow at the institute.
586 **The next day Louis:** Louis Fischer to Michael Josselson, 2/25/67, folder 3a, box 5, Fischer Papers, PUL, courtesy of George Dyson.
587 **"he mumbled so badly"** *and subsequent quotes:* Ibid.
587 **"I walked him":** Francis Fergusson, interview by Sherwin, 7/7/79, p. 19, and 6/23/79, p. 10.
587 **"His death was":** JRO death certificate, no. 08006, State Department of Health of New Jersey; Dyson, *Disturbing the Universe,* p. 81; Sabra Ericson, interview by Sherwin, 1/13/82, p. 20. According to Dr. Stanley Bauer, director of pathology at Princeton Hospital, Oppenheimer's autopsy report indicated that his liver showed signs of necrosis due to an external toxic substance, presumably the chemotherapy. It also seems that the radiation treatment had completely eradicated Oppenheimer's throat cancer—in which case, he died as a result of the chemotherapy.
587 **"grieved at the news":** Strauss to Kitty Oppenheimer, cable, 2/20/67, Strauss Papers, HHL.
588 **"Renaissance man":** Ferenc M. Szasz, "Great Britain and the Saga of J. Robert Oppenheimer," *War in History,* vol. 2, no. 3 (1995), p. 320.
588 **"The world has lost":** NYT, 2/20/67.
588 **"a man of exceptional":** "Talk of the Town," *The New Yorker,* 3/4/67.
588 **"Let us remember":** *Congressional Record,* 2/19/67.
588 **"In Oppenheimer," he wrote:** Rabi, et al., *Oppenheimer,* p. 8.
588 **"That's where he wanted":** John and Irva Green, and Irva Claire Denham, interview by Sherwin, 2/20/82, pp. 1–2.

Epilogue: "There's Only One Robert"

589 **Within a year or two:** Charlotte Serber committed suicide in 1967.
589 **In 1972, Kitty:** Serber, *Peace and War,* pp. 218–19.
589 **Kitty died of an embolism:** Serber, *Peace and War,* p. 221; Pais, *The Genius of Science,* p. 285.
590 **"The whole point":** Hilde Hein, *The Exploratorium,* pp. ix–x, xiv–xv, 14–21.

590   **"Toni always felt":** Robert Serber, interview by Sherwin, 3/11/82, p. 20.

590   **"She could shift":** Sabra Ericson, interview by Sherwin, 1/13/82, p. 9.

590   **The FBI opened:** "Letter to Newark," 12/22/69, sect. 59, JRO FBI files (declassified 6/23/99).

590   **"She made the mistake":** Serber, interview by Sherwin, 3/11/82, p. 18; June Barlas, interview by Sherwin, 1/19/82, pp. 1–7.

591   **"But when she did mention":** June Barlas interview by Sherwin, 1/19/82, p. 1; Ellen Chances interview by Sherwin, 5/10/79.

591   **"her resentment toward":** Inga Hiilivirta interview by Sherwin, 1/16/82, p. 20.

591   **She swam for a long time:** Ed Gibney interview by Bird, 3/26/01.

591   **On a Sunday afternoon:** June Barlas interview by Sherwin, 1/19/82, p. 5; Fiona St. Clair interview by Sherwin, 2/17/82, p. 4; Sabra Ericson interview by Sherwin, 1/13/82, p. 12.

BIBLIOGRAPHY

Acheson, Dean. *Present at the Creation: My Years in the State Department.* New York: Norton, 1969.

Albright, Joseph, and Marcia Kunstel. *Bombshell: The Secret Story of America's Unknown Atomic Spy Conspiracy.* New York: Times Books, 1997.

Allen, James S. *Atomic Imperialism.* New York: 1952.

Alperovitz, Gar. *Atomic Diplomacy: Hiroshima and Potsdam: The Use of the Atomic Bomb and the American Confrontation with Soviet Power.* New York: Simon & Schuster, 1965.

———. *The Decision to Use the Atomic Bomb.* New York: Alfred A. Knopf, 1995.

Ambrose, Stephen E. *Eisenhower: The President, 1952–1969.* London: George Allen & Unwin, 1984.

Alsop, Joseph and Stewart. *We Accuse: The Story of the Miscarriage of American Justice in the Case of J. Robert Oppenheimer.* New York: Simon & Schuster, 1954.

Alvarez, Luis W. *Alvarez: Adventures of a Physicist.* New York: Basic Books, 1987.

Barrett, Edward L., Jr. *The Tenney Committee: Legislative Investigation of Subversive Activities in California.* Ithaca, NY: Cornell University Press, 1951.

Badash, Lawrence, Joseph O. Hirschfelder, and Herbert P. Broida, eds. *Reminiscences of Los Alamos, 1943–45.* Dordrecht, Holland: D. Reidel Publishing Company, 1980.

Bartusiak, Marcia. *Einstein's Unfinished Symphony: Listening to the Sounds of Space-Time.* New York: Berkeley Books, 2000.

Baruch, Bernard. *Baruch: My Own Story.* New York: Henry Holt & Co., 1957.

———. *The Public Years.* New York: Holt, Rinehart & Winston, 1960.

Belfrage, Cedric. *The American Inquisition, 1945–1960.* Indianapolis and New York: Bobbs-Merrill Co., 1973.

Bell, Daniel. *The Coming of Post-Industrial Society: A Venture in Social Forecasting.* New York: Basic Books, 1973.

Benson, Robert Louis, and Michael Warner. *Venona: Soviet Espionage and the American Response, 1939–1957.* Washington, DC: National Security Agency and Central Intelligence Agency, 1996.

Bernstein, Barton J., ed. *The Atomic Bomb: The Critical Issues.* Boston: Little, Brown & Co., 1976.

Bernstein, Jeremy. *Experiencing Science.* New York: Basic Books, 1978.

———. *Hans Bethe: Prophet of Energy.* New York: Basic Books, 1980.

———. *Quantum Profiles.* Princeton, NJ: Princeton University Press, 1991.

———. *The Merely Personal: Observations on Science and Scientists.* Chicago: Ivan R. Dee, 2001.

———. *The Life It Brings: One Physicist's Beginnings.* New York: Penguin Books, 1987.

———. *Oppenheimer: Portrait of an Enigma.* Chicago: Ivan R. Dee, 2004.

Berson, Robin Kadison. *Marching to a Different Drummer: Unrecognized Heroes of American History.* Westport, CT: Greenwood Press, 1994.

Bird, Kai. *The Chairman: John J. McCloy and the Making of the American Establishment.* New York: Simon & Schuster, 1992.

———. *The Color of Truth: McGeorge Bundy and William Bundy, Brothers in Arms.* New York: Simon & Schuster, 1992.

Bird, Kai, and Lawrence Lifschultz, eds. *Hiroshima's Shadow: Writings on the Denial of History and the Smithsonian Controversy.* Stony Creek, CT: Pamphleteer's Press, 1998.

Birmingham, Stephen. *Our Crowd.* New York: Future Books, 1967.

———. *The Rest of Us: The Rise of America's Eastern European Jews.* Boston: Little, Brown & Co., 1984.

Blackett, P. M. S. *Fear, War, and the Bomb: Military and Political Consequences of Atomic Energy.* New York: McGraw-Hill, 1948, 1949.

Blum, John Morton, ed. *The Price of Vision: The Diary of Henry A. Wallace, 1942–1946.* Boston: Houghton Mifflin, 1973.

Bohlen, Charles E. *Witness to History: 1929–1969.* New York: Norton, 1973.

Born, Max. *My Life: Recollections of a Nobel Laureate.* New York: Charles Scribner's Sons, 1975.

Boyer, Paul. *By Bomb's Early Light: American Thought and Culture at the Dawn of the Atomic Age.* Chapel Hill, NC: University of North Carolina Press, 1994 (Pantheon, 1985).

Brechin, Gray. *Imperial San Francisco: Urban Power, Earthly Ruin.* Berkeley: University of California Press, 1999.

Brian, Denis. *Einstein: A Life.* New York: John Wiley & Sons, 1996.

Brode, Bernice. *Tales of Los Alamos: Life on the Mesa, 1943–1945.* Los Alamos, NM: Los Alamos Historical Society, 1997.

Brome, Vincent. *The International Brigades: Spain, 1936–1939.* New York: William Morrow & Co., 1966.

Brown, John Mason. *Through These Men: Some Aspects of Our Passing History.* New York: Harper & Brothers, 1956.

Bruner, Jerome Seymour. *In Search of Mind.* New York: Harper & Row, 1983.

Bundy, McGeorge. *Danger and Survival: Choices About the Bomb in the First Fifty Years.* New York: Random House, 1988.

Burch, Philip H., Jr. *Elites in American History.* Vol. 3, *The New Deal to the Carter Administration.* New York: Holmes & Meier, 1980.

Bush, Vannevar. *Pieces of the Action.* New York: William Morrow & Co., 1970.

Byrnes, James F. *Speaking Frankly.* New York: Harper & Brothers, 1947.

Calaprice, Alice, ed. *The Expanded Quotable Einstein.* Princeton, NJ: Princeton University Press, 2000.

Calvovoressi, Peter, and Guy Wint. *Total War: The Story of World War II.* New York: Pantheon Books, 1972.

Carroll, Peter N. *The Odyssey of the Abraham Lincoln Brigade: Americans in the Spanish Civil War.* Stanford, CA: Stanford University Press, 1994.

Cassidy, David. *J. Robert Oppenheimer and the American Century.* Indianapolis, IN: Pi Press, 2004.

———. *The Uncertainty Principle: The Life and Science of Werner Heisenberg.* New York: W. H. Freeman, 1992.

Chambers, Marjorie Bell, and Linda K. Aldrich. *Los Alamos, New Mexico: A Survey to 1949.* Los Alamos, NM: Los Alamos Historical Society, monograph 1, 1999.

Chevalier, Haakon. *The Man Who Would Be God.* New York: G. P. Putnam's Sons, 1959.

———. *Oppenheimer: The Story of a Friendship.* New York: George Braziller, 1965.

Childs, Herbert. *An American Genius: The Life of Ernest Orlando Lawrence.* New York: E. P. Dutton & Co., 1968.

Christman, Al. *Target Hiroshima: Deke Parson and the Creation of the Atomic Bomb.* Annapolis, MD: Naval Institute Press, 1998.

Church, Peggy Pond. *Bones Incandescent: The Pajarito Journals of Peggy Pond Church.* Lubbock, TX: Texas Tech University Press, 2001.

————. *The House at Otowi Bridge: The Story of Edith Warner and Los Alamos.* Albuquerque, NM: University of New Mexico Press, 1959.

Clark, Ronald W. *Einstein: The Life and Times.* New York: HarperCollins, Avon Books, 1971, 1984.

Cohen, Sam. *The Truth About the Neutron Bomb.* New York: William Morrow, 1983.

Coleman, Peter. *The Liberal Conspiracy: The Congress for Cultural Freedom and the Struggle for the Mind of Postwar Europe.* New York: The Free Press, 1989.

Compton, Arthur H. *Atomic Quest.* New York: Oxford University Press, 1956.

Cook, Fred J. *The FBI Nobody Knows.* New York: Macmillan Co., 1964.

————. *The Nightmare Decade: The Life and Times of Senator Joe McCarthy.* New York: Random House, 1971.

Corson, William R. *The Armies of Ignorance: The Rise of the American Intelligence Empire.* New York: Dial, 1977.

Crease, Robert P., and Charles C. Mann. *The Second Creation: Makers of the Revolution in 20th Century Physics.* New York: Macmillan Co., 1986.

Curtis, Charles P. *The Oppenheimer Case: The Trial of a Security System.* New York: Simon & Schuster, 1955.

Dallet, Joe. *Letters from Spain.* New York: Workers Library Publishers, 1938.

Davis, Nuel Pharr. *Lawrence and Oppenheimer.* New York: Simon & Schuster, 1968.

Dawidoff, Nicholas. *The Catcher Was a Spy: The Mysterious Life of Moe Berg.* New York: Pantheon, 1994.

Dean, Gordon E. *Forging the Atomic Shield: Excerpts from the Office Diary of Gordon E. Dean.* Ed. Roger M. Anders. Chapel Hill, NC: University of North Carolina Press, 1987.

Donaldson, Scott. *Archibald MacLeish: An American Life.* Boston: Houghton Mifflin, 1992.

Dyson, Freeman. *Disturbing the Universe.* New York: HarperCollins, 1979.

————. *From Eros to Gaia,* New York: Pantheon, 1992.

————. *Weapons and Hope.* New York: Harper & Row, 1984.

Eisenberg, Carolyn. *Drawing the Line: The American Decision to Divide Germany, 1944–1949.* New York: Cambridge University Press, 1996.

Else, Jon, *The Day After Trinity: J. Robert Oppenheimer and the Atomic Bomb* (documentary film). Image Entertainment DVD, 1980. Transcript and supplemental files. Courtesy of Jon Else.

Eltenton, Dorothea. *Laughter in Leningrad: An English Family in Russia, 1933–1938.* London: Biddle Ltd., 1998.

Feynman, Richard. *"Surely You're Joking, Mr. Feynman!"* New York: Norton, 1985.

Fine, Reuben. *A History of Psychoanalysis.* New York: Columbia University Press, 1979.

Fölsing, Albrecht. *Albert Einstein.* New York: Viking Penguin, 1997.

Foreign Relations of the United States (FRUS), 1950, vol 1.

Friedan, Betty. *Life So Far: A Memoir.* New York: Simon & Schuster, 2000.

Friess, Horace L. *Felix Adler and Ethical Culture: Memories and Studies.* New York: Columbia University Press, 1981.

Gell-Mann, Murray. *The Quark and the Jaguar: Adventures in the Simple and the Complex.* New York: W. H. Freeman & Co., 1994.

Gilpin, Robert. *American Scientists and Nuclear Weapons Policy.* Princeton, NJ: Princeton University Press, 1962.

Giovannitti, Len, and Fred Freed. *The Decision to Drop the Bomb.* London: Methuen & Co., 1965, 1967.

Gleick, James. *Genius: The Life and Science of Richard Feynman.* New York: Vintage, 1992.

Goldstein, Robert Justin. *Political Repression in Modern America.* Cambridge, MA: Schenkman Publishing Co., 1978.

Goodchild, Peter. *J. Robert Oppenheimer: Shatterer of Worlds*. Boston: Houghton Mifflin Co., 1981.

Goodman, Walter. *The Committee*. New York: Farrar, Straus & Giroux, 1968.

Gowing, Margaret. *Britain and Atomic Energy, 1939–1945*. New York: St. Martin's Press, 1964.

Greene, Brian. *The Elegant Universe: Superstrings, Hidden Dimensions, and the Quest for the Ultimate Theory*. New York: Random House, 1999; Vintage, 2003.

Grew, Joseph C. *Turbulent Era: A Diplomatic Record of Forty Years*. Vol. 2. Boston: Houghton Mifflin, 1952.

Gribbin, John. *Q Is for Quantum: An Encyclopedia of Particle Physics*. New York: Simon & Schuster, 1998.

Grigg, John. *1943: The Victory That Never Was*. London: Eyre Methuen, 1980.

Groves, Leslie M. *Now It Can Be Told: The Story of the Manhattan Project*. New York: Harper, 1962; Da Capo Press, 1983.

Guttmann, Allen. *The Wound in the Heart: America and the Spanish Civil War*. New York: 1962.

Haynes, John Earl, and Harvey Klehr. *In Denial: Historians, Communism and Espionage*. San Francisco: Encounter Books, 2003.

———. *Venona: Decoding Soviet Espionage in America*. New Haven CT: Yale University Press, 1999.

Healey, Dorothy. *Dorothy Healey Remembers*. New York: Oxford University Press, 1990.

Hein, Hilde. *The Exploratorium: The Museum As Laboratory*. Washington, DC: Smithsonian Books, 1991.

Herken, Gregg. *Brotherhood of the Bomb: The Tangled Lives and Loyalties of Robert Oppenheimer, Ernest Lawrence, and Edward Teller*. New York: Henry Holt & Co., 2002.

———. *Cardinal Choices: Presidential Science Advising from the Atomic Bomb to SDI*. New York: Oxford University Press, 1992.

———. *Counsels of War*. New York: Alfred A. Knopf, 1985.

———. *The Winning Weapon: The Atomic Bomb in the Cold War, 1945–1950*. New York: Alfred A. Knopf, 1980.

Hershberg, James. *James B. Conant: Harvard to Hiroshima and the Making of the Nuclear Age*. New York: Alfred A. Knopf, 1993.

Hewlett, Richard G., and Oscar E. Anderson, Jr. *The New World, 1939–1946*. Vol. 1, *A History of the United States Atomic Energy Commission*. University Park, PA: Pennsylvania State University Press, 1962.

Hewlett, Richard G., and Francis Duncan. *Atomic Shield, 1947–1952*. Vol. 2, *A History of the United States Atomic Energy Commission*. University Park, PA: Pennsylvania State University Press, 1969.

Hewlett, Richard G., and Jack M. Holl. *Atoms for Peace and War, 1953–1961: Eisenhower and the Atomic Energy Commission*. Berkeley, CA: University of California Press, 1989.

Hinckle, Warren, and William W. Turner. *The Fish Is Red: The Story of the Secret War Against Castro*. New York: HarperCollins, 1981.

Hixson, Walter L. *George F. Kennan: Cold War Iconoclast*. New York: Columbia University Press, 1989.

Hoddeson, Lillian, Laurie M. Brown, Michael Riordan and Max Dresden, eds. *The Rise of the Standard Model: A History of Particle Physics from 1964 to 1979*. New York: Cambridge University Press, 1983.

Hoddeson, Lillian, Paul W. Henriksen, Roger A. Meade and Catherine Westfall. *Critical Assembly*. New York: Cambridge University Press, 1993.

Hollinger, David A. *Science, Jews, and Secular Culture*. Princeton, NJ: Princeton University Press, 1996.

Holloway, David. *Stalin and the Bomb: The Soviet Union and Atomic Energy, 1939–1956.* New Haven, CT: Yale University Press, 1994.

Holton, Gerald. *Einstein, History, and Other Passions.* Woodbury, NY: American Institute of Physics Press, 1995.

Horgan, Paul. *A Certain Climate: Essays in History, Arts, and Letters.* Middletown, CT: Wesleyan University Press, 1988.

Isserman, Maurice. *Which Side Were You On? The American Communist Party During the Second World War.* Middletown, CT: Wesleyan University Press, 1982.

James, Henry. *The Beast in the Jungle and Other Stories.* New York: Dover Publications, 1992.

Jenkins, Edith A. *Against a Field Sinister: Memoirs and Stories.* San Francisco: City Lights, 1991.

Jette, Eleanor. *Inside Box 1663.* Los Alamos, NM: Los Alamos Historical Society, 1977.

Jones, Ernest. *The Life and Work of Sigmund Freud.* New York: Basic Books, 1957.

Jones, Vincent C. *Manhattan: The Army and the Atomic Bomb.* Washington, DC: Center of Military History, United States Army, 1985.

Jungk, Robert. *Brighter Than a Thousand Suns: A Personal History of the Atomic Scientist.* New York: Harcourt, Brace & Co., 1958.

Kamen, Martin D. *Radiant Science, Dark Politics: A Memoir of the Nuclear Age.* Berkeley: University of California Press, 1985.

Kaplan, Fred. *Gore Vidal.* New York: Doubleday, 1999.

Kaplan, Fred. *The Wizards of Armageddon.* New York: Simon & Schuster, 1983.

Kaufman, Robert G. *Henry M. Jackson: A Life in Politics.* Seattle: University of Washington Press, 2000.

Keitel, Wilhelm. *Mein Leben. Pflichterfüllung bis zum Untergang. Hitler's Generalfeldmarschall und Chef des Oberkommandos der Wehrmacht in Selbstzeugnissen.* Berlin: Quintessenz Verlags, 1998.

Kempton, Murray. *Rebellions, Perversities, and Main Events.* New York: Times Books, 1994.

Kevles, Daniel J. *The Physicists: A History of a Scientific Community in Modern America.* New York: Vintage Books, 1971.

Kipphardt, Heinar. *In the Matter of J. Robert Oppenheimer.* Translated by Ruth Speirs. New York: Hill and Wang, 1968.

Klehr, Harvey. *The Heyday of American Communism: The Depression Decade.* New York: Basic Books, 1984.

Klehr, Harvey, John Earl Haynes, and Fridrikh Igorevich Firsov, *The Secret World of American Communism.* New Haven, CT: Yale University Press, 1995.

Kragh, Helge. *Quantum Generations: A History of Physics in the Twentieth Century.* Princeton, NJ: Princeton University Press, 1999.

Kraut, Benny. *From Reform Judaism to Ethical Culture: The Religious Evolution of Felix Adler.* Cincinnati, OH: Hebrew Union College Press, 1979.

Kunetka, James W. *City of Fire: Los Alamos and the Birth of the Atomic Age, 1943–1945.* Englewood Cliffs, NJ: Prentice-Hall, 1978.

———. *Oppenheimer: The Years of Risk.* Englewood Cliffs, NJ: Prentice-Hall, 1982.

Kuznick, Peter. *Beyond the Laboratory: Scientists as Political Activists in 1930s America.* Chicago: University of Chicago Press, 1987.

Lamont, Lansing. *Day of Trinity.* New York: Atheneum, 1985.

Lanouette, William, with Bela Silard. *Genius in the Shadows: A Biography of Leo Szilard, the Man Behind the Bomb.* New York: Charles Scribner's Sons, 1992.

Larrowe, Charles P. *Harry Bridges: The Rise and Fall of Radical Labor in the U.S.* New York: Independent Publications Group, 1977.

Lawren, William. *The General and the Bomb: A Biography of General Leslie R. Groves, Director of the Manhattan Project.* New York: Dodd, Mead & Co., 1988.

Leffler, Melvyn P. *A Preponderance of Power: National Security, the Truman Administration, and the Cold War.* Stanford, CA: Stanford University Press, 1992.

Lewis, Richard, and Jane Wilson, eds. *Alamogordo Plus Twenty-five Years.* New York: Viking Press, 1971.

Libby, Leona Marshall. *The Uranium People.* New York: Crane, Russak & Co., 1979.

Lieberman, Joseph I. *The Scorpion and the Tarantula: The Struggle to Control Atomic Weapons, 1945–1949.* New York: Houghton Mifflin, 1970.

Lilienthal, David E. *The Journals of David E. Lilienthal.* Vol. 2, *The Atomic Energy Years, 1945–1950.* New York: Harper & Row, 1964.

——. *The Journals of David E. Lilienthal.* Vol. 3, *Venturesome Years, 1950–1955.* New York: Harper & Row, 1966.

——. *The Journals of David E. Lilienthal.* Vol. 4, *The Road to Change, 1955–1959.* New York: Harper & Row, 1969.

——. *The Journals of David E. Lilienthal.* Vol. 5, *The Harvest Years, 1959–1963.* New York: Harper & Row, 1971.

——. *The Journals of David E. Lilienthal.* Vol. 6, *Creativity and Conflict, 1964–1967.* New York: Harper & Row, 1976.

Madsen, Axel. *Malraux: A Biography.* New York: William Morrow & Co., 1976.

Marbury, William L. *In the Catbird Seat.* Baltimore: Maryland Historical Society, 1988.

Mayers, David. *George Kennan and the Dilemmas of US Foreign Policy.* New York: Oxford University Press, 1988.

McGrath, Patrick J. *Scientists, Business, and the State, 1890–1960:* Chapel Hill, NC: University of North Carolina Press, 2002.

McMillan, Priscilla J. *The Ruin of J. Robert Oppenheimer and the Birth of the Modern Arms Race.* New York: Viking, 2005.

Merriman, Marion, and Warren Lerude. *American Commander in Spain: Robert Hale Merriam and the Abraham Lincoln Brigade.* Reno, NV: University of Nevada Press, 1986.

Merry, Robert W. *Taking on the World: Joseph and Stewart Alsop—Guardians of the American Century.* New York: Viking Press, 1996.

Michelmore, Peter. *The Swift Years: The Robert Oppenheimer Story.* New York: Dodd, Mead & Co., 1969.

Miller, Barbara Stoler, trans. *Bhartrihari: Poems.* New York: Columbia University Press, 1967.

Miller, Merle. *Plain Speaking: An Oral Biography of Harry S. Truman.* New York: G. P. Putnam's Sons, 1973.

Mills, Walter, ed. *The Forrestal Diaries.* New York: Viking Press, 1951.

Mitford, Jessica. *A Fine Old Conflict.* New York: Alfred A. Knopf, 1977.

Morgan, Ted. *Reds: McCarthyism in Twentieth-Century America.* New York: Random House, 2003.

Moynahan, Lt. Col. John F. *Atomic Diary.* Newark, NJ: Barton Publishing Co., 1946.

Nasar, Sylvia. *A Beautiful Mind.* New York: Simon & Schuster, 1998.

Navasky, Victor. *Naming Names.* New York: Viking Press, 1980.

Nelson, Cary, and Jefferson Hendricks, eds. *Madrid 1937: Letters of the Abraham Lincoln Brigade from the Spanish Civil War.* New York: Routledge, 1996.

Nelson, Steve, James R. Barrett, and Rob Ruck. *Steve Nelson: American Radical.* Pittsburgh, PA: University of Pittsburgh Press, 1981.

Nichols, Kenneth D. *The Road to Trinity.* New York: William Morrow and Co., 1987.

Norris, Robert S. *Racing for the Bomb: General Leslie R. Groves, the Manhattan Project's Indispensable Man*. South Royalton, VT: Steerforth Press, 2002.

Offner, Arnold A. *Another Such Victory: President Truman and the Cold War, 1945–1953*. Stanford, CA: Stanford University Press, 2002.

Oppenheimer, J. Robert. *The Flying Trapeze: Three Crises for Physicists*. London: Oxford University Press, 1964.

———. *The Open Mind*. New York: Simon & Schuster, 1955.

Paine, Jeffery. *Father India: How Encounters with an Ancient Culture Transformed the Modern West*. New York: HarperCollins, 1998.

Painter, David S. *The Cold War: An International History*. London and New York: Routledge, 1999.

Pais, Abraham. *The Genius of Science: A Portrait Gallery of Twentieth-Century Physicists*. Oxford: Oxford University Press, 2000.

———. *Inward Bound: Of Matter and Forces in the Physical World*. New York: Oxford University Press, 1986.

———. *Niels Bohr's Times in Physics, Philosophy, and Polity*. Oxford: Clarendon Press, 1991.

———. *A Tale of Two Continents: A Physicist's Life in a Turbulent World*. Princeton, NJ: Princeton University Press, 1997.

Pais, Abraham, Robert P. Crease, Ida Nicolaisen and Joshua Pais. *Shatterer of Worlds: A Life of J. Robert Oppenheimer*. New York: Oxford University Press, 2005.

Pais, Abraham, Maurice Jacob, David I. Olive and Michael F. Atiyah. *Paul Dirac: The Man and His Work*. Cambridge: Cambridge University Press, 1998.

Palevsky, Mary. *Atomic Fragments: A Daughter's Questions*. Berkeley, CA: University of California Press, 2000.

Pash, Boris T. *The Alsos Mission*. New York: Award House, 1969.

Pearson, Drew. *Diaries 1949–1959*. Ed. Tyler Abell. New York: Holt, Rinehart & Winston, 1974.

Peat, F. David. *Infinite Potential: The Life and Times of David Bohm*. Reading, MA: Helix Books, Addison-Wesley, 1997.

Pettitt, Ronald A. *Los Alamos Before the Dawn*. Los Alamos, NM: Pajarito Publications, 1972.

Pfau, Richard. *No Sacrifice Too Great: The Life of Lewis L. Strauss*. Charlottesville, VA: University Press of Virginia, 1985.

Polenberg, Richard, ed. *In the Matter of J. Robert Oppenheimer: The Security Clearance Hearing*. Ithaca, NY: Cornell University Press, 2002.

Polmar, Norman, and Thomas B. Allen. *Rickover: Controversy and Genius*. New York: Simon & Schuster, 1982.

Powers, Thomas. *Heisenberg's War: The Secret History of the German Bomb*. New York: Alfred A. Knopf, 1993.

Prochnau, William W., and Richard W. Larsen. *A Certain Democrat: Senator Henry M. Jackson, A Political Biography*. Englewood Cliffs, NJ: Prentice-Hall, Inc., 1972.

Rabi, I. I., Robert Serber, Victor F. Weisskopf, Abraham Pais and Glenn T. Seaborg. *Oppenheimer*. New York: Charles Scribner's Sons, 1969.

Reeves, Thomas C. *The Life and Times of Joe McCarthy: A Biography*. New York: Stein & Day, 1982.

Regis, Ed. *Who Got Einstein's Office?* Reading, MA: Addison-Wesley, 1987.

Reston, James. *Deadline: A Memoir*. New York: Random House, 1991.

Rhodes, Richard. *Dark Sun: The Making of the Hydrogen Bomb*. New York: Simon & Schuster, 1995.

———. *The Making of the Atomic Bomb*. New York: Simon & Schuster, 1986.

Rigden, John S. *Rabi: Scientist and Citizen*. Cambridge, MA: Harvard University Press, 1987.

Robertson, David. *Sly and Able: A Political Biography of James F. Byrnes*. New York: Norton, 1994.

Roensch, Eleanor Stone. *Life Within Limits*. Los Alamos, NM: Los Alamos Historical Society, 1993.

Romerstein, Herbert, and Eric Breindel. *The Venona Secrets: Exposing Soviet Espionage and America's Traitors*. Washington, DC: Regnery, 2000.

Rosenstone, Robert A. *Crusade of the Left: The Lincoln Battalion in the Spanish Civil War*. New York: Pegasus, 1969.

Royal, Denise. *The Story of J. Robert Oppenheimer*. New York: St. Martin's Press, 1969.

Saunders, Frances Stonor. *The Cultural Cold War: The CIA and the World of Arts and Letters*. New York: The New Press, 2000.

Schrecker, Ellen. *Many Are the Crimes: McCarthyism in America*. Boston: Little, Brown & Co., 1998.

———. *No Ivory Tower: McCarthyism and the Universities*. New York: Oxford University Press, 1986.

Schwartz, Stephen I., ed. *Atomic Audit: The Cost and Consequences of U.S. Nuclear Weapons Since 1940*. Washington, DC: Brookings Institution Press, 1998.

Schwartz, Stephen. *From West to East: California and the Making of the American Mind*. New York: The Free Press, 1998.

Schweber, S. S. *In the Shadow of the Bomb: Bethe, Oppenheimer and the Moral Responsibility of the Scientist*. Princeton, NJ: Princeton University Press, 2000.

Seaborg, Glenn T. *A Chemist in the White House*. Washington, DC: American Chemical Society, 1998.

Segrè, Emilio. *Enrico Fermi: Physicist*. Chicago: University of Chicago Press, 1970.

———. *A Mind Always in Motion: The Autobiography of Emilio Segrè*. Berkeley: University of California Press, 1993.

Serber, Robert. *The Los Alamos Primer*. Berkeley: University of California Press, 1992.

———. with Robert P. Crease. *Peace and War: Reminiscences of a Life on the Frontiers of Science*. New York: Columbia University Press, 1998.

Sevareid, Eric. *Small Sounds in the Night: A Collection of Capsule Commentaries on the American Scene*. New York: Alfred A. Knopf, 1956.

Sherwin, Martin. *A World Destroyed: Hiroshima and Its Legacies* (3rd ed.). Stanford, CA: Stanford University Press, 2003. Originally published as *A World Destroyed: The Atomic Bomb and the Grand Alliance*. New York: Alfred A. Knopf, 1975.

Shirer, William L. *Twentieth-Century Journey: A Native's Return, 1945–1988*. Boston: Little, Brown & Co., 1990.

Simpson, Christopher. *Blowback: America's Recruitment of Nazis and Its Effect on the Cold War*. New York: Weidenfeld & Nicolson, 1988.

Singer, Gerald, ed. *Tales of St. John and the Caribbean*. St. John, VI: Sombrero Publishing Co., 2001.

Smith, Alice Kimball. *A Peril and a Hope: The Scientists' Movement in America: 1945–47*. Cambridge, MA: MIT Press, 1965.

———. and Charles Weiner, eds. *Robert Oppenheimer: Letters and Recollections*. Stanford, CA: Stanford University Press, 1995. Originally published in 1980 by Harvard University Press.

Smith, Richard Norton. *The Harvard Century: The Making of a University to a Nation*. New York: Simon & Schuster, 1986.

St. John People: Stories About St. John Residents by St. John Residents. St. John, VI: American Paradise Publishing, 1993.

Steeper, Nancy Cook. *Gatekeeper to Los Alamos: The Story of Dorothy Scarritt McKibbin.* Los Alamos, NM: Los Alamos Historical Society, 2003.

Stern, Philip M., with Harold P. Green. *The Oppenheimer Case: Security on Trial.* New York: Harper & Row, 1969.

Strauss, Lewis L. *Men and Decisions.* Garden City, NY: Doubleday, 1962.

Szasz, Ferenc Morton. *The Day the Sun Rose Twice: The Story of the Trinity Site Nuclear Explosion, July 16, 1945.* Albuquerque, NM: University of New Mexico Press, 1984.

Tanenhaus, Sam. *Whittaker Chambers: A Biography.* New York: Random House, 1997.

Taubman, William. *Khrushchev: The Man and His Era.* New York: Norton, 2000.

Teller, Edward, and Allen Brown. *The Legacy of Hiroshima.* New York: Doubleday, 1962.

Teller, Edward, with Judith Shoolery. *Memoirs: A Twentieth-Century Journey in Science and Politics.* Cambridge, MA: Perseus Publishing, 2001.

Terkel, Studs. *The Good War: An Oral History of World War Two.* London: Hamish Hamilton, 1985.

Thomas, Hugh. *The Spanish Civil War.* New York: Harper & Brothers, 1961.

Truman, Harry S. *Memoirs by Harry S. Truman.* Vol. 1, *Year of Decisions,* Garden City, NY: Doubleday & Co., 1955.

———. *Off the Record: The Private Papers of Harry S. Truman.* Robert H. Ferrell, ed. New York: Penguin, 1982.

Trumpbour, John, ed. *How Harvard Rules: Reason in the Service of Empire.* Boston: South End Press, 1989.

United States Atomic Energy Commission, *In the Matter of J. Robert Oppenheimer: Transcript of Hearing Before Personnel Security Board and Texts of Principal Documents and Letters.* Foreword by Philip M. Stern. Cambridge, MA: MIT Press, 1971 (referred to in endnotes as "JRO hearing").

Vidal, Gore. *Palimpsest: A Memoir.* New York: Random House, 1995.

Voros, Sandor. *American Commissar.* Philadelphia: Chilton Company, 1961.

Wang, Jessica. *American Science in an Age of Anxiety: Scientists, Anticommunism, and the Cold War.* Chapel Hill: University of North Carolina Press, 1999.

Weisskopf, Victor. *The Joy of Insight: Passions of a Physicist.* New York: Basic Books, 1991.

Werth, Alexander. *Russia at War, 1941–1945.* New York: Carroll & Graf, 1964.

Wheeler, John Archibald, with Kenneth Ford. *Geons, Black Holes, and Quantum Foam: A Life in Physics.* New York: W. W. Norton, 1998.

Weinstein, Allen. *Perjury: The Hiss-Chambers Case.* New York: Alfred A. Knopf, 1978.

———. and Alexander Vassiliev. *The Haunted Wood: Soviet Espionage in America—The Stalin Era.* New York: Random House, 1999.

Wigner, Eugene. *The Recollections of Eugene P. Wigner as Told to Andrew Szanton.* New York: Plenum Press, 1992.

Williams, Robert Chadwell. *Klaus Fuchs: Atomic Spy.* Cambridge, MA: Harvard University Press, 1987.

Wilson, Edmund. *The Fifties: From the Notebooks and Diaries of the Period,* ed. Leon Edel. New York: Farrar, Straus & Giroux, 1986.

Wilson, Jane S. *All in Our Time.* Chicago: Bulletin of the Atomic Scientists, 1974.

Wilson, Jane S., and Charlotte Serber, eds. *Standing By and Making Do: Women of Wartime Los Alamos.* Los Alamos, NM: Los Alamos Historical Society, 1988.

Wirth, John D., and Linda Harvey Aldrich. *Los Alamos: The Ranch School Years, 1917–1943.* Albuquerque, NM: University of New Mexico Press, 2003.

Ybarra, Michael J. *Washington Gone Crazy: Senator Pat McCarran and the Great American Communist Hunt.* Hanover, NH: Steerforth Press, 2004.

Yoder, Edwin M., Jr. *Joe Alsop's Cold War: A Study of Journalistic Influence and Intrigue.* Chapel Hill, NC: University of North Carolina Press, 1995.

694 *Bibliography*

York, Herbert. *The Advisors: Oppenheimer, Teller, and the Superbomb.* Stanford, CA: Stanford University Press, 1976, 1989.

Zubok, Vladislav, and Constantine Pleshakov. *Inside the Kremlin's Cold War: From Stalin to Khrushchev.* Cambridge, MA: Harvard University Press, 1996.

Major Articles and Dissertations

Alperovitz, Gar, and Kai Bird. "The Centrality of the Bomb," *Foreign Policy,* Spring 1994.

Barnett, Lincoln. "J. Robert Oppenheimer," *Life,* 10/10/49.

Bernstein, Barton J. "Eclipsed by Hiroshima and Nagasaki: Early Thinking about Tactical Nuclear Weapons," *International Security,* vol. 15, Spring 1991.

———. "Four Physicists and the Bomb: The Early Years, 1945–1950," *Historical Studies in Physical Sciences,* vol. 18, no. 2, 1988.

———. "Interpreting the Elusive Robert Serber: What Serber Says and What Serber Does Not Explicitly Say," *Studies in History and Philosophy of Modern Physics,* vol. 32, no. 3, 2001, pp. 443–86.

———. "In the Matter of J. Robert Oppenheimer," *Historical Studies in the Physical Sciences,* vol. 12, part 2, 1982.

———. "The Oppenheimer Loyalty-Security Case Reconsidered," *Stanford Law Review,* July 1990.

———. "Oppenheimer and the Radioactive-Poison Plan," *Technology Review,* May–June 1985.

———. "Reconsidering the Atomic General: Leslie R. Groves," *The Journal of Military History,* July 2003.

———. "Seizing the Contested Terrain of Early Nuclear History: Stimson, Conant, and Their Allies Explain the Decision to Use the Atomic Bomb," *Diplomatic History* 17, Winter 1993.

Bernstein, Jeremy. "Profiles: Physicist," *The New Yorker,* 10/13/75 and 10/20/75.

Birge, Raymond T. "History of the Physics Department," vol. 4, "The Decade 1932–1942," unpublished manuscript, University of California, Berkeley.

Boulton, Frank. "Thomas Addis (1881–1949): Scottish Pioneer in Haemophilia Research," *Journal of the Royal College of Physicians of Edinburgh,* no. 33, 2003, pp. 135–42.

Bundy, McGeorge. "Early Thoughts on Controlling the Nuclear Arms Race." *International Security,* Fall 1982.

———. "The Missed Chance to Stop the H-Bomb," *New York Review of Books,* 5/13/82.

Coughlan, Robert. "The Tangled Drama and Private Hells of Two Famous Scientists," *Life,* 12/13/63.

———. "The Equivocal Hero of Science: Robert Oppenheimer," *Life,* February 1967.

Davis, Harry M. "The Man Who Built the A-Bomb," *New York Times Magazine,* 4/18/48.

Day, Michael A. "Oppenheimer on the Nature of Science." *Centaurus,* vol. 43, 2001.

"The Eternal Apprentice," *Time,* 11/8/48.

Galison, Peter, and Barton J. Bernstein. "In Any Light: Scientists and the Decision to Build the Superbomb, 1952–54," *Historical Studies in Physical Sciences,* vol. 19.

Gidney, Nancy. "Finding Out Different," in *St. John People: Stories about St. John Residents by St. John Residents.* St. John V. I.: American Paradise Publishing, 1993.

Green, Harold P. "The Oppenheimer Case: A Study in the Abuse of Law," *Bulletin of the Atomic Scientists,* September 1977.

Gundel, Jeremy. "Heroes and Villains: Cold War Images of Oppenheimer and Teller in Mainstream American Magazines," July 1992, Occasional Paper 92-1, Nuclear Age History and Humanities Center, Tufts University.

Hershberg, James G. "The Jig Was Up: J. Robert Oppenheimer and the International Control of Atomic Energy, 1947–49." Paper presented at Oppenheimer Centennial Conference, Berkeley, CA, 4/22–24/04.

Hijiya, James A. "The Gita of J. Robert Oppenheimer," *Proceedings of the American Philosophical Society,* vol. 144, no. 2, June 2000.

Holton, Gerald. "Young Man Oppenheimer," *Partisan Review,* vol. XLVIII, 1981.

Kempton, Murray. "The Ambivalence of J. Robert Oppenheimer," *Esquire,* December 1983.

Leffler, Melvyn. "Inside Enemy Archives: The Cold War Re-Opened," *Foreign Affairs,* Summer 1996.

Lemley, Kevin V., and Linus Pauling. "Thomas Addis," *Biographical Memoirs,* vol. 63. Washington, DC: National Academy of Sciences, 1994.

Morgan, Thomas B. "A Visit with J. Robert Oppenheimer," *Look,* 4/1/58.

———. "With Oppenheimer, on an Autumn Day: A Thoughtful Man Talks Searchingly About Science, Ethics, and Nuclear War on a Quiet Afternoon During a Bad Time," *Look,* 12/27/66.

Newman, Steven Leonard. "The Oppenheimer Case: A Reconsideration of the Role of the Defense Department and National Security." Dissertation, New York University, February 1977.

Oppenheimer, Robert. "Niels Bohr and Atomic Weapons," *New York Review of Books,* 12/17/64.

———. "On Albert Einstein," *New York Review of Books,* 3/17/66.

Preuss, Paul. "On the Blacklist," *Science,* June 1983, p. 35.

Rhodes, Richard. "I Am Become Death . . ." *American Heritage,* vol. 28, no. 6, 1987, pp. 70–83.

Rosenberg, David Alan. "The Origins of Overkill: Nuclear Weapons and American Strategy, 1945–60," *International Security,* no. 7, Spring 1983.

Sanders, Jane A. "The University of Washington and the Controversy Over J. Robert Oppenheimer," *Pacific Northwest Quarterly,* January 1979.

Stern, Beatrice M. *A History of the Institute for Advanced Study, 1930–1950,* p. 613, unpublished manuscript, archives, Institute for Advanced Studies.

Szasz, Ferenc M. "Great Britain and the Saga of J. Robert Oppenheimer," *War in History,* vol. 2, no. 3, 1995.

Thorpe, Charles Robert. "J. Robert Oppenheimer and the Transformation of the Scientific Vocation," Dissertation, UC–San Diego, 2001.

———. and Steven Shapin. "Who Was J. Robert Oppenheimer?" *Social Studies of Science,* August 2000.

Trilling, Diana. "The Oppenheimer Case: A Reading of the Testimony," *Partisan Review,* November–December 1954.

Wilson, Robert. "Hiroshima: The Scientists' Social and Political Reaction," *Proceedings of the American Philosophical Society,* September 1996.

Manuscript Collections

Acheson, Dean (YUL)
Barnard, Chester (Harvard Business School Library)
Baruch, Bernard (PUL)
Bethe, Hans (CUL)
Bohr, Niels (AIP)
Bush, Vannevar (LC and MIT)
Byrnes, James F. (CU)
Clark, Grenville (Dartmouth College)
Clayton, William (HSTL)

Clifford, Clark (HSTL)

Committee to Frame a World Constitution (University of Chicago)

Compton, Arthur (Washington University)

Compton, Karl (MIT)

Conant, James B. (HU)

DuBridge, Lee (Caltech)

Dulles, John Foster (PUL and DDEL)

Eisenhower, Dwight D., Presidential Papers collections (DDEL)

Federation of Atomic Scientists and numerous associated manuscript collections such as Atomic Scientists of Chicago, Fermi papers and Hutchins papers (University of Chicago)

Forrestal, James (PUL)

Frankfurter, Felix (LC and Harvard Law School)

Groves, Leslie, Record Group (RG) 200, National Archives (NA)

Harriman, Averell (LC and Kai Bird personal archive)

Lamont, Lansing (HSTL)

Lawrence, E. O. (UCB)

Lilienthal, David (PUL)

Lippmann, Walter (YUL)

McCloy, John J. (Amherst College archives)

Niebuhr, Reinhold. (LC)

Oppenheimer, J. Robert (LOC and IAS)

Osborn, Frederick (HSTL)

Patterson, Robert (LC)

Peters, Bernard (Niels Bohr Archive, Copenhagen)

Roosevelt, Franklin D., Presidential Papers collection (Roosevelt Library)

Stimson, Henry L. (YUL)

Strauss, Lewis L. (HHL)

Szilard, Leo (UCSDL)

Tolman, Richard (Caltech)

Truman, Harry S., Presidential Papers collections (HSTL)

University of Michigan records of the theoretical physics summer schools during the 1930s

Urey, Harold (UCSDL)

Wilson, Carroll (MIT)

Government Document Collections

Atomic Energy Commission, National Archives

Manhattan Engineering District, Harrison-Bundy files, RG 77, NA

National Defense Research Council and Office of Scientific Research and Development, RG 227, NA

Federal Bureau of Investigation records on J. Robert Oppenheimer, FBI Headquarters, Washington, DC (Name files: J. Robert Oppenheimer, Katherine Oppenheimer, Frank Oppenheimer, Haakon Chevalier, and Klaus Fuchs)

Los Alamos National Laboratory Archives, numerous files

Secretary of Defense Papers, RG 330, NA

Secretary of War Papers, RG 107, NA

Joint Committee on Atomic Energy, RG 128, NA

Special Committee on Atomic Energy, RG 46, NA

Department of State, AEC files and the records of the Special Assistant to the Secretary of State for atomic energy matters, RG 50, NA

Interviews

The interviews listed here were conducted by Martin Sherwin (MS), Kai Bird (KB), Jon Else (JE), Alice Kimball Smith (AS), and Charles Weiner (CW). The Sherwin and Bird interview transcripts are in the possession of the authors. The Jon Else interviews were conducted for use in Else's 1980 documentary film *The Day After Trinity*—and we are grateful for his permission to quote from them. The Smith and Weiner interviews were conducted in connection with their edited collection of Oppenheimer's letters, *Robert Oppenheimer: Letters and Recollections*. While Smith and Weiner graciously gave us copies of these interviews for use in this biography, most of their interview transcripts are on file with MIT's Oral History Program in Cambridge, MA.

Anderson, Carl, 3/31/83 (MS)
Bacher, Jean, 3/29/83 (MS)
Bacher, Robert, 3/29/83 (MS)
Barlas, June, 1/19/82 (MS); 3/28/01 (KB)
Bernheim, Frederic, 10/27/75 (CW)
Bethe, Hans, 7/13/79 (JE); 5/5/82 (MS)
Bohm, David, 6/15/79 (MS)
Boyd, William, 12/21/75 (AS)
Bradbury, Norris, 1/10/85 (MS)
Bundy, McGeorge, 12/2–3/92 (KB)
Chance, Ellen, 5/10/79 (MS)
Cherniss, Harold F., 4/21/76 (AS); 5/23/79 (MS); 11/10/76 (AS)
Chevalier, Haakon, 6/29/82, 7/15/82 (MS)
Chevalier, Haakon, Jr., 3/9/02 (MS)
Christy, Robert, 3/30/83 (MS)
Colgate, Sterling, 11/12/79 (JE)
Compton, Margaret, 4/3/76 (AS)
Crane, Horace Richard, 4/8/83 (MS)
Dale, Betty, 1/21/82 (MS)
Denham, Irva Claire, 1/20/82 (MS)
Denham, John, 1/20/82 (MS)
DeWire, John, 5/5/82 (MS)
DuBridge, Lee, 3/30/83 (MS)
Duffield, Priscilla Greene, 1/2/76 (AS)
Dyer-Bennett, John, 5/15/01 (KB phone interview)
Dyson, Freeman, 12/10/79 (JE); 2/16/84 (MS)
Ecker, Allan, 7/16/91 (MS)
Edsall, John, 7/16/75 (CW)
Edwards, Steve, 1/18/82 (MS)
Ericson, Sabra, 1/13/82 (MS)
Fergusson, Francis, 4/23/75, 4/21/76 (AS); 6/8/79, 6/18/79, 6/23/79, 7/7/79 (MS)
Fontenrose, Joseph, 3/25/83 (MS)
Fowler, William A., 3/29/83 (MS)
Frank, Sis, 1/18/82 (MS)
Freier, Phyllis, 3/5/83 (MS)
Friedan, Betty, 1/24/01 (KB)
Friedlander, Gerhart, 4/30/02 (MS)
Garrison, Lloyd, 1/31/84 (KB)

Geurjoy, Edward, 6/26/04 (KB)
Gibney, Ed, 3/26/01 (KB)
Gibney, Eleanor, 3/27/01 (KB)
Green, John and Irva, 2/20/82 (MS)
Goldberger, Marvin, 3/28/83 (MS)
Goldberger, Mildred, 3/3/83 (MS)
Gordon, Lincoln, 5/18/04 (KB phone interview)
Hammel, Edward, 1/9/85 (MS)
Hawkins, David, 6/5/82 (MS)
Hempelmann, Louis, MD, 8/10/79 (MS)
Hein, Hilde Stern, 3/11/04 (KB)
Hiilivirta, Inga, 1/16/82 (MS); 3/26/01 (KB)
Hobson, Verna, 7/31/79 (MS)
Horgan, Paul, 3/3/76 (AS)
Jadan, Doris and Ivan, 1/18/82, 3/26/01 (MS); 3/28/01 (KB)
Jenkins, Edith Arnstein, 5/9/02 (interview by Gregg Herken); 7/25/02 (KB phone interview)
Kamen, Martin D., 1/18/79 (MS)
Kayser, Jane Didisheim, 6/4/75 (CW)
Kelman, Dr. Jeffrey, 2/3/01 (KB)
Kennan, George F., 5/3/79 (MS)
Langsdorf, Babette Oppenheimer, 12/1/76 (AS)
Lilienthal, David E., 10/14/78 (MS)
Lomanitz, Rossi, 7/11/79 (MS)
Manfred, Ken Max (Friedman), 1/14/82 (MS)
Manley, John, 1/9/85 (MS)
Mark, J. Carson, 12/19/79 (JE)
Marks, Anne Wilson, 3/5/02, 3/14/02, 5/9/02 (KB)
Marquit, Irwin, 3/6/83 (MS)
McCloy, John J., 7/10/86 (KB)
McKibbin, Dorothy, 1/1/76 (AS); 7/20/79, 12/10/79 (JE)
Motto, Dr. Jerome, 3/14/01 (KB phone interview)
Mirsky, Jeanette, 11/10/76 (AS)
Morrison, Philip, 6/21/02 (MS); 10/17/02 (KB phone interview)
Nedelsky, Leo, 12/7/76 (AS)
Nelson, Steve and Margaret, 6/17/81 (MS)
Nier, Alfred, 3/5/83 (MS)
Oppenheimer, J. Robert, 11/18/63 (interview by T. S. Kuhn), AIP, APS
Oppenheimer, Frank, 2/9/73 (CW); 3/17/75, 4/14/76 (AS); 12/3/78 (MS)
Oppenheimer, Peter, 7/79 (MS); 9/23–24/04 (KB)
Peierls, Sir Rudolph, 6/5–6/79 (MS)
Phillips, Melba, 6/15/79 (MS)
Pines, David, 6/26/04 (KB)
Plesset, Milton, 3/28/83 (MS)
Pollak, Inez, 4/20/76 (AS)
Purcell, Edward, 3/5/79 (MS)
Rabi, I. I., 3/12/82 (MS)
Rosen, Louis, 1/9/85 (MS)
Rotblat, Joseph, 10/16/89 (MS)
St. Clair, Fiona and William, 2/17/82 (MS)
Serber, Robert, 3/11/82 (MS); 12/15/79 (JE)
Sherr, Patricia, 2/20/79 (MS)

Silverman, Albert, 8/9/79 (MS)

Silverman, Judge Samuel, 7/16/91 (MS)

Smith, Alice Kimball, 4/26/82 (MS)

Smith, Herbert, 8/1/74 (CW); 7/9/75 (AS)

Stern, Hans, 3/4/04 (KB phone interview)

Stratchel, John, 3/19/80 (MS)

Strunsky, Robert, 4/26/79 (MS)

Smyth, Henry DeWolf, 3/5/79 (MS)

Tatlock, Hugh, 2/01 (MS)

Teller, Edward, 1/18/76 (MS)

Uehling, Edwin and Ruth, 1/11/79 (MS)

Uhlenbeck, Else, 4/20/76 (AS)

Ulam, Stanislaw L., 7/19/79 (MS)

Ulam, Stanislaw and Françoise, 1/15/80 (JE)

Voge, Hervey, 3/23/83 (MS)

Wallerstein, Dr. Robert S., 3/19/01 (KB phone interview)

Weinberg, Joseph, 8/11/79; 8/23/79 (MS)

Weisskopf, Victor, 3/23/79; 4/21/82 (MS)

Whidden, Georgia, 4/25/03 (KB)

Wilson, Robert, 4/23/82 (interview by Owen Gingrich)

Wyman, Jeffries, 5/28/75 (CW)

Yedidia, Avram, 2/14/80 (MS)

Zorn, Jans, 4/8/83 (MS)

INDEX

chin in hand, JROMC. Page 3: JRO on horseback, JROMC; JRO as young man, AIP; young JRO and Frank, AIP. Page 4: Paul Dirac, NA; Max Born, NA; JRO with Kramers, AIP; JRO and others on a boat, AIP. Page 5: Serber, Fowler, JRO, and Alvarez, AIP; JRO in Caltech courtyard, Caltech; Serber at blackboard, Berkeley. Page 6: Lawrence with JRO leaning on car, AIP; JRO with horse, LANL; The authors at Perro Caliente, BS. Page 7: JRO with Fermi and Lawrence, Berkeley; Joe Weinberg, Lomanitz, Bohm, and Freidman, NA; Niels Bohr, AP. Page 8: Jean Tatlock at Vassar, Vassar; Jean Tatlock facing camera, Tatlock; Dr. Thomas Addis, NAS; FBI document, FBI. Page 9: Hoke Chevalier, Johan Hagemeyer Portrait Collection, Bancroft; George Eltenton, Voge; Col. Boris Pash, NA; Martin Sherwin with Chevalier, BS. Page 10: Kitty in jodphurs, BS; Kitty passport photo, BS; Kitty in lab, BS; JRO's lab pass, BS. Page 11: Kitty smoking on couch, JROMC; Kitty's Los Alamos pass, BS; Kitty smiling, JROMC. Page 12: Kitty and Peter, JROMC; JRO feeding baby Peter, JROMC. Page 13: JRO at Los Alamos party, LANL; Dorothy McGibbin, JRO, and Victor Weisskopf, LANL. Page 14: JRO et al. at a lecture, LANL; Hans Bethe portrait, NA; Frank Oppenheimer inspecting instrument, Berkeley; Groves with Stimson, NA. Page 15: JRO pouring coffee, AIP; JRO silhouetted, LANL; Trinity test explosion, LANL. Page 16: Panorama of Hiroshima, NA; Mother and child survivors in Nagasaki, Yamahata. Page 17: JRO et al. at machine, AIP-PTC; Physics Today cover, UPI; JRO, Conant, and Vannevar Bush in tuxedos, Harvard. Page 18: Frank Oppenheimer in lab, NA; Frank and cow, AP; Anne Wilson Marks in boat, Marks; Richard and Ruth Tolman, BS. Page 19: Cover of *TIME,* Getty; JRO et al. with airplane, LANL; JRO et al. at Harvard, Harvard. Page 20: Olden Manor, BS; Kitty, Toni, and Peter outside Olden Manor, Whitehead; JRO, Toni, and Peter in grass, Sonnenberg. Page 21: Kitty in greenhouse, Eisenstadt. Page 22: JRO and Neumann in Princeton, Richards; JRO teaching class, Eisenstadt. Page 23: JRO with Eleanor Roosevelt and others, Getty; JRO portrait, NA; JRO with Greg Breit, NA. Page 24: Herblock cartoon, Herblock; Lewis Strauss portrait, NA; JRO walking with cigarette, Getty. Page 25: Ward Evans, Northwestern; Gordon Gray, UNC; Henry DeWolf Smyth, NA; Eugene Zuchert, NA; Roger Robb, Getty. Page 26: Toni on horse, BS; Kitty and JRO, BS; Peter in coat and tie, JROMC. Page 27: Kitty sailing, BS; JRO sailing, BS; Oppenheimer family on beach, JROMC. Page 28: Neils Bohr and JRO on couch, Bohr; Kitty and JRO in Japan, JROMC. Page 29: Oppie smoking pipe, Steltzer; JRO and Jackie Kennedy, Getty; Frank Oppenheimer at Exploratorium, Exploratorium. Page 30: JRO with Kitty receiving Fermi prize, JROMC; JRO with LBJ, Berkeley; JRO shaking hands with Teller, Getty. Page 31: JRO at beach house, Bukowski; Toni on floor, BS; Toni, Inga, Kitty, and Doris on swing, Hiilivirta. Page 32: Portrait of JRO, Steltzer. Part Title I: JRO as young man, AIP-BAS. Part Title II: JRO at blackboard, JROMC. Part Title III: JRO and Groves at Trinity site, AP. Part Title IV: Einstein and JRO, Eisenstadt. Part Title V: JRO in profile, Karsh.